Introduction to

Statistics and SPSS in Psychology

Introduction to Statistics and SPSS in Psychology

Andrew Mayers

PEARSON

Harlow, England • London • New York • Boston • San Francisco • Toronto • Sydney • Auckland • Singapore • Hong Kong
Tokyo • Seoul • Taipei • New Delhi • Cape Town • São Paulo • Mexico City • Madrid • Amsterdam • Munich • Paris • Milan

PEARSON EDUCATION LIMITED
Edinburgh Gate
Harlow CM20 2JE
Tel: +44 (0)1279 623623
Fax: +44 (0)1279 431059
Website: **www.pearson.com/uk**

First published 2013 (print and electronic)

ISBN: 978-0-273-73101-6 (print)
978-0-273-73102-3 (PDF)
978-0-273-78689-4 (eText)

British Library Cataloguing-in-Publication Data
A catalogue record for the print edition is available from the British Library

Library of Congress Cataloging-in-Publication Data
A catalog record for the print edition is available from the Library of Congress

10 9 8 7 6 5 4 3 2
16 15 14 13

Getty images

Print edition typeset in 9/12 and GiovanniStd-Book
Print edition printed and bound by L.E.G.O. S.p.A., Italy

NOTE THAT ANY PAGE CROSS REFERENCES REFER TO THE PRINT EDITION

Contents

About the author

Dr Andrew Mayers gained his PhD at Southampton Solent University, and has held a number of academic, teaching and research positions. Previously at London Metropolitan University and the University of Southampton, he is now a Senior Lecturer in psychology at Bournemouth University. He teaches statistics and clinical psychology to undergraduate and postgraduate students, and has received a number of teaching awards. His research focuses on mental health, particularly in children and families. Currently, that work focuses on postnatal depression, children's sleep, community mental health, care farming, and behavioural, emotional and emotional problems in children. He has frequently appeared on national television and radio about children's sleep problems, and has published widely on factors relating to sleep and mood. He is passionate about reducing mental health stigma and supports a number of groups, such as the national Time to Change campaign. He is a board member of Barnardo's Bournemouth Children's Centres and is Patron for Bournemouth and District Samaritans.

Companion Website

For open-access **student resources** specifically written to complement this textbook and support your learning, please visit **www.pearsoned.co.uk/mayers**

Lecturer Resources

For password-protected online resources tailored to support the use of this textbook in teaching, please visit **www.pearsoned.co.uk/mayers**

Acknowledgements

I would like to thank my friends and family for sticking with me during this mammoth project. I particularly dedicate this book to my wife Sue, to whom I will always be grateful for her support. I also thank our Sammy, Holly, Katy and Simon for all their encouragement over the years. Finally, I would like to thank all of my students who have used the draft versions of this book and have given me such valuable feedback.

Publisher's acknowledgments

We are grateful to the following for permission to reproduce copyright material:

Figures

Figure 1.2 from screenshots from IBM SPSS Statistics, copyright © IBM SPSS Statistics Software; Figure 2.11 from Microsoft Excel screenshots and icons, Microsoft product screenshots reprinted with permission from Microsoft Corporation; Figure 4.5 from Normal Distribution Calculator, http://stattrek.com/Tables/Normal.aspx, copyright © 2012 StatTrek.com. All Rights Reserved; Figure 4.7 from G*Power opening screen, http://www.psycho.uni-duesseldorf.de/abteilungen/aap/gpower3/, Letzte Änderung: 20.08.2012. Reproduced with kind permission from Professor Dr Axel Buchner; Figure 4.8 from G*Power outcome, http://www.psycho.uni-duesseldorf.de/abteilungen/aap/gpower3/, Letzte Änderung: 20.08.2012. Reproduced with kind permission from Professor Dr Axel Buchner; Figure 4.9 from G*Power outcome for calculating required sample size, http://www.psycho.uni-duesseldorf.de/abteilungen/aap/gpower3/, Letzte Änderung: 20.08.2012. Reproduced with kind permission from Professor Dr Axel Buchner; Figure 10.9 from G*Power data input screen for repeated-measures one-way ANOVA http://www.psycho.uni-duesseldorf.de/abteilungen/aap/gpower3/, Letzte Änderung: 20.08.2012. Reproduced with kind permission from Professor Dr Axel Buchner

In some instances we have been unable to trace the owners of copyright material, and we would appreciate any information that would enable us to do so.

Guided Tour

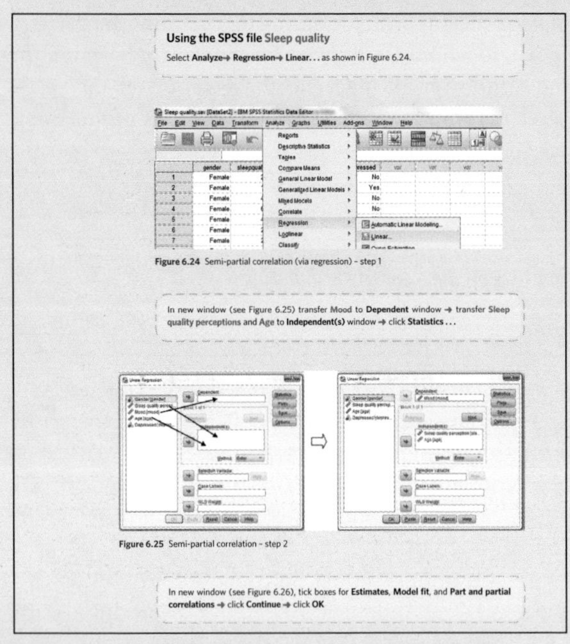

Using the SPSS file Sleep quality

Select **Analyze → Regression → Linear...** as shown in Figure 6.24.

Figure 6.24 Semi-partial correlation (via regression) – step 1

In new window (see Figure 6.25) transfer Mood to **Dependent** window → transfer Sleep quality perceptions and Age to **Independent(s)** window → click **Statistics...**

Figure 6.25 Semi-partial correlation – step 2

In new window (see Figure 6.26), tick boxes for **Estimates**, **Model fit**, and **Part and partial correlations** → click **Continue** → click **OK**

SPSS screenshots and accompanying step-by-step instructions guide you through the processes you need to carry out, using datasets provided on the companion website.

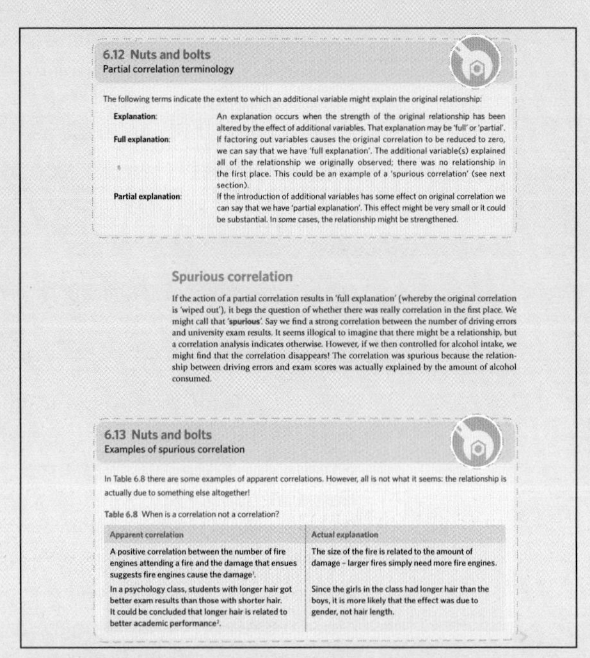

6.12 Nuts and bolts
Partial correlation terminology

The following terms indicate the extent to which an additional variable might explain the original relationship:

Explanation: An explanation occurs when the strength of the original relationship has been altered by the effect of additional variables. That explanation may be 'full' or 'partial'.

Full explanation: If factoring out variables causes the original correlation to be reduced to zero, we can say that we have 'full explanation'. The additional variable(s) explained all of the relationship we originally observed; there was no relationship in the first place. This could be an example of a 'spurious correlation' (see next section).

Partial explanation: If the introduction of additional variables has some effect on original correlation we can say that we have 'partial explanation'. This effect might be very small or it could be substantial. In *some* cases, the relationship might be strengthened.

Spurious correlation

If the action of a partial correlation results in 'full explanation' (whereby the original correlation is 'wiped out'), it begs the question of whether there was really correlation in the first place. We might call that '**spurious**'. Say we find a strong correlation between the number of driving errors and university exam results. It seems illogical to imagine that there might be a relationship, but a correlation analysis indicates otherwise. However, if we then controlled for alcohol intake, we might find that the correlation disappears! The correlation was spurious because the relationship between driving errors and exam scores was actually explained by the amount of alcohol consumed.

6.13 Nuts and bolts
Examples of spurious correlation

In Table 6.8 there are some examples of apparent correlations. However, all is not what it seems: the relationship is actually due to something else altogether!

Table 6.8 When is a correlation not a correlation?

Apparent correlation	Actual explanation
A positive correlation between the number of fire engines attending a fire and the damage that ensues suggests fire engines cause the damage[1].	The size of the fire is related to the amount of damage – larger fires simply need more fire engines.
In a psychology class, students with longer hair got better exam results than those with shorter hair. It could be concluded that longer hair is related to better academic performance[2].	Since the girls in the class had longer hair than the boys, it is more likely that the effect was due to gender, not hair length.

Nuts and Bolts boxes help you to understand the conceptual issues and to go beyond the basics.

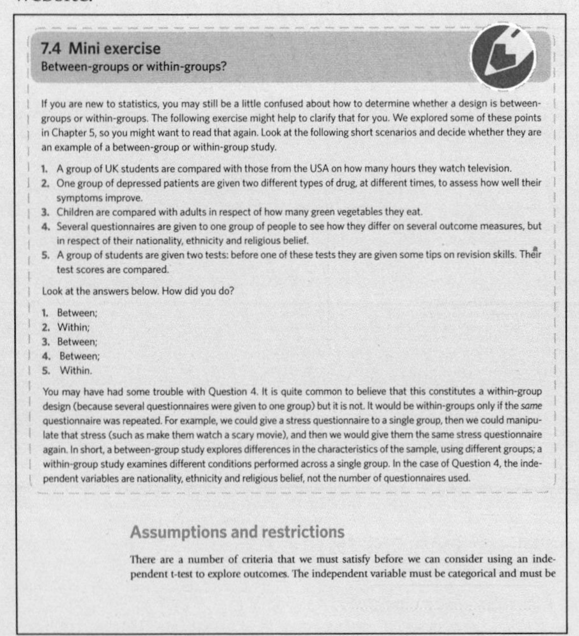

7.4 Mini exercise
Between-groups or within-groups?

If you are new to statistics, you may still be a little confused about how to determine whether a design is between-groups or within-groups. The following exercise might help to clarify that for you. We explored some of these points in Chapter 5, so you might want to read that again. Look at the following short scenarios and decide whether they are an example of a between-group or within-group study.

1. A group of UK students are compared with those from the USA on how many hours they watch television.
2. One group of depressed patients are given two different types of drug, at different times, to assess how well their symptoms improve.
3. Children are compared with adults in respect of how many green vegetables they eat.
4. Several questionnaires are given to one group of people to see how they differ on several outcome measures, but in respect of their nationality, ethnicity and religious belief.
5. A group of students are given two tests: before one of these tests they are given some tips on revision skills. Their test scores are compared.

Look at the answers below. How did you do?

1. Between;
2. Within;
3. Between;
4. Between;
5. Within.

You may have had some trouble with Question 4. It is quite common to believe that this constitutes a within-group design (because several questionnaires were given to one group) but it is not. It would be within-groups only if the *same* questionnaire was repeated. For example, we could give a stress questionnaire to a single group, then we could manipulate that stress (such as make them watch a scary movie), and then we would give them the same stress questionnaire again. In short, a between-group study explores differences in the characteristics of the sample, using different groups; a within-group study examines different conditions performed across a single group. In the case of Question 4, the independent variables are nationality, ethnicity and religious belief, not the number of questionnaires used.

Assumptions and restrictions

There are a number of criteria that we must satisfy before we can consider using an independent t-test to explore outcomes. The independent variable must be categorical and must be

Mini exercises are practical things you can do to improve your understanding of new concepts.

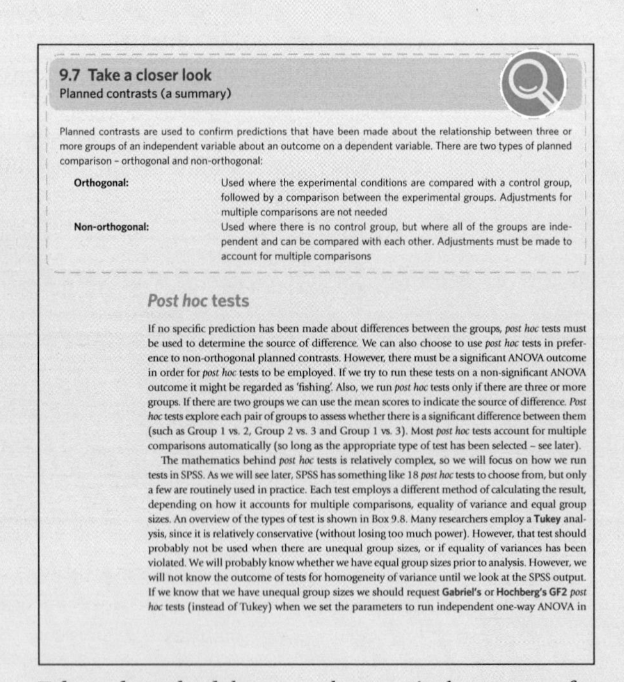

9.7 Take a closer look
Planned contrasts (a summary)

Planned contrasts are used to confirm predictions that have been made about the relationship between three or more groups of an independent variable about an outcome on a dependent variable. There are two types of planned comparison – orthogonal and non-orthogonal:

Orthogonal: Used where the experimental conditions are compared with a control group, followed by a comparison between the experimental groups. Adjustments for multiple comparisons are not needed

Non-orthogonal: Used where there is no control group, but where all of the groups are independent and can be compared with each other. Adjustments must be made to account for multiple comparisons

Post hoc tests

If no specific prediction has been made about differences between the groups, *post hoc* tests must be used to determine the source of difference. We can also choose to use *post hoc* tests in preference to non-orthogonal planned contrasts. However, there must be a significant ANOVA outcome in order for *post hoc* tests to be employed. If we try to run these tests on a non-significant ANOVA outcome it might be regarded as 'fishing'. Also, we run *post hoc* tests only if there are three or more groups. If there are two groups we can use the mean scores to indicate the source of difference. *Post hoc* tests explore each pair of groups to assess whether there is a significant difference between them (such as Group 1 vs. 2, Group 2 vs. 3 and Group 1 vs. 3). Most *post hoc* tests account for multiple comparisons automatically (so long as the appropriate type of test has been selected – see later).

The mathematics behind *post hoc* tests is relatively complex, so we will focus on how we run tests in SPSS. As we will see later, SPSS has something like 18 *post hoc* tests to choose from, but only a few are routinely used in practice. Each test employs a different method of calculating the result, depending on how it accounts for multiple comparisons, equality of variance and equal group sizes. An overview of the types of test is shown in Box 9.8. Many researchers employ a **Tukey** analysis, since it is relatively conservative (without losing too much power). However, that test should probably not be used when there are unequal group sizes, or if equality of variances has been violated. We will probably know whether we have equal group sizes prior to analysis. However, we will not know the outcome of tests for homogeneity of variance until we look at the SPSS output. If we know that we have unequal group sizes we should request **Gabriel's** or **Hochberg's GF2** *post hoc* tests (instead of Tukey) when we set the parameters to run independent one-way ANOVA in

Take a closer look boxes explore particular aspects of the topics in more detail.

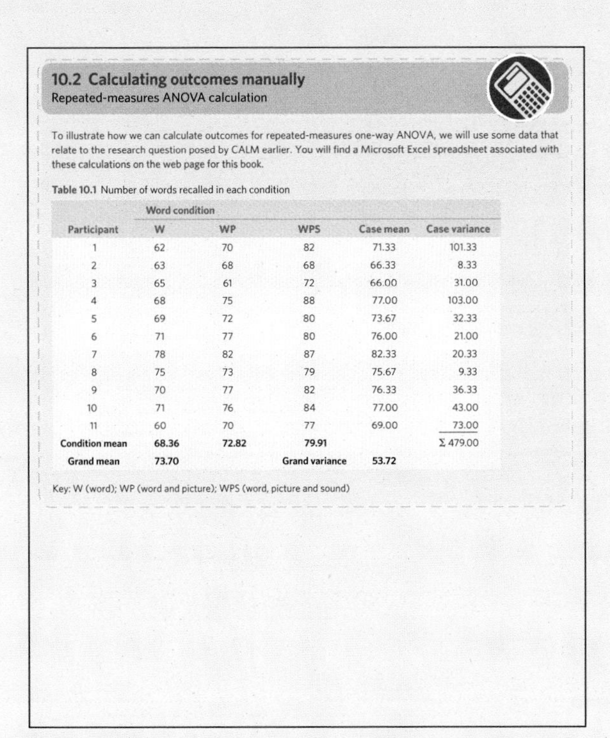

10.2 Calculating outcomes manually
Repeated-measures ANOVA calculation

To illustrate how we can calculate outcomes for repeated-measures one-way ANOVA, we will use some data that relate to the research question posed by CALM earlier. You will find a Microsoft Excel spreadsheet associated with these calculations on the web page for this book.

Table 10.1 Number of words recalled in each condition

	Word condition				
Participant	W	WP	WPS	Case mean	Case variance
1	62	70	82	71.33	101.33
2	63	68	68	66.33	8.33
3	65	61	72	66.00	31.00
4	68	75	88	77.00	103.00
5	69	72	80	73.67	32.33
6	71	77	80	76.00	21.00
7	78	82	87	82.33	20.33
8	75	73	79	75.67	9.33
9	70	77	82	76.33	36.33
10	71	76	84	77.00	43.00
11	60	70	77	69.00	73.00
Condition mean	**68.36**	**72.82**	**79.91**		Σ 479.00
Grand mean	**73.70**		**Grand variance**	**53.72**	

Key: W (word); WP (word and picture); WPS (word, picture and sound)

Calculating outcomes manually boxes show you how to do the calculations by hand so that you understand how they work.

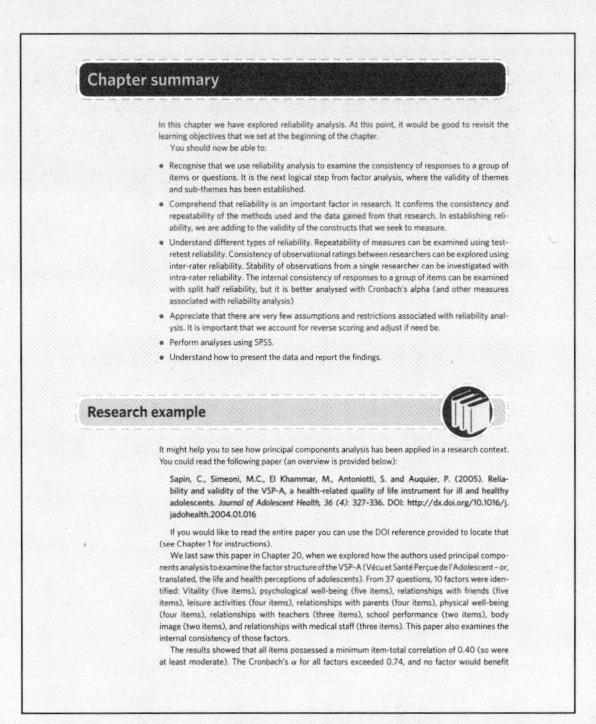

Chapter summary

In this chapter we have explored reliability analysis. At this point, it would be good to revisit the learning objectives that we set at the beginning of the chapter.

You should now be able to:

- Recognise that we use reliability analysis to examine the consistency of responses to a group of items or questions. It is the next logical step from factor analysis, where the validity of themes and sub-themes has been established.
- Comprehend that reliability is an important factor in research. It confirms the consistency and repeatability of the methods used and the data gained from that research. In establishing reliability, we are adding to the validity of the constructs that we seek to measure.
- Understand different types of reliability. Repeatability of measures can be examined using test-retest reliability. Consistency of observational ratings between researchers can be explored using inter-rater reliability. Stability of observations from a single researcher can be investigated with intra-rater reliability. The internal consistency of responses to a group of items can be examined with split half reliability, but it is better analysed with Cronbach's alpha (and other measures associated with reliability analysis)
- Appreciate that there are very few assumptions and restrictions associated with reliability analysis. It is important that we account for reverse scoring and adjust if need be.
- Perform analyses using SPSS.
- Understand how to present the data and report the findings.

Research example

It might help you to see how principal components analysis has been applied in a research context. You could read the following paper (an overview is provided below):

Sapin, C., Simeoni, M.C., El Khammar, M., Antoniotti, S. and Auquier, P. (2005). Reliability and validity of the VSP-A, a health-related quality of life instrument for ill and healthy adolescents. *Journal of Adolescent Health, 36 (4)*: 327-336. DOI: http://dx.doi.org/10.1016/j.jadohealth.2004.01.016

If you would like to read the entire paper you can use the DOI reference provided to locate that (see Chapter 1 for instructions).

We last saw this paper in Chapter 20, when we explored how the authors used principal components analysis to examine the factor structure of the VSP-A (Vécu et Santé Perçue de l'Adolescent – or, translated, the life and health perceptions of adolescents). From 37 questions, 10 factors were identified: Vitality (five items), psychological well-being (five items), relationships with friends (five items), leisure activities (four items), relationships with parents (four items), physical well-being (four items), relationships with teachers (three items), school performance (two items), body image (two items), and relationships with medical staff (three items). This paper also examines the internal consistency of those factors.

The results showed that all items possessed a minimum item-total correlation of 0.40 (so were at least moderate). The Cronbach's α for all factors exceeded 0.74, and no factor would benefit

Research examples put the statistical tests in the context of real-world research, while the **chapter summaries** bring everything together and recap what you've read.

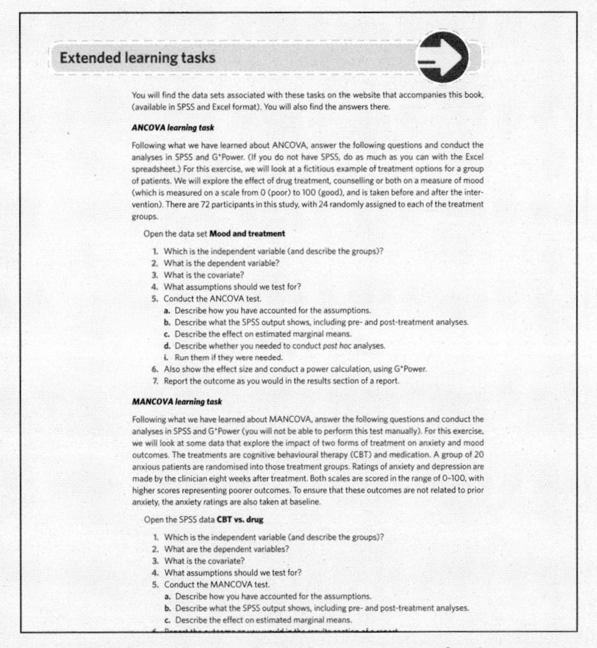

Extended learning tasks

You will find the data sets associated with these tasks on the website that accompanies this book, (available in SPSS and Excel format). You will also find the answers there.

ANCOVA learning task

Following what we have learned about ANCOVA, answer the following questions and conduct the analyses in SPSS and G*Power. (If you do not have SPSS, do as much as you can with the Excel spreadsheet.) For this exercise, we will look at a fictitious example of treatment options for a group of patients. We will explore the effect of drug treatment, counselling or both on a measure of mood (which is measured on a scale from 0 (poor) to 100 (good), and is taken before and after the intervention). There are 72 participants in this study, with 24 randomly assigned to each of the treatment groups.

Open the data set **Mood and treatment**

1. Which is the independent variable (and describe the groups)?
2. What is the dependent variable?
3. What is the covariate?
4. What assumptions should we test for?
5. Conduct the ANCOVA test.
   a. Describe how you have accounted for the assumptions.
   b. Describe what the SPSS output shows, including pre- and post-treatment analyses.
   c. Describe the effect on estimated marginal means.
   d. Describe whether you needed to conduct *post hoc* analyses.
   i. Run them if they were needed.
6. Also show the effect size and conduct a power calculation, using G*Power.
7. Report the outcome as you would in the results section of a report.

MANCOVA learning task

Following what we have learned about MANCOVA, answer the following questions and conduct the analyses in SPSS and G*Power (you will not be able to perform this test manually). For this exercise, we will look at some data that explore the impact of two forms of treatment on anxiety and mood outcomes. The treatments are cognitive behavioural therapy (CBT) and medication. A group of 20 anxious patients are randomised into those treatment groups. Ratings of anxiety and depression are made by the clinician eight weeks after treatment. Both scales are scored in the range of 0–100, with higher scores representing poorer outcomes. To ensure that these outcomes are not related to prior anxiety, the anxiety ratings are also taken at baseline.

Open the SPSS data **CBT vs. drug**

1. Which is the independent variable (and describe the groups)?
2. What are the dependent variables?
3. What is the covariate?
4. What assumptions should we test for?
5. Conduct the MANCOVA test.
   a. Describe how you have accounted for the assumptions.
   b. Describe what the SPSS output shows, including pre- and post-treatment analyses.
   c. Describe the effect on estimated marginal means.

Extended learning tasks help you to go further, using the datasets provided on the website to carry out extra data analysis.

Visit **www.pearsoned.co.uk/mayers** for datasets to use for the exercises in the text, answers to the all learning exercises, revision questions and much more.

1 INTRODUCTION

Why I wrote this book — what's in it for you?

There are a lot of statistics books around, so why choose this one? I have been teaching research methods and statistics in psychology for many years, in several universities. When I recently set about writing my lecture notes, I had to choose a course book to recommend. When I looked at what was available I noticed a number of things. Some books explain when to use a statistical test, and give a broad overview of the theory and concepts, but don't show you how to run it using statistical analysis software. Others show you just how to run the test in that software, but don't explain how and when to use the test, nor do they tell you very much about the theory behind the test. There are several that are very complicated, with loads of maths and formulae – and take themselves far too seriously. Others still are less serious in their approach. I wanted to find something in between all of that; I hope this is it.

In this book you should find sufficient theory and rationale to tell you when you should use a test, why you should use it and how to do so. I will also explain when it is probably not so good to use the test, if certain assumptions are not met (and what to do instead). Then there's the maths thing. I know that most people hate maths, but there is good reason for learning this. When I started studying psychology and statistics, computers and statistical analysis software were all pretty new. It took so long for the valves on the computer to warm up that, by the time it was ready, the data were too old to use. So we had to use maths. Once using a computer was viable, statistical analysis software became the thing to use and it was all very exciting. There seemed little need to ever go back to doing it by hand, I thought. Press a few buttons and off you go. However, when I started teaching statistics, I had another go at doing it all manually and was surprised how much it taught me about the rationale for the test. Therefore, I have decided to include some sections on maths in this book. I really do recommend that you try out these examples (I have attempted to make it all quite simple) – you may learn a lot more than you imagined.

For many, statistics is their very idea of hell. It need not be that way. As you read this book, you will be gently led and guided through whole series of techniques that will lay the foundations for you to become a confident and competent data analyst. How can I make such a bold claim? Well, you only need to ask my students, who have read various iterations of this book. Their feedback has been one of the most motivating aspects of writing it. Over the past few years, several hundred psychology students have used draft versions of this book as part of their studies. They have frequently reported on how the book's clarity and humour really helped them. Many have told me that the friendly style has helped them engage with a subject that had always troubled them before. They also like the unique features of the book that combine theory, rationale, step-by-step guides to performing analyses, relevant real-world research examples, and useful learning exercises and revision.

Above all, I want to make this fun. There will be occasional (hopefully appropriate) moments of humour to lighten the mood, where points may be illustrated with some fun examples. I hope you enjoy reading this book as much as I enjoyed writing it. If you like what you see, tell your friends; if you hate it, don't tell them anything.

Why do psychologists need to know about statistics?

Much of what we explore in psychological research involves people. That much may seem obvious. But because we are dealing with people, our investigations are different to other scientific methods. All the same, psychology remains very much a science. In physical science, 'true experiments' manipulate and control variables; in psychology, we can do that only to a certain extent. For example, we cannot induce trauma in a group of people, but we can compare people who have experienced trauma with those who have not. Sometimes, we can introduce an intervention, perhaps a new classroom method, and explore the effect of that. All of this is still scientific, but there will always be some doubt regarding how much trust we can put in our observations.

A great deal of the time a psychology researcher will make predictions and then design studies to test their theory. We may observe children in a classroom, or investigate attitudes between two groups of people, or explore the risk factors for depression. When we design our experiments and research studies, we will be pleased when our predicted outcomes have been demonstrated. However, we need to be confident that what we have observed is due to the factors that we predicted to be 'responsible' for that outcome (or that might illustrate a relationship) and not because of something else. The observed outcome could just possibly have occurred because of chance or random factors. We are dealing with people, after all. Try as we might, we cannot control for all human factors or those simply down to chance. That's where statistics come in.

Throughout this book you will encounter a whole series of different statistical techniques. Some will be used to explore differences between groups, others examine changes across time, while some tests may simply look at relationships between outcomes. Whatever the focus of that investigation, we need to find some way to measure the likelihood that what we observed did not happen by chance, thus increasing our confidence that it probably occurred because of the factors that we were examining. The statistical analyses in this book have one thing in common: they express the likelihood that the outcome occurred by chance. We will see how to apply that to the many contexts that we are likely to encounter in our studies.

1.1 Take a closer look
Who should use this book?

- This book is aimed at anyone who needs some direction on how to perform statistical analyses.
- The main target audience is probably psychology students and academics, but I hope this book will be equally useful for those working in medicine, social sciences, or even natural sciences.
- Most students are likely to be undergraduates, but this book should also be a valuable resource to postgraduates, doctoral students, lecturers and researchers.
- You may be new to all of this statistics stuff, or an old lag in need of a refresher.
- Whatever your reason for picking up this book, you are most welcome.

How this book is laid out - what you can expect

Introductory chapters: the basics

Chapter 2 will introduce you to some of the basic functions of **SPSS** (a software package designed for analysing research data). In this book, we are using SPSS version 19. You will be shown how to create data sets, how to define the variables that measure the outcome, and how to input those data You will learn how to understand the main functions of SPSS and to navigate the menus. You will see how to investigate, manipulate, code and transform data The statistical chapters will explain how to use SPSS to perform analyses and interpret the outcome.

1.2 Nuts and bolts

I don't have SPSS! Is that a problem?

One of the central features of this book is the way in which it will guide you through using SPSS. The web page resources for this book include SPSS data sets for all of the worked examples and learning exercises. If you are a psychology student at university, it is quite likely that you will have access to the latest version of SPSS during the course of your studies. The licence is renewed each year, so once you leave, the program may stop working. If that happens, you may feel a little stuck. Alternatively, you may not have access to SPSS at all. Either way, it is extraordinarily expensive to buy a single-user copy of SPSS. To address that, all of the data sets are also provided in spreadsheet format, which can be opened in more commonly available programs such as Microsoft Excel.

Chapter 3 explores the concept of normal distribution. This describes how the scores are 'distributed' across a data set, and how that might influence the way in which you can examine those data. We will explore why that is important, and we will learn how to measure and report normal distribution. If the outcome data are not 'normally distributed' we may not be able to rely on them to represent findings. We will also see what we can do if there is a problem with normal distribution.

Chapter 4 examines three ways in which we can measure the impact of our results: statistical significance, effect size and power. We will not explore what those concepts mean here, as that would involve exposing you to factors that you have not learned yet. Most importantly, we will learn about how probability is used in statistics to express the likelihood that an observed outcome happened due to chance factors. We will discuss effect size and power briefly a little later in this chapter.

Chapter 5 provides an overview of experimental methods and guidance on how to choose the correct statistical test. We will learn how to understand and interpret the key factors that determine which procedure we can perform. Using that information, we will explore an overview of the statistical tests included in this book, so that we can put all of it into context.

1.3 Take a closer look

Icons

A common feature throughout the chapters in this book relates to the use of 'boxes'. This 'Take a closer look' box will be employed to explore aspects of what you have just learned in a little more detail, or will summarise the main points that have just been made.

Nuts and bolts 

Within the chapter text, you should find all you need to know to perform a test. However, it is important that you also learn about conceptual issues. You can do the tests without knowing such things, but it is recommended that you read these 'Nuts and bolts' sections. The aim is to take you beyond the basic stuff and develop points a little further.

Calculating outcomes manually 

In all of the statistical chapters, you will be shown how to run a test in SPSS. For most readers, this will be sufficient. However, some of you may want to see how the calculations are performed manually. In some cases your tutor will expect you to be able to do this. To account for those situations, most of the statistical tests performed in SPSS will also be run manually. These mathematical sections will be indicated by this calculator icon. While these sections are optional, I urge you to give them a go – you can learn so much more about a test by taking it apart with maths. Microsoft Excel spreadsheets are provided to help with this.

Statistical chapters (6–21)

Each of the statistical chapters presents the purpose of that procedure, the theory and rationale for the test, the assumptions made about its use, and the restrictions of using the measure. In many cases we will explore how to calculate the test manually, using mathematical examples. Before learning how to perform the test in SPSS, you will see how to set up the variables in the data set and how to enter the data. You will then be guided gently through data entry and analysis with a series of screenshots and clear instructions. You will learn about what the output means and how to interpret the statistics (often with use of colour to highlight the important bits). You will be shown how to report the outcome appropriately, including graphical displays and correct presentation of statistical terminology. You will also be able to read about some examples of how those tests have been reported in published studies, to give you a feel for their application in the real world (and sometimes how not to do it). Finally, you have the opportunity to practise running the tests for yourself with a series of extended learning exercises.

Statistical chapter features

The format of the statistical chapters has been standardised to help give you a better understanding of each test. Certain features will be common across the chapters.

Learning objectives

At the start of each chapter you will be given an overview of what you can expect to learn.

Research question

Throughout each chapter, a single research theme will be used to illustrate each statistical test. This will help maintain some consistency and you will get a better feel for what that procedure is intended to measure.

Theory and rationale

In order to use a test effectively, it is important that you understand why it is appropriate for the given context. You will learn about the theoretical assumptions about the test and the key factors that we need to address. Much of this will focus on the arguments we explore in Chapter 5, relating to the nature of the variables that you are exploring. Sometimes you will be shown how the test compares to other statistical procedures. This will help you put the current test into context, and will give you a better understanding of what it does differently to the others.

Assumptions and restrictions

Related to the last section, each test will come with a set of assumptions that determines when it can be legitimately used. Often this will relate to factors that we explore in Chapters 3 and 5 regarding whether the data are normally distributed and the nature of the data being measured. We will explore the importance of those assumptions and what to do if they are violated.

Performing manual calculations

Although a main feature of this book focuses on the use of SPSS, wherever possible there will be instructions about how to calculate the outcomes manually. There are several reasons for doing this. As we saw earlier, witnessing how to explore the outcome using maths and formulae can reinforce our understanding of the analyses. Also, some of you simply may not have SPSS. Most of these calculations are provided just prior to the SPSS instructions. However, some are a little more complex, so they are safely tucked away at the end of the chapter to protect the faint-hearted, or those of a more nervous disposition. Where appropriate, those calculations are supported by a Microsoft Excel spreadsheet that is provided on the web page for this book. These could also be used as a template to analyse other Excel-formatted data sets (such as those provided for the learning exercises). In some cases, those data can also be used to perform the complete statistical test in Microsoft Excel.

Creating the SPSS data set

Many statistical books show you how to perform a test in SPSS; this book is quite unique in the way that it shows you how to set up the data set in the first place. Data analysis can be so much easier if we create data sets that are appropriate for the type of analysis that we need to conduct. Using procedures that we learned in Chapter 2, we will explore the best way to create a data set, suitable for your analysis.

Conducting tests in SPSS

Each statistical chapter includes full instructions about how to perform the test using SPSS. These include easy-to-follow boxes that will guide you on how to undertake each stage of the statistical analyses. An example is shown in Figure 1.1.

Open the SPSS data set Sleep 2

Select **Analyze → Compare Means → Independent-Samples T Test...** (in new window) transfer Sleep Quality to **Test Variable List** → transfer **HADS cut-off depression** to **Grouping Variable** → click **Define Groups** button → (in the window) enter **1** in box for **Group 1** → enter **2** in box for **Group 2** → click **Continue** → click **OK**

Figure 1.1 An example of SPSS procedure instructions

You will also be shown screenshots of the SPSS displays that you will encounter during the process. You can refer to these to ensure that you are using the recommended method. An example of this is shown in Figure 1.2.

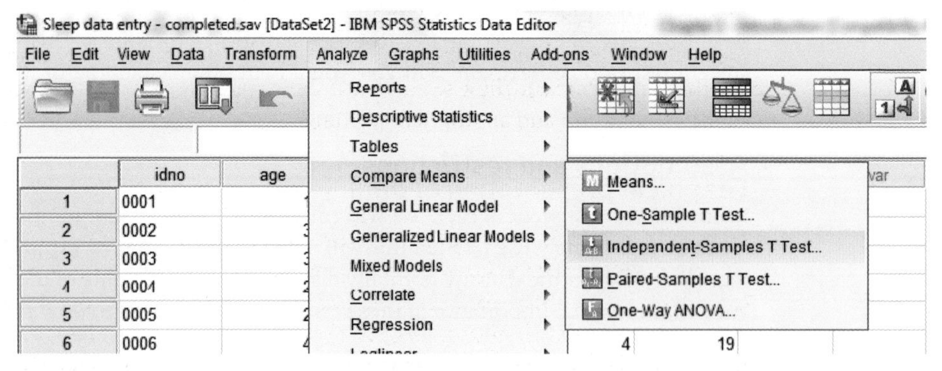

Figure 1.2 An example of SPSS screenshot

Interpretation of output

Once each test has been run, you will be taken through the SPSS output more thoroughly, so that you understand what each table of results shows and what the implications are. In some cases, this output is relatively easy to follow – there may be just one line of data to read. In other cases, there may be several lines of data, some of which are not actually that important. Where there may be some doubt about what part of the output to read, colour and font will be used to illustrate where you should be focusing your attention. An example of this is shown in Figure 1.3.

Tests of between-subject effects

Dependent variable: HADS anxiety score

Source	Type III sum of squares	df	Mean square	F	Sig.	Partial eta squared
Corrected model	1102.107[a]	5	220.421	16.688	.000	.476
Intercept	5193.219	1	5193.219	393.177	.000	.810
HADSDbase	336.365	2	168.183	**12.733**	**.000**	.217
hxinsom	74.793	1	74.793	**5.663**	**.019**	.058
HADSDbase * hxinsom	61.310	2	30.655	2.321	.104	.048
Error	1215.169	92	13.208			
Total	8799.000	98				
Corrected total	2317.276	97				

a. R squared = .476 (adjusted R squared = .447)

Figure 1.3 An example of annotated SPSS output

Effect size and power

In addition to reporting statistics, it is important that you state the effect size and power of the outcome. You will learn more about what that means in Chapter 4. Briefly, effect size represents the actual magnitude of an observed difference or relationship; power describes the probability that we will correctly find those effects.

Writing up results

Once you have performed the statistical analyses (and examined effect size and power where appropriate), you need to know how to write up these results. It is important that this is done in a standardised fashion. In most cases you will be expected to follow the guidelines dictated by the British Psychological Society (BPS) (although those rules will vary if you are presenting data in other subject areas). These sections will show you how to report the data using tables, graphs, statistical notation and appropriate wording.

Graphical presentation of data

You will be shown how to draw graphs using the functions available in SPSS, and you will learn when it is appropriate to use them. Drawing graphs with SPSS is much easier than it used to be (compared with earlier versions of the program). In many cases, you can simply drag the variables that you need to measure into a display window and manipulate the type of graph you need. In other cases, you will need to use the menu functions to draw the graphs.

Research example

To illustrate the test you have just examined, it might help to see how this has been applied in real-world research. At the end of each chapter you will find a summary of a recently published research article that uses the relevant statistical tests in its analyses. The papers focus on topics that may well be related to your own research. While those overviews should be very useful in extending your understanding, you are encouraged to read the full version of that paper. For copyright reasons, we cannot simply give these to you. However, each paper is provided with a link that you can enter into an internet browser. In most cases this will be the '**DOI code**'. These initials stand for 'Digital Object Identifier'. It is an internationally recognised unique character string that locates electronic documents. Most published articles provide the DOI in the document description. Leading international professional bodies, such as the BPS, dictate that the DOI should be stated in reference lists. A typical DOI might be http://dx.doi.org/10.1080/07420520601085925 (they all start with 'http://dx.doi.org/').

Once you enter the DOI into an internet browser, you are taken directly to the publisher's web page, where you will be given more details about the article, usually including the Abstract (a summary of that paper). If you want to access the full article you will have a series of choices. If you, or your educational institution/employer, have a subscription with that publisher you can download a PDF copy. If not, you can opt to buy a copy. Alternatively, you can give those details to your institutional librarian and ask them to get you a copy. Wherever possible, the DOI will be provided alongside the citation details for the summarised paper; when that is not available an alternative web link will be presented.

Extended learning task

To reinforce your learning, it is useful to undertake some exercises so that you can put this into practice. You will be asked to manipulate a data set according to the instructions you would have learned earlier in the chapter. You will find these extended learning examples at the end of each chapter (or in some cases within the chapter when there are several statistical tests examined). You will be able to check your answers on the web page for this book.

1.4 Take a closer look
Chapter layout

Each statistical chapter will follow a similar pattern, providing you with consistency throughout. This might help you get a better feel of what to expect each time. A typical running order is shown below:

- Learning objectives
- Research question
- Theory and rationale
- Assumptions and restrictions
- Performing manual calculations
- Setting up the data set in SPSS
- Conducting test in SPSS
- Interpretation of output
- Effect size and power
- Writing up results
- Presenting data graphically
- Chapter summary
- Research example
- Extended learning task

Online resources

A series of additional resources is provided on the web page for this book, which you can access at www.pearsoned.co.uk/mayers. These resources are designed to supplement and extend your learning. The following list provides a guide to what can expect to find there:

- Data sets:
  - to be used with worked examples and learning exercises
  - available in SPSS and Excel formats.
- Multiple-choice revision tests.
- Answers to all learning exercises.
- Excel spreadsheets for manual calculations of statistical analyses.
- Supplementary guides to SPSS (tasks not covered in the book).
- More extensive versions of distribution tables.

2 SPSS – THE BASICS

Learning objectives

By the end of this chapter you should be able to:

- Understand the way in which data and variables can be viewed in SPSS
- Recognise how to define variables and set parameters
- Enter data into SPSS and navigate menus
  - How to use them to enhance, manipulate and alter data
  - How to transform, recode, weight and select data
- Understand basic concepts regarding syntax

Introduction

SPSS® is one of the most powerful statistical programs available, and probably the most popular. Originally called the 'Statistical Package for the Social Sciences', SPSS has evolved to be much more than a program for social scientists, but the acronym remains. Many published studies, in a very wide variety of research fields, include statistics produced with SPSS. To the uninitiated, the program appears daunting and is associated with the horrors of maths and statistics. However, it need not be that scary; SPSS can be easy to learn and manipulate. Most of the tasks are available at the press of a button, and it is a far cry from the days when even the most basic function had to be activated by using programming code. The trick is learning what button to press. Many books report on how to use the functions, but very few provide even the most basic understanding. Some of you may be experienced enough not to need this chapter, in which case, you can happily pass on to the next chapter. However, even if you have been using SPSS for several years, you may benefit from learning about some of the newer functions now available.

We will start by looking at some of the most basic functions of SPSS, such as how to set up new data sets and how to use the main menus. To create a data set, we need to define **variables** – we will learn how to set the parameters according to the type of test we need to perform. We will see that there are two ways that we can view a data set: a '**variable view**', where we define those variable parameters and a '**data view**', where we enter data and manipulate them. Once we have created a data set, it would be useful if we learned how to use important menu functions such as 'Save' and 'Edit'. Then we will proceed to some slightly more advanced stuff. Now, it's quite likely that some bits about data editing and manipulation will be beyond you at this stage, particularly if you are new to statistics. If that happens, don't worry. This chapter is not designed to be read in one go; you can return to it again later when you have learned more about statistical analyses themselves. The rationale for this approach is a simple one: it keeps all of the instructions for performing the main functions in one place. In many cases we will revisit the procedures in later chapters, when they become appropriate. However, it is useful to have the most basic instructions all together. We will not explore the data analysis and graphical functions in this chapter, as it is better that we see how to do that within the relevant statistical chapters. But we will (briefly) consider how SPSS uses '**syntax**' language to perform tasks. You will rarely have to use this programming language, but it may be useful for you to see what it is used for. Throughout this book we will be using SPSS for Windows version 19.

Viewing options in SPSS

One of the first things to note is that there are two editing screens for SPSS (called '**Data Editors**'): 'Variable View' and 'Data View'. Variable View is used to set up the data set parameters (such as variable names, type, labels and constraints). Data View is used to enter and manipulate actual data. An example of each is shown in Figure 2.1 and Figure 2.2. Before you enter any data, you should set up the parameters and limits that define the variables (in Variable View). Once you have those variables set up, you can proceed to enter the data; you will do that via Data View.

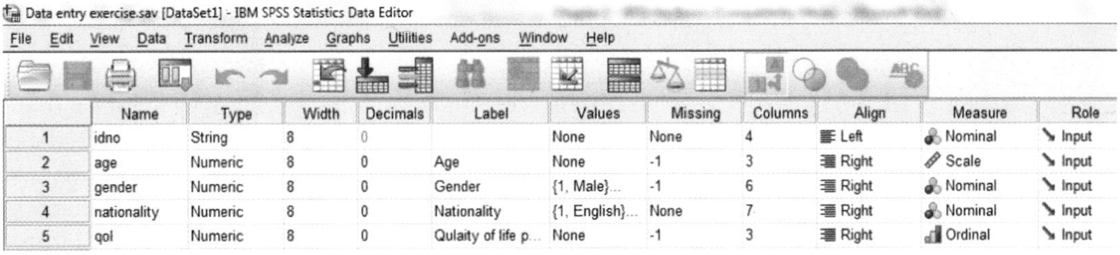

Data entry exercise.sav [DataSet1] - IBM SPSS Statistics Data Editor

File Edit View Data Transform Analyze Graphs Utilities Add-ons Window Help

	Name	Type	Width	Decimals	Label	Values	Missing	Columns	Align	Measure	Role
1	idno	String	8	0		None	None	4	Left	Nominal	Input
2	age	Numeric	8	0	Age	None	-1	3	Right	Scale	Input
3	gender	Numeric	8	0	Gender	{1, Male}...	-1	6	Right	Nominal	Input
4	nationality	Numeric	8	0	Nationality	{1, English}...	None	7	Right	Nominal	Input
5	qol	Numeric	8	0	Qulaity of life p...	None	-1	3	Right	Ordinal	Input

Figure 2.1 SPSS Variable View

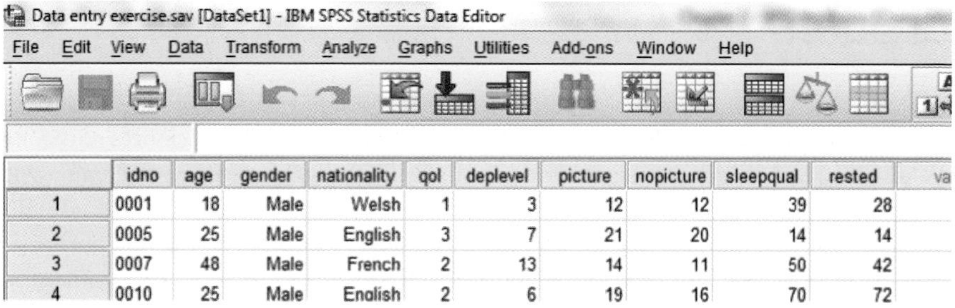

Data entry exercise.sav [DataSet1] - IBM SPSS Statistics Data Editor

File Edit View Data Transform Analyze Graphs Utilities Add-ons Window Help

	idno	age	gender	nationality	qol	deplevel	picture	nopicture	sleepqual	rested
1	0001	18	Male	Welsh	1	3	12	12	39	28
2	0005	25	Male	English	3	7	21	20	14	14
3	0007	48	Male	French	2	13	14	11	50	42
4	0010	25	Male	English	2	6	19	16	70	72

Figure 2.2 SPSS Data View

Variable View is arranged in columns that relate to the parameters that we will set for each variable. Each row relates to a single variable in the data set.

Data View is arranged in columns that show each of the variables included in the data set (these are the same as the rows in Variable View). Each row represents a single participant or case. Now we should see how we define and enter the information, so that we get the information that is displayed in Figures 2.1 and 2.2.

Defining variable parameters

It might help you understand the functions of SPSS by defining some variables and then entering data. To help us, we are going to use a small data set that will examine participants' age, gender, nationality, perceived quality of life and current level of depression. We will also examine how many words the group can recall (with or without a picture prompt). Finally, we will record the participants' perceptions of sleep quality and how well rested they felt when they woke up that morning.

Starting up a new SPSS data file

Before we start, we need to open a new (blank) SPSS data file. When SPSS is open for the first time, you may be presented with a range of screens. The default view (shown in Figure 2.3) requests options of how to proceed:

Open SPSS 19 from your program menu, or click on the SPSS icon.

In this case, we do not need any of those options, so just click on Cancel; a blank window will open (similar to Figures 2.3 or 2.4). On other occasions, you may wish to perform one of the other functions, but we will look at that later. In Figure 2.3, you will notice that there is a tick-box option saying 'Don't show this dialog in the future'. If that has previously been selected, you will not see Figure 2.3 at all; the program will just open straight into a blank window. Once you have opened a new data file, click on the Variable View button (at the bottom of the page). An example of a brand new Variable View page is shown in Figure 2.4.

Defining variable parameters: rules and limits

When we open Variable View we see a range of parameter descriptions across the column headings. Before we define those we should explore what each of the descriptors means:

Name: Give your variable a name that is relevant to what it measures, but try to keep it short. The limit for SPSS 19.0 is 64 characters, but it is advisable to make it more manageable (you can always provide a fuller description in the 'Label' column). The name should start with a letter; subsequent characters can be any combination of letters, numbers and almost any other character. There are some exceptions, and you will get an error message should you select any of those. You cannot use blanks: 'age of participant' is not acceptable, but 'age_of_participant' is fine. This field is not case sensitive.

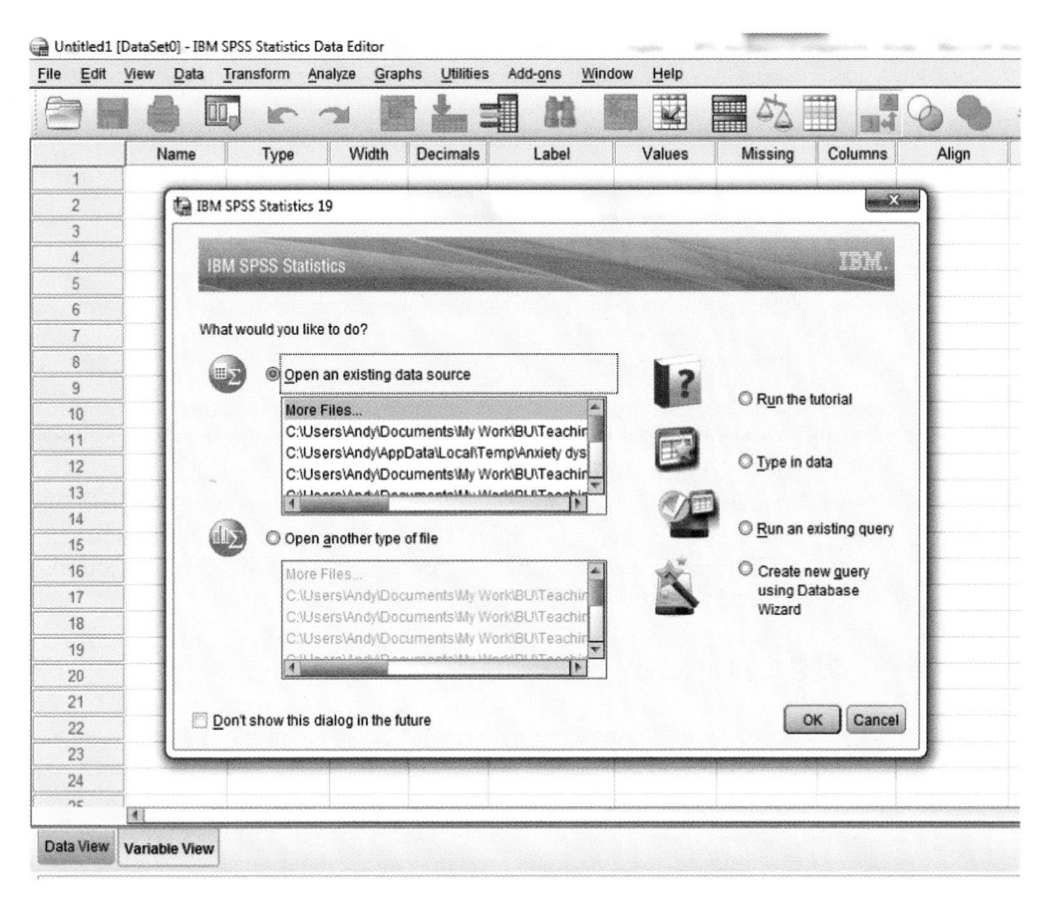

Figure 2.3 SPSS opening view

Figure 2.4 Blank Variable View

Type: If you click on the cell for this parameter you will be presented with a row of dots (. . .). Click on that and you will see a list of options (see Figure 2.4). The default is 'Numeric', which you will use most often. The most likely alternative is 'String', which you could use for participant identification. 'Numeric' can be used even when the variable is categorical, such as gender. This is because 'numbers' can be allocated to represent the groups of the variable (see 'Values').

Width: It is unlikely that you will need to change the default on this, unless you expect to require more digits than the default (eight characters). You may need to extend that if you want very large numbers, or if you need to display numbers with several decimal places (see below).

Decimals: Setting decimals applies only when using numeric data. You can use this to determine how many decimal places you show (in the data set). The default setting is for two decimal places. For something like age, you may want to change this to '0' (use the arrows to the right of the cell to make changes). For more specific data (such as reaction times) you may want any number of decimal places. This option has no effect on the number of decimal places shown in the results.

Label: This is where you can enter something more specific about the nature of the variable, so you can include a longer definition (and there are no limits). For instance, the 'Label' could be 'Depression scores at baseline', while the 'Name' parameter might be 'depbase'. Always put something here, as that label is shown in some parts of the SPSS **output**.

Values: As we will see in later chapters, a categorical variable is one that measures groups (such as gender). So SPSS understands that we are dealing with categorical variables, we need to allocate 'numbers' to represent those groups. For example, we cannot expect SPSS to differentiate between the words 'male' and 'female', but can use the values facility to indicate that '1' represents male and '2' is female. If there are no groups, you would leave the Values cell as 'None' (the default). If you do have groups, you must set these values (you will see how later).

Missing: It is always worth considering how you will handle missing data. If there is a response absent from one of your variables, SPSS will count that empty cell as '0'. This will provide a false outcome. For example, the mean (average) score is based on the sum of scores divided by the number of scores. If one of those scores is incorrectly counted as 0, the mean score will be inaccurate. You should include '0' only if it actually represents a zero score. If the data are missing, you can define a specific 'missing variable value'. This will instruct SPSS to skip that cell (a mean score will be based on the remaining values). The missing 'value' indicator must be sensible; it must not be in the range of numbers you might be expecting (otherwise a real number might be ignored). The same applies to numbers used to define groups. A good choice for missing values is -1: it should cover most scenarios. We will see how to do this later.

Columns: This facility determines the width of the column reserved for that variable in the Data View. So long as you can see the full range of digits in the cell, it does not really matter. Set this to be your preference.

Align: Data can appear to the left of a cell, the middle, or to the right – the choice is yours.

Measure: You need to define what type of variable you are measuring. Click on the arrow ▼ in the Measure cell. The options for Numeric data are **Scale**, **Ordinal** or **Nominal**. For String the options are Ordinal or Nominal. Select the appropriate one from the pull-down list.

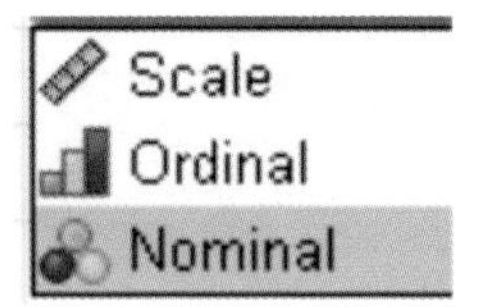

The Scale measure is ruler – representing a range of scores.

The Ordinal measure is step – representing an order of groups.

The Nominal measure is distinct circles – representing categories.

With numeric data, 'Scale' refers to scores such as age, income or numbers that represent ranges and magnitude. These numbers are what we would normally categorise as interval or ratio data. 'Ordinal' data are also 'numerical' but only in the sense that the number represents a range of abstract groups; you will typically find ordinal data in attitude scale (where 1 = strongly agree, through to 5 = strongly disagree). You will learn more about interval, ratio and ordinal data in Chapter 5, so don't worry if that's all a bit confusing right now. 'Nominal' refers to distinct categories such as gender (male or female).

Role: Just use 'Input' for now; you can learn about the rest another time.

Creating new variable parameters

At this stage it would be useful to set up an example set of variables. You will recall that we are creating a data set that examines the participants' age, gender (male or female), nationality (English, Welsh or French), perceived quality of life, current level of depression, how many words they can recall (with and without a picture prompt), perceived sleep quality and how rested the participants felt when they woke up. We will also have a variable called 'participant identifier' (the usefulness of that will become apparent later). Table 2.1 shows the information we are about to enter into our new SPSS data set.

Table 2.1 Data set

SPSS variable									
idno	**age**	**gender**	**Nationality**	**qol**	**deplevel**	**picture**	**nopicture**	**sleepqual**	**rested**
0001	18	Male	Welsh	1	3	12	12	39	28
0002	38	Female	English	4	18	21	20	14	14
0003	30	Female	French	4	?	14	11	50	42
0004	22	Female	English	5	20	19	16	70	72
0005	25	Male	French	3	7	12	12	63	62
0006	40	Female	Welsh	4	19	11	11	39	39
0007	48	Male	English	2	13	21	22	59	39
0008	35	Female	Welsh	5	20	24	20	55	54
0009	45	Female	Welsh	3	10	17	21	39	42
0010	25	Male	English	2	6	18	12	57	60
0011	50	Male	French	5	24	18	11	59	57
0012	35	Male	English	2	11	18	9	74	78
0013	?	Female	Welsh	4	28	14	11	17	27
0014	32	Female	English	5	25	19	14	24	24
0015	40	Male	Welsh	1	12	23	18	50	47
0016	53	Female	Welsh	4	23	15	15	57	61
0017	35	Male	French	3	16	21	12	57	46
0018	30	Male	English	2	13	24	19	61	58
0019	20	Female	French	4	16	17	14	31	24

2.1 Nuts and bolts
SPSS instruction boxes

We will be using instruction boxes throughout this book to show how we perform a function in SPSS. To maintain consistency, fonts will be employed to indicate a specific part of the process:
Black bold: this represents a command or menu options shown within the data window.
Green bold: this indicates the item to select from a list within the menu or variable.
Blue bold: this refers to words and/or numbers that you need to type into a field.

We will now set up the parameters for the variables in this data set. Remember we need a new row for each variable. Go to the blank Variable View window for the new data set.

Participant identifier (Row 1):

We will start with a 'variable' that simply states the participant's identification number. This can be useful for cross-referencing manual files.

In **Name** type **idno** → in **Type** click on the dots... (you will be presented with a new window as shown in Figure 2.5) → select **String** radio button → everything else in this row can remain as default

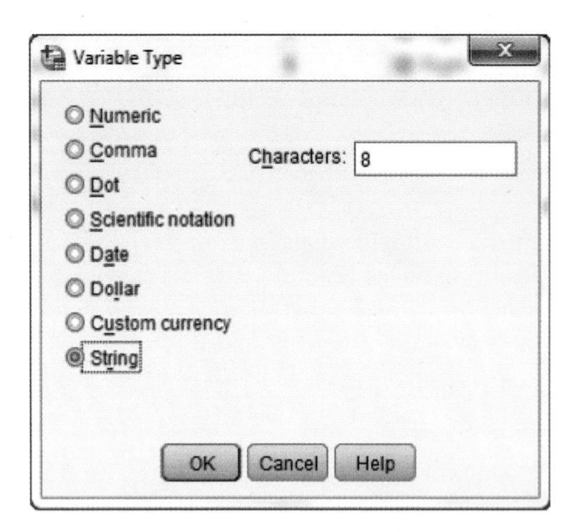

Figure 2.5 Setting type

Age (Row 2):

In **Name** type **age** → set **Type** to **Numeric** → ignore **Width** → change **Decimals** to **0** → in **Label** type **Age** → ignore **Values**
To set the parameter for **Missing** values, click on that cell and then the dots ... (you will be presented with a new window, as shown in Figure 2.6) → select **Discrete missing values** radio button → type **-1** in first box → click **OK** → back in original window, ignore **Columns** → ignore **Align** → click **Measure** → click arrow ▼ → select **Scale**

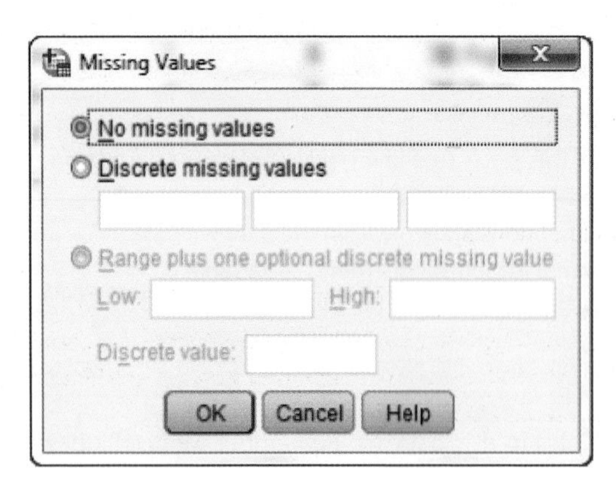

Figure 2.6 Missing values

Gender (Row 3):

In **Name** type **gender** → set **Type** to Numeric → ignore **Width** → change **Decimals** to 0 → in **Label** type **Gender**
Gender is a 'group' (categorical) variable, so we have to set some **Values** → click on that cell and then the dots ... (you will be presented with a new window, as shown in Figure 2.7) → in **Value** type **1** → in **Label** type **Male** → click **Add** → in **Value** type **2** → in **Label** type **Female** → click **Add** → click **OK** → back in original window, set **Missing** to **−1** → ignore **Columns** → ignore **Align** → set **Measure** to Nominal

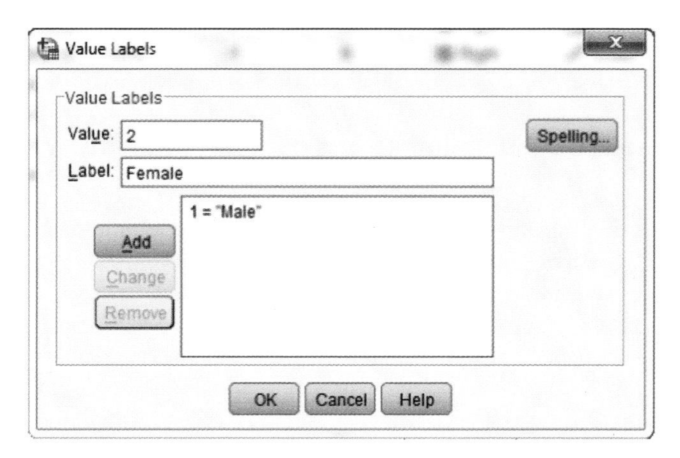

Figure 2.7 SPSS value labels

Nationality (Row 4):

In **Name** type **nationality** → set **Type** to Numeric → ignore **Width** → change **Decimals** to 0 → in Label type **Nationality** → set Values as **1 = English**, **2 = Welsh**, and **3 = French** respectively (you saw how just now) → set **Missing** to **−1** → ignore **Columns** → ignore **Align** → set Measure to Nominal

Quality of life perception (Row 5):

In **Name** type **qol** → set **Type** to Numeric → ignore **Width** → change **Decimals** to 0 → in **Label** type **Quality of life perception** → ignore **Values** → set **Missing** to **−1** → ignore **Columns** → ignore **Align** → set **Measure** to Ordinal

Current level of depression: (Row 6):

In **Name** type **deplevel** → set **Type** to Numeric → ignore **Width** → change **Decimals** to 0 → in **Label** type **Current level of depression** → ignore **Values** → set **Missing** to **−1** → ignore **Columns** → ignore **Align** → set **Measure** to Scale

Picture: (Row 7):

In **Name** type **picture** → set **Type** to Numeric → ignore **Width** → change **Decimals** to 0 → in **Label** type **Words recalled with picture** → ignore **Values** → set **Missing** to −**1** → ignore **Columns** → ignore **Align** → set **Measure** to Scale

No picture: (Row 8):

In **Name** type **nopicture** → set **Type** to Numeric → ignore **Width** → change **Decimals** to 0 → in **Label** type **Words recalled without picture** → ignore **Values** → set **Missing** to −**1** → ignore **Columns** → ignore **Align** → set **Measure** to Scale

Sleep quality: (Row 9):

In **Name** type **sleepqual** → set **Type** to Numeric → ignore **Width** → change **Decimals** to 0 → in **Label** type **Sleep quality** → ignore **Values** → set **Missing** to −**1** → ignore **Columns** → ignore **Align** → set **Measure** to Scale

Rested: (Row 10):

In **Name** type **rested** → set **Type** to Numeric → ignore **Width** → change **Decimals** to 0 → in **Label** type Rested on waking → ignore **Values** → set **Missing** to −**1** → ignore **Columns** → ignore **Align** → set **Measure** to **Scale**

Entering data

To start entering data, click on the Data View tab and you will be presented with a window similar to the one in Figure 2.8. Remember, each row in Data View will represent a single participant.

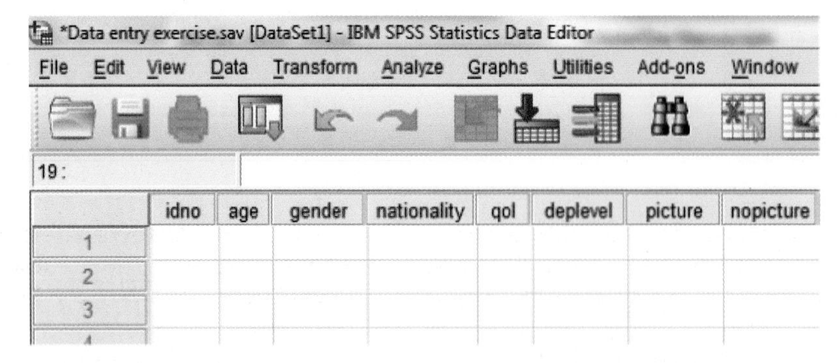

Figure 2.8 Blank Data View

To enter the data, we will use the information from Table 2.1. To get some practice you should enter these data, following the instructions shown below (note that there are some data 'missing').

Using the SPSS data set that we have just created, enter the following information:

Row 1: In **idno** type **0001** → in **age** type **18** → in **gender** type **1** → in **nationality** type **2** → in **qol** type**1** → in **deplevel** type **3** → in **picture** type **12** → in **nopicture** type **12** → in **sleepqual** type **39** → in **rested** type **28**

Row 2: In **idno** type **0002** → in **age** type **38** → in **gender** type **2** → in **nationality** type **1** → in **qol** type **4** → in **deplevel** type **18** → in **picture** type **21** → in **nopicture** type **20** → in **sleepqual** type **14** → in **rested** type **14**

Row 3: In **idno** type **0003** → in **age** type **30** → in **gender** type **2** → in **nationality** type **3** → in **qol** type **4** → in **deplevel** type **-1** (the 'depression score' is missing; so we enter the 'missing value' indicator instead) → in **picture** type **14** → in **nopicture** type **11** → in **sleepqual** type **50** → in **rested** type **42** . . . **and so on**

Perhaps you would like to enter the remaining data (from Table 2.1); there will some further exercises at the end of this chapter.

SPSS menus (and icons)

Now we have created our first data set, we should explore how we use the 'menus' (refer to Figure 2.8 to see the range of menu headings). You will need to use only some of the functions found within these menus, so we will look at the most commonly used. In some cases, a menu function has an icon associated with it (located at the top of the view window). You can click on an icon to save time going through the menus; we look at the most useful of those icons (these are displayed below the menu headings, as shown in Figure 2.9). There are actually many more icons that could be included. You can add and remove the icons that are displayed, but we will not look at how to do that in this section. You can see how to do this in the supplementary facilities supplied in the web features associated with this chapter. The menu structure is the same in Data View and Variable View screens.

Figure 2.9 SPSS menus and icons

File menu

2.2 Nuts and bolts
SPSS files

SPSS uses two main file types: one for data sets (these are illustrated by files that have the extension '.sav') and one for saving the output (the tables of outcome that report the result of a procedure) - these are indicated by files that have the extension '.spv'. A file extension is the letters you see after the final dot in a filename. It determines which program will open the file, and what type of file it is within that program. For example, word-processed files often have the file extension '.doc'. There are other file types in SPSS, such as those used for the syntax programming language. However, most of the time you will use only .sav and .spv files.

When the 'File' menu is selected, a series of options will appear (see Figure 2.10). The file menu is pretty much the same as you will find in most popular software programs, with some exceptions. There are several functions available here. Some of these are more advanced than we need, so we will focus on those that you are most likely to use for now.

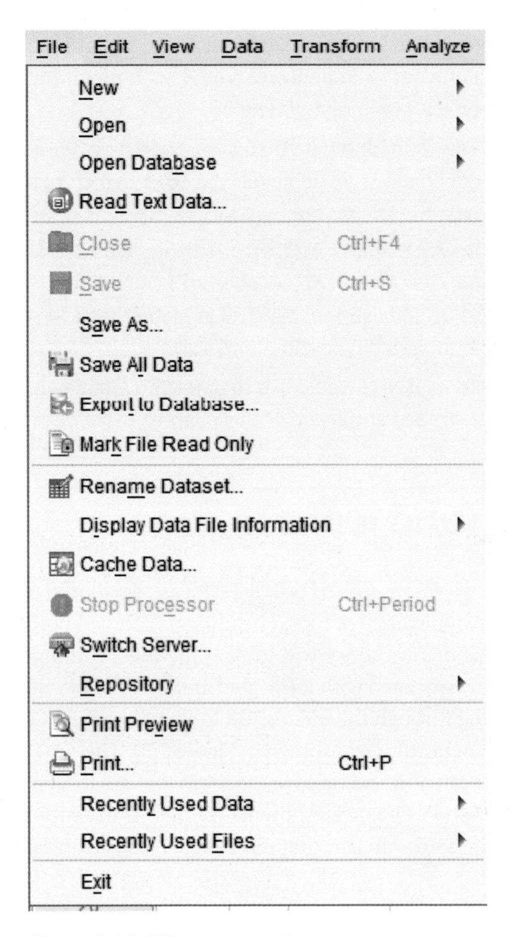

Figure 2.10 File menu options

New: Use this to start a new data file. It is most likely that this will be a new data set, in which case you would follow the route: (click on) **File ➔ New ➔ Data**. However, you might equally choose to start a new **Syntax** or **Output** file.

Open:

Use this to open an existing file, perhaps one that you have worked on previously. If you want to open a data file, perform **File ➔ Open ➔ Data**. To open a saved output file, perform **File ➔ Open ➔ Output**. There is an icon associated with this function, which you can use just by clicking on it (saving a little time from selecting the menu route):

You can also open a file by clicking on it directly from your own folders (see Figure 2.11).

Save:

It is good practice to save data sets and output files frequently, not just when you have finished. If your computer crashes, you might lose everything. To save the file, select **File ➔ Save** (regardless of whether you are saving a data set or an output file). Alternatively, you can click on the icon shown here. If the file has not been saved before, you will be asked to create a name and indicate where you want the file saved. If it is an existing file, it will save any new changes.

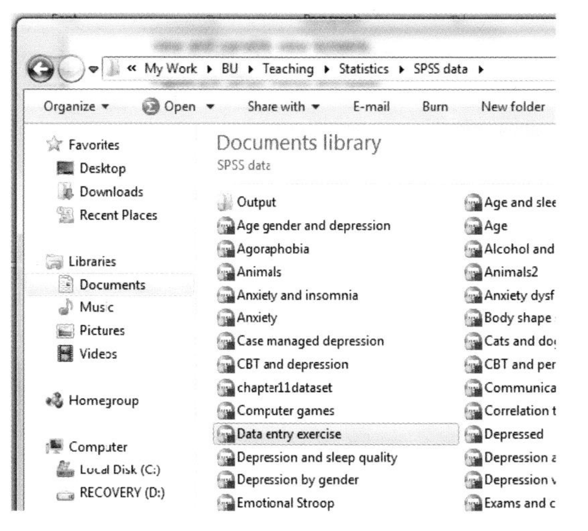

Figure 2.11 Opening a file from general folders

Double click on the required file and it will open in the SPSS program.

Save As: If you make changes to a file but want to keep the original file, use this function to save the changed version to a different file. Select **File ➔ Save As** (regardless of whether you are saving a data set or an output file). Do *not* use the 'file save' function: the details in the file prior to the changes will be overwritten.

Mark File Read Only: You can protect your file from any further changes being made; new changes can be made to a new file using 'Save As'. Select **File ➔ Mark File Read Only** (you will be reminded to save current unsaved changes).

Print Preview: You may want to see what a printed copy of your file will look like, without actually printing it (for example, you may want to change margins to make it fit better) – this saves printing costs. Select **File ➔ Print Preview.**

Print: If you are happy to print the file, send this to a printer of your choice by selecting **File ➔ Print.** You will be given a list of printers that this can be sent to. If you use the 'Print' icon the print will be automatically sent to your default printer.

Exit: As the name implies, this closes down the file. You will be warned if data have not been saved. You also get a warning if the file is the last SPSS data set still open (it closes the whole program). You can also click on the cross in the top right-hand corner to close the file. Make sure you save before you close anything.

Recently Used Data: This provides a similar function to 'Open' but will locate the most recently used data sets. This is often quicker because, using 'Open', you may need to trawl through several folders before you find the file you are after. However, this function remembers file names only. If you have moved the file to another folder since it was last used, you will get an error message. Select **File ➔ Recently Used Data** and choose the file you want to open.

Recently Used Files: This is the same as 'Recently Used Data' but it locates all other files that are not data sets (output files, for example).

Edit menu

The Edit menu also shares properties with other software programs that you may be more familiar with. When this menu is selected, a number of options are displayed (see Figure 2.12).

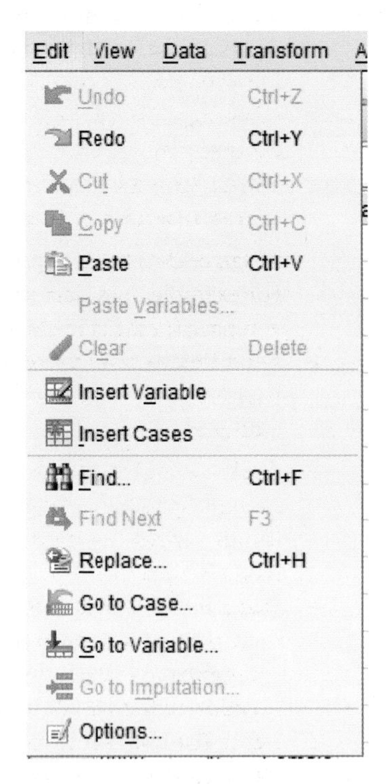

Figure 2.12 Edit menu options

We will explore some of the more common functions here. Where an icon is displayed, this can be selected instead of using the full menu function:

Undo: Sometimes you may enter data incorrectly, or make some other error that you want to 'undo'. Use this function to do that by selecting **Edit → Undo**.

Redo: Having undone what you believed to be incorrect, you may decide it was OK after all and want to put the information back in again. You can redo what was undone by selecting **Edit → Redo**.

Cut: If you want to move information from a current cell and put it somewhere else, you need to use this 'Cut' facility. It's rather like deleting, but the information is saved in a memory cache until you find somewhere else to put it (see 'Paste'). To do this, select **Edit → Cut.**

Copy: If you want to copy information from the current cell (to somewhere else) but also keep the current information where it is, you need this 'Copy' function. To do this, select **Edit → Copy**. You will need the 'Paste' function to complete the task.

Paste: Use this to paste information that has been cut or copied from somewhere else into a cell by selecting **Edit → Paste.**

Insert Variable: You can use this function to insert a new variable in Variable View. In many cases, we would simply start a new row (rather like we did earlier). However, sometimes you might decide to include a new variable but would like to have it placed next to an existing one (perhaps because it measures something similar). To do this, go to Variable View and click on the row above which you want to insert the new variable. Then select **Edit ➜ Insert Variable**. You would then need to set the parameters as you have been shown.

Insert Case:

You can use this function to insert a new 'case'. In most data sets, a case will be a participant. It is quite likely that it will not matter what order you enter data, but sometimes you may want to keep similar participants together (such as all of the depressed people in one place). In that scenario, you may want to insert a participant into a specific row of your data set. To do that, go to Data View and click on the row above which you want to insert the new case. Then select **Edit ➜ Insert Case**. You can then enter the data for your new participant.

Find:

In larger datasets it can be time consuming to look for specific bits of data. For example, in a data set of 1,000 people you may want to find cases where you have (perhaps mistakenly) used '99' to indicate a missing variable. You can select **Edit ➜ Find** to locate the first example of 99 in your data set. Once you have found the first example, you can use the 'Next' button to locate subsequent examples.

Replace:

Having found the items you are looking for, you may wish to replace them. For example, you have originally chosen to use 99 as your missing value indicator for all variables, including age. Later, you discover that one of your participants is aged 99! If you kept 99 as the missing variable it would not count that person. So you decide to change the missing value indicator to – 1. If there were 50 missing values in all variables across the data set, it would take some time to change them and you might miss some. However, the 'Replace' function will do that for you all at once. Go to **Edit ➜ Replace ➜** enter 99 in the **Find** box **➜** enter – 1 in the **Replace with** box **➜** click on **Replace All**. However, do be careful that there are not other (valid) cases of 99 – you might replace true data with an invalid missing value. If you are not sure, use the **Find Next** button instead of 'Replace All'.

Options:

This function enables you to change a whole series of default options, including the font display, how tables are presented, how output is displayed, and so on. Much of this is entirely optional and will reflect your own preferences.

View menu

The View menu offers fewer features than the others, but those that are there are very useful.

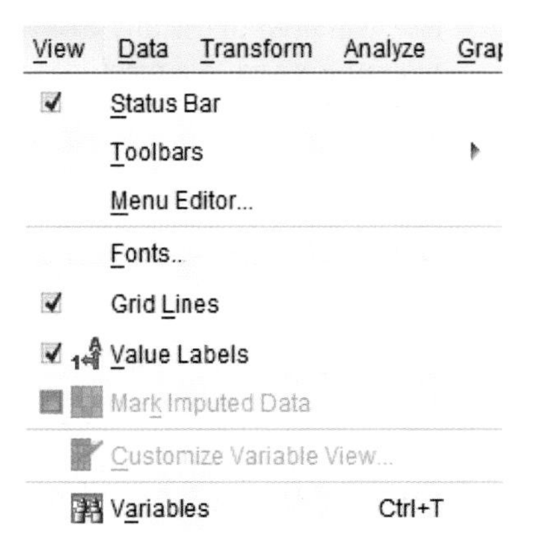

Figure 2.13 View menu options

Three functions can be selected via tick boxes:

Status bar: This function confirms current functions at the foot of the display window. This can be quite reassuring that the process is working, so it is a good idea to leave this ticked.

Grid lines: This function allows you to show grid lines between cells, or to remove them; it is entirely optional.

Value Labels: This is a very useful function. Earlier, we saw how to set up categorical variables that represent groups. For example, we created a Gender variable and used codes of 1 and 2 for 'male' or 'female' respectively. When we display the data set, we can choose whether to show the numbers (such as 1, 2) or the value labels (such as male, female) by ticking that box. Alternatively, you can click on the icon in the toolbar – if you are currently showing numbers it will switch to value labels, and vice versa.

Other functions are selected by clicking on that option and following additional menus:

Toolbars: You can use this function to choose which icons to include on the toolbar. Select **View → Toolbars → Customize** (a new window opens) → click **Edit**. From the operations window you can select a menu and choose which icons you can drag onto the toolbar.

Fonts: You can use this facility to change the way in which fonts are displayed in the data set. This is entirely your choice. Select **View → Fonts** if you want to change anything.

The next three menus are used to manipulate data. To fully illustrate these functions, we will undertake some of the procedures as well as explain what the menu aims to do.

Data menu

The data menu examines and arranges the data set so that specific information can be reported about those data. In some cases this has an impact on the way in which data are subsequently analysed. There are many functions in this menu, so we will focus on those that are probably most useful to you for the moment.

We can perform these functions on the data set that we created earlier. If you want to see the completed data set, you will find it in the online resources for this book. The file is called '**Data entry exercise**'.

Define Variable Properties: This function confirms how a variable has been set up and reports basic outcomes, such as the number of cases meeting a certain value. To perform this task, select **Data → Define Variable Properties.**

Copy Data Properties: This function enables you to copy the properties of one variable onto another by selecting **Data → Copy Data Properties.**

Sort cases: This useful facility allows you to 'sort' one of the columns in the data set in ascending or descending order. For example, using the data set we created, we could sort the 'Current level of depression' column from lowest score to highest score. To illustrate this important function, we will perform that task without data:

Using the SPSS data set Data entry exercise

Select **Data** → **Sort Cases** (see Figure 2.14) → (in new window) transfer **Current level of depression** to the **Sort by** window (by clicking on the arrow, or by dragging the variable to that window) → select radio button by **Ascending** → click **OK** (as shown in Figure 2.15).

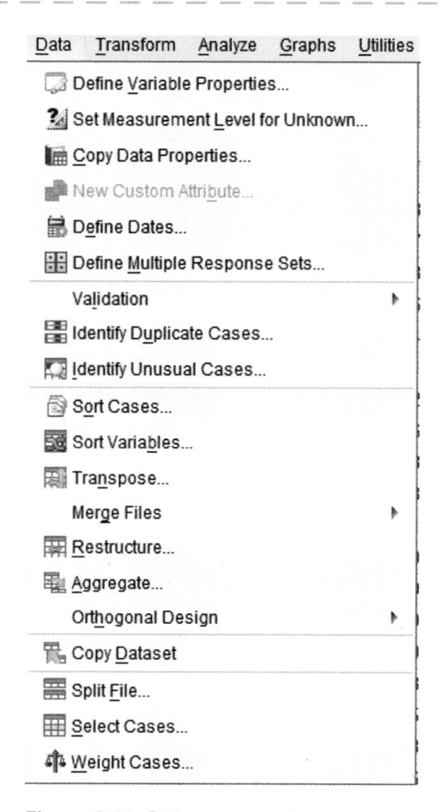

Figure 2.14 Data menu options

Return to the data set and you will notice the column for 'Current level of depression' is now in order, from the lowest to the highest.

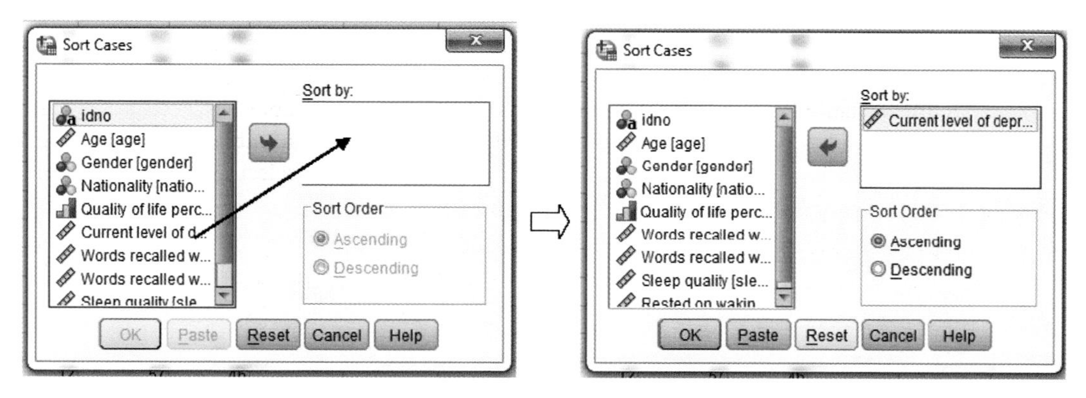

Figure 2.15 Sort cases function

Split File: This is another extremely useful facility. It enables you to split the data set according to one of the (categorical variable) groups. This can be used to report outcomes across remaining variables but separately in respect of those groups.

We will use this function in very important analyses later in the book, notably for multi-factorial ANOVAs (Chapters 11 and 13). However, we can illustrate this function with a simple example now. In the data set that we created, we have two variables that measure 'word recall'. These measure how words can be recalled by the participants when they are given a picture prompt to aid recall (' Words recalled with picture') and when they are not ('Words recalled without picture'). If we examine our entire sample across those two variables, we can compare the outcomes. We call that a within-group study (we will encounter these often throughout the book). We might find that people recall more when they are given the picture prompt. This is all very well, but we might also want to know whether that outcome differs according to gender. We can do this with the split file.

Before we split the file, we should look at some basic outcome regarding the word recall across the group.

		Mean	N	Std. deviation	Std. error mean
Pair 1	Words recalled with picture	17.79	19	4.008	.920
	Words recalled without picture	14.74	19	4.067	.933

Figure 2.16 Mean number of words recalled in each condition

Figure 2.16 appears to show that more words are recalled when the group are given the picture prompt (mean [average] words remembered = 17.79) than when no picture is given (mean = 14.74). We should analyse that statistically, but we will leave all of that for later chapters, when you have learned more about such things. For now, let's see what happens when we 'split the file' by gender:

Select **Data** → **Split File** (see Figure 2.14) → (in new window) click radio button for **Compare groups** → transfer **Gender** to **Groups Based on:** window → click **OK** (as shown in Figure 2.17). Choosing the 'Compare Groups' option here will result in output that directly compares the groups. This is probably better than selecting the 'Organize output by groups' option, which would produce separate reports for each group.

Now we can examine the difference in word recall across the picture conditions, now according to gender. We can see some fundamental differences between the groups on these outcomes. Figure 2.18 suggests that there is very little difference in mean words recalled between conditions for men, but women appear to recall far more words when prompted with the picture than with no picture.

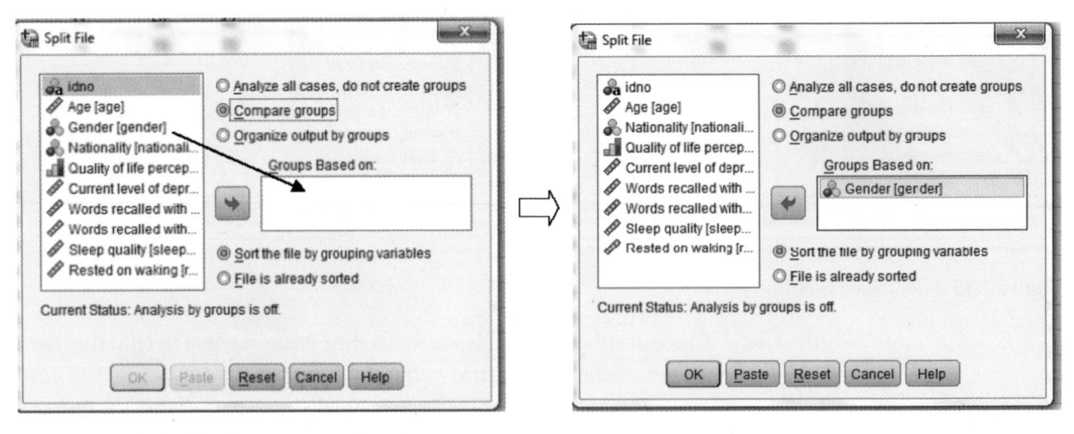

Figure 2.17 Split File function, step 2

Gender			Mean	N	Std. deviation	Std. error mean
Male	Pair 1	Words recalled with picture	16.78	9	4.738	1.579
		Words recalled without picture	16.11	9	4.676	1.559
Female	Pair 1	Words recalled with picture	18.70	10	3.199	1.012
		Words recalled without picture	13.50	10	3.171	1.003

Figure 2.18 Mean number of words recalled in each condition (by gender)

You must remember to return the data set to a state where there is no 'split' – otherwise all subsequent analyses will be affected.

Select **Data** → **Split File** → click radio button for **Analyze all cases, do not split groups** → click **OK**

Select Cases: This function allows you to explore certain sections of the data. In some respects it is similar to what we saw for the 'Split File' facility, but there are several more options. For example, you can exclude a single group from the data set and report outcomes on the remaining groups. In our data set, we could decide to analyse only English and Welsh participants, excluding French people. In effect, we 'switch off' the French participants from the data. This is how we do it:

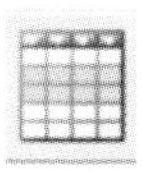

Select **Data** → **Select Cases** (see Figure 2.14) → (in new window) select **If condition is satisfied** radio button → click on **If . . .** box (as shown in Figure 2.19)

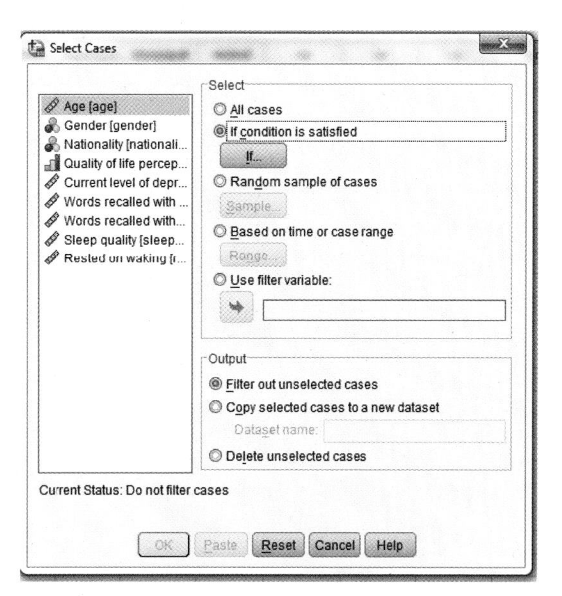

Figure 2.19 Select cases function, step 1

In that new window (see Figure 2.20), transfer **Nationality** to blank window to the right ('Nationality' will now appear in that window) → click on ~=(this means 'does not equal') → Type **3** (because 'Nationality = 3' represents French people (who we want to deselect) → click **Continue** → click OK (see Figure 2.21 to see completed action)

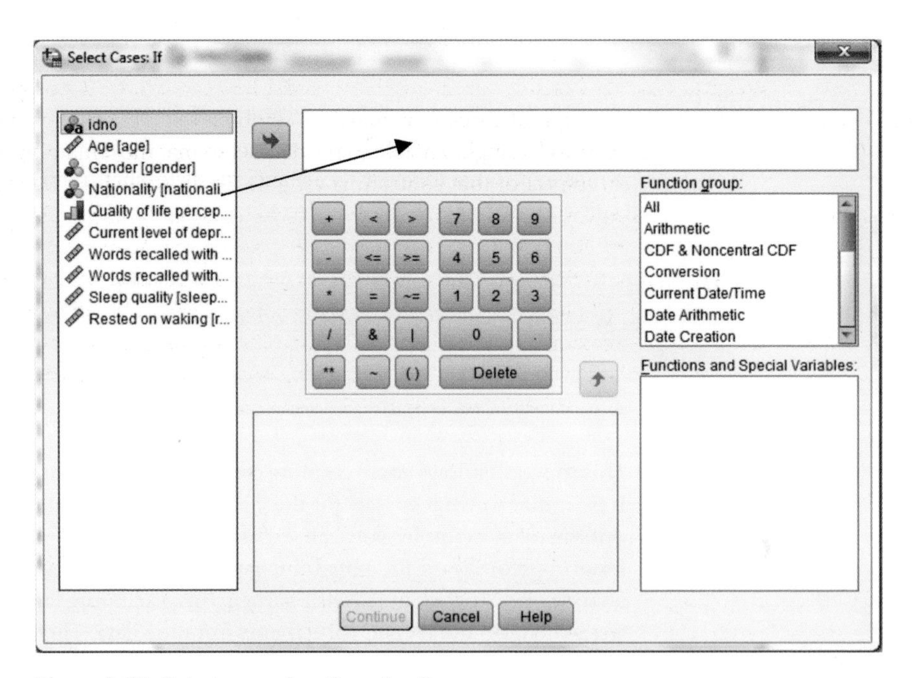

Figure 2.20 Select cases function, step 2

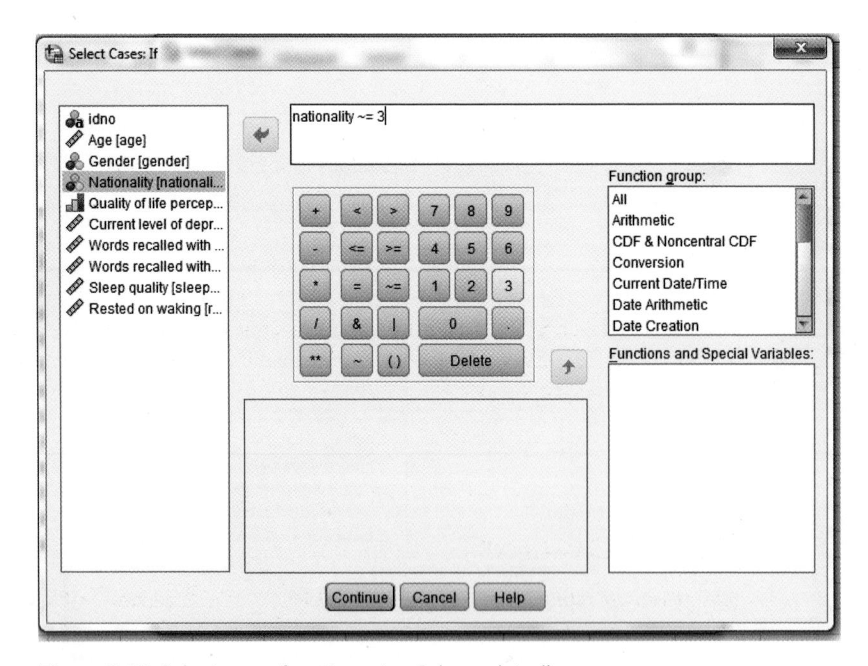

Figure 2.21 Select cases function, step 2 (completed)

When you return to Data View you will notice that all of the cases referring to French people are now crossed out. You would now be able to perform your analyses just based on English and Welsh people. Before you can use the data for other functions, you will need to remove the selected cases and return to the full data set:

Select **Data** → **Select Cases** → click All cases → click **OK**

Weight cases: This facility has a couple of useful functions. First, it can be used to count the number of cases that match a combination of scenarios. Or, second, we can 'control' a single variable in the data set so that the remaining variables are 'equal' in respect of that controlling variable. To illustrate how we can use this function to count cases we need a much larger data set. In this scenario, we have a sample of 200 people, for whom we measure two variables: gender (males/females) and whether they watch football on TV (yes/no). Now imagine how long it would take to enter data for 200 participants. Thankfully, there is a shortcut. We can count the number of times we find the combination of the following: males who watch football on TV, males who do not, females who do and females who do not. The data set might look something like Figure 2.22.

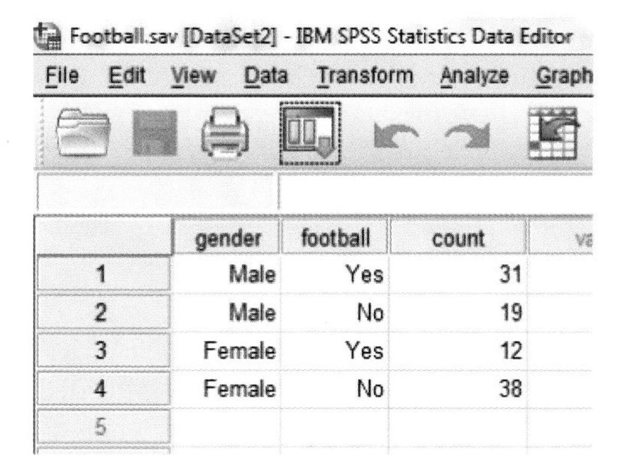

	gender	football	count
1	Male	Yes	31
2	Male	No	19
3	Female	Yes	12
4	Female	No	38
5			

Figure 2.22 SPSS data set: watching TV by gender

However, as it stands, the 'count' is simply another variable. To use it to count the number of cases that match the scenarios in the first two columns, we need to use the 'weight' function.

Open SPSS data set Football

Select **Data** → **Weight Cases** (see Figure 2.14) → select **Weight cases by** radio button → transfer **Count** to **Frequency Variable** window → click **OK** (see Figure 2.23)

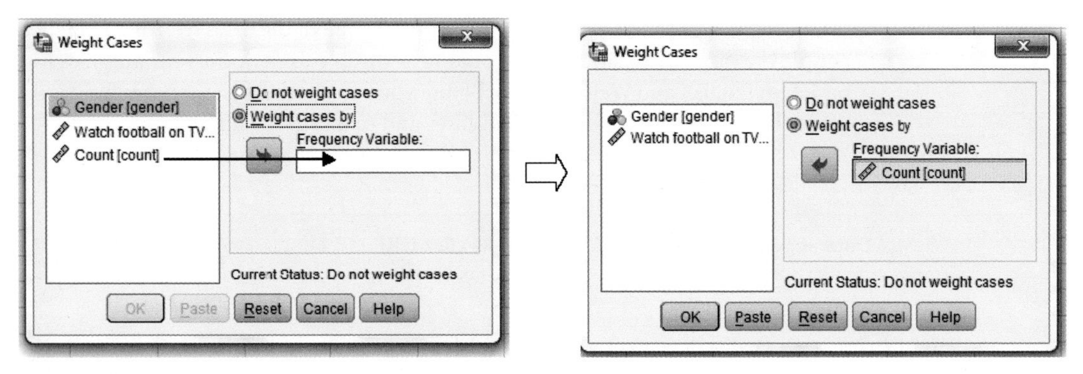

Figure 2.23 Weight Cases function

Now that the data are 'weighted' by count, analyses can be performed to explore how men and women differ in watching football.

We can also use the 'weight' function to 'normalise' data. In social science research (including psychology) it is difficult to control all of the variables. Using the data set that we created earlier, we might choose to explore 'current level of depression' by gender. We might find that women score more highly (poorly) on depression scores than men. However, what if we also notice that depression scores increase with age? How can we be sure that the observed outcome is not the result of age rather than gender? To be confident that we are measuring just depression scores by gender, we need to 'control' for age. By using the 'weight' function, we can adjust the depression scores so that everyone is equal in age. As we will see later in this book, there are more sophisticated tests that can do this (see ANCOVA, Chapter 15). However, the weight function provides one fairly easy way of exploring a simple outcome. This is how we do it:

Select **Data** → **Weight Cases** → select **Weight cases by** radio button → transfer **Age** to **Frequency Variable** window → click **OK**

Before you can use the data for other functions, you will need to remove the weighting function:

Select **Data** → **Weight Cases** → click on **Do not weight cases** → click **OK**

Transform menu

The transform menu undertakes a series of functions that can change the properties of variables, or create new variables based on the manipulation of existing variables. Once again, we will focus on the ones that you are most likely to use. To illustrate those important facilities, we will perform the functions using example data.

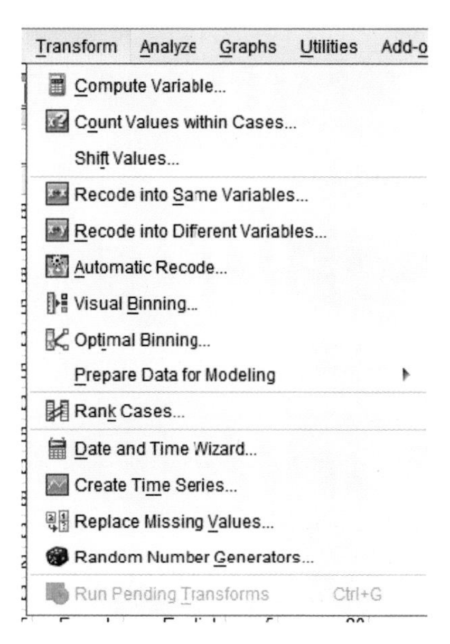

Figure 2.24 Transform menu options

Compute Variable: You can use this to perform calculations on your variables, perhaps to adjust them or create new variables. For example, you might have several variables that measure similar concepts, so you decide to create a new variable that is the sum of those added together. In the data set that we created earlier, we had one variable for 'Sleep quality' and one for 'Rested on waking'. We could combine those into a new variable called 'Sleep_perceptions'. Here's how we do that:

Using the SPSS data set Data entry exercise

Select **Transform** → **Compute variable** (see Figure 2.24) → (in new window, as shown in Figure 2.25), for**Target Variable** type **Sleepperceptions** → transfer **Sleep quality** to **Numeric Expression** window → click on **+** (the 'plus' sign shown in keypad section below the **Numeric Expression** window) → transfer **Rested on waking** to **Numeric Expression** window → click **OK** (see Figure 2.26 for completed action)

Go back to the data set. You will see that a new variable (sleepperceptions) has been included.

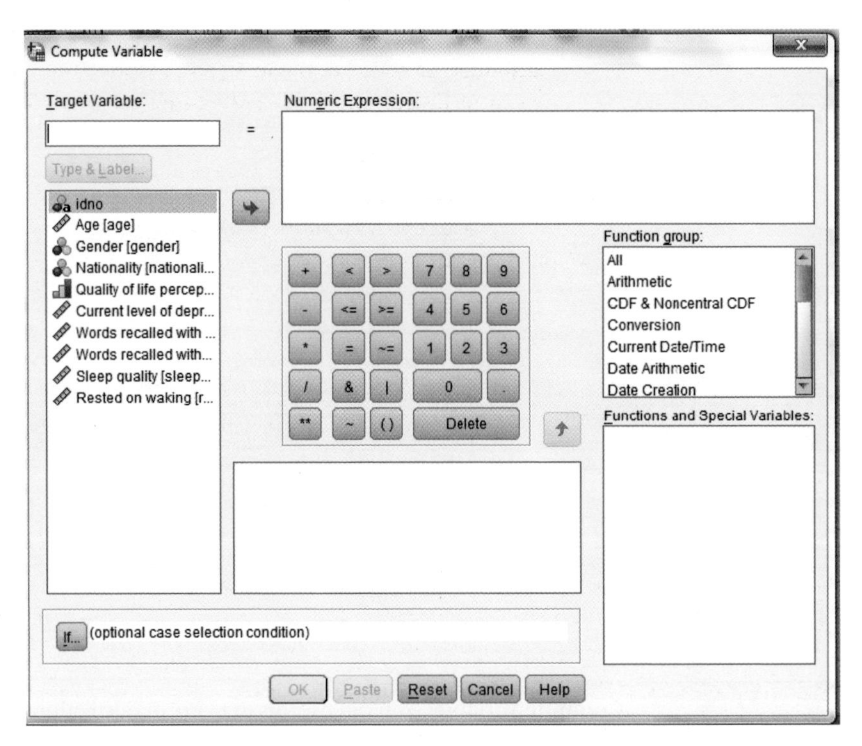

Figure 2.25 Transform Compute Variable

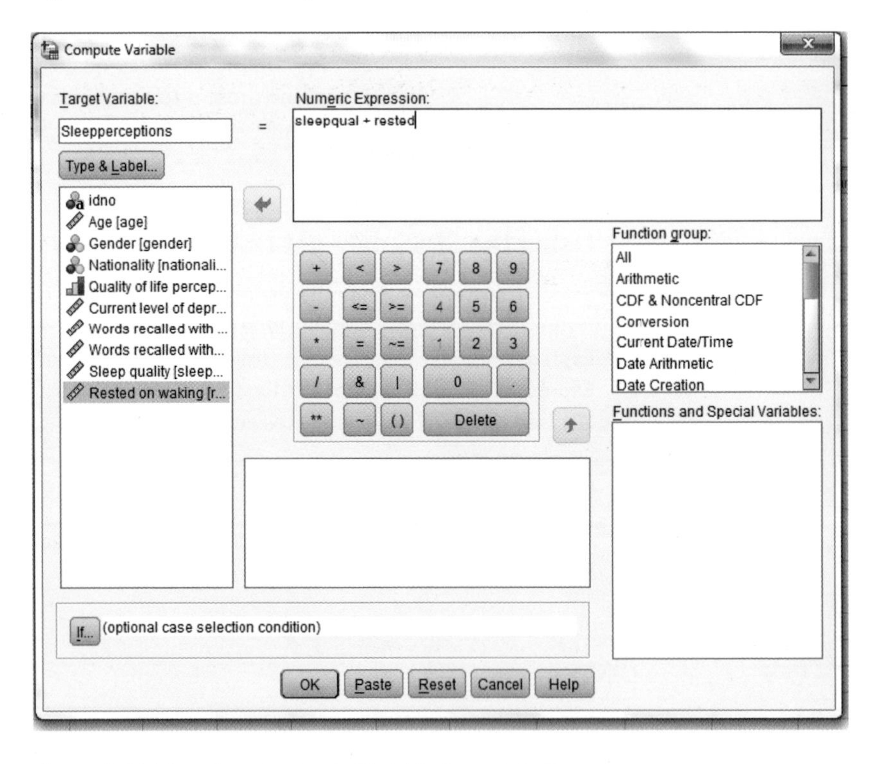

Figure 2.26 Transform Compute Variable (completed)

Recode into Same Variables: Sometimes you may need to recode the values of your variables. For example, when we created our data set, we input the values for gender as 1 (male) and 2 (female). However, as we will see in later chapters, some statistical procedures (such as linear regression – Chapter 16) require that categorical variables can have only two groups and must be coded as 0 and 1 (don't worry about why for the moment). This is how we make those changes (this procedure will overwrite the values that we set up before):

> Select **Transform** ➔ **Recode into Same Variables** (see Figure 2.24) ➔ in new window (as shown in Figure 2.27) transfer **Gender** to **Variables** window (which becomes renamed as '**NumericVariables**') ➔ click **Old and New Values ...**

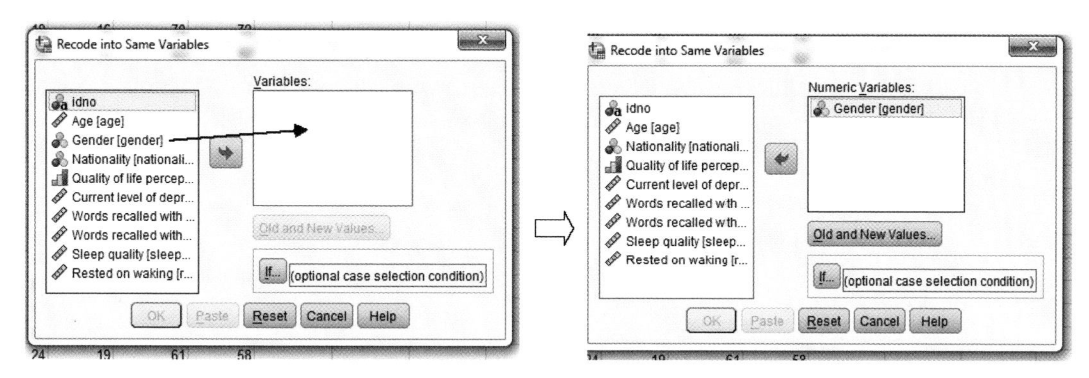

Figure 2.27 Recode into Same Variables function – step 1

> In new window (as shown in Figure 2.28), under **Old Value**, select **Value** radio button ➔ type **1** in box ➔ under **New Value**, select **Value** radio button ➔ type **0** in box ➔ click **Add** (1 --> 0 appears in **Old --> New** box) ➔ for **Old Value**, type **2** ➔ for **New Value**, type **1** ➔ click **Add** (2 --> 1 appears in **Old --> New** box) ➔ click **Continue** ➔ (in original window) click **OK**

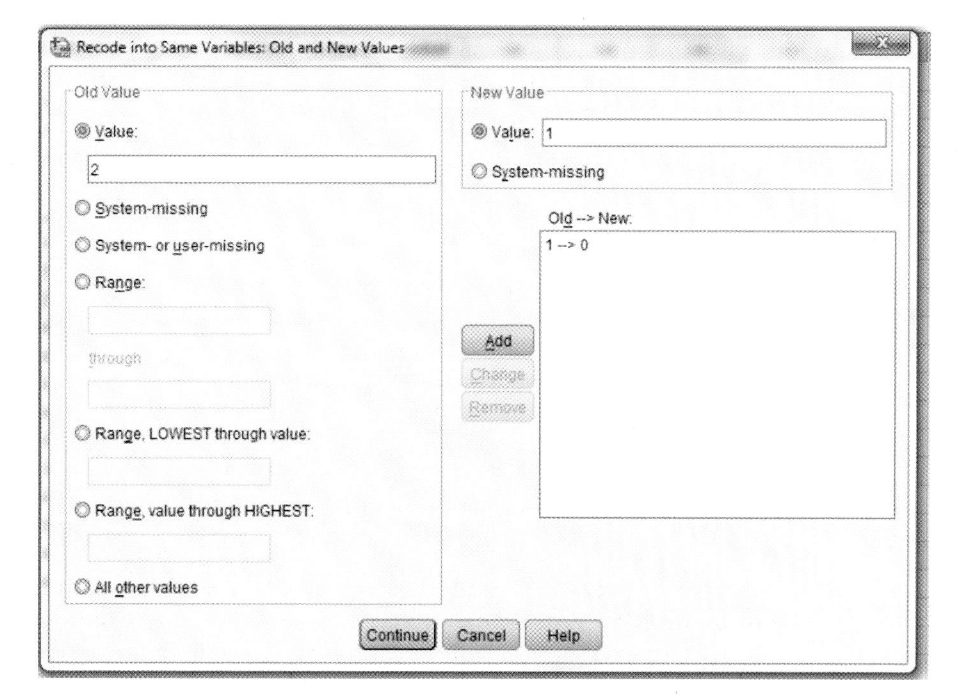

Figure 2.28 Recode into Same Variables function – step 2

If you look at the data set you will see that the gender data now show '0 and 1' where '1 and 2' used to be. But now, the variable is coded incorrectly. You must go to Variable View and change the value codes to show males = 0, females = 1.

Recode in Different Variables: This is the same as what we have just seen, but a new variable is created rather than changing the existing one (it will not overwrite the original variable information).

Analyze menu

This menu contains the statistical techniques that we can use to analyse and manipulate data. We will be exploring how to analyse data in the statistical chapters later, so we do not need to look at this in too much detail here. This menu permits a wide range of statistical analyses, each with different rules of operation so we will leave that for now.

Direct Marketing menu

This menu is more likely to be useful for market researchers. According to SPSS, it 'provides a set of tools designed to improve the results of direct marketing campaigns by identifying demographic, purchasing, and other characteristics that define various groups of consumers and targeting specific groups to maximize positive response rates'.

Graphs menu

Once you have reported your results, you may want to represent the outcome graphically. This menu provides a wide range of graphs that can be used. However, we will explore that in more detail when we get to the statistical chapters.

Utilities menu

We will not dwell on this menu – the facilities are more likely to be attractive to advanced users. It offers further opportunities to view the properties of variables (how they are defined in the program, including the programming language parameters). Perhaps the most useful facility is one where you can change the output format so that it can be sent to another medium (such as Word, PDF, etc.). You can append comments to SPSS files, which may be useful if you are sharing data Other facilities are much more advanced and might be useful only to those who understand the more technical aspects of programming (so, not me then).

Add-ons menu

This menu highlights a number of additional products that SPSS would like you to be aware of, such as supplementary programs or books about using SPSS. There is then a link to a website that invites you to buy these products. Enough said.

Window menu

This is simply a facility whereby you change the way in which the program windows are presented, such as splitting the screen to show several windows at once.

Help menu

This does exactly as it says on the tin: it helps you find stuff. You can search the index for help on a topic, access tutorials on how to run procedures, and scan contents of help files. This can be very useful even for the most experienced user.

Syntax

Syntax is the programming language that SPSS uses (mostly in the background). For the most part, you will not need to use this, as the functions are performed through menus and options. However, there are times when using syntax is actually much quicker than entering all of the required information using the main menus. We may need to run the same statistical test many times, particularly if we are collecting data on an ongoing basis. Running tests in SPSS can be relatively straightforward (such as an independent t-test), while others are rather more complex (such as a mixed multi-factorial ANOVA or a multiple regression). Using syntax can save a lot of time and energy in setting up the parameters for those tests. As you will see when you run each statistical test, the SPSS output includes a few lines of syntax code (just before the main outcome tables). If we want to run a subsequent test on this data set, we can cut and paste the code into the syntax operation field. The test will run without having to redefine the test parameters. There may also be some occasions when you will need to write some syntax to perform a task that is not available through the normal menus (see Chapter 11 for an example).

Chapter summary

In this chapter we have explored some of the basic functions of SPSS. At this point, it would be good to revisit the learning objectives that we set at the beginning of the chapter.

You should now be able to:

- Recognise that SPSS presents data sets in two 'views': the Variable View where variables are defined and parameters are set, and the Data View where the raw data are entered.
- Understand that there are a number of limits that we must observe when setting up those parameters.
- Appreciate the need to correctly define 'missing variables' so that blank spaces in the data set are not treated as '0'.
- Perform basic data entry in SPSS.
- Understand the purpose of the SPSS menus, and the function of the more popular sub-menus, including basic data manipulation and transformation.

Extended learning task

Following what we have learned about setting up variables, data input and data manipulation, perform the following exercises. Your task will be to create an SPSS data set that will explore outcome regarding mood, anxiety and body shape satisfaction in respect of gender and age group. The variable parameters are as follows:

Gender: male (1), female (2)
Age group: under 25 (1), 25–40 (2), 41–55 (3)

Outcome measures: anxiety, mood (measured on an interval scale), body shape satisfaction (measured on an ordinal scale)

We have some raw data in respect of eight participants, shown in Table 2.2.

Table 2.2 Raw data

Gender	Age group	Anxiety	Mood	Body shape satisfaction
Male	<25	87	74	11
Female	25-40	54	61	23
Female	41-55	31	38	?
Male	25-40	43	39	34
Male	<25	69	82	8
Female	41-55	18	12	51
Female	25-40	38	77	29
Male	<25	74	65	16

Open a new SPSS data set.

1. Create the variables, using the parameters shown above.
2. Enter the data, using the raw data from Table 2.2.
3. Create a new variable that measures a combination of 'Anxiety' and 'Mood' scores added together (called 'Affect').
   a. Format the variable parameters for the new variable.
4. Recode the Gender variable (using values of 0 = male and 1 = female).
   a. Format the variable parameters for the new variable.

3 NORMAL DISTRIBUTION

Learning objectives

By the end of this chapter you should be able to:

- Understand the importance of normal distribution
- Recognise the effects of skew and kurtosis, and what we mean by 'outliers'
- Appreciate how to measure and interpret normal distribution graphically and statistically
- Recognise ways in which we can deal with potential violations in normal distribution
- Understand how to adjust outliers and transform variables
- Recognise what we mean by homogeneity and sphericity of variances

What is normal distribution?

Normal distribution describes the way in which data are 'spread'. Imagine that we collected some information about the age for a group of 30 people, aged between 18 and 50. Some of those people would be younger, some older, others somewhere in between. **Probability** statistics describe the likelihood of something happening based on what we know about previous outcomes. In probability, we expect things to happen in a predictable, uniform way. If our group was representative of the general population, we would expect the ages of our group to be pretty evenly spread out. However, there may be circumstances that might cause those ages to be not so even. If this were a group of university students, we might expect most of the ages to tend towards being younger; if the group were members of a crown green bowls club, the ages might be somewhat older. In normal distribution, we start with the assumption that the data we collect represent something close to the general population. In our example, we could plot the ages in a graph: the range of ages would be placed in ascending order along the horizontal (x) axis and we would count the number of people matching that age along the vertical (y) axis. The graph might look something like the one shown in Figure 3.1.

The bars in Figure 3.1 represent a group of age bands, with the height of the bar showing how many people are in that group of ages. We have added a curve that shows the trend of the ages (we will see how to draw this graph, including adding the curve, later on). That curve is useful in two respects: it shows how 'evenly spread' the ages are across the group and it provides some information on what the average age of the group is likely to be. The peak of that curve approximately indicates the average age (around 35 in this example). Overall, we appear to have a gentle 'bell-shaped' curve where the distribution of ages is roughly equal either side of the mean. It is this type of 'normal distribution' that we should be aiming for. In this chapter we will explore exactly how we quantify that.

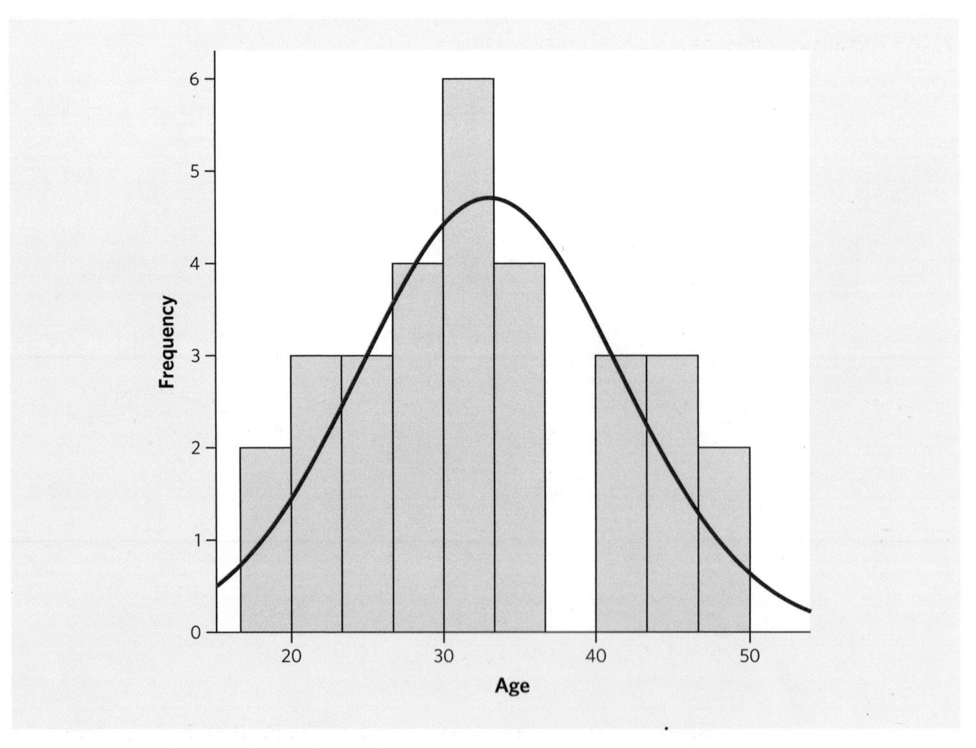

Figure 3.1 Distribution of ages (n = 50)

What does normal distribution look like?

Looking at Figure 3.1, we have some idea of what normal distribution looks like. For data to be 'normally distributed' we expect them to be 'evenly' distributed either side of the mean, illustrated by a smooth, bell-shaped pattern, and where the 'peak' of that distribution is neither 'too pointed' nor 'too flat'. Graphically, we often draw a curve through the data to indicate the trend in those data; we call these '**histograms**'. To illustrate a good example of normal distribution, compared with examples where normal distribution may have been compromised, we need to look at a series of histograms. We need to compare the curves in these histograms to appreciate how they differ. However, before we start, we need to learn some basic terms about how we measure data (see Box 3.1).

3.1 Nuts and bolts
Basic units of measurement

Mean: This is the average number in a data set. We add up all of the numbers in the data set and divide the answer by the number of cases (or people).

Median: This is the middle number in a data set, when those numbers have been ordered numerically from lowest to highest (or vice versa).

Mode: This is the most common number in a data set.

We will start with an example of a normal distribution. Table 3.1 shows what some normally distributed data might look like.

Table 3.1 Example data for normal distribution

Ages															Mean	Median	Mode
20	23	28	28	32	32	35	35	35	38	38	42	42	47	50	35.0	35	35

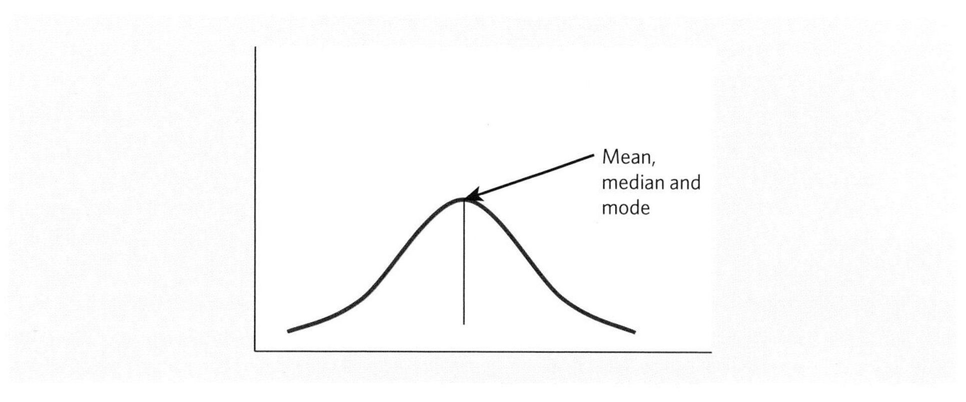

Figure 3.2 Normal distribution

Figure 3.2 is an example of a normal distribution. It is signified by a smooth, bell-shaped curve. The mean and median are identical.

Skewed data

By definition, normal distribution describes a range of data where the scores at either end of the distribution are the same distance to the mean. In our example, the eldest person is 15 years older than the mean age; the youngest is 15 years younger than the mean age. If there are extreme scores at one end of the distribution it is likely to '**skew**' the mean score away from the median. We call those extreme scores 'outliers'. If the data are skewed, this can distort the mean score and can bias any test that depends on it (as we will see later).

Positively skewed data

When the data are positively skewed, there are extreme (outlier) scores at the higher end of the range of data. This might cause the mean score to be overstated (see Table 3.2).

Table 3.2 Example of positively skewed data

Ages															Mean	Median	Mode
20	23	28	28	32	32	35	35	35	38	38	42	42	55	60	36.2	35	35

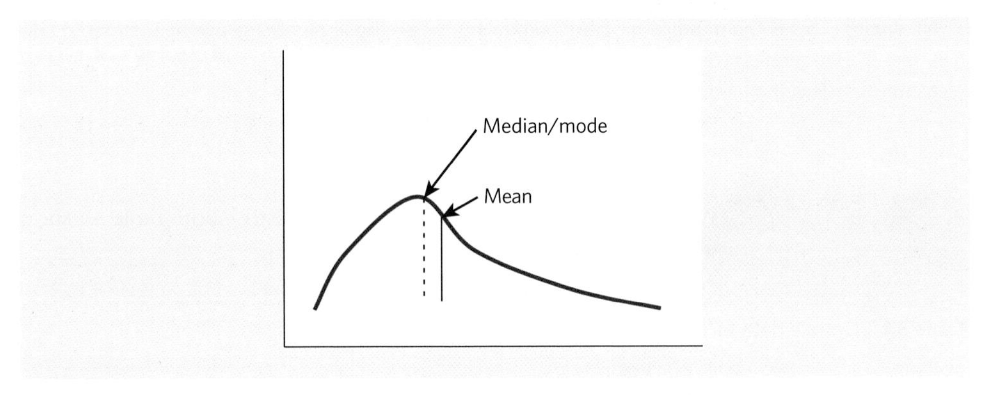

Figure 3.3 Positively skewed distribution

Figure 3.3 shows data that are positively skewed. One tip of the curve points towards the right-hand side of the distribution. The mean is drawn to the right of the median and mode. The high extreme scores may have artificially inflated the mean score.

Negatively skewed data

When the data are negatively skewed, there are extreme scores at the lower end of the range of data. This might cause the mean score to be understated (see Table 3.3).

Table 3.3 Example of negatively skewed data

Ages															Mean	Median	Mode
9	10	28	28	32	32	35	35	35	38	38	42	42	47	50	33.4	35	35

Figure 3.4 presents an example of negative skew. One tip of the curve points towards the left-hand side of the distribution. The mean is drawn to the left of the median and mode. The low extreme scores may have artificially deflated the mean score.

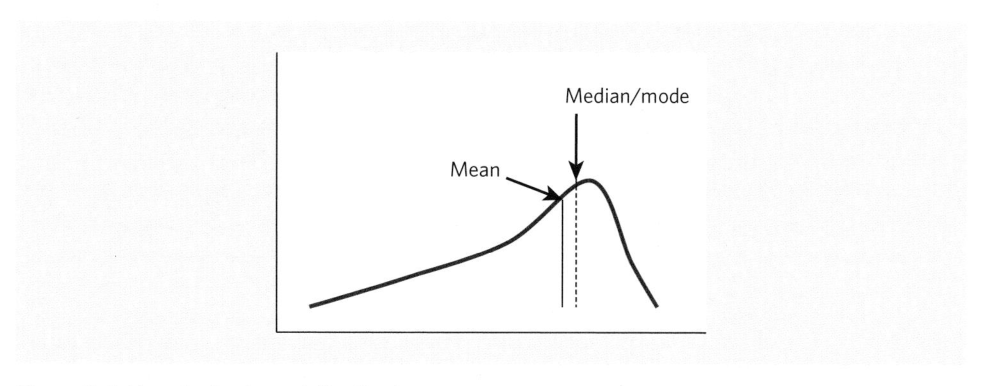

Figure 3.4 Negatively skewed distribution

Kurtosis

In addition to skew, we need to measure **kurtosis**. This describes the 'peakedness' of the curve. A normal distribution is often referred to as being 'mesokurtic', which is another reference to the 'bell shape' that we are aiming for. However, we may encounter problems with curves that are too 'peaked', or ones that are too 'flat'.

Leptokurtic distributions

A **leptokurtic** distribution describes a curve that is 'peaked', like a pointed hat (see Table 3.4).

Table 3.4 Example of leptokurtic data

Ages															Mean	Median	Mode
31	31	32	32	34	34	35	35	35	36	36	38	38	39	39	35	35	35

Although the mean and median are the same, there is very little variation in the data, making analyses difficult. Graphically, the data distribution might look like Figure 3.5.

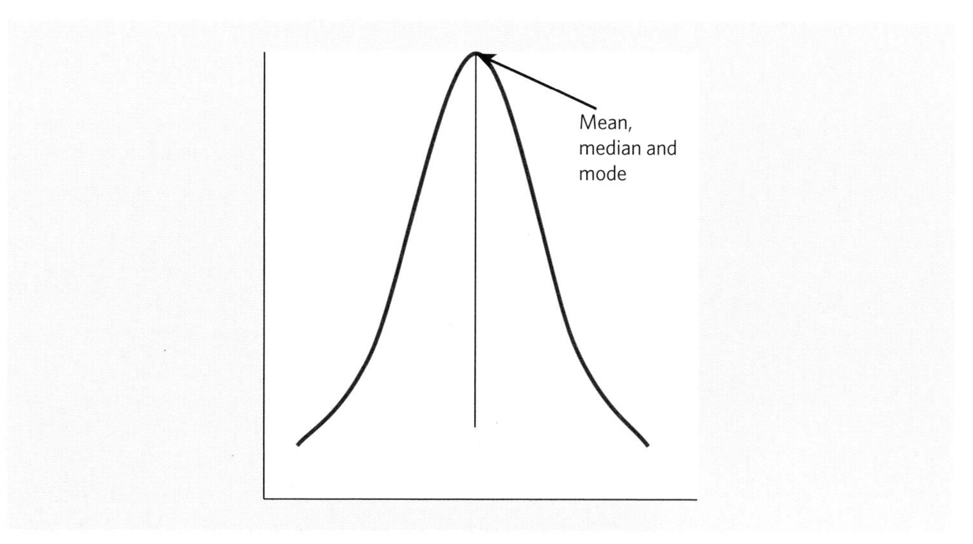

Figure 3.5 Leptokurtic distribution

Platykurtic distributions

A **platykurtic** distribution describes a curve that is flat (see Table 3.5).

Table 3.5 Example of platykurtic data

Ages															Mean	Median	Mode
20	22	24	26	28	30	34	35	36	40	42	44	46	48	50	35	35	None

Once again, the mean and median are the same, but now there is too much variation in the data to make analyses viable. Graphically, the data distribution might look like Figure 3.6.

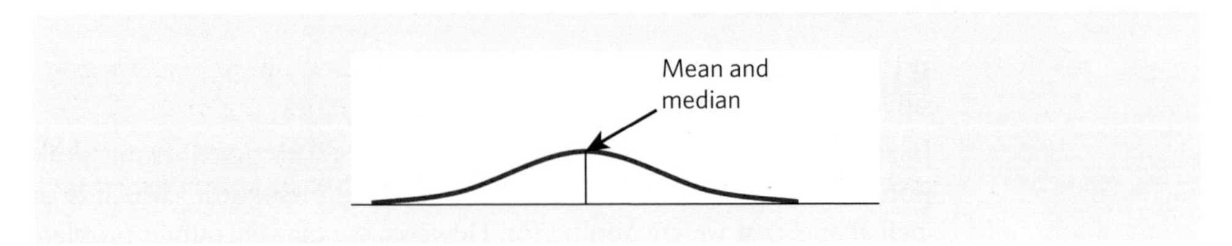

Figure 3.6 Platykurtic distribution

3.2 Take a closer look
Terms used in measuring normal distribution

Normally distributed: data are evenly distributed either side of the mean (see Figure 3.2)
Positive skew: where there are outliers at the higher end of a data set (see Figure 3.3)
Negative skew: where there are outliers at the lower end of a data set (see Figure 3.4)
Kurtosis: describes the peakedness of a normal distribution curve
Mesokurtic: a 'normal' curve, as demonstrated by the bell shape (see Figure 3.2)
Leptokurtic: very 'peaked' distribution, with little variation in the data (see Figure 3.5)
Platykurtic: very 'flat' distribution, with data widely dispersed across the data set (see Figure 3.6)

What happens when data are not normally distributed?

As we have just seen, data may not be normally distributed if there are problems with skew and kurtosis. Data that are positively skewed may cause the mean score to be artificially inflated. This may have occurred because there are some extreme high scores. Without those outliers, a more realistic mean score might have been somewhat lower. Similarly, data that are negatively skewed might lead to an artificially deflated mean because of some extreme low scores. Either way, the mean score in skewed data may not be reliable. We also saw that deviations in kurtosis may cause a problem. Leptokurtic distributions may offer too little variation in the data, while platykurtic distributions may have too much variation. But why might all of this be a problem? Many of the statistical procedures that we will explore in this book depend on measuring differences in mean scores. We will come to know these as parametric tests (we will explore this in more depth in Chapter 4). Normal distribution is a major determinant in deciding whether we can classify our data as parametric. If normal distribution has been compromised, we may no longer be able to trust the mean score as truly reflecting the data. If we cannot trust the mean score, we may have less confidence in the outcome produced by parametric tests. In short, if we lack normal distribution we may need to choose alternative tests (such as those examined in Chapter 18).

Measuring normal distribution

So how can we check that our data are normally distributed? We can get SPSS to help us here. This can be achieved through the production of graphs (such as histograms, **box plots** or **stem-and-leaf plots**), or we can employ statistical procedures. We will look at each of these in turn.

Graphical procedures

Histograms

In Figure 3.1, we saw a graphical representation of normal distribution. This type of graph is called a histogram. It is a bar chart, where bars represent individual cases or groups, and where the height of the bar indicates the frequency of that outcome. We can add a curve to the display to illustrate normal distribution. We can get SPSS to draw this histogram:

Open the SPSS file Age and sleep quality

Select **Analyze → Descriptive Statistics → Frequencies** (as shown in Figure 3.7)

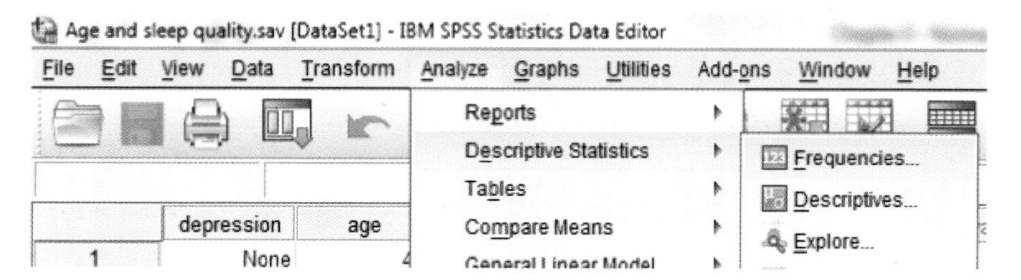

Figure 3.7 Creating histograms - step 1

In new window (see Figure 3.8) transfer Age to **Variable(s)** window (by clicking on the arrow to the left of that window, or by 'dragging' the variable there) → click **Statistics**

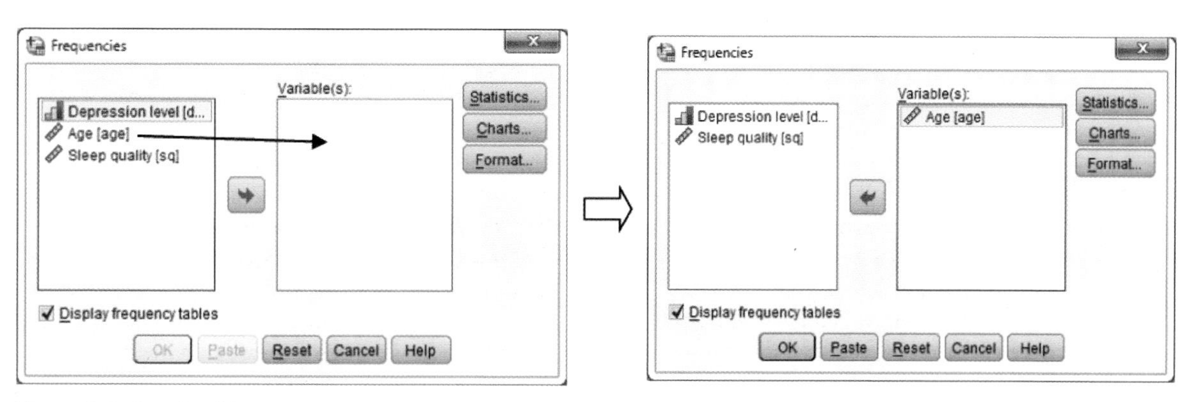

Figure 3.8 Creating histograms - step 2

In new window (see Figure 3.9) select **Mean**, **Median**, and **Mode** radio buttons → click **Continue** → (in original window) click **Charts**

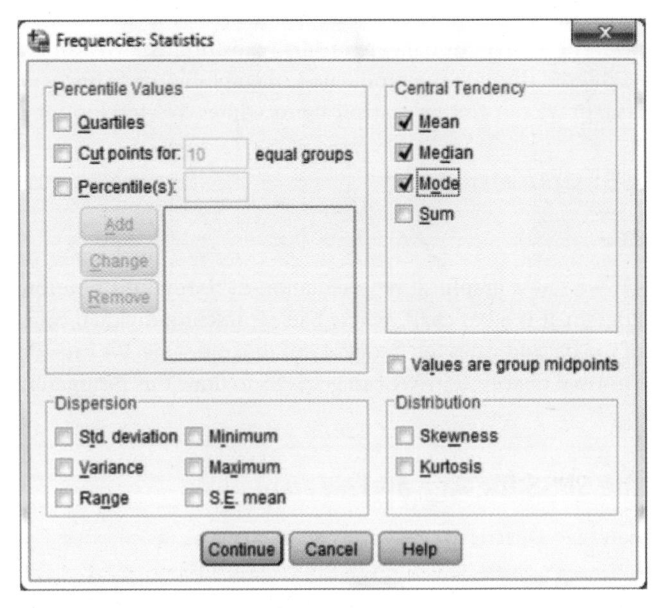

Figure 3.9 Creating histograms – step 3

In new window (see Figure 3.10) click **Histogram** radio button → tick **Show normal curve on histogram** box → click **Continue** → (in original window) click **OK**

If you need further guidance on these procedures, you can visit the website for this book and follow the video guides for SPSS

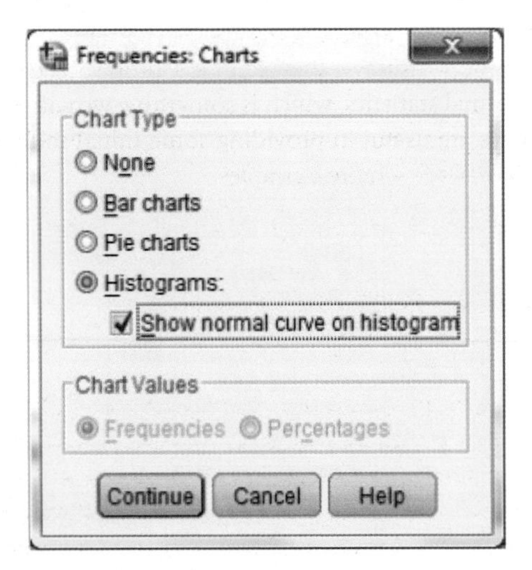

Figure 3.10 Creating histograms – step 4

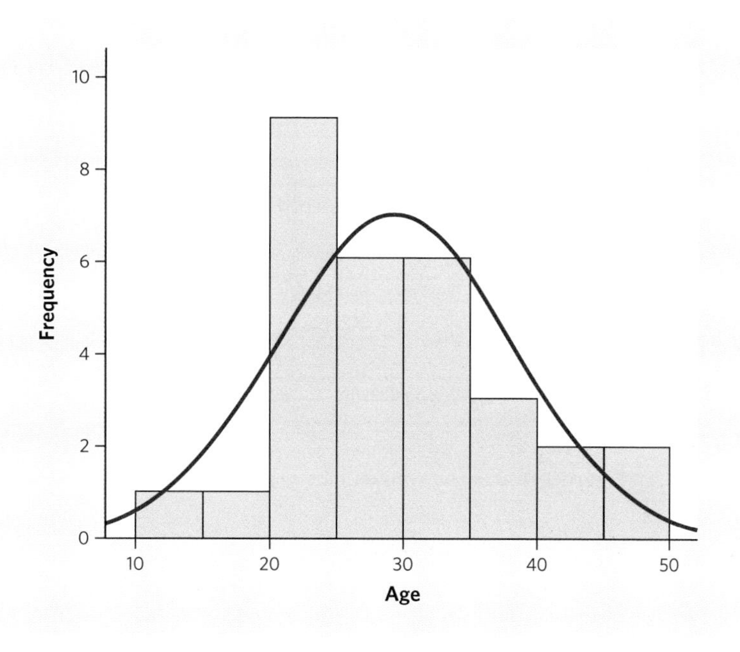

Figure 3.11 Completed histogram

This *appears* to be a pretty good example of a normal distribution, at least according to the curve that has been added to the graph (see Figure 3.11). However, we may feel that the bars suggest slightly positively skewed data. To help us here, we can refer to the descriptive statistics that we asked for (see Table 3.6).

Table 3.6 Descriptive data

	Mean	Median	Mode
Age	29.30	27	24

Table 3.6 suggests that there are some differences in the mean, median and mode. These differences might cause us to question whether the data are normally distributed after all. This illustrates a drawback of graphical displays: they can be a little subjective. However, we can supplement the graphs with formal statistics, which is something we will look at shortly. Nevertheless, these graphical displays are useful in providing some initial indications about normal distribution, so we should look at a few more examples.

Box plots

Another graphical display that we can use is called a box plot (also known as a box and whisker plot, for reasons that are about to become obvious). Some examples of box and whisker plots are shown in Figure 3.12.

Box plots show how the data are spread around the median (the thick line through the box, representing the middle point of the data). The inter-quartile ranges are represented by the 'hinges' at either end of the box. The bottom hinge is equivalent to the (lower) 25 per cent data point; the higher hinge symbolises the (upper) 75% data point. The 'whiskers', either side of the boxes, approximately represent the lowest and highest scores (unless there are outliers – see later).

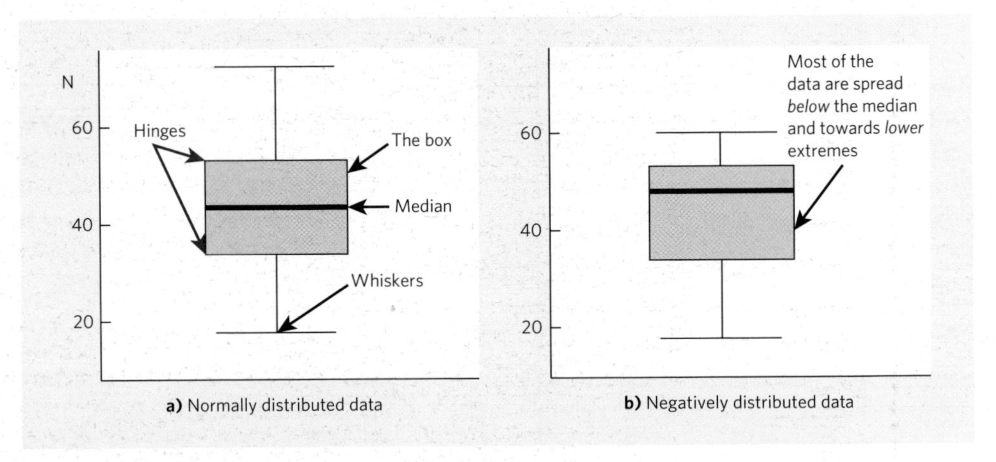

Figure 3.12 Box and whisker plot

Figure 3.12a shows an example of a normal distribution – data are evenly spread either side of the median, with whiskers at equal length above and below the box. Figure 3.12b illustrates some negatively skewed data – there is a larger shaded area below the median line than above it, and there is a disproportionately longer whisker below the box than above it. Positively skewed data will show the opposite of this. This is how we can request a box plot in SPSS (using the same data as we examined with a histogram):

Select **Analyze** → **Descriptive Statistics** → **Explore** (see Figure 3.7) → (in new window, as shown in Figure 3.13) transfer **Age** to **Dependent List** window → select **Plots** radio button → click **Plots** box

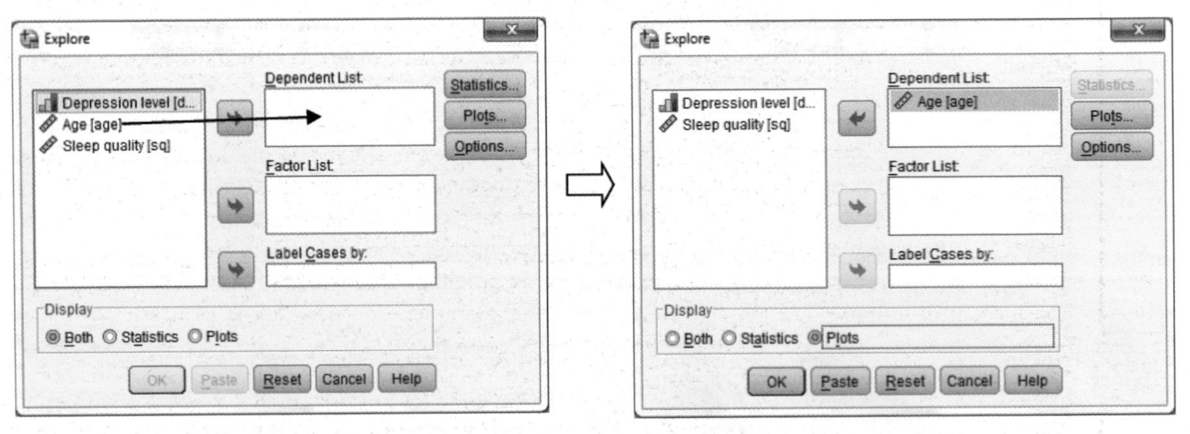

Figure 3.13 Creating box plots - step 1

In new window (as shown in Figure 3.14), click **Factor levels together** radio button (under **Boxplot)** → make sure that **Stem-and-leaf** and **Histogram** (under **Descriptive**) are <u>unchecked</u> (for now) → click **Continue** → (in original window) click **OK**

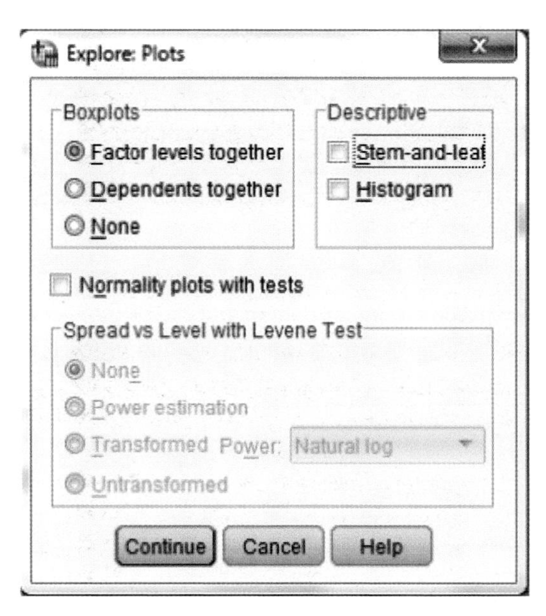

Figure 3.14 Creating box plots – step 2

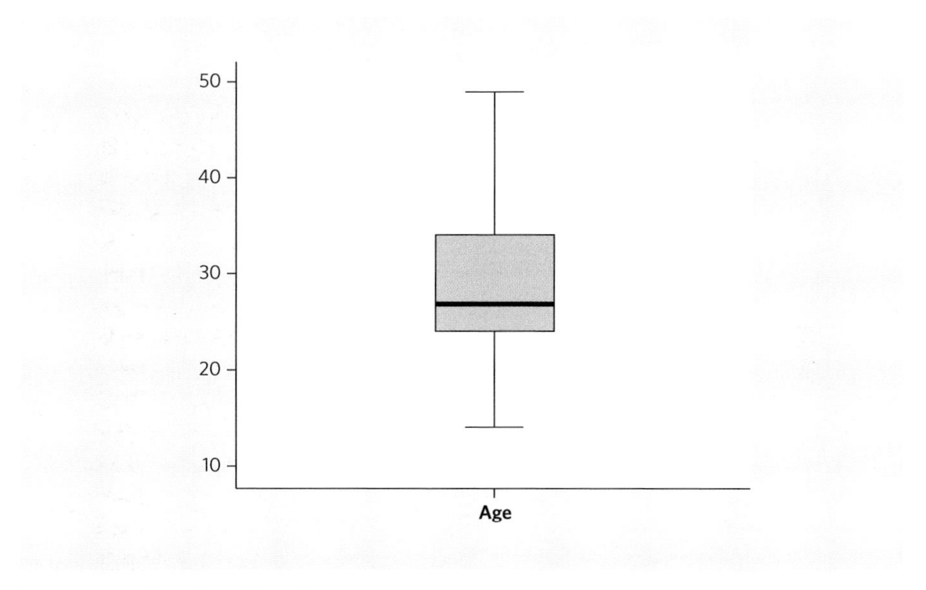

Figure 3.15 Completed box plot

Given what we saw in Figure 3.12b, we might conclude that the data appear to be positively skewed (an outcome potentially supported by the data in Table 3.6).

Stem-and-leaf plots

Stem-and-leaf plots are another way in which we can present data to visually examine normal distribution. The style of presentation is similar to histograms but has the added advantage of retaining the actual numbers within the graphical display. The 'stem' refers to a group of data (usually tens, hundreds, thousands, etc.) and the 'leaf' refers to units within that group. An example is shown in Figure 3.16.

Stem	Leaf	
Tens	Units	
0	3 3 3 5	The red bold number in this row represents 3
1	2 4 6	The red bold number in this row represents 16
2	0 0 2 4	The red bold number in this row represents 20

Figure 3.16 Simple stem-and-leaf plot

Larger data sets are arranged in a similar fashion, but can be more easily assessed to establish whether those data are normally distributed. A larger set of numbers is shown in Figure 3.17.

Stem	Leaf
Tens	Units
0	
1	
2	
3	1,3,3
4	2,3,3,4,4,8
5	0,0,2,4,5,6,8
6	1,1,4,4,4,4,5,8,8,9
7	0,2,3,3,4,6,7
8	1,1,4,4,5,6
9	4,4,6
10	
11	

Figure 3.17 Normally distributed stem-and-leaf plot

The data in Figure 3.17 appear to be normally distributed, because the numbers are evenly spread either side of those in the 60s range. If we rotated the display 90° (anticlockwise), we would see the bell-shaped curve typical of normal distributions presented by histograms (as shown by Figure 3.2). However, this 'histogram' has actual numbers in it.

Figure 3.18 presents a stem-and-leaf plot where the data may be positively skewed. If we were to rotate this 90° (anticlockwise), we would see a distribution similar to the positively skewed histogram we saw in Figure 3.3. The tail tends towards the higher numbers. A negatively skewed

Stem	Leaf
Tens	Units
0	
1	
2	
3	1,3,3,5
4	2,3,3,4,4,8,8
5	0,0,2,4,5,6,7,9
6	1,1,4,4,4,4,5,8,8
7	0,2,3,3,4,6
8	1,1,4,4,6
9	4,4,6
10	1,7
11	3
12	5

Figure 3.18 Positively distributed stem-and-leaf plot

stem-and-leaf plot would show the opposite of this. We can also produce stem-and-leaf plots in SPSS (we will use the same data again):

Select **Analyze** → **Descriptive Statistics** → **Explore** (see Figure 3.7) → (in new window) transfer **Age** to **Dependent List** window → select **Plots** radio button → click **Plots** box → (in new window) select **None** radio button for (under **Boxplots**) → check **Stem-and-Leaf** box (under **Descriptive**) → click **Continue** → (in original window) click **OK**

```
Frequency    Stem &  Leaf

    1.00        1 .  4
    1.00        1 .  8
    9.00        2 .  112224444
    6.00        2 .  555778
    6.00        3 .  001444
    3.00        3 .  779
    2.00        4 .  04
    2.00        4 .  79

Stem width:         10
Each leaf:        1 case(s)
```

Figure 3.19 Completed stem-and-leaf plot

The pattern displayed in Figure 3.19 is actually quite similar to Figure 3.3, further suggesting that we might have positive skew. However, as we said earlier, we additionally need some formal statistics to be more confident about the outcome.

Statistical assessment of normal distribution

Graphical information provides some very useful guidance about normal distribution, but it might be more useful to have some formal statistics to illustrate the outcome. However, there are different views about this. There are some statisticians who argue that graphical displays tell us all we need to know, while others wholly advocate statistics. When we analyse normal distribution statistically, there are several methods that we can use. We will focus on the most commonly used: the **Kolmogorov-Smirnov** and **Shapiro-Wilk** tests (we will look at these together because they involve the same method, but different interpretation), z-score analyses tests for skew and kurtosis, and counting outliers. Throughout this book, we will mostly focus on statistical procedures to examine normal distribution rather than use graphical analyses.

Kolmogorov-Smirnov and Shapiro-Wilk tests

The Kolmogorov-Smirnov (KS) and Shapiro-Wilk (SW) tests are reported in many published studies. These tests are obtained in the same way through SPSS. However, there is some debate about which statistic we should report once we are given the results. Several sources suggest that the KS test is less powerful than the SW test (Eadie *et al.*, 1971). Others suggest that both tests tend to falsely reject normal distribution in larger samples (www.basic.northwestern.edu/statguidefiles/n-dist_exam_res.html) and that graphical displays are better after all. A common suggestion is that the KS test should be used in samples greater than 50, while the SW test is better for samples smaller than that. We will apply this criterion throughout this book.

The method that we use to perform these tests depends on the nature of our data and the statistical test that we are likely to use to examine the main outcome. Before we explore some of those methods, we need to understand some basic concepts about the definition of variables (see Box 3.3).

3.3 Nuts and bolts
Variable types and research methods

Dependent variable: The outcome measure being investigated. It is the variable that is expected to change (as a result of factors such as groups or conditions).

Independent variable: A factor (such as groups or conditions) that is thought to be responsible for changes in an outcome measure.

Between-group studies: Where the independent variable is measured between two or more distinct groups of people or cases.

Within-group studies: Where the independent variable is measured across one group, in respect of two or more conditions.

In some cases, we need to explore normal distribution across single variables (in correlation, for example). In between-group studies, we examine whether the dependent variable scores are normally distributed across each independent variable group. In within-group studies, we investigate whether the dependent variable scores are normally distributed at each of the conditions. As we progress through this book, we will see that there are slight variations in the method of measuring normal distribution for each statistical test. However, so that we can examine some basic methods, we will now look at examples for a single variable, between-group data and within-group data.

Using KS/SW tests across single variables

When we explore normal distribution across single variables in SPSS, we use the methods shown below. We will explore whether 'anxiety scores' are normally distributed in a group of 60 people (Figure 3.20):

Open the SPSS file Mood and gender

Select **Analyze** ➜ **Descriptive Statistics** ➜ **Explore** ➜ (in new window) transfer **Anxiety scores** to **Dependent List** window ➜ check **Plots** radio button ➜ click **Plots** box ➜ (in new window) check **Normality plots with tests** radio button ➜ select **None** under **Boxplots** ➜ un-tick all boxes under **Descriptive** ➜ click **Continue** ➜ (in original window) click **OK**

	Kolmogorov–Smirnov[a]			Shapiro–Wilk		
	Statistic	df	Sig.	Statistic	df	Sig.
Anxiety scores	.075	60	.200*	.980	60	.427

Figure 3.20 Kolmogorov-Smirnov and Shapiro-Wilk test for anxiety scores

This function also produces a series of graphs; ignore those and focus on the statistical outcome. The KS and SW tests examine whether the data are significantly different to a normal distribution. We explore **statistical significance** in depth in Chapter 4, so it might be better to leave fuller explanations about that until then. However, for now, we just need to know that if the outcome shown in the 'Sig.' column is less than .050, it suggests that the data are 'significantly different' to a normal distribution. In other words, there is less than 5% probability that the data are normally distributed. If that is the case, we cannot be confident that these data are normally distributed. If the output shows that 'Sig.' is greater than (or equal) to .050, it suggests that the data are probably not different to a normal distribution. Therefore, we can be more confident that the data are normally distributed. For reasons that do not matter here, KS outcomes are reported using the letter 'D' and SW outcomes with the letter 'W'. So, which test do we report? Earlier, we proposed that we should use the KS test in samples of 50 or more and the SW test in smaller samples. Since we have 60 participants, we should choose the former. Therefore, we can see that anxiety scores are (probably) normally distributed, D(60) = .075, p = .200.

Using KS/SW tests in between-group studies

When we explore normal distribution for data in between-group studies, we need to follow a similar method to that we have just seen, but we must account for how the data are distributed across each independent variable group. We will measure whether mood scores are normally distributed across gender.

Using the SPSS file Mood and gender

Select **Analyze** → **Descriptive Statistics** → **Explore** → (in new window) transfer **Mood scores** to **Dependent List** window → transfer **Gender** to **Factor List** window → select **Plots** radio button → click **Plots** box → (in new window) click **Normality plots with tests** radio button → click **Continue** → (in original window) click **OK**

	Gender	Kolmogorov–Smirnov[a]			Shapiro–Wilk		
		Statistic	df	Sig.	Statistic	df	Sig.
Mood scores	Male	.131	39	.089	.940	39	.038
	Female	.113	21	.200*	.982	21	.950

Figure 3.21 Kolmogorov-Smirnov and Shapiro-Wilk test for mood scores by gender

Figure 3.21 indicates that there may be some inconsistency in the normal distribution of mood scores when examined across gender groups. Since there are less than 50 people in each group, we should report the SW outcome. Mood scores appear to be normally distributed for women, W(21) = .982, p = .950, but may not be for men, W(39) = .940, p = .038. Is this a problem? It depends on how severe you want to be. In most tests, we are looking for *reasonable* normal distribution. The outcome for males is only just below the cut-off point for significance. However, if you are more cautious, you might like to additionally test for **z-scores** of the skew and kurtosis (we will see how to do this shortly).

Using KS/SW tests for within-group studies

When we examine normal distribution for within-group data, we explore the dependent variable scores at each condition. We will use the same data set as the last two examples, but look at some variables that are more suited to within-group analysis. This means that we explore outcomes across a single group in respect of two or more conditions. In our data set, we have two variables that measure fatigue: one for 'fatigue week 1' and one for 'fatigue week 4'. These examine the extent of fatigue reported by the entire group at two different time points. This is how we would check normal distribution in that scenario:

Using the SPSS file Mood and gender

Select **Analyze** → **Descriptive Statistics** → **Explore** → (in new window) transfer **Fatigue week 1** and **Fatigue week 4** to **Dependent List** window (we do NOT select anything for **Factor List**) → select **Plots** radio button → click **Plots** box → (in new window) click **Normality plots with tests** radio button → click **Continue** → (in original window) click **OK**

	Kolmogorov–Smirnov[a]			Shapiro–Wilk		
	Statistic	df	Sig.	Statistic	df	Sig.
Fatigue week 1	.089	60	.200*	.953	60	.022
Fatigue week 4	.095	60	.200*	.962	60	.058

Figure 3.22 Kolmogorov-Smirnov and Shapiro–Wilk tests for fatigue reports across time

Since we have 60 participants at each time point, we can report the KS outcome. Figure 3.22 indicates that fatigue scores are normally distributed at week 1, D(60) = .089, p = .200, and week 4, D(60) = .095, p = .200.

Z-score tests of skew and kurtosis

When we explored an example of normal distribution in a between-group study, Figure 3.21 suggested that we might have a problem with the mood scores for men. Once we have employed a KS or SW test, there are additional statistical measures that can be undertaken if there is still any uncertainty. We can calculate something called a 'z-score' of the skew and kurtosis. In fact, some statisticians prefer this method to KS and SW tests.

Skew is measured in terms of whether it is positive or negative. Data that are potentially negatively skewed will be indicated in the SPSS output by a minus sign, so the absence of that minus sign will suggest possible positive skew (unless the outcome is 0, which suggests no skew). Kurtosis is also measured either side of 0 (mesokurtic – normal), with positive scores representing leptokurtic (peaked) data and negative scores representing platykurtic (flattened) curves. But how do we determine whether the outcome violates limits for skew and kurtosis? In addition to the main outcome for skew and kurtosis, SPSS reports something called 'standard error'. We will see more about this in Chapter 4 but, in short, it is an estimate of how much the data vary either side of the mean, relative to the sample size.

A rough guide suggests that the skew or kurtosis should not be more than two times greater than its 'standard error' (Coolican, 2009). More specifically, we can convert the skew and kurtosis scores to a z-score. This is obtained by dividing an actual value of the skew or kurtosis by its respective standard error (we call this 'standardisation'). Once we have done that we have a 'standardised' score that can be viewed within a normal distribution. In a normal distribution the mean score is 0. Either side of 0, scores are evenly distributed as positive and negative scores; these are known as z-scores because they lie within a 'z-distribution' (you don't need to know why it's 'z'). Because we know how these z-scores should be distributed, we know when those scores are so high (or low) that they are beyond the bounds of normal distribution. Once again, this is based on probability and statistical significance (we will revisit this in Chapter 4). Data are seen to be significantly outside the bounds of normal distribution when their probability is less than 5%. Statisticians have calculated that we reach the limits of normal distribution when z-scores are greater than ± 1.96 (plus *or* minus 1.96); this approximates to the 'two' times the standard error suggested by Coolican. A z-score greater than (+)1.96 indicates significant positive skew; a negative number greater than −1.96 suggests significant negative skew. Similar interpretations can be made for kurtosis. These 'limits' can be viewed in a z-score (or normal distribution) table–see Appendix 1.

3.4 Take a closer look

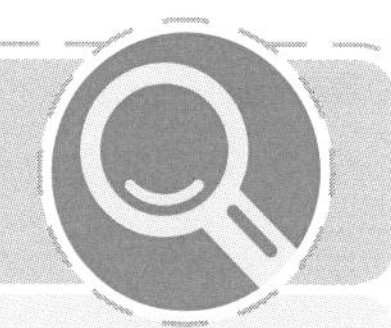

Guidelines for z-score cut-off points

Sample size	z-score cut-off
< 50	±1.96
51 - 100	±2.58
> 100	±3.29

Setting the z-score limits at ±1.96 is probably good enough for smaller sample sizes, but we can use additional cut-off points for larger samples. The initial cut-off point represents the outermost 5% of the data (where $p < .05$). In larger samples we can be more lenient and set a cut-off point of 2.58 (placing outliers in the outer 1% of our data, where $p < .01$). In larger samples still, we can be even more relaxed and use a cut-off point of 3.29; outliers are now deemed to be in the outer 0.1% of data (where $p < .001$). That might all seem useful, but there are few guidelines to tell us what a larger sample is! The suggestions offered in Box 3.4 are a basic guide only and should be used in conjunction with other considerations (such as graphical displays). We should now see how to explore normal distribution, using z-scores of skew and kurtosis (Figure 3.23), by examining those data we investigated earlier:

Using the SPSS file Mood and gender

Select **Analyze** ➜ **Descriptive Statistics** ➜ **Explore** ➜ (in new window) transfer Mood scores to **Dependent List** window ➜ transfer Gender to **Factor List** window ➜ select **Statistics** radio button ➜ click **OK**

To examine the data for normal distribution we need to focus on the skew and kurtosis data, along with the relevant standard error outcome. Strictly speaking, we need to do this only for the male data because that's where our potential problem lies. If we divide the main scores (highlighted in blue for the male data) by the standard error (green) we get the z-scores that will help determine whether the data are normally distributed – those calculations are shown in Table 3.7.

This shows that the z-scores for skew and kurtosis are within limits (±1.96), except for males. There still appears to be positive skew for the men in this sample. However, all is not lost; there are still some procedures that we could employ. Besides, some people might feel that there is sufficient evidence for *reasonable* normal distribution here in any case.

Table 3.7 *z*-scores for skew and kurtosis, in respect of mood scores by gender

DV: Mood scores		Statistic	SE	z-score
Male	Skewness	0.909	.378	2.40
	Kurtosis	0.908	.741	1.23
Female	Skewness	0.016	.501	0.03
	Kurtosis	−0.194	.972	−0.20

Descriptives

	Gender			Statistic	Std. error
Mood scores	Male	Mean		10.85	.803
		95% confidence interval for mean	Lower bound	9.22	
			Upper bound	12.47	
		5% trimmed mean		10.55	
		Median		10.00	
		Variance		25.134	
		Std. deviation		5.013	
		Minimum		2	
		Maximum		24	
		Range		22	
		Interquartile range		5	
		Skewness		**.909**	**.378**
		Kurtosis		**.908**	**.741**
	Female	Mean		9.95	.674
		95% confidence interval for mean	Lower bound	8.55	
			Upper bound	11.36	
		5% trimmed mean		9.95	
		Median		10.00	
		Variance		9.548	
		Std. deviation		3.090	
		Minimum		4	
		Maximum		16	
		Range		12	
		Interquartile range		5	
		Skewness		.016	.501
		Kurtosis		−.194	.972

Figure 3.23 Example of skew and kurtosis data from SPSS

Count the number of outliers

Another way in which we can examine normal distribution involves counting the number of outliers in a data set. An outlier is any data point that is found beyond certain limits. We express outliers through z-scores. Just now, we said that a z-score greater than ±1.96 is located within the outer 5% of a normal distribution, 2.58 within the outer 1%, and 3.29 within the outer 0.1%. We can use those limits to set targets for determining outliers. Table 3.8 shows the cut-off points for a range of sample sizes.

Table 3.8 z-score outlier limits by sample size

z-score	Limit	n = 50	n = 100	n = 200	n = 1000
1.96	5%	2-3	5	10	50
2.58	1%	0	1	2	10
3.29	0.1%	0	0	0	1

We can use SPSS to help us count the number of **outliers** we have in a data set. We said that a z-score is any number that is divided by its standard error; we call this process '**standardisation**'. We can ask SPSS to standardise a variable (we will use the same data set that we have used throughout this section):

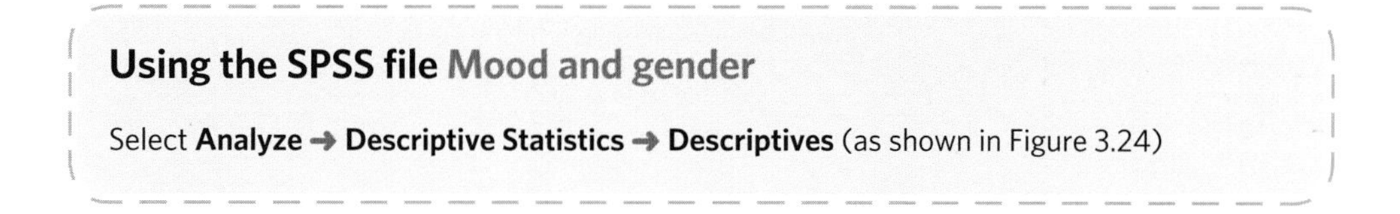

Using the SPSS file Mood and gender

Select **Analyze → Descriptive Statistics → Descriptives** (as shown in Figure 3.24)

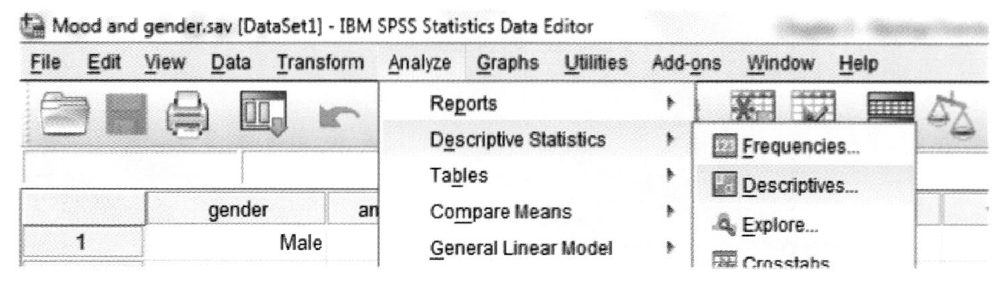

Figure 3.24 Standardising variables - step 1

In new window (see Figure 3.25), transfer Mood scores to **Variable(s)** window → check **Save standardized values as variables** box → click **OK**

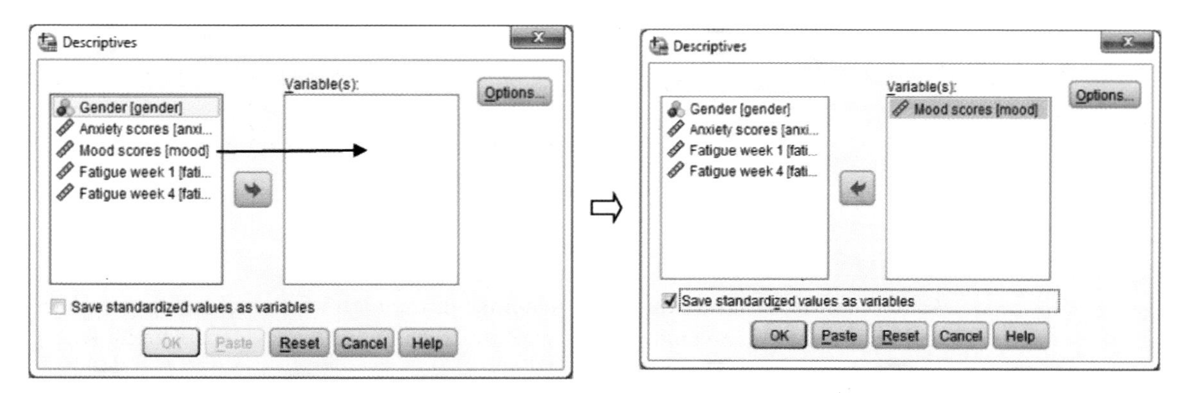

Figure 3.25 Standardising variables - step 2

For this test we do not need to refer to output tables. Instead, we can go back to the data set where we will see that a new variable has been created called 'Zmood' (see Figure 3.26). The values in this new variable represent the original scores converted into z-scores. Using that new variable, we need to count how many scores exceed the limits we have set for outliers. In our example, we have a relatively small data set, so it is quite easy to look for these. In larger data sets we might need to sort the variable in ascending or descending order (refer to Chapter 2 to see how).

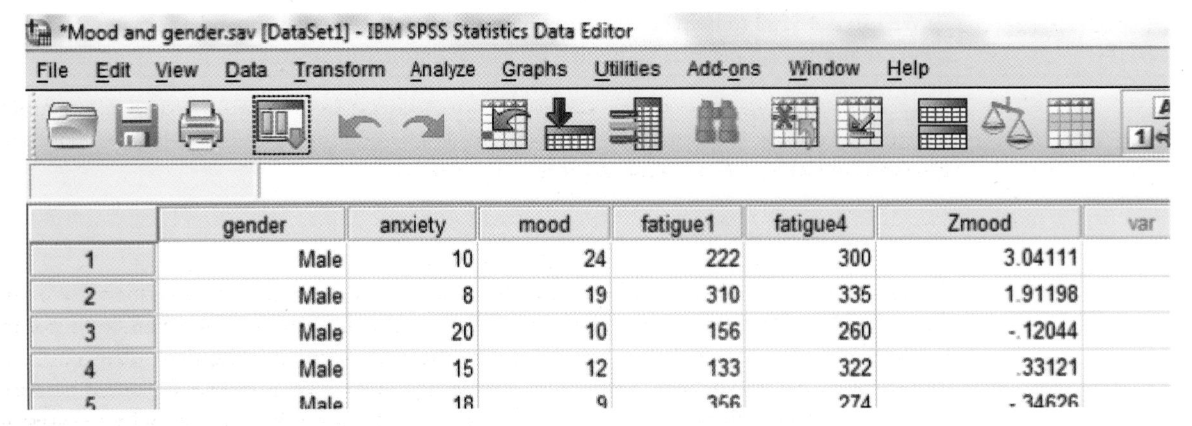

Figure 3.26 Data set with standardised variables

If we look through the values in the 'Zmood' variable we can count the number of z-scores that exceed each limit. We will see that we have two scores greater than 1.96. Both of these z-scores are 3.04, so they also exceed the 2.58 limits, but not the upper cut-off point (3.29). The outcome for our variable is shown in Table 3.9.

Table 3.9 Outliers in 'mood and gender' data set

z-score	Limit %	Limit n = 60	Actual
1.96	5%	3	2
2.58	1%	0	2
3.29	0.1%	0	0

Table 3.9 suggests that, although we satisfied the lower cut-off point for outliers for mood scores, we did have two cases (both men) that exceed the 2.58 z-score limit (ideally, we should have had none). Once again, there is conflicting evidence about normal distribution. More cautious researchers may want to account for these outliers – we will see how to do that shortly.

I don't have normal distribution. What can I do?

You may have noticed throughout these analyses of normal distribution that we have had some potential problem with some of the mood scores, particularly for a couple of male participants whose scores might represent outliers. If we feel that we have violated normal distribution, we have a number of options open to us. We could do nothing, but report the outcomes with caution. This strategy will depend on how much the data have deviated from a normal distribution. In our example, there may be more justification for cautiously accepting normal distribution; we are only looking for *reasonable* outcomes after all. In more extreme cases, we may need to make some adjustments; we will look at some of the ways in which we can do that in the next section. Ultimately, if none of these procedures helps and we are left with data that are clearly not normally distributed, we will probably need to abandon parametric tests that rely on the mean score to determine outcome. We will look at these non-parametric tests in Chapter 18.

Adjusting non-normal data

If our investigations suggest that normal distribution might have been violated, we could consider a number of adjustments to those data (such as removing and adjusting outliers, and transforming data). We will look at those options now.

Removing outliers

As we saw earlier, outliers are data points that exceed certain limits. Those outliers are often identified through z-scores, where we express the number in relation to the standard error of that variable. We saw that we can expect a certain number of potential outliers in any data set; the extent of that will depend on the sample size (see Table 3.8). A data set is skewed because of outliers; if the outliers were not there the data would not be skewed. We had two cases in our example data set where some mood scores for men appeared to be outliers; we might be tempted to remove these. However, there must be a really good reason for doing so. It is always a good idea to identify where the outliers are – there may be data-entry errors. We saw a method for identifying outliers earlier (when we standardised the data into a new variable). For example, we could return to the raw data for the male participants that show a mood score of 24, only to find that we should have entered a score of 20. If we examined normal distribution for the variable again, we might find that it is fine now.

Adjusting outliers

In certain circumstances it might be appropriate to 'adjust' outliers. There are several ways that we can do this, but perhaps the most common method is to replace the outlier with a score that represents the 'mean score plus two standard deviations'. We already know that the mean is the average score of all of the data points in the variable. Within that variable, all of the scores will vary either side of the mean score; standard deviation measures average of that variation. We saw how to obtain descriptive data earlier on (when we requested mean, median, mode outcomes). To ask for standard deviation we simply tick that box as well (see Figure 3.9).

In our example data, we had some potential trouble with apparent mood scores for two males. Table 3.10 presents some information on the mean score and standard deviation for the mood variable, and shows how to adjust the outliers accordingly.

Table 3.10 Adjusting an outlier (with two standard deviations from mean)

Original score (outlier)	24
Mean score	10.53
Standard deviation	4.43
Calculation	10.53 + (2 × 4.43)
Adjusted score	19.39

We could replace the original (outlier) score of 24 with the adjusted score (19.39). However, we would have to report the outcome with some caution. There are many statisticians who report extreme reservations about this method of adjustment, advocating formal **transformation** instead.

Transforming data

Another frequently used method of adjusting skewed data is through transformation. We will explore some of the methods that can be used, but it might pay you to read more advanced books on this subject (such as Howell, 2010, pp. 338–342). To illustrate transformation, we will use the example data set that we have been using throughout, focusing on the potential problem posed by normal distribution of mood scores for men.

Some of the more popular methods of transformation are discussed in Box 3.5. We will explore how to use SPSS to perform transformation shortly.

3.5 Nuts and bolts

Common methods of transformation

Logarithmic: This method is particularly useful for positively skewed distribution (as it compresses higher scores). It does not matter what logarithm base we use, but most researchers choose base 10 logs ($\log_{10}$) or linear logs ($\log_e$). Further adjustments are needed if we have '0' scores in our data, as there is no log of zero (see Box 3.6). For the purposes of this exercise you do not need to know what logarithms are, but if you are curious, you can find out more about them in Chapter 17. We will use this type of transformation to illustrate the methods used in performing the task in SPSS.

Square root: This method is often used when the data represent a count (rather than a continuous scale). For example, we might count the number of hospital admissions someone has (count data); this can be contrasted with the length of time that they may stay in hospital (continuous data). The data points in the variable are converted into the square root of that number, thus reducing the variance (the variance is equivalent to the standard deviation squared).

Reciprocal: This method is helpful when there is no specific upper limit to the values in the variable. The numbers are measured on a (potentially) infinite range (unlike questionnaires, where the limit is defined). Reciprocal transformation might be used when the actual magnitude of difference is not important. If we measured how long it took people to read this chapter we might expect an upper limit of 20 minutes, but find that someone took 2 hours (120 minutes). The extent to which they took longer than 20 minutes might be irrelevant in the study that we choose to conduct. A reciprocal score is anything where 1 is divided by 'x'. For example, $1 \div 20 = 0.05$; $1 \div 200 = 0.005$. Further adjustments are also needed with this method if we have '0' scores in the data, because numbers cannot be divided by zero (see Box 3.6).

Logarithmic transformation

Since there appears to be some positive skew in our example data, we will use logarithmic transformation to illustrate how to perform the procedure in SPSS (the other transformation procedures are fundamentally the same, we just change the method description). Before using this method we must check that there are no zero scores in our data because there is no log of '0' and it would mess up the calculations (see Box 3.6). In our example data we do not have any.

Using the SPSS file Mood and gender

Select **Transform → Compute** (as shown in Figure 3.27)

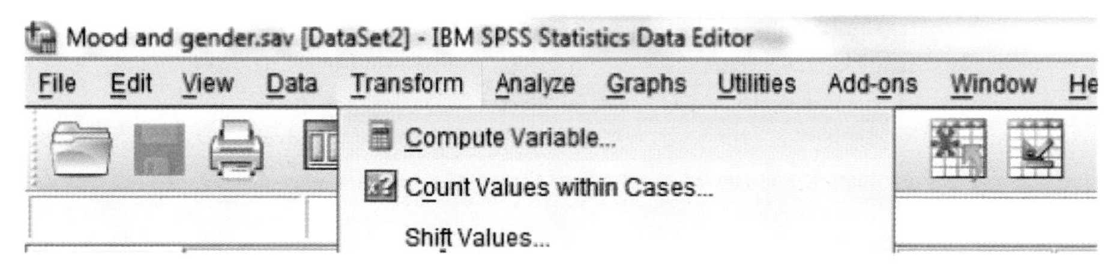

Figure 3.27 Transforming data - step 1

In new window (see Figure 3.28) type **moodlog** in **Target Variable** → click on **Type & Label** button → (in new window) type **Mood scores log transformed** in **Label** → tick **Numeric** radio button → click **Continue** → (back in original window) select **Arithmetic** from **Function group** → scroll and select **Lg10** from **Functions and Special Variables** → click on 'up' arrow (" LG10(?)" will appear in **Numeric Expression** window) → transfer **Mood scores** to **Numeric Expression** window using the arrow (it should now read "LG10(mood)" → click **OK** (see Figure 3.29 for completed action)

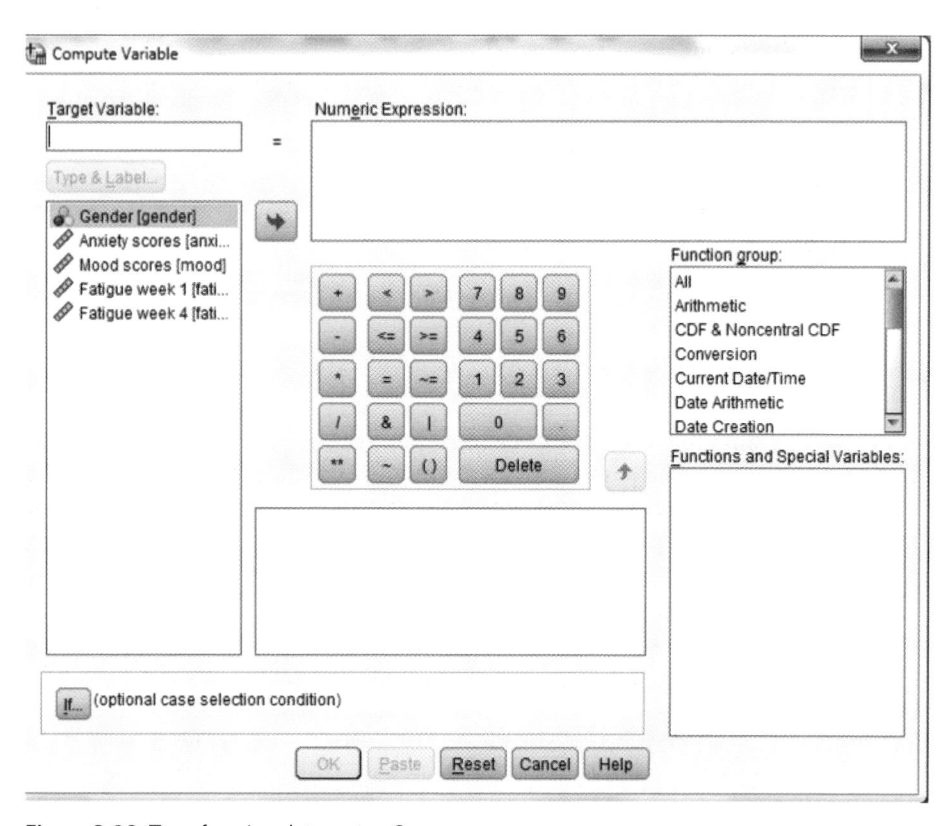

Figure 3.28 Transforming data - step 2

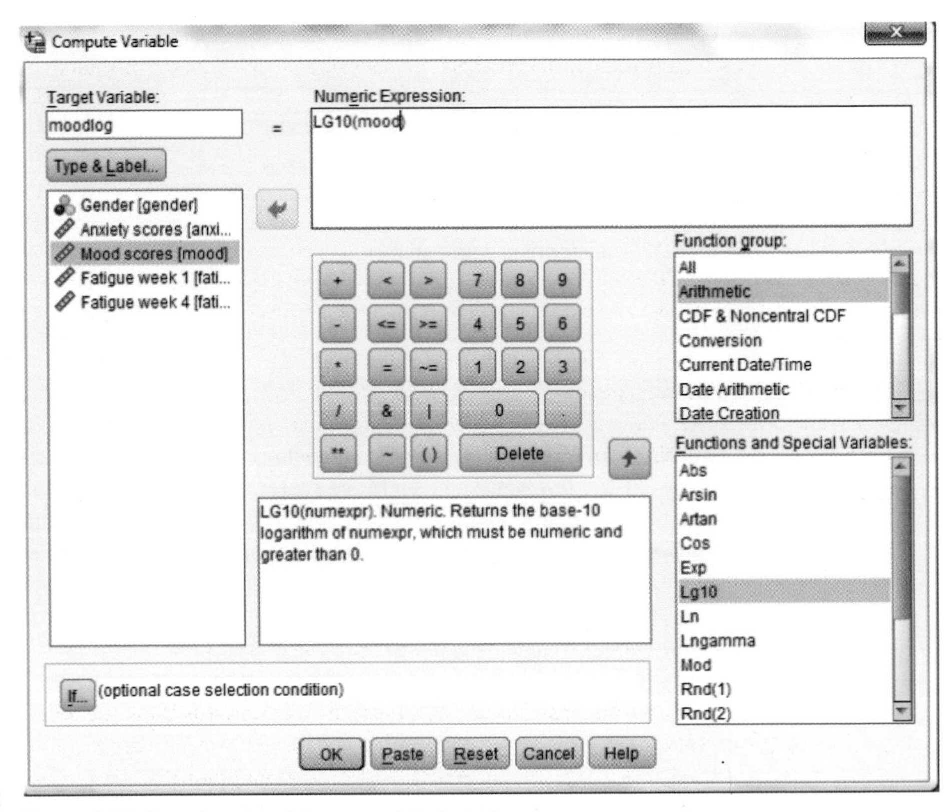

Figure 3.29 Transforming data – completed window

This action does not produce an output, but does add a new variable ('moodlog") to our data set (as shown in Figure 3.30).

*Mood and gender.sav [DataSet2] - IBM SPSS Statistics Data Editor

File Edit View Data Transform Analyze Graphs Utilities Add-ons Window Help

	gender	anxiety	mood	fatigue1	fatigue4	moodlog	var
1	Male	10	24	222	300	1.38	
2	Male	8	19	310	335	1.28	

Figure 3.30 Data set showing transformed variable

We can now use that new variable in the methods shown earlier to examine whether the transformed data are normally distributed (see Table 3.11).

Table 3.11 z-scores for skew/kurtosis for log-transformed mood scores by gender

DV: Mood scores		Statistic	SE	z-score
Male	Skewness	−0.715	.378	−1.89
	Kurtosis	1.556	.741	2.10
Female	Skewness	−0.862	.501	−1.72
	Kurtosis	0.746	.972	0.77

We can compare the outcome shown in Table 3.11 with what we saw in Table 3.7. The z-score for male mood scores now appears to be normally distributed.

3.6 Nuts and bolts

Getting rid of zero scores ahead of transformation

In some forms of transformation we must not have zero scores. We can use the Data - Transform function to eliminate them - we simply add '1' to all of the scores. If we had any zeros values in our example data, this is how we would have adjusted that:

Using the SPSS file **Mood and gender**
Select **Transform** → **Compute** → (in new window) in **Target Variable** type **moodplus1** →
transfer **Mood scores** to **Numeric Expression** window → type **+ 1** → click **OK**

We could then perform transformation on the new variable 'moodplus1'.

Homogeneity of between-group variance

When we explore specific statistical tests throughout this book, we will see that there are many other assumptions that we need to address beyond normal distribution. In between-group studies we need to account for something called (between-group) '**homogeneity of variance**'. Variance is the amount that scores vary around the mean score. When we are examining groups of data in respect of an outcome, the variance should be similar between the groups – we call that homogeneity of variances ('homogeneity' means 'sameness'). Although we might expect mean scores to differ between the groups, we need the extent that the scores vary either side of each mean to be similar. If they are not similar, it might affect the validity of the outcome. As we will learn as we progress through this book, statistical significance is often based on how much the scores vary. If we are comparing group means, we make false assumptions about differences if we do not account for how much the scores have varied within the group. The smaller the variance, the more likely we will have a significant outcome. This is a particular problem if the group sizes are unequal. If larger groups have greater proportional variance than smaller groups, we run the risk of understating significant outcomes; if larger groups have the smaller variance, we may be overstating the likelihood of significance. This may become clearer after you have read Chapter 4.

Because there are other things we need to know about measuring between-group differences, we will not explore how to investigate homogeneity of variance here. Instead we leave that until we get to those chapters where we will also look at how to interpret outcome and how to deal with violations in homogeneity.

Sphericity of within-group variance

As homogeneity of variance applies to between-group studies, sphericity is related to within-group variances. As we saw earlier, within-group studies are used to examine outcomes across a single group, but across several conditions. In some respects, the problem of individual differences that we encounter in between-group studies is reduced in within-group analyses. However, when we explore three or more conditions across the group we can have a problem if variance differs between pairs of conditions. We measure this with something called **sphericity**. Violations in sphericity may change the way in which we interpret the outcome. Once again, it would make sense to explore how we measure and interpret sphericity when we get to those chapters, later in this book.

Chapter summary

In this chapter we have explored normal distribution. At this point, it would be good to revisit the learning objectives that we set at the beginning of the chapter.

You should now be able to:

- Understand that normal distribution is achieved when data are evenly distributed either side of the mean score. This is illustrated graphically by a smooth, bell-shaped curve.
- Appreciate that normal distribution is important because it determines how much trust we can place in the mean score.
- Recognise that skewed data might be caused by extreme scores (outliers). High outliers may artificially inflate the mean score; low outliers may deflate it. Should that happen we may no longer be able to use the mean score to report outcome. Kurtosis describes the peakedness of the curve. A leptokurtic distribution is shown by an abnormally 'peaked'curve where there is too little variation in the data; a platykurtic distribution is short and flat, with too much variation.
- Appreciate that we can measure normal distribution graphically and statistically. Graphs can include histograms, box plots and stem-and-leaf plots. These can be used in conjunction with statistical evidence. We can analyse normal distribution through the Kolmogorov–Smirnov and Shapiro–Wilk tests and by examining the z-scores of the skew and kurtosis. We can also count the number of outliers to ascertain whether we have normal distribution.
- Recognise ways in which we can deal with potential violations in normal distribution. If those violations are minor we can simply report the outcome cautiously. When deviations from normal distributions are more serious we can make some adjustments to the outliers, or (preferably) we can transform the variable. Where there are serious violations we may need to abandon parametric tests and examine outcome with procedures that do not rely on the mean score.
- Understand that we could remove outliers (but only if there appear to be errors), or replace them with a score that represents the mean score plus two standard deviations. Preferably, we can undertake transformation. There are many methods that we can use to transform the data. The most popular are log-transformation, square root transformation and reciprocal transformation.
- Recognise some basic concepts regarding homogeneity of between-group variances and sphericity of within-group variances. We will explore these more fully in subsequent chapters.

Extended learning task

You will find the SPSS data associated with this task on the website that accompanies this book. You will also find the answers there.

Following what we have learned about normal distribution, answer the following questions and conduct the analyses in SPSS. For this exercise, we examine responses from 350 participants regarding quality of life, depression, anxiety, sleep and relationship perceptions. These are explored in respect of gender.

Open the SPSS data set **QOL and Gender**.

1. Conduct tests for normal distribution, using appropriate graphical and statistical analyses for the following variables:
   a. Depression as the dependent variable, in respect of gender (the independent variable).
   b. Relationship satisfaction as the dependent variable, in respect of gender (the independent variable).
2. If there remains a problem with normal distribution, transform the data using an appropriate method.

4 SIGNIFICANCE, EFFECT SIZE AND POWER

Learning objectives

By the end of this chapter you should be able to:

- Understand the principles of establishing significant differences or relationships
- Recognise the role that probability plays in examining significance
- Understand the definitions of null and alternative hypotheses, one-tailed and two-tailed tests, and Type I vs. Type II errors
- Appreciate the importance of variance, standard deviation, standard error and confidence intervals in measuring significance
- Recognise the principles of sampling distributions and central limit theorem
- Appreciate how we put all of this together to estimate statistical significance
- Understand the importance of effect size and statistical power

Introduction

In this chapter we explore some of the most important factors in statistical analyses of research data. We begin with the key concept of statistical significance. As we will see, this indicates the probability that an observed outcome has occurred by chance. Alternatively, the result may be due to fundamental differences between the groups or conditions that we are measuring (or because of the **association** between variables). To fully understand the principles of significance, we will need to explore the laws of probability. We can then use this to test hypotheses. We will also look at effect size and statistical power. Effect size describes the strength of the relationship in relation to sample size and average variation – it is a very useful supplement to statistical significance. Statistical power describes the extent to which the data are robust enough to find that effect.

Statistical significance

In statistical analyses, we should use the word 'significance' with caution. It is a common error to call any 'big' difference 'significant'. It may make the difference sound more convincing, but 'significance' can be used only once the differences have been subjected to rigorous statistical testing. In short, statistical significance examines the likelihood that an outcome happened by chance. It is measured from 0% (it could not have happened by chance) to 100% (it must have happened by chance) – the former is very unlikely in psychology research.

In psychological studies we often aim to support a prediction about some kind of outcome – we call this prediction a '**hypothesis**'. We will learn more about these hypotheses a little later, but we need a brief overview now to underpin an important issue. You might be forgiven for believing that much of psychological research is about proving hypotheses. However, statistical analyses are actually about 'rejecting the null hypotheses'. An '**experimental hypothesis**' may predict that observed differences in an outcome between groups of people was due to the factor that we are examining (we usually call this the **alternative hypothesis**, for reasons that will become clearer later). In contrast, the '**null hypothesis**' states that there are no differences, or that observed differences were due to chance (and not because of the factors being measured). That's where statistical analyses and probability come in. We usually say that an outcome is 'statistically significant' if there is a less than 5% probability that it happened by chance or (more precisely) that there is a less than 5% probability that the null hypothesis is true. If that chance likelihood is less than 5%, we report that in terms of probability (p). We say that an outcome is (statistically) significant if 'p' is less than 0.05 (which is 5% written as a decimal); we usually report that $p < .05$ (but more of that later).

In Chapter 5 we will explore an overview of the statistical procedures that we cover in this book. The type of statistical test we employ to investigate data will depend on a number of factors, relating to the nature of the data and the method of collection. We will not go into too much detail about all of that here, but we can summarise some key points. In very general terms, statistical outcomes in psychological research will fall into one of four categories: between-group studies, within-group studies, associations and **mixed designs** (which combine any of these types). Between-group studies explore dependent variable outcomes across two or more distinct groups (the independent variable). We introduced the terms dependent and independent variable in Chapter 3 (see Box 3.3). For example, we might investigate how mood scores differ between men and women. Within-group studies examine dependent variable outcomes for a single group, but do so across a series of conditions (such as time points). For example, we might measure stress levels in a group of people before and after they watch a scary movie. Associations measure the relationship between variables. For example, we might measure the relationship between income in a group of people and the amount of money they spend on luxury goods. In all of these examples, when we measure significance we are exploring the probability that the observed difference or relationship occurred due to chance factors.

Significance and probability

In probability statistics, we expect events to happen in a uniform, predictable manner. If we toss a coin, we know that there is a one-in-two chance of getting heads. If we tossed that coin ten times, we might be justified in believing that it would be unlikely to get heads on each occasion. But when does something become so unlikely that it is statistically significant? Up to which point do we remain confident that our observation happened by chance? We need some kind of measure that provides an objective way of making that judgement. The laws of probability play a very large part in how we determine a difference or relationship to be 'significant'.

4.1 Nuts and bolts
Probability in action

The following example demonstrates how we can use probability to predict the likelihood of getting heads when tossing a coin:

If we toss one coin, we have a one-in-two chance that we will get 'heads'. We can write that in fraction form as $\frac{1}{2}$, in decimal form as 0.50, or as a percentage (50%).

If we tossed two coins in succession, the chances of both coming up heads is:

$\frac{1}{2} \times \frac{1}{2} = \frac{1}{4} = 0.25 = 25\%$

If we tossed three coins in succession, the chances of all of them coming up heads is:

$\frac{1}{2} \times \frac{1}{2} \times \frac{1}{2} = \frac{1}{8} = 0.125 = 12.5\%$

So, the more times we toss the coin, the less likely it is that we will get heads on each occasion. Let's take that to the extreme and measure the likelihood of getting heads every time following ten tosses:

$$\frac{1}{2} \times \frac{1}{2} \times \frac{1}{2} \times \frac{1}{2} \times \frac{1}{2} \times \frac{1}{2} \times \frac{1}{2} \times \frac{1}{2} \times \frac{1}{2} \times \frac{1}{2} = \frac{1}{1024} = 0.001 = 0.1\%$$

The example shown in Box 4.1 illustrates how we use probability in predicting outcome. Each time we toss the coin, we might have some opinion about the likelihood of getting heads. If we get three, four or even five heads in a row, we might think that is unusual, but perhaps not beyond the bounds of probability. But how might we consider the likelihood of getting seven or eight consecutive heads? Or even nine or ten? When would we start thinking that the coin might be biased? It might be reasonable to say that we might question the bias of the coin after six consecutive heads. But what do the laws of probability say about that?

We can illustrate how probability works by using the same coin example. Let's say we take a coin and toss it ten times and record the number of heads we get. We then repeat that on several more occasions, recording the number of heads we get in each batch of ten throws. If we do this enough times, we will see a pattern emerge that reflects probability factors. On most occasions, we might reasonably expect to get five heads in a batch of ten throws.

Table 4.1 Probability factors for getting heads in ten tosses of a coin

No. of heads	Probability
0	.001
1	.010
2	.044
3	.117
4	.205
5	.246
6	.205
7	.117
8	.044
9	.010
10	.001

However, sometimes we will get four heads from ten throws, other times we might get six; on a few occasions we might even get ten or none. The data from a block of several trials are presented in Table 4.1 – the outcome is shown in terms of probability factors. As might be expected, the most common outcome is five heads from ten tosses, followed by four and six, then three and seven, and so on. The least likely outcome was ten heads and no heads. We could also plot the outcome in a histogram, as shown in Figure 4.1.

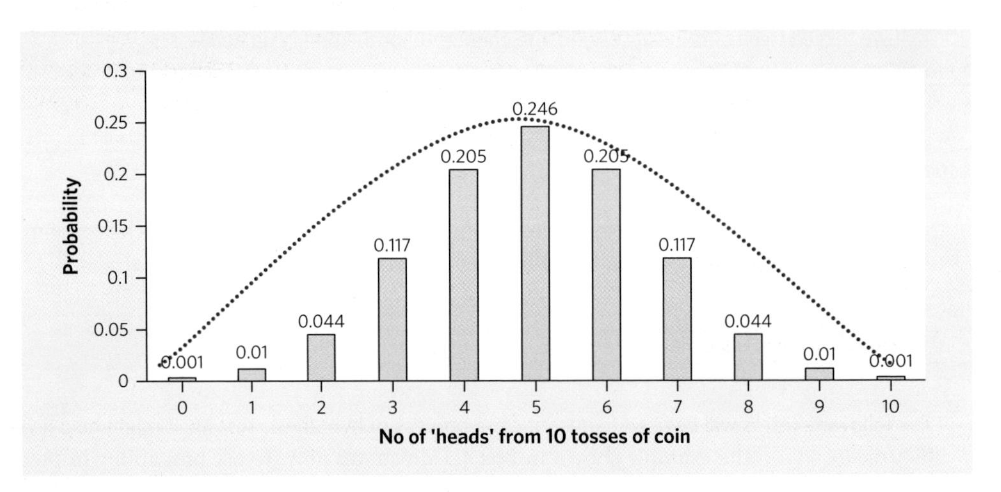

Figure 4.1 Distribution of probability of getting heads from ten tosses

You might have noticed that the data in Figure 4.1 look like something we saw in Chapter 3 when we explored normal distribution, especially with the curve added to show the trend of outcome. Much of what we do in significance testing is related to looking at data within a normal distribution. In this example, if the coin is 'normal' and we perform enough blocks of tosses, the laws of probability dictate that the outcome will tend towards a normal distribution. The most likely outcome (five heads) will be the most common and will represent the mean outcome. The outer tails of the distribution then become important in determining where we might consider the outcome to be statistically outside the bounds of normality.

Earlier, we said that we usually say that an outcome is statistically significant when there is a less than 5% probability of it happening by chance. We should put this in context of what we have just seen with the coin example. If we refer to Figure 4.1, we can see that there is a 0.1% probability of getting ten heads in ten tosses (where $p = .001$). To find the probability of getting nine or more heads, we add the probability of getting nine heads and the probability of getting ten heads ($0.01 + 0.001 = 0.011$). So, the probability of nine or more heads is still comfortably beyond chance likelihood, if we use 0.05 as the cut-off point. The probability of getting eight or more heads from ten throws is 0.055 ($0.044 + 0.01 + 0.001$). This is just about at the limits of chance. This tells us a great deal about the chance likelihood when we employ $p = .05$ is the cut-off point for significance. In our coin example, only an outcome of getting eight or more heads from ten tosses is (almost) statistically outside chance factors. Now imagine we observed that women scored significantly poorer mood scores than men ($p < .05$). This suggests that there is a less than 5% probability the observed difference in mood scores between men and women happened by chance. How unlikely is that? It is about as unlikely as getting eight or more heads every time you toss ten coins. The outcome is very likely to have occurred because there is a very real difference between men and women in respect of mood (at least in that sample in any case).

Significance and hypotheses

As we saw briefly earlier, a key aspect of research involves making predictions about what we expect an outcome is likely to be. We call these predictions experimental hypotheses. We specify these hypotheses on the basis that there will either be no difference or that there will be one. Statistical significance is used to examine those hypotheses. Throughout this book we will encounter a series of statistical procedures that aims to test our predictions. We will explore a summary of the most common experimental research methods in Chapter 5. However, it is important that we understand the concept of hypothesis testing here, and how we use statistical significance to explore that. Before we start, it might help if we define some of the key terms that we use when testing hypotheses (see Box 4.2).

4.2 Nuts and bolts
Terminology in hypothesis testing

The following terms will be used to describe the process of hypothesis testing. It might help if we understood what they mean:

Null hypothesis	There is *no* difference (or there is *no* relationship) between the variables. Or, the observed difference between 'X' and 'Y' is *not* because of 'Z'.
Alternative hypothesis	There *is* a difference (or there *is* an association) between the variables. Or, the observed difference between 'X' and 'Y' *is* because of 'Z'.
One-tailed hypothesis	A specific prediction regarding the direction of an outcome, stating *how* the variables will differ (e.g. that 'A' will be higher than 'B').
Two-tailed hypothesis	A non-specific prediction just stating there will be a difference or relationship (e.g. that 'A' will differ from 'B').
Type I error	Where the null hypothesis is rejected when it should have been accepted.
Type II error	Where we fail to reject the null hypothesis when we should have done so.

Null hypothesis vs. the alternative hypothesis

We can illustrate the process of hypothesis testing with an example. Let's say we collect some data from 40 men and 40 women about their current mood. We do this by giving everyone a questionnaire that asks all sorts of questions about happiness and satisfaction. Each questionnaire is assessed by scoring the answers, where a higher score indicates poorer mood. Based on previous evidence, we might predict that women will report poorer mood scores than men. That prediction would be our (alternative) hypothesis. By contrast, the null hypothesis would be that there will be no difference in mood scores between men and women. To test our prediction, we must investigate whether we can reject the null hypothesis (or not) before we can say anything about the alternative hypothesis. Why? Well, it goes back to the point we were making about probability in statistical significance. By stating that there is less than 5% probability that an outcome occurred by chance, we are actually saying that there is a less than 5% probability that the null hypothesis is 'true' (that there is no difference).

Once we have collected the data, we might observe that women have indeed reported higher mood scores than men. Statistical analyses might show that there is a 3% probability that the outcome occurred by chance. Because this is lower than the 5% cut-off point that we usually set for significance, it would appear that our prediction is correct. However, this is only half of the picture – the process of testing hypothesis testing must start with the null hypothesis. According to our results here, we can reject the null hypothesis because there is not enough evidence to support that it is true (because the outcome was significant at $p = .03$). As a result, we can say that the null hypothesis is rejected *in favour of* the alternative hypothesis. Strictly speaking, we *cannot* say that we have '*accepted* the alternative hypothesis' (although many people do this, even in the most prestigious journals).

Similarly, we might still find that women reported higher mood scores than men, but statistical analyses suggest that there is a 6% probability that the outcome occurred by chance (where $p = .06$, or $p > .05$). Because this is greater than the 5% cut-off point, we cannot reject the null hypothesis. This does not mean that the null hypothesis *is* true, but simply that there is not evidence that it is a false. Once again, strictly speaking, we should not say that the alternative hypothesis is rejected (although, again, many researchers do say that), we should always phrase the outcome in terms of the null hypothesis.

When we *make* predictions, we should state this in our reports in terms of what we *expect* to find. However, when we report the findings we should say one of two things: 'statistical analyses suggest that we can reject the null hypotheses, in favour of the alternative hypotheses' or 'statistical analyses suggest that we cannot reject the null hypothesis'.

One-tailed vs. two-tailed hypotheses

When we make predictions we could be specific and state that females will report poorer depression scores than men, or we could be more general and say that mood reports will differ between men and women. The first statement is an example of a **one-tailed hypothesis** – a specific, directional prediction. In another research study, we might predict that patients' anxiety scores will improve after undergoing cognitive therapy. Or we might posit that the sale of ice creams will increase as temperature increases. All of these are examples of a one-tailed hypothesis. In contrast, a two-tailed hypothesis is a general, non-directional prediction. For example, we might speculate that anxiety scores will be different before and after cognitive therapy. Or we could suggest that there will be a relationship between ice cream sales and temperature. The use of one-tailed or **two-tailed hypotheses** has an impact on how we interpret significance. So why we do we refer to these predictions in terms of tails?

When we first encountered normal distribution in Chapter 3, we saw that there are two 'tails' at either end of the curve. These tails relate to portions of the distribution of scores where values are least likely. In Figure 4.1, we saw the probability distribution of tossing a coin ten times. The least likely outcomes were zero heads and ten heads, followed by one heads and nine

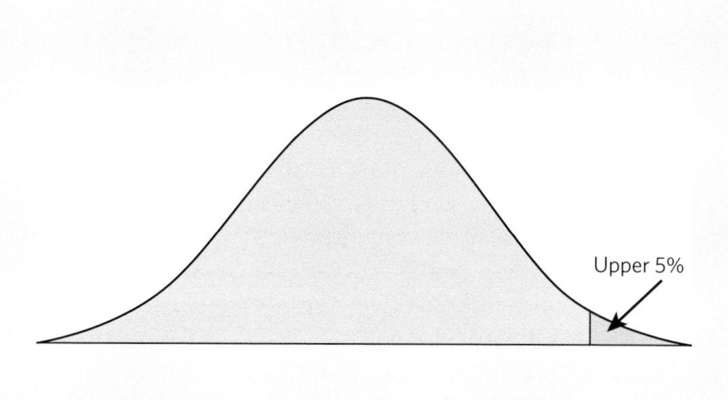

Figure 4.2 One-tailed (positive) test

heads, and so on. We said that the probability of getting eight or more heads in ten tosses of the coin was (roughly) in the outer 5% of the distribution (so was significantly unlikely). To be more precise, this outcome was located within the upper 2.5% of that distribution. We equally could have said that getting two or fewer heads in ten tosses is located in the lower 2.5% of that distribution.

Significance with one-tailed tests

When we test hypotheses we will (usually) set the significance level at 5%. If we employ a one-tailed test, we are predicting that our 'outcome' will reside in the outer 5% of one end of the sampling distribution (we will see more about sampling distributions later). If we predict that A will be greater than B, we would expect to find the outcome in the upper 5% of the sampling distribution (see Figure 4.2). For example, we might predict that mood scores will be higher for women than for men. If we find that women *do* report higher mood scores than men *and* statistical analyses indicate that there is a less than 5% probability that this happened by chance, we can reject the null hypothesis (in favour of the alternative hypothesis). If men score more highly than women (even if there is a less than 5% probability that this occurred by chance), we cannot reject the null hypothesis (because the outcome contradicts our prediction).

However, if we predict that X will be less than Y, we would expect to find the outcome in the lower 5% of the sampling distribution (as shown in Figure 4.3). For example, we might predict that IQ scores of cats might be less than for dogs. If we find that cats present lower IQ scores than dogs, and statistical analyses indicate that there is a less than 5% probability that this happened by chance, we can reject the null hypothesis.

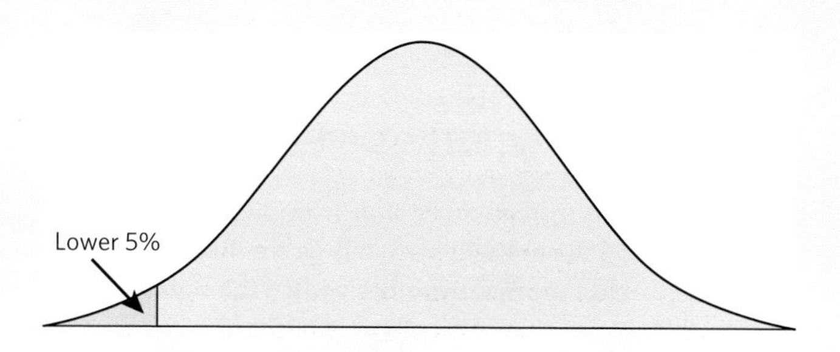

Figure 4.3 One-tailed (negative) test

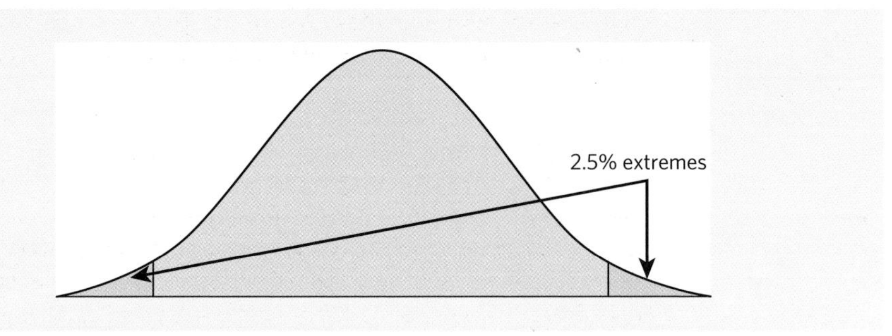

Figure 4.4 Two-tailed test

Significance with two-tailed tests

Sometimes, we may not have enough evidence to make a specific prediction. However, we might be able to suggest that there will be a difference, without specifying the direction of that difference. For example, we could predict that there will be a difference in the hours spent in lectures across the student groups, but not predict *which* group will spend more time in lectures than the other. In this instance, we have made a two-tailed hypothesis. In a non-directional test, we still (usually) set the significance level at 5%, but we have to share that between the two tails of the distribution because the difference could reside at either end. Our significance level at either end is now 2.5%, as shown in Figure 4.4. If we find that there is a difference between the groups in respect of hours spent in lectures, and statistical analyses indicate that there is a less than 2.5% probability that this happened by chance, we can reject the null hypothesis.

4.3 Nuts and bolts

Don't move the goal posts!

As we saw just now, when we state a two-tailed hypothesis we must divide the significance cut-off between the tails of the distribution. If we use the traditional cut-off point of 5% to determine significance, we must share that between the two tails. Therefore, we can reject the null hypothesis only if the significance is less than 2.5% (where $p < .025$). For example, you might predict that income will differ between doctors and nurses.

When the data are collected you find that income does indeed differ between the groups (doctors earn more than nurses). However, statistical analyses indicate that there was a 3% probability that the difference was due to chance (where $p = .03$). Because the prediction was two-tailed, the null hypothesis cannot be rejected because the significance was greater than 2.5%. You might be annoyed that you did not demonstrate your prediction. Had you stated a one-tailed hypothesis in the first place (that doctors will earn more than nurses), you would have been able to reject the null hypothesis (because the significance is less than 5%). However, you cannot simply change a two-tailed hypothesis into a one-tailed prediction just to fit the statistics.

Errors in hypothesis testing

A key factor to remember with hypothesis testing is that we are dealing with probability, not certainty. Statistics will only tell us the likelihood that the outcome occurred by chance. We usually reject the null hypothesis when the significance is less than 5%. Earlier, we compared the probability of chance factors explaining significant outcomes as being as likely as getting eight or more heads in a series of ten tosses of the coin. But no matter how compelling the evidence might be, there is still a chance that we have made a false assumption. We might reject the null hypothesis when we should not have done so, or we might decide not to reject the null hypothesis when we should have. We refer to the outcomes as Type I and Type II errors respectively.

Type I error

A **Type I** error occurs when we incorrectly reject the null hypothesis in favour of the alternative hypothesis. We may have had good reason for rejecting the null hypothesis, the most likely one being that we had a significant outcome (where $p < .05$). However, we might find other evidence that causes us to question that initial assumption. In the example we used earlier, we might find that there was no difference between men and women in respect of mood scores after all (despite what the statistics tell us). There might be several reasons for this:

- We have set the significance level too high. When we use the traditional 5% cut-off point for significance, there can be up to a 1-in-20 likelihood that the outcome happened by chance. This means that if we were to repeat our test 20 times, we could get a significant outcome just by probability factors alone. To try to avoid this, we could be more cautious and set significance at $p < .01$; we would reject the null hypothesis only if there was less than 1% probability that the outcome occurred by chance.
- Related to this last point, we may undertake several analyses of a single data set. The more tests we do, the more likely it is that we will get a significant outcome. To account for this we should adjust the significance cut-off point accordingly. For example, a Bonferroni correction divides the significance cut-off ($p < .05$) by the number of tests undertaken, So, if we performed three tests on the same data, we should divide '.05' by 3; now we will have only a significant outcome when $p < .016$.
- The method of data collection was biased. For example, the questions inadvertently might have led women to report poorer mood, rather than reflect actual mood.
- We might be using inappropriate statistical analyses. For example, we might be using parametric statistics to analyse data that are not normally distributed.
- Sometimes, we might think that our data are measuring a concept when they are actually measuring something else. In our example, we suggested that we were measuring mood via a questionnaire. Women may well have reported higher scores than men, so we conclude that women were probably reporting poorer mood. However, what if the questions were actually asking about happiness with factors such as sleep? What if sleep was poorer in the female group because most of them were either pregnant or new mums? Although there may have been significant difference in the 'scores', it may be measuring sleep satisfaction and not mood after all.

Type II error

A **Type II** error occurs when we do not reject the null hypothesis when we should have done so. The most likely reason for not rejecting the null hypothesis is that the outcome was non-significant. In our example we are suggesting that there is no difference between men and women in respect of mood scores. However, we might subsequently find evidence that there is a difference and we were wrong not to reject the null hypothesis. There might be several reasons for this, but the most likely one is that our study and/or sample lacked power. We will explore statistical power later in this chapter. We should be designing our studies in such a way that we reduce the likelihood of making Type II errors – we may be missing important effects. Cohen (1992) said that we should avoid getting Type II errors on more than 20% of occasions (we should aim to find at least 80% of true effects). Here some reasons why we might lack power:

- We may not have a sufficient sample to find the true 'effect'. When writing up research reports, it is quite common for students to say that they might have achieved a significant outcome if they had recruited more people. On its own, that statement is probably a bit lame (and should be avoided). However, it is true to say that the study would have achieved more power with a more robust sample. If we know what sort of outcome we are looking for, we can use 'power statistics' to estimate how many people we need to recruit to make it more likely that we will find that effect (we will see more about power and effect size later in this chapter).

- There are too many outliers in the sample, relative to the sample size. While an outlier may be legitimate, it might also be an anomaly that has occurred just in this sample. Replications of the study with new samples may reinforce that the outlier is not representative. If we 'exclude' the outlier we might be able to reject the null hypothesis and eliminate the Type II error.
- The study design might be inappropriate. It is vital that the data-collection materials are sensitive enough to find the effect being sought. Vague questions and poor definitions can lead to inconsistent responses.

Replication

The very presence of Type I and Type II errors reinforces the need for studies to be replicatable. If we conduct a study and reject the null hypothesis, we need to make sure that we have not committed a Type I error. We can get other researchers to investigate the study using our methods, but they can do that only if we provide enough information about how we did it. If several researchers repeat our study and also get positive results, the likelihood of a Type I error is very small indeed. Replication is also useful in the reduction of Type II errors. We may have failed to reject the null hypothesis because of some outliers. If other researchers repeat our work and achieve positive outcomes, it strengthens our claim that our outliers were an anomaly (and that we would have rejected the null hypotheses had they been absent).

Measuring statistical significance

The method for calculating significance varies with each type of statistical test, so it is better that we leave the precise techniques until we explore those tests. However, we can look at some general matters. As a rule, significance calculations will be based on one or more of three key determinants: **variance**, **standard deviation** and **standard error** (we will define these terms shortly). **Parametric** tests base outcomes on mean scores; significance often focuses on how mean scores differ between groups or across conditions. Significance in **non-parametric** studies is more likely to focus on median scores and on how ranked scores differ between groups or across conditions. We will explore the concept of parametric data in more depth throughout Chapter 5. We defined 'mean' and 'median' scores in Chapter 3 (see Box 3.1).

To estimate significance, an outcome score is often compared to a 'known' distribution of scores. The actual distribution that is used varies according to the type of statistical test being employed (some of the most commonly used include the t-distribution, F-distribution, z-score distribution and chi-squared distribution). In any distribution, probabilities have been calculated for a range of scores for every possible sample size. Significance is established by whether the observed outcome exceeds cut-off points within the known distribution. Those cut-off points vary according to sample size, the level of significance being set (usually $p = .05$) and, for some tests, whether we are employing a one-tailed or two-tailed test. We should now explore how key factors such as variance, standard deviation and standard error play a role in determining significance.

Variance

Variance (σ^2) is demonstrated by the extent that scores vary around the mean score. The mean is the average score and is calculated by dividing the sum of all the scores in the data by the number of scores. Each score in the distribution will vary from that mean: some will be less than the mean; others will be greater than the mean. We need to know the 'average' variation, as this will tell us something about how the data are spread. However, if we used the pure data, that average would be zero (add up the red numbers in Box 4.4), so we 'square' the variation to get a whole number. You can see how to calculate variance in Box 4.4.

4.4 Calculating outcomes manually
Variance

To illustrate how to calculate variance in a sample, we will use the data from Table 4.2, which refer to the distribution of 'mood scores' across a sample of 11 people.

Table 4.2 Mood scores

Scores (x_i)	9	12	14	15	15	16	18	18	19	24	30
$x_i - \bar{x}$	−8.3	−5.3	−3.3	−2.3	−2.3	−1.3	0.7	0.7	1.7	6.7	12.7
$(x_i - \bar{x})^2$	68.4	27.8	10.7	5.2	5.2	1.6	0.5	0.5	3.0	45.3	162.0

Mood scores are represented by x_i; the mean of that range is 17.3. We deduct the mean from each score to get $x_i - \bar{x}$ (shown in red font). We square that to get $(x_i - \bar{x})^2$ We need to 'sum' all of the outcomes to obtain $\Sigma(x_i - \bar{x})^2$.

The formula for variance is $\sigma^2 = \dfrac{\Sigma(x_i - \bar{x})^2}{N - 1}$ $\Sigma(x_i - \bar{x})^2 = 330.20$ and $N = 11$ so $\sigma^2 = \dfrac{330.20}{10} = 33.02$

Standard deviation

Standard deviation (s) is the average variation in that sample. As we saw just now, all values in a distribution will vary from the mean score, being either higher or lower. Because of that, the pure average would be zero, so we square the differences to find the variance (σ^2). To get the standard deviation, we simply find the square root of the variance. We could use the example data from Table 4.2 to show how we can calculate the standard deviation of a sample.

$$\text{Standard deviation } (s) = \sqrt{\sigma^2} = \sqrt{33.02} = 5.75$$

Confusingly, some sources allocate the symbol σ to represent standard deviation. In reality that is better used to denote standard error (see below). In this book we will use the symbol s to represent standard deviation.

Standard error

In short, standard error (σ) is an estimation of standard deviation in the entire population. When we analyse a sample of people in respect of an outcome we know that this sample is only a very small proportion of the population. The mood scores shown in Box 4.4 are taken from only 11 participants. It would be unrealistic to propose that these mood scores represent everyone in the world (often referred to as the ***population***). To get a better representation of that population, we could collect data from many samples. This task would be onerous, so we can use statistics to 'model' those theoretical samples. We call this a *sampling distribution*. Had we actually collected all possible samples, each sample would have a different mean and standard deviation. In a sampling distribution, we assume that the mean is the same as it is in the entire population, so long as that population is normally distributed. In repeated collections of samples, the mean would automatically tend to the population mean in such cases. However, the sampling distribution also has a standard deviation. To make any assessment of probability regarding statistical significance, we must know the mean and standard deviation of the sample and the sampling distribution. The standard deviation of the sampling distribution is called the *standard error of the mean* (often just referred to as standard error).

We can calculate the standard error of the mean in a sampling distribution ($\sigma_{\bar{x}}$) from the standard deviation of the population (S), the size of the population (N) and the sample size (n):

$$\sigma_{\bar{x}} = S\sqrt{\frac{1}{n} - \frac{1}{N}}$$

4.5 Calculating outcomes manually

Standard error of mean (when the population is known)

When we know the mean and standard deviation of the population, we can calculate the standard error based on what we know. We can illustrate this with some data. We will stay with the mood scores example that we used earlier. Let's say that the 'population' refers to all 25-year-old people in a town in the UK, representing 500 people (N). The mean mood score is 18 and the standard deviation of the population is 2. If we collect some new data from 15 people (n), what is the standard error of this sampling distribution?

$$\sigma_{\bar{x}} = S\sqrt{\frac{1}{n} - \frac{1}{N}} = 2 \times \sqrt{\frac{1}{15} - \frac{1}{500}} = 2 \times \sqrt{0.0647} = 0.509$$

Standard error for infinite populations

The calculations that we have just seen are all very well, so long as we know enough about the overall population, namely the size and standard deviation. In reality we rarely know that. In theory, the sample size is likely to be infinite. Because of that, we need to adjust the calculation of the standard error; we make an estimate of the standard deviation of the entire population. The formula for that is probably more familiar. We divide the sample standard deviation by the square root of the sample size:

$$\text{Standard error } (\sigma) = \frac{s}{\sqrt{n}} \text{ where } s = \text{ the sample standard deviation}$$

$$\text{So, using our data, } \sigma = \frac{5.75}{\sqrt{11}} = 1.73$$

Standard deviation in significance testing

When we examine probability in statistical significance we often use standard deviation or (more likely) standard error somewhere in that process. We will begin this illustration with a simple examination of probability of outcomes within a sample of scores. As we saw earlier, any score that is located within the outer 5% limits of a distribution can be considered to be an outlier (with 2.5% at either end of the distribution). We can use the mean score and standard deviation of that sample to calculate the probability that a score is an outlier. The mathematics involved in calculating probability within any distribution is complex, but there are several very good probability calculators available online (e.g. http://stattrek.com/Tables/Normal.aspx) – see Figure 4.5.

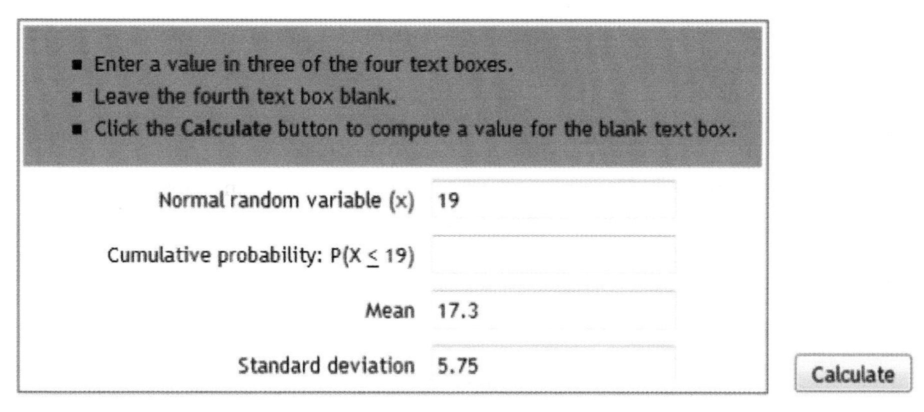

Figure 4.5 Normal distribution calculator

We could use the data from Box 4.4 as an example. Let's see whether some of those mood scores represent outliers in the distribution of data shown in Table 4.2. We said that outliers occur in the outer 5% extremes of a distribution (2.5% in either tail). If the score is within the range of normally distributed data, the probability will be between 2.5% and 97.5%.

We will start with a value of 19: Enter 19 in **Normal random variable**
Enter 17.3 in **Mean**
Enter 5.75 in **Standard deviation**
Click on **Calculate**

The 'answer' will appear in **Cumulative probability: P (X ≤ 19)**; in this case: 0.616. It means that there is a 61.6% probability that scores will be less than 19. This is within the range 2.5% to 97.5% – it is probably not an outlier.

Now we will try 30 (so do the same again, but enter 30 in 'Normal random variable' instead of 19, before clicking on Enter again).

This time the 'answer' is 0.986 – there is a 98.6% probability that scores will be less than 30. This is outside the range 2.5% to 97.5% – it is probably an outlier.

To fully illustrate what we are trying to demonstrate, we will try a low number (5). This was not in our original range, but it is useful for showing what happens at the other end of a distribution.

Using the procedures that we have seen before, the new answer is 0.016 – there is a 1.6% probability that scores will be less than 5. This is outside the range 2.5% to 97.5% – it is also probably an outlier, but at the lower end of the distribution.

Standard error in significance testing

We often use standard error to examine probability in significance testing. We can use the standard error to create a z-score, from which we can estimate the likelihood of a significant outcome. As we saw in Chapter 3, a z-score is a value that resides within the normally distributed z-score distribution. In that distribution, we have a mean score of 0 and standard deviation of 1. We can obtain a z-score by dividing a value by the standard error of means for the sampling distribution. We can use these principles in significance testing. For example, we might want to examine whether mean scores differ significantly between groups or across conditions. We will focus on group differences to illustrate the point.

To assess whether observed differences in scores between the groups are statistically significant, we need to know the mean score and standard deviation of each group. From this we can calculate variance and standard error for each group. Crucially, we can also calculate something called the ***standard error of differences***. To demonstrate how this works, we will maintain the focus on mood scores, but use some new data based on 32 men and 32 women. The male group has a mean mood score of 17.34, with a standard deviation of 5.78, while the female group mean is 20.91, with a standard deviation of 6.81.

4.6 Mini exercise
Calculate standard error and variance

Based on what we learned earlier, calculate standard error and variance for men and women in respect of mood scores, using the information that you have just been given about the mean score and standard deviation.

Answer: *Male*: mean 17.34; standard deviation (*s*) 5.78; n = 32:

standard error $= s \div \sqrt{n} = 5.78 \div \sqrt{32} = 1.02$; variance $= \sigma^2 = 5.78^2 = 33.46$

Female: mean 20.91; standard deviation (*s*) 6.81; *n* = 32:

standard error $= 6.81 \div \sqrt{32} = 1.21$; variance $= 6.81^2 = 46.47$

To estimate the probability that there is a significant difference in mood scores between men and women (in these samples) we use the mean difference and the standard error of differences. The mean difference is found simply by deducting one mean score from the other (male, 17.34; female, 20.91: mean difference $= 20.91 - 17.34 = 3.57$). To estimate the standard error of differences ($\sigma\,\bar{x}$), we need to refer to the variance (σ^2) for each group. The formula is shown below:

$$\sigma\bar{x} = \sqrt{\left(\frac{\sigma_a^2}{n_a}\right) + \left(\frac{\sigma_b^2}{n_b}\right)} = \sqrt{\left(\frac{33.46}{32}\right) + \left(\frac{46.47}{32}\right)} = 1.58$$

In the same way that we can divide any value by the standard error to get a z-score, we can divide the mean difference by the standard error of differences:

$$3.57 \div 1.58 = 2.26$$

We can apply this to the z-score distribution (which we know has a mean score of 0); we need to examine how this z-score (2.26) differs from 0. We can use the normal distribution calculator to help us here again (using a mean score of 0 and a standard deviation of 1).

Enter 2.26 in Standard score
Enter 0 in Mean
Enter 1 in Standard deviation
Click on Calculate
You will get the 'answer' as 0.988. It means that there is a 98.8% probability that remaining z-scores will be less than 2.26. This is outside the range 2.5% to 97.5%, so it is probably an outlier. It *suggests* that there may be a significant difference between the mean mood scores of men and women in this example, where $p < .05$.

Let's look at this in more depth. When we examined z-scores for normal distribution in Chapter 3, we said that a z-score greater than ± 1.96 indicated an outlier, where $p < .05$. We should put that in context, using the normal distribution calculator.

Enter 1.96 in **Standard score** (keeping mean and standard deviation as 0 and 1). We get an answer of .975, indicating that we are at the 97.5% limit for the upper tail of the normal distribution. Now enter −1.96 and repeat the process. We get an answer of .025, indicating that we are at the 2.5% limit for the lower tail of the normal distribution.

What we have just seen is a useful indication of how we might gauge significance in a two-tailed test, where the 5% significance parameters are shared between the tails of normal distribution.

If we are examining outcomes in a one-tailed test, we can use the 5% limits in either tail to illustrate significance. We can see what that means for z-scores by using the normal distribution calculator again:

Leave **Standard score** blank this time
Enter 0.95 in **Cumulative probability** (leave mean and standard deviation as 0 and 1)
Click on **Calculate**
The answer (in the Standard score box) shows 1.645
Now enter 0.05 in **Cumulative probability** and follow the same procedure as above
Now the outcome shows −1.645

What we have just seen provides us with some valuable information about significance and z-scores for two-tailed and one-tailed tests. We will have a significant outcome in a two-tailed test if the z-score is greater than ± 1.96. In a one-tailed test, significant outcomes are confirmed when z-scores exceed ± 1.645 (but only if the outcome is in the predicted direction – see earlier section on one-tailed vs. two-tailed tests).

Central limit theorem and sampling distributions

In statistics, **central limit theorem** states that the mean of the sampling distribution equals the mean of the population and that the standard error of the mean equals the standard deviation of the population. Where the population is infinite, standard error is found by dividing the sample standard deviation by the square root of the sample size. This much we have already seen. However, the impact of central limit theorem cannot be understated. In social science research we often use samples that are used as an approximation of what might occur in the entire population. So long as distributions are relatively normal, we can use the principles of central limit theorem to make inferences about probability and statistical significance with relatively small samples. But how small is small? In general terms, a sample of 30 or more will probably suffice. However, we can be more precise if we make the effort to find out more about the distribution of our sample. According to central limit theorem, the sample is large enough if any of the following holds true:

1. Where the sample size is 15 or less:
    a. The distribution must be normally distributed,
    b. have no outliers and
    c. must be unimodal (have one peak in the curve).
2. Where the sample size is between 16 and 40:
    a. the distribution must be no more than moderately skewed,
    b. have no outliers and
    c. must be unimodal.
3. Where the sample size is greater than 40:
    a. the distribution must have no outliers.

Confidence intervals

Using **confidence intervals** we can estimate a range of values that is likely to be included within a given proportion of a sampling distribution. We use what we know about the mean score and standard error of the mean to calculate those values, according to the significance limits that we have set. Usually, we describe these parameters in terms of 95% confidence intervals, where we have set significance as $p < .05$. The values within this range represent an estimation of scores within a distribution, excluding the extreme scores (as they might

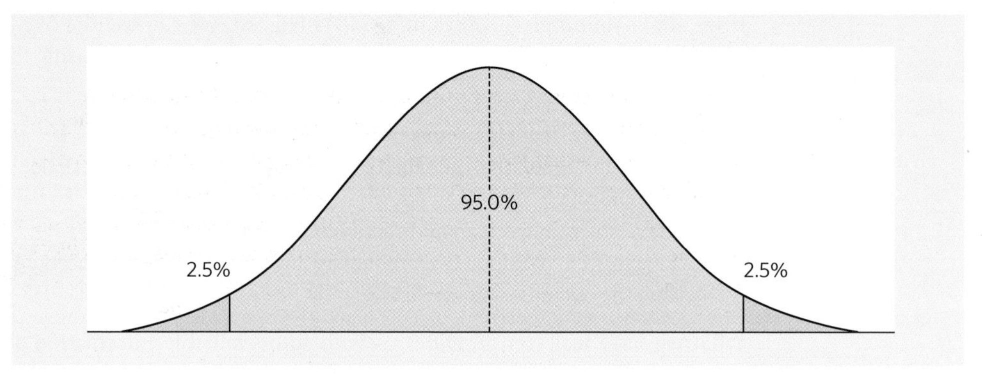

Figure 4.6 Graphical representation of 95% confidence intervals

represent outliers). In the previous sections, we learned how to estimate the probability of certain ranges of scores within a distribution. We can apply this to what we now know about mean and standard error of the sampling distribution. As we saw earlier, the population is potentially infinite, so we need to estimate the standard deviation of the population from the standard error. To calculate the range of values within the 95% confidence intervals we use the sample mean and the standard error (we will see how shortly). Confidence intervals have an upper and lower boundary, beyond which we find the outer 5% of the distribution (represented by lower and upper 2.5% tails). An illustration of confidence intervals is shown in Figure 4.6.

Although it can be useful to show confidence intervals graphically, we are more likely to see the range of values written out numerically (with respect to the lower and upper boundaries of those values). As we saw earlier, we can express the outer 5% of a normal distribution in terms of a z-score, representing the two tails of 2.5%, where values exceed ± 1.96. Based on that information, we calculate the upper and lower boundaries for 95% confidence intervals in the following way:

Lower boundary: Mean − (1.96 × standard error)
Upper boundary: Mean + (1.96 × standard error)

Let's put that in context of the descriptive data we found in respect of mood scores for men and women.

95% confidence intervals for mood scores (men)

Lower boundary: $17.34 - (1.96 \times 1.02) = 15.34$
Upper boundary: $17.34 + (1.96 \times 1.02) = 19.34$

95% confidence intervals for mood scores (women)

Lower boundary: $20.91 - (1.96 \times 1.21) = 18.54$
Upper boundary: $20.91 + (1.96 \times 1.21) = 23.28$

We can now present some important descriptive data about our sample (see Table 4.3).

Table 4.3 Mean, SE and 95% CI for mood scores

	Mood scores		
	Mean	**SE**	**95% CI**
Male (n = 32)	17.34	1.02	15.34 − 19.34
Female (n = 32)	20.91	1.21	18.54 − 23.28

Confidence intervals of difference

We can produce confidence intervals for any sampling distribution. When we compare two samples (or two conditions from the same sample) we might like to know whether they differ from each other. To explore that we could create a distribution of scores that represents the range of differences between them. We could estimate confidence intervals for that distribution of differences; we call that the ***confidence intervals of difference***. We could illustrate this with our example data once more, but return to the scenario where the data are displayed in Table 4.3. The calculation for confidence intervals of difference is a little more complex – you can see how this is done in Box 4.7.

4.7 Calculating outcomes manually
Estimating 95% confidence intervals of difference

The calculations for 95% confidence intervals of difference involve a little more work and require us to look up some values in distribution tables. The formula is shown below:

95% CI of differences $= M_d \pm t \times S_{Md}$

where M_d = mean difference

t = relevant score from the t-distribution, according to the sample size

S_{Md} = estimate of standard error of differences ($\sigma\bar{x}$)

We will illustrate this with the example that we have been using in respect of mood scores for men and women. We already know the mean difference (3.57) and the standard error of differences (1.58) from what we did earlier. The 't-score' is something we need to look up in distribution tables.

To find the t-score, we need to consult cut-off values in something called a t-distribution (see Appendix 2). Throughout this book we will be referring to a number of distribution tables that will guide us in determining significance; this is just one of those. To use the t-distribution, we need to locate the cut-off point that relates to the degrees of freedom (*df*) for the data. Degrees of freedom describe the sample sizes, but allow one score to remain constant. The *df* in this case is calculated from the sample sizes for both groups, minus the constant: males 32 − 1 (31) *plus* females = 32 − 1(31); *df* in our example is 62. We now go to the t-distribution table and look up the cut-off value for *df* = 62 and where p = .05 (for two-tails, because we are looking for a difference, and not saying which will be higher, as we saw earlier). Using those criteria, cut-off t value = 2.0.

Now we can apply that to the upper and lower boundaries for 95% CI of differences:

Lower boundary $3.57 - (2.0 \times 1.58) = 0.41$

Upper boundary $3.57 + (2.0 \times 1.58) = 6.73$

Once we know the confidence intervals of difference we can plot them in a table of descriptive data as we did before (see Table 4.4). Notice how we present the data ('95% CI of diffs') between the rows for male and female outcomes; this is because it represents the ranges of difference in values between the groups.

We know from our earlier exercise that there was a significant difference in the mean mood scores between these groups. However, the 95% confidence intervals of difference provide an additional clue that would lead us to suspect that this would be the case. The range of scores does not cross zero; all of the values in the range are positive. (It would also be OK if all of the values in the range were negative.) It suggests that differences *may* be significant. However, if the 95% confidence

Table 4.4 Mean, SE, 95% CI and 95% CI of differences for mood scores by gender

	Mood scores				
	Mean	SE	95% CI	SE of diffs	95% CI of diffs
Male (n = 32)	17.34	1.02	15.34-19.34	1.58	0.41-6.73
Female (n = 32)	20.91	1.21	18.54-23.28		

intervals of difference cross zero (there are positive and negative values in the range) there will *not* be a significant difference in mean scores between the groups or conditions (see Box 4.8).

4.8 Nuts and bolts
What happens when 95% CI of differences cross zero?

The range of values described by the 95% confidence intervals of difference is very important when estimating whether two sets of data might differ significantly. The keyword is 'consistency'. In our example the range was found to be 0.41 to 6.73. This suggests that, within this range of values, the largest difference was represented when female mood scores exceed male scores by 6.73; the lowest when female scores exceed males by 0.41. These scores are consistently positive. It would be equally consistent if all of the scores within the 95% confidence intervals of difference were negative; it would just mean that all of the scores in that range are lower than the scores from the second range.

However, if that range were to include positive and negative numbers, it would not be consistent. Let's say that the 95% confidence intervals of difference for our example was −1.06 to 7.34. This would mean that (at one extreme) females exceed males on mood scores by 7.34; but now it would also mean that (at the other extreme) males exceed females by 1.06. When this happens, we say that the range has 'crossed zero' – in these cases we will not find a significant difference.

4.9 Take a closer look
Key factors in significance testing

In the previous sections, we have been exploring a range of factors that may be involved in statistical significance testing. Here is a summary of the key terms:

Variance:	the extent that scores vary around the mean
Standard deviation:	the average variation of scores, in relation to the sample mean
Standard error:	the average variation of scores in a sampling distribution, or the estimated variation in the population
Sampling distribution:	a theoretical calculation of all possible samples in a population
Standard error of difference:	an estimate of the standard deviation in the sampling distribution representing differences between two samples (or two conditions of the same sample)
Confidence intervals:	an estimate of the range of values likely to be included within a given proportion of a sampling distribution
Confidence intervals of difference:	an estimate of the range of values within a given proportion that represents differences between two samples or two conditions of the same sample

Effect size

Up until this point we have been stressing the importance of statistical significance when examining differences and relationships. For example, we have seen that we cannot take observed between-group differences at face value. Even if there were an apparently large difference in mean mood scores between men and women, it is statistically significant only if there is a less than 5% probability that the outcome happened by chance. Now, that might tell us a great deal about how much we can trust the result, and that it is likely to happen if we repeat the methods used. However, it does not tell us much about the actual size of the difference, relative to the number of cases used to measure that difference. For that we need ***effect size***.

Effect size indicates the actual magnitude of the difference between scores, without considering how that relates to an overall population. It is based on the sample mean and sample standard deviation; it does not account for standard error. When we find a significant difference, we use that as evidence to reject the null hypothesis in favour of the alternative hypothesis; we may feel pleased about that. However, significance should not be taken in isolation. Even the smallest differences can be significant if the sample is large enough. For example, say we want to test a new antidepressant to examine whether it provides better improvements in mood than previous drugs. We could test the new drug in a randomly controlled trial (see Chapter 5 for more details on experimental methods). Once we have collected the data we might find that the new drug produces significantly better outcomes than the old one ($p<.001$). However, now let's say that we conducted this trial with 2,000 patients to find that effect. On closer inspection, we see that the improvement represents 1% change on illness rating scores. The difference may be statistically significant, but it is hardly clinically relevant given the small change (especially as there may be side effects). You can see an even more spectacular (real-life) example in Box 4.11.

Effect size also allows us to compare more easily between studies carried out by different researchers. Reporting effect size is becoming more established in published studies, so comparison is often quite easy. However, even when this is not done, there should be enough information reported from which to infer effect sizes. For example, we might find two studies that report outcomes that appear to suggest that older people (aged 60 or above) experience fewer hours' sleep than younger people. One study reports a non-significant result ($p = .065$), while the other indicates a significant one ($p = .004$). It would appear that there is some inconsistency here. However, we may notice that the first study recruited 50 participants, while the second study observed 500 people. Once we account for actual differences, sample size and standard deviation, we might find very similar outcomes.

Measuring effect size

There are several ways to measure effect size, but the most commonly used are Pearson's *r* and Cohen's *d*. Pearson's effect size focuses on associations between samples and is often used in correlation (see Chapter 6). Cohen's methods explore effect size by examining differences relative to sample sizes and pooled standard deviation (you can see how to calculate Cohen's effect size in a simple two-sample between-group example in Box 4.10). Throughout this book we will mostly use Cohen's methods, largely because there is a very good (and utterly free) software program available to help us perform effect size calculations for almost every statistical procedure (at least the ones you are likely to use). The program is called G*Power, which we will explore in a little more depth later on. In the meantime, since you will probably come across both types of effect size in your reading, you can see a general overview of how to interpret effect sizes in Table 4.5 (but do remember that these values can vary between statistical tests). Note how Pearson's *r* effect size ranges from 0 to 1, while Cohen's *d* effect size can exceed 1.

Table 4.5 Effect size guidelines

Size	Pearson's *r*	Cohen's *d*
Small	0.1-0.3	<0.25
Medium	0.3-0.5	0.25-0.4
Large	0.5-1.0	0.4-∞

Note: ∞ infinity

4.10 Calculating outcomes manually

Cohen's effect size calculation (for two sample means)

Formula for calculating Cohen's (d) effect size:

$$d = \frac{\text{Mean a} - \text{Mean b}}{\text{Pooled standard deviation } (S_p)} \quad S_p = \sqrt{\frac{(n_1 - 1)S_1^2 + (n_2 - 1)S_2^2 \ldots + (n_k - 1)S_k^2}{n_1 + n_2 - k}}$$

k = no. of conditions; S^2 = variance

We can illustrate this with the mood score data that we used earlier:

Male: mean score, 17.34; variance, 33.46, n = 32; Female: mean score, 20.91; variance, 46.47, n = 32

$$S_p = \sqrt{\frac{(31 \times 33.46) + (31 \times 46.47)}{32 + 32 - 2}} = 6.321 \text{ so, } d = \frac{17.34 - 20.91}{6.321} = 0.564$$

4.11 Take a closer look

Aspirin and heart attack - high statistical significance, but what about effect size?

A recent study reported important results from a large clinical trial proclaiming how aspirin might reduce heart attacks. The findings were headline news across the world. We should explore what was found in light of what we have been learning about significance and effect size. In this longitudinal study, the investigators demonstrated that 104 (out of 11,037) people taking aspirin subsequently had heart attacks, compared with 189 (out of 11,034) people in the placebo group who went on to present heart attacks. This means that nearly half as many people taking aspirin experienced heart attacks than those taking the placebo. Is this impressive? The statistics appear to say so.

The null hypothesis (that there would be no difference in heart attacks for those taking aspirin and placebo) was rejected, where $p < .00001$; this means that there is a less than 1 in 100,000 probability that the outcome happened by chance. Surely this is enough evidence to suggest that we should all take aspirin to reduce the likelihood of heart attacks?

But let's look at these findings a little more closely. Aspirin was associated with a 0.94% chance of heart attack, while the risk with placebo was 1.71% - a difference of 0.77%. This equates to a difference of fewer than 8 people in every 1,000. When the data are examined according to mean difference, standard deviation and sample size, calculations show the effect size to be 0.06 (very small). So, what appeared to be a really impressive outcome was actually very small in terms of real effect. This shows why it is always good to explore significance and effect size.

Statistical power

Another key measurement in reporting outcomes is represented by **statistical power**, which measures the probability of correctly rejecting the null hypothesis. You may have read about research being 'underpowered' and wondered what that meant. You may also have asked yourself (or more likely your statistics tutor) about how many participants will be needed in a study. These questions can be answered using power calculations. There are four factors in a power calculation: the effect size (which we have just seen); the probability or significance level (also known as α, usually set at .05); the statistical power; and the number of participants that need to be recruited to achieve that effect size and power.

Earlier we said that a Type II error occurs when we incorrectly fail to reject the null hypothesis. Cohen (1992) said that we should avoid getting Type II errors too often. He said that we should aim to correctly reject the null hypothesis on at least 80% of occasions. If we present 80% as a decimal, we get 0.80; most power calculations are based on that. To be able to achieve a power of 0.80, we need to make sure we recruit enough participants to obtain a large enough effect size, using an appropriate level of significance. The calculation for that is complex, so we will not explore that here. Instead, we will be using G*Power to calculate outcomes.

4.12 Take a closer look
Effect size and statistical power

These summary definitions might be useful:

Effect size: a measure of the actual size of differences between two variables, in relation to the sample mean and sample standard deviation. It makes no assumption about the population mean

Statistical power: the probability that a test will correctly reject the null hypothesis. We should aim to achieve this on at least 80% of occasions (thus avoiding too many Type II errors)

Measuring effect size and power using G*Power

G*Power is an extremely useful program that enables you to calculate outcomes in a power analysis. Typically, we would use this software to do one of two things (although it will do other stuff, too). We can calculate (or have G*Power calculate it) the statistical power of a completed study, since we know the effect size, the sample size and the significance level. We can also estimate the number of participants we need to recruit for a study, assuming that we are aiming for a power of 0.80, and based on an estimate of the expected effect size, and the level of significance that we have decided to set. At the time of going to press, the latest version of this program is G*Power 3.1.3 and you can download it, free of charge, from the Internet at www.psycho.uni-duesseldorf.de/abteilungen/aap/gpower3/. Follow the instructions for downloading the program (according to the operating system for your PC or laptop) and then how to activate it. Once installed, you can open the program from the Programs menu by clicking on the shortcut on your desktop (if you asked for that). When you first open the program, you will get a screen like the one shown in Figure 4.7.

Using G*Power to examine power of a completed study

In this first example we will demonstrate how we can calculate the achieved statistical power, based on outcomes from a completed study. We will illustrate this with the mood scores data

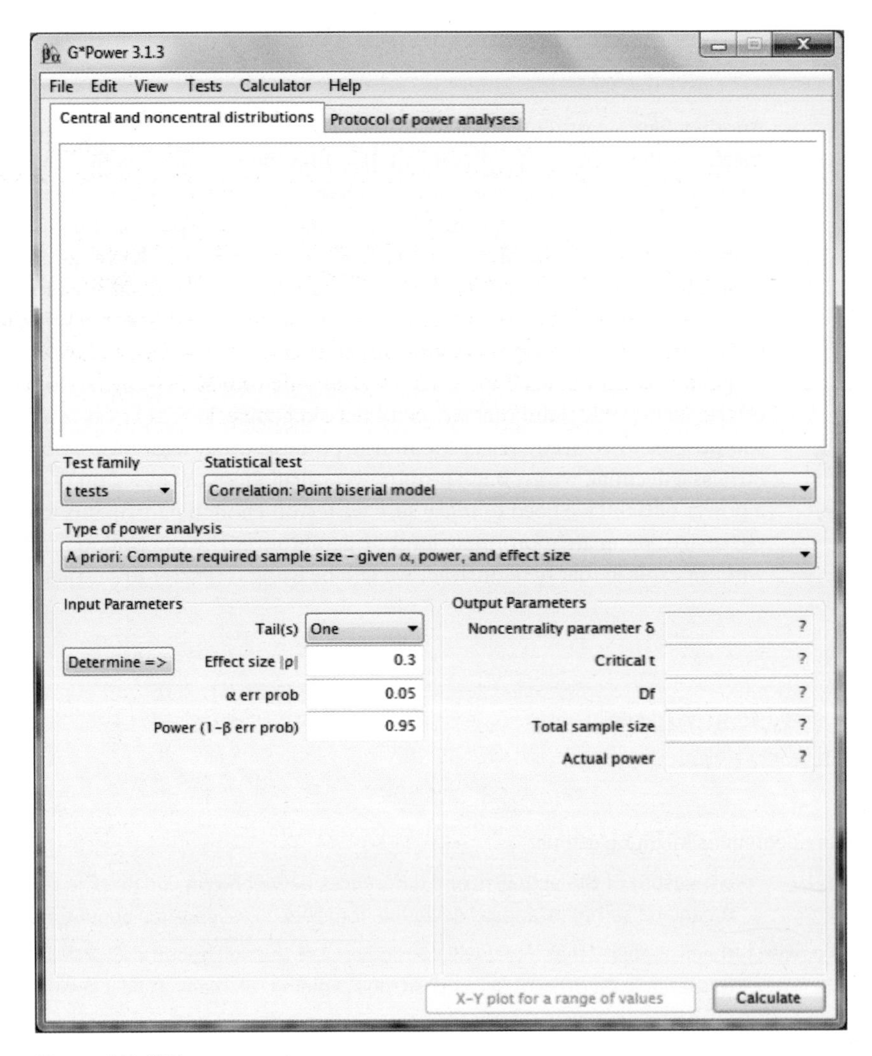

Figure 4.7 G*Power opening screen

that we explored in respect of gender when we were learning about significance testing. This is an example of a between-group analysis. Assuming that the data are appropriate, we would probably explore these outcomes formally using an independent t-test. We will learn more about that procedure in Chapter 7, but we can examine the statistical power and effect size of this outcome without having to know too much about the rules of performing that t-test. The first action we need to take is to define the type of test we have used (we said that we would probably have examined the outcome using an independent t-test).

From **Test family** select **t-tests**
From **Statistical test** select **Means: Difference between two independent means (two groups)**
From **Type of power analysis** select **Post hoc: Compute achieved – given α, sample size and effect size power**

Now we enter the outcome data into the **Input Parameters**:

From the **Tails** box, select **Two** (we did not predict which group would be higher)

To calculate the **Effect size d**, click on the **Determine** button (a new box appears).
In that new box, for **Mean group 1** type **17.34** ➜ for **Mean group 2** type **20.91** ➜ for **SD** σ **group 1** type **5.78** ➜ for **SD** σ **group 2** type **6.81** ➜ click on **Calculate and transfer to main window**

Back in original display, for α **err prob** type **0.05** (the significance level) ➜ for **Sample size group 1** type **32** ➜ for **Sample size group 2** type **32** ➜ click on **Calculate**

There are two outcome measures that we are interested in: *Effect size* (d) **0.565** (according to the limits shown in Table 4.5 it demonstrates a medium effect, and confirms what we calculated manually earlier); and *Power* (1- β err prob) **0.604** (which is not so good, as it is below the desired 0.80 level; we have achieved 'poor power').

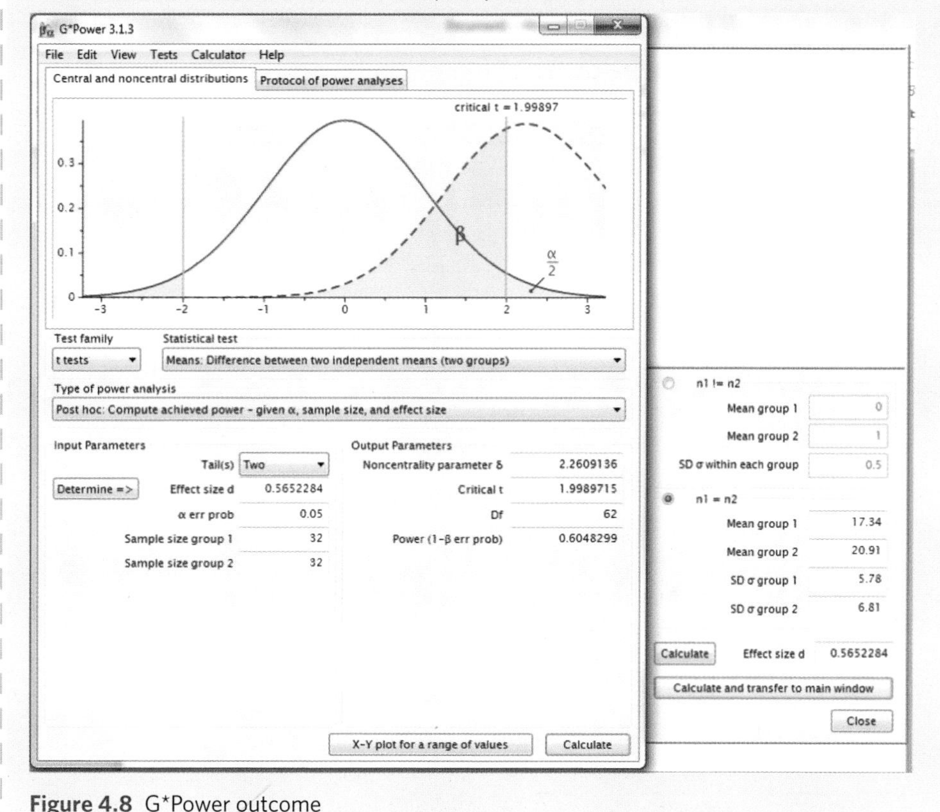

Figure 4.8 G*Power outcome

Using G*Power to estimate sample size for a future study

Although the effect size in our last example was good, it could be said that the study was 'underpowered' as we achieved a statistical power of only 0.604. We said that we should aim to achieve power of at least 0.80. The low power might have been because there were not sufficient participants to find the effect that we achieved. So how many would have been enough? G*Power can be used to calculate how many participants we should recruit to achieve a power of 0.80, where significance is $p = .05$. We also need to estimate the effect size that we are trying to find. In this next example, we will assume that we would like to repeat the effect size that we found (0.565), but want to ensure that we recruit enough participants to achieve sufficient power. This is how we examine that:

From **Test family** select t-tests
From **Statistical test** select **Means: Difference between two independent means (two groups)**
From **Type of power analysis** select **A priori: Compute required sample size - given α, power, and effect size**

Now enter the **Input Parameters**:

For **Effect size d** type **0.565** → for **α err prob** type **0.05**; for **Power (1-β err prob)** type **0.80** (this is the optimal power we are seeking) → for **Allocation ratio N2/N1** type **1** (assuming we want equal group sizes) → click on **Calculate**
We are interested in one outcome: Total sample size **102** (51 participants in each group)

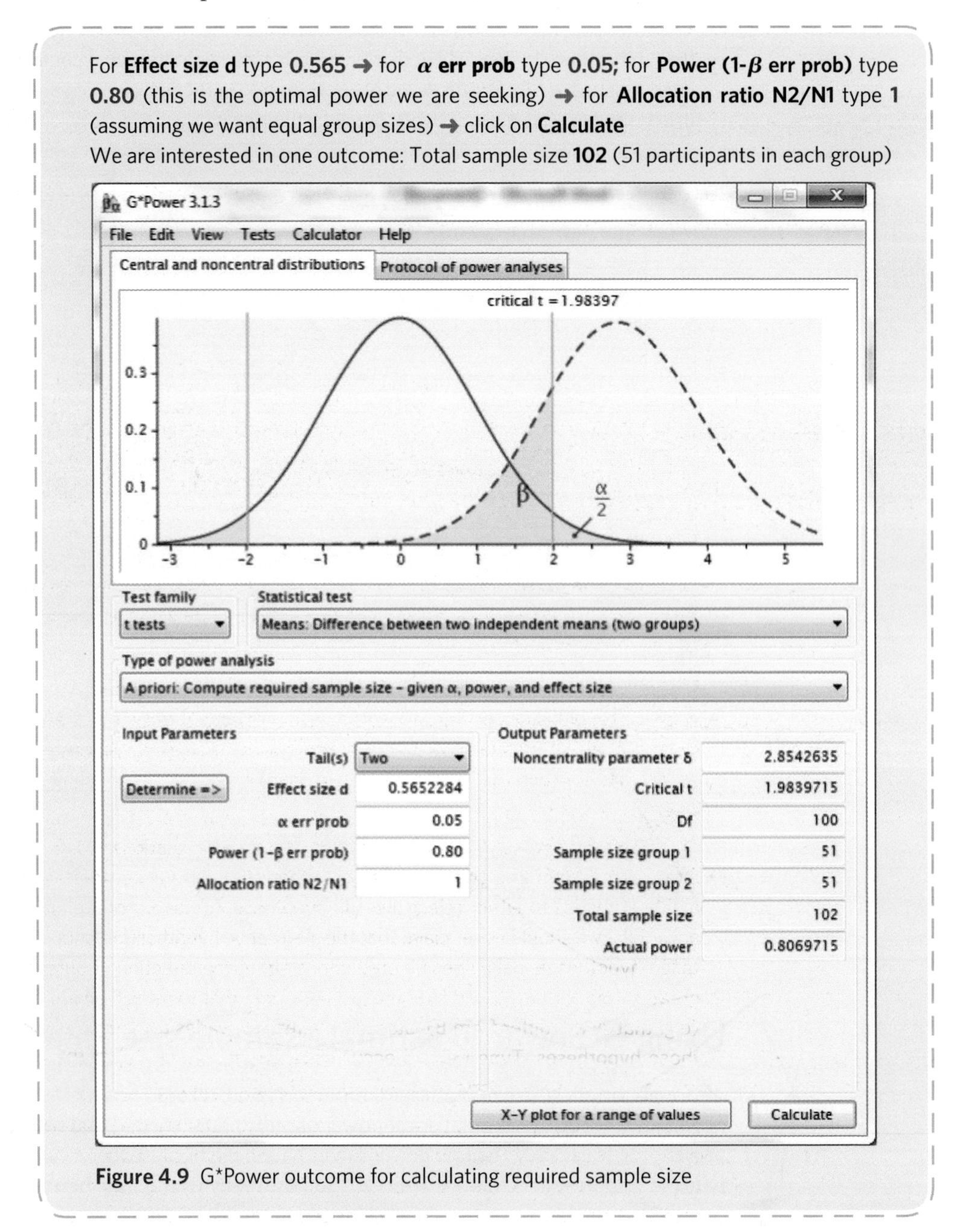

Figure 4.9 G*Power outcome for calculating required sample size

If you need to write a statement of power calculation (for a project proposal, for instance), you would write something like this:

We need to recruit at least 102 participants to a medium effect of 0.565, using a significance level of 0.05 and a power of 0.8 to detect that effect.

4.13 Take a closer look
Calculate sample size requirements for different effect sizes

Power calculations are very useful. We based the last sample size estimate on a medium effect size of 0.565. However, we might have evidence to suggest that we could expect a very strong effect size (such as 0.85). How many people would we need to find that effect (assuming significance and power target remain constant)?

We would use the same methods as we have just seen, but enter 0.85 into Effect size d (instead of 0.565). Now we are told that we need only 52 participants (26 in each group). The greater the differences that we believe we are likely to find between the groups, the fewer people we need to find that effect.

But what if evidence suggests that we are likely to find only a small effect at best (such as 0.20)? How many people do we need to find a much smaller effect?

Now we enter 0.20 into Effect size d. Now it would appear that we need 900 participants (450 in each group)! We need an awful lot more people to find smaller differences.

Chapter summary

In this chapter we have explored statistical significance, effect size and power. At this point it would be good to revisit the learning objectives that we set at the beginning of the chapter.
You should now be able to:

- Understand that we use statistical significance to express the probability that observed differences or relationships occurred by chance. In most cases, we say that an outcome is 'significant' if the probability of chance factors is less than 5%. It suggests that there is a very strong probability that the outcome 'supports' the predicted event.
- Recognise the definitions in testing hypotheses for statistical significance. We begin with the null hypothesis (which states there is no difference or relationship).This is compared with the alternative hypothesis (which states that there is a difference or relationship). We use probability statistics to either reject the null hypothesis (in favour of the alternative hypothesis) or accept it - we should never claim that the alternative hypothesis is 'supported' or 'rejected'. One-tailed hypotheses make specific predictions about the direction of outcome (e.g. that A will be greater than B); two-tailed hypotheses state only that there will be a difference or relationship (e.g. that A will differ from B). Statistical inference varies according to the nature of the tails in those hypotheses. Type I errors occur when we incorrectly reject the null hypothesis; Type II errors happen if we incorrectly fail to reject the null hypothesis.
- Appreciate the importance of key measures used to estimate the probability of significance. Variance measures the extent that scores vary around the mean score. Standard deviation describes the average variation around the mean within the sample. Standard error is an estimate of the standard deviation within sampling distributions, or within the overall population. Sampling distributions are a statistical model of all of the possible samples that can be drawn from a population. Probability statistics use variance and/or standard error to explore the likelihood of significance (in relation to known distributions). Confidence intervals describe a range of values that is likely to be included within a given proportion of a sampling distribution. Confidence intervals

of difference describe those values in a sampling distribution that represent differences between two samples (or two conditions within the same sample).

- Understand the importance of effect size and statistical power. Significance can be misleading when there are large sample sizes. Even the smallest difference can look important (as the aspirin and heart attack example demonstrated in Box 4.11). Effect size reflects actual differences in relation to sample mean and standard deviation (but not in context of the overall population). This should always be stated alongside significance. Statistical power measures the probability of correctly rejecting the null hypothesis. We should aim to correctly reject the null hypothesis (and avoid Type II errors) on at least 80% of occasions.

Extended learning task

Following what we have learned about statistical significance, effect size and power, answer the following questions. You will find the answers on the web page associated with this book.

1. Describe the null and alternative hypotheses.
2. How do we use probability to make decisions about those hypotheses?
3. What are standard deviation and standard error? Why do we need to examine them when assessing significance?
4. Briefly explain the following:
   a. One-tailed vs. two-tailed hypotheses.
   b. Type I and Type II errors.
5. What are confidence intervals?
6. What is the implication if the 95% confidence intervals of difference cross zero?
7. Why is it important that we measure effect size in addition to significance?
8. How do we use statistical power?

5 EXPERIMENTAL METHODS – HOW TO CHOOSE THE CORRECT STATISTICAL TEST

Learning objectives

By the end of this chapter you should be able to:

- Understand the basic features of experimental methods
- Define data types and appreciate whether they are parametric
- Recognise the factors that determine which test to perform:
  - Data type and parametric assumptions
  - The number of variables being measured
  - Measuring differences or relationships
  - Examining outcome between groups or within groups

Introduction

Before we explore the statistical tests covered in this book, it would be useful to understand a little about experimental methods. We will examine why it is difficult to undertake more traditional experiments in psychology, and seek to use alternative methods. We will consider the impact of exploring differences compared with measuring relationships. We will investigate how to examine cause and effect using longitudinal research. Other factors will also help us decide which statistical tests we should employ. For example, we will look at how to define and measure independent and dependent variables, we will compare between-group studies and within-group designs, we will decide whether the data are parametric, and we will seek to understand the nature of the data measured by those variables. We will discover the importance of **validity** and **reliability**, and explore ways in which we can measure these. We will then conduct an overview of the statistical tests that you will encounter in subsequent chapters.

Conducting 'experiments' in psychology

When we examine outcomes in 'pure' science (such as chemistry, physics and biology) experimental methods are generally quite straightforward. In these experiments we speak in terms of initial states, upon which we perform some kind of manipulation that produces an effect. For instance, if we take a small piece of potassium (initial state) and drop it into water (manipulation) it fizzes and burns furiously around the water (effect). I always remember that from my school days, partly because it is really cool and also because my chemistry teacher was somewhat more than eccentric (we said that he was more volatile than most of his experiments). The science example illustrates how we can measure an outcome, based on something that we did, to change the properties of the initial state.

It's not that easy in psychology. In fact, many purists say that psychology cannot be a science precisely because of that. There are many reasons why we cannot 'manipulate' a variable when dealing with humans. For instance, we might want to examine the effect of children living in one-parent families, compared with those in traditional families. In doing so, it would be completely unethical to remove one of the parents from some of the families to examine the outcome. Instead, we would need to conduct what we call a quasi-experiment. In this context, we could examine naturally occurring differences between children who are in two-parent families compared with those in one-parent families. However, because psychology (usually) employs the rigour of scientific methods to its explorations, many feel justified in considering psychology a natural science.

Problems with psychological experimental methods

However, we may still encounter problems. In the example we saw with potassium, we are likely to observe that outcome every time, so long as we controlled other conditions. In psychology, controlling for 'other factors' can be more problematic than with traditional science. Let's say we examine a group of students on their scores from a social psychology exam. We might aim to investigate whether females perform better than males. We may support that prediction, but what else should we consider? It may be that exam scores can change according to mood, IQ, examination anxiety, age, experience and so on. The observed difference might be due to any one (or more) of those factors in addition to, or instead of, gender. We could try to control for all of those things by recruiting equal numbers of men and women, who are also equally matched on mood, IQ, examination anxiety, age, experience and so on. However, the more controls we place on recruitment, the harder it will be to find participants (which is usually the trickiest part of conducting psychological research). One way around this is to use statistics. As we will see later, we can use measures to control for one or more factor, while focusing on the variable we actually want to measure. In effect, additional factors are held constant.

Despite that, there are manipulations that we can undertake (subject to ethical approval). We could pilot a new drug on one group of depressed patients and compare the outcome to patients receiving a placebo (a pill that looks like the real drug but has no effect – only the patient does not know that). In that case, we can still say that we are investigating cause and effect, so it is very similar to pure science. We could also examine the effect of different teaching methods on students in respect of attendance at lectures. Or we could observe naturally occurring events to generate theory or confirm hypotheses. However, whichever method we use, how do we go about measuring the outcome? Do we look at differences between separately recruited groups of people? Do we examine those differences across the same group of people, but simply vary the **conditions** at different time points? Or do we just look at the **relationship** between one event and another? These questions and more will be addressed in the following sections.

Factors that determine the appropriate statistical test

Before we can make any decisions about which statistical test to perform, we need to know several important aspects about the nature of our data and how they will be examined. These relate to the type of data that we are seeking to investigate, whether those data are parametric, whether differences are being measured or relationships explored, and whether the examination is being conducted as a between-group or within-group study. We will explore those factors briefly here, before considering the steps we need to take to select the appropriate statistical test.

What type of data are being measured?

In simple terms, your data will be either numerical (relate to numbers and counting) or **categorical** (relate to descriptions and groups). In reality it is a lot more complex than that. **Discrete** data represent distinct units. Typically, they will be specific categories or groups, such as gender (male and female) or nationality (British, French and American). However, discrete data can also relate to numbers, but only if this represents a specific count (rather than a range). For example, if you count the *number* of females who are depressed (this is discrete), while the depression *scores* associated with a group of people is '**continuous**' (this is not discrete). Categorical data are an example of discrete data that relate to homogeneous groups, such as animal type (cat, dog or hamster). To aid statistical analysis, these groups can be 'defined' by values when we set them up in SPSS (such as 1 = cat, 2 = dog and 3 = hamster – we saw how to do that in Chapter 2). No numerical weight can be inferred from those values; they are simply used to differentiate the groups. SPSS calls these 'nominal' variables. If we have only two groups or categories (as we might with gender) we call this a **dichotomous** (or binary) variable.

'Numerical' data can be further categorised into sub-types: **ordinal**, **interval** and **ratio**. Ordinal data refer to those which can be ordered by rank. Final position in a race (first – second – third) is one example. A more common example is found in **Likert scales**. These are used to measure attitudes, opinions and satisfaction, where numbers can be allocated to those perceptions (such as 5 = 'very satisfied', 4 = 'satisfied', 3 = 'neither satisfied nor dissatisfied', 2 = 'dissatisfied' and 1 = 'very dissatisfied'). Equally, we could ask someone to rate their current satisfaction on a scale of 1 to 10 (with 10 representing the most satisfaction). Those numbers carry more weight than those allocated to nominal variables, but still cannot be inferred in the same way as we can do for interval or ratio data. For example, the numbers used to define gender (1 = male, 2 = female) are arbitrary; a score of 2 is not 'higher' than a score of 1 in this instance. Meanwhile, an ordinal score (such as 5 = 'very satisfied') could be considered to be 'higher' than another score in the scale (such as 2 = 'dissatisfied'). However, little inference can be given about the 'distances' between those numbers. As we will see shortly, interval data relate to numbers where such inference *can* be made. The difference between an age of 50 and an age of 25 is objective, measurable and undeniable. Yet one person's satisfaction rating of 4 might be very different to the next person's. Such 'numbers' might be seen as subjective differences; in the values used in those ratings is less clear. Because of that we often rank ordinal scores rather than treat them as a number that can be manipulated.

Interval and ratio data refer to numbers that are measurable; differences between numbers in a range are more obvious. Some good examples are age, income and temperature. We can compare sets of numbers using descriptive data, such as the average score (something we cannot do with ordinal data). Although interval data are usually confined to a range of scores, they can also include discrete data that refer to counts, such as the number of people attending lectures. Interval data may be represented by numbers in a range where there are 'equal distances' – the difference between the ages 8 and 6 is the same as the difference between 25 and 23. Those differences are clearly objective. Another good example of interval data is temperature. There are clear differences between 70° and 60° Fahrenheit, as there are between 20° and 10°Celsius.

Interval data may also be described as ratio data, but only when values can be compared with each other in relative terms. Someone who is 50 years of age is twice as old as someone who is 25. To qualify as ratio data, the range of scores must include an absolute 0 (age does). Temperature could not be considered as ratio data (although it is interval); there is no absolute zero for Fahrenheit or Celsius (0° is arbitrary). We cannot say that 70° Fahrenheit is twice as hot as 35° Fahrenheit. Think about what happens if we convert Fahrenheit to Celsius (21.1° and 1.6° respectively). While the number 70 is twice as high as 35, 21.1 is not twice as high as 1.6. Time is a good example of ratio data – it has a zero point, so we can say that 20 minutes is twice as long as 10 minutes. Despite those differences, interval and ratio data tend to be grouped together as 'interval' data (SPSS calls these data 'scale').

5.1 Take a closer look

Glossary of data types

Having explored data types in the last section, you may find this summary useful:

Discrete: data that are distinct, or separate, entities. This can include groups (such as gender) or a *count* of numbers (but not a *range* of numbers)

Categorical: discrete data that are distinct homogeneous (descriptive) entities. These can represent distinct groups, such as gender (male vs. female), or within-group conditions, such as time points (before test vs. after test)

Nominal: another term for categorical data

Dichotomous (binary): where there are only two categories for a descriptive variable (such as with gender)

Continuous data: any non-discrete data (i.e. not categorical)

Ordinal: where numerical data can be ordered by rank. The numbers used to define the ranks have more numerical meaning than nominal data, but inferences cannot be made about the distances between numbers

Interval: where numerical data have objective differences between numbers in the scale (unlike ordinal data, where those differences could be seen as subjective)

Ratio: where interval data can be related to each other in terms of relative amounts. Time is a good example of ratio data; temperature is not. To fit this criterion, there must an absolute zero within the data

Are the data parametric?

Determining whether data are parametric is pivotal to choosing the correct statistical test. Understanding this concept, and how to measure it, is probably one of the most important things you can learn about in data analysis. To qualify as parametric, the dependent variable data should be (reasonably) normally distributed and must be interval or ratio (not ordinal, and definitely not categorical). We explored normal distribution in Chapter 3; we examined interval and ratio data just now.

Normal distribution is achieved when data are evenly distributed either side of the mean (average) score. If there are extreme scores at either end of the range of numbers (outliers), this can cause the distribution to skew. If we have high extreme scores, we can get positive skew; if the outliers are low, we can get negative skew. We saw graphical examples of these distributions in Figures 3.2 – 3.4. Positive skew can cause the mean score to be artificially increased; negative skew can understate the mean score. This is important because parametric tests depend on the mean score to determine outcome. Such methods are typically employed in t-tests, ANOVAs, Pearson's correlation, and linear regression. If the mean scores have been biased by outliers, we should not rely on statistical tests that use that to examine outcome – it could produce false outcomes (see Type I and Type II errors in Chapter 4). We should probably consider using non-parametric tests (see Chapter 18).

We have a similar problem if we fail to meet the requirements for interval or ratio data (even if those data are normally distributed). As we saw earlier, although ordinal data have 'some' numerical value, it is questionable whether we can infer scores in the same way as we can with interval or ratio data. For example, if we examine Likert scale scores (where 1 = 'very dissatisfied' through to 5 = 'very satisfied'), what does a mean score of 4.35 suggest? It is a little convoluted to say that it reflects a perception somewhere in between 'satisfied' and 'very satisfied'. If we ask participants to rate their own satisfaction on a scale of (say) 1 to 10 (where 10 is the most satisfaction), we may have a little more faith in a mean score of 7.62. Many opinion scales are rated that way. Also, one person's satisfaction rating of 7 may be very different to someone else's rating of 7. How much can we trust a mean score that relates to subjective ratings of several different people? In those circumstances, it might be more appropriate to compare groups on how those ratings are ranked by using a non-parametric test. That way the absolute value of the rating has less impact. Many researchers argue that ordinal data are not suitable for tests that rely on the mean score to determine outcome. However, as we will see throughout this book, ordinal data still tend to be used in some parametric tests.

5.2 Take a closer look
What is a parametric test?

The following summary might be useful for determining whether data are parametric:

Parametric data:	Where interval or ratio data are normally distributed. Parametric tests tend to use the mean score to evaluate differences or relationships
Normal distribution:	Where the data are evenly distributed either side of the mean, with no outliers at either end of the distribution. We saw how to measure that in Chapter 3
Non-parametric:	Where the data fail to meet one or more of the requirements for parametric data. Differences or relationships tend to focus on the ranking of the data

How many variables are there?

Once you know what type of data are included in your variables, you will need to know how many you will be examining. The number of variables that you measure will determine the type of test you can use. Of those variables, you will also need to know how many of them are dependent variables and how many are independent variables. The dependent variable (DV) is the outcome and is often represented by a range of scores. For example, we could measure examination marks from a group of students. Sometimes the dependent variable might be a categorical outcome. For example, we might explore whether a person is depressed (or not). The independent variable (IV) is the factor that we believe will have an effect on the outcome; it is usually categorical. An IV could be represented by specific groups (such as gender: male or female) or it might be conditions that are examined across a single group. We might expect scores on the dependent variable to vary between the groups or across the conditions (we will explore the difference in those examples shortly).

5.3 Take a closer look
Dependent variable vs. independent variable

One of the first things we have to remember when learning about statistics is differentiating between the dependent variable and the independent variable; the following summary might help.

Dependent variable:	The outcome measure being investigated that is expected to change (as a result of factors such as groups or conditions)
Independent variable:	A factor (such as groups or conditions) that is thought to be responsible for changes in an outcome measure

Differences vs. relationships

Another important factor in selecting a statistical test focuses on whether we are measuring 'differences' or exploring 'relationships'. When we examine differences, we will often investigate how dependent variable scores vary across distinct groups, or over several conditions for a single group. These groups or conditions represent the independent variable; occasionally, they may be something that we have 'manipulated'. For example, we could randomly split participants into groups before exploring how they differ on a given outcome. Or we could investigate a single group of people and measure how their mood differs on various days of the week. Both are examples of exploring differences. Sometimes we measure cause and effect. For example, we might measure illness severity in a group of people, according to the dose of medication that we give them. Other times, we simply measure how outcomes differ naturally (such as mood scores between men and women). In contrast, when we measure relationships we are not concerned with differences. Instead, we are observing how outcomes on one variable change as outcomes vary on another variable. For example, we could explore how the ice-cream sales vary as temperature changes – such examinations rarely focus on cause and effect.

Between-group vs. within-group

If we are exploring a difference, we need to know whether we are going to use a between-group or within-group approach. In between-group studies, we investigate differences in dependent variable scores in respect of distinct groups that represent the independent variable. These groups must be wholly independent from each other – no person or case can appear in more than one group. In within-group studies, we explore a series of conditions across a single group. For example, let's say we want to examine the effect that a new antidepressant has on illness scores in depressed patients (perhaps compared with an existing antidepressant). We might predict that we expect to find a difference in illness rating scores, according to the type of antidepressant. If we chose to examine this in a between-group study, we could divide our depressed patients into groups, where we give one group the new antidepressant and the other group the old one. We could then compare those groups on illness scores and see which group shows the most improvement. Conversely, if we decided to use a within-group approach, we would measure outcomes across a single cohort and give all of the depressed patients the new antidepressant (and measure the illness score) and then give them the old antidepressant (and measure the illness score again). We can measure the difference in illness scores across those two conditions (but within a single cohort).

5.4 Take a closer look
Between-group vs. within-group

Students new to statistics often have trouble identifying when they have a between-group study and when it is a within-group study. These (brief) definitions might help:

Between-group:	Where the independent variable is measured between two or more distinct groups of people or cases
Within-group:	Where the independent variable is measured across one group, in respect of two or more conditions

Are within-group designs better than between-group studies?

Before we explore statistical methods that investigate between-group and within-group studies, perhaps we should pause to consider whether one method is 'better' than the other. No matter how hard we try, controlling for all possible variables in a between-group study is problematic. In within-group designs, individual differences are reduced because it is the same person in each condition. There is less likely to be unexpected variations in the outcome measure (whereas this is more likely to happen in between-group designs, because different people are represented in the groups; additional individual differences may explain the variation). Another big advantage of within-group studies is that you need fewer people to conduct them with. In between-group studies, you need to recruit participants into each of the study groups (often there will be three or more groups); in a within-group study you need only one participant to represent all of the conditions.

Despite these benefits, the 'repeated-measures' design does have its limitations. On the downside, within-group studies are prone to something called **order effects**. While individual differences are less likely to occur, those potential **confounding variables** are not completely eradicated. Once participants have conducted a test once, they might be more familiar with the procedures by the second presentation. Would that make them quicker on the second test? Might they recall their previous answer? They may get bored having to do the same test again. Might they pay less attention and make more errors on the second test? Furthermore, in some within-group studies, the purpose of the study may become apparent to the participant. This may influence them to respond in a way that might please the experimenter. The participant is more likely to remain naive in between-group studies. These order effects can interfere with the outcome (although they can be overcome by **counterbalancing** the conditions). Using counter-balancing, the order in which conditions are presented can be shuffled between participants. Allocation to the order of presentation can be managed using established procedures (see Chapter 10). To overcome recall effects, we could leave a longer gap between trials. However, this could mean that data collection takes much longer than it might have done if a between-group study had been employed.

Another problem for within-group studies is time: by their very nature the conditions are conducted over several time points. This is not a problem with between-group studies, where the conditions can be examined concurrently across the groups, saving a great deal of time. Also by default, within-group studies mean that participants must be present in every condition. The statistics are calculated in respect of how each participant responds across the conditions. If a participant misses a condition they must be excluded from the study. That can be tiresome, given the difficulties of participant recruitment. In between-group studies it is preferable to have equal numbers in each group, but statistics can adjust for missing participants. The impact of losing a participant from a condition is less serious in this context. You can read an extension of this debate in Chapter 8 (see Box 8.4).

Exploring differences

5.5 Nuts and bolts
Cause and effect

When measuring differences we often talk in terms of **cause and effect**. We might believe that the independent variable will cause a change in the dependent variable. This is more likely to occur when we explore differences than when we look at relationships.

Research methods

There are several statistical tests that we can use to explore differences. To help us decide which one to choose, we need to consider all those factors that we examined earlier. What sort of data do we have? Are those data parametric? Will we be exploring those differences across several groups? Or will we choose to explore the differences within one group over several conditions?

Once we have decided that, we need to know how we collect the data. There are several options open to us. We might choose experimental methods (or **quasi-experimental** methods, if that is not possible). We might opt for a **cross-sectional** approach, or decide that a longitudinal study is better. Alternatively, we may collect our data retrospectively. We will explore what is involved in each of these methods now.

Using a traditional experimental method, we directly manipulate the independent variable. In a between-group study we would decide which group to allocate people to. For example, if we were to examine the effect of a new antidepressant on a group of depressed patients, we could allocate those patients to the 'new antidepressant' or the 'placebo' group as we chose. However, it is more likely that we would randomly allocate group selection. Indeed, to ensure objective evaluation, we would probably use a 'blind' method so that we did not know which group the patient had been allocated to (the tablets would look identical). This might remove subjectivity, but it might not be considered as a **'true' experimental** method. Similarly, if we want to compare outcomes on something like gender, such groups are naturally occurring. On other occasions, ethical guidelines will determine whether we can use group allocation to assess outcomes (such as exploring the effect of children being in one-parent families). On those occasions we might use **observational** research, where we would record what happens over time and compare outcomes between groups. Alternatively, we could trawl through historical records – we call this **retrospective** research.

True experiments are conducted within 'laboratory' conditions, where variables can be controlled more easily. When we conduct research with people, such controls are more problematic. First of all, laboratory studies might not reflect 'real life'. It is often better to explore outcomes in contexts that represent what the participants normally encounter. It is also very difficult to control all variables. If we want to measure mood reports according to gender, ideally the participants would differ on that only – age, income, education, housing and a whole load of other individual differences should be identical (otherwise, we cannot be certain that we are measuring only gender differences). We could 'control' that by matching participants on all of those factors. However, this can make recruitment very difficult. As an alternative, we can use statistical procedures to control such things (as we will see later).

Much of what we have just described explores outcomes 'here and now'. These between-group studies are usually best for cross-sectional research, but it often says very little about cause and effect. If we want to be more confident about that, we may be better off choosing a prospective longitudinal study, conducted under within-group conditions. For example, to explore the effect of a new antidepressant in a single group of depressed patients, we could examine illness severity scores at a series of time points. This is actually a variation on the experimental method, because we are still manipulating the independent variable. Only this time we are doing this (perhaps more ethically) in the form of conditions performed on a single group.

5.6 Take a closer look
Research methods in psychology

We have a number of options open to us when we investigate differences. This summary might help you remember the terminology:

Experimental: Where the independent variable is directly manipulated, just as it is in traditional science

Quasi-experiments: Quite often we cannot manipulate the independent variable, perhaps because of ethical constraints. Retrospective studies and those which explore naturally occurring events are good examples of a quasi-experiment

Observational: Where we observe events, rather than intervene

Cross-sectional: A quasi-experimental design that focuses on measuring groups

Longitudinal: Where outcome is measured sequentially over a series of time points

Retrospective: Where we examine historical data to investigate outcome in respect of an independent variable, rather than seek to manipulate it

Sampling methods

Earlier, we said that there are a number of ways in which we can allocate participants to groups in quasi-experimental studies. When we choose some non-specific method to do this, it is often (mistakenly) referred to as **random** allocation. In student projects it is quite likely that they will approach willing volunteers in cafes, libraries and bars. Despite claims, this is not a random form of participant recruitment. At best it is a systematic approach. In reality it is probably an **opportunity or convenience** sample. True random methods are completely impartial. They often use random number generators. They always use methods that are blind to the researcher. To be truly random, the researcher cannot have any influence over who is recruited, or to which experimental condition they are allocated.

A good example of randomised allocation often occurs in clinical drug trials. These studies can explore a range of drugs and doses, many including placebo conditions. To improve objectivity, the clinicians treating the patient (and rating their illness) are 'blind' to whether they are taking the placebo or the real drug (and dose of that drug). Equally, to ensure that the patient rates their response objectively, they, too, are unaware of what they have been given. For this to work, once patients have been recruited to the study, the clinician requests a code from the pharmaceutical company sponsoring the trial. This code, randomly generated from their computer system, is presented to the hospital pharmacy, where the trial drugs are stored according to those numbers. The relevant drug pack is administered to the patient. Neither the patient nor the clinician knows what is in the tablets (all of the pills and packaging are identical). The identity of the tablet is revealed only once the trial is over. An exception would be if a patient develops a serious problem – clinical intervention might dictate needing to know what the patient was taking (the patient would be withdrawn from the study).

Systematic sampling occurs when participants are recruited according to a specific number or order. For example, every tenth patient from a list of current outpatients could be invited to take part in a study. Opportunity sampling is probably the most common form of recruitment in student studies: this is where participants are recruited by availability through being in the right place (such as the student bar) at the right time. We also call this convenience sampling. Sometimes we select participants using **quota** sampling. This is when we recruit our groups in proportions that reflect the ratios seen in the general population. We may know quite a lot about the typical profile of depressed people in the population: age, gender, income, housing, type of job, etc. If we are recruiting a large sample of depressed patients, we might want to recruit them in the proportions known to exist in each age group, job type, etc. A **stratified** sample is similar to a quota sample, but differs in the fact that participants within each cluster are recruited at random.

5.7 Take a closer look
Summary of sampling methods

When we recruit participants to between-group studies, we can use a series of methods to allocate them to groups. Here is a summary of the points that we have just made:

Random:	Participants selected by random number generators, or some other way that is blind to the researchers
Systematic:	Participants recruited by choosing the nth person available from a specified group
Opportunity:	Participants recruited on the basis of availability; also known as convenience sampling
Cluster:	Participants chosen at random, but from very specific groups assumed to be representative of the population of interest
Quota:	Participants recruited into groups in proportions that reflect how they are represented in the general population
Stratified:	Same as quota sampling, but the participants within the groups are randomly selected

Measuring between-group differences

If we know that we are examining our data using between-group methods, the choice of statistical test will then depend on three further factors: Are the data parametric? How many dependent variables are being explored? How many independent variables are involved? We will not explore each test in any great depth, as that analysis can be reviewed in the relevant chapters. Many statistical books provide flow diagrams and other such charts to guide the researcher to the most appropriate statistical test. We will not seek to reinvent the wheel by adding to those. However, an overview of between-group statistical tests is shown in Table 5.1, indicating in which chapter the test is explored and how the procedure matches the criteria for data type, number of variables and whether this is a parametric or non-parametric test.

Table 5.1 Statistical tests for between-group studies

		DV		IV			
Main tests	Ch	Type	No	Type	No	Groups	P/N
Independent t-test	7	Con	1	Cat	1	2	P
Mann Whitney U	18	Con	1	Cat	1	2	NP
Independent one-way ANOVA[1]	9	Con	1	Cat	1	2+	P
Kruskal Wallis	18	Con	1	Cat	1	3+	NP
Independent multi-factorial ANOVA	11	Con	1	Cat	2+	2+	P
MANOVA	14	Con	2	Cat	1+	2+	P
ANCOVA[2]	15	Con	1	Both	1+	2+	P

Key Con: continuous variable; Cat: categorical variable; P: parametric; NP: non-parametric

Notes [1]: Independent one-way ANOVA can be performed with two groups, but t-test usually performed in that context; [2]: ANCOVA is used to examine effect of 'controlling' variables.

Measuring within-group differences

If we know that we will be exploring differences within-groups, we will then need to make choices about which test to perform so that we can analyse the data. As usual, those options will based on the factors relating to the nature of the data and variables (what type, how many, parametric issues, etc.). Table 5.2 provides a summary of within-group tests.

Table 5.2 Statistical tests for within-group studies (and mixed models)

		DV		IV			
Main tests	**Ch**	**Type**	**No**	**Type**	**No**	**Conditions**	**P/N**
Related t-test	8	Con	1	Cat	1	2	P
Wilcoxon signed ranks	18	Con	1	Cat	1	2	NP
Repeated-measures one-way ANOVA[1]	10	Con	1	Cat	1	2+	P
Friedman's ANOVA	18	Con	1	Cat	1	3+	NP
Repeated-measures multi-factorial ANOVA	12	Con	1	Cat	2+	2+	P
Mixed models							
Mixed multi-factorial ANOVA[2]	13	Con	1	Cat	2+	1+ WG	P
						1+ BG	
Repeated-measures MANOVA	14	Con	2	Cat	1+	1+ BG	P
MANCOVA	15	Con					P

Key Con: continuous variable; Cat: categorical variable; P: parametric; NP: non-parametric

Notes [1]: Repeated-measures one-way ANOVA can be performed with 2 groups, but t-test usually performed in that context;
[2]: Mixed multi-factorial ANOVA examines at least one within-group (WG) IV and at least one between-group (BG) IV.

Examining relationships

Sometimes, rather than explore differences, we might want to look at the relationship between variables. We might examine how the scores on one variable change in relation to the scores on another variable (as we do with correlation). Or we may investigate how much variance in the scores for an outcome can be explained by variations in 'predictor' variables (as we do with regression). **Correlation** examines the relationship between two variables and represents the extent that one variable changes, the other variable changes accordingly. For example, we could measure how ice-cream sales vary as temperature changes.When we conduct correlation we are less likely to talk in terms of dependent and independent variables. Furthermore, we cannot measure cause and effect (unlike some measures of difference). There are several types of correlation that we can use, dependent on the nature of the data (see Table 5.3). There is little point addressing the exact pre-requisites for these tests here, so we will leave that until we get to Chapter 6.

Regression examines how much 'variance' can be 'explained' in an outcome, and which variables are responsible for contributing to that outcome. **Linear regression** focuses on numerical dependent (outcome) variables; independent (predictor) variables are examined to see how well they explain that outcome. Simple linear regression involves a single predictor variable; multiple linear regression has several predictors. For example, we could explore how much variance in mood scores can be explained by variations in sleep satisfaction, age and gender. **Logistic regression** explores categorical outcome variables; predictor variables are investigated with respect to how much they explain the 'likelihood' of that outcome. For example, we could explore how the likelihood of passing an exam is explained by variations in revision time, lecture attendance, time of day and amount of lecturer support. The choice of test and rules of engagement are somewhat more complex than these simplistic overviews, but we will explore that when we get to the relevant chapters. Table 5.3 provides a summary of factors that help us decide which test to use.

Table 5.3 Statistical tests for measuring relationships

Correlation	Ch	Variable type	No	P/N
Pearson's	6	Continuous	2	Parametric
Spearman's or Kendall's tau	6	Continuous	2	Non-parametric
Partial[1]	6	Continuous	3	Either
Biserial or point biserial[2]	6	Both	2	Either
Regression	**Ch**	**Outcome**	**Predictors**	**P/N**
Simple linear[3]	16	Continuous	1	Parametric
Multiple linear[3]	16	Continuous	2+	Parametric
Logistic	17	Categorical	1+	Non-parametric

Notes [1:] Partial correlation explores relationship between two variables, controlling for a third variable;
[2:] Biserial correlation explores relationship between one continuous variable and one categorical variable;
[3:] It is *preferable* that linear regression examine interval outcomes

Additional tests of association

There are further tests of relationships that do not quite fit the examples that we have just seen – see Table 5.4. These relate to cases where all of the variables are categorical, but that measure frequency data. By definition these tests are non-parametric. Typically, we measure such outcome using **Chi-square** (χ^2) tests and loglinear analysis (see Chapter 19 for more details).

Table 5.4 Other tests

	No. variables	Groups
Pearson's χ^2, Yates' continuity correction, Fishers exact test	2	2+
Layered χ^2, loglinear analysis	3+	2+

Validity and reliability

When we conduct research and collect data, we must make sure that we are actually measuring what we claim to be, and that we are doing so in a consistent way. We examine these important factors through validity and reliability. If we can demonstrate that we have accounted sufficiently for these, we can have more confidence in our outcome – and others are more likely to trust our data. For example, these outcomes are often used to describe the robustness of a questionnaire. Good published studies will report the validity and reliability of the scales that they have used in their study. It is these concepts that the final two chapters of this book focus on. In those chapters we are introduced to some statistical procedures that cannot be described in terms of the parameters we have been using so far.

Validity examines whether we are actually measuring what we think we are. For example, we could ask someone to report their IQ. If it is high we might claim that the person is intelligent. However, such an assumption might lack validity because we cannot be certain that IQ really does measure intelligence. Related to that, we also might want to explore the component structure of a questionnaire, to assess what 'factors' are present. In Chapter 20 we will explore a statistical procedure called factor analysis (more specifically principal components analysis) that can help us do just that.

Reliability measures the consistency and repeatability of an outcome. Once we observe a specific outcome, we would expect to see a similar result if we were to repeat the procedure (or if someone else used our methods). Consistency can be examined over time, between several researchers, for single researchers (in respect of their own consistency of ratings), and to measure the internal consistency of concepts within a questionnaire (to ensure that they appear to be measuring the same theme). In Chapter 21 we will focus on the last of those examples, with a statistical test called reliability analysis. It would be pointless to go into any more detail about these procedures at this stage, so we will leave that until we get to the relevant chapters.

5.8 Take a closer look
Validity and reliability

We can summarise the points that we have just made:

Validity: The extent that we are measuring what we claim to be.

Reliability: Describes the consistency of our data, across items, over time and between researchers.

Chapter summary

In this chapter we have explored experimental methods and have applied this to selecting the correct test to examine data in given contexts. At this point, it would be good to revisit the learning objectives that we set at the beginning of the chapter.

You should now be able to:

- Understand that, due to various restrictions posed by research with people, psychological research can rarely use true experimental methods. Laboratory conditions may not reflect real life; it is difficult to control for all possible variables. Instead, we tend to use quasi-experimental and correlational methods.
- Define data types: discrete data are distinct entities, represented by categorical groups or 'counts' of numbers; continuous data are numerical ranges (anything that is not discrete); categorical are discrete data that are represented by groups or conditions; ordinal data are numerical data that can be ordered by rank, but little inference can be taken in the magnitude of numbers in these ranges; interval data are more obviously numerical, where measurement can be based on magnitude and distance between numbers; ratio data are interval values that can be related to each other relatively.
- Appreciate that data are parametric if they are (reasonably) normally distributed and at least interval in nature. Parametric tests rely on the mean score to determine outcome. If we cannot trust the mean score, we should use non-parametric tests.
- Recognise the factors that determine which statistical test to perform: data types (whether the variables are continuous or categorical); whether the data are parametric; how many variables are being measured (including the number of dependent and independent variables, and the

number of groups and/or conditions being measured within the independent variables); *whether* differences are being examined, or relationships are being observed; and whether the data are being measured between-groups or within-groups (or a mixed model is being used).

- Recognise how to differentiate between differences and relationships.
- Appreciate how to specify between-group and within-group designs (and understand the relative benefits of each of them).
- Understand different types of sampling methods and research designs.
- Know how to define and measure validity and reliability.

Extended learning task

Following what we have learned about research methods and statistical tests, answer the following questions.

1. Describe each of the following variables in respect of the characteristics of the data that they aim to measure. Refer to terms such as discrete, categorical, dichotomous, ordinal, interval and ratio data (some answers may include more than one term):
   a. A diagnosis of depression (yes or no)
   b. Anxiety groups (none, mild or moderate)
   c. Position in a race (first, second, third... fifteenth)
   d. Subjective rating of mood (on a Likert scale where 1 = very happy through to 5 = very unhappy)
   e. Children's IQ
   f. Height of participants (in centimetres)
   g. Number of goals scored by each striker in a football season

2. Look at the following research summaries:
   a. We believe that as age increases, anxiety scores increase. However, we notice that the age variable is not normally distributed.
   b. We want to see if anxiety scores increase proportionately with sleep disturbance scores; both variables are normally distributed.
   c. We want to examine if women spend more money on clothes than men; the amount spent by both groups is not normally distributed.
   d. We want to measure quality-of-life scores among some participants who have been categorised according to their depression scores: no depression, mild depression and moderate depression. The quality-of-life scores are normally distributed.
   e. We want to measure the effect of a new teaching method across a group of students, to whom we present both techniques and measure them on satisfaction with the teacher.

For each of the examples in Question 2, answer the following questions:

i. Is this examining a relationship or exploring a difference?
ii. Describe the dependent and independent variables.
iii. Describe the levels on the independent variable (if appropriate).
iv. Indicate whether there is evidence that relevant variables are parametric.
v. Suggest a suitable statistical test to investigate the research, or describe the range of options if some information is missing.

6 CORRELATION

Learning objectives

By the end of this chapter you should be able to:

- Recognise when it is appropriate to use correlation
- Appreciate the different types of correlation and the factors that determine which type should be performed
- Understand the theory, rationale and assumptions associated with each test
- Calculate outcomes manually (using maths and equations)
- Perform analyses using SPSS
- Understand how to present the data and report the findings

What is correlation?

The term correlation represents a series of statistical tests that measures the relationship between two variables. Usually, both variables will be represented by ordinal or interval data As we saw in Chapter 5, both of those data types have a numerical form in one way or another (we will have a reminder about the distinction between them later). In these cases, correlation explores the way in which the values in the two variables vary with each other (involving the same cases or participants). These changes may occur in the same direction, they could operate in opposition to each other, or there may be no relationship at all. For example, we might find that as salaries increase, the amount spent on luxury goods also increases. Or we might observe that as unemployment increases, the amount spent on luxury goods decreases. Meanwhile, there is probably no relationship between the amount spent on luxury goods and hair colour. Less commonly, correlation can also be conducted between one continuous variable and one categorical variable. For example, we might choose to examine whether there is a *relationship* between gender and the amount spent on clothes. In this chapter, we will explore a range of correlation tests. The choice of test type will depend on several factors, such as whether the data are parametric (see Chapter 5 for a definition of parametric data).

Research questions for correlation

To illustrate the various types of correlation, we will pose a series of research questions set by the (fictitious) Mood, Anxiety and Sleep research group (MOANS). They decide to investigate whether there is a relationship between sleep quality perceptions and mood. They examine data from two questionnaires that they present to their participants. MOANS predict that as participants' perceptions of sleep quality worsen, their reports of mood scores will get poorer. To extend their analyses the researchers also record the age and gender of the participants, and whether they have a formal diagnosis of depression.

Theory and rationale

Correlation: the basics

Correlation describes the relationship between variables. We assess that relationship (or association) in terms of a 'correlation **coefficient**', which measures the way in which the 'values' in one variable change in relation to 'values' in a second variable. A **positive correlation** coefficient occurs when values change in the same direction. For example, we might expect the sale of ice creams to increase as temperature increases. A **negative correlation** will exist when

6.1 Take a closer look
Correlation: size and direction

Correlation is measured in terms of magnitude and direction. Here is a summary of the key factors in interpreting correlation coefficient:

+1 Perfect positive correlation
−1 Perfect negative correlation
0 No correlation

values change in opposite directions. We might predict that the sale of overcoats will decrease as temperature increases. Alternatively, there might be no correlation whatsoever, as might happen if we measured the relationship between temperature and the sale of hamsters. The correlation coefficient is measured on a scale of 0 (no correlation) to +1 (perfectly positive correlation) or −1 (perfectly negative correlation). When we report correlation we do so in terms of the letter r. For example, the correlation between ice-cream sales and temperature might indicate $r = .75$, while the correlation between temperature and the sale of overcoats might show $r = -.66$. If we examined the relationships between temperature and the sale of hamsters we might find a correlation of $r = .04$. We will see how to interpret the magnitude of correlation shortly.

Conventional interpretation of the magnitude of correlation was set by Cohen (1988). However, as is often the case in statistics, there are others who hold a slightly different view, such as the guidelines suggested by Brace *et al.* (2006). There is little to choose from between these guidelines, although Cohen's interpretation is directly related to effect size, which may be useful (see Chapter 4). A summary of those interpretations are shown in Table 6.1.

Table 6.1 Correlation coefficients - two different interpretations

Coefficient	Cohen, 1988	Brace *et al.*, 2006
Weak	± 0.1	$\leq \pm 0.2$
Moderate	± 0.3	$\pm 0.3 - 0.6$
Strong	± 0.5	$\geq \pm 0.7$

Viewing correlation graphically

Correlation between two variables is often presented graphically. A data point is plotted for each participant (or case). Each axis represents a variable; data points are placed along those axes according to the value for each variable. For example, we could assign 'sleep quality perceptions' along the horizontal (x) axis and 'mood scores' along the vertical (y) axis. If a participant reported a sleep quality perception score of 15 and a mood score of 62, a data point would be drawn 15 units along the x axis and 62 units up the y axis. Once all of the data have been plotted, the cluster of data points indicates the magnitude and direction of correlation. This type of graph is called a **scatterplot**. We can draw a line through the cluster that illustrates the trend of those data points, which we call a line of best fit. Some examples are shown in Figure 6.1.

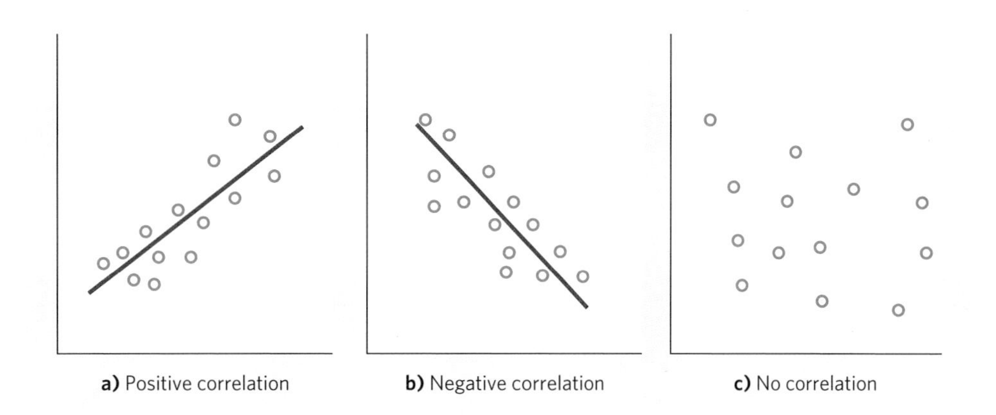

Figure 6.1 Correlation scatterplots

Figure 6.1a shows an example of positive correlation. It is represented by a line that slopes upwards from left to right – the relationship between ice-cream sales and temperature might be presented like this. Figure 6.1b shows an example of negative correlation. The line slopes downwards from left to right – the relationship between temperature and overcoat sales may look like that. Figure 6.1c shows an example of no correlation. There is no pattern to the cluster of data (such as we might find with the relationship between temperature and the sale of hamsters).

Correlation: common myths

There are a number of myths associated with correlation that we should dispel. To begin with, there will rarely be perfect correlation (positive or negative), even in the most obvious relationships. Other factors might interfere, such as personal preferences or individual differences. For example, some people may buy overcoats in hot weather because they may be cheaper than during cold spells. Another myth is that negative correlation is 'bad'. Students often think that they have failed if they find anything other than positive correlation. This is simply not true: both are equally important. Often, the direction of correlation depends on how a questionnaire is scored. For example, you may want to measure the relationship between participants' age and their sleep quality perception scores. You might expect that sleep quality perceptions worsen with age. If poorer sleep quality is indicated by lower scores, then your prediction is correct *if* you find a negative correlation.

Even if there is a strong relationship between variables, it can *never* mean that we can infer cause and effect. There may be some cases where there may *appear* to be cause (perhaps in the relationship between temperature and ice-cream sales), but correlation cannot measure that. We can say only that there is a relationship; we will not be able to say that changes in one variable *cause* variations in the other. For example, evidence suggests that there is a correlation between inner-city dwelling and rates of schizophrenia. From that, you might argue that living in inner cities *causes* schizophrenia. In reality, at best, it is only one of many risk factors. In any case, it is also quite common for people with schizophrenia to drift towards inner cities. Therefore, there appears to a relationship, but we cannot be clear about cause and effect.

Another common error is to put too much emphasis on the statistical *significance* of a relationship. We saw some guidelines for interpreting the *magnitude* in Table 6.1. The significance of the relationship suggests how unlikely it is that the observed coefficient occurred by chance factors (in that sample). More precisely, it shows the *improbability* that the null hypothesis is true (that there is no relationship – see Chapter 4). For most of the other statistical procedures that we encounter in this book, we will see that statistical significance is of primary importance. For example, if we find a *difference* between men and women on sleep quality, it may mean nothing if that difference is not statistically significant. With correlation, such assumptions are rather simplistic. Earlier, we saw how we can measure the magnitude of the correlation from effect sizes suggested by Cohen (1988). The effect size says more about the strength of the association than the significance. In very large samples it is possible for a small effect size (such as $r = .10$) to be highly significant. Equally, in small samples, it is possible for a large effect size (such as $r = .70$) to be non-significant. With larger samples, we should not make too much of the significance – we need to focus on the magnitude of the correlation to illustrate whether the outcome is meaningful. In smaller samples, we must pay more attention to the significance of the relationship, as well as to the magnitude. A summary is provided in Box 6.2.

Applications of correlation

So far, we have seen a number of examples where we could use correlation to explore the relationship between variables. However, there are a number of other applications of correlation In Chapter 5 we briefly explored validity and reliability. Validity describes how well we are actually

6.2 Take a closer look

Important things you should know about correlation

Here are some key factors worth remembering about correlation:

1. You will rarely find 'perfect correlation'. Individual differences and other factors may interfere with observed relationships.
2. Negative correlation is not bad, it simply describes the direction of the relationship. It does not mean there is 'no relationship' because it is 'less than 0'.
3. Correlation *never* implies cause and effect.
4. Don't put too much emphasis on significance in larger samples; focus on the correlation coefficient. Always report significance with the coefficient in smaller samples.

measuring what we claim to be. For example, let's say we have designed a new questionnaire to measure self-esteem. We could give this to some participants and measure their responses. We could also ask them to complete an established self-esteem questionnaire. If our new questionnaire truly measures aspects of self-esteem, the responses on that should be similar to those given on the established scale. We can examine the strength of this validity by comparing the scores (for each participant) across both questionnaires. A high correlation would indicate good validity.

Reliability measures the consistency of our data. For example, to assure ourselves of the reliability of our new self-esteem questionnaire, we need to know that people will respond in the same way each time that they complete the scale (all else being equal). To do this we can use something called **test-retest reliability** (we will see more about this in Chapter 21). We can give the questionnaire to a group of people on one day and record the responses. Two weeks later, we could give the same people the same questionnaire and record those responses again. To assess reliability we simply compare the two sets of responses using correlation – the higher the coefficient, the higher the reliability.

Correlation also plays a large part in many other statistical tests, as we will see as we venture through this book. It is an integral part of **linear regression** (Chapter 16), factor analysis (Chapter 20) and **reliability analysis** (Chapter 21). Correlation is also an important consideration for weighing up assumptions and restrictions of other tests, such as MANOVA (Chapter 14) and ANCOVA (Chapter 15).

Types of correlation

We have been using the word 'correlation' quite liberally so far. In reality, there are several types – the choice depends on a number of factors relating to the nature of the variables being measured. In the subsequent sections we explore six methods of correlation: Pearson's correlation, Spearman's correlation, Kendall's Tau-b, partial (and semi-partial) correlation, biserial correlation and point-biserial correlation. We will look at the theory behind each of these now, and will explore how to perform the tests in the remaining sections.

Pearson's correlation is probably the most commonly used of these tests (although some would argue that it is the most commonly misused). We should employ Pearson's correlation only when both variables are parametric. This is because Pearson's correlation is based on how case scores vary from (variable) mean scores across the respective variables. We saw how to determine whether our data are parametric in Chapter 5. In short, the variables must be at least interval and should be (reasonably) normally distributed (we explored normal distribution in Chapter 3). **Spearman's correlation** should be used if the data for at least one of the variables is not parametric. It might be that some data are represented by ordinal data, or that one (or both) of the variables are not normally distributed. Rather than rely on mean scores, the outcome is based on how scores are ranked across a variable. **Kendall's Tau-b** is very similar to Spearman's correlation, in that it is used for non-parametric data. However, it might be employed if there are too many 'ties' in the ranked scores. In some cases we can measure correlation where one of the variables is categorical (such as gender: male vs. female). The second variable must be ordinal or interval. To measure correlation in this context we need to use something called **biserial correlation** (or **point-biserial correlation**, depending on the nature of the categorical variable – as we will see later).

We will also explore **partial correlation** and **semi-partial correlation** in this chapter, but it is better that we leave the explanation of that until later – you need to understand the fundamentals of correlation before we address slightly more complex issues.

6.3 Take a closer look
Basic types of correlation: a summary

We have just explored several types of correlation. Here is a summary of those points:

Pearson's correlation:	Used where both variables are parametric
Spearman's correlation:	Used when at least one of the variables is not parametric
Kendall's Tau-b:	Used instead of Spearman's correlation if there are too many tied ranks
(Point) Biserial correlation:	Used when one of the variables is categorical

Pearson's correlation

The magnitude of the coefficient is reported using **Pearson's *r*** (e.g. $r = .75$). As the correlation coefficient cannot exceed 1, it is normal convention to omit the leading zero before the decimal point. To illustrate how we can use Pearson's correlation to examine relationships, we will return to the research question posed by MOANS (our research group). They are seeking to explore the relationship between sleep quality perceptions and mood. In this example, 15 participants have been given two questionnaires: one that measures a series of factors about perceived sleep quality (feeling refreshed, sleep satisfaction, having enough sleep, etc.) and a scale that measures current mood. Both questionnaires are measured on a scale of 0–100. Higher values on the sleep quality scale represent better perceptions, while higher mood scores represent poor perceptions. MOANS predict that there will be a *negative* correlation between sleep quality perceptions and mood.

6.4 Take a closer look
Hypothesis for research question

There will be a negative correlation between sleep quality perceptions and mood.

Assumptions and restrictions

Before we proceed, we should examine the assumptions and restrictions for Pearson's correlation. Generally, Pearson's correlation is conducted between two parametric variables. The data should be interval (or ratio) and reasonably normally distributed (we explored normal distribution in Chapter 3). Interval data are represented by meaningful, objective, numerical values. Pearson's correlation outcomes are based on how case scores vary from the mean score for each variable. If the data are not parametric, this might compromise the mean score, making it unreliable. We could be reporting inaccurate outcomes if we fail to recognise this; it increases the likelihood of Type I and Type II errors (see Chapter 4).

However, in reality, things are never quite as simple as that. A great deal of psychological research is conducted using questionnaires. As these frequently explore subjective data, such as perceptions, we might rarely use Pearson's correlation to examine quasi-experimental research data (see Chapter 5 for a review of psychological research methods). A quick scan of published studies will reveal that ordinal data are frequently examined with parametric tests. Our variables explore subjective sleep perceptions and self-assessments of mood, using Likert scales. These are questionnaires that elicit responses such as '1 = strongly agree' through to '5 = strongly disagree'. Some sources claim that these are the very essence of ordinal data, so should not be measured with parametric tests (Jamieson, 2004). Others argue that a well-designed Likert scale that has been highly validated can approximate interval scores (Reips and Funke, 2008). For the purposes of illustration, we will assume that our data come from questionnaires like that. Furthermore, it would be useful to compare outcomes from a series of correlation methods – using the same data set. However, we should be careful to check that we have (reasonable) normal distribution (which we will do shortly).

6.5 Take a closer look
Summary of assumptions and restrictions

- Both variables should be parametric
- The data (on both variables) should be at least interval
- Those data should be reasonably normally distributed

Establishing the magnitude of Pearson's correlation coefficient

The magnitude of the coefficient for Pearson's correlation is based on how much the data (within each variable) vary according to their respective means, and how much those scores vary for each participant across both variables (for non-human research, we would explore outcomes across cases, rather than participants). You can see how this is done manually in Box 6.6.

6.6 Calculating outcomes manually
Pearson's correlation calculation

To illustrate how to calculate Pearson's correlation manually, we will refer back to the research question set by MOANS (the data are presented in Table 6.2). MOANS are examining the relationship between sleep quality perceptions (SQ) and mood, using a sample of 15 participants. The outcome from both variables is scored from 0–100 – higher sleep quality scores represent 'better' perceptions, while higher mood scores are poorer. **You will find a Microsoft Excel spreadsheet associated with these calculations on the web page for this book.**

Table 6.2 Sleep quality perceptions and mood data

Participant	SQ (x)	$A(x_i - \bar{x})$	Mood (y)	$B(y_i - \bar{y})$	A × B
1	48	−23.67	26	−8.27	195.64
2	80	8.33	30	−4.27	−35.56
3	78	6.33	23	−11.27	−71.36
4	87	15.33	34	−0.27	−4.09
5	66	−5.67	40	5.73	−32.49
6	70	−1.67	25	−9.27	15.44
7	67	−4.67	28	−6.27	29.24
8	62	−9.67	64	29.73	−287.42
9	85	13.33	33	−1.27	−16.89
10	43	−28.67	73	38.73	−1110.36
11	79	7.33	20	−14.27	−104.62
12	62	−9.67	37	2.73	−26.42
13	79	7.33	20	−14.27	−104.62
14	83	11.33	40	5.73	64.98
15	86	14.33	21	−13.27	−190.16
Mean x ($\bar{x}$)	**71.67**	**Mean y ($\bar{y}$)**	**34.27**	**Sum (A × B):**	
SD x (S_x)	**13.60**	**SD y (S_y)**	**15.54**	$\sum(x_i - \bar{x})(y_i - \bar{y})$	**−1678.67**

To find Pearson's correlation we need the following equation: $r = \dfrac{\sum(x_i - \bar{x})(y_i - \bar{y})}{(N - 1)S_x S_y}$

N = sample size (15); SD = standard deviation (S); we saw how to calculate SD in Chapter 4 (but also see Excel spreadsheet).

We take each participant's score in variable x and deduct the mean of x. We put that answer in column A:

e.g. Participant 1: x_i (48) − $\bar{x}$ (71.67) = −23.67

We repeat that for each participant

Then we do the same for variable y, putting the answer to that in column B.

Then we multiply column A by column B and put the answer in column 'A × B'.

e.g. Participant 1: −23.67 × −8.27 = 195.64 (allow for rounding)

We add all of the answers in column 'A × B' to get $\sum(x_i - \bar{x})(y_i - \bar{y})$

We put all of that into the Pearson's correlation equation:

So, $r = \dfrac{-1678.67}{14 \times 13.60 \times 15.54} = -.567$ (a strong negative correlation, using Brace, *et al.*'s (2006) guide)

Correlation and significance

The calculations that we have just performed suggest a strong relationship. However, as we saw earlier, it is important to assess the significance of the relationship in samples as small as this (in larger samples we pay less attention to significance). In Chapter 4 we discovered that statistical significance examines the likelihood that the null hypothesis is true (in this case, that there is no relationship between sleep quality perceptions and mood). If that likelihood is less than 5%, we can reject the null hypothesis in favour of the alternative hypothesis (that there will be negative correlation between the variables). We state significance in terms of the probability (p); a significant outcome is observed when $p < .05$ (although we tend to report the full value of that significance, as we will see later).

To determine significance for Pearson's correlation, we can refer to Pearson's *r* tables (see Appendix 3). We look up the *r* value in that table, according to the '**degrees of freedom**', the target significance level (usually $p = .05$), and depending on whether we have a one-tailed or two-tailed test. In correlation, a one-tailed test represents a specific prediction (that there will be a positive correlation between the variables *or* that there will be a negative correlation). A two-tailed test relates to non-specific predictions (simply that there will a *relationship* between the variables). The degrees of freedom are related to the sample size. We will encounter these throughout this book. Degrees of freedom (often shown as *df*) refer to the number of values that are 'free to vary' in the calculation, while everything else is held constant. Usually, *df* represents the number of values being measured (N) *minus* the number of parameters being used to measure it. In this case we have 15 numbers (the sample size) minus 2 variables (the parameters), so $df = 15-2 = 13$.

Those parameters direct us to a value in the *r* value that represents a cut-off point – if the observed *r* value exceeds that, we can say that the relationship is significant. In our example, we have a one-tailed test,where $df = 13$ and where significance is $p < .05$; the cut-off point for that is $r = .441$. Our correlation coefficient was $r = -.567$, so we have a significant (negative) relationship (we can reject the null hypothesis).

We can also use Microsoft Excel to calculate the critical value of *r* and to provide the actual p value. You can see how to do that on the web page for this book. In this case we find that $p = .0138$. On that spreadsheet, you will also see how to perform the entire test in Excel.

How SPSS performs Pearson's correlation

We can get SPSS to perform Pearson's correlation. However, I do urge you to try those manual calculations that we explored in Box 6.6 – you can learn so much more about statistics when you do that. To illustrate how we perform the analysis in SPSS, we will maintain our focus on the MOANS research example. You will recall that the research group sought to explore the relationship between sleep quality perceptions and mood, predicting that there will be a negative correlation. You can see how to create a data set in SPSS for Pearson's correlation in Box 6.7.

6.7 Nuts and bolts
Setting up the data set in SPSS

When we create the SPSS data set for Pearson's correlation, we simply need to set up columns for the two variables that we are seeking to measure (we saw how to create data sets in Chapter 2).

Sleep quality & mood.sav [DataSet1] - IBM SPSS Statistics Data Editor

File Edit View Data Transform Analyze Graphs Utilities Add-ons Window Help

	Name	Type	Width	Decimals	Label	Values	Missing	Columns	Align	Measure
1	Sleepqual	Numeric	8	0	Sleep quality p...	None	None	8	Right	Scale
2	Mood	Numeric	8	0	Mood	None	None	8	Right	Scale

Figure 6.2 Variable View for 'sleep quality and mood' data

Figure 6.2 shows how the SPSS Variable View should be set up. The first variable is called 'sleepqual', which will be used to record sleep quality perception. The second variable is called 'mood', which will be used to report mood perceptions. Both variables are classed as 'Scale' in the 'Measure' column.

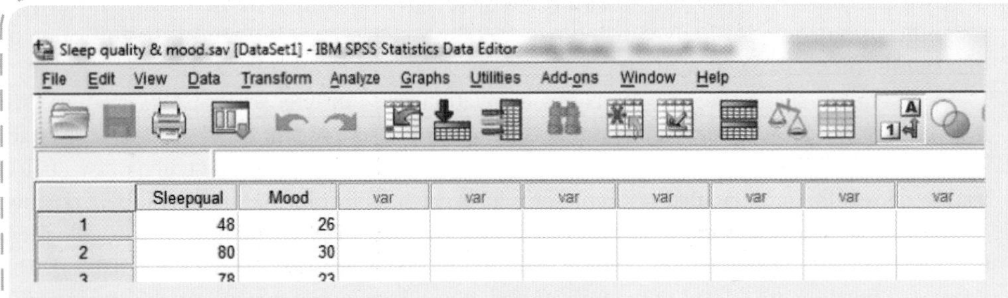

Figure 6.3 Data View for 'sleep quality and mood' data

Figure 6.3 illustrates how this will appear in the Data View. Each row represents a participant; values represent the scores reported for each participant.

Testing for normal distribution

Earlier, we said that we can perform Pearson's correlation on parametric data only. Putting aside the arguments about ordinal data for the moment, we still need to check that the data are normally distributed for both variables. Initially, we will run the Kolmogorov–Smirnov and Shapiro–Wilk tests to examine this (we saw the full instructions for this test in Chapter 3):

Open the SPSS file Sleep quality and mood

Select **Analyze** → **Descriptive Statistics** → **Explore** → (in new window) transfer **Sleep quality perceptions** and **Mood** to **Dependent List** window (by clicking on arrow, or by dragging the variables there) → select **Plots** radio button → click on **Plots** box → (in new window) select **None** radio button (under **Boxplot)** → make sure that **Stem-and-leaf** and **Histogram** (under **Descriptive**) are unchecked → select **Normality plots with tests** radio button → click **Continue** → click **OK**

	Kolmogorov–Smirnov[a]			Shapiro–Wilk		
	Statistic	df	Sig.	Statistic	df	Sig.
Sleep quality perceptions	.213	15	.067	.899	15	.092
Mood	.223	15	.044	.810	15	.005

Figure 6.4 Kolmogorov-Smirnov/Shapiro-Wilk test for sleep quality perceptions and mood

Since we have a sample of 15 participants, we should refer to the Shapiro–Wilk test (in Chapter 3 we saw that we should use the Kolmogorov–Smirnov outcome only when we have samples greater than 50). Figure 6.4 indicates that sleep quality perceptions appear to be OK, $W(15), = .899$, $p = .092$ (if the significance [Sig.] is greater than .05 the data are probably normally distributed). However, mood may not be normally distributed, $W(15), = .810$, $p = .005$. On the basis of that outcome, perhaps we should inspect normal distribution in respect of mood scores a little further, by examining z-scores for skew and kurtosis (Figure 6.5) (we learned about this in Chapter 3):

Select **Analyze** → **Descriptive Statistics** → transfer **Sleep quality perceptions** and **Mood** to **Dependent List** window → select **Statistics** radio button → click **OK**

Descriptives

			Statistic	Std. error
Mood	Mean		34.27	4.012
	95% confidence interval for mean	Lower bound	25.66	
		Upper bound	42.87	
	5% trimmed mean		32.91	
	Median		30.00	
	Variance		241.495	
	Std. deviation		15.540	
	Minimum		20	
	Maximum		73	
	Range		53	
	Interquartile range		17	
	Skewness		1.615	.580
	Kurtosis		2.220	1.121

Figure 6.5 Skew and kurtosis data for mood scores

To assess normal distribution, we divide the skew and kurtosis data by the respective standard error (see Chapter 3 for more details). This produces z-scores; these can be used to make judgements about normal distribution (see Table 6.3). We only need to do this for the mood scores because we know that the sleep quality perception data are probably normally distributed.

Table 6.3 z-scores for skew and kurtosis, in respect of mood scores

	Statistic	SE	z-score
Skewness	1.615	0.580	2.78
Kurtosis	2.220	1.121	1.98

In a sample this small, we do not want z-scores to be noticeably greater than $\pm$ 1.96 (see Chapter 3). It could be argued that there is significant positive skew in the mood data, suggesting that normal distribution has been compromised. Earlier, we said that we need 'reasonable' normal distribution. It could be argued that these data are not parametric. We will continue with the Pearson's correlation analysis for now (mostly so that you can see how to run it). However, it is quite likely that we should employ a non-parametric test of correlation (such as Spearman's). We will do that later.

Running Pearson's correlation in SPSS

Using the SPSS file Sleep quality and mood

Select **Analyze → Correlate → Bivariate...** as shown in Figure 6.6

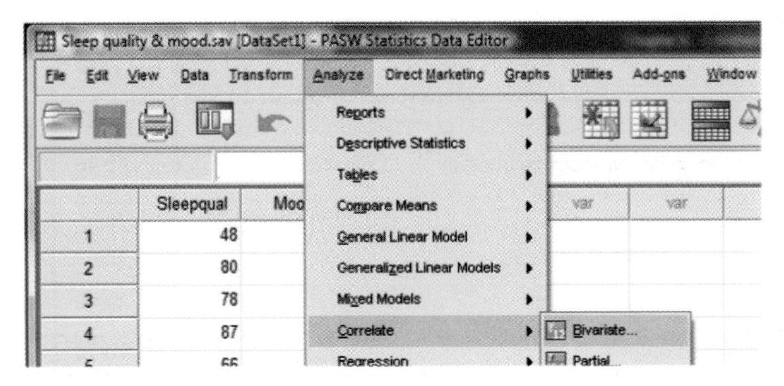

Figure 6.6 Pearson's correlation – step 1

In new window (see Figure 6.7), transfer **Sleep quality perceptions** and **Mood** to **Variables** window ➜ tick boxes for **Pearson** and **One-tailed** ➜ click **OK**

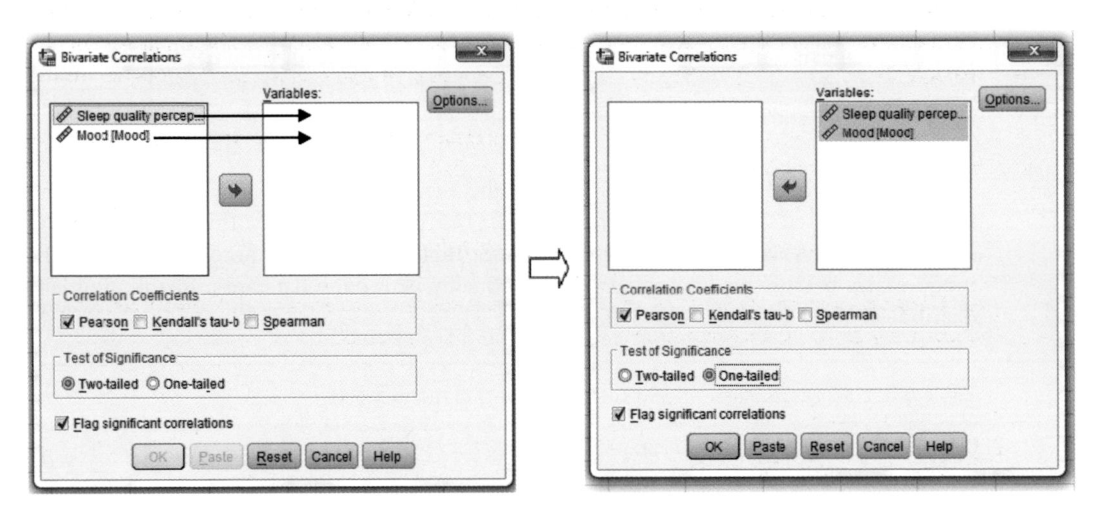

Figure 6.7 Pearson's correlation – step 2

Interpretation of output

Figure 6.8 indicates that we have a 'strong' negative (significant) correlation between sleep quality perceptions and mood. In our write-up, we report the correlation coefficient (stating

		Sleep quality perceptions	Mood
Sleep quality perceptions	Pearson correlation	1	–.567*
	Sig. (1-tailed)		.014
	N	15	15
Mood	Pearson correlation	–.567*	1
	Sig. (1-tailed)	.014	
	N	15	15

Figure 6.8 Pearson's correlation output

direction and 'magnitude' – see Box 6.2 for guidelines on interpreting coefficients). Then we present the **statistical notation**; we start with *r* (the sign for Pearson's correlation), followed by the degrees of freedom (*df*, which is presented within brackets; remember, $df = N - 2$), the correlation 'value', and the *full* significance value (unless SPSS has shown that as .000, in which case we present that as $p < .001$). In our case, the full notation is $r = (13) = -.567$, $p = .014$.

Variance

From these data, we can also report how much variance is explained by the relationship. We do this by squaring the *r* value; in our example $.567^2 = .321$. This suggests that 32.1% of variance in mood is explained by sleep quality perceptions.

Writing up results

Throughout these chapters, we will see how to write up our results as if we were doing so for a report. We might include a table of data (or a graph) and write something appropriate that describes the outcome and displays the statistical notation. However, it is bad form to include tabulated data *and* a graph that effectively show the same thing. Also, you should *never* 'cut and paste' SPSS output into your results. The write-up for our current results is pretty straightforward. We probably only need to write something like this:

There was a strong negative (Pearson's) correlation: poorer sleep quality perceptions were associated with poorer mood: $r(13) = -.567$, $p = .014$.

Using SPSS to draw correlation scatterplot

In Figure 6.1 we saw a range of graphical representations for correlation. We can get SPSS to draw a graph for us, showing how the data points are 'clustered'.

Using the SPSS file Sleep quality and mood

Select **Graphs → Chart Builder → . . .** as shown in Figure 6.9

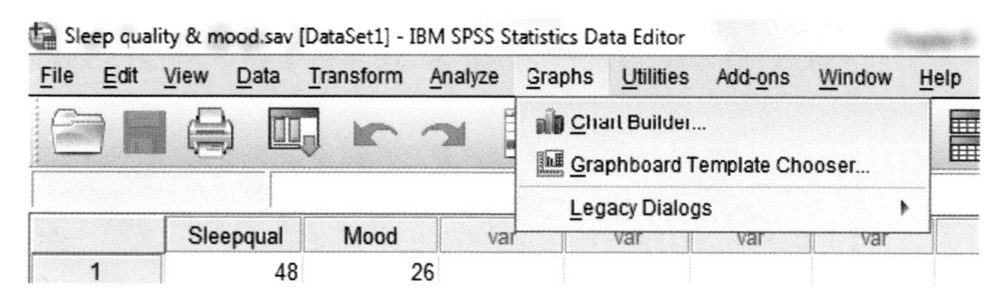

Figure 6.9 Drawing a scatterplot – step 1

In new window (see Figure 6.10), select **Scatter/dot** (from list under **Choose from:**) → drag **Simple Scatter** graphic (top left corner) into **Chart Preview** window

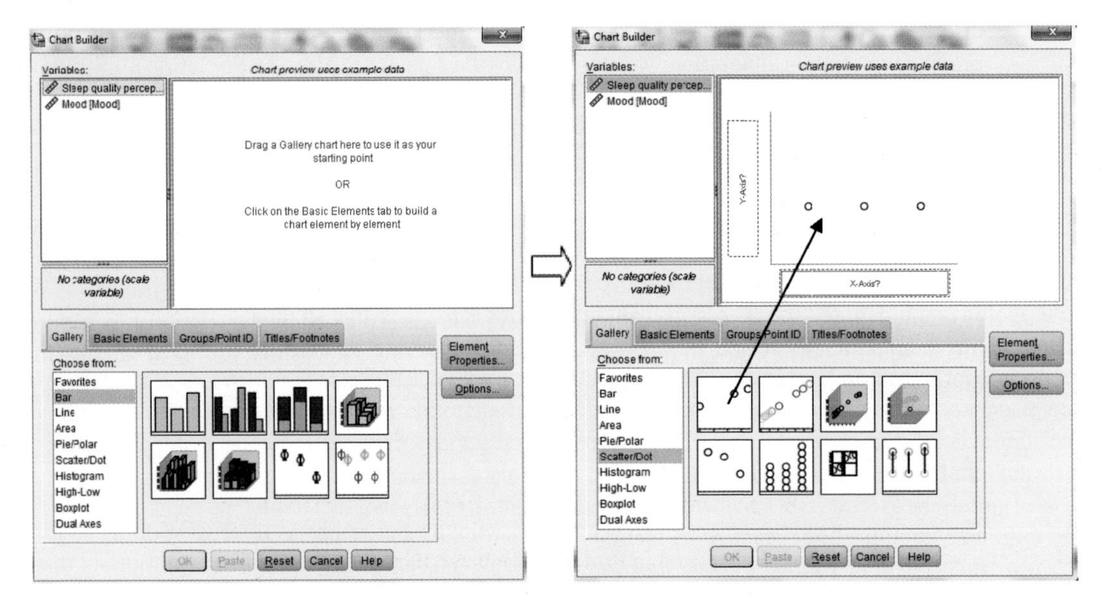

Figure 6.10 Drawing a scatterplot – step 2

Transfer **Mood** to **Y-Axis** box (to left of new graph – see Figure 6.11) ➜ transfer **Sleep quality perceptions** to **X-Axis** box (under graph) ➜ click **OK**
The action will produce a scatterplot, as shown in Figure 6.12.

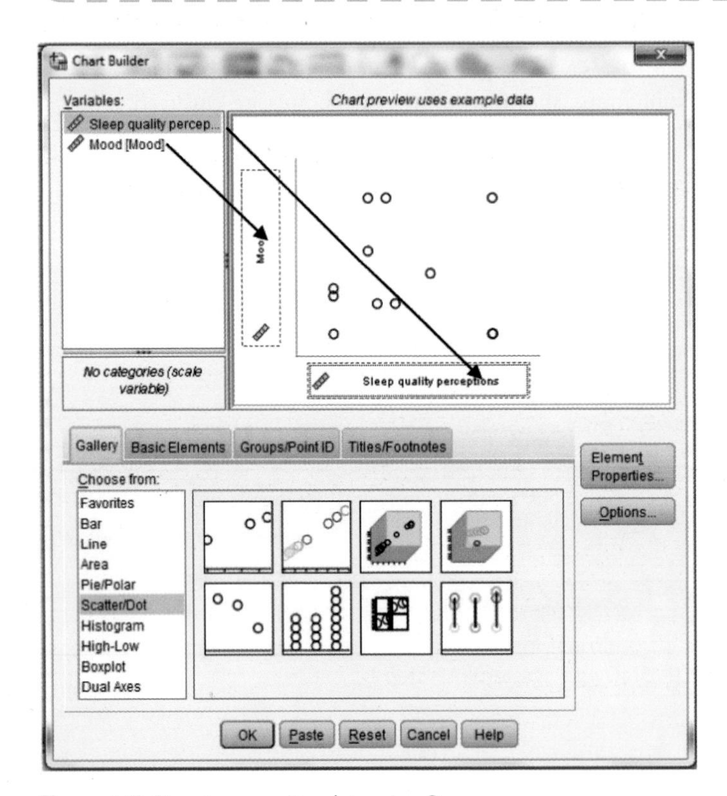

Figure 6.11 Drawing a scatterplot – step 3

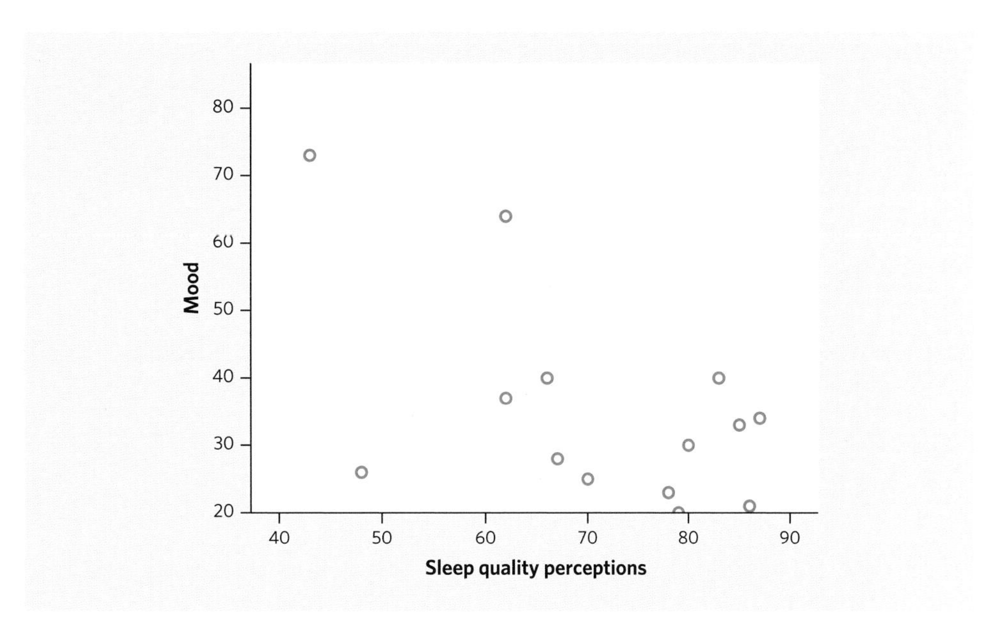

Figure 6.12 Scatterplot: sleep quality perceptions vs. mood

It is also useful to draw a line of regression through the data that describes the 'best fit' (the most typical trend of data) – it illustrates the '**regression' line** (or **gradient**), which tells us how values in 'mood' change for each unit change in 'sleep quality perceptions'. We learn more about the implications of that when we explore linear regression in Chapter 16.

In the SPSS output, double click on the graph (it will open in a new window, and will display some additional options) → click on the icon '**Add Fit Line at Total**' (in the icons displayed above the graph - see Figure 6.13) → click on **Close** → click on cross in top right hand corner of window showing adjusted graph

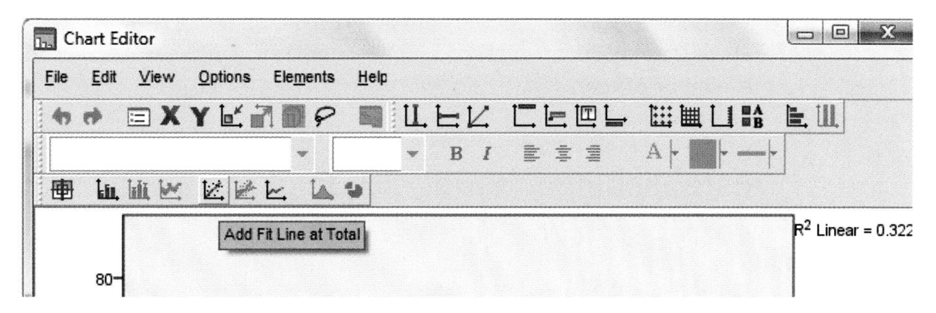

Figure 6.13 Adding the Fit Line

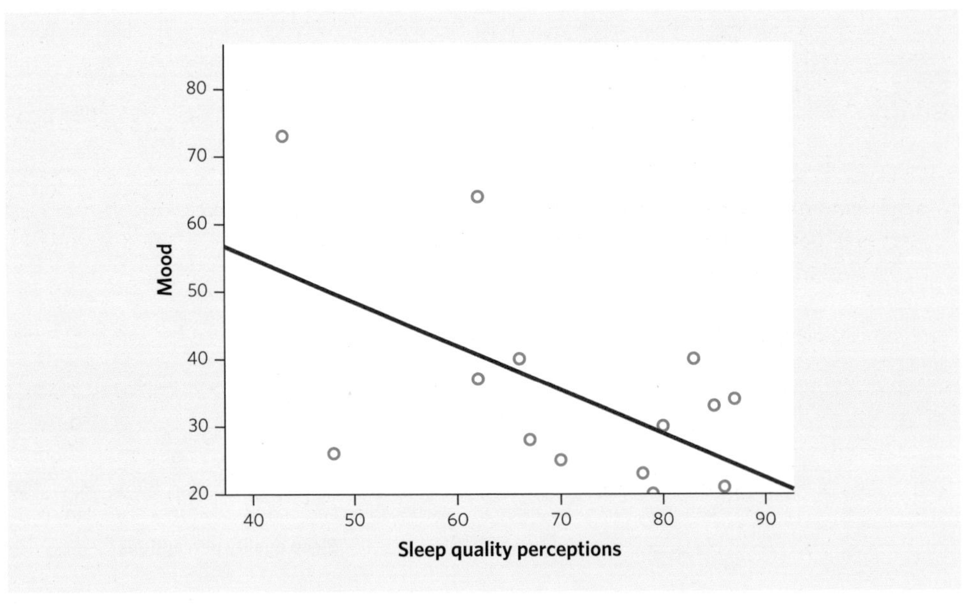

Figure 6.14 Scatterplot: sleep quality perceptions vs. mood (with line of fit)

Figure 6.14 suggests a negative correlation between sleep quality perceptions and mood.

Spearman's rank correlation

Spearman's rank correlation is used if one (or both) of the variables are non-parametric. When we performed Pearson's correlation earlier, we had some doubts over normal distribution. Furthermore, the subjective ratings of sleep quality and mood were almost certainly ordinal data. Either of these factors suggest that we ought to explore this relationship using a non-parametric test, such as Spearman's correlation (refer to Chapter 5 to see an extended account of these arguments).

Assumptions and restrictions

The most notable feature of Spearman's correlation is that there are fewer restrictions on its use. There is no requirement for data to be parametric, but the scores must be at least ordinal (they cannot be categorical).

Establishing the magnitude of Spearman's correlation coefficient

Because the data are not parametric with Spearman's correlation, outcomes cannot be based on mean scores. Instead, coefficients are calculated based on how much the **ranked scores** within each variable deviate from the mean rank. These rank variations are assessed for each participant (or case) across both variables. Once this has been done, the formula for calculating correlation is the same as it is for Pearson's. To illustrate these procedures we focus on the same data that we used for Pearson's (it will serve as a useful comparison). You will recall that MOANS (our research group) are seeking to examine the relationship between sleep quality perceptions and mood – they have predicted a negative correlation. You can see how this is done manually in Box 6.8.

6.8 Calculating outcomes manually

Spearman's correlation calculation

To illustrate how we calculate Spearman's correlation manually, we will reanalyse the MOANS data (see Table 6.4). The initial data are the same (outcomes in both variables are still scored from 0-100; higher sleep quality (SQ) scores represent 'better' perceptions, while higher mood scores are poorer. **You will find a Microsoft Excel spreadsheet associated with these calculations on the web page for this book.**

Table 6.4 Sleep quality perceptions and mood data

Case no	SQ (×)	Rank x	A: Rank x dev to mean	Mood (y)	Rank y	B: Rank y dev to mean	A × B
1	48	2	−6.0	26	6	−2.0	12.00
2	80	11	3.0	30	8	0.0	0.00
3	78	8	0.0	23	4	−4.0	0.00
4	87	15	7.0	34	10	2.0	14.00
5	66	5	−3.0	40	12.5	4.5	−13.50
6	70	7	−1.0	25	5	−3.0	3.00
7	67	6	−2.0	28	7	−1.0	2.00
8	62	3.5	−4.5	64	14	6.0	−27.00
9	85	13	5.0	33	9	1.0	5.00
10	43	1	−7.0	73	15	7.0	−49.00
11	79	9.5	1.5	20	1.5	−6.5	−9.75
12	62	3.5	−4.5	37	11	3.0	−13.50
13	79	9.5	1.5	20	1.5	−6.5	−9.75
14	83	12	4.0	40	12.5	4.5	18.00
15	86	14	6.0	21	3	−5.0	−30.00
Mean rank $\bar{x}$		8.0		$\bar{y}$	8.0	**Sum A × B**	**−98.50**
Rank sd* S_x		4.46		S_y	4.46		

*Rank sd = 'rank standard deviation', which is calculated like any other sample standard deviation (see Chapter 4)

Spearman's correlation takes each variable (x and y) and ranks the scores *within each variable*, from the smallest number (which receives rank #1) to the largest (which receives the highest rank). Tied ranks are shared (e.g. a 'sleep quality' score of 62 is shared between two participants; these scores occupy ranks 3 and 4; so we average that: 3 + 4 ÷ 2 = 3.5). The ranks are shown in columns 'Rank x' and 'Rank y'. Once those scores are ranked, the mean rank is calculated for each variable. Each score is then assessed with regard to how much it deviates from the mean rank - this is shown in columns A and B. Then, for each participant, we multiply columns A and B to provide outcomes in column 'A × B'. This is repeated for all participants and summed (Sum A × B).

Once we have Sum A × B, we can use Pearson's equation to find Spearman's rank correlation (r_s):

$$r_s = \frac{\sum(x_i - \bar{x})(y_i - \bar{y})}{(N - 1)S_x S_y}$$

$$\text{So } r_s = \frac{Sum\ A \times B}{(N-1)S_x S_y} = \frac{-98.50}{14 \times 4.46 \times 4.46} = -.353 \text{ (a moderate negative correlation)}$$

Estimating significance

Once again, we can use Pearson's *r* tables to estimate significance (Appendix 3). As we saw earlier, we can also use Excel to calculate the critical value of r and to provide the actual p value (see associated web page). In this case, we find that the p-value = .098. On that spreadsheet, you will also see how to perform the entire test in Excel.

Running Spearman's correlation in SPSS

Once again, we can ask SPSS to perform Spearman's correlation. We will keep the focus on the MOANS data, exploring the relationship between sleep quality perceptions and mood. The data would be created in SPSS as shown in Box 6.7. MOANS predicted that there will be a negative correlation, so we will be performing a one-tailed test.

Using the SPSS file Sleep quality and mood

Select **Analyze** → **Correlate** → **Bivariate...** (as shown in Figure 6.6) → transfer **Sleep quality perceptions** and **Mood** to **Variables list** → tick boxes for **Spearman** and **One-tailed** → click **OK**

Interpretation of output

Figure 6.15 confirms a moderately negative, but non-significant, correlation: $r_s(13) = -.353$, $p = .098$. If we compare this to the outcome we found for Pearson's correlation, we can see that the observed relationship is weaker using non-parametric methods. This is also a classic example of where we must pay attention to significance – the relationship might have been moderate, but it is not significant. We cannot reject the null hypothesis.

			Sleep quality perceptions	Mood
Spearman's rho	Sleep quality perceptions	Correlation coefficient	1.000	–.353
		Sig. (1-tailed)	.	.098
		N	15	15
	Mood	Correlation coefficient	–.353	1.000
		Sig. (1-tailed)	.098	.
		N	15	15

Figure 6.15 Spearman's correlation output

Variance

As we saw earlier, variance is the square of the correlation. So, $-.353^2 = .125$. This suggests that only 12.5% of variance in mood scores is explained by sleep quality perceptions in this sample, when we explore the outcome using non-parametric methods.

Writing up results

In our report we should write:

There was a moderate, but non-significant, correlation between sleep quality perceptions and mood: $r_s(13) = -.353$, $p = .098$.

Kendall's Tau-b

Kendall's Tau-b is very similar to Spearman's rank correlation in that it is used where some data are not parametric. However, this test is often employed in preference to Spearman's correlation when there are too many tied ranks. In Box 6.8, we saw how a data set is ranked according to values in each variable. There were several instances where scores received tied ranks. For example, using our data, there are two ties for sleep quality perception scores and two for mood scores. It might be considered that this represents too many ties, so Kendall's Tau-b may be more appropriate.

In your reading you may also come across a test called Kendall's Tau-a. In case you were wondering how that test differs to what we are doing (although I suspect you were not), Kendall's Tau-a is just another way of calculating correlation, but without adjusting for ties.

Assumptions and restrictions

There is nothing we can add here to what we have already said about the assumptions and restrictions for Spearman's correlation. This test is still used for cases where at least one of the variables includes non-parametric data. The main difference is that Kendall's Tau-b should be used if there are too many tied ranks. How many is too many? There is no golden rule, although some would say that you should not use Spearman's correlation if there are any ties.

Running Kendall's Tau-b in SPSS

We will not look at how to run manual calculations for Kendall's Tau-b as these are quite complex. Instead, we will go straight to performing the test in SPSS. We will use the same data again, based on the MOANS research question:

Using the SPSS file Sleep quality and mood

Select **Analyze** → **Correlate** → **Bivariate...** (as shown in Figure 6.6) → transfer **Sleep quality perceptions** and **Mood** to **Variables** window → tick boxes for **Kendall's Tau-b** and **One-tailed** → click **OK**

Interpretation of output

Figure 6.16 confirms that we have a moderately negative, but non-significant, correlation between sleep quality perceptions and mood: *Tau-b* (13) = −.272, p = .082. This is even weaker than the correlation that we found using Spearman's correlation.

			Sleep quality perceptions	Mood
Kendall's Tau_b	Sleep quality perceptions	Correlation coefficient	1.000	−.272
		Sig. (1-tailed)	.	.082
		N	15	15
	Mood	Correlation coefficient	−.272	1.000
		Sig. (1-tailed)	.082	.
		N	15	15

Figure 6.16 Kendall's Tau-b output

Biserial (and point-biserial) correlation

For the final examples of correlation, we look at situations where one of the variables is measured by categorical data. Any categorical variable we use must be dichotomous – it can have only two categories (such as we find with gender, male vs. female). Ideally, when we create the data set in SPSS, we should code categorical variables using 0 and 1 (perhaps 0 = Male, 1 = Female). We can use biserial (or point-biserial) correlation in these situations.

6.9 Nuts and bolts

Biserial vs. point-biserial correlation

There are two types of correlation that can be used when one of the variables is categorical: biserial and point-biserial. We summarise the difference between them in Table 6.5.

Table 6.5 Definitions and examples of biserial and point-biserial correlation

Definition	Examples
Biserial correlation: Where two 'poles' of the categorical variable are considered to be on a 'continuum' between 0 and 1.	Sleep quality perceptions (continuous variable) could be compared with depression diagnosis (categorical: depressed or not depressed).
Point-biserial correlation: Where the two categories are distinct groups, represented by value codes of 0 and 1.	Mood perceptions (continuous variable) could be correlated to gender (categorical: male or female).

Comparing biserial to point-biserial correlation

There are some subtle differences between biserial and point-biserial correlation (see Box 6.9). The categories for gender are (arguably) quite straightforward: someone is (usually) either male or female. It could be said that a diagnosis of depression is less clear, particularly if that diagnosis is based on mood scale thresholds: there are degrees of depression severity. These distinctions determine which type of 'biserial' correlation we can employ. Clearly, categorical data (such as gender) are examined using point-biserial correlation. Where those boundaries are somewhat fuzzier, we use biserial correlation. We can use SPSS to perform point-biserial correlation (although we will see how to calculate that manually in Box 6.10). Biserial correlation can only be calculated manually – see Box 6.11.

Establishing the magnitude of point-biserial correlation coefficient

We will see how to perform point-biserial correlation in SPSS shortly. In the meantime, we should explore how to do this manually. We will use a new MOANS data set, where information was collected from 98 participants regarding sleep quality perceptions, mood perceptions, age, gender, and current diagnosis of depression Similar to the first data set that we examined, sleep quality and mood are measured on a scale of 0–100, with higher sleep quality scores representing better perceptions and higher mood scores indicating poorer perceptions. Gender (male vs. female) is the categorical variable. We have established that gender is discrete, so we can examine the relationship with mood using point-biserial correlation. MOANS predict that there will be a 'relationship' between mood and gender, but do not specify the direction (which means that we have a two-tailed test). Before we do that, we need to report some descriptive data about the mean scores and standard deviation for mood scores, in respect of gender. We can get that from simple 'descriptive' analyses in SPSS:

Using the SPSS file Sleep quality

Select **Analyze** → **Compare means** → **Means...** → transfer Mood to **Dependent List** window → select Gender from list → click on arrow by **Independent List** → click **OK**

Table 6.6 Mood scores and standard deviation (SD) by gender

	Mean	SD
Male (n = 21)	57.05	18.36
Female (n = 77)	63.92	21.25

6.10 Calculating outcomes manually

Point-biserial calculation

Point-biserial correlation is used when one of the variables is represented by distinct, dichotomous, categorical data. Using the MOANS data we have been given (see Table 6.6), we will explore the relationship between mood perceptions and gender. Gender is coded as: males = 0, females = 1.

The formula for point-biserial correlation (r_{pb}) is: $\frac{(Y_1 - Y_0) \times \sqrt{pq}}{SY}$

To find Y_1 and Y_0 we need the data from Table 6.6. Y_1 is the mean sleep quality score for women (x = 1) : 63.92; Y_0 is the mean for men (x = 0): 57.05.

p is the proportion of the sample represented when x = 0: 21 ÷ 98 = .2143

q is the proportion of the sample represented when x = 1: 77 ÷ 98 = .7857

SY represents the pooled standard deviation of Y, which we derive using the standard deviation data for men (18.36) and women (21.25), as shown in Table 6.6, and apply that to yet another formula!

$$SY = \sqrt{\frac{(n_0 - 1)S_0^2 + (n_1 - 1)S_1^2}{n_0 + n_1 - 2}} = \sqrt{\frac{(20 \times 18.36^2 + 76 \times 21.25^2)}{21 + 77 - 2}} = 20.68$$

$$\text{So, } r_{pb} = \frac{(63.92 - 57.05) \times \sqrt{.2143 \times .7857}}{20.68} = .137\text{, a very weak correlation}$$

The direction of the correlation will depend on how 'x' and' y' are coded, so we must be careful when interpreting outcomes. In our analysis, the correlation was positive, where 'women' were coded as 1 – we *could* say that mood perceptions are *marginally* more likely to be poorer for women. However, the effect was very small indeed. We can calculate variance in a relationship by squaring the correlation co-efficient ($.137^2 = .019$). This suggests that only 1.9% of all variance in mood scores is explained by gender. Using the Pearson's *r* calculator that we referred to earlier, we find that we have a non-significant relationship: r_{pb} (96) = .137, p = .179.

Running point-biserial correlation in SPSS

We can use SPSS to perform point-biserial correlation (so long as we have indicated that one of the variables is categorical (we set'nominal' codes in SPSS as '0' and '1') :

Using the SPSS file Sleep quality

Select **Analyze** → **Correlate** → **Bivariate...** → transfer **Mood** and **Gender** to **Variables** window → tick boxes for **Pearson** and **Two-tailed** → click **OK**

		Mood	Gender
Mood	Pearson correlation	1	.137
	Sig. (2-tailed)		.180
	N	98	98
Gender	Pearson correlation	.137	1
	Sig. (2-tailed)	.180	
	N	98	98

Figure 6.17 Point-biserial output

Figure 6.17 confirms what we calculated manually in Box 6.10. We have a weak, non-significant correlation between mood and gender: r_{pb} (96) = .137, p = .180.

Establishing the magnitude of biserial correlation coefficient

We cannot perform biserial correlation in SPSS, but you can see how to calculate the outcome manually in Box 6.11. As we said earlier, biserial correlation is performed when the categorical variable is more likely to be on a continuum (such as a diagnosis of depression). To examine this, we still use the latest MOANS data set, but will focus on two different variables: sleep quality perceptions and depression diagnosis. Before we undertake those analyses, we need to find the mean and standard deviation for those variables. You saw how to request descriptive statistics in SPSS just now, so we do not need to repeat those instructions. The outcome is shown in Table 6.7.

Table 6.7 Sleep quality scores and standard deviation (SD) by depression diagnosis

	Mean	SD
Depressed (n = 55)	33.97	16.78
Not depressed (n = 43)	44.76	17.59

6.11 Calculating outcomes manually
Biserial calculation

We can calculate biserial correlation from the following formula (using much of what we learned in Box 6.10).

The formula for biserial correlation (r_b) is: $\dfrac{(Y_1 - Y_0) \times \dfrac{pq}{Y}}{SY}$

Using the data from Table 6.7, the values of Y_1 and Y_0 are 44.76 and 33.97 respectively.

Using the rationale from Box 6.10, we can see that p = 55 ÷ 98 = .5612, and q = 43 ÷ 98 = .4388

We can also calculate (*SY*)based on standard deviations of 17.59 and 16.78 (see Table 6.7).

$$SY = \sqrt{\frac{(n_0 - 1)S_0^2 + (n_1 - 1)S_1^2}{n_0 + n_1 - 2}} = \sqrt{\frac{(54 \times 16.78^2 + 42 \times 17.59^2)}{55 + 43 - 2}} = 17.14$$

We know that the area under a normal distribution curve is 1 (see Chapter 3). That area can be divided into 'larger' and 'smaller' portions (but their sum will always be 1). The values of p and q represent those portions (the larger portion is .5612; the smaller portion is .4388). Using normal distribution tables, we can estimate the height of the normal distribution curve (Y) where that distribution of portions is observed. We need to find Y when the portions under the curve are in the ratio of .5612 to .4388. This can be found in specially adapted normal distribution tables, such as the one published in Field (2009, pp. 797–802). Using that table, we look for a distribution similar to our ratio. The nearest we can find is .5596/.4404; at that point, y = .3945.

$$\text{So, } \frac{(Y_1 - Y_0) \times \dfrac{pq}{Y}}{SY} = \frac{(44.76 - 33.97) \times \dfrac{.5612 \times .4388}{.3945}}{17.14} = .393\text{, a moderate correlation}$$

Using the Pearson's *r* calculator that we referred to earlier, we find that we have a significant relationship: $r_b(96) = .393$, $p < .001$. Given that we have a positive correlation, and we coded 'depressed' as 0 and 'not depressed' as 1 (where higher sleep quality scores are 'better'), we might observe that sleep quality perceptions appear to be better for non-depressed participants (although the relationship is still only moderate, despite the highly significant outcome).

Partial correlation

Partial correlation can be used to examine how a relationship between two variables might be 'explained' by one or more additional (potentially confounding) variables. The original relationship is compared with the new outcome to see if there are important changes in that relationship once other variables are included. Until now in this chapter, we have been using the MOANS data to explore the relationship between sleep quality perceptions and mood. We have found conflicting outcomes depending on the way in which we perform the correlation analysis. Nevertheless, in all cases, there was at least a moderate relationship. The significance was compromised in non-parametric correlation tests, but this is probably due to the small sample. If we used a more robust sample, we might find a stronger, significant, relationship. However, even when we have more convincing outcomes, there may be more to the relationship than we first observe. We may find that there are additional variables that are interfering with the observed relationship.

Let's say that MOANS explore outcomes from a new, larger data set (still investigating the relationship between sleep quality perceptions and mood). From that, they might observe a strongly positive (significant) correlation. However, what happens if they are provided with evidence that suggests how mood perceptions worsen with increasing age, and then notice that most of the people in the MOANS data set who were reporting poorer sleep quality were older? How can they be sure that the relationship between sleep quality and mood is not reflecting age? That's where partial correlation can help: we can investigate the relationship between two variables while 'factoring out' the effect of other variables.

Effect, explanation and suppression

The outcome of partial correlation is illustrated by a series of key factors. These factors demonstrate the extent of the 'interference', from 'no effect at all' to 'cancelling out' the relationship altogether. We measure the effect in terms of how much the additional factor 'explains' the relationship that we thought we had observed. There will be full, partial, or no explanation – these all relate to what we call 'suppression'.

We could illustrate the action of effect and suppression graphically. Figures 6.18 and 6.19 shows how partial suppression might look like in two contexts. Let's say that the original relationship between sleep quality perceptions (S) and mood (M) is $r = .60$; the variance (r^2) is .36. Next, we could show the extent to which the relationship with age (A) 'overlaps' with both of those variables. Figure 6.18. shows an example where the relationship might be slightly compromised (partial explanation); Figure 6.19 presents a scenario where there is potentially greater interference (full explanation).

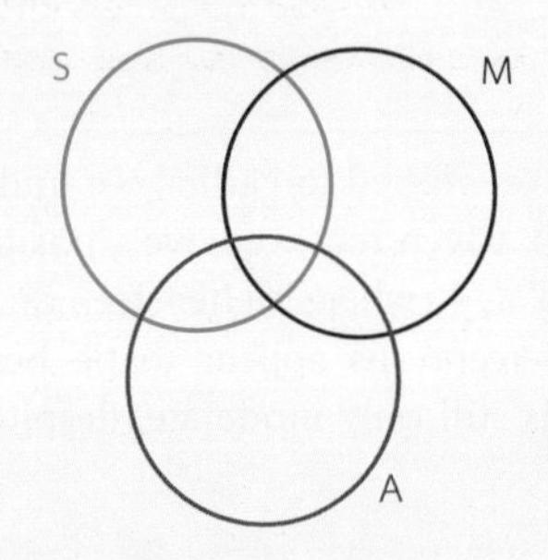

The 'effect' of suppression from the additional variable (age) upon the relationship between sleep quality perceptions and mood is shown by the intersection of the three circles. Using a visual assessment, we could say that about one-quarter of the relationship between sleep quality perceptions and mood is 'overlapped' with the relationship between sleep quality perceptions and age, and between mood and age.

Figure 6.18 Partial suppression (small effect)

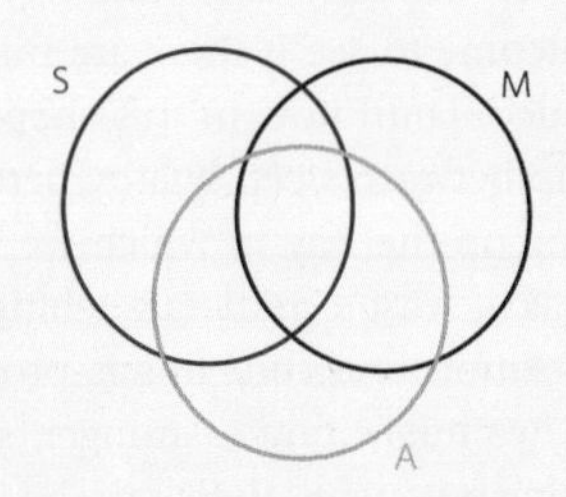

In this example, the relationship between sleep quality perceptions and mood is still the same, but now we have a much larger 'interference' from age. The intersection between the three circles appears to account for about two-thirds of the variance between sleep quality perceptions and mood. It is almost a 'full explanation' (see Box 6.12). It may be that the relationship between sleep quality perceptions and mood is a 'spurious' one.

Figure 6.19 Partial suppression (large effect)

6.12 Nuts and bolts
Partial correlation terminology

The following terms indicate the extent to which an additional variable might explain the original relationship:

Explanation: An explanation occurs when the strength of the original relationship has been altered by the effect of additional variables. That explanation may be 'full' or 'partial'.

Full explanation: If factoring out variables causes the original correlation to be reduced to zero, we can say that we have 'full explanation'. The additional variable(s) explained all of the relationship we originally observed; there was no relationship in the first place. This could be an example of a 'spurious correlation' (see next section).

Partial explanation: If the introduction of additional variables has some effect on original correlation we can say that we have 'partial explanation'. This effect might be very small or it could be substantial. In *some* cases, the relationship might be strengthened.

Spurious correlation

If the action of a partial correlation results in 'full explanation' (whereby the original correlation is 'wiped out'), it begs the question of whether there was really correlation in the first place. We might call that '**spurious**'. Say we find a strong correlation between the number of driving errors and university exam results. It seems illogical to imagine that there might be a relationship, but a correlation analysis indicates otherwise. However, if we then controlled for alcohol intake, we might find that the correlation disappears! The correlation was spurious because the relationship between driving errors and exam scores was actually explained by the amount of alcohol consumed.

6.13 Nuts and bolts
Examples of spurious correlation

In Table 6.8 there are some examples of apparent correlations. However, all is not what it seems: the relationship is actually due to something else altogether!

Table 6.8 When is a correlation not a correlation?

Apparent correlation	Actual explanation
A positive correlation between the number of fire engines attending a fire and the damage that ensues suggests fire engines cause the damage[1].	The size of the fire is related to the amount of damage – larger fires simply need more fire engines.
In a psychology class, students with longer hair got better exam results than those with shorter hair. It could be concluded that longer hair is related to better academic performance[2].	Since the girls in the class had longer hair than the boys, it is more likely that the effect was due to gender, not hair length.

Apparent correlation	Actual explanation
During the summer, someone notices that there is a negative correlation between ice-cream sales and the amount of clothes people wear. It could be suggested that eating ice cream causes people to remove their clothes.	It is simply a factor of the hot weather. The hotter it is, the more people buy ice cream. The heat would also explain the tendency to wear less.
Hospital statistics suggest that higher rates of radiotherapy are associated with greater death rates. It could be argued that radiotherapy is responsible for the deaths.	People who undergo radiotherapy are more likely to have cancer. More virulent cancer may need more treatment. More serious forms of cancer pose a greater risk of death.

Sources: 1. Burns (1996); 2. Vogt (2005)

Controlling for more than one variable

There is no limit to how many controlling variables we can use in partial correlation. A **zero-order correlation** is one without controls; it is just basic correlation. A **first-order correlation** has one controlling variable (for simplicity's sake, we will focus on that in this chapter), a second-order correlation has two controlling variables, while a third-order correlation has three, and so on.

However, if you need to conduct larger analyses, you would be better off using more robust statistical methods such as loglinear analysis for categorical data (see Chapter 19), **path analysis** or **structural equation modelling** for interval data (we do not deal with these methods in this book).

Assumptions and restrictions

In theory, to perform partial correlation we need to follow the guidelines for parametric data that we have seen so far. In reality, SPSS makes no allowances in respect of which form of correlation is used. Also, it is possible to conduct partial correlation where the controlling variable is categorical. So you could examine the relationship between sleep quality perceptions and mood, controlling for gender. However, the controlling categorical variable must be dichotomous (there must be only two levels, such as male vs. female). Furthermore, one of the variables in the zero-order correlation can also be dichotomous. So you could explore the correlation between mood and gender, while factoring out age (but see biserial correlation later).

Measuring the effect of partial correlation

Partial correlation is calculated by pairing up each of the correlations involved. The first 'pair' is the relationship representing the main analysis. Each variable in that correlation is then paired with the 'additional' variable. Those two correlations are multiplied together and deducted from the original correlation This outcome is then divided by a multiple of the 'square root of 1 minus the squared correlation' representing the relationships between the additional variable and *each* of the original variables. Does that sound a little complicated? Perhaps the manually calculated example shown in Box 6.14 will help to clarify the process. We will see how to perform partial correlation in SPSS shortly. To examine the effect of partial correlation, the MOANS research group collect data from a larger sample (98 participants). We are measuring the relationship between sleep quality perceptions and mood, taken from questionnaires (both scored 0–100, with higher sleep quality scores representing 'better' perceptions and higher mood scores being poorer). MOANS predict that there will still be a negative correlation. However, they also decide to examine whether the participants' ages have any effect on the observed outcome.

6.14 Calculating outcomes manually

Partial correlation calculation

To illustrate the effect of partial correlation, we will perform a 'standard' (Pearson's) correlation between each of the variables using SPSS (you saw how to perform this manually earlier, so we do not need to do that again). From the reported outcomes, we can apply some equations to calculate the remaining analyses of partial correlation manually. The data set (sleep quality) include the 98 participants that we have just referred to in the previous section. The outcome of Pearson's correlation is shown in Table 6.9. The main analysis is between mood (variable A in Table 6.9) and sleep quality perceptions (variable B). We then explore the effect of 'partialling out' age (variable C).

Table 6.9 Correlation (*r*)

Pair	Relationship	*r*	r^2
AB	Mood vs. sleep quality perceptions*	−.324	.105
AC	Mood vs. age	.645	.416
BC	Sleep quality perceptions vs. age	−.314	.099

Table 6.9 shows that there is a moderately negative correlation between sleep quality perceptions and mood ($r_{AB} = -.324$). But how will that relationship appear once we factor in the age of the participants? To calculate partial correlation we use the data from Table 6.9 and apply them to the following formula for partial correlation:

$$\frac{r_{AB} - (r_{AC})(r_{BC})}{\left(\sqrt{1-r_{AC}^2}\right) \times \left(\sqrt{1-r_{BC}^2}\right)} = \frac{-.324 - (.645)(-.314)}{\left(\sqrt{1-.416}\right) \times \left(\sqrt{1-.099}\right)} = -.167$$

The adjusted correlation ($r = -.167$) is dramatically reduced from the original outcome ($r = -.324$). The relationship between sleep quality perceptions and mood initially appeared to be moderate, but it is very much weaker once we account for the age of the participants.

How SPSS performs partial correlation

SPSS performs partial correlation using Pearson's formula. This assumes that the data are parametric. If the data are clearly non-parametric, we may have initially explored a relationship using Spearman's correlation. If we feel that our partial correlation should be undertaken using non-parametric methods, we must run analyses for all of the variables in SPSS, using the Spearman's option. Once that has been done, we can manually calculate partial correlation using the methods shown in Box 6.14. In the meantime, we will see how SPSS performs partial correlation (assuming it's OK to use Pearson's assumptions).

Running partial correlation in SPSS

To illustrate how SPSS performs partial correlation, we use the revised MOANS data set we referred to in Box 6.14. These data explore the relationship between sleep quality perceptions and mood in 98 participants. We will explore that correlation initially. Then we will explore how that relationship is affected by running a partial correlation, accounting for the potential effect of the participants' ages. We should know how to run Pearson's correlation by now, but here is a summary of those methods again. MOANS predict that there will be a negative correlation, so we have another one-tailed test.

Open the SPSS file Sleep quality

Select **Analyze → Correlate → Bivariate...** (as shown in Figure 6.6) → transfer **Sleep quality perceptions** and **Mood** to **Variables** window → tick boxes for **Pearson** and **One-tailed** → click **OK**

		Sleep quality perception	Mood
Sleep quality perception	Pearson correlation	1	–.324**
	Sig. (1-tailed)		.001
	N	98	98
Mood	Pearson correlation	–.324**	1
	Sig. (1-tailed)	.001	
	N	98	98

Figure 6.20 Pearson's correlation output

Figure 6.20 shows that there is a moderately negative, highly significant, correlation between sleep quality perceptions and mood: r (96) = −.324, p = .001. But what happens when we account for age? This is how we run partial correlation in SPSS:

Using the SPSS file Sleep quality

Select **Analyze → Correlate → partial...** as shown in Figure 6.21

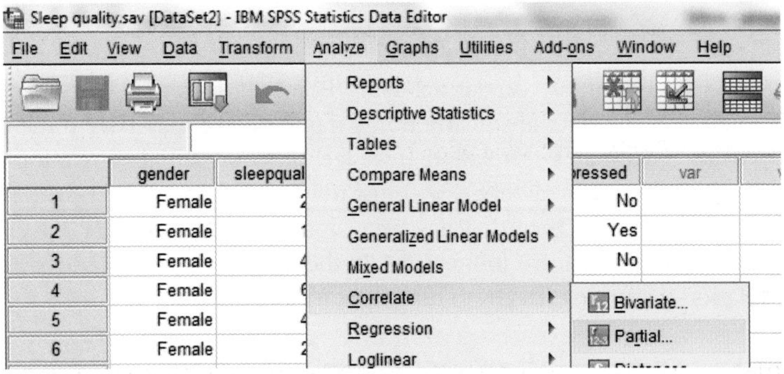

Figure 6.21 Partial correlation – step 1

In new window (see Figure 6.22), transfer **Sleep quality perceptions** and **Mood** to **Variables** window → transfer **Age** to **Controlling for** window → tick **One-tailed** box (we will assume that we have predicted that there will be an effect) → click **OK**

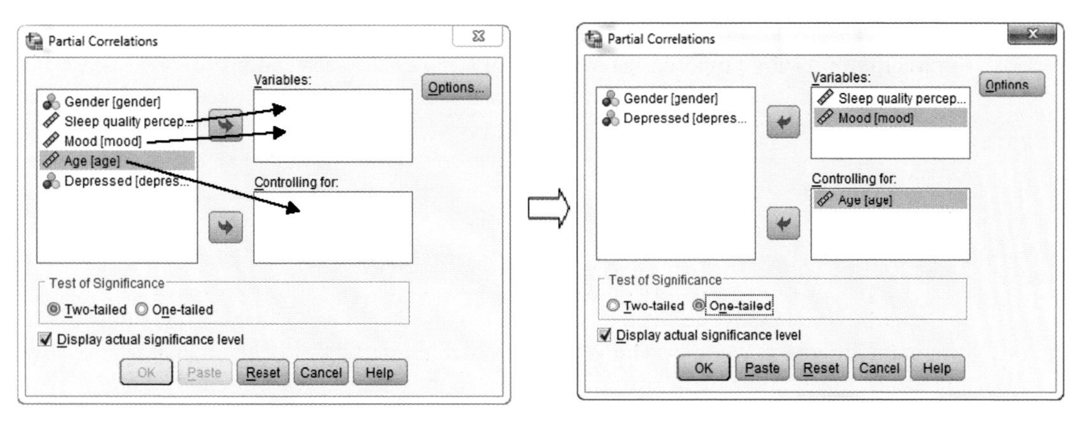

Figure 6.22 Partial correlation - step 2

Interpretation of output

Figure 6.23 shows that, following partial correlation, we now have a very much weaker negative correlation between sleep quality perceptions and mood, once we have accounted for age. Furthermore, the relationship is no longer significant: $r\,(95) = -.167$, $p = .051$ (this time, SPSS actually presents the degrees of freedom (*df*) rather than the sample size). We initially thought that the relationship was moderate and significant. It would appear that there was no relationship after all: the observed 'relationship' was actually a reflection of the participants' ages.

Control variables			Sleep quality perception	Mood
Age	Sleep quality perception	Correlation	1.000	–.167
		Significance (1-tailed)	.	.051
		df	0	95
	Mood	Correlation	–.167	1.000
		Significance (1-tailed)	.051	.
		df	95	0

Figure 6.23 Partial correlation output

Semi-partial correlation

Partial correlation explores correlation between two variables, while holding a third variable constant in respect of both original variables. We saw that just now, when we examined the correlation between sleep quality perceptions and mood, after holding age constant (for sleep quality perceptions and mood). In our case we saw that the original, significantly moderate, correlation was no longer significant when accounting for that third variable. Given that outcome we might feel that the relationship between sleep quality and age is potentially more important. We can explore that prediction with something called semi-partial correlation. With this analysis, we can explore the relationship between two variables, while holding a third variable constant but only in respect of one of the original variables. So, using semi-partial correlation, we can investigate the relationship between sleep quality perceptions and mood, while holding age constant for sleep quality perceptions (only). This type of analysis is central to linear regression (as we will see in Chapter 16). It actually makes more sense to use semi-partial correlation in a regression context (and is rarely used outside of that). However, we should explore the effect on correlation here, before we examine the wider aspects in the later chapter.

Measuring the effect of semi-partial correlation

Semi-partial correlation is calculated in much the same way as partial correlation, but with one important difference. We still examine pairs of variables, starting with the original relationship, and deduct the multiple of the correlations between the additional variable and each of the two original variables. However, this time, we divide that product by only *one* of the squared correlation calculations (see Box 6.15).

6.15 Calculating outcomes manually

Semi-partial correlation calculation

To illustrate the effect of semi-partial correlation, we return to the example we explored in Box 6.14. You may recall that we initially explored the relationship between sleep quality perceptions and mood, finding a moderately negative (significant) correlation of $r = -.324$. Then, we decided to explore the impact of adding age to relationship. The data we saw earlier are repeated in Table 6.10 (to save you trawling back).

Table 6.10 Correlation (r)

Pair	Relationship	r	r^2
AB	Mood vs. sleep quality perceptions*	−.324	.105
AC	Mood vs. age	.645	.416
BC	Sleep quality perceptions vs. age	−.314	.099

To calculate semi-partial correlation we use the data from Table 6.10 and apply it to the following formula. It is similar to the one we saw for partial correlation, except that we lose one of the items in the denominator (because we are only controlling the additional variable against one of the original variables). In this case, we might expect 'mood' to be affected by sleep quality perceptions and/or age. Since we suspect the relationship sleep quality perceptions vs. age is more important than it is for sleep quality perceptions vs. mood, we will hold age constant only for sleep quality perceptions:

$$\frac{r_{AB} - (r_{AC})(r_{BC})}{\left(\sqrt{1 - r_{BC}^2}\right)} = \frac{-.324 - (.645)(-.314)}{\left(\sqrt{1 - .099}\right)} = -.128$$

The semi-partial correlation between sleep quality perceptions and mood ($r = -.128$) is even more dramatically reduced from the original outcome ($r = -.324$), when we hold age constant for sleep quality perceptions only.

How SPSS performs semi-partial correlation

Actually, it doesn't, at least not directly! We cannot obtain semi-partial correlation from the correlation menus, but instead we can get the information we need from regression analyses.

It would be pointless explaining too much about that now, as we will explore such things in Chapter 16. So, for now, simply follow these instructions:

Using the SPSS file Sleep quality

Select **Analyze→ Regression→ Linear. . .** as shown in Figure 6.24.

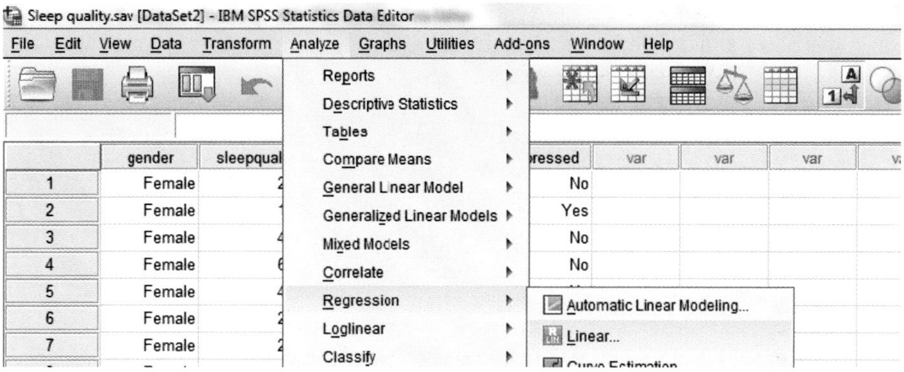

Figure 6.24 Semi-partial correlation (via regression) – step 1

In new window (see Figure 6.25) transfer **Mood** to **Dependent** window → transfer **Sleep quality perceptions** and **Age** to **Independent(s)** window → click **Statistics . . .**

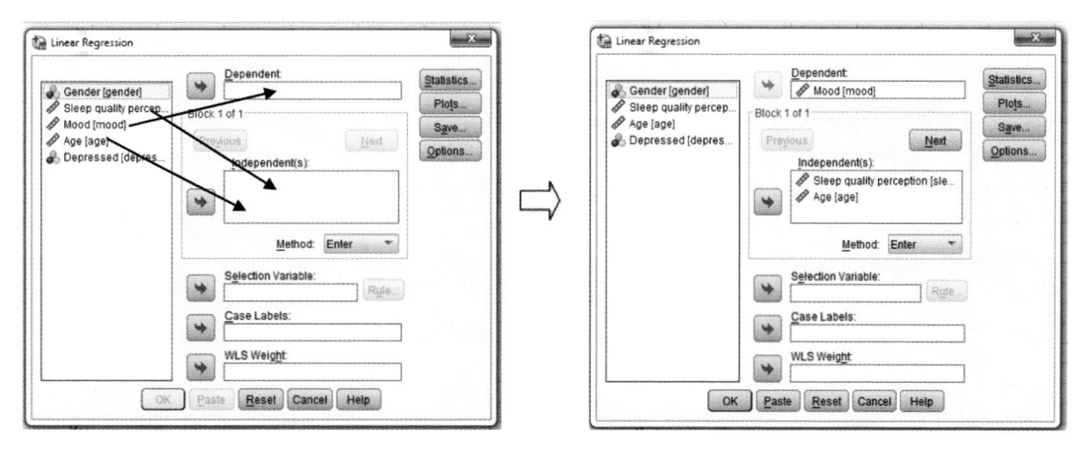

Figure 6.25 Semi-partial correlation – step 2

In new window (see Figure 6.26), tick boxes for **Estimates**, **Model fit**, and **Part and partial correlations** → click **Continue** → click **OK**

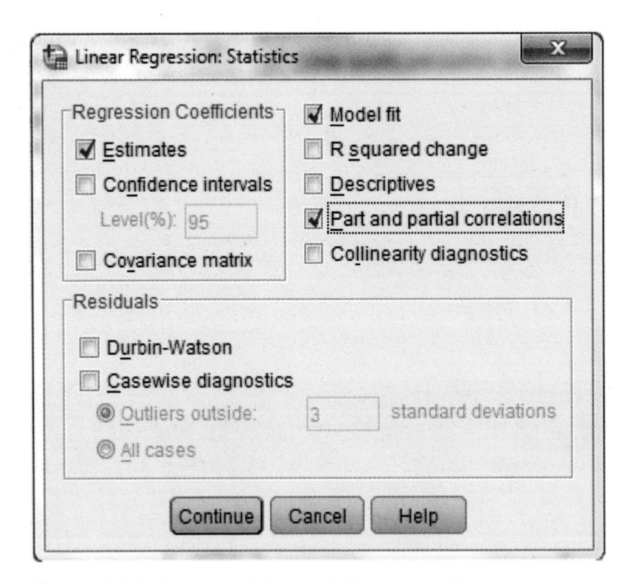

Figure 6.26 Semi-partial correlation – step 3

Interpretation of output

The semi-partial correlation is shown by the 'Part' data in Figure 6.27. This confirms what we calculated manually in Box 6.15: the semi-partial correlation between sleep quality perceptions and mood, while holding age constant just for sleep quality perceptions is low, $r = -.128$. We will explore the implications of this further when we examine linear regression in Chapter 16.

Coefficients[a]

Model		Unstandardised coefficients		Standardised coefficients	t	Sig.	Correlations		
		B	Std. error	Beta			Zero-order	Partial	Part
1	(Constant)	29.539	7.465		3.957	.000			
	Sleep quality perception	–.156	.095	–.135	–1.654	.101	–.324	–.167	–.128
	Age	.930	.126	.603	7.405	.000	.645	.605	.572

a. Dependent variable: mood

Figure 6.27 Semi-partial correlation output (via regression coefficients)

Chapter summary

In this chapter we have explored several forms of correlation. At this point, it would be good to revisit the learning objectives that we set at the beginning of the chapter.

You should now be able to:

- Recognise that we use correlation to measure the relationship between two variables. The relationship is measured on a scale of 0 (no correlation) to ±1 (perfect correlation). A positive correlation reflects how the 'values' of both variables change in the same direction; a negative correlation occurs when values change in opposite directions. Correlation never indicates cause and effect.

- Appreciate the different types of correlation. Pearson's correlation is used when both variables are parametric; outcomes are calculated according to mean scores. Spearman's correlation is used when at least one of the variables is non-parametric; outcomes are determined by ranking scores. Kendall's Tau-b is used in preference to Spearman's correlation if there are too many ties in those ranked scores. Biserial (and point-biserial) correlation examine the relationship between two variables when one of these is represented by categorical data.
- Understand that partial correlation enables us to examine the relationship between two variables while 'factoring out' the potential effect of additional variables.
- Recognise that semi-partial correlation also factors out an additional variable, but only controls that variable against one of the original variables.
- Calculate outcomes manually for each test (using maths and equations).
- Perform analyses for each test using SPSS (where appropriate).
- Understand how to present the data and report the findings.

Research example

It might help you to see how correlation has been applied in published research. As we saw earlier, correlation can be used to measure the validity of questionnaires; this paper does just that. In this section you can read an overview of the following paper. If you would like to read the entire paper you can use the DOI reference provided to locate that (see Chapter 1 for instructions).

Bush, S.H., Parsons, H.A., Palmer, J.L., Li, Z., Chacko, R. and Bruera, E. (2010). Single- vs. multiple-item instruments in the assessment of quality of life in patients with advanced cancer. ***Journal of Pain and Symptom Management*****, 39 (3): 564–571. DOI: http://dx.doi.org/10.1016/j.jpainsymman.2009.08.006**

The study investigated the merit of using single-item measurements to examine quality of life (QoL) in cancer patients, as opposed to employing traditional multi-dimensional approaches. The authors argued that QoL in cancer is so complex that even multi-dimensional methods fail to capture all of the facets, and take too long to complete. Single-item measures may be more effective and completion is quicker. An existing (validated) multi-dimensional QoL scale, the Functional Assessment of Cancer Therapy-General (FACT-G; Cella *et al.*, 1993), was used as the comparison scale. The authors proposed that QoL in cancer patients could be sufficiently captured by the 'feeling of well-being' item of the Edmonton System Assessment System (ESAS; Bruera *et al.*, 1991) single-item scale (i.e. the ESAS WB). They compared the ESAS WB to the FACT-G total score and the sub-domains of the FACT-G relating to physical well-being (PWB), social well-being (SWB), emotional well-being (EWB) and functional well-being (FWB). This was conducted at baseline (T1) and at the time of primary outcome (T2). An assessment was also undertaken in respect of change in scores between T1 and T2. Data from 218 cancer patient records were assessed across six clinical trials (although only 146 patients were available for the T2 stage). Analyses were performed using Pearson's and Spearman's correlations.

Spearman's correlation indicated that ESAS WB scores were moderately correlated with FACT-G total scores at T1: r_s (218) = −.48, $p < .0001$; and T2: r_s (146) = −.47, $p < .001$. Similarly moderate correlations were found across most of the sub-scales, except SWB, where weak correlations were observed (the full range of results is not shown here, to avoid repetition). Pearson's correlation showed that change-scores between T1 and T2 were moderately correlated between ESAS WB FACT-G in respect of total scores: $r(146) = -.36$, $p < .0001$; PWB: $r(146) = -.31$, $p = .0001$; and FWB: $r(146) = -.30$, $p = .0002$. The ESAS WB holds up pretty well against FACT-G total scores and all domains (other than social well-being). The authors argue this is justification for the single-item ESAS WB in measuring QoL in cancer patients.

Extended learning task

You will find the data set associated with this task on the website that accompanies this book (available in SPSS and Excel format). You will also find the answers there.

Following what we have learned about correlation, answer these questions and conduct the analyses in SPSS (if you do not have SPSS, do as much as you can with the Excel spreadsheet). For this exercise, 60 participants have been examined in a series of tests that records reaction times to a sequence of events on a driving simulator. This is measured in relation to how much alcohol the participant has consumed prior to the tasks. Average alcohol consumption across a series of recording sessions is measured in 'units'. The researchers also record the participants' mood with a questionnaire. Responses are measured on a scale 0–100, with higher scores representing better perceptions. Gender is also noted.

Open the data set **Alcohol and reaction time**

1. Run the appropriate correlation analysis for each of the following:
   a. Average units of alcohol with reaction time.
   b. Mood with reaction time.
   c. Mood with reaction time, controlling for average units of alcohol.
   d. Mood with gender.

7 INDEPENDENT T-TEST

Learning objectives

By the end of this chapter you should be able to:

- Recognise when it is appropriate to use an independent t-test
- Understand the theory, rationale, assumptions and restrictions associated with each test
- Calculate outcomes manually (using maths and equations)
- Perform analyses using SPSS
- Explore effect size and statistical power
- Understand how to present the data and report the findings

What is a t-test?

A t-test examines differences in the mean scores of a parametric dependent variable across two groups or conditions (the independent variable). As we saw in Chapter 5, data are parametric if they are represented by interval values and are *reasonably* normally distributed. The t-test outcome is based on differences in mean scores between groups and conditions, in relation to the 'standard error' of the differences (we saw what that meant in Chapter 4, but we will look at this again later). We explore two types of t-test in this book: the independent t-test (for between-group analyses) and related t-test for within-group studies. This chapter focuses on the former.

What is an independent t-test?

An independent t-test measures differences between two distinct groups. Those differences might be directly manipulated (e.g. drug treatment group vs. placebo group), they may be naturally occurring (e.g. male vs. female), or they might be beyond the control of the experimenter (e.g. depressed people vs. healthy people). In an independent t-test mean dependent variable scores are compared between the two groups (the independent variable). For example, we could measure differences in the amount of money spent on clothes between men and women.

Research question for independent t-test

We can illustrate the rationale and outcome for an independent t-test with a research question. A group of researchers called FRET (Federation of Research into Emotion and Threat) decide to explore the effect of anxiety on attention to threatening stimuli. A number of research groups have shown that anxious people often ruminate on situations and objects that make their stress worse. Patients diagnosed with Generalised Anxiety Disorder (GAD) tend to worry about a whole series of events throughout the day, to the extent that it significantly interferes with their ability to conduct a normal life. It is common for GAD patients to see threat in almost any context, often because they scan their environment looking for those threats. The way in which GAD patients unconsciously focus their thoughts is often illustrated by these attentional biases.

Emotional Stroop tests are a common way to measure attentional biases. Participants are shown a series of words (on a computer screen) presented in various font colours. The computer keyboard is adapted so that certain keys represent those font colours. The participants are told that they must select the correct colour. If the word 'biscuit' is presented on the screen in blue font, the participant should hit the 'blue' key. A computer program measures how quickly the key is selected (the reaction time). The word list is mostly made of 'neutral' words (such as house, plate or newspaper). However, other words are interspersed within the list, which some people might find more threatening (such as money, death or deadline). Because anxious people focus more on threat than non-anxious people, they may spend more time 'reading' the words that cause them stress; this may cause a delay in participants choosing which key to hit. To investigate this, FRET decide to conduct an experiment, using

7.1 Take a closer look
Variables for independent t-test

Dependent variable: Reaction times (on Emotional Stroop test).
Independent variable: Anxiety group (anxious vs. not anxious).

the emotional Stroop, comparing reaction times between anxious people and non-anxious people. 'Reaction times' represent the dependent variable, while 'anxiety group' (anxious vs. non-anxious) is the independent variable. FRET predict that reaction times will be slower for anxious people than for non-anxious people.

Theory and rationale

Measuring differences

In Chapter 6 we learned how to investigate *relationships* between variables; we sought to discover how the values of two variables were related to each other. In this chapter we are focusing on *differences* – we will explore how outcomes in one variable differ between two groups. But how do we measure differences? As with all statistical analyses, we need a starting point from which we can make and test predictions. In Chapter 4 we learned that the null hypothesis states that there will be no difference between groups in respect of an outcome. Using our research example, this means that there would be no difference between anxious people and non-anxious people in respect of reaction times on the emotional Stroop test. FRET suggested that reaction times would be slower for the anxious group than for the non-anxious group – this is the alternative hypothesis.

To investigate attentional biases in anxiety, we could perform a series of emotional Stroop tests and measure reaction times in selecting the correct font colour key. If reaction times were identical between the anxious and non-anxious groups, the '**mean difference**' between them would be zero. In that case, the null hypothesis could not be rejected as there would be no evidence to counter the prediction that there would be no difference between the groups. However, in most cases there probably *will* be a 'difference'. The task of the independent t-test is to establish whether that difference is statistically significant, whereby we could reject the null hypothesis. In Chapter 4, we saw that statistical significance measures the likelihood that the outcome occurred by chance. If the probability is less than 5% ($p < .05$), we can be confident that the outcome did not happen by chance. Before we go further, have a look at the between-group differences displayed in Box 7.2. These examples show how difficult it is to make assumptions about differences on face value.

7.2 Nuts and bolts
When is a difference big enough?

Table 7.1 shows the average amount of money spent on clothes between 40 men and 40 women at selected shops over the course of one year.

Table 7.1 Average amount spent on clothes per year (£)

	Shop 1	Shop 2	Shop 3
Male (40)	150	250	350
Female (40)	150	235	200
Mean diff:	**0**	**15**	**150**

There is no difference in average spending between men and women at shop 1, but there are differences in the other two shops. How do we decide when a difference is large enough to be significant?

The answer is that we need to run an independent t-test to find out.

We can use the examples in Box 7.2 to illustrate between-group differences. In Shop 1, there are no differences in the average spending on clothes between men and women. In Shops 2 and 3 there *appears* to be some difference. However, evaluations like this are highly subjective – that's why we need to test the differences for statistical significance. Just because we think we can see a difference in clothes spending between men and women, it does not mean that the difference was indeed due to gender differences – it may have been a random (chance) difference. In simple terms, a difference is more likely to be 'significant' if it is large enough, relative to the magnitude in the variation of those scores. Those differences also need to be 'consistent'. Indeed, it is possible that the difference in average spending in Shop 2 is significant, but not in Shop 3. Why? If the pattern of spending in Shop 2 showed that all of the male shoppers spent more than the female shoppers, that would be consistent. Furthermore, the average variation of the spending patterns in Shop 2 might be much less than it is in Shop 3. If outcomes vary too much, relative to the mean score, the likelihood of significance is reduced.

An independent t-test assesses whether the observed difference in the mean dependent variable scores between the two groups is large enough, relative to the standard error of differences. We explored standard error in Chapter 4. In short, it is an estimation of the standard deviation in the overall population. The sample standard deviation measures the average variation of scores either side of the mean. To obtain the standard error, we divide the sample standard deviation by the square root of the sample size. The standard error of differences is calculated in a similar way, but outcomes are based on the variance of the scores within each group (see Box 7.7). So, two factors will determine whether a difference is significant: how much larger the mean difference is than zero and the standard error of differences. This 'error' represents how much the observed differences are explained by random chance factors. To calculate significance, we divide the observed 'mean difference' by that error. The higher the outcome, the more it will be that the difference is significant. Because of that, we do not want the error statistic to be too high.

7.3 Nuts and bolts
Other names for the independent t-test

If you read other books, or have spoken to other tutors, or have taken other courses in statistics, you may have come across other names for the independent t-test. Some of the more common ones are listed below. However rest assured, they all measure the same thing.

Between-groups t-test
Unrelated t-test
Independent measures t-test
Independent samples t-test
Student's t-test

On that last point, have you ever wondered why this test is sometimes called the Student's t-test? Ever really cared? You would be forgiven for thinking that it is because students often use the test, but it is not that at all. The explanation actually goes back to the time when William Gosset invented the t-test in the first place. Early in the 20th century Gosset sought to examine the quality of his employer's product (Guinness stout) in relation to other beers. The Guinness brewery in Dublin often recruited high-quality graduates. When Gosset came to publish his findings, he was prevented by Guinness from using his real name (so rivals companies would not poach him). So, Gosset used the pseudonym 'Student' instead. No one outside the company knew his real name, not even his statistician colleagues and friends. Gosset's statistical procedure became known as Student's t-test.

Comparison with other tests

The independent t-test is an example of a between-group test; differences are measured in respect of distinct groups. If we needed to measure differences in a single group, across several conditions, we would employ a within-group design For example, we could explore the effect of two types of teaching method performed across one cohort of students. So long as the data are parametric, we would use a related t-test to explore the outcome (see Chapter 8). The defining aspect of the independent t-test is that there are just two groups. If we want to measure differences across three or more groups (and the data are parametric) we would need to use an independent one-way ANOVA (see Chapter 9). For example, we could investigate differences in mean income between university lecturers, college lecturers and school teachers. Finally, if the data in either of our two groups are not parametric (perhaps reaction times for anxious people are not normally distributed), we need to employ a Mann–Whitney U test, the non-parametric equivalent to the independent t-test.

7.4 Mini exercise
Between-groups or within-groups?

If you are new to statistics, you may still be a little confused about how to determine whether a design is between-groups or within-groups. The following exercise might help to clarify that for you. We explored some of these points in Chapter 5, so you might want to read that again. Look at the following short scenarios and decide whether they are an example of a between-group or within-group study.

1. A group of UK students are compared with those from the USA on how many hours they watch television.
2. One group of depressed patients are given two different types of drug, at different times, to assess how well their symptoms improve.
3. Children are compared with adults in respect of how many green vegetables they eat.
4. Several questionnaires are given to one group of people to see how they differ on several outcome measures, but in respect of their nationality, ethnicity and religious belief.
5. A group of students are given two tests: before one of these tests they are given some tips on revision skills. Their test scores are compared.

Look at the answers below. How did you do?

1. Between;
2. Within;
3. Between;
4. Between;
5. Within.

You may have had some trouble with Question 4. It is quite common to believe that this constitutes a within-group design (because several questionnaires were given to one group) but it is not. It would be within-groups only if the *same* questionnaire was repeated. For example, we could give a stress questionnaire to a single group, then we could manipulate that stress (such as make them watch a scary movie), and then we would give them the same stress questionnaire again. In short, a between-group study explores differences in the characteristics of the sample, using different groups; a within-group study examines different conditions performed across a single group. In the case of Question 4, the independent variables are nationality, ethnicity and religious belief, not the number of questionnaires used.

Assumptions and restrictions

There are a number of criteria that we must satisfy before we can consider using an independent t-test to explore outcomes. The independent variable must be categorical and must be represented by two distinct, exclusive, groups (it can only be possible to be a member of one group at a time). A good example would be gender (male vs. female). When we create the variable in SPSS we allocate 'numbers' to represent groups when creating variables in SPSS (such as 1 = male; 2 = female) – we saw how to do that in Chapter 2.

The data on the dependent variable should be parametric. We explored what that meant in Chapter 5. In short, the dependent variable data must be interval or ratio, and should be *reasonably* normally distributed. Interval data represent numbers that can be objectively measured (such as age, income or reaction times). Although ordinal data have 'numerical' form, they tend to be measured in terms of relative rank, rather than magnitude; they cannot be examined with an independent t-test. This is often quite obvious when applied to measurements such as finishing position in a race; on other occasions the boundaries are fuzzier. This point is probably best exemplified by attitude questionnaires, such as Likert scales (see Box 7.6). If the dependent variable data are not normally distributed, it might be evidence of the presence of extreme scores. If that is the case, the mean score might be artificially inflated or deflated. The independent t-test relies on the mean score to determine outcome. If we cannot trust the mean score, we should not use a parametric test to explore the outcome – non-parametric tests may be more appropriate. We can check normal distribution through a number of statistical techniques, as we saw in Chapter 3. Our data (reaction times) are clearly interval, but we will need to check whether the scores are normally distributed across the anxiety groups (see later).

7.5 Take a closer look

Summary of assumptions and restrictions

- The independent variable must be categorical
    - It must consist of two distinct groups
    - Group membership must be independent and exclusive
        - No person (or case) can appear in more than one group
- There must be one parametric dependent variable
    - The dependent variable data must be interval or ratio
        - And should be reasonably normally distributed (across both groups)
    - We should check for homogeneity of variances
    - If these assumptions are not met the non-parametric Mann–Whitney U test could be considered (see Chapter 18)

7.6 Mini exercise

Do attitude scales measure ordinal or interval data?

Look at these examples of attitude measurement:

1. My statistics tutor is the best ever. Rate this statement on the following scale: 1 = strongly agree; 2 = agree; 3 = neither agree nor disagree; 4 = disagree; 5 = strongly disagree.
2. Indicate how happy you are with this book, using the following scale: 1 = it rocks; 2 = it is very good; 3 = it is pretty good; 4 = it's OK; 5 = it is quite poor; 6 = it is poor; 7 = it sucks.
3. Rate how satisfied you are with the usefulness of this book on a scale of 0–100, where 0 is 'useless - I would not even use it as a door stop' and 100 represents 'incredibly useful - I can't live without it'.

Which of those examples might be best described as ordinal data? Which are more likely to be classed as interval? Is it determined by the magnitude of the scores? Is Question 3 a better example of interval data because scores are measured on a scale of 0–100?

Strictly speaking, they are all ordinal and probably should not be used in parametric tests. Some statisticians feel that the difference between 'strongly agree' and 'agree' is subjective, and that this varies between people. The same could be said when rating an opinion between 1 and 100 – the difference between one person's rating of 85 and 65 may be very different to the next person's rating (and might also be considered subjective). By contrast, the difference between an age of 50 and an age of 40 is obvious, objective and measurable. Having said that, as we saw in Chapter 6 (assumptions and restrictions), there are some arguments to support the measurement of Likert scales with parametric tests.

We also need to check for homogeneity of variance across the independent variable groups (we also explored what that meant in Chapter 3). As we will see later, the independent t-test makes an adjustment if homogeneity of variance is violated (using **Levene's test**). We will explore this issue in more depth in Chapter 9. If we violate parametric assumptions we should use a non-parametric test. In this case we would need to choose the Mann – Whitney U test (see Chapter 18).

Establishing significant differences

We can use the independent t-test to examine differences in mean dependent variable scores, between two groups, and establish whether that difference is statistically significant. The method used in the independent t-test is relatively simple: the 'mean difference' between the groups is divided by the 'standard error of differences'. We will explore how to do this manually in Box 7.7, and will see how to perform the test in SPSS later. To illustrate how we

7.7 Calculating outcomes manually
Independent t-test calculation

The data in Table 7.2 represent reaction times in responding to tasks on an Emotional Stroop test, as presented to ten anxious people and ten non-anxious people. You will find a Microsoft Excel spreadsheet associated with these calculations on the web page for this book.

Table 7.2 Reaction times to an Emotional Stroop test, by anxiety group

	Reaction times (milliseconds)	
	Anxious[a]	Not anxious[b]
	821	742
	763	642
	726	640
	701	585
	707	561
	680	503
	644	501
	641	480
	625	486
	520	441
Mean ($\bar{x}$)	682.80	558.10
Variance (Var)	*6824.40*	*8844.99*

The formula for the independent t-test is: $t = \dfrac{\bar{x}_a - \bar{x}_b}{\sqrt{\dfrac{Var_a}{N_a} + \dfrac{Var_b}{N_b}}}$

The top line in the equation is the difference between the mean scores; the lower line represents the standard error of differences. In our case males are represented by 'a' and females by 'b'. We calculate our outcome in a number of stages.

First, we calculate the mean scores for each group and then deduct one from the other to get the mean difference:

$$\text{Mean difference} = \text{Mean}_a(\bar{x}_a) - \text{Mean}_b(\bar{x}_b) = 682.80 - 558.10 = 124.70$$

To calculate the standard error of differences, we need to establish the variance for each group - we saw how to calculate that in Chapter 4 (Box 4.4). The variance (Var) for each group is shown in Table 7.2.

The calculation for standard error of differences is shown in the lower part of the t-test equation:

$$\text{Standard error of differences} = \sqrt{\frac{6824.40}{10} + \frac{8844.99}{10}} = \sqrt{1566.94} = 39.58$$

We can now put all of this into the original equation: $t = \dfrac{\bar{x}_a - \bar{x}_b}{\sqrt{\dfrac{Var_a}{N_a} + \dfrac{Var_b}{N_b}}} = \dfrac{124.70}{39.58} = 3.15$

To assess whether a *t* score is significant, we compare to a *t* score distribution (see Appendix 2) to find a 'cut-off' point where p = .05 for the relevant degrees of freedom (*df*)in a one-tailed test. If the *t* score is higher than that, we have a significant outcome. As we saw in Chapter 6, degrees of freedom relate to the number of values that are 'free to vary' in the calculation, while everything else is held constant. The *df* for an independent t-test is ($N_a + N_b$) − 2; in our case (10 + 10) − 2 = 18). In the *t* score distribution the cut-off for p = .05, where *df* = 18, in a one-tailed test = 1.734; we have a significant outcome because our *t* score is higher than that. In some cases, the *t* score will be negative. This has no impact on significance (it simply indicates which way round you presented group scores).

We can also use Microsoft Excel to calculate the critical value of t and to provide the actual p value. You can see how to do that on the web page for this book. In our case, p = .003 (one-tailed). In that spreadsheet, you can also see how to perform the whole test in Excel.

calculate the outcome for an independent t-test, we will refer to the research question posed by FRET. They are seeking to investigate attentional biases in anxiety. They have predicted that reaction times on an Emotional Stroop test will be significantly slower for anxious people than for non-anxious people (which means that we have a one-tailed test, because the prediction is specific – see Chapter 4). We will explore this in a small sample of 20 people. Usually, we should aim for larger samples (but it is easier to show the calculations in these smaller groups).

How SPSS performs an independent t-test

We can perform an independent t-test in SPSS (see Box 7.8). However, I would urge you to work through the manual calculations shown in Box 7.7: you will learn so much more that way. To explore outcomes, we will use the data shown in Table 7.2. We are examining reaction times (in milliseconds) to Emotional Stroop test tasks, according to anxiety group in a sample of 20 people (10 anxious and 10 non-anxious). FRET have predicted that reaction times for anxious people will be significantly slower for anxious people than for non-anxious people. This is a one-tailed test (a fact that will be important when interpreting the outcome). Reaction times are objective measures, so we can consider them to be interval, meeting requirements for parametric data However, as yet, we do not know whether the data are normally distributed.

7.8 Nuts and bolts
Setting up the data set in SPSS

When we create the SPSS data set for an independent t-test, we need to set up one column for the dependent variable (which will have a continuous score) and one column for the independent variable (which will have a categorical coding).

Emotional Stroop.sav [DataSet1] - IBM SPSS Statistics Data Editor

	Name	Type	Width	Decimals	Label	Values	Missing	Columns	Align	Measure
1	anxiety	Numeric	8	0	Anxiety group	{1, Anxious}...	None	8	Right	Nominal
2	RT	Numeric	8	2	Reaction time (...	None	None	8	Right	Scale

Figure 7.1 Variable View for 'Emotional Stroop' data

Figure 7.1 shows how the SPSS Variable View should be set up (you should refer to Chapter 2 for more information on the procedures used to set the parameters). The first variable is called 'anxiety'. This is the categorical independent variable representing the anxiety groups. In the Values column, we include '1 = Anxious' and '2 = Not anxious'; the Measure column is set to Nominal. The second variable is called 'RT'. This is the continuous dependent variable representing reaction times. We do not need to adjust anything in the Values column; the Measure column is set to Scale.

Emotional Stroop.sav [DataSet2] - IBM SPSS Statistics Data Editor

	anxiety	RT	var	var	var	var	var	var
1	Anxious	821						
2	Anxious	763						
3	Not anxious	742						

Figure 7.2 Data View for 'Emotional Stroop' data

Figure 7.2 illustrates how this will appear in the Data View. Each row represents a participant. When we enter the data for 'depressed', we input 1 (to represent Anxious) or 2 (to represent Not anxious). The 'anxiety' column will display the descriptive categories (Anxious or Not anxious) or it will show the value numbers, depending on how you choose to view the column (you can change that using the Alpha Numeric button in the menu bar - see Chapter 2).

Testing for normal distribution

We saw how to investigate normal distribution statistically in Chapter 3. We examine the outcomes from a Kolmogorov–Smirnov and Shapiro–Wilk test (the sample size determines which outcome you should report). If the outcome casts any doubt regarding normal distribution, we could also explore z-scores for the skew and kurtosis (but do remember that we only need 'reasonable' outcomes). Normal distribution of data to be used in an independent t-test needs to be explored across both groups of the independent variable:

Open the SPSS file Emotional Stroop

Select **Analyze** → **Descriptive Statistics** → **Explore** → (in new window) transfer **Reaction times** to **Dependent List** window (by clicking the arrow, or by dragging the variable to that window) → transfer **Anxiety group** to **Factor** window → select **Plots** radio button → click **Plots** box → (in new window) check **None** radio button (under **Boxplot)** → make sure that **Stem-and-leaf** and **Histogram** (under **Descriptive**) are unchecked → check **Normality plots with tests** radio button → click **Continue** → click **OK**

		Kolmogorov–Smirnov[a]			Shapiro–Wilk		
	Anxiety group	Statistic	df	Sig.	Statistic	df	Sig.
Reaction time (msecs)	Anxious	.142	10	.200*	.976	10	.943
	Not anxious	.221	10	.182	.927	10	.418

Figure 7.3 Kolmogorov-Smirnov/Shapiro-Wilk test for reaction times, in respect of anxiety

Because we have a sample size of ten (for each group), we should refer to the Shapiro–Wilk outcome (in Chapter 3 we saw that we should use the Kolmogorov–Smirnov outcome only when we have samples greater than 50). Figure 7.3 confirms that the data are probably normally distributed for both groups (because the significance [Sig.] is *greater* than .05). Remember, the Kolmogorov–Smirnov and Shapiro–Wilk tests investigate whether the data are significantly *different* to a normal distribution–we don't want that to happen. Therefore, we probably have normal distribution in reaction times for the anxious group: $W(10) = .976$, $p = .943$, and for the non-anxious group: $W(10) = .927$, $p = .418$.

Running independent t-test in SPSS

Using the SPSS file Emotional Stroop

Select **Analyze** → **Compare Means** → **Independent-Samples T Test...** as shown in Figure 7.4

In new window (see Figure 7.5), transfer **Reaction times** to **Test Variable(s)** window → transfer **Anxiety group** to **Grouping Variable** window → click on **Define Groups**

In new window (see Figure 7.6), enter **1** in **Group 1** box → enter **2** in **Group 2** box → click **Continue** → (in original window) click **OK** (these values are those set up for 'anxious' and 'not anxious' - you will need to check this when defining groups for the independent t-test)

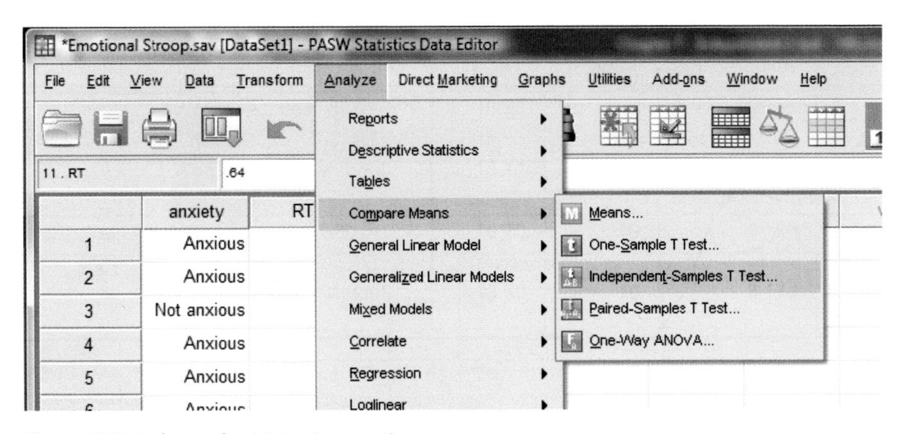

Figure 7.4 Independent t-test - step 1

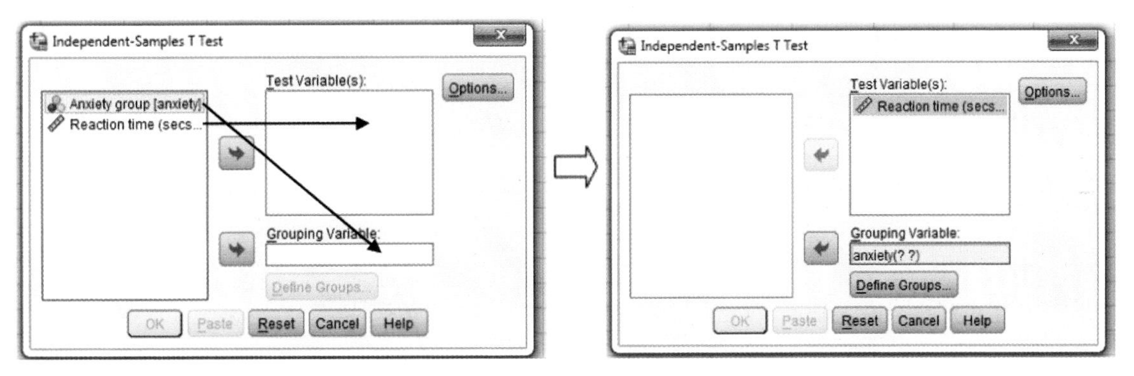

Figure 7.5 Independent t-test - step 2

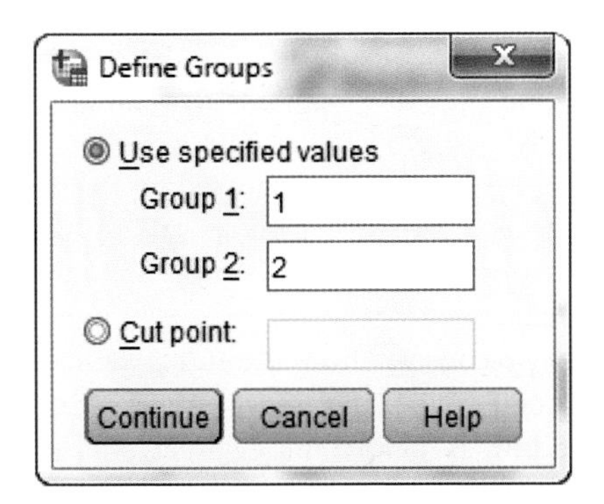

Figure 7.6 Independent t-test - step 3

Interpretation of output

Figure 7.7 shows the descriptive statistics, including mean reaction times, standard deviation and standard error in respect of the anxious and non-anxious groups. It would appear that reaction times are slower for the anxious group. However, we need to refer to the statistical outcome before we can make any assumptions about that.

	Anxiety group	N	Mean	Std. deviation	Std. error mean
Reaction time (msecs)	Anxious	10	682.80	82.610	26.124
	Not anxious	10	558.10	94.048	29.741

Figure 7.7 Descriptive statistics

		Levene's test for equality of variances		t-test for equality of means						
									95% confidence interval of the difference	
		F	Sig.	t	df	Sig. (2-tailed)	Mean difference	Std. error difference	Lower	Upper
Reaction time (msecs)	Equal variances assumed	.442	.515	3.150	18	.006	124.700	39.585	41.536	207.864
	Equal variances not assumed			3.150	17.706	.006	124.700	39.585	41.437	207.963

Figure 7.8 Independent t-test statistics

Figure 7.8 shows two lines of data: one each for 'Equal variances assumed' and 'Equal variances not assumed'. Levene's test (as shown to the left of this output) examines the equality of variances. Earlier, we stated that one of the assumptions that had to be met for the independent t-test was that there must equality (homogeneity) of variances between the groups. If the Levene's test shows a significant difference ('Sig.' $< .05$), it means that the variances are significantly different from each other, therefore not equal. If the outcome is not significant ($> .05$), the variances are not significantly different, so they are assumed to be equal.

In this case we did have equal variances ($F = 0.442$, $p = .515$), so we should read the line for 'Equal variances assumed' (highlighted in red font in Figure 7.8). The t-test outcome indicates that we have a significant difference in reaction times between anxious people and non-anxious people ($p = .006$). However, SPSS has calculated the outcome according to a two-tailed outcome (which would be based on a non-specific hypothesis). The FRET prediction was specific (that reaction times would be slower for anxious people than for non-anxious people). This is a one-tailed hypothesis, so we can divide the reported significance by two, so now we can say that $p = .003$. You can refer back to Chapter 4 if you want to read more about tails and hypotheses.

Using the mean score data from Figure 7.7 and the (adjusted) statistical outcome from Figure 7.8, we now know that reaction times are significantly slower for anxious participants than for non-anxious ones. When we write up the results, we should state that and show the statistical notation. In this case, we start with *t*, followed by the degrees of freedom (*df*) in brackets (see Box 7.7), the *t* value, and then the *full* 'p value' (followed by a statement saying whether this was a one-tailed or two-tailed outcome). So, our statistical notation is: $t(18) = 3.150$, $p = .003$ (one-tailed).

It is generally considered better practice to show the full p value (rather than $p < .05$) unless SPSS shows the outcome as '.000' (in which case we write $p < .001$). It is also common convention to not present the leading '0' before the decimal point (because p cannot be greater than 1).

Effect size and power

In Chapter 4, we learned how we should not just focus on significance when reporting outcomes. We saw examples of how very small 'effects' can be significant in larger samples; sometimes the effect is so small the outcome is a little meaningless. To ensure that we have a balanced view, we can also report the magnitude of the outcome through effect size. We also learned about how important it is that our studies have sufficient power. An underpowered study increases the likelihood that we make Type II errors (when we fail to reject the null hypothesis when we should have done so). We saw that we should aim to find true effects on at least 80% of occasions. To help achieve this we set a target power of 0.80 for our studies. We can use a software program called G*Power to calculate effect size and statistical power for us. You will find details of how to obtain and install this program in Chapter 4.

Open G*Power:

From **Test family** select **t-tests**

From **Statistical test** select **Means: Difference between two independent means (two groups)**

From **Type of power analysis** select **Post hoc: Compute achieved power - given α, sample size and effect size**

Now we enter the **Input Parameters**:

From the **Tails** box, select **One**

To calculate the **Effect size**, click on the **Determine** button (a new box appears).

In that new box, for **Mean group 1** type **682.80** → for **Mean group 2** type **558.10** → for **SD σ group 1** type **82.61** → for **SD σ group 2** type **94.05** (we get that from Figure 7.7) → click on **Calculate and transfer to main window**

Back in original display, for **α err prob** type **0.05** (the significance level) → for **Sample size group 1** type **10** → for **Sample size group 2** type **10** → click on **Calculate**

From this, we can observe two outcomes: Effect size (d) **1.41** (which is very strong); and Power (1-β err prob) **0.92** (which is very good, and higher than the desired 0.80 level).

Writing up results

We need to consider how to report and display mean data and other factors. More recently, it has become convention to show confidence intervals and standard error or standard deviation (or both). An example of some tabulated data is shown in Table 7.3.

Table 7.3 Reaction times (seconds) to Emotional Stroop test tasks, according to anxiety group

	Mean	Standard deviation	Standard error	95% CI of difference
Anxious (n = 10)	682.80	82.60	26.12	41.54 to 207.86
Not anxious (n = 10)	558.10	94.05	29.74	

Once the tabular data have been presented, we report the statistics and significance in the following format, using the statistical data from Figure 7.8 and the G*Power data:

> Using an independent t-test, it was confirmed that reaction times were significantly slower for anxious people than for non-anxious individuals, $t(18) = 3.150$, p = .003 (one-tailed). This represented a very strong effect, $d = 1.41$.

Presenting data graphically

You might also want to present those data graphically (a picture can show a lot more than words sometimes). However, never replicate data that has already been shown in tables. Nevertheless, there are times when graphs are very helpful. In this case, we will see how to draw a **bar chart** (with **error bars** to show 95% confidence intervals):

> Select **Graphs** ➜ select **Chart Builder**. . . as shown in Figure 7.9
>
> In new window, select **Bar** (from list under **Choose from**) ➜ drag **Simple Bar** graphic (top left corner) into empty chart preview area (as shown in Figure 7.10)

Figure 7.9 Creating a bar chart - step 1

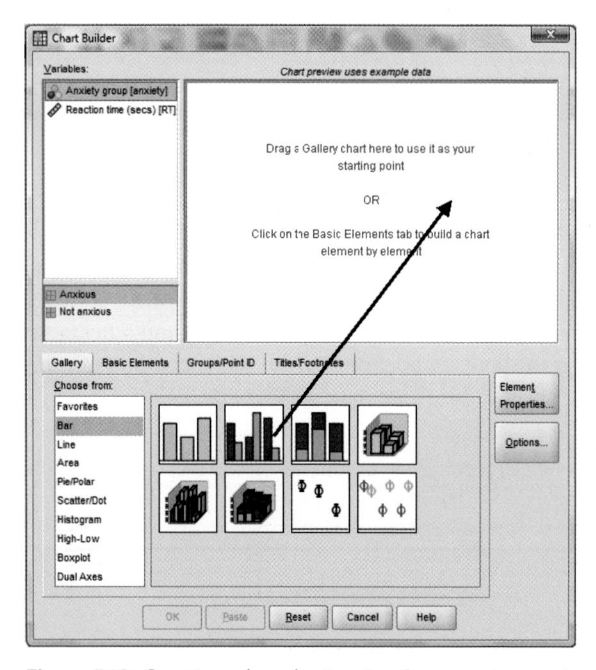

Figure 7.10 Creating a bar chart - step 2

Now we can add our data:

> In new window, transfer **Reaction times** to **Y-Axis** box (to left of graph) ➜ transfer **Anxiety group** to **X-Axis** box (under graph) - as shown in Figure 7.11.
>
> These actions will be confirmed inside the axis boxes, as shown in Figure 7.12.

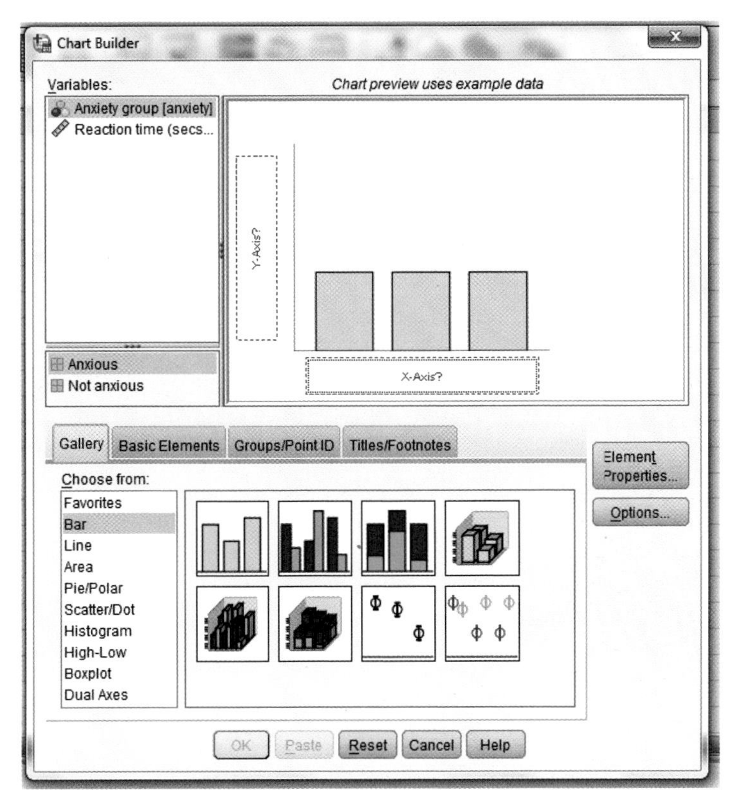

Figure 7.11 Creating a bar chart – step 3

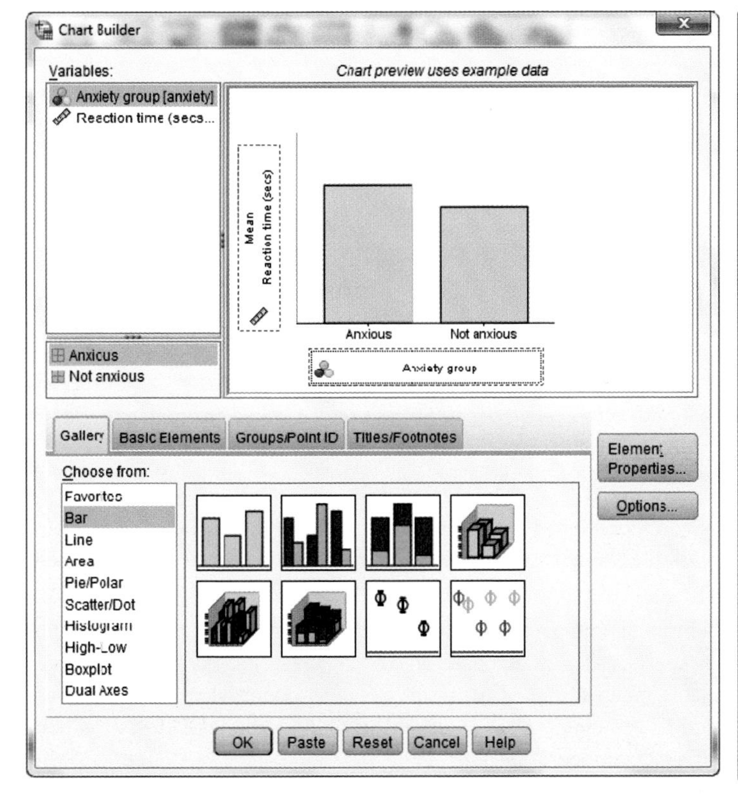

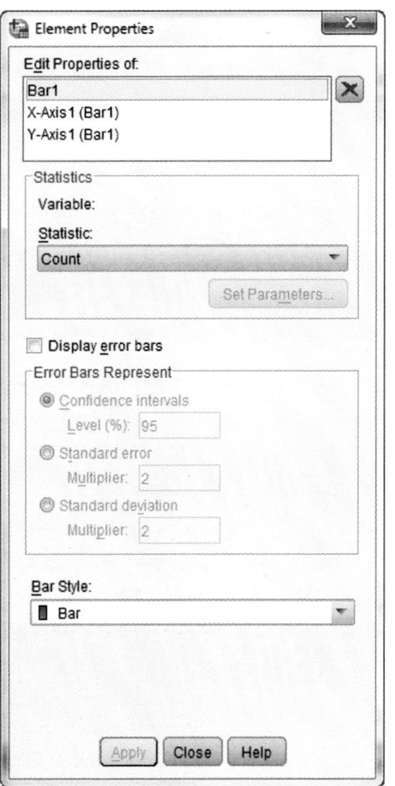

Figure 7.12 Creating a bar chart – step 3: completed

We also need to add some 'error bars' to represent the 95% confidence intervals (we learned about the importance of confidence intervals in Chapter 4). We can request error bars via the **Element Properties** box (to the right of the main display):

In the **Element Properties** box, to the right of the graph window (see Figure 7.12), tick box for **Display error bars** → ensure that it states 95% confidence intervals in box below → click on **Apply** (the error bars appear) → click **OK**.

Those actions will produce the completed bar graph, as shown in Figure 7.13.

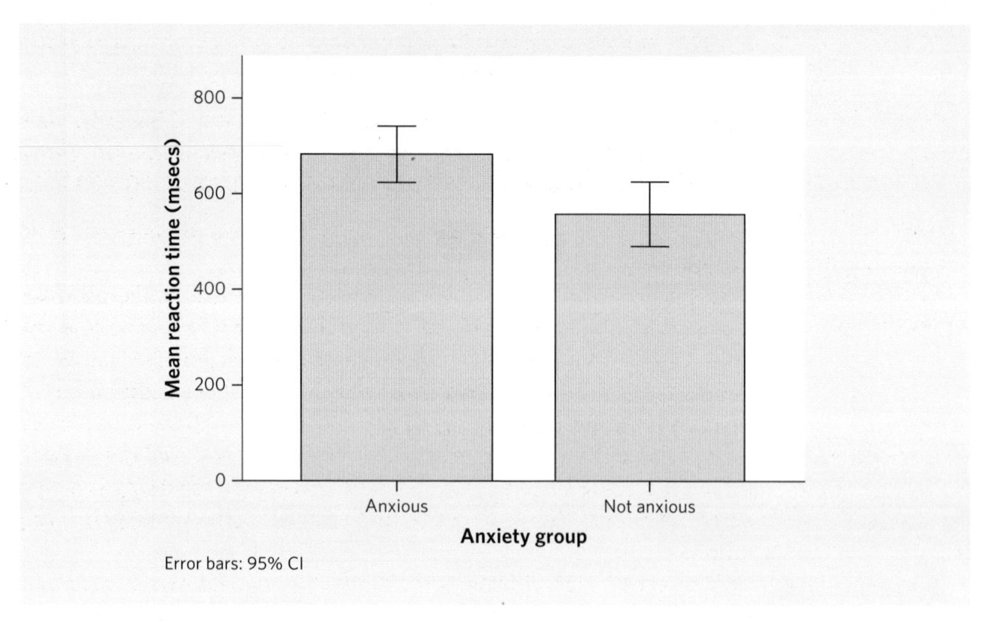

Figure 7.13 Completed bar chart

Chapter summary

In this chapter we have explored the independent t-test. At this point, it would be good to revisit the learning objectives that we set at the beginning of the chapter.

You should now be able to:

- Recognise that we use an independent t-test to examine differences in mean scores of a parametric dependent variable, across two distinct groups of a categorical independent variable.
- Understand that the data must be interval or ratio, and should be reasonably normally distributed. Outcomes for the independent t-test are based on mean scores. Non-parametric data can be associated with unreliable mean scores. If we have reason to doubt whether the data are parametric, we might need to examine outcomes using the Mann-Whitney U test (the non-parametric equivalent of the independent t-test).

- Calculate outcomes for an independent t-test manually (using maths and equations).
- Perform an independent t-test using SPSS.
- Explore effect size and statistical power.
- Understand how to present the data and report the findings.

Research example

It might help you to see how the independent t-test has been applied in published research. In this section you can read an overview of the following paper:

> **Schmidt, P. J., Murphy, J. H., Haq, N., Rubinow, D. R. and Danacea, M. A. (2004). Stressful life events, personal losses, and perimenopause-related depression.** ***Archives of Women's Mental Health*****, *7*: 19–26. DOI: http://dx.doi.org/10.1007/s00737-003-0036-2**

If you would like to read the entire paper you can use the DOI reference provided to locate that (see Chapter 1 for instructions).

In this research the authors investigated the number of life events and the quality of those events experienced by women approaching menopause - otherwise known as the perimenopausal period. These events were compared between perimenopausal women with and without depression (aged 44–55). Diagnosed depression had to be associated with the perimenopausal period (recent previous history of depression was excluded); healthy controls were matched for age and (menstrual) clinical history. The incidence (and quality) of life events focused on school, work, love, health, legal, financial, childbirth, family, residence, personal and death of someone close. The events were measured in respect of observed frequency and subjective reports of impact.

The results showed that depressed perimenopausal women reported significantly more life events than non-depressed controls: $t(98) = 2.3$, $p = .02$ (although this was not significant when accounting for multiple comparisons). However, there were a number of other significant aspects. Depressed perimenopausal women reported more negative events than non-depressed controls: $t(98) = 3.9$, $p < .001$, while they did not differ on positive events. Depressed women reported significantly more life events that decreased their self-esteem: $t(98) = 4.1$, $p < 0.001$, that were not anticipated: $t(98) = 3.6$, $p = .003$, and that were felt to be out of the respondent's control: $t(98) = 3.8$, $p < .001$.

Extended learning task

You will find the data set associated with this task on the website that accompanies this book (available in SPSS and Excel format). You will also find the answers there.

Following what we have learned about the independent t-test, answer these questions and conduct the analyses in SPSS and G*Power. (If you do not have SPSS, do as much as you can with the Excel spreadsheet.) The fictitious data explored the number of hours young people played on computer games, according to their gender and nationality. It was predicted that boys would play for longer than girls, but no predictions were made about nationality.

Open the data set **Computer games**

1. There are two independent variables in this data set, but which one is more appropriate for an independent t-test? Why is that?
2. Using the appropriate variables:
   a. Check for normal distribution across hours played on computer games.
   b. Conduct an independent t-test.
3. Describe what the SPSS output shows.
4. Explain how you accounted for equal variances.
5. State the effect size and power, using G*Power.
6. Report the outcome as you would in the results section of a report.

8

RELATED T-TEST

Learning objectives

By the end of this chapter you should be able to:

- Recognise when it is appropriate to use a related t-test
- Understand the theory, rationale, assumptions and restrictions associated with each test
- Calculate outcomes manually (using maths and equations)
- Perform analyses using SPSS
- Explore effect size and statistical power
- Understand how to present the data and report the findings

What is the related t-test?

A related t-test examines differences in mean (parametric) dependent variable scores across two within-group conditions (the independent variable), measured across a single group. We examined the criteria for parametric data in Chapter 5. For example, we could measure heart rate in a group of participants before and after 15 minutes of exercise. The *t* score is found by examining the mean difference between the conditions in relation to the extent that scores vary in the sample. In this chapter we will explore how to measure differences using the related t-test.

Research question for the related t-test

We can illustrate the rationale and outcome for a related t-test with a research question. A fictitious group of researchers, Centre for Advanced Learning and Memory (CALM), seek to discover what factors aid more effective memory recall. They advocate that people can remember things more readily if they are associated with meaningful information. For example, words might be easily recalled if they are paired with graphical associations. To illustrate this, CALM recruit a group of 12 people to whom they present a series of words, one at a time, on a computer screen. One hour later, the participants are asked to recall as many of those words as they can. The group undergo two memory trials: in one condition they simply see the words on a computer screen; in a second condition they see the words, but are also shown a picture that illustrates each word. The dependent variable is the number of (correct) words recalled; the independent variable is the presentation type, for which there are two conditions ('with picture' and 'without picture'). CALM predict that the group will recall more words in the 'with picture' condition than in the 'without picture' condition.

8.1 Take a closer look
Variables for related t-test

Dependent variable: Number of words recalled
Independent variable: Presentation type ('with picture' and 'without picture')

Theory and rationale

Problems with between-group studies

We often use within-group methods to conduct research because of the difficulties that we can encounter with between-group methods. For example, it is difficult to completely eradicate individual differences in between-group studies. The only sure way to control for potentially **extraneous variables** is to 'match' the groups on everything but the independent variable. If we do not control individual differences properly, we cannot be confident that the observed differences are not actually due to the extraneous variables. There are some statistical procedures that we can use to help control for these (see ANCOVA in Chapter 15), but even this cannot account for potential interfering variables. Because of that, many researchers choose within-group studies to overcome this dilemma. This is one reason why within-group methods, such as those measured by the related t-test, are popular.

8.2 Nuts and bolts
When can within-group studies be used?

Within-group studies can be used in a wide variety of contexts. They are employed in longitudinal research, where outcomes are measured at various time points; they can explore the effect of an intervention; or they can be used to measure attitudes and opinions under various conditions. These are just a few general scenarios. The key point is that within-group studies explore differences in dependent variable scores, in respect of independent variable conditions that are measured across one group. These are called **repeated-measures** studies, because the outcome variable is repeated in various situations, but with the same people. The number of repetitions of the outcome variable determines the type of test we must use. If we have parametric data with two repetitions we use the related t-test (as explored in this chapter); if we have more than two repetitions we use one of the repeated-measures ANOVAs (see Chapters 10, 12 and 13). If the data are not parametric, we should use Wilcoxon or Friedman's ANOVA respectively (see Chapter 18 for a review of non-parametric tests). Some specific examples of where we might use a within-group study are shown below:

1. A group of patients are measured in respect of the effectiveness of a new medication. The outcome measure is illness severity, which is examined at baseline (before the new treatment is given) and at the fourth week after the new treatment is given.
2. A group of young people complete a questionnaire that measures attitudes towards smoking before and after watching a video that shows the effects of nicotine on the lungs.
3. A large group of children are measured in a longitudinal study that explores the number of words they can produce at ages 1, 3, 5, 7 and 9.
4. A group of university students are asked to rate their satisfaction with their course at the end of their first, second and third years of study.

These are just a few examples. The first two would be suitable for a related t-test; the second two would need to be examined via a repeated-measures one-way ANOVA.

Resolving extraneous variables using within-group studies

We can reduce extraneous variables by employing within-group methods to study the outcome. By exposing one group to all conditions at different times, and measuring the outcome at each event, we 'lose' the differences observed in between-group studies. Rather than explore differences across groups, we examine those differences over conditions that are represented by the same person. Extraneous variables are then 'absorbed'.

However, despite these advantages, within-group studies can be prone to errors related to repeating a measure to one group of people. Typically, these are demonstrated by **practice or boredom effects** between tests, or by changes in an individual's characteristics between testing (perhaps motivation, attitude, mood, etc.). This can be reduced by altering the order of tests (counterbalancing). For example, CALM might present six of the participants with the 'picture' condition followed by the 'no picture' condition, while the remaining six participants receive the presentation in reverse order.

8.3 Nuts and bolts

Other names for the related t-test

There are several names for the related t-test which can be very confusing if you are just getting used to these tests. Here are just a few of those names:

Within-group t-test
Repeated-measures t-test
Related samples t-test
Paired samples t-test

8.4 Nuts and bolts

Are within-group studies better than between-group studies?

This is not an easy question to answer, since both have relative merits and restrictions. Table 8.1 summarises the key points in this debate.

Table 8.1 Advantages/disadvantages of within-group vs. between-group studies

Study design	Advantages	Disadvantages
Within group	Need fewer participants	Order effects
	Effect stronger	Lost participants at follow up
	Control for extraneous variables	Participants can learn aim of experiment
	No need to test homogeneity of variances	May have to wait to allow participants to forget stimulus
	Participants can act as own control	
Between-group	Naive participants	Larger samples needed
	Only need to run test once	Extraneous variables may not be controlled
	No order effects	Need to examine homogeneity of variance
	Unequal samples OK	Less powerful
	No need to wait for participants to forget stimulus	May need control group

One of the most useful advantages of within-group studies is that we need fewer participants to examine the outcome. This is because the same participants are used for each condition. We can recruit 40 participants to explore an outcome over three conditions within one group. A between-group study would require different participants in each of the three groups, which would require a sample of 120. Another advantage of within-group studies is that they are more powerful statistically; this is related to the fact that fewer participants are needed to find the desired effect. As we discussed earlier, one major problem with between-group studies is that we can never be certain that we have accounted for all potential extraneous variables. We might want to examine gender differences, only to find that the groups also differed on other factors, such as mood, motivation or IQ. This may reduce the validity of any assumptions that we make. Extraneous variables are much more easily controlled for in within-group studies. However, as Table 8.1 indicates, within-group studies are not without limitations.

Measuring differences

When we examine differences statistically we start with the assumption of the null hypothesis (that there will be no difference in dependent variable scores across the two within-group conditions). Using our research example, this means that there would be no difference in the number of words that a single group of people can recall between when they are given a picture prompt and when they are not. If we counted the number of words remembered in each condition, the mean difference between those conditions would be zero. CALM predicted that people would recall more words when they are provided with additional information (to aid memory) – this is the alternative hypothesis. We can reject the null hypothesis in favour of the alternative hypothesis only if significantly more words are recalled in the 'picture' condition. We can use a related t-test to examine the mean difference between the two conditions and establish whether the difference is significant. We learned about statistical significance in Chapter 4. Using a related t-test, a significant difference occurs if there is a less than 5% probability that any observed differences occurred by chance (where $p < .05$). As we saw with the independent t-test, the likelihood of significance will depend on the size of the differences between the two conditions and the extent of variation seen in the sample. We will explore how to do all of that a little later.

Comparison with other tests

The related t-test shares a number of similarities with the independent t-test (see Chapter 7). Most notably, both are used to measure two 'levels' of an independent variable in respect of differences in a single parametric dependent variable. However, the tests differ in key ways, too. In an independent t-test we explore the differences between two distinct groups (such as gender: male vs. female); in a related t-test those differences are measured as conditions across a single group (such as before and after an intervention). If we want to maintain a within-group design but measure differences across three or more conditions we would need to use repeated-measures one-way ANOVA, so long as the data are parametric (see Chapter 10). For example, we could investigate differences in mood scores, from a single group of people, on three different days of the week; Monday, Wednesday and Friday. Finally, if we have two conditions but find that the dependent variable data (at either condition) are not parametric, we need to employ a Wilcoxon signed-rank test, the non-parametric equivalent to the related t-test.

Assumptions and restrictions

There are several assumptions that need to be met for the related t-test. The dependent variable scores must be parametric, across both within-group conditions (the independent variable). Those conditions must be measured across a single group. We saw what we meant by parametric data in Chapter 5. In short, the data must be represented by interval or ratio scores and should be reasonably normally distributed (we explored normal distribution in Chapter 3). Previously, we have seen that ordinal data are not considered to be parametric. We probably should not use these data in a related t-test, but researchers often do so. Typically, ordinal data might be represented by subjective ratings – differences between these should be based on how the scores are ranked rather than focus on the magnitude. Parametric tests use the mean score to determine outcome – if data are ordinal or are not normally distributed, we may not be able to trust that mean score. Non-parametric tests may be more appropriate in those circumstances. In this case we would use a Wilcoxon signed-rank test (see Chapter 18). Also, each person (or case) being measured must be present at both conditions. The whole point of the related t-test is that it measures the difference in the scores across the conditions for each individual.

8.5 Take a closer look

Summary of assumptions and restrictions

- The independent variable must be represented by two conditions:
  - Measured across one group
  - Each person must be present in both conditions
- Dependent variable data must be parametric:
  - The data must be interval or ratio (not ordinal or categorical)
  - Data should be reasonably normally distributed (within each condition)
- If assumptions are not met, the Wilcoxon signed-rank test could be considered (see Chapter 18)

Establishing significant differences

We can use the related t-test to examine differences in mean dependent variable scores, across two within-group conditions, and establish whether that difference is statistically significant. Similar to the methods that we saw for the independent t-test, the mean difference between the conditions is divided by the standard error of differences. We will explore how to do this manually in Box 8.6, and will see how to perform the test in SPSS later. We will use the research question set by CALM. The group predicted that more words would be remembered in the 'with picture' condition than

8.6 Calculating outcomes manually

Related t-test calculation

To explore outcomes for a related t-test manually, we will use some data that reflect the research question that we have been posing. The data are presented in Table 8.2. You will find a Microsoft Excel spreadsheet associated with these calculations on the web page for this book.

Table 8.2 Words recalled in 'no picture' and 'with picture' conditions

Participant	No picture	With picture	Diff (d)
1	14	19	−5
2	16	16	0
3	21	23	−2
4	19	26	−7
5	23	30	−7
6	16	26	−10
7	26	23	3
8	12	16	−4
9	14	14	0
10	21	23	−2
11	19	28	−9
12	23	18	5

Mean difference
(D) −3.17

Standard deviation of difference
(σD) 4.69

To examine outcomes in the related t-test, we start by finding the 'mean difference' across the conditions. We do this by calculating the difference in outcome values for each participant (d) and calculate the average of those differences. Then we calculate the standard deviation of those differences (we saw how to calculate standard deviation in Chapter 4). Then we divide that outcome by the square root of the sample size (to find the standard error of differences). Finally, we divide the mean difference by that standard error (to find the *t* score).

$$\text{The formula for the related t-test: } t = \overline{D} \div \frac{\sigma\overline{D}}{\sqrt{n}} = -3.17 \div \frac{4.69}{\sqrt{12}} = -2.34$$

The *t* score can be compared with a *t*-score distribution table (see Appendix 2), at the given degrees of freedom (*df*; in this case, the sample size *minus* 1: 12 − 1 = 11 [As we saw in Chapter 6, degrees of freedom refer to the number of values that are 'free to vary' in the calculation, while everything else is held constant]), at the agreed level of significance (usually p = .05), according to whether a one-tailed or two-tailed hypothesis has been made. If the *t* score exceeds that cut-off point, we have a significant difference.

In our example, the cut-off point for *df* = 11, where p = .05, in a one-tailed test =1.796. Our *t*-score (2.34) is higher; we have a significant difference in the number of words recalled across the conditions.

We can also use Microsoft Excel to calculate the critical value of t and to provide the actual p value. You can see how to do that on the web page for this book. In our case, p = .020 (one-tailed). In that spreadsheet, you can also see how to perform the entire test in Excel.

in the 'without picture' condition. This specific prediction represents a one-tailed hypothesis. Had CALM simply predicted that 'there would be a difference' in correct recall between the conditions, it would have been a two-tailed hypothesis (see Chapter 4 for more information on this).

How SPSS performs the related t-test

We can also perform a related t-test in SPSS. To explore outcomes, we will use the data shown in Table 8.2. We are examining how a picture prompt might aid word recall. CALM predicted that participants would recall more words when presented with a relevant picture than when not given the picture. The number of words that are recalled can be regarded as interval data, so we

8.7 Nuts and bolts
Setting up the data set in SPSS

When we create the SPSS data set for the related t-test, we need to set up two columns: each represents the dependent variable score for each condition (the scores are 'continuous' - see Chapter 5 for more information on data types).

Word recall.sav [DataSet1] - IBM SPSS Statistics Data Editor

File Edit View Data Transform Analyze Graphs Utilities Add-ons Window Help

	Name	Type	Width	Decimals	Label	Values	Missing	Columns	Align	Measure
1	none	Numeric	8	0	No Picture	None	None	8	Right	Scale
2	picture	Numeric	8	0	Picture	None	None	8	Right	Scale

Figure 8.1 Variable View for 'Word recall' data

Figure 8.1 shows how the SPSS Variable View should be set up (you should refer to Chapter 2 for more information on the procedures used to set the parameters). The variables are 'none' and 'picture', representing the two within-group conditions. Although these illustrate the independent variable, continuous dependent variable scores will be recorded here. We do not need to adjust anything in the Values column; the Measure column is set to Scale.

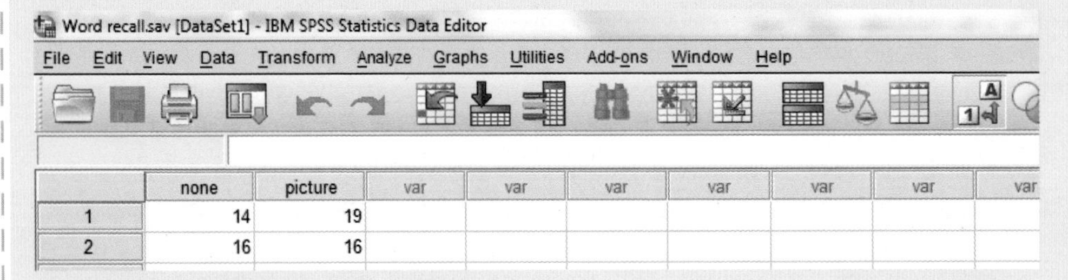

Figure 8.2 Data View for 'Word recall' data

Figure 8.2 illustrates how this will appear in the Data View. Each row represents a participant. Dependent variable scores (number of words recalled) will be entered for each participant in respect of each condition.

have satisfied that part of the requirements for parametric data. However, we need to examine whether the 'recalled words' are normally distributed across both conditions.

Testing for normal distribution

To examine normal distribution we will request Kolmogorov–Smirnov and Shapiro–Wilk tests through SPSS as we have done in previous chapters. On this occasion, we need to explore the outcome across the two within-group conditions. If the outcome indicates that normal distribution may have been compromised, we can employ z-score tests of skew and kurtosis (as we saw in Chapter 3). However, it is worth remembering that we are seeking *reasonable* normal distribution.

Open the SPSS file Word recall

Select **Analyze** → **Descriptive Statistics** → **Explore** → (in new window) transfer **No Picture** and **Picture** to **Dependent List** window, by clicking on arrow or by dragging the variable to that window (nothing is entered in **Factor List** as there are no between-group factors) → (in new window) click on **Plots** radio button → click on **Plots** button → tick **Normality with plots box** → click **Continue** → click **OK**

	Kolmogorov–Smirnov^a			Shapiro–Wilk		
	Statistic	df	Sig.	Statistic	df	Sig.
No picture	.148	12	.200*	.961	12	.798
Picture	.172	12	.200*	.947	12	.598

Figure 8.3 Kolmogorov-Smirnov/Shapiro-Wilk test for word recall conditions

Since we have a sample size of ten (for each condition), we should refer to the Shapiro–Wilk outcome (see Figure 8.3). We need the significance [Sig.] to be *greater* than .05 (because these tests investigate whether the data are significantly *different* to a normal distribution). Figure 8.3 shows that word recall is probably normally distributed for both conditions: No picture, W(12) = .961, p = .798; and Picture, W(12) = .947, p = .598.

Running the related t-test in SPSS

Using the SPSS file Word recall

Select **Analyze → Compare Means → Paired-Samples T Test...** as shown in Figure 8.4

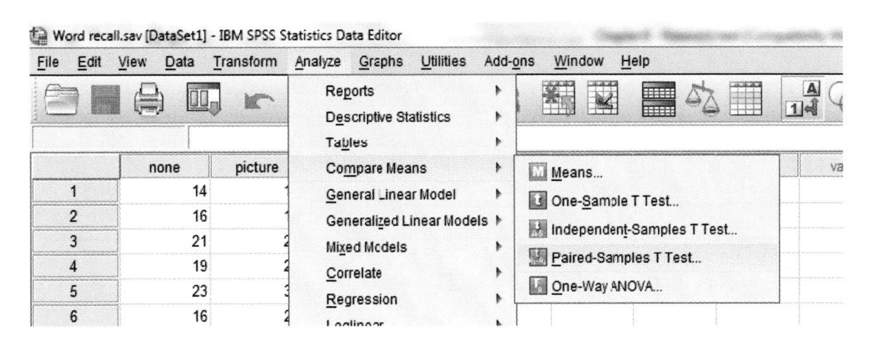

Figure 8.4 Related t-test: procedure 1

In new window (see Figure 8.5), transfer **No Picture** to **Pair 1, Variable 1** in **Paired Variables** window → transfer **Picture to Pair 1, Variable 2** → click **OK**

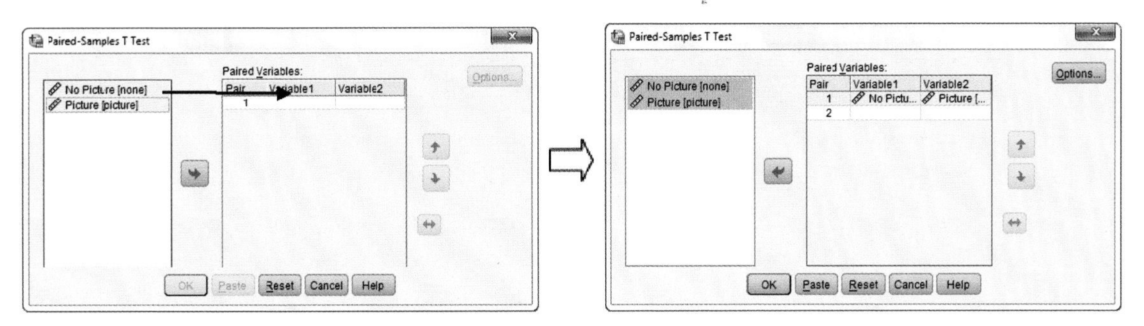

Figure 8.5 Related t-test: procedure 2

Interpretation of output

		Mean	N	Std. deviation	Std. error mean
Pair 1	No picture	18.67	12	4.313	1.245
	Picture	21.83	12	5.184	1.497

Figure 8.6 Descriptive statistics

Figure 8.6 suggests that more words were recalled in the 'picture' condition than the 'no picture' condition (as expected). However, we need to refer to the statistical outcome before we can declare whether that difference is significant.

Paired samples correlations

		N	Correlation	Sig.
Pair 1	No picture & picture	12	.526	.079

Figure 8.7 Correlation

Figure 8.7 shows some correlation statistics. This is of little use to us right now, but it will become useful when we examine effect size and power, which we will see later.

		Paired differences					t	df	Sig. (2-tailed)
					95% confidence interval of the difference				
		Mean	Std. deviation	Std. error mean	Lower	Upper			
Pair 1	No picture – picture	–3.167	4.687	1.353	–6.145	–.189	–2.340	11	.039

Figure 8.8 Related t-test statistics

Figure 8.8 indicates that there was a significant difference in word recall across the conditions ($p = .039$). However, as we saw with the independent t-test, SPSS calculates the significance based on a two-tailed outcome. This is fine for a non-specific hypothesis, but CALM specifically predicted that more words would be recalled in the 'picture' condition. Since this is a one-tailed hypothesis, we can divide that significance outcome by 2 so now we can say that $p = .020$ (see Chapter 4 for more information about tails and hypotheses).

Using the mean score data from Figure 8.6 and the (adjusted) statistical outcome from Figure 8.8, we now know that significantly more words were recalled in the 'picture' condition than for 'no picture'. When we write up the results, we should include a sentence that suggests that and show the statistical notation. As we saw with the independent t-test, we start with *t*, followed by the degrees of freedom (*df*) in brackets (see Box 8.6), the *t* value, and then the ***full*** p value (followed by a statement saying whether this was a one-tailed or two-tailed outcome). So, our statistical notation is: $t\,(11) = -2.340$, $p = .020$ (one-tailed).

Effect size and power

As we saw in Chapter 4, we can use G*Power to help us find Cohen's *d* effect size. This is how we explore these outcomes for a related t-test:

Open G*Power (for a screen shot, see Chapter 4, Figure 4.7):

From **Test family** select **t-tests**
From **Statistical test** select **Means: Difference between two dependent pairs (matched groups)**
From **Type of power analysis** select **Post hoc: Compute achieved power - given α, sample size and effect size**

Now we enter the **Input Parameters**:

From the Tails box, select One

To calculate the **Effect size**, click on the **Determine** button (a new box appears).
In that new box, for **Mean group 1** type **18.67** → for **Mean group 2** type **21.83** → for **SD σ group 1** type **4.31** → for **SD σ group 2** type **5.18** → for **correlation** type **0.526** (we got that from Figure 8.7) → click **Calculate and transfer to main window**
Back in original display, for **α err prob** type **0.05** (the significance level) → for **Sample size** type **12** → click on **Calculate**
From this, we can observe two outcomes: Effect size (d) **0.67** (which is very strong); and Power (1-β err prob) **0.71** (which is a little weak, given our target power of 0.80).

Writing up results

Table 8.3 Words recalled across the conditions (n = 12)

Words recalled	Mean	Standard deviation	Standard error	95% CI of difference
No picture	18.67	4.31	1.25	−6.15 to −0.19
Picture	21.83	5.18	1.50	

We report the statistics and significance in the following format, using the statistical data from Figure 8.8 and the G*Power data:

A related t-test confirmed that more words were recalled when there was a picture prompt than when there was no prompt, $t(11) = -2.340$, p = .020 (one-tailed). This represented a very strong effect, $d = 0.67$.

Presenting data graphically

You might also want to provide a graph for this result (but remember, you should never repeat data in figures that are also shown in tables). The method for drawing graphs for within-group studies is little different to what we saw for the independent t-test:

Select **Graphs** → **Legacy Dialogs** → **Bar...** as shown in Figure 8.9

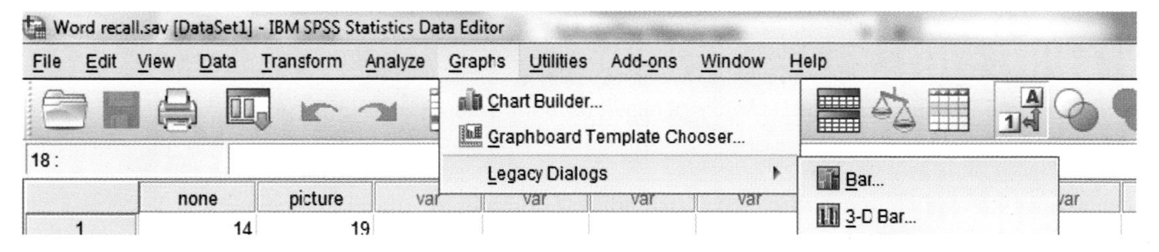

Figure 8.9 Selecting the type of graph

In new window (as shown in Figure 8.10) click on **Simple** box → tick **Summaries for separate variables** radio button → click **Define**

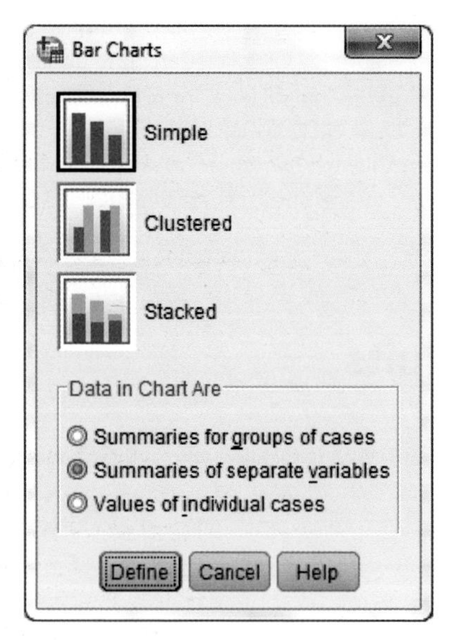

Figure 8.10 Selecting the options for related t-test graph

In new window (see Figure 8.11), transfer **No Picture** and **Picture** to **Bars Represent** window (make sure that data displayed in the window is preceded by 'MEAN'; if it is not, click on **Change Statistic...** button and select **Mean of values)** → click **Options**
Click **Options** → tick **Display error bars** box → select **Confidence intervals** radio button → ensure that **Level (%)** is set to **95** → click **Continue** → click **OK**

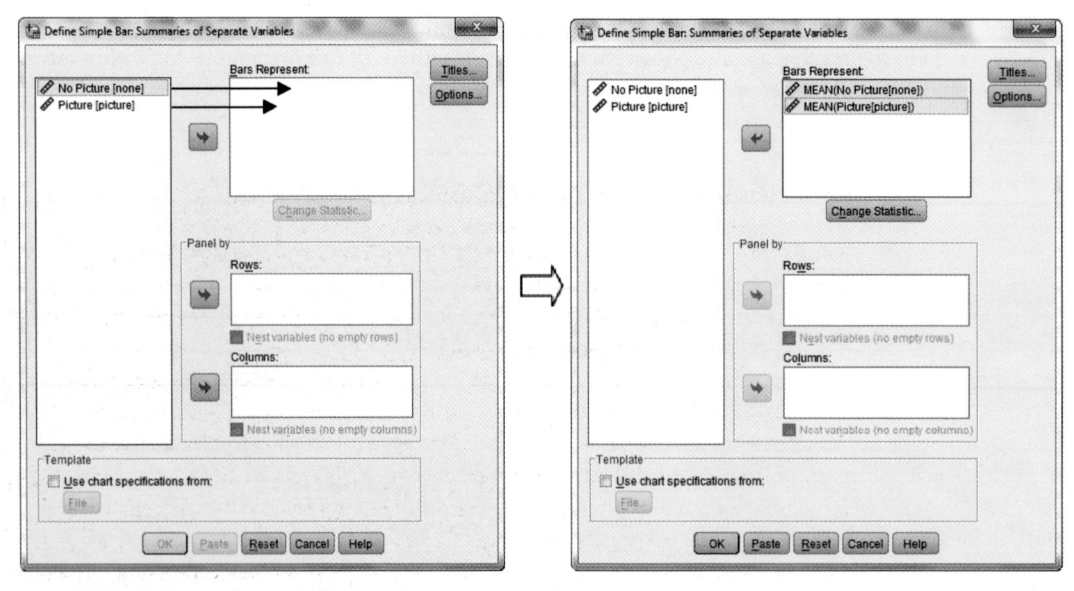

Figure 8.11 Selecting the variables for related t-test graph

In new window (see Figure 8.12), tick **Display error bars** box ➜ select **Confidence intervals** radio button ➜ ensure that **Level (%)** is set to **95** ➜ click **Continue** ➜ click **OK**.
These actions will produce the completed, as shown in Figure 8.13.

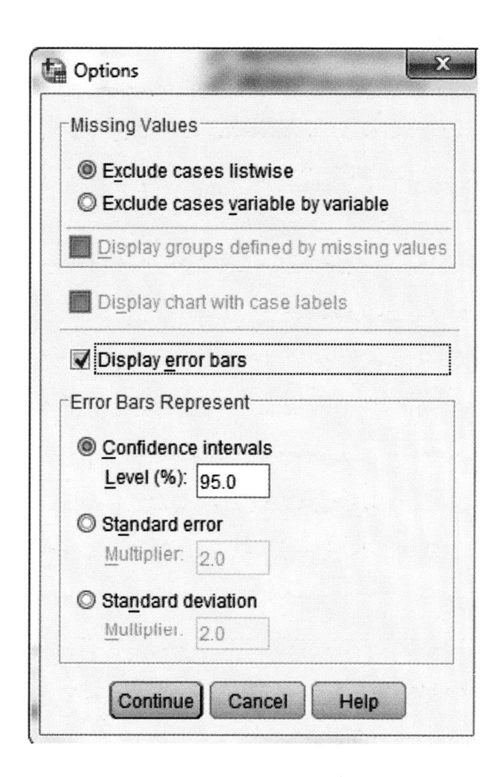

Figure 8.12 Options

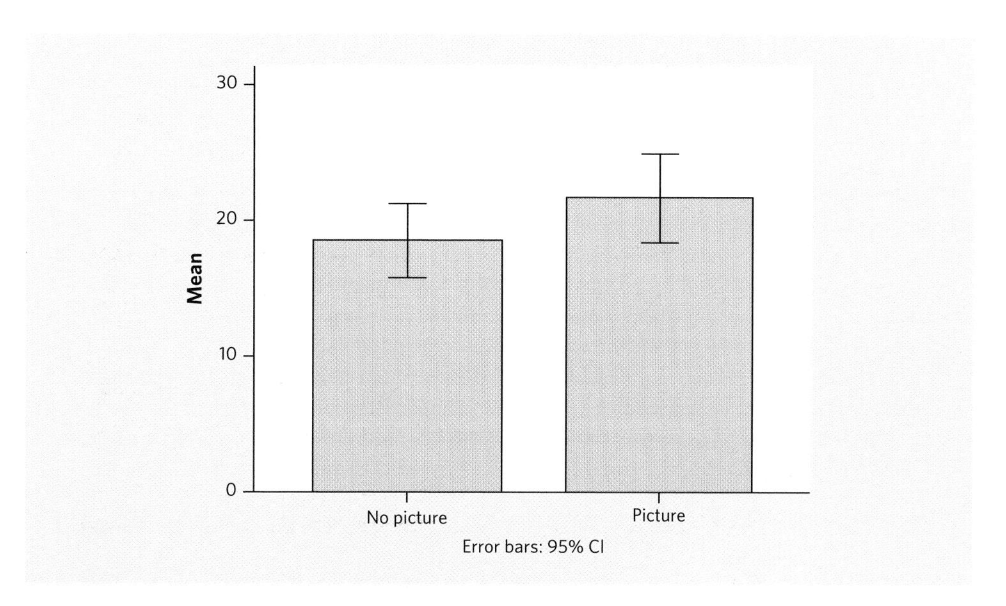

Figure 8.13 Completed bar chart

Chapter summary

In this chapter we have explored the related t-test. At this point, it would be good to revisit the learning objectives that we set at the beginning of the chapter.

You should now be able to:

- Recognise that we use a related t-test to examine differences in mean scores of a parametric dependent variable, across two within-group conditions (the independent variable) across a single sample.
- Understand that the data must be interval or ratio, and should be reasonably normally distributed. Outcomes for the related t-test are based on mean scores. Non-parametric data can be associated with unreliable mean scores. If we have reason to doubt whether the data are parametric, we might need to examine outcomes using the Wilcoxon signed-rank test (the non-parametric equivalent of the related t-test). Every person must be present in both conditions.
- Calculate outcomes for a related t-test manually (using maths and equations)
- Perform analyses using SPSS.
- Explore effect size and statistical power.
- Understand how to present the data and report the findings.

Research example

It might help you to see how the related t-test has been applied in published research. In this section you can read an overview of the following paper:

Rosenbloom, T. (2006). Driving performance while using cell phones: an observational study. *Journal of Safety Research. 37 (2)*: **207–12. DOI: http://dx.doi.org/10.1016/j.jsr.2005.11.007**

If you would like to read the entire paper you can use the DOI reference provided to locate that (see Chapter 1 for instructions). In this research the author investigated the extent that using a mobile phone (hands-free) might interfere with driving safely. Several studies have been conducted using simulators, but such research may lack ecological validity. This study examined drivers' attention while actually driving the vehicle on a busy road. Observation was undertaken by the front-seat passenger (without the knowledge of the driver) and was performed on 23 male drivers. Telephone calls were made to the driver by an associate of the researcher and lasted for between 5 and 15 minutes. Observations were made regarding speed travelled and distance from the car in front (two separate dependent variables), in respect of when the driver was engaged in a telephone conversation and when he was not (independent variable).

The results showed that there was no significant difference in driving speed between when the driver was engaged in conversation on the hands-free mobile phone and when he was not: $t(22) = 0.54$, $p = .59$. However, the distance from the car in front was significantly less (6.3 metres) when speaking on the phone than when not (8.3 metres): $t(22) = 4.56$, $p < .001$. There would appear to be safety concerns for drivers using mobile phones, even when the hands-free function is employed. The researchers were able to show this using a related t-test.

Extended learning task

You will find the data set associated with this task on the website that accompanies this book (available in SPSS and Excel format). You will also find the answers there.

Following what we have learned about the related t-test, answer these questions and conduct the analyses in SPSS and G*Power. (If you do not have SPSS, do as much as you can with the Excel spreadsheet.) For this exercise, we examine factors relating to the possible relationship between chronic illness and depression. This (fictitious) example represents a group of 50 patients who attended two follow-up sessions after an operation. At the first appointment, they received no information about their treatment. At the second appointment, they received clear information. On both occasions an independent rater observed the patients and completed a scale that indicated how satisfied the patients appeared to be with their treatment. Scores ranged from 0 (very satisfied) to 100 (very unsatisfied). The same rater was used in both conditions. In this example, we are exploring differences rather than making predictions. Since the ratings are being undertaken by one person, it is quite likely that the data are interval.

Open the data set **Satisfaction with treatment**

1. Using available data:
   a. Check for normal distribution at both time points.
   b. Conduct a related t-test.
2. Describe what the SPSS output shows.
3. State the effect size and power, using G*Power.
4. Report the outcome as you would in the results section of a report.

9 INDEPENDENT ONE-WAY ANOVA

Learning objectives

By the end of this chapter you should be able to:

- Understand the principles of analysis of variance (ANOVA)
- Recognise when it is appropriate to use an independent one-way ANOVA
- Understand the theory, rationale, assumptions and restrictions associated with independent one-way ANOVA
- Calculate outcomes manually (using maths and equations)
- Perform analyses using SPSS and interpret outcomes
- Explore effect size and statistical power
- Understand how to present the data and report the findings

Setting the scene: what is ANOVA?

ANOVA is an acronym for **Analysis of Variance**. It covers a series of tests that explores differences between groups or across conditions. The type of ANOVA employed depends on a number of factors: how many independent variables there are; whether those independent variables are explored between- or within-groups; and the number of dependent variables being examined. A summary of those tests is shown in Box 9.1. ANOVA explores the amount of variance that can be 'explained'. We first encountered variance in Chapter 4; in these next few chapters we will be seeing a great deal more of it. Any distribution of scores will usually vary. For example, we could give a questionnaire to a group of people. Responses to the questions can be allocated scores according to how they have been answered. Because people are different, it is quite likely that the response scores will differ between them. To examine that, we find the **overall (grand) mean** score and see how much the scores vary either side of that. The amount that the scores vary is called the variance.

When using ANOVA tests, we can partition that variance into separate pots. We call those pots the '**sum of squares**' (for reasons that will become clear in Box 9.3). The overall variance is found in the '**total sum of squares**': this illustrates how much the scores have varied overall to that grand mean. Sometimes the scores will vary a lot, other times they will vary a little, and occasionally the scores will not vary at all. When the scores vary, we need to investigate the cause of the variation. The scores may have varied because of some fundamental differences between the people answering the questions. For example, income may vary across a sample of people according to the level of education. If those differences account for most of the variation in the scores, we could say that we have 'explained' the variance. At the other extreme, the scores may have varied simply due to random or chance factors: this is the variance that we cannot explain. ANOVA tests seek to explore how much variance can be explained – this is found in the '**model sum of squares**'. Any variance that cannot be explained is found in the '**residual sum of squares**' (or error). We express the sum of squares in relation to the relevant degrees of freedom (we will see how later). This produces the model **mean square** and residual mean square. If we divide the **model mean square** by the **residual mean square**, we are left with something called an '**F ratio**'– this illustrates the proportion of the overall variance that has been explained (in relation to the unexplained variance). The higher the F ratio, the more likely it will be that there is a statistically significant difference in mean scores between groups (or conditions for within-group ANOVAs).

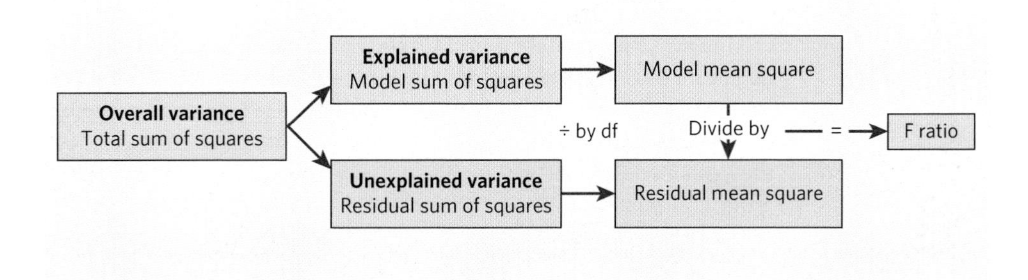

What is independent one-way ANOVA?

An independent one-way ANOVA explores differences in mean scores from a single (parametric) dependent variable (usually) across three or more distinct groups of a categorical independent variable (the test *can* be used with two groups, but such analyses are usually undertaken with an independent t-test). We explored the requirements for determining whether data are parametric in Chapter 5, but we will revisit this a little later in this chapter. As we saw just now, the variance in dependent variable scores is examined according to how

9.1 Nuts and bolts
Types of ANOVA

The ANOVA type depends on the number of independent variables (IV), whether those are measured between-groups (BG) or within-groups (WG), and the number of dependent variables (DV).

Independent one-way ANOVA: Scores from one DV, compared across one BG IV (with three or more groups*).
Repeated-measures one-way ANOVA: Scores from one DV, compared across one WG IV (with three or more conditions*).
Independent multi-factorial ANOVA: Scores from one DV, compared across two or more BG IVs (each with two or more groups); two-way ANOVA - two IVs; three-way ANOVA - three IVs, etc.
Repeated-measures multi-factorial ANOVA: Scores from one DV, compared across two or more WG IVs (each with two or more conditions).
Mixed multi-factorial ANOVA: Scores from one DV, compared across one or more BG IV (with two or more groups) *and* one or more WG IV (with two or more conditions).
MANOVA: Scores from two or more DVs, compared across one or more BG IV (each with two or more groups).
ANCOVA: Scores from one DV, compared across one or more BG IV (each with two or more groups), while controlling (factoring out) one or more additional variables.

* One-way ANOVA *can* be performed on just two groups or conditions (but see main text).

much of that can be explained (by the groups) in relation to how much cannot be explained (the error variance). The main outcome is illustrated by the '**omnibus**' (overall) ANOVA test, but this will indicate only whether there is a difference in mean scores across the groups. Additional tests *may* be needed to explore the source of difference (see later).

Research question for independent one-way ANOVA

We can illustrate independent one-way ANOVA by way of a research question. A group of higher education researchers, FUSS (Fellowship of University Student Surveys), would like to know whether contact time varies between university courses. To explore this they collect data from several universities and investigate how many hours are spent in lectures, according to three courses (law, psychology and media). FUSS expect that there is a difference but they have not predicted which group will spend more time attending lectures. In this example, the explained variance will be illustrated by how mean lecture hours vary across the student groups, in relation to the grand mean. For the record, all outcomes in this chapter are based on entirely fictitious data!

9.2 Take a closer look
Variables for independent one-way ANOVA

Dependent variable:	Lecture hours attended per week (hours)
Independent variable:	Student group (law, psychology and media)

Theory and rationale

Identifying differences

Essentially, all of the ANOVA methods have the same method in common: they assess **explained (systematic) variance** in relation to **unexplained (unsystematic) variance** (or 'error'). With an independent one-way ANOVA, the explained variance is calculated from the **group means** in comparison to the grand mean (the overall average dependent variable score from the entire sample, regardless of group). The unexplained (error) variance is similar to the standard error that we encountered in Chapter 4. In our example, if there is a significant difference in the number of hours spent in lectures between the university groups, the explained variance must be sufficiently larger than the unexplained variance. We examine this by partitioning the variance into the model sum of squares and residual sum of squares. We can see how this is calculated in Box 9.3. It is strongly recommended that you try to work through this manual example as it will help you understand how variance is partitioned and how this relates that to the statistical outcome in SPSS.

9.3 Calculating outcomes manually

Independent one-way ANOVA calculation

Table 9.1 No. of lecture hours attended per week

	Law (L)	Psychology (P)	Media (M)
	15	14	13
	10	13	12
	14	15	11
	15	14	11
	17	16	14
	13	15	11
	13	15	10
	19	18	9
	16	19	8
	16	13	10
Group mean	14.80	15.20	10.90
Standard deviation	2.486	1.990	1.792
Group variance	6.18	3.96	3.21
Grand mean	13.63	**Grand variance**	8.03

To illustrate how we can calculate the outcome of an independent one-way ANOVA, Table 9.1 presents some (fictitious) data based on the FUSS research question. The overall variance is measured by the total sum of squares. We need to 'partition' this into the model sum of squares and the residual sum of squares. The model sum of squares is found by squaring the difference between the group mean and the grand mean (the average score, regardless of group), multiplying that by the group size and summing the answer for each group (that is why it is called the sum of squares). The residual sum of squares is found from the variance of each group (which is the sum of the squared differences between each score and the group mean). These are expressed in terms of 'degrees of freedom' (the number of values that are 'free to vary' in the calculation, while everything else is held constant - see Chapter 6). This produces the model mean square and residual mean square. From these outcomes we find the F ratio, which can be compared with cut-off points to determine whether the between-group differences are significant.

You will find a Microsoft Excel spreadsheet associated with these calculations on the web page for this book.

Total sum of squares (SS_T):

$SS_T = S^2\text{grand}\,(N - 1) =$ grand variance $\times$ sample size (30) *minus* 1 $= 8.03 \times 29 =$ **232.97**

You will need to allow for slight 'rounding errors' due to decimal places throughout these calculations.

Grand variance: Deduct grand mean from each score, square it, repeat for all scores, add these up, divide by number of scores minus 1: $([15 - 13.63]^2 + [14 - 13.63]^2 + \ldots [10 - 13.63]^2) \div (30 - 1) = 8.03$

Model sum of squares (SS_M): The formula for model sum of squares: $SS_M = \sum n_k(\bar{x}_k - \bar{x}_{grand})^2$

Deduct grand mean from group mean, square it, multiply by no. of scores in group (10)

So $SS_M = 10 \times (14.80 - 13.63)^2 + 10 \times (15.20 - 13.63)^2 + 10 \times (10.90 - 13.63)^2 =$ **112.87**

We have three groups, so degrees of freedom (*df*) for $SS_M(df_M) = 3 - 1 = 2$ (this is the numerator *df*)

Residual sum of squares (SS_R): Formula for the residual sum of squares: $(SS_R) = \sum s_k^2(n_k - 1)$

Multiply group variances by group size *minus* 1 ($10 - 1 = 9$):

$SS_R = (6.18 \times 9) + (3.96 \times 9) + (3.21 \times 9) =$ **120.15** (allow for decimal place rounding)

Group variance: Deduct group mean from each group score, square it, repeat for all group scores, add these up, divide by group size minus 1: e.g. for Gender: $([15 - 14.80]^2 + \ldots [16 - 14.80]^2) \div (10 - 1) = 6.18$

df for SS_R = sample size minus 1(30 − 1) − $df_M(2)$: so $df_R = 29 - 2 = 27$ (this is the denominator *df*)

Mean squares: This is found by dividing model sum of squares and residual sum of squares by the relevant *df*:

Model mean square (**MS_M**): $SS_M \div df_M = 112.87 \div 2 =$ **56.43**

Residual mean square (**MS_R**): $SS_R \div df_R = 120.15 \div 27 =$ **4.45**

F ratio: This is calculated from model mean square divided by residual mean square:

$$F = \frac{MS_M}{MS_R} = 56.43 \div 4.45 = \mathbf{12.68}$$

We compare that F ratio to **F-distribution tables** (Appendix 4), according to **numerator** and **denominator degrees of freedom**, at the agreed level of significance (usually $p = .05$). In our example, we have a numerator (2) and denominator (27) degrees of freedom ($df = 2, 27$) and we set significance as $p = .05$, so the F cut-off point (**critical value**) is 3.35. Our F ratio (12.68) is greater than that, so we have a significant difference in lecture hours across student groups.

We can also use Microsoft Excel to calculate the critical value of F and to provide the actual p value. You can see how to do that on the web page for this book. In our example, $p < .001$, You can also see how to perform the entire test in Excel.

Statistical significance: putting it into context

The outcome shown in Box 9.3 suggests that there is a significant difference in the number of hours spent in lectures across the university groups. When we perform the test in SPSS later, we can refer to the F ratio and significance outcome to confirm these findings. However, although it would be easy for you to simply look at that outcome, it would help your understanding of what the result means to view this in terms of explained vs. unexplained variance and relate that to the sum of squares and mean squares outcome from Box 9.3. The explained variance is that which relates to differences between the groups; it is illustrated by the model sum of squares (SS_M). The unexplained variance is the error, or that which is not related to the differences between the groups; it is shown by the residual sum of squares (SS_R). To take account of the number of groups and sample size, we need to express the model and residual sum of squares in terms of the degrees of freedom (*df*) – see Box 9.3. This produces the model mean square (MS_M) and residual mean square (MS_R). We divide MS_M by MS_R to get the F ratio. In our case, MS_M is larger than MS_R, which means that we have 'explained' most of the overall variance in lecture hours (the between-group differences will never be significant if there is more unexplained than explained variance). To assess whether the outcome is statistically significant, we compare the F ratio to F-distribution tables – if we exceed the relevant cut-off point we know that the observed between-group differences are significant.

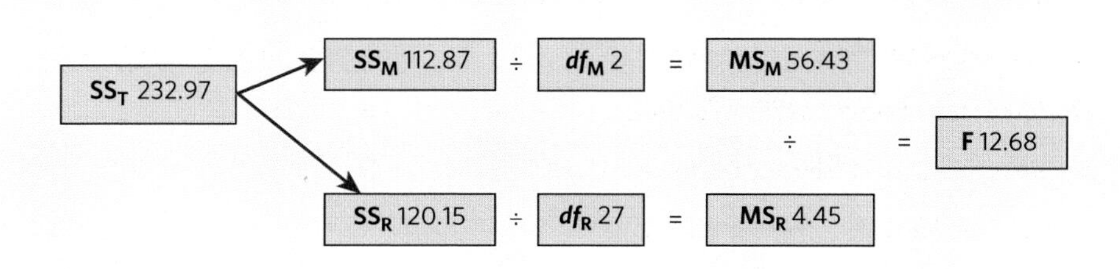

9.4 Exercise

Partitioning sum of squares mini-exercise

Using the data set shown below, calculate the sum of squares, mean squares and F ratio. State whether there is a significant difference in quality of life scores between the town residents.

Table 9.2 Quality of life scores

Town A	Town B	Town C
36	94	26
18	56	67
81	20	83
30	53	94
36	56	69
76	56	67
31	76	77
29	24	66
55	69	38
16	59	42

Exercise outcome

To assess whether there is a significant difference in quality of life scores between the towns, you should have started by calculating the group means, group variance (you may have needed some help from Chapter 4 there) and grand mean. Those outcomes are shown in Table 9.3.

Table 9.3 Mean and variance outcomes for quality of life scores

	Quality of life scores		
	Town A	Town B	Town C
Group mean	40.80	56.30	62.90
Group variance	509.96	483.34	452.10
Grand mean 53.33			

Using the information in Table 9.3 and the guidelines in Box 9.3, you should have found the following:

Model sum of squares (SS_M): 2574.07
Residual sum of squares (SS_R): 13008.60
Model degrees of freedom (*df*): 3 groups *minus* 1=2
Residual *df*: sample size *minus* 1 *minus* model *df* =27
Model mean square (MS_M): $SS_M \div 2 = 1287.03$
Residual mean square (MS_R): $SS_R \div 27 = 481.80$
F ratio: $MS_M \div MS_R = 2.67$

Using F-distribution tables (or the Excel function shown in the associated spreadsheet) we see that the cut-off point for F when $df = 2, 27$ is 3.35. The calculated F ratio (2.67) is *less* than that, so there is *not* a significant difference in quality of life scores between the towns (exact p value = .087).

Finding the source of difference

The F ratio will indicate whether we have a significant difference in mean scores between the groups – we call this the 'omnibus' ANOVA. However, *if* we have three or more groups, this tells us only that most of the overall variance is explained by differences across those groups; it will not indicate *where* those differences are. If we have a significant difference across two groups, we can simply compare the mean scores to tell where the differences are. If we have significant difference across three or more groups, we need more information.

When we explored the FUSS data just now, we saw that there was a significant difference in the number of lecture hours attended between three university course groups. The mean data suggest that media students attended for fewer hours than psychology and law students (who appear to be similar in the number of hours attending lectures). However, we need to do more than make visual comparisons: we must explore the differences statistically. To do this, we need to perform additional analyses, using either **planned contrasts** or ***post hoc* tests**. We use planned contrasts if we have predicted a *specific* outcome about differences between the groups; otherwise we must use *post hoc* tests.

Planned contrasts, *post hoc* tests and multiple comparisons

The key difference between planned contrasts and *post hoc* tests rests on the way in which they account for multiple comparisons. Usually, we base significance testing on the probability that there is less than a 5% likelihood that the observed differences occurred by chance (see Chapter 4). The more tests we run, the greater the possibility we have of finding a significant outcome simply by chance factors alone (known as the, '**familywise error**'). To account for that possibility, we may need to adjust the level at which we declare statistical significance. For every additional test that we run, we may need to divide the significance cut-off according to how many of those tests we undertake. As we will see throughout the next few sections, the way in which we handle **multiple comparisons** will depend on what type of test we employ to examine those differences.

But remember, neither planned contrasts nor *post hoc* tests are needed if, a) there are only two groups, or b) the overall outcome is not significant.

Planned contrasts

Planned contrasts can be used only if *specific* (one-tailed) predictions have been made about outcomes between the groups. For example, using the FUSS research example, a specific prediction might state that psychology students will spend more time in lectures than the other two groups, while there will be no difference in attendance between law students and media students. In that scenario we can use planned contrasts. If FUSS predicts only that there will be difference between the groups, planned contrasts cannot be employed; *post hoc* tests must be undertaken instead. The type of planned contrast used depends on whether one of the groups being analysed represents a **control group** (this is the group to which outcome across **experimental groups** are compared). If we do have a control group, we explore between-group differences using 'orthogonal' planned contrasts; if there is no control group, a non-orthogonal test must be employed. Outcomes from orthogonal planned contrasts can be reported without adjusting for multiple comparisons, otherwise (with non-orthogonal analyses) we must adjust the significance cut-off point by the number of additional tests that we run.

Orthogonal planned contrasts

When we have a control group, we need to know how the other experimental groups compare with that control group (we call that Contrast 1). We also need to know how the experimental groups compare with each other. If we have two experimental groups we explore that in one further test (Contrast 2); if we have more than two additional groups we would need several extra contrasts. To enter the data into SPSS we need to assign values to the contrasts. For every positive value entered we need corresponding negative values, so that the overall values sum to zero. The weight of the values will depend on how many groups there are. Box 9.5 shows how we would allocate values to some example planned contrasts.

9.5 Nuts and bolts
Value allocation in orthogonal planned contrast

Say we examine one control group (Group 1) against two experimental groups (Groups 2 and 3). In this scenario, we allocate −2 for the control group (because there are two experimental groups to compare with). To balance back to 0, we allocate +1 to each of the experimental groups. Then we compare the experimental groups with each other. The control group is now redundant, so is given the value 0. The remaining experimental groups receive values of −1 and +1 respectively (so balance to 0 once again).

Value allocation: one control group and two experimental groups

	Compared groups	Redundant group	Contrast values		
Contrast 1	Control vs. Groups 1 and 2	None	−2	+1	+1
Contrast 2	2 vs. 3	Control	0	−1	+1

If we have three experimental groups (Groups 2-4) to compare to the single control group (Group 1), we have additional comparisons to undertake. We allocate −3 to the control group and +1 to each of the experimental groups. We then perform three additional contrasts allocating −1 and +1 to each pair of groups that we need to compare, leaving the control group and remaining experimental group redundant.

Value allocation: one control group and three experimental groups

	Compared groups	Redundant groups	Contrast values			
Contrast 1	Control vs. Groups 1, 2, and 3	None	−3	+1	+1	+1
Contrast 2	2 vs. 3	Control, 4	0	−1	+1	0
Contrast 3	2 vs. 4	Control, 3	0	−1	0	+1
Contrast 4	3 vs. 4	Control, 2	0	0	−1	+1

Non-orthogonal planned contrasts

If we have made a specific prediction about the outcomes between the groups, but none of the groups is being used as a control group, we must use a non-orthogonal planned contrast. This means that the contrasts are no longer independent. The method for calculating values in a non-orthogonal planned contrast is shown in Box 9.6. Effectively, it is the same as the three-group example in Box 9.5, but without the control condition. Once we have created the contrast values, and calculated outcomes across the contrasts, we must adjust the significance level to account for multiple comparisons. If we have three groups, we have three pairwise comparisons, so we divide the significance cut-off point by three. Assuming overall significance is set at $p < .05$, contrast outcomes will be significant only where $p < .016$ ($0.05 \div 3 = 0.016$). If we have four groups, there will be six contrasts; significance occurs only where $p < .008$, and so on. It could be argued that the additional work undertaken to run non-orthogonal contrasts is not worth the effort. We gain nothing in terms of improving chances of finding significant outcomes. As we will see soon, *post hoc* tests (in SPSS) are performed with little fuss and (most) automatically adjust for multiple comparisons.

9.6 Nuts and bolts
Value allocation in non-orthogonal planned contrast

In this scenario we examine pairs of groups, such as Group 1 vs. Group 2 (Contrast 1), 1 vs. 3 (Contrast 2), and 2 vs. 3 (Contrast 3). We must assign values for each of the groups in each contrast, using −1 and +1 for the comparison pair and 0 for the redundant group. We will see how to enter this into SPSS later.

	Compared groups	Redundant group	Contrast values		
Contrast 1	1 vs. 2	3	−1	+1	0
Contrast 2	1 vs. 3	2	−1	0	+1
Contrast 3	2 vs. 3	1	0	−1	+1

Standard contrasts

SPSS has a set of pre-defined methods that we could use. The rationale for their use is complex and beyond the scope of this book. However, one of the standard contrasts may be quite useful. The polynomial contrast can be employed to confirm whether the data have a linear or quadratic trend. A **linear trend** happens when mean scores consistently increase (or decrease) across groups in a straight line. For example, if we were measuring quality of life scores at various levels of depression severity (mild, moderate and severe), we might expect quality of life scores to worsen with increasing severity of depression – that would be linear. A **quadratic trend** occurs when scores increase from one point to another, but then decrease thereafter (or vice versa). For example, quality of life scores might worsen between mild and moderate depression, but improve between moderate and severe depression. Or quality of life scores might improve between mild and moderate depression, but worsen between moderate and severe depression. This contrast can be used only if specific predictions have been made about trends; otherwise *post hoc* tests must be employed.

9.7 Take a closer look
Planned contrasts (a summary)

Planned contrasts are used to confirm predictions that have been made about the relationship between three or more groups of an independent variable about an outcome on a dependent variable. There are two types of planned comparison – orthogonal and non-orthogonal:

Orthogonal: Used where the experimental conditions are compared with a control group, followed by a comparison between the experimental groups. Adjustments for multiple comparisons are not needed

Non-orthogonal: Used where there is no control group, but where all of the groups are independent and can be compared with each other. Adjustments must be made to account for multiple comparisons

Post hoc tests

If no specific prediction has been made about differences between the groups, *post hoc* tests must be used to determine the source of difference. We can also choose to use *post hoc* tests in preference to non-orthogonal planned contrasts. However, there must be a significant ANOVA outcome in order for *post hoc* tests to be employed. If we try to run these tests on a non-significant ANOVA outcome it might be regarded as 'fishing'. Also, we run *post hoc* tests only if there are three or more groups. If there are two groups we can use the mean scores to indicate the source of difference. *Post hoc* tests explore each pair of groups to assess whether there is a significant difference between them (such as Group 1 vs. 2, Group 2 vs. 3 and Group 1 vs. 3). Most *post hoc* tests account for multiple comparisons automatically (so long as the appropriate type of test has been selected – see later).

The mathematics behind *post hoc* tests is relatively complex, so we will focus on how we run tests in SPSS. As we will see later, SPSS has something like 18 *post hoc* tests to choose from, but only a few are routinely used in practice. Each test employs a different method of calculating the result, depending on how it accounts for multiple comparisons, equality of variance and equal group sizes. An overview of the types of test is shown in Box 9.8. Many researchers employ a **Tukey** analysis, since it is relatively conservative (without losing too much power). However, that test should probably not be used when there are unequal group sizes, or if equality of variances has been violated. We will probably know whether we have equal group sizes prior to analysis. However, we will not know the outcome of tests for homogeneity of variance until we look at the SPSS output. If we know that we have unequal group sizes we should request **Gabriel's** or **Hochberg's GF2** *post hoc* tests (instead of Tukey) when we set the parameters to run independent one-way ANOVA in

SPSS (we will see how to do that later). Since we do not know about homogeneity of variance, we should take the precaution of selecting the **Games-Howell** option *in addition* to the main test. With these alternative tests, adjustments are made to account for differences in group size and/or homogeneity of variance. If we have equal group sizes and homogeneity of variance, we should be safe choosing Tukey, although some researchers prefer **Scheffé**, **Bonferonni** and **REGWQ**. These *post hoc* tests are OK, but some feel they are too conservative.

9.8 Take a closer look

Summary of common *post hoc* tests

Although there are several *post hoc* tests, Table 9.4 provides some information on the ones you are most likely to encounter and the conditions under which you might use them.

Table 9.4 *Post hoc* tests

Condition	Examples
Equal groups; equal variances assumed	Tukey is the most commonly used (it is conservative, but with good power).
	Scheffé, Bonferonni and REGWQ can be used, but some might consider them a little too conservative.
	REGWQ might be more useful where there are more than four groups.
	LSD (Least Significant Difference) is not recommended because it does not control for multiple comparisons.
Non-equal groups	Gabriel's or Hochberg's GF2 are most favoured here.
	Hochberg's GF2 may be better if there is a larger difference in group sizes.
Equal variances not assumed	Games-Howell is the most commonly used here. It is probably wise to select it when setting up SPSS, since you may not know whether you have equality of variances until other test outcomes are known.

There are many more tests available than this in SPSS. Other texts may recommend other choices. However, the suggestions here represent an overview of the most commonly reported views.

9.9 Nuts and bolts

One-way ANOVA: one-tailed or two-tailed?

With t-tests in SPSS, outcomes are reported assuming a two-tailed test. If the results relate to a one-tailed hypothesis, we can divide the significance by two. We cannot do that with ANOVA - the outcome is based on differences across all groups, not on specific ones between pairs of groups.

Assumptions and restrictions

We need to satisfy a number of assumptions to perform an independent one-way ANOVA. No person (or case) can appear in more than one group at any one time (the groups must be independent). The dependent variable scores should be parametric. As we saw in Chapter 5, this means that the data should be interval or ratio and should be reasonably normally distributed (we explored normal distribution in Chapter 3). However, it is quite common for independent one-way ANOVA to be used with Likert scales (self-report opinion and attitude questionnaires). Although there is no question that these produce ordinal scores, many researchers claim highly validated, well-designed Likert scales can elicit interval-style outcome (we explored that argument in Chapter 6). If we violate parametric assumptions we should employ a non-parametric test instead (Kruskal–Wallis test in this case – see Chapter 18). However, it is worth remembering that ANOVA is robust enough to overcome modest violations of parametric requirements.

Homogeneity of variance

We must also check that we have 'homogeneity of between-group variance' across the groups. This examines the extent that scores vary either side of the mean. When we examined outcomes across two groups with an independent t-test (Chapter 7), SPSS provided an adjusted outcome that we can use if homogeneity of variance has been violated. SPSS produces a Levene's test for independent one-way ANOVA (but we must request it). The test investigates whether the variances vary significantly between the groups. We don't want that, so we need the outcome to be non-significant. Violation of this assumption can be serious. If we have equal group sizes, ANOVA is robust enough to withstand unequal variances. However, this is much more of a problem when we do not have equal group sizes. If larger groups have proportionally larger variance than the smaller groups, the F ratio tends to be understated. This means that a significant outcome is less likely and we risk making a Type II error. If the larger groups have the smaller variance, the F ratio tends to be overstated, making significant outcomes more likely and risking Type I errors. Either way, it's not good and we should account for that. There are two methods by which we can adjust outcomes if there are unequal variances: **Brown–Forsythe F** and **Welch's F**. You probably do not need to know how they are calculated (at least for most undergraduate and postgraduate studies), but you could consult other sources if you feel the need. We should always request these tests if we know we have unequal group sizes (since we will not know whether we have equal variances until we see the results). SPSS will then produce an adjusted outcome. If the Levene's test suggests that we have equal variances, we can consult the main SPSS outcome. If there are unequal variances, we should refer to the outcome from the Brown–Forsythe F or Welch's F adjustments. There is very little to choose between the two tests, although most sources tend to suggest that Welch's F is more powerful (in most cases).

9.10 Take a closer look
Summary of assumptions and restrictions

- The IV must be categorical, with at least two distinct groups (but usually three or more)
- Membership of a group must be independent
- There must be one numerical dependent variable
  - DV data should be interval or ratio, and reasonably normally distributed
- If these assumptions are not met, the non-parametric Kruskal-Wallis test could be considered
- There should be homogeneity in the variances between the groups
  - If this assumption is violated we should consult adjusted outcomes (such as Welch's F)

How SPSS performs independent one-way ANOVA

We can perform an independent one-way ANOVA in SPSS. To illustrate, we will use the same data that we examined manually earlier. We are investigating whether there is a significant difference in the number of lecture hours attended according to student groups (law, psychology and media). There are ten students in each group. This means that we will not need to consult

Brown–Forsythe F or Welch's F outcomes (see assumptions and restrictions). However, because we do not know whether we have homogeneity of variance between the groups, we still need to potentially account for that in choosing the correct *post hoc* test (so we should request the outcome for Games–Howell in addition to Tukey – see Box 9.8). We can be confident that a count of lecture hours attendance represents interval data, so that the parametric criterion is OK However, we do not know whether the data are normally distributed; we will look at that shortly.

9.11 Nuts and bolts
Setting up the data set in SPSS

When we create the SPSS data set for an independent one-way ANOVA, we need to set up one column for the dependent variable (which will have a continuous score) and one column for the independent variable (which will have a categorical coding).

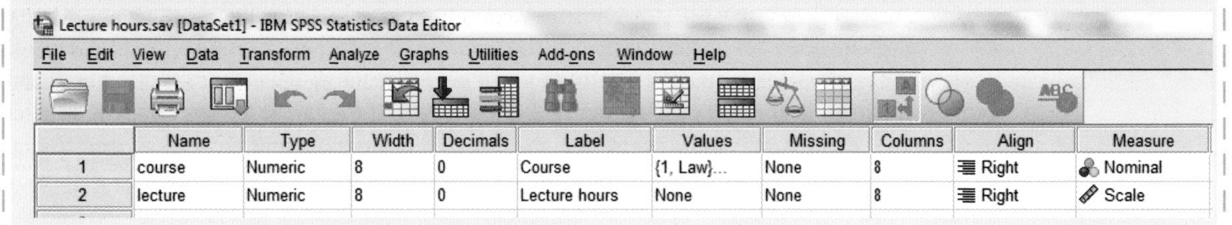

	Name	Type	Width	Decimals	Label	Values	Missing	Columns	Align	Measure
1	course	Numeric	8	0	Course	{1, Law}...	None	8	Right	Nominal
2	lecture	Numeric	8	0	Lecture hours	None	None	8	Right	Scale

Figure 9.1 Variable View for 'Lecture hours' data

Figure 9.1 shows how the SPSS Variable View should be set up. The first variable is called 'course'. This is the categorical independent variable representing the student groups. In the Values column, we include '1 = Law', '2 = Psychology' and '3 = Media'. The Measure column is set to Nominal. The second variable is 'lecture'. This is the continuous dependent variable representing the number of hours spent in lecture. We do not need to adjust anything in the Values column. The Measure column is set to Scale.

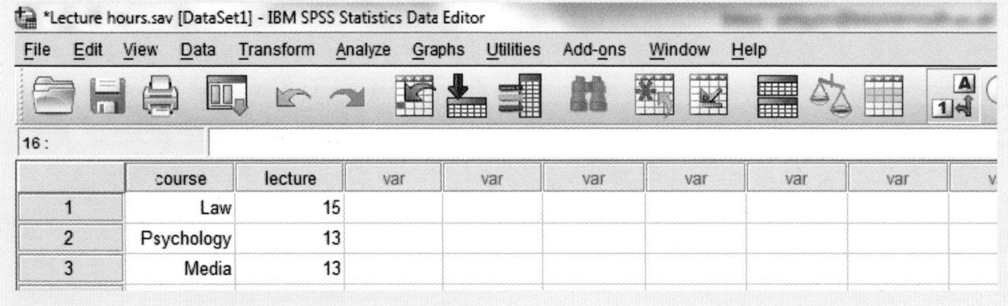

	course	lecture	var	var	var	var	var	var
1	Law	15						
2	Psychology	13						
3	Media	13						

Figure 9.2 Data View for 'Lecture hours' data

Figure 9.2 illustrates how this will appear in the Data View. Each row represents a participant. When we enter the data for 'course', we input 1 (to represent Law), 2 (to represent Psychology) or 3 (to represent Media). The 'course' column will display the descriptive categories (Law, Psychology or Media) or it will show the value numbers, depending on how you choose to view the column. The numbers of hours attended will be entered into the 'lecture' column.

Testing for normal distribution

To examine normal distribution, we start by running the Kolmogorov–Smirnov/Shapiro–Wilk tests (we saw how to do this for between-group studies in Chapter 7). If the outcome indicates that the data may not be normally distributed, we could additionally run z-score analyses of skew and kurtosis, or look to 'transform' the scores (see Chapter 3). We will not repeat the instruction for performing Kolmogorov–Smirnov/Shapiro–Wilk tests in SPSS (but refer to Chapter 3 for guidance). However, we should look at the outcome from those analyses (see Figure 9.3).

	Course	Kolmogorov–Smirnov[a]			Shapiro–Wilk		
		Statistic	df	Sig.	Statistic	df	Sig.
Lecture hours	Law	.134	10	.200*	.976	10	.943
	Psychology	.240	10	.107	.893	10	.184
	Media	.178	10	.200*	.976	10	.937

Figure 9.3 Kolmogorov-Smirnov/Shapiro-Wilk test for lecture hours according to course

Because we have a sample size of ten (for each group), we should refer to the Shapiro–Wilk outcome. Figure 9.3 indicates that lecture hours appear to be normally distributed for all courses: Law, W (10) = .976, p = .943); Psychology, W (10) = .893, p = .184); and Media, W (10) = .976, p = .937). You will recall that KS and SW tests investigate whether the data are significantly *different* to a normal distribution, so we need the significance ('Sig.') to be greater than .05.

Running independent one-way ANOVA in SPSS

Now we can perform the main analysis:

Using the SPSS file **Lecture hours**
Select **Analyze → Compare Means → One-Way ANOVA...** as shown in Figure 9.4

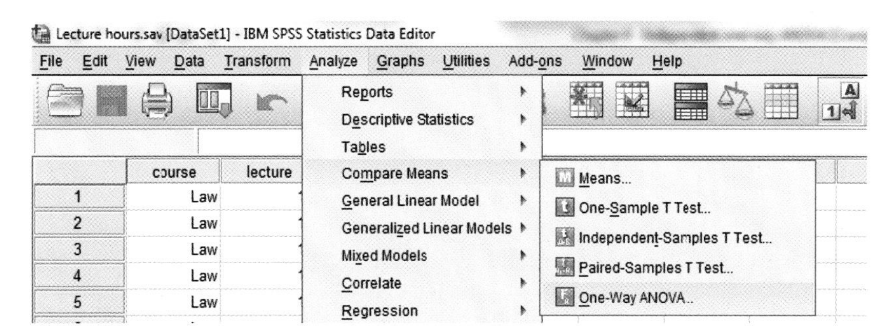

Figure 9.4 Independent one-way ANOVA: procedure 1

In new window (see Figure 9.5), transfer **Lecture hours** to **Dependent List** window (by clicking on arrow, or by dragging the variable to that window) → transfer **Course** to **Factor window** → click **Post Hoc...** button (because FUSS did not specify where differences may be, we must employ *post hoc* tests to explore the source of difference)
Note: we will explore how to run planned contrasts later

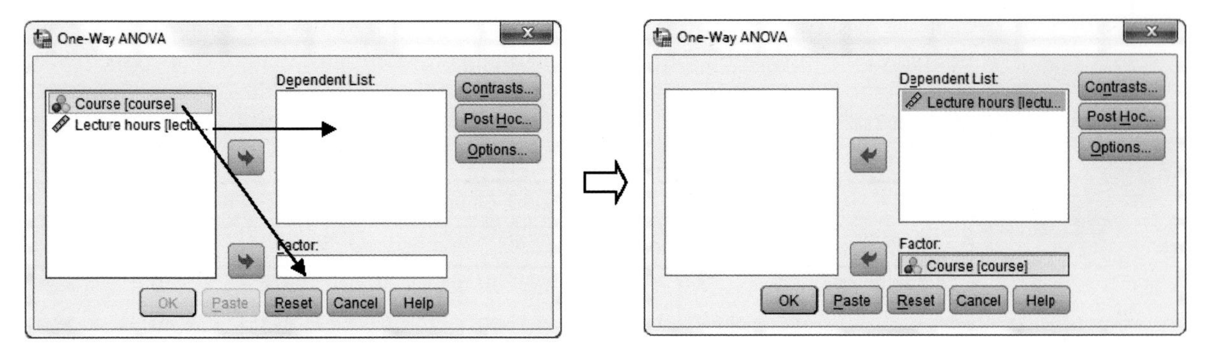

Figure 9.5 Independent one-way ANOVA: procedure 2

We explored the various *post hoc* options earlier. We know that we have equal group sizes (ten in each group), so we are pretty safe in selecting Tukey. However, we do not yet know whether we have equality of variances, so we should also select Games–Howell just in case.

In Post Hoc window (see Figure 9.6) tick **Tukey** and **Games-Howell** boxes ➜ click on **Continue** ➜ click **Options...** button

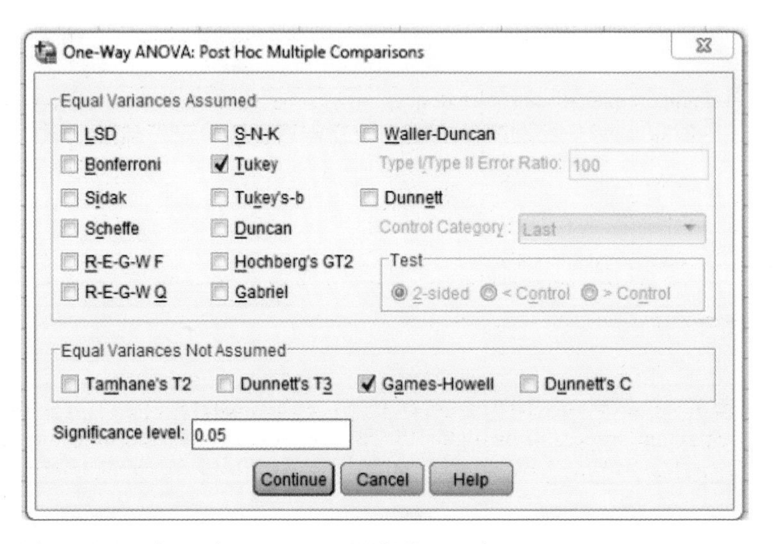

Figure 9.6 Independent one-way ANOVA: *post-hoc* options

In new window (see Figure 9.7), tick boxes for **Descriptives, Homogeneity of variance test, Brown-Forsythe** and **Welch** (as we do not yet know the outcome for homogeneity of variance, we need to run analyses for Brown-Forsythe and Welch's adjustments, just in case we need them) ➜ click **Continue** ➜ (back in original window) click **OK**

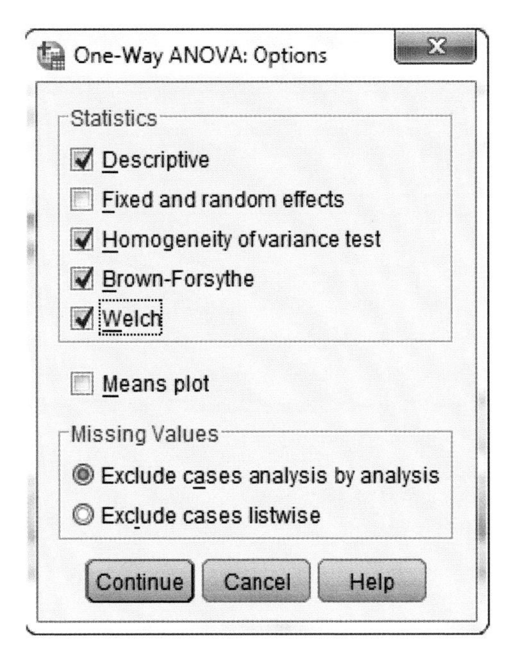

Figure 9.7 Independent one-way ANOVA: statistics options

Interpretation of output

					95% confidence interval for mean			
	N	Mean	Std. deviation	Std. error	Lower bound	Upper bound	Minimum	Maximum
Law	10	14.80	2.486	.786	13.02	16.58	10	19
Psychology	10	15.20	1.989	.629	13.78	16.62	13	19
Media	10	10.90	1.792	.567	9.62	12.18	8	14
Total	30	13.63	2.834	.517	12.57	14.69	8	19

Figure 9.8 Descriptives for independent one-way ANOVA

Figure 9.8 shows the lecture hours appear to be higher for psychology than for the other two courses, and appear to be higher for law than media However, we cannot make any inferences about that until we have assessed whether these differences are significant.

Levene statistic	df1	df2	Sig.
.408	2	27	.669

Figure 9.9 Test for homogeneity of variances

We need to check homogeneity of (between-group) variances. Figure 9.9 indicates that significance is greater than .05. Since the Levene statistic assesses whether variances are significantly *different* between the groups we can be confident that the variances are equal.

	Sum of squares	df	Mean square	F	Sig.
Between groups	112.867	2	56.433	12.687	.000
Within groups	120.100	27	4.448		
Total	232.967	29			

Figure 9.10 ANOVA statistics

Figure 9.10 indicates that there is a significant difference in mean lecture hours across the courses. We report this as follows: F (2, 27) = 12.687, p < .001. The between-groups line in SPSS is equivalent to the model sum of squares we calculated in Box 9.3. The within-groups line equates to the residual sum of squares (or error).

There are conventions for reporting statistical information in published reports, such as those suggested by the American Psychological Association (APA). The British Psychological Society also dictates that we should adhere to those conventions. When reporting ANOVA outcomes, we state the degrees of freedom (*df*), which we present (in brackets) immediately after 'F'. The first figure is the between groups (numerator) *df*; the second figure is the within groups (denominator) *df*. The 'Sig' column represents the significance (p). It is generally accepted that we should report the actual p value (e.g. p = .002, rather than p < .05). When a number cannot be greater than 1 (as is the case with probability values) we drop the 'leading' 0 (so we write 'p = 002', rather than 'p = 0.002'). The only exception is when the significance is so small that SPSS reports it as .000. In this case we report the outcome as p < .001. We cannot say that p = 0, because it almost certainly is not (when we explored the ANOVA outcome manually earlier, we saw that p = .0001; this is less than .001, but it is not 0).

Welch/Brown-Forsythe adjustments to ANOVA outcome

If we find that homogeneity of variance has been violated, we should refer to the Welch/Brown–Forsythe statistics to make sure that the outcome has not been compromised. The tests make adjustment to the degrees of freedom (*df*) and/or F ratio outcomes (shown under 'Statistic'–Figure 9.11). We do not need to consult this outcome, as we did satisfy the assumption of homogeneity of variance. However, it is useful that you can see what the output would look like if you did need it.

	Statistic[a]	df1	df2	Sig.
Welch	14.792	2	17.714	.000
Brown–Forsythe	12.687	2	24.994	.000

a. Asymptotically F distributed

Figure 9.11 Adjusted outcome for homogeneity of variance

Post hoc outcome

As we stated earlier, if we have three or more groups (as we do), the ANOVA outcome tells us only that we have a difference; it does not indicate where the differences are. We need to refer to *post hoc* tests to help us with that.

	(I) Course	(J) Course	Mean difference (I-J)	Std. error	Sig.	95% confidence interval	
						Lower bound	Upper bound
Tukey HSD	Law	Psychology	–.400	.943	.906	–2.74	1.94
		Media	**3.900***	.943	**.001**	1.56	6.24
	Psychology	Law	.400	.943	.906	–1.94	2.74
		Media	**4.300***	.943	**.000**	1.96	6.64
	Media	Law	–3.900*	.943	.001	–6.24	–1.56
		Psychology	–4.300*	.943	.000	–6.64	–1.96
Games–Howell	Law	Psychology	–.400	1.007	.917	–2.98	2.18
		Media	3.900*	.969	.003	1.41	6.39
	Psychology	Law	.400	1.007	.917	–2.18	2.98
		Media	4.300*	.847	.000	2.14	6.46
	Media	Law	–3.900*	.969	.003	–6.39	–1.41
		Psychology	–4.300*	.847	.000	–6.46	–2.14

*. The mean difference is significant at the 0.05 level

Figure 9.12 *Post-hoc* statistics

Figure 9.12 shows two *post hoc* tests: one for Tukey and one for Games–Howell. We selected both because, at the time, we were not certain whether we had homogeneity of variances. As we do, we can refer to the Tukey outcome. The first column confirms the *post hoc* test name. The second column is split into three blocks: Law, Psychology and Media (our groups). The third column shows how each group compares to the two remaining groups (in respect of lecture hours). For example, Law vs. Psychology (row 1) and Law vs. Media (row 2). The fourth column shows the difference in the mean lecture hours between that pair of course groups. To assess the source of difference we need to find between-pair differences that are significant (as indicated by cases where 'Sig.' is less than .05). Those that are significant will also be shown by an asterisk next to the mean difference. In our example, we have two such instances, which are highlighted in red in Figure 9.12. It shows that mean lecture hours are significantly higher for Law than they are for Media, and significantly higher for Psychology than for Media; there are no significant differences elsewhere. You should also note that the *post hoc* table presents the differences twice (where A is greater than B, B is also greater than A, but we need to report each between-pair only once).

Performing planned contrasts

We cannot conduct *post hoc* tests and planned contrasts together: we must choose one or the other. On this occasion we (correctly) opted for *post hoc* tests. If FUSS had made specific predictions about outcomes between the groups, we may have chosen to run planned contrasts. There is no control group in this example, so we must use a non-orthogonal method. We use the method that we saw in Box 9.6, which is revised to illustrate our groups in Box 9.12.

9.12 Nuts and bolts

Planned contrast value allocation for FUSS data

	Compared groups	Redundant group	Contrast values		
Contrast 1	Law vs. Psychology	Media	−1	+1	0
Contrast 2	Law vs. Media	Psychology	−1	0	+1
Contrast 3	Psychology vs. Media	Law	0	−1	+1

This is how we create the values in SPSS:

Select **Analyze** → **Compare Means** → **One-Way ANOVA...** (the variable settings will still be there from what you did just now - if not, follow the instructions from earlier) → click **Contrasts...** button → (in new window - see Figure 9.13) type **-1** in **Co-efficients**box → click on **Add** → type **1** → click on **Add** → type **0** → click on **Add** → click on **Next** → type **-1** → click on **Add** → type **0** → click on **Add** → type **1** → click on **Add** → click on **Next** → type **0** → click on **Add** → type **-1** → click on **Add** → type **1** → click on **Add** → click on **Next** → click on **Continue** → (back in original window) click **OK**

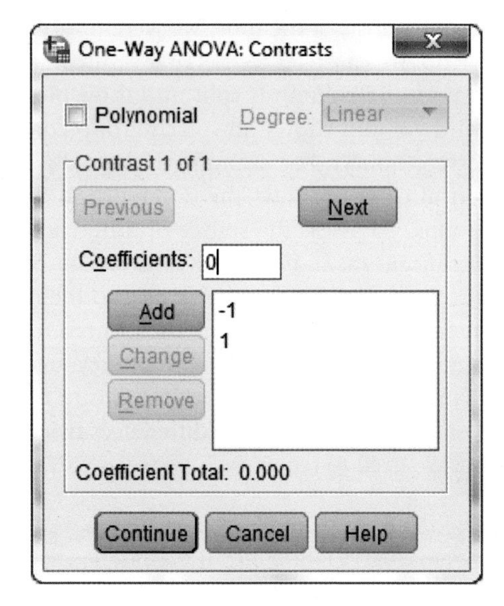

Figure 9.13 Independent one-way ANOVA: planned contrasts options

Contrast coefficients

Contrast	Course		
	Law	Psychology	Media
1	−1	1	0
2	−1	0	1
3	0	−1	1

Figure 9.14 Independent one-way ANOVA: planned contrasts output 1

Figure 9.14 confirms the code values that we created. So, Contrast 1 is Law vs. Psychology; Contrast 2 is Law vs. Media; and Contrast 3 is Psychology vs. Media.

		Contrast	Value of contrast	Std. error	t	df	Sig. (2-tailed)
Lecture hours	Assume equal variances	1	.40	.943	.424	27	.675
		2	−3.90	.943	−4.135	27	.000
		3	−4.30	.943	−4.559	27	.000
	Does not assume equal variances	1	.40	1.007	.397	17.174	.696
		2	−3.90	.969	−4.025	16.366	.001
		3	−4.30	.847	−5.079	17.808	.000

Figure 9.15 Independent one-way ANOVA: planned contrasts output 2

Figure 9.15 indicates the significance of the contrasts. There are two outcomes: one where equal variances are assumed and one where they are not. We know from earlier that we did have homogeneity of variance, so we can choose the first option. However, we must account for multiple comparisons. We had three pairings, so we divide the significance cut-off by 3 ($p = .05 \div 3 = .016$). We will have a significant outcome only if $p < .016$. We can see that Contrasts 2 and 3 are significant, but Contrast 1 is not (so there are differences between Law and Media, and between Psychology and Media, but not between Law and Psychology). The outcome is very similar to what we found with *post hoc* tests (but we had to work much harder to get there!). This perhaps reinforces the suggestion that planned contrasts are best kept for studies with specific predictions that include a control group.

Effect size and power

We can use G*Power to calculate effect size and to show how much power was achieved (see Chapter 4 to read more about the rationale for effect sizes).

Open G*Power:

From **Test family** select **F tests**
From **Statistical test** select **ANOVA: Fixed effects, omnibus, one-way**
From **Type of power analysis** select **Post hoc: Compute achieved power - given α, sample size and effect size**

Now we enter the **Input Parameters**:

To calculate the **Effect size**, click on the **Determine** button (a new box appears).
In that new box, for **Number of groups** type **3** for **SD σ → within each group** type **2.83** (the overall standard deviation from Figure 9.8) → for **Mean group 1** type **14.80** → **Mean group 2** type **15.20** → **Mean group 3** type **10.90** → for **size group 1** type **10** → **size group 2** type **10** → **size group 3** type **10** → click on **Calculate and transfer to main window**

Back in original display, for **α err prob** type **0.05** (the significance level) → for **Total sample size** type **30** → click on **Calculate**

From this, we can observe two outcomes: Effect size (d) **0.68** (which is strong); and Power (1- β err prob) **0.90** (which is very good, easily above the target level of 0.80).

Writing up results

We can present the mean data through tables, as in Table 9.5, and report the outcome in words, presenting the statistical notation.

Table 9.5 Mean lecture hours by course

Course	Mean lecture hours	Standard deviation	Standard error	95% CI to mean
Law (n = 10)	14.80	2.49	0.79	13.02 to 16.58
Psychology (n = 10)	15.20	1.99	0.63	13.78 to 16.62
Media (n = 10)	10.90	1.79	0.57	9.62 to 12.18

We should report the statistics and significance in the following format, using the statistical data from Figure 9.10 and the G*Power data:

An independent one-way ANOVA indicated that lecture hours attended differed significantly according to course, F (2, 27) = 12.687, p < .001. A *post hoc* Tukey indicated that media students attended significantly fewer lecture hours than law students (p = .001) and than psychology students (p < .001). This was represented by a strong effect, *d* = 0.68.

Never tabulate the *post hoc* data, but do report it in the narrative statistics (using the data from Figure 9.12).

Presenting data graphically

We can also draw a graph for this result. However, never just repeat data in a graph that has already been shown in tables. The illustration is shown in Figure 9.16, in case you ever need to run the graphs. We can use SPSS to 'drag and drop' variables into a graphic display with between-group studies (it is not so straightforward for within-group studies).

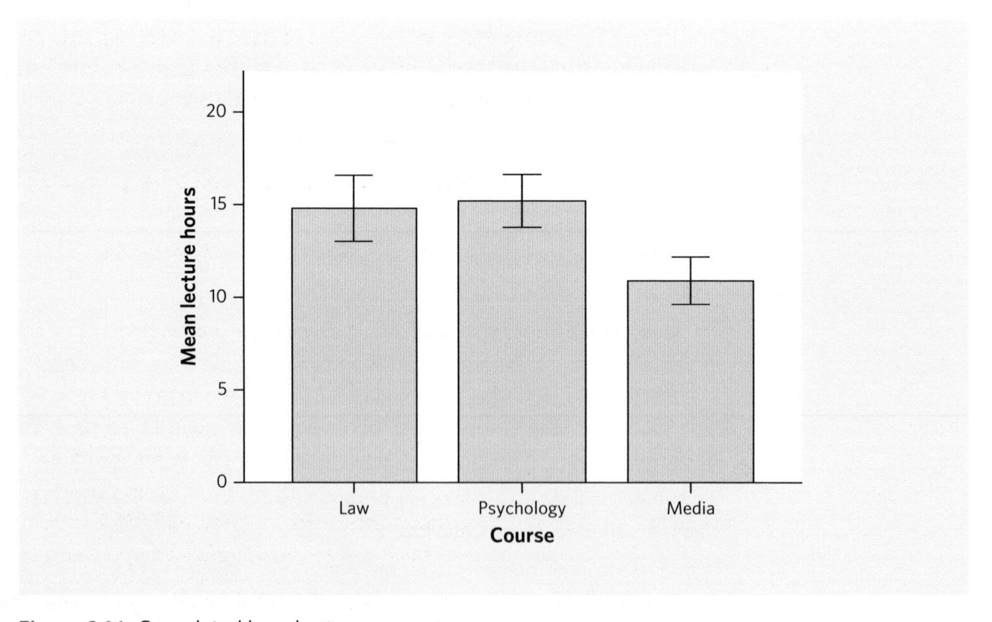

Figure 9.16 Completed bar chart

Select **Graphs** → **Chart Builder** → select **Bar** from list under **Choose from:** → drag **Simple Bar** graphic (top left corner) into empty chart preview area → transfer **Lecture hours** to **Y-Axis** box → transfer **Course** to **X-Axis** box → to include error bars, tick box for **Display error bars** in **Element Properties** box (to right of main display box) → ensure that it states 95% confidence intervals in the box below → click **Apply** (the error bars appear) → click **OK**

Chapter summary

In this chapter we have explored the independent one-way ANOVA. At this point, it would be good to revisit the learning objectives that we set at the beginning of the chapter.

You should now be able to:

- Understand that the term analysis of variance (ANOVA) refers to a series of tests that explores differences between groups, or across conditions. The type of ANOVA employed depends on the number of dependent and independent variables being examined and whether the data are being explored between- or within-groups.
- Recognise that we use an independent one-way ANOVA to explore differences in mean scores from a single parametric dependent variable, across three or more groups from a categorical independent variable (the test can be performed on two groups, but we generally use an independent t-test to do that).
- Appreciate that once we find a significant outcome with an independent one-way ANOVA, we need additional tests to locate the source of the difference. Planned contrasts can be used if we have made specific predictions about the outcomes between each group; *post hoc* tests must be applied in all other cases.
- Understand that the data should be interval or ratio, and be reasonably normally distributed. There should also be homogeneity of variances between the groups. If we have reason to doubt whether the data are parametric, we might need to examine outcomes using Kruskal–Wallis (the non-parametric equivalent of independent one-way ANOVA). Group membership must be exclusive: no person (or case) can appear in more than one group.
- Calculate outcomes manually (using maths and equations).
- Perform analyses using SPSS, and know how to select the appropriate planned contrast or *post hoc* test.
- Examine significance of outcomes and the source of significance.
- Explore effect size and statistical power.
- Understand how to present the data and report the findings.

Research example

It might help you to see how the independent one-way ANOVA has been applied in published research. In this section you can read an overview of the following paper:

Lim, J., Kim, M., Chen, S.S. and Ryder, C.E. (2008). An empirical investigation of student achievement and satisfaction in different learning environments. *Journal of Instructional Psychology.*

35 (2): 113–119. Web link (no DOI): www.eric.ed.gov/ERICwebportal/search/detailmini.jsp?_nfpb=true&_&ERICExtsearch_SearchValue_0=EJ813314&ERICExtsearch_SearchType_0=no&accno=EJ813314

If you would like to read the entire paper you can use the web link show above.

In this research the authors examined the effect of three teaching methods (online instruction, traditional face-to-face instruction and a combination of online and traditional instruction) on undergraduate student achievement and satisfaction levels. A total of 153 students from an American university were included in the study. The students chose the form of instruction that they preferred to receive. Those selections produced three groups: online - 31 students (14 men, 17 women); face-to-face - 82 students (42 men, 40 women); and combination - 40 students (15 men, 25 women). There were age differences between the groups (although we are not told the nature of those differences). However, all students were examined on course content knowledge before and after instruction. There were no significant differences between the groups prior to instruction. Student satisfaction (with the course and tutors) was measured using the Online Education Survey, while a Student Evaluation on Teaching Survey examined other aspects of student experience, including grading.

The results showed a significant difference in post-instruction student achievement across the groups: $F(2,150) = 5.60$, $p < .01$ (that's how the authors presented the outcome - see later). *Post hoc* Scheffé tests indicated that students in the combined learning and online learning groups had significantly higher achievement than the face-to-face group; there were no significant differences between the combined and online groups. A further independent one-way ANOVA showed that student satisfaction significantly differed between the instruction groups: $F(2,150) = 4.8$, $p < .05$. *Post hoc* Scheffé tests indicated that students in the combined learning experienced significantly greater satisfaction with their course than the face-to-face group; there were no significant differences between the combined and online groups. Other factors relating to satisfaction were also reported.

This paper demonstrates the use of independent one-way ANOVA in a familiar context. However, there are a number of problems with the conclusions that can be drawn from the outcome. Did you spot any of the inconsistencies? The group 'allocation' was uneven, and the gender ratio and age differed across those groups. Although we were told that there were no pre-existing differences on 'knowledge' prior to instruction, we have no way of knowing the effect of age and gender on the outcome measures. It may be that student performance (after any instruction) is better for women, and improves with age - we were not given any of those data. Although very few papers actually report homogeneity of variance, it would have been useful in this case because of the unequal group sizes. The use of Scheffé as a *post hoc* measure might be considered inappropriate. It is generally used only where there are equal groups; we know that there were not. Gabriel's or Hochberg's GF2 might have been a better choice.

Extended learning task

You will find the data set associated with this task on the website that accompanies this book (available in SPSS and Excel format). You will also find the answers there.

Following what we have learned about independent one-way ANOVA, answer these questions and conduct the analyses in SPSS and G*Power. (If you do not have SPSS, do as much as you can with the Excel spreadsheet.) The fictitious data explored the number of hours young people played computer games, according to their gender and nationality. It was predicted that boys would play for longer than girls, but no predictions were made about nationality.

Open the data set **Computer games**

1. There are two independent variables in this data set, but which one is *normally* better suited for an independent one-way ANOVA? Why is that?

2. Using that independent variable:
   a. Check for normal distribution across relationship satisfaction.
   b. Conduct an independent one-way ANOVA.
3. Describe what the SPSS output shows.
4. Explain how you accounted for homogeneity of variance.
5. Conduct appropriate additional analyses to indicate where the differences are. State the effect size and power, using G*Power.
6. Report the outcome as you would in the results section of a report.
7. Draw a bar chart to display the answer graphically.

Hint: There are unequal group sizes – how will this affect the way you analyse the outcome?

10 REPEATED-MEASURES ONE-WAY ANOVA

Learning objectives

By the end of this chapter you should be able to:

- Recognise when it is appropriate to use repeated-measures one-way ANOVA
- Understand the theory, rationale, assumptions and restrictions associated with the test
- Calculate outcomes manually (using maths and equations)
- Perform analyses using SPSS
- Examine significance of outcomes and the source of significance
- Explore effect size and statistical power
- Understand how to present the data and report the findings

What is repeated-measures one-way ANOVA?

A repeated-measures one-way ANOVA examines differences in mean (parametric) dependent variable scores across two or more (usually at least three) within-group conditions of categorical independent variable (we saw what we mean by parametric in Chapter 5). The key factor is that differences are explored across one group rather than between distinct groups (which we explored with independent one-way ANOVA in Chapter 9). Differences are assessed on how individuals differ across conditions rather than on how people (or cases) differ from each other.

Research question for repeated-measures one-way ANOVA

We can illustrate repeated-measures one-way ANOVA with a research example. When we explored the related t-test in Chapter 8, we encountered a group of researchers called CALM (Centre for Advanced Learning and Memory). They conduct studies that investigate what factors are associated with better memory recall. Part of their work focuses on whether facts are more easily remembered if they are associated with meaningful information. In Chapter 8, CALM sought explore whether more words are recalled when paired with relevant pictures. They now choose to extend the research. Will audio information aid recall still further? To explore this, CALM once again present words on a computer screen and ask participants to recall as many words as possible one hour later. The single sample is exposed to three conditions: in one condition they are presented with a word only (with no additional prompts); in another condition they are presented with a word and a relevant picture; and in a third condition they are presented with the word and relevant picture, but also hear that word being spoken. CALM predict that the more information that is presented, the greater the number of words that will be recalled.

10.1 Take a closer look
Variables for repeated-measures one-way ANOVA

Dependent variable: Number of words correctly recalled
Independent variable: Presentation: picture only, picture and word, and picture, word and sound

Theory and rationale

Identifying differences

Testing for significance in a repeated-measures one-way ANOVA is similar to what we did with the between-group version of this test. Any range of scores will usually vary for one reason or another. When we measure differences across conditions within a single group, the scores might vary due to fundamental differences between those conditions, or they may vary due to random or chance factors. For repeated-measures one-way ANOVA, the overall variance (total sum of squares) still describes how much the scores varied overall. This includes a proportion that varies *between* the participants and a portion that varies *within* each participant (the within-participant sum of squares). We focus on the latter for this test. The within-participant sum of squares needs to be partitioned to the portion that can be explained (model sum of squares) and that which cannot (residual sum of squares).

To calculate sum of square outcomes, we need to find the mean score for each condition, the grand mean (the average of all scores, regardless of condition) and the **case variance** (the variance across the conditions for each participant). The within-participant sum of squares is the case variance multiplied by the number of conditions being measured *minus* 1. The model sum of squares (the explained variance) represents the squared difference between the mean score for each condition and the grand mean, multiplied by the number of participants. The residual sum of squares (the unexplained variance, or error) is whatever is left over (within-participant sum of squares = model sum of squares + residual sum of squares). We express the sum of squares in relation to the relevant degrees of freedom (*df*; the number of values that are 'free to vary' in the calculation, while everything else is held constant – see Chapter 6). This produces the model mean square and residual mean square. The F ratio is found by dividing the model mean square by the residual mean square. This is the final expression that tells us the ratio of explained to unexplained variance. The higher the F ratio, the more likely that there is a significant difference in mean dependent variable scores between the conditions. We will use some example data from the CALM research question to show how the calculations and partitioning are undertaken manually (see Box 10.2). We can compare this to the SPSS output that we will obtain later.

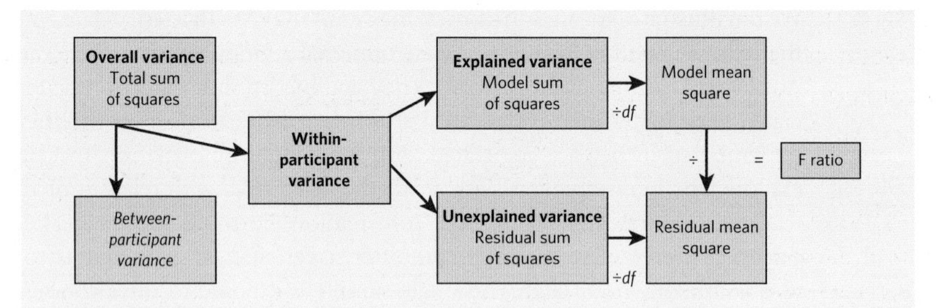

10.2 Calculating outcomes manually

Repeated-measures ANOVA calculation

To illustrate how we can calculate outcomes for repeated-measures one-way ANOVA, we will use some data that relate to the research question posed by CALM earlier. You will find a Microsoft Excel spreadsheet associated with these calculations on the web page for this book.

Table 10.1 Number of words recalled in each condition

	Word condition				
Participant	**W**	**WP**	**WPS**	**Case mean**	**Case variance**
1	62	70	82	71.33	101.33
2	63	68	68	66.33	8.33
3	65	61	72	66.00	31.00
4	68	75	88	77.00	103.00
5	69	72	80	73.67	32.33
6	71	77	80	76.00	21.00
7	78	82	87	82.33	20.33
8	75	73	79	75.67	9.33
9	70	77	82	76.33	36.33
10	71	76	84	77.00	43.00
11	60	70	77	69.00	73.00
Condition mean	**68.36**	**72.82**	**79.91**		Σ 479.00
Grand mean	**73.70**		**Grand variance**	**53.72**	

Key: W (word); WP (word and picture); WPS (word, picture and sound)

Total sum of squares (SS_T):

$SS_T = s^2_{grand}(N - 1) =$ grand variance × number of scores size *minus* 1 = 53.72 × 32 = **1718.96**

We saw how to calculate grand variance in Chapter 9 (but see MS Excel spreadsheet for this chapter).

Within-participant sum of squares (SS_W):

$SS_W =$ Case variance × number of conditions *minus* 1 = 479.00 × 2 = **958.00** *(a case is a 'participant')*

Case variance: Deduct case mean from each case score (for a single participant), square it, repeat for each case score, add these up, divide by number of conditions minus 1:

e.g. for Participant 1: $([62 - 71.33]^2 + [70 - 71.33]^2 + [70 - 71.33]^2) \div (3 - 1) = 101.33$

Degrees of freedom (*df*) for $SS_W(df_w)$ = (number of conditions *minus* 1) × no. of participants

$df_w = (3 - 1) \times 11 = 22$

Between-participant sum of squares (SS_B): $SS_T - SS_W = 760.96$ (this is not needed in final calculations)

Model sum of squares (SS_M) The formula for model sum of squares: $SS_M = \Sigma n_k (\overline{X}_k - \overline{X}_{grand})^2$

Deduct grand mean from condition mean, square it, multiply by number of participants (11)

$SS_M = 11 \times (68.36 - 73.70)^2 + 11 \times (72.82 - 73.70)^2 + 11 \times (79.91 - 73.70)^2 =$ **745.88**

df for SS_M (df_M) = (number of conditions *minus* 1) = 3 − 1 = 2 (this is the numerator *df*)

Residual sum of squares (SS_R):

Whatever is left over from within-participant sum of squares: $SS_R = SS_w - SS_M = 958.00 - 745.88 =$ **212.12**

df for SS_R (df_R) = $df_W - df_M = 22 - 2 = 20$ (this is the denominator *df*)

Mean squares:

Found by dividing model sum of squares and residual sum of squares by the relevant *df*

Model mean square (MS_M): $SS_M \div df_M = 745.88 \div 2 =$ **372.94**

Residual mean square (MS_R): $SS_R \div df_R = 212.12 \div 20 =$ **10.61**

F ratio:

$$F = \frac{MS_M}{MS_R} = 372.94 \div 10.61 = \mathbf{35.16}$$

The F ratio is compared with cut-off points in a F-distribution table (Appendix 4). The cut-off points are determined by the numerator (model) and denominator (residual) degrees of freedom (*df*), and the agreed level of significance (usually $p = 0.05$). In our example, the cut-off point for *df* (2,20) = 3.49. As our F ratio (35.16) is higher than that, we can say that there is a significant difference in the number of words recalled across the presentation conditions.

We can also use Microsoft Excel to calculate the critical value of F and to provide the actual p value. You can see how to do that on the web page for this book. In our example, $p < .001$.

Finding the source of difference

When we find a significant outcome repeated-measures one-way ANOVA, we may need additional tests to locate the source of difference, rather like we saw with independent one-way ANOVA (Chapter 9). If we have three or more conditions, the overall (omnibus) ANOVA outcome only tells us that most of the overall variance is explained by differences across the conditions. If we have two conditions only, we can consult the mean scores to explain the difference. When there are three or more conditions, we cannot rely on the mean scores to explain differences between pairs of conditions. When we explored the CALM data just now, the mean scores appeared to suggest that more words were recalled in the 'word, picture and sound' condition than in the 'word and picture' and 'word only' conditions. The mean scores also *suggested* that more words were recalled for 'word and picture' than 'word only'. But we cannot be certain without further analyses. Similar to independent one-way ANOVA, the choice of additional tests focuses on planned contrasts and *post hoc* tests. The rationale for their use is the same, although the method of performing the tests is different. Planned contrasts can be used if specific (one-tailed) predictions have been made about the relationship between each pair of conditions; otherwise, *post hoc* tests must be used. As we have seen before, these types of analyses differ in the way that they account for multiple comparisons.

Planned contrasts

We can perform planned contrasts in SPSS for repeated-measures one-way ANOVA, but there is no specific method that will allow us to pre-define those contrasts (unlike independent one-way ANOVA). There are a number of 'standard contrasts', but we can use those only if we have made specific predictions about the differences between pairs of conditions. In that scenario, it could be argued that we do not need to adjust the significance cut-off points to account for multiple comparisons. You will recall from previous chapters that the more additional tests that we run subsequent to the main analysis, the greater the likelihood that we will find a significant outcome by chance factors alone. If we have predicted outcomes about each subsequent analysis, we may be justified in keeping the significance cut-off point unadjusted (usually $p < .05$). In independent one-way ANOVA this remained true only for orthogonal planned contrasts. Such an option is available among the standard contrasts in SPSS (it is called 'Simple' contrast), but this can be employed only when there is a control group. By default, there is no control group in repeated-measures one-way ANOVA (as there is only one group). We could add a control group, but then we would need to analyse the outcome with a mixed ANOVA (see Chapter 13). The remaining planned contrasts that are available in SPSS for repeated-measures one-way ANOVA are non-orthogonal. On that basis, we probably do need to adjust for multiple comparisons. In the case of our CALM data, we have three pairs of conditions to analyse, so any one of those pairs will be significant only where $p < .016$ (because $0.05 \div 3 = 0.016$. Although there are several options of standard contrast in SPSS, the most likely one that we would use is called 'Repeated'– we will see how to request that and learn how to interpret the outcome later.

Post hoc tests

Repeated-measures *post hoc* options are less easy to find in SPSS than they are for between-group studies. We do not use the given 'post hoc' function here; this can be used only for

10.3 Take a closer look

Planned contrasts and *post hoc* tests in repeated-measures ANOVA

One of these methods must be used to determine the source of difference between pairs of conditions.

Table 10.2 Methods for finding the source of difference in repeated-measures one-way ANOVA

Method	Comments
Planned contrasts	Can be used if specific predictions were made about the outcome on pairs of conditions. Only 'standard' contrasts available in SPSS.
	Repeated contrast option most appropriate in this context, but probably still need to account for multiple comparisons.
***Post hoc* tests**	Chosen via 'estimated marginal means' adjustments, rather than specific *post hoc* option.
LSD	No option at all, given that no adjustment is made for multiple comparisons.
Bonferroni	The favoured *post hoc* test – makes appropriate adjustments.
Sidak	Also makes adjustments, but probably too conservative.

between-group portions of mixed ANOVAs (see Chapter 13). Instead, we 'adjust' the confidence intervals via the Options facility (we will see how to do that later). Also, there are fewer *post hoc* tests options (there is no option for Tukey or Scheffé, for example). We have just three options: LSD, Bonferroni and Sidak. The LSD (least squares difference) option is really no choice at all, as it offers no adjustment for multiple comparisons. The Bonferroni option is probably the most powerful and it automatically adjusts for multiple comparisons. The Sidak does that too, but some feel that it is somewhat conservative. Alternatively, we could perform separate related t-tests for each pair of conditions and then manually adjust for multiple comparisons.

Assumptions and restrictions

As usual, there are a number of assumptions that we need to meet before we can perform this test. All of the conditions on the independent variable must be measured across one group. Furthermore, each person (or case) must be present in all conditions – the outcome of the test depends on comparing people (or cases) across each condition. If any person is missing at any time point, all of that person's data should be excluded. There are exceptions, particularly in clinical trials (you can read about that in Box 10.7). The dependent variable scores should be parametric, ideally with interval or ratio scores that are *reasonably* normally distributed. It is argued that self-rated rating scores (such as Likert scales) are subjective, so should be considered as ordinal data. However, that argument is perhaps less convincing for within-group analyses. Outcomes are based on how scores change across conditions within each participant. Any subjectivity is confined to each case and has no effect on the outcome for other participants. On that basis, it could be said that such scores are more acceptable in repeated-measures ANOVAs.

Sphericity of within-group variance

In previous chapters we have encountered the term 'homogeneity (or equality) of variance'. This is an important measurement that examines how groups (or conditions) vary in the extent to which scores vary either side of respective means. We must measure this in all between-group studies (see Chapter 9), but we also need to explore that in within-group studies where there are three or more conditions, such as we find with repeated-measures ANOVA. (We do not measure equality of variances where there are only two conditions.) Within-group equality of variance is measured by something called 'sphericity'. The problem is illustrated with example data in Box 10.4.

Mauchly's test of sphericity

The assessment we made in Box 10.4 is a little subjective – we must use statistical analyses to examine whether the within-group variances differ significantly between pairs of conditions. **Mauchly's test** provides a chi-square (χ^2) score that determines whether the variances are significantly *different* to each other. We do not want that, so we need that χ^2 score to be small and non-significant ($p > .05$). Mauchly's test is produced automatically in SPSS for all repeated-measures ANOVA tests. The outcome shows the χ^2 outcome and significance. If that outcome is non-significant, we can say that sphericity is assumed. SPSS will show this outcome in terms of the Mauchly's value and whether that is significant. However, we need to read that outcome only if there are three or more conditions. If we have only two conditions, we assume sphericity automatically. This is important because the Mauchly's test output in SPSS for two condition outcomes can be misread (see Box 10.9).

As we will see later, when we perform repeated-measures one-way ANOVA in SPSS, we are presented with four lines of outcome, each reporting potentially different F ratio and

10.4 Nuts and bolts

Sphericity of within-group variance

To illustrate the need to examine whether there is sphericity between pairs of within-group conditions, we will revisit the CALM data that we explored manually earlier. We were investigating the number of words recalled in word (W), word and picture (WP), and word, picture and sound (WPS) conditions.

Table 10.3 Number of words recalled in each condition

	W	WP	WPS	$A - \bar{A}$	$B - \bar{B}$	$C - \bar{C}$
Participant	A	B	C	a	b	c
1	62	70	82	−6.36	−2.82	2.09
2	63	68	68	−5.36	−4.82	−11.91
3	65	61	72	−3.36	−11.82	−7.91
4	68	75	88	−0.36	2.18	8.09
5	69	72	80	0.64	−0.82	0.09
6	71	77	80	2.64	4.18	0.09
7	78	82	87	9.64	9.18	7.09
8	75	73	79	6.64	0.18	−0.91
9	70	77	82	1.64	4.18	2.09
10	71	76	84	2.64	3.18	4.09
11	60	70	77	−8.36	−2.82	−2.91
Mean	68.4	72.8	79.9			
	Variance			30.45	31.36	35.49

The key information is shown in columns a, b, and c. These indicate the variance for each condition (we saw how to calculate variance in Chapter 4). Sphericity is measured comparing the variance across pairs of conditions ('ab', 'ac' and 'bc'). We do not want the variance across the pairs to differ significantly. They look fairly similar here, but we need formal statistical calculations to confirm that via SPSS.

significance results. It is important that we choose the appropriate one. If Mauchly's test is non-significant (or if we have only two conditions), we can refer to the line that reads 'Sphericity Assumed'. If Mauchly's outcome is significant, sphericity cannot be assumed, so we must choose from one of the other three lines: **Greenhouse-Geisser**, **Huynh-Feldt** or '**Lower-bound**'. Each of those remaining options adjusts the F ratio to account for the lack of sphericity, but each does so in a different way. As usual, there is some debate about which option should be selected. All of the outcomes report something called **epsilon** (ε). If sphericity is assumed (ε) = 1. When sphericity has been violated the F ratio is adjusted based on the magnitude of ε. This outcome will vary between 1 and $1 \div k$ (where k is the number of conditions). If there are three conditions, 'ε' will range between 1 and 0.33. The closer the result is to 1, the more equal the variance is assumed to be. So which one should we choose? Well, most researchers choose Greenhouse–Geisser or Huynh–Feldt – there is little to choose between the arguments, so either is fine. However, Field (2009) recommends taking an average of the p values for Greenhouse–Geisser and Huynh–Feldt where there is a large difference between their outcomes (there usually is not much difference).

Mauchly's outcome with two conditions

Most of the time, the independent variable being analysed in repeated-measures one-way ANOVA will have three or more conditions; Mauchly's test works just fine here. However, occasionally, there may be only two conditions. In that case, we will still get a Mauchly's outcome, but it may look a little odd. It will confirm a maximum sphericity of 1, but the significance will show just a dot (see Figure 10.12). This simply means that the significance is also 1 (definitely not significant). Actually, because there were only two conditions, you can ignore Mauchly's altogether.

The CALM data that we will examine in SPSS using repeated-measures one-way ANOVA has three conditions. However, it is worth pursuing a little further why sphericity matters only when there are three or more conditions. When we assess within-group variance we take account of something called the **variance-covariance matrix**. Variance measures how much values vary either side of the mean score in a single condition. **Covariance** measures the extent that these values vary in pairs of conditions (it is also an integral part of calculating the correlation between two variables – see Chapter 6). Box 10.5 shows how we calculate the covariance in our CALM data.

10.5 Nuts and bolts
Covariance

To calculate covariance we need to assess how each condition score varies to the condition mean. Then we multiply those outcomes across each pair of conditions. You will recall that we had three conditions: word only (W), word and picture (WP), and word, picture and sound (WPS).

Table 10.4 Number of words recalled in each condition

	W	WP	WPS	$A - \bar{A}$	$B - \bar{B}$	$C - \bar{C}$	Covariance		
Participant	A	B	C	a	b	c	ab	ac	bc
1	62	70	82	−6.36	−2.82	2.09	17.93	−13.31	−5.89
2	63	68	68	−5.36	−4.82	−11.91	25.84	63.88	57.38
3	65	61	72	−3.36	−11.82	−7.91	39.75	26.60	93.47
4	68	75	88	−0.36	2.18	8.09	−0.79	−2.94	17.65
5	69	72	80	0.64	−0.82	0.09	−0.52	0.06	−0.07
6	71	77	80	2.64	4.18	0.09	11.02	0.24	0.38
7	78	82	87	9.64	9.18	7.09	88.48	68.33	65.11
8	75	73	79	6.64	0.18	−0.91	1.21	−6.03	−0.17
9	70	77	82	1.64	4.18	2.09	6.84	3.42	8.74
10	71	76	84	2.64	3.18	4.09	8.39	10.79	13.02
11	60	70	77	−8.36	−2.82	−2.91	23.57	24.33	8.20
Mean	68.36	72.82	79.91	0.00	0.00	0.00	20.16	15.94	23.44
Variance	30.45	31.36	35.49	30.45	31.36	35.49			

Compare this outcome to the two-condition example we explored with a related t-test (Chapter 8):

Table 10.5 Number of words recalled in each condition

	W	WP	A − $\bar{A}$	B − $\bar{B}$	Covar
Participant	A	B	a	b	ab
1	14	19	−4.27	−3.18	13.60
2	16	16	−2.27	−6.18	14.05
3	21	23	2.73	0.82	2.23
4	19	26	0.73	3.82	2.78
5	23	30	4.73	7.82	36.96
6	16	26	−2.27	3.82	−8.68
7	26	23	7.73	0.82	6.32
8	12	16	−6.27	−6.18	38.78
9	14	14	−4.27	−8.18	34.96
10	21	23	2.73	0.82	2.23
11	19	28	0.73	5.82	4.23
12	23	18	4.73	−4.18	−19.77
Mean	18.27	22.18	0.00	0.00	13.40
Variance	18.61	26.88	18.61	26.88	

30.45	20.16	15.94
20.16	**31.36**	23.44
15.94	23.44	**35.49**

Figure 10.1a 3-condition variance-covariance matrix

4.31	10.78
10.78	**5.18**

Figure 10.1b 2-condition variance-covariance matrix

Figure 10.1a shows the variance for the three conditions down the diagonal (bold font), with the covariance shown in the off-diagonals. These are repeated either side of the diagonals, so we need to refer to only one of these. The shaded blocks show that there may be some differences in the covariance (Mauchly's test examines whether these are significantly different). Figure 10.1b shows the variance for two conditions (bold), but there is only one covariance outcome. Although there may be some apparent difference in the variance of those within-group conditions, there is nothing to compare the covariance data to, so sphericity must be assumed.

10.6 Take a closer look
Summary of assumptions and restrictions

- IV must be categorical, with at least two conditions (usually three or more) measured across one group:
  - Each person (or case) must be present in all conditions
- There must be one numerical dependent variable:
  - DV data should be interval or ratio, and reasonably normally distributed
- If these assumptions are not met, the non-parametric Friedman test could be considered
- We need to account for sphericity of within-group variances

10.7 Nuts and bolts

Clinical trials: last observation carried forward

A key assumption of repeated-measures ANOVAs is that every person must be present in all conditions. Longitudinal clinical trials often have several follow-up points to measure clinical effectiveness. Occasionally, a patient drops out of the study. Their data are important and should not be lost. As a result, some studies will employ a technique called 'last observation carried forward (LOCF)'. In this case, the last score taken for the patient is copied across the remaining time points. LOCF might seem an attractive solution, but there are problems. We cannot be confident that we are reporting a true mean, as we cannot know for sure how the patient would have been likely to progress. Table 10.6 shows some fictitious data from an antidepressant trial, where the mean scores represent depression severity rating.

Table 10.6 Mean depression severity scores

	Follow-up week					
Patient #	4	6	12	16	26	52
1	58	46	36	18	22	14
2	50	41	30	22	**15**	15
3	49	40	33	23	15	11
4	50	38	35	**12**	12	12
5	59	43	**38**	38	38	38
6	53	45	40	35	30	27

Three patients dropped out of the study. Their final recorded score is shown in bold, followed by LOCF scores shown in red. The impact might well be different for each one. For Patient 2, we already have most of their data, so the final missing time point might not severely affect outcomes. However, Patient 4's depression severity scores dropped dramatically at week 16, and then they withdrew from the study. It would appear that this patient responded very well, but we cannot be sure that the week 16 score was not just an unusual response; there is nothing to say that scores might not increase again. In this case, we might be understating the mean score. Patient 5 dropped out quite early (at week 12), so the last score we have for them is 38. Carrying their relatively high score forward might not reflect the general trend for scores to fall across the time points. In this case, we might be overstating the mean score.

How SPSS performs repeated-measures one-way ANOVA

To show how we can perform repeated-measures one-way ANOVA in SPSS, we will use the CALM data that we examined manually earlier. We are investigating whether there is a significant difference in the number of words that are recalled by a single group of 11 people, who experience three presentation conditions (word only, word and picture, and word, picture and sound). We can be confident that a count of words recalled represents interval data, so that part of parametric testing is OK However, we do not know whether the data are normally distributed – we will look at that shortly.

10.8 Nuts and bolts

Setting up the data set in SPSS

When we create the SPSS data set for repeated-measures one-way ANOVA (with three conditions), we need to set up three columns: each column represents the dependent variable score for that condition (the scores are 'continuous' - see Chapter 5 for more information on data types).

Word recall 2.sav [DataSet1] - IBM SPSS Statistics Data Editor

File Edit View Data Transform Analyze Graphs Utilities Add-ons Window Help

	Name	Type	Width	Decimals	Label	Values	Missing	Columns	Align	Measure
1	word	Numeric	8	0	Word only	None	None	8	Right	Scale
2	wordpic	Numeric	8	0	Word & picture	None	None	8	Right	Scale
3	wordpicsound	Numeric	8	0	Word, picture &...	None	None	8	Right	Scale

Figure 10.2 Variable View for 'Word recall 2' data

Figure 10.2 shows how the SPSS Variable View should be set up. The variables are 'word' (for the word only condition), 'wordpic' (for word and picture) and 'wordpicsound' (for word, picture and sound). These represent the three within-group conditions for the independent variable. Continuous dependent variable scores will be recorded here; they will be the 'dependent variable scores at each condition'. We do not need to adjust anything in the Values column; the Measure column is set to Scale.

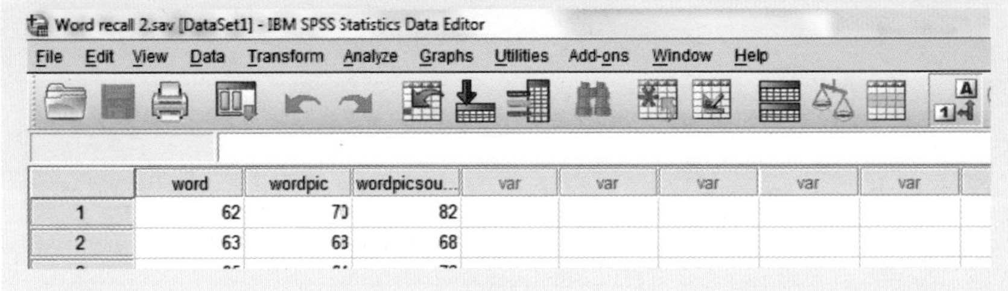

Word recall 2.sav [DataSet1] - IBM SPSS Statistics Data Editor

File Edit View Data Transform Analyze Graphs Utilities Add-ons Window Help

	word	wordpic	wordpicsou...	var	var	var	var	var
1	62	70	82					
2	63	63	68					

Figure 10.3 Data View for 'Word recall 2' data

Figure 10.3 illustrates how this will appear in the Data View. Each row represents a participant. Dependent variable scores (number of words recalled) will be entered for each participant in respect of each condition.

Testing for normal distribution

To examine normal distribution, we start by running the Kolmogorov–Smirnov/Shapiro–Wilk tests (we saw how to do this for within-group studies in Chapter 8). If the outcome indicates that the data may not be normally distributed, we could additionally run z-score analyses of skew and kurtosis, or look to 'transform' the scores (see Chapter 3). We will not repeat the instruction for performing KS/SW tests in SPSS (but refer to Chapter 3 for guidance). However, we should look at the outcome from those analyses (see Figure 10.4).

	Kolmogorov–Smirnov[a]			Shapiro–Wilk		
	Statistic	df	Sig.	Statistic	df	Sig.
Word only	.135	11	.200*	.971	11	.898
Word & picture	.137	11	.200*	.965	11	.830
Word, picture & sound	.167	11	.200*	.946	11	.595

Figure 10.4 Kolmogorov-Smirnov/Shapiro-Wilk test for recall by condition

Because we have a sample size of 11 (for each condition), we should refer to the SW outcome. Figure 10.4 suggests that the data are normally distributed at each condition: word only W (11), = .971, p = .898; word and picture W (11), = .965, p = .830; and word, picture and sound W (11), = .946, p = .595 (because 'Sig.' outcomes are greater than .05).

Running repeated-measures one-way ANOVA in SPSS

Using the SPSS file Word recall 2

Select **Analyze → General Linear Model → Repeated Measures...** as shown in Figure 10.5

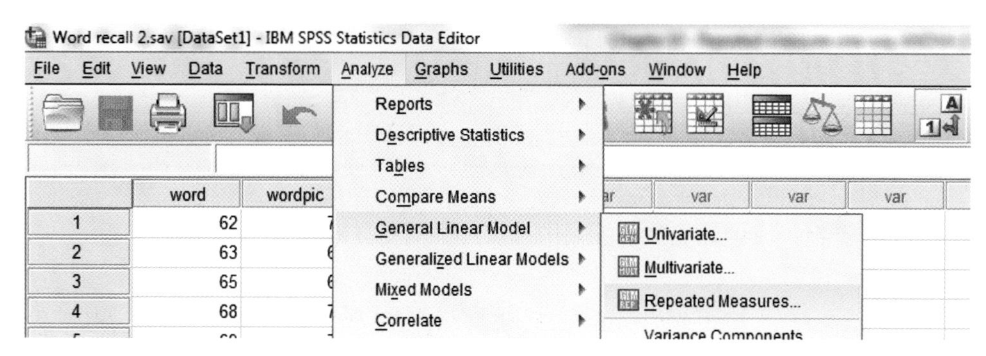

Figure 10.5 Repeated-measures one-way ANOVA: procedure 1

In new window (see Figure 10.6), Type **Recall** in **Within-Subject Factor Name** box → type **3** for **Number of Levels** → click on **Add** → click on **Define**

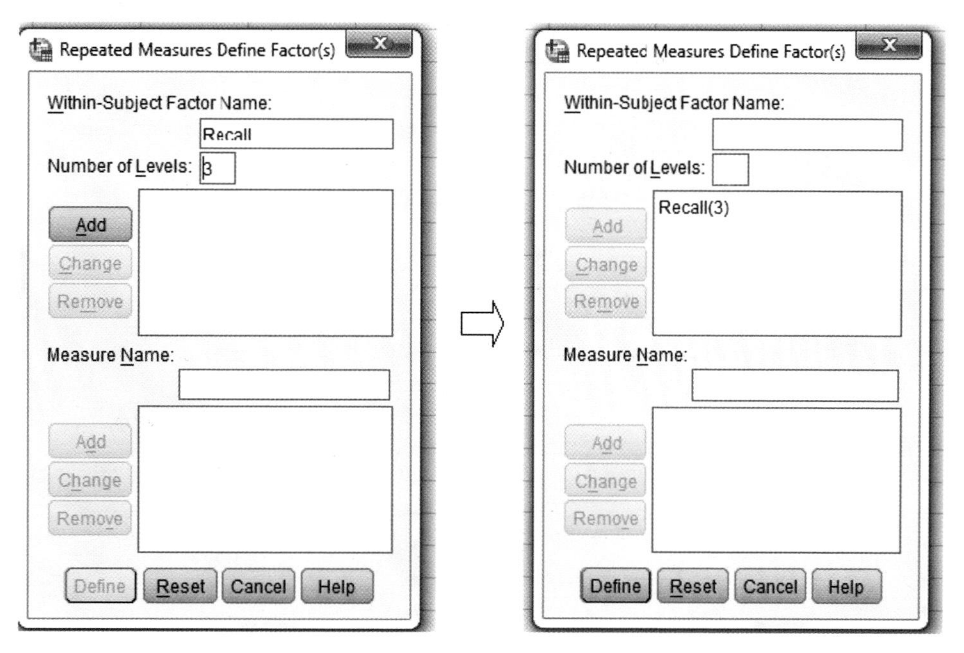

Figure 10.6 Repeated-measures one-way ANOVA: procedure 2

In next window (see Figure 10.7), transfer **Word only**, **Word & picture**, and **Word, picture & sound to Within-Subjects Variables** window - in that same order - (by clicking on arrow, or by dragging the variables to that window) → click **Options...** button

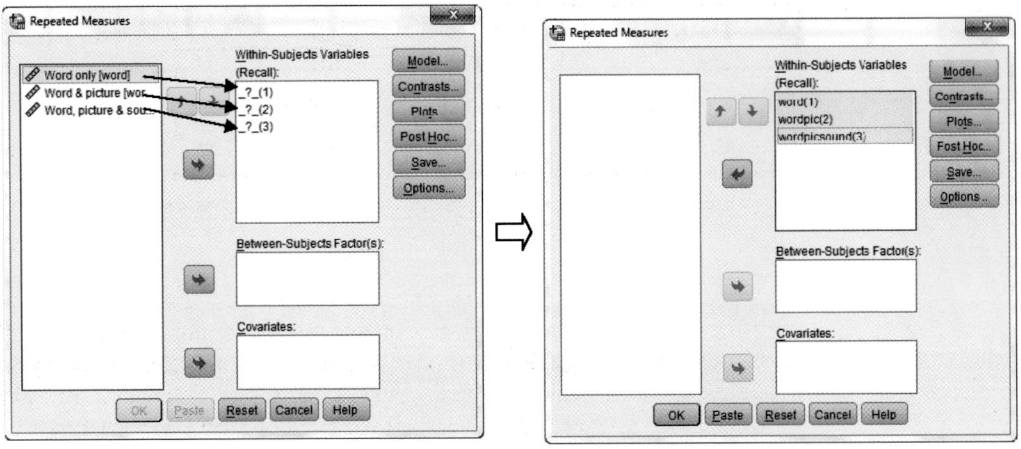

Figure 10.7 Repeated-measures one-way ANOVA: procedure 3

This is where we set up the options for *post hoc* test for repeated-measures one-way ANOVA. We do not use the Post Hoc button; this is for between-group analyses only.

In the Options window (see Figure 10.8), transfer **Recall** to **Display Means for** window → the options (for adjusting confidence intervals) now become visible → tick box for **Compare main effects** → click pull-down arrow (under **Confidence interval adjustment**) → select **Bonferroni** (we looked at why we choose that one earlier) → tick **Descriptive statistics** and **Estimates of effect size** boxes (under **Display**) → click **Continue** → click **OK**

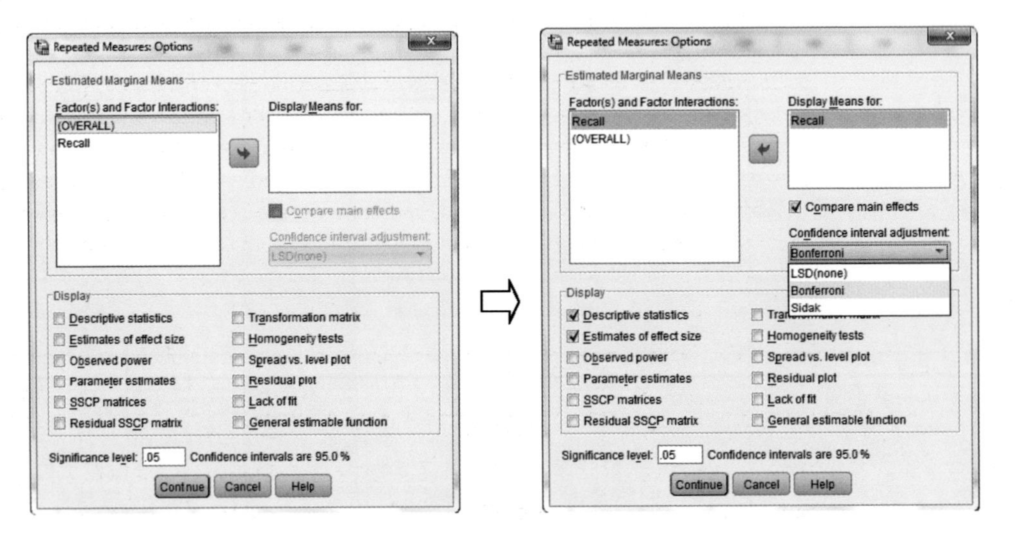

Figure 10.8 Repeated one-way ANOVA: *post hoc* and options selections

Interpretation of output

Figure 10.9 presents the mean scores and standard deviation for the numbers of words recalled in each condition It would appear that highest recall is in the 'word, picture and sound' condition, but we need to check if this outcome is significant.

	Mean	Std. deviation	N
Word only	68.36	5.519	11
Word & picture	72.82	5.600	11
Word, picture & sound	79.91	5.957	11

Figure 10.9 Descriptives for repeated-measures one-way ANOVA

Figure 10.10 shows the outcome for sphericity of within-group variances. We can be confident that sphericity can be assumed as there was a non-significant outcome ('Sig.' [p] > .05 – we explored the rationale behind this earlier). We will be able to select the 'Sphericity Assumed' line of data when we examine the main ANOVA outcome. If Mauchly's test had been significant we would need to choose one of the sphericity adjustment options (shown in the final three columns of Figure 10.10).

Within subjects effect	Mauchly's W	Approx. chi-square	df	Sig.	Epsilon[a]		
					Greenhouse–Geisser	Huynh–Feldt	Lower bound
Recall	.789	2.132	2	.344	.826	.968	.500

Figure 10.10 Mauchly's Test of Sphericity

Figure 10.11 presents the statistical outcome. We need to refer to the first block of data (labelled Recall in this instance) and choose one of the four rows to determine outcome. Notice how 'F' remains the same throughout, while the *df* and Mean Square data change. This is because the various methods make an adjustment when calculating the F ratio. Since we demonstrated that sphericity was assumed, we can select that line. There is a significant difference in the number of items recalled across the conditions: $F(2,20) = 35.163$, $p < .001$ (highlighted in red). We can compare that line of outcome with what we calculated manually in Box 10.2. However, the omnibus ANOVA outcome tells us only that there is a significant difference in the mean number of words recalled across the conditions; we need additional tests to examine differences between pairs of conditions. The final column provides the **partial eta squared** (η^2 highlighted in orange). We do not need that yet, but will do when we come to calculate the effect size later.

Source		Type III sum of squares	df	Mean square	F	Sig.	Partial eta squared
Recall	Sphericity assumed	745.879	2	372.939	35.163	.000	.779
	Greenhouse–Geisser	745.879	1.652	451.608	35.163	.000	.779
	Huynh–Feldt	745.879	1.937	385.144	35.163	.000	.779
	Lower bound	745.879	1.000	745.879	35.163	.000	.779
Error (recall)	Sphericity assumed	212.121	20	10.606			
	Greenhouse–Geisser	212.121	16.516	12.843			
	Huynh–Feldt	212.121	19.366	10.953			
	Lower bound	212.121	10.000	21.212			

Figure 10.11 Tests of within-subjects effects

10.9 Nuts and bolts

What does Mauchly's outcome look like when there are only two conditions?

Earlier, we said that we need to consult Mauchly's outcome only when we have three or more conditions. We do not need this test when there are only two conditions because there are no between-pair differences to explore. If we were to run a repeated-measures one-way ANOVA where there were only two conditions, we would see something like the output shown in Figure 10.12.

Within subjects effect	Mauchly's W	Approx. chi-square	df	Sig.	Epsilon[a]		
					Greenhouse–Geisser	Huynh–Feldt	Lower bound
Recall	1.000	.000	0	.	1.000	1.000	1.000

Figure 10.12 Mauchly's outcome for two conditions

Sphericity is confirmed as 1.000 (maximum), which is highly non-significant. The epsilon adjustments make no difference to the outcome.

Locating source of difference

Because we have three conditions, we need additional tests to locate the source of difference. We saw a number of methods to do this earlier. For within-group studies, it is probably better to employ Bonferroni *post hoc* tests as a matter of course (due to restrictions on other methods – see earlier). We run these *post hoc* tests only if we have a significant main outcome, and if we have three or more conditions (if there are just two conditions, we can use the mean scores to illustrate where the differences are).

(I) Recall	(J) Recall	Mean difference (I-J)	Std. error	Sig.[a]	95% confidence interval for difference[a]	
					Lower bound	Upper bound
1	2	−4.455*	1.260	.016	−8.072	−.837
	3	−11.545*	1.675	.000	−16.354	−6.737
2	1	4.455*	1.260	.016	.837	8.072
	3	−7.091*	1.179	.000	−10.475	−3.707
3	1	11.545*	1.675	.000	6.737	16.354
	2	7.091*	1.179	.000	3.707	10.475

Figure 10.13 *Post hoc* data (pairwise comparisons)

Figure 10.13 presents those data. We are looking for instances where the significance outcome ('Sig.') is less than .05 (highlighted in red font here). This is also confirmed by an asterisk next to the number in the 'Mean Difference' column. The output produced by SPSS can be a little confusing; labelling the conditions as numbers does not help. To help interpret the condition numbers, we need to refer to the table shown in Figure 10.14.

Recall	Dependent variable
1	word
2	wordpic
3	wordpicsound

Figure 10.14 Condition numbers

Using the data from Figure 10.13 (with assistance from Figure 10.14) we can see that recall is significantly higher in the 'Word & picture condition' than 'Word only' ($p = .016$), and that recall is significantly higher in the 'Word, picture & sound' condition than 'Word only' ($p < .001$) and 'Word & picture' ($p < .001$). You probably would have noticed that the differences are shown twice (Condition 3 > Condition 1 and Condition 1 < Condition 3); we need only one of these. Remember, although SPSS presents the outcome as .000, we must show this as < .001.

Linear vs. quadratic outcome

Source	Recall	Type III sum of squares	df	Mean square	F	Sig.	Partial eta squared
Recall	Linear	733.136	1	733.136	47.494	.000	.826
	Quadratic	12.742	1	12.742	2.206	.168	.181
Error (recall)	Linear	154.364	10	15.436			
	Quadratic	57.758	10	5.776			

Figure 10.15 Tests of (polynomial) within-subjects contrasts

Figure 10.15 shows whether the data fit a linear or quadratic trend when plotted on a graph. A significant linear trend suggests that the scores change in a straight line, which is what happened here (highlighted in red). That makes sense, because word recall has increased incrementally through the conditions (where additional prompts were added to aid recall). A significant quadratic line would be presented as a 'U' shape on a graph – this might happen if scores drop from the first time point to the second, but increase from the second to the third. Note that we will get this output in SPSS by default if we leave the Contrast option unchanged (at polynomial). If we select another option (such as 'Repeated') we will get a different output (see Figure 10.18).

Planned contrasts

As we saw earlier, we can run planned contrasts only if we have made specific predictions about the outcome between pairs of conditions. It could be argued that CALM did state a one-tailed hypothesis in their research example by saying 'the more information that is presented, the greater the number of words that will be recalled'. That being the case, perhaps we should look at some planned contrast outcome to see what that might tell us about between-pair differences. To do this, we follow the same procedure as we undertook just now, but also select the 'Contrasts' button (we will choose the 'repeated' option (for the reasons that we stated earlier). Although there was a specific prediction, none of the conditions represents a control group. Therefore, strictly speaking, we have a non-orthogonal planned contrast and should adjust for multiple comparisons.

Using the SPSS file Word recall 2

Select **Analyze → General Linear Model → Repeated Measures... →** (the variable settings will still be there from what you did just now - if not, follow the instructions from earlier) → click **Contrasts... →** (in new window, see Figure 10.16), click pull-down arrow by **Contrast: Polynomial →** select **Repeated →** click **Change** button → click **Continue →** click **OK**

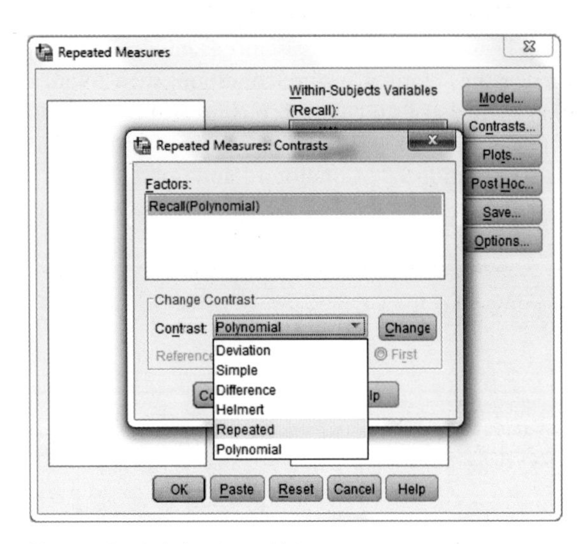

Figure 10.16 Selecting within-group contrasts

We encountered contrast coding in Chapter 9 and the principle is the same here. Comparison pairs are indicated by the codes 1 and −1, while the remaining condition is ignored. Figure 10.17 shows that in the first contrast, the pair 'word only' and 'word and picture' are compared. In the second contrast, 'word and picture' and 'word, picture and sound' are compared. Curiously, 'word only' and 'word, picture and sound' are not compared.

Dependent variable	Recall	
	Level 1 vs. Level 2	Level 2 vs. Level 3
Word only	1	0
Word & picture	−1	1
Word, picture & sound	0	-1

Figure 10.17 Contrast value coding

Figure 10.18 suggests that difference in mean word recall appears to be significantly different between conditions for both pairs. However, we need to adjust for multiple comparisons. We have three pairs of conditions (despite what SPSS is showing here), so we must divide the significance cut-off point accordingly (.05 ÷ 3 = .016). Subsequent to that adjustment, the outcome is still significant, but we are interpreting it appropriately. Equally, we could multiply the given significance by three to get a clearer picture. If we do that for the pair 'Level 1 vs. Level 2' we would get a revised significance of .015. Compare that to the Bonferroni *post hoc* outcome in Figure 10.15 – the significance levels are now very similar. Therefore, it is probably

Source	Recall	Type III sum of squares	df	Mean square	F	Sig.	Partial eta squared
Recall	Level 1 vs. Level 2	218.273	1	218.273	12.492	.005	.555
	Level 2 vs. Level 3	553.091	1	553.091	36.171	.000	.783
Error (recall)	Level 1 vs. Level 2	174.727	10	17.473			
	Level 2 vs. Level 3	152.909	10	15.291			

Figure 10.18 Tests of (repeated) within-subjects contrasts

more efficient to simply report the Bonferroni outcome and not bother with planned contrasts (especially since we are given all of the contrasts in any case).

Effect size and power

You may recall from Chapter 4 that we can use G*Power to calculate effect size for us (based on Cohen's *d* formula) and to show how much power our study had.

Open G*Power:

From **Test family** select **F tests**
From **Statistical test** select **ANOVA: Repeated measures, within factors** From **Type of power analysis** select **Post hoc: Compute achieved power - given α, sample size and effect size power**

You will get a new window that looks rather different to the previous examples (see Figure 10.19).

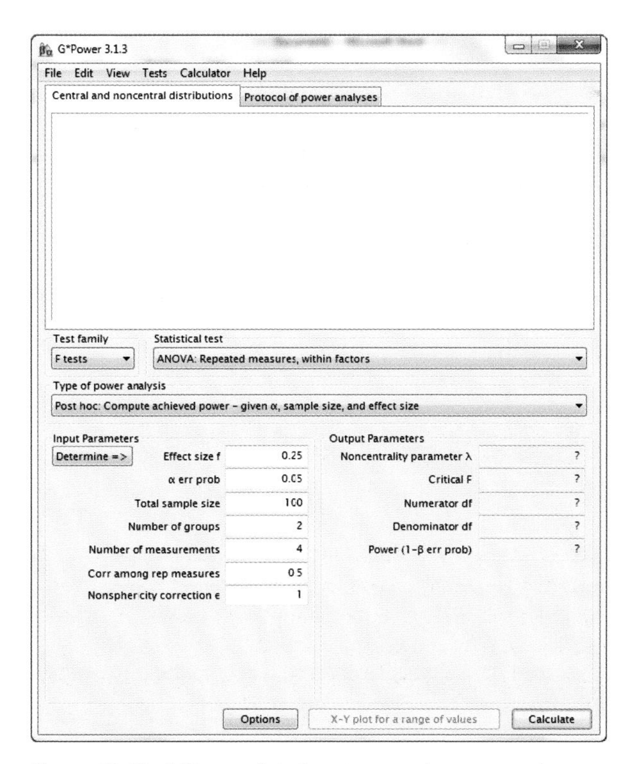

Figure 10.19 G*Power data input screen for repeated-measures one-way ANOVA

Before we proceed with entering the data, we need to make one further calculation. Most of the information we need can be gleaned from the repeated-measures analysis. However, we were not given anything for 'correlation between repeated measures'. The only way we can get that is to run a correlation between the three conditions, treating them like variables. We saw how to perform correlation in SPSS in Chapter 6. The output for this is shown in Figure 10.20.

		Word only	Word & picture	Word, picture & sound
Word only	Pearson correlation	1	.717*	.533
	Sig. (2-tailed)		.013	.091
	N	11	11	11
Word & picture	Pearson correlation	.717*	1	.773**
	Sig. (2-tailed)	.013		.005
	N	11	11	11
Word, picture & sound	Pearson correlation	.533	.773**	1
	Sig. (2-tailed)	.091	.005	
	N	11	11	11

Figure 10.20 Correlation between repeated-measures conditions

To find the overall estimate of correlation within repeated-measures we can find the average of the correlation co-efficients shown in Figure 10.20 (so, $[.717 + .533 + .773] \div 3 = .675$).

Now we enter the **Input Parameters**:

For **α err prob** type **0.05** (the significance level) → for **Total sample size** type **11** → **Number of groups** select 1 (we had one group of participants) → for Repetitions type **3** (there were 3 conditions for the IV) → for **Corr among rep measures** type **0.675** (we saw why just now) → for **Nonsphericity correction ϵ** type **1** (because there was no correction; had we needed to use one of the non-sphericity corrections, such as Huynh-Feldt, we would use the epsilon figure from Figure 10.10).

To calculate the **Effect size**, click on the **Determine** button (a new box appears).

In that new box, click on the radio button for **Direct**; in **Partial η^2** type **0.779** (this is the eta-squared parameter from Figure 10.11, highlighted in orange) → Click on **Calculate and transfer to main window**

Click on **Calculate**

From this, we can observe two outcomes: Effect size (r) **1.88** (which is very strong); and Power (1−β err prob) **1.00** (which is excellent - clearly above the target of .80).

Writing up results

We can present the mean data through tables and report the outcome in words, presenting the statistical notation (Table 10.7).

Table 10.7 Mean number of words recalled and standard deviation (SD) across prompt conditions

Recall	Mean	SD
Word only	68.36	5.52
Word & picture	72.82	5.60
Word, picture & sound	79.91	5.96

Once you have presented tabulated data, you should report the statistics and significance in the following format, using the statistical data from Figures 10.9 and 10.11 and the G*Power data. You do not tabulate the *post hoc* data, but you must report it in the narrative statistics:

> A repeated-measures one-way ANOVA indicated that there was a significant difference in the number of words recalled according to the type of memory prompt given, $F(2, 20) = 35.163$, $p < .001$. A *post hoc* Bonferroni analysis indicated that significantly more words were recalled in the 'word & picture' condition than word only ($p = .016$), and significantly more words recalled in the 'word, picture and sound' condition than 'word & picture ($p < .001$) and 'word only' ($p < .001$). This was represented by a very strong effect, $d = 1.88$.

Presenting data graphically

You could also provide a graph for this result. Given the way that these scores change over time, it may be better to present the data in **line graphs**, with error bars to indicate overlap between scores:

> Select **Graphs → Legacy Dialogs → Line...** as shown in Figure 10.21

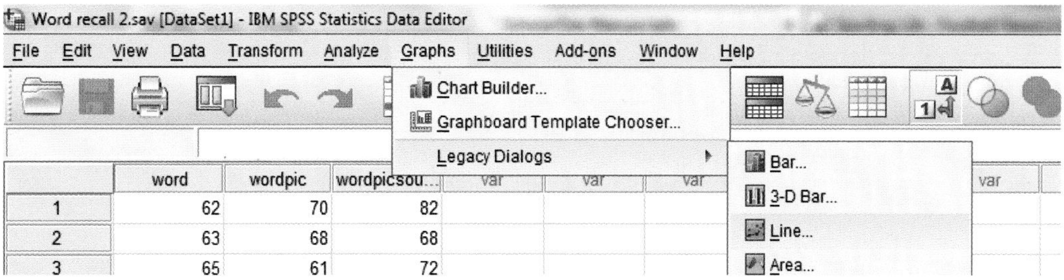

Figure 10.21 Creating a line chart – step 1

> In new window (see Figure 10.22), select **Simple** box → tick **Summaries for separate variables** radio button → click **Define**

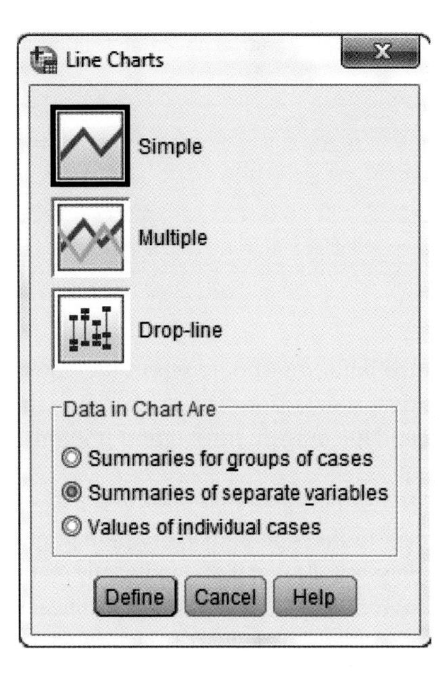

Figure 10.22 Creating a line chart – step 2

In next window (see Figure 10.23), transfer **Word, Word & picture**, and **Word, picture & sound** to **Line Represents** window ➜ click **Options**

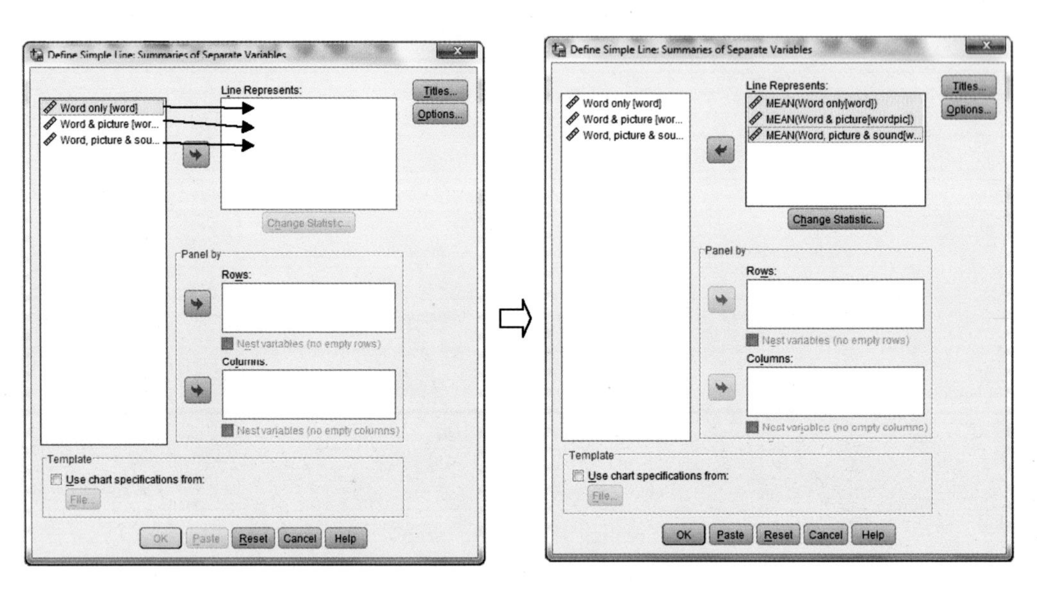

Figure 10.23 Creating a line chart – step 3

In next window (see Figure 10.24), tick box for **Display error bars** ➜ select **Confidence intervals** radio button ➜ ensure **Level (%)** is set to **95** ➜ click **Continue** ➜ click **OK**

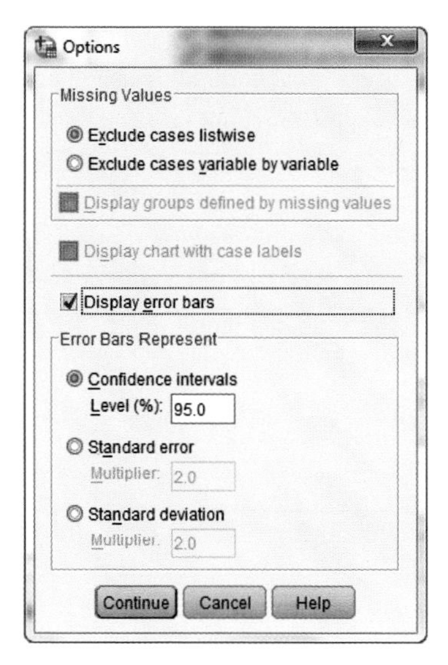

Figure 10.24 Options

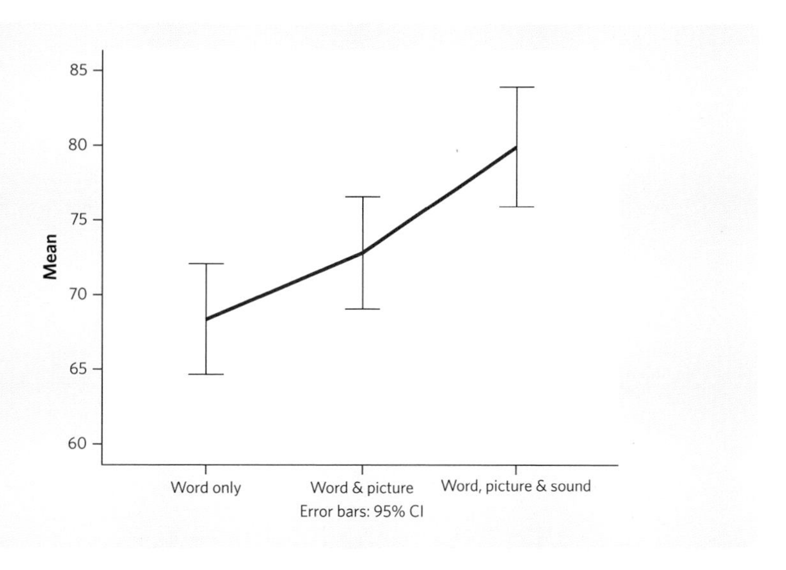

Figure 10.25 Completed line graph

Chapter summary

In this chapter we have explored the repeated-measures one-way ANOVA. At this point, it would be good to revisit the learning objectives that we set at the beginning of the chapter.

You should now be able to:

- Recognise that we use repeated-measures one-way ANOVA to explore differences in mean scores from a single parametric dependent variable, across three or more within-group conditions from a categorical independent variable (the test can be performed with two conditions, but we generally use a related t-test to do that).

- Appreciate that once we find a significant outcome, we need additional tests to locate the source of the difference. Bonferroni *post hoc* tests are most commonly used to perform those analyses.
- Understand that the data should be interval or ratio, and be reasonably normally distributed. If we have reason to doubt whether the data are parametric, we might need to examine outcomes using Friedman's ANOVA (the non-parametric equivalent of repeated-measures one-way ANOVA). We need to account for sphericity of within-group variances; these determine how we interpret the ANOVA outcome. Every person (or case) must appear in all conditions.
- Calculate outcomes manually (using maths and equations).
- Perform analyses using SPSS, including *post hoc* tests where appropriate.
- Examine significance of outcomes and the source of significance.
- Explore effect size and statistical power.
- Understand how to present the data and report the findings.

Research example

It might help you to see how repeated-measures one-way ANOVA has been applied in published research. In this section you can read an overview of the following paper:

> **Bernstein, G.A., Carroll, M.E., Dean, N.W., Crosby, R.D., Perwien, A.R. and Benowitz, N.L. (1998). Caffeine withdrawal in normal school-age children.** ***Journal of the American Academy of Child & Adolescent Psychiatry. 37 (8):*** **858–865. DOI: http://dx.doi.org/10.1097/00004583-199808000-00016**

If you would like to read the entire paper you can use the DOI reference provided to locate that (see Chapter 1 for instructions).

In this research the authors examined the effect of caffeine, and its withdrawal, on attention, anxiety and motor performance in children. Previous evidence suggests that caffeine is associated with increased arousal, jitteriness, sleep disturbance, decreased fatigue and faster reaction time. Caffeine withdrawal has been linked with increased depression, anxiety, fatigue, headaches and poor motor control. However, most of that research has focused on adults; very few studies have examined younger participants. The effects of caffeine were investigated on 30 children (17 boys, 13 girls, mostly white, with an average age of 10.1 years). Effects were measured at **baseline** (no caffeine), administration of 120–145 mg caffeine daily (for 13 days), during 24 hours of withdrawal, and back to baseline (no caffeine again). Caffeine was consumed via 'popular carbonated drinks'. It was predicted that the children would show withdrawal effects immediately after the caffeine-drinking period; no other predictions were made about the conditions.

Attention was measured using the Test of Variables of Attention (TOVA; Greenburg and Waldmen, 1993). Caffeine withdrawal symptoms were examined using the self-report Caffeine Rating Scale (CRS; created by one of the authors). State and trait anxiety were investigated using State-Trait Anxiety Inventory for Children (STAIC; Spielberger, 1973). Depression symptoms were ascertained via the Children's Depression Inventory (CDI; Kovacs, 1981). Motor performance was assessed with two tasks: a pegboard test involving placing pegs into a pegboard, using the dominant and non-dominant hands and a finger-tapping test involving pressing a lever with the index finger (of both hands) as often as possible in ten seconds. The same 30 children were examined throughout (data from nine children were excluded because they did not take part in all four conditions correctly). Thus a repeated-measures one-way ANOVA was clearly the appropriate measure (although we were not told about normality of distribution, but then we seldom are). The authors reported that they used paired t-tests to examine *post hoc* data, but used a Bonferroni adjustment to account for multiple comparisons.

The authors reported a range of results, a summary of which is given here. There were significant differences in response times (on the TOVA): F = 24.13, *df* = 3,84, p < .001 (we are not told about sphericity outcomes). The children were faster in the baseline and caffeine conditions than they were in the withdrawal and return to baseline conditions. There were also significant differences across the conditions for reported withdrawal symptoms from the CRS (F = 6.07, *df* = 3,87, p < .001), state anxiety (F = 8.01, *df* = 3,87, p < .001), and trait anxiety (F = 10.21, *df* = 3,87, p < .001). In all cases, baseline scores were poorer than 'return to baseline', and there were no differences between the caffeine and withdrawal conditions. It was reported that there were no significant differences in depression scores across the conditions (although I feel that ANOVA outcome should be shown for that, too). Motor performance was significantly different across the conditions in respect of the pegboard task (dominant hand: F = 20.32, *df* = 3,87, p < .001; non-dominant hand: F = 10.41, *df* = 3,87, p < .001) and the finger-tapping task (dominant hand: F = 15.52, *df* = 3,87, p < .001; non-dominant hand: F = 12.30, *df* = 3,87, p < .001).

This is a good example of how a repeated-measures one-way ANOVA has been used in published research. There may have been a need for a little more detail about the statistics in places, but this may reflect the fact that this was published in 1998. For example, the reporting conventions for degrees of freedom (*df*) have changed a lot since then.

Extended learning task

You will find the data set associated with this task on the website that accompanies this book (available in SPSS and Excel format). You will also find the answers there.

Following what we have learned about repeated-measures one-way ANOVA, answer these questions and conduct the analyses in SPSS and G*Power. If you do not have SPSS, do as much as you can with the Excel spreadsheet. In this example we have a group of participants who have sought help with gaining control of their lives. They are given the same questionnaire at four stages of treatment: baseline (prior to treatment), after watching a video about mastery, after receiving a short course of cognitive-behavioural therapy (CBT), and after receiving a combined treatment of CBT and video. The questionnaire asks the participants to rate their perceived control, which is scored from 0 (low) to 100 (high). No specific predictions are made about outcomes between conditions.

Open the data set **CBT and perceived control**

1. Which is the independent variable?
2. Which is the dependent variable?
3. Using the correct independent variable:
   a. Check for normal distribution across the conditions.
   b. Conduct a repeated-measures one-way ANOVA.
4. Describe what the SPSS output shows.
5. Explain how you accounted for sphericity.
6. Conduct appropriate *post hoc* analyses to indicate where the differences are (if there were any).
7. State the effect size and power, using G*Power.
8. Report the outcome as you would in the results section of a report.

11 INDEPENDENT MULTI-FACTORIAL ANOVA

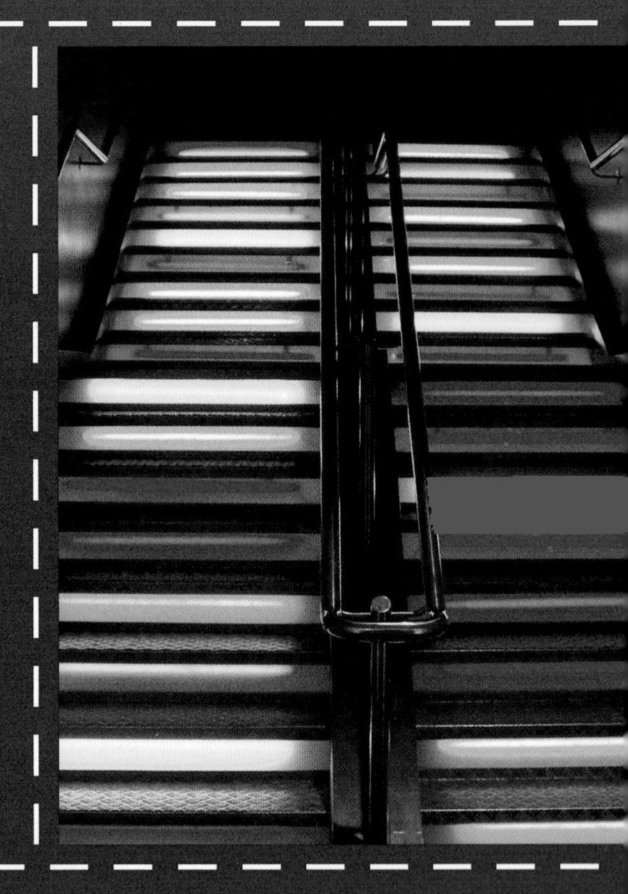

Learning objectives

By the end of this chapter you should be able to:

- Recognise when it is appropriate to use an independent multi-factorial ANOVA
- Understand the theory, rationale, assumptions and restrictions associated with this test
- Understand what is meant by 'main effects' and 'interactions'
- Calculate outcomes manually (using maths and equations)
- Perform analyses using SPSS, and examine the significance of outcomes, the source of significance and source of interaction
- Explore effect size and statistical power
- Understand how to present the data and report the findings

What is independent multi-factorial ANOVA?

An independent multi-factorial ANOVA examines differences in mean scores from a single (parametric) dependent variable across two or more categorical independent variables, each represented by two or more distinct groups. We explored the criteria for parametric data in Chapter 5, but we revisit these a little later. A key feature of this test is that we can measure outcomes for each independent variable and the extent that the independent variables 'interact'. An interaction occurs when the outcome across one independent variable differs across the groups or conditions of another independent variable. The interaction is an essential part of this test – it is more than likely that you would choose to perform an independent multi-factorial ANOVA *because* you had predicted that there would be an interaction.

Research question for independent multi-factorial ANOVA

We can illustrate independent multi-factorial ANOVA by way of a research question, posed by a group of sleep researchers called SNORES (Sleep and Nocturnal Occurrences Research Group). They decide to investigate how sleep quality perceptions change according to current depression status. Evidence suggests that sleep is poorer for people with depression. SNORES are also interested in how these sleep reports vary according to gender, since further evidence suggests that women report poorer sleep than men. Given that prior evidence, SNORES design a study that will explore the extent that sleep satisfaction reports change with depression status, and particularly whether those changes are more pronounced in women. Questionnaires are used to capture reports of sleep satisfaction. These are scored on a scale of 0 – 100, with higher scores indicating poorer satisfaction. Equal numbers of men and women are recruited to the study. Within the gender groups, participants are rated according to their current depression status. There are three groups (none, mild and severe depression). These groups also have equal numbers.

11.1 Take a closer look
Variables for independent multi-factorial ANOVA

Dependent variable: Sleep satisfaction scores
Independent variable 1: Depression status (none, mild, severe)
Independent variable 2: Gender (male, female)

Theory and rationale

Terminology with multi-factorial ANOVA

The SNORES research is an example of an independent two-way ANOVA. This is because there are two independent variables: gender and depression status. The term 'two-way' simply describes the number of independent variables used in the multi-factorial ANOVA. For example, we could extend the example we have been using and add 'personality type' as another between-group independent variable. We might now examine sleep satisfaction scores in respect of depression status, gender and personality type. This would be an example of an independent three-way ANOVA. In your further reading, you may see studies reported as being of '2 × 2 design', or that a '2 × 2 independent multi-factorial ANOVA' was used. This simply indicates how many groups or conditions there are within each independent variable. The SNORES example we used earlier would be called a '2 × 3 independent multi-factorial ANOVA'; there are two independent variables, one with two groups (gender) and one with three groups (depression status).

Multi-factorial ANOVA is often confused with multivariate ANOVA. The latter is used to describe cases when we have more than one dependent variable; we will explore that in more depth when we look at MANOVA (Chapter 14). It is a **multivariate ANOVA** because there are several outcomes (or variates) that might vary. An example of a MANOVA might be where we measure gender (the independent variable) against sleep satisfaction scores *and* anxiety scores (two dependent variables). With **multi-factorial ANOVA** there is only one dependent variable (it is often referred to as univariate). However, there are several independent variables (another name for an independent variable is a factor, so perhaps that might help clarify the distinction).

11.2 Take a closer look

One-way, multi-factorial ANOVA and multivariate ANOVA: some clarification

It is very common for students to confuse these terms. The following explanations may help:

One-way ANOVA: Where there is one independent variable (IV)
Multi-factorial ANOVA: Where there are two or more IVs
Two-way ANOVA, three-way ANOVA (etc.): Describes the number of IVs in a multi-factorial ANOVA
Multivariate ANOVA: Where there are two or more dependent variables

Main effects and interactions

Perhaps the most important feature of independent multi-factorial ANOVA is the way in which it examines the effect of each independent variable (in respect of outcome on the dependent variable), and the way in which those independent variables interact with each other. A **main effect** describes whether there is a significant difference in the dependent variable scores across the groups – we investigate this for each of the independent variables. An **interaction** occurs when the outcome across one independent variable differs significantly across the groups of another independent variable. The interaction effect should never be seen as a by-product of the analysis. Such an effect should have been predicted in the first place; it would have been why an independent multi-factorial ANOVA was chosen to analyse the data in the first place (see later).

We can illustrate this with an example using the scenario posed by the SNORES research question. The researchers are examining sleep satisfaction reports according to depression status (none, mild or severe) and gender (male or female). We can see some of the possible main effect and interaction outcomes below:

Possible main effect outcomes

Between-group independent variable 1: depression status

Sleep satisfaction scores may be poorer for those with severe depression, compared with those without depression.

Between-group independent variable 2: gender

Women might report poorer sleep satisfaction than men.

Possible (two-way) interaction

Between-group interaction: 'depression status vs. gender'

Women might report poorer sleep satisfaction than men, but only if they are severely depressed. There may be no gender differences in sleep satisfaction with mild or no depression.

That last example is a **two-way interaction** because there were two independent variables compared in respect of the dependent variable outcome. If we add another independent variable (such as personality type), things get rather more complex. We would examine three main effects (depression status, gender and personality type) and would also investigate three two-way interactions ('depression status vs. gender', 'depression status vs. personality type' and 'gender vs. personality type'), but we would also need to investigate whether we had a **three-way interaction** ('depression status vs. gender vs. personality type'). In the case of the SNORES research, we might find that severely depressed women report poorer sleep satisfaction scores than severely depressed men, but only when they have a neurotic personality type.

Graphical representation of main effects and interaction

We can use line graphs to get a visual illustration of main effects and interactions. To put this into context, we can refer to the SNORES research example concerning how depression status and gender (two independent variables) impact on sleep satisfaction scores (dependent variable). We plot one independent variable (such as 'depression status') along the bottom (x) axis, while the lines represent the second independent variable (e.g. 'gender'); 'sleep satisfaction' scores are shown on the side (y) axis. The slope of the lines indicates whether there might be a main effect for depression status (the greater the slope, the more likely there may be an effect), while the distance between the lines suggests whether there may be a main effect for gender (larger gaps may indicate the presence of an effect). Any potential interaction is determined by whether the lines are parallel. If so, there is no interaction – the gap between men and women remains the same at all depression levels. If the lines are not parallel, there may be one – it suggests that the gap between males and females differs across depression severity groups.

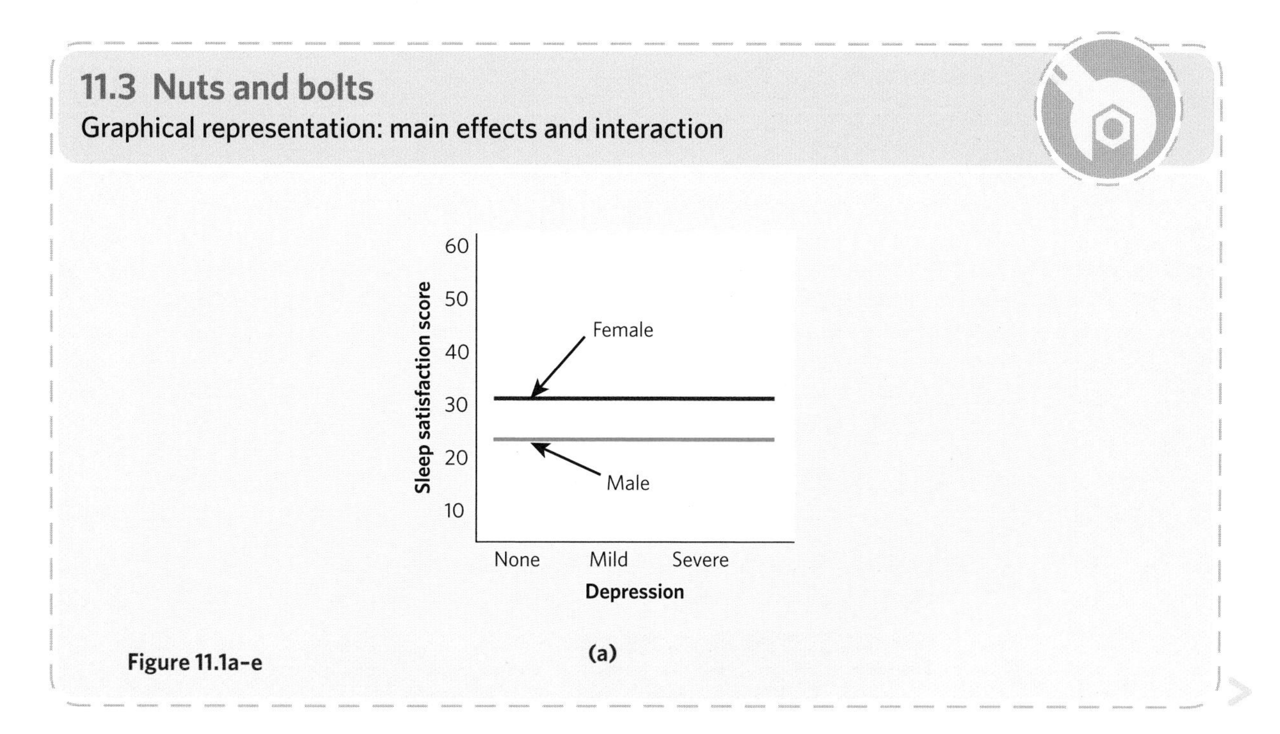

Figure 11.1a–e

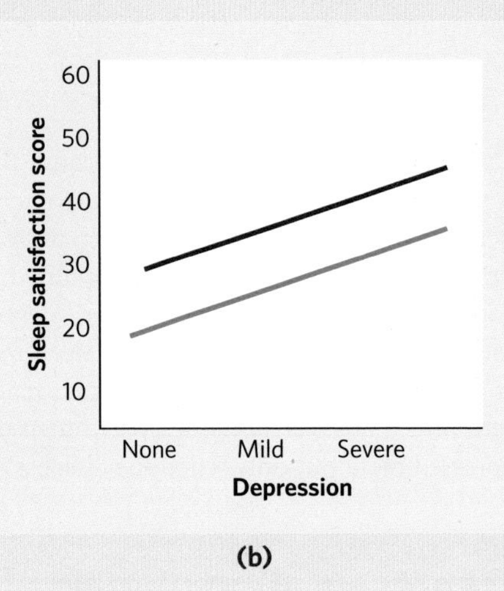
60
50
40
30
20
10
Sleep satisfaction score
None
Mild
Severe
Depression

(b)

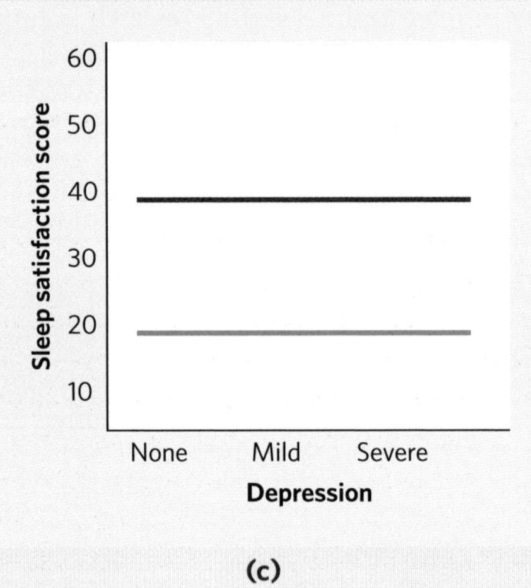
60
50
40
30
20
10
Sleep satisfaction score
None
Mild
Severe
Depression

(c)

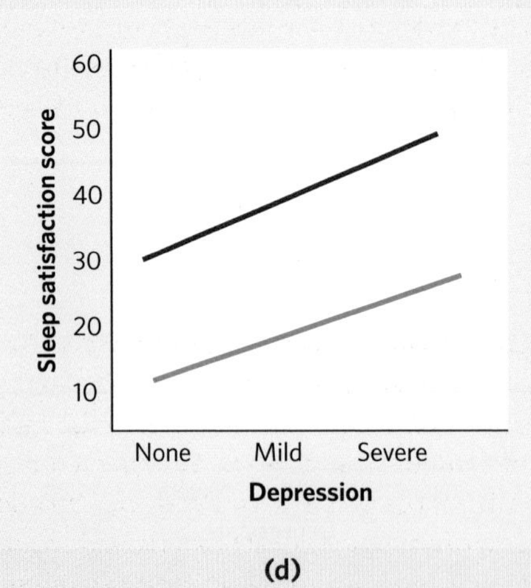
60
50
40
30
20
10
Sleep satisfaction score
None
Mild
Severe
Depression

(d)

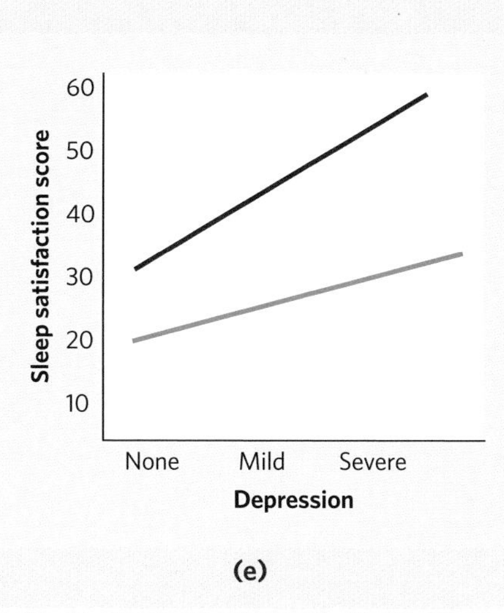

(e)

Figure 11.1a indicates that there is no main effect for depression status – sleep satisfaction scores do not change across depression severity (both lines are level). There is probably no main effect for gender either – the lines representing men and women are close together, suggesting little difference between them. There is no interaction because the lines are parallel. Figure 11.1b indicates that there may be a main effect for depression – sleep satisfaction scores increase with worsening depression severity (both lines are sloped).There is probably still no main effect for gender – the gap between the lines is small. There is no interaction – the lines are parallel. Figure 11.1c indicates that there is no main effect for depression status (both lines are level), but there may be a main effect for gender – the gap between the lines for male and female is quite large. There is still no interaction. Figure 11.1d indicates that there are probably main effects for both depression status and gender – both lines are sloped and quite far apart. There is still no interaction. Figure 11.1e indicates that there are probably main effects for both depression status and gender, but now there may be an interaction – the line sare not parallel. Sleep satisfaction scores appear to increase more dramatically with increasing depression severity for women rather than men – that is a *potential* interaction. Figure 11.1e is just an example of what an interaction might look like (some others are shown in Box 11.4). The key point is that the lines are clearly not parallel in any of the examples. It is probably a good idea to see a graph of the main effects and interactions before analysing the data formally with statistics (we will see how to do that later). Graphs can provide a 'feel' for the outcome.

Identifying differences

The methods used to identify between-group differences for independent multi-factorial ANOVA are much the same as we saw with one-way ANOVA. Mean scores and variance are calculated in respect of dependent variable scores across each of the independent variable groups. This information is used to 'partition' the variance. We explored the concept of partitioning in depth in Chapter 9, so we will not repeat that here. However, with multi-factorial ANOVA, we need to partition the total sum of squares into several model sums of squares – one for each of the independent variables (main effects) and interaction(s). These will illustrate how much of the variance we can 'explain'. The unexplained variance (or error) is captured by the residual sum of squares. The sums of squares are each expressed in terms of the relevant degrees of freedom (the number of values that

11.4 Nuts and bolts

Other examples of interaction

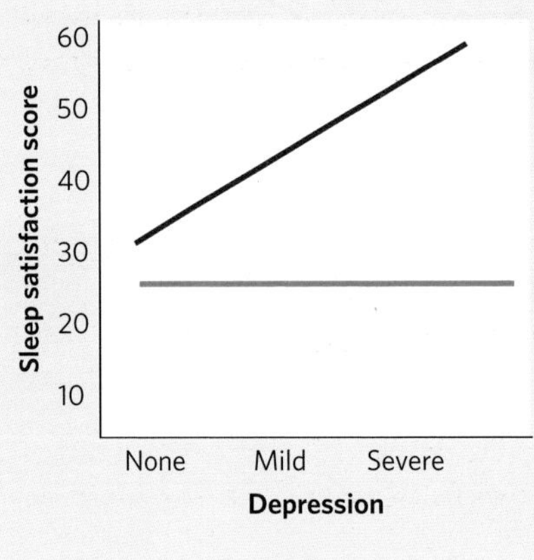

(f)

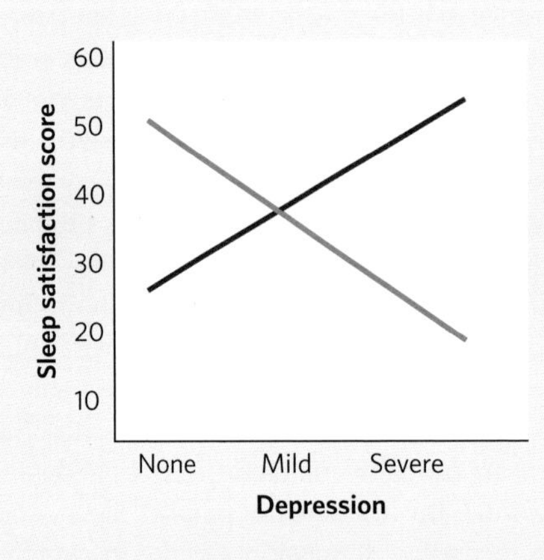

(g)

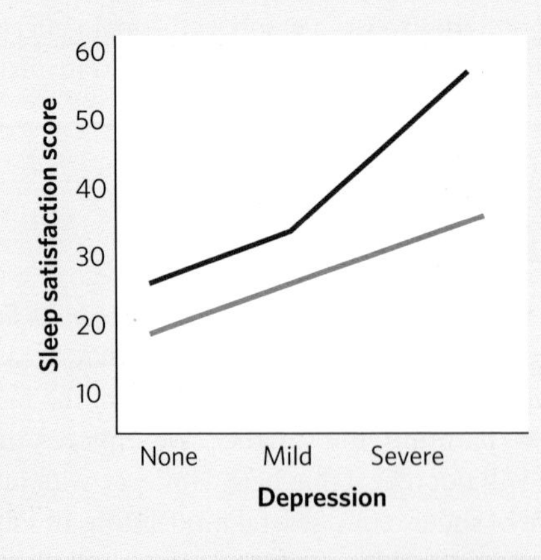

(h)

Figure 11.1f–h

are 'free to vary' in the calculation, while everything else is held constant – see Chapter 6). These produce model mean squares for each independent variable and interaction, and the residual mean square (we will see how to do this in Box 11.5). F ratios are calculated for each main effect and interaction (by dividing the respective model mean square by the residual mean square). Those F ratios are compared with cut-off points to assess whether they are statistically significant. If a significant main effect relates to three or more groups, we will need additional tests to locate the source of difference. If there are significant interactions, we will need to locate the source of that, too. Box 11.5 illustrates how we can partition the sums of squares, using data from the SNORES research example.

11.5 Calculating outcomes manually

Independent multi-factorial ANOVA calculation

To illustrate how we can calculate the outcome of an independent multi-factorial ANOVA, we will use some data that relate to the research question posed by SNORES earlier. The data are shown in Table 11.1. You will find a Microsoft Excel spreadsheet associated with these calculations on the web page for this book.

Table 11.1 Sleep satisfaction scores by depression status and gender

Depression:	**None**		**Mild**		**Severe**	
Gender	**Male**	**Female**	**Male**	**Female**	**Male**	**Female**
	42	42	50	76	48	50
	44	64	55	48	65	71
	67	42	56	37	58	94
	50	35	66	45	58	66
	54	38	49	77	66	76
	45	35	56	50	45	77
	45	38	20	48	67	59
	65	67	53	51	70	67
	46	65	56	54	72	69
	45	37	56	48	69	92
	56	41	76	66	45	87
	38	53	24	70	65	71
	36	44	69	74	76	67
	51	53	59	86	47	77
	33	46	59	46	42	97
Sub-group mean	47.80	46.67	53.60	58.40	59.53	74.67
Variance	93.89	123.52	215.40	222.40	129.12	174.52
Gender mean	**Male**	53.64	**Female**	59.91		
Dep mean	**None**	47.23	**Mild**	56.00	**Severe**	67.10
Grand mean	56.78		**Grand variance**		239.01	

Total sum of squares SS_T

$SS_T = s^2_{grand}(N - 1)$ = grand variance × sample size *minus* 1 = 239.01 × 89 = 21271.56

We saw how to calculate grand variance in Chapter 9 (but also see Excel spreadsheet).

Model sums of squares:

Formula for model sum of squares: $\sum n_k(\bar{x}_k - \bar{x}_{grand})^2$

(Overall) model sum of squares (SS_M):

Deduct grand mean from 'sub-group' mean, square it, multiply by no. of scores in sub-group (15)

$SS_M = 15 \times (47.80 - 56.78)^2 + 15 \times (46.67 - 56.78)^2 + 15 \times (53.60 - 56.78)^2$

$+ 15 \times (58.40 - 56.78)^2 + 15 \times (59.53 - 56.78)^2 + 15 \times (74.67 - 56.78)^2 = 7847.56$

We have six groups, so degrees of freedom (*df*) for $SS_M(df_M) = 6 - 1 = 5$

Model sum of squares - gender main effect (SS_G)

Deduct grand mean from gender mean, square it, multiply by no. of scores in group (45)

$SS_G = 45 \times (53.64 - 56.78)^2 + 45 \times (59.91 - 56.78)^2 = 883.60$

We have two groups, so *df* for $SS_G(df_G) = 2 - 1 = 1$

Model sum of squares - depression status main effect (SS_D)

Deduct grand mean from depression mean, square it, multiply by no. of scores in group (30)

$SS_D = 30 \times (47.23 - 56.78)^2 + 45 \times (56.00 - 56.78)^2 + 45 \times (67.10 - 56.78)^2 = 5947.49$

We have three groups, so *df* for $SS_D(df_D) = 3 - 1 = 2$

Model sum of squares - interaction effect ($SS_{G \times D}$)

What is left from (overall) model sum of squares:

$SS_M - SS_G - SS_D = 7847.56 - 883.60 - 5947.49 = 1016.47$

df for $SS_{G \times D}(df_{G \times D}) = \; = df_M - df_G - df_D = 2$

Residual sum of squares:

Formula for the residual sum of squares: $(SS_R) = \sum s_k^2(n_k - 1)$

Multiply sub-group variances by sub-group size *minus* 1 (15 − 1 = 14):

$SS_R = (93.89 \times 14) + (123.52 \times 14) + (215.40 \times 14) + (222.40 \times 14)$

$+ (129.12 \times 14) + (174.52 \times 14) = 13424.00$

We saw how to calculate group variance in Chapter 9 (but also see Excel spreadsheet).

df for SS_R (df_R) = group size (15) *minus* 1 (14), multiplied by no. of groups (6)

$df_R = 14 \times 6 = 84$

Alternatively, SS_R can be calculated from what is left over from total sum of squares:

$SS_R = SS_T - SS_M = 21271.56 - 7847.56 = 13424.00$

Mean squares:

Found by dividing sums of squares by relevant degrees of freedom:

Gender effect $MS_G = \dfrac{SS_G}{df_G} = 883.60 \div 1 = 883.60$

Depression status effect $MS_D = \frac{SS_D}{df_D} = 5947.49 \div 2 = 2973.74$

Interaction $MS_{G\times D} = \frac{SS_{G\times D}}{df_{G\times D}} = 1016.47 \div 2 = 508.23$

Residual sum $MS_R = \frac{SS_R}{df_R} = 13424.00 \div 84 = 159.81$

F ratios:

Found by dividing mean square by residual mean square:

Gender $F_G = \frac{MS_G}{MS_R} = 883.60 \div 159.81 = 5.529$

Depression status $F_D = \frac{MS_D}{MS_R} = 2973.74 \div 159.81 = 18.608$

Interaction $F_{G\times D} = \frac{MS_{G\times D}}{MS_R} = 508.23 \div 159.81 = 3.180$

Each F ratio is compared with the relevant cut-off point in the F-distribution table (see Appendix 4), according to the *df* and significance level. In using that table, we need to refer to the cross-section of the numerator *df* (model sum degrees of freedom) and the denominator *df* (residual sum degrees of freedom). We express that as (numerator *df*, denominator *df*):

Gender (1, 84): cut-off = 3.95; F = 5.53 ➪ significant main effect across gender

Depression status (2, 84): cut-off = 3.11; F = 18.61 ➪ significant main effect across depression status

Interaction (2, 84): cut-off = 3.11; F = 3.18 ➪ significant interaction between gender and depression status (but only just significant)

We can also use Excel to calculate the critical value of F and to provide the actual p value. You can see how to do this on the web page for this book. Using those calculations we find the following p values for our outcome: Gender: $p = .021$ Depression Status: $p < .001$ Interaction: $p = .047$. You can also see how to perform the entire test in Excel.

We have now partitioned the variance into the parts that are explained by between-group differences, and the interactions between them, and the part that is left unexplained (the error). There are three portions within the (overall) model sum of square (SS_M): gender (SS_G), depression status (SS_D) and the interaction between gender and depression status (SS_{GxD}). The residual sum of squares (SS_R) is calculated from the group variances, in relation to group size (or it can simply be found by deducting the model sum of squares from the total sum of squares).

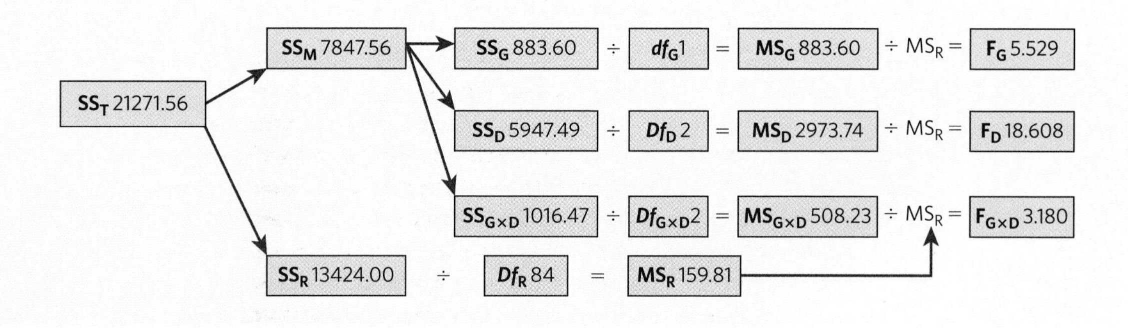

Locating source of main effects

If we find that we have a significant (main) effect for an independent variable, we need to declare the source of that difference. If there are two independent variable groups (as is the case with gender in our example), we can use the mean scores to indicate where the differences are. If there are three or more groups (as we have with the 'depression status' independent variable), we need additional tests to show the source of difference (just as we saw with independent one-way ANOVA). We do not investigate the source of difference if the main ANOVA outcome is not significant. In our example, we need to compare sleep satisfaction scores across three pairs of groups: 'severe depression' vs. 'mild depression', 'severe depression' vs. 'no depression', and 'mild depression' vs. 'no depression'. We do so with the aid of 'planned contrasts' or '*post hoc* tests'.

As we saw in Chapter 9, we can use a planned contrast only if a specific (one-tailed) hypothesis has been made about the outcome between the groups; otherwise a *post hoc* test must be employed. The key difference between these tests is the way that they account for multiple comparisons. Remember, the more additional tests that we undertake, the greater the likelihood that we will find a significant outcome by chance factors alone; we risk making Type I errors (see Chapter 4). If we make a specific prediction about the outcome and one of the groups represents a control group, we can undertake orthogonal planned contrasts. Then we are justified in leaving the significance cut-off point unadjusted (see Chapter 9). In all other cases, we must divide that cut-off point by the number of additional tests that have been undertaken. In our example, because we need three additional tests, mean sleep satisfaction scores are significantly different between pairs of groups only if $p < .016$ (usual cut-off $p < 0.05 \div 3 = 0.016$). Since *post hoc* tests are easier to perform than planned contrasts, it is probably sensible that we focus on the former. The guidelines for choosing the appropriate *post hoc* test remain the same as we saw in Chapter 9 (see Box 9.8). However, SPSS does not provide an option for violations of homogeneity of variance.

Locating the source of interaction

If we find that we have a significant interaction, we need to explore the source of that. In the SNORES research example, it may be that sleep satisfaction scores are only significantly poorer for women than men when they are severely depressed; there may be no significant difference for gender in respect of sleep satisfaction across other depression severity groups. The mean scores might provide a clue, but we need something more tangible. A line graph might help to see what the interaction might look like. We could compare that to the examples that we saw in Box 11.3. However, to be certain, we need statistical confirmation. There are a couple of ways that we can do this. First, we can use the 'Spilt File' facility in SPSS to explore the outcome according to each group (we will see how later). For example, we could split the data file into depression severity groups. Then we can look at sleep satisfaction scores in respect of gender, but report separate independent t-test outcomes for each level of depression severity. Alternatively, we could perform a '**simple effect**' test – this is somewhat more complex, but potentially more accurate (because analyses explore each main effect without the 'bias' of the other independent variables). Undertaking simple effects involves the use of syntax (the basic language that SPSS speaks). As this is a little more advanced, those procedures are shown at the very end of this chapter.

Interactions: predicting the outcome

The interaction is central to multi-factorial ANOVA. It should never be seen as a by-product of the analyses. Indeed, an expectation of an interaction between two independent variable factors, in respect of a dependent variable outcome, is why this test would have been chosen to analyse the data. Let's say you are a psychologist working in the SNORES group. During your clinical practice (and research) you notice that your patient reports of sleep quality get poorer when they are more depressed, and that this appears to be worse for women. To examine this observation, you would need to collect data on sleep satisfaction scores from a group of people, where gender and depression status were also recorded. The data could then be analysed with independent multi-factorial ANOVA. The test would have been chosen *because* the interaction was expected.

Assumptions and restrictions

We need to satisfy a number of assumptions before we can perform an independent multi-factorial ANOVA. There must be at least two independent variables, each with at least two distinct categorical groups, measured against one dependent variable. The groups must be independent of each other; no one can appear in more than one group. The dependent variable data should be parametric (interval or ratio dependent variable scores should be *reasonably* normally distributed, as we saw in Chapter 5). But don't be overly cautious about normal distribution, as ANOVA is quite robust to modest violations. Also, it is common for researchers to use these tests to examine outcomes from attitude and satisfaction questionnaires, such as Likert scales (even though they are considered to represent original scores). We have considered that debate in previous chapters, so we will not extend it here. However, it *is* important that there is homogeneity of variances across the independent variable groups. Violations are particularly a problem if there are unequal group sizes, as we saw in Chapter 9. Larger groups with variances that are greater than those seen in the smaller groups can reduce the likelihood of finding a significant outcome, while larger groups with smaller variances can inflate that chance.

Minor problems with normal distribution may warrant a word of caution when interpreting outcomes: more serious problems may justify using transformation (see Chapter 3). However, despite what was said there, transformation of data is not without problems. Another way around this involves a method called 'bootstrapping'. We will not dwell on this in depth here, as the principles are too advanced for this relatively introductory book. In essence, bootstrapping will assess and adjust for violations, with the aim of making the data sample 'more robust'. You can read about this in more depth in Wilcox (2005), where you will also find information about add-on programs for SPSS. IBM has also recently introduced bootstrapping plug-ins for SPSS. In the meantime, we could try non-parametric tests, but there is no direct equivalent for an independent multi-factorial ANOVA. We could perform separate Mann–Whitney U tests and/or Kruskal–Wallis tests for each independent variable, but this would not allow us to examine the potential interaction between the variables.

Violations of homogeneity of variance are even less easy to address. In Chapter 9, we learned about adjustments that can be made using Brown–Forsythe F or Welch's F statistics. These are quite complex to run manually (especially with several independent variables with three or more groups). Although we can select these tests in SPSS for independent one-way ANOVA, they are not available for multi-factorial ANOVA (nor are options for appropriate *post hoc* tests in this respect). If we violate homogeneity of variance where there are equal groups (or even relatively small differences in those group sizes) we are probably OK to proceed (at least with caution). If there are noticeably unequal groups, we could try running independent one-way ANOVAs for each independent variable and see what happens to ANOVA outcomes subsequent to Brown–Forsythe F or Welch's F adjustments (you can see how to do that in Chapter 9). If there is no major difference, we can at least be confident about the main effects. Equally, we could use Games–Howell *post hoc* tests in those one-way ANOVA analyses. However, none of this investigates the impact on the interaction. Perhaps this illustrates why it is important to aim to have equal group sizes.

11.6 Take a closer look
Summary of assumptions and restrictions

- There must be at least two categorical independent variables:
  - Each of which must have at least two distinct groups
- Membership of a group must be independent
- There must be one numerical dependent variable
  - DV data should be interval or ratio, and reasonably normally distributed
- There should be homogeneity in the variances between the groups

11.7 Nuts and bolts
Setting up the data set in SPSS

When we create the SPSS data set for an independent multi-factorial ANOVA, we need to set up one column for the dependent variable (which will have a continuous score) and a column for each of the independent variables (each of which will have a categorical coding).

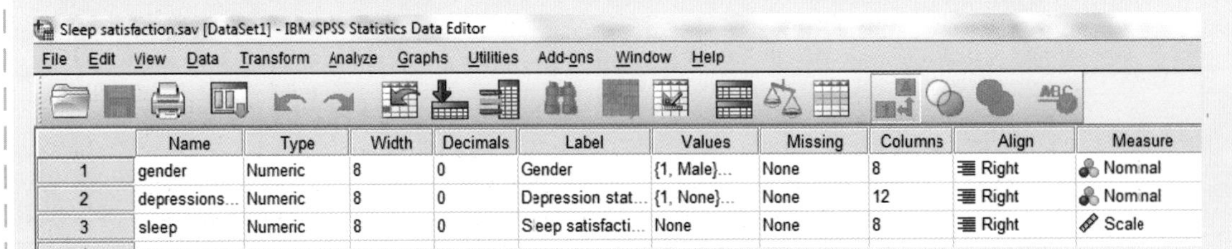

	Name	Type	Width	Decimals	Label	Values	Missing	Columns	Align	Measure
1	gender	Numeric	8	0	Gender	{1, Male}...	None	8	Right	Nominal
2	depressions...	Numeric	8	0	Depression stat...	{1, None}...	None	12	Right	Nominal
3	sleep	Numeric	8	0	Sleep satisfacti...	None	None	8	Right	Scale

Figure 11.2 Variable View for 'Sleep satisfaction' data

Figure 11.2 shows how the SPSS Variable View should be set up (you should refer to Chapter 2 for more information on the procedures used to set the parameters). The first variable is called 'gender' - this is the first independent variable. In the Values column, we include '1 = Male' and '2 = Female'; the Measure column is set to Nominal. The second variable is 'depression status' - this is the second independent variable. In the Values column, we include '1 = None', '2 = Mild', and '3 = Severe'; the Measure column is set to Nominal. The third variable is 'sleep' - this is the continuous dependent variable representing the sleep satisfaction scores. We do not adjust anything in the Values column; the Measure column is set to Scale.

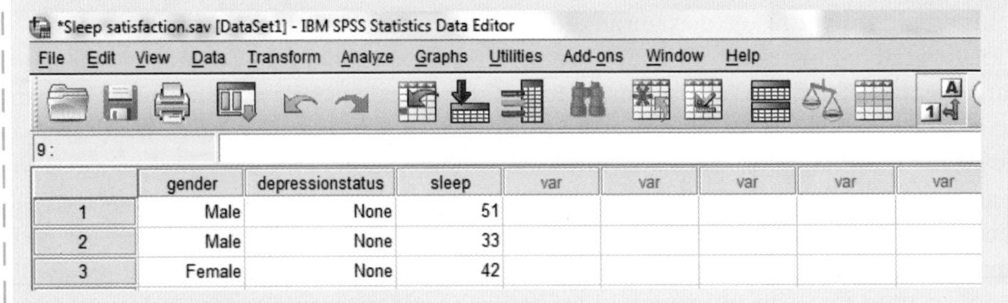

	gender	depressionstatus	sleep	var	var	var	var	var
1	Male	None	51					
2	Male	None	33					
3	Female	None	42					

Figure 11.3 Data View for 'Sleep satisfaction' data

Figure 11.3 illustrates how this will appear in the Data View. Each row represents a participant. When we enter the data for 'gender', we input 1 (to represent 'Male') or 2 (to represent 'Female'), and for 'depressionstatus', we input 1 (for 'None'), 2 (for 'Mild'), or 3 (for 'Severe'). Those columns will display the descriptive categories or the value numbers, depending on how you choose to view the column (you can change that using the Alpha Numeric button in the menu bar - see Chapter 2). Sleep satisfaction scores will be entered into the 'sleep' column.

How SPSS performs independent multi-factorial ANOVA

We can perform an independent multi-factorial ANOVA in SPSS. To do that, we will refer once again to the SNORES research question, using the same data set that we explored manually earlier. There are equal numbers in each of the groups in both independent variables. When we select a *post hoc* test to explore the source of difference (if there is one), we know that we can safely choose the Tukey option (see Box 9.8). There are no *post hoc* test options in SPSS to account for violation of homogeneity of variance. It *could* be argued that the dependent variable data (sleep satisfaction reports) are ordinal, because they *might* be considered to be subjective. However, there are probably just as many arguments to suggest that the range of potential responses is linear, and therefore interval. We should still check for normal distribution (which we will do shortly), although ANOVA is pretty robust to minor violations.

Producing a line graph

As we said earlier, it might be useful to have a look at a graphical display of the main effects and interactions before we analyse the statistics. We could draw a line and compare that with the graphs that we saw in Boxes 11.3 and 11.4. Here's how we do that.

Open the SPSS file Sleep satisfaction

Select **Graphs** → **Chart Builder** → select **Line** from list under **Choose from:** → drag **Multiple Line** graphic into empty chart preview area → transfer **Sleep satisfaction scores** to **Y-Axis** box → transfer **Depression status** to **X-Axis** box → transfer **Gender** to **Set Color** box → click **OK**

Figure 11.4 suggests that there may be a significant effect for depression status because the lines are sloped (for men and women); sleep satisfaction scores appear to worsen with increasing depression severity. There may also be an effect for gender, as the lines are quite far apart (although, only in the presence of depression); women (green line) appear to be reporting poorer sleep satisfaction than men (blue line). There could well be an interaction, as it seems that the rate of deterioration in sleep satisfaction across increasing depression severity is more dramatic for women. This graph is quite similar to Figure 11.1e.

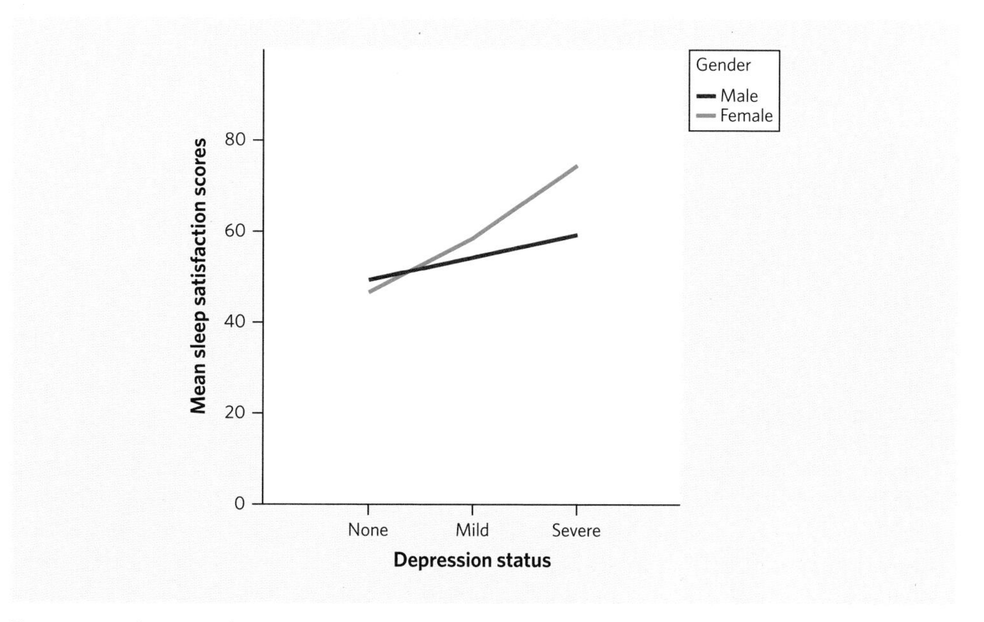

Figure 11.4 Sleep satisfaction in respect of gender vs. depression severity

Testing for normal distribution

As we have seen in previous chapters (refer to Chapter 7), to examine normal distribution we start by running the Kolmogorov–Smirnov/Shapiro–Wilk tests. If the outcome indicates that the data may not be normally distributed, we could additionally run z-score analyses of skew and kurtosis, or look to 'transform' the scores (see Chapter 3).

	Gender	Kolmogorov–Smirnov[a]			Shapiro–Wilk		
		Statistic	df	Sig.	Statistic	df	Sig.
Sleep satisfaction scores	Male	.102	45	.200*	.971	45	.317
	Female	.122	45	.089	.946	45	.036

Figure 11.5 Kolmogorov-Smirnov/Shapiro-Wilk test sleep satisfaction vs. gender

Because we have group sizes of less than 50 (for both independent variables) we should refer to the Shapiro–Wilk outcome. Figure 11.5 indicates that sleep satisfaction scores appear to be normally distributed for men (W (45) = .971, p = .317), but may be more of a problem for female scores (W (45) = .946, p = .036). Figure 11.6 suggests that sleep satisfaction scores *may* not be normally distributed for those with no depression (W (30) = .909, p = .014), but appear fine for mild depression (W (30) = .957, p = .252) and severe depression (W (30) = .954, p = .210). ANOVA is robust to minor violations to normal distribution, so these outcomes are probably OK. However, if you are the more cautious type, you could run additional z-score tests of skew and kurtosis (we saw how to do that in Chapter 3). As it happens, should you choose to run those tests, you will find that outcomes confirm normal distribution in any case.

	Depression status	Kolmogorov–Smirnov[a]			Shapiro–Wilk		
		Statistic	df	Sig.	Statistic	df	Sig.
Sleep satisfaction scores	None	.181	30	.013	.909	30	.014
	Mild	.133	30	.183	.957	30	.252
	Severe	.142	30	.127	.954	30	.210

Figure 11.6 Kolmogorov-Smirnov/Shapiro-Wilk test sleep satisfaction vs. depression status

Running independent multi-factorial ANOVA in SPSS

Using the SPSS file Sleep satisfaction

Select **Analyze → General Linear Model → Univariate . . . →** as shown in Figure 11.7

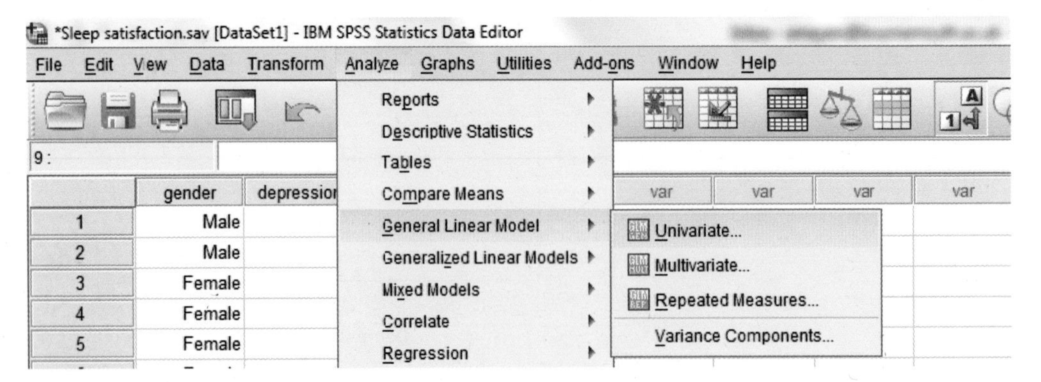

Figure 11.7 Independent multi-factorial ANOVA selection

In new window (see Figure 11.8), transfer **Sleep satisfaction scores** to **Dependent Variable** (by clicking on arrow, or by dragging the variable to that window) ➜ transfer **Gender** and **Depression status** to **Fixed Factors** ➜ click **Post Hoc...**

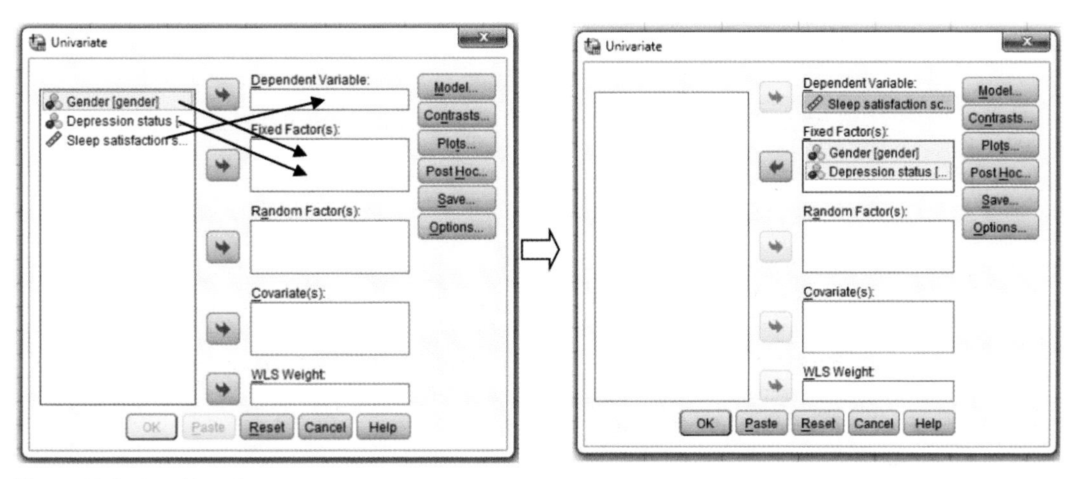

Figure 11.8 Variable selection

In new window (see Figure 11.9), transfer **depression status** to **Post Hoc Tests for** (we do not include 'gender', because that independent variable only has two groups) ➜ tick **Tukey** box (see Box 9.8 for the criteria for choosing the correct *post hoc* test, but note that options for unequal homogeneity of variance are not available in SPSS for independent multi-factorial ANOVA) ➜ click **Continue** ➜ click **Options...** ➜ (in new window) tick **Descriptives**, **Estimates of effect size** and **Homogeneity of variance test** ➜ click **Continue** ➜ click **OK**

To see graphical displays for Options functions, refer to Chapter 9 for independent one-way ANOVA

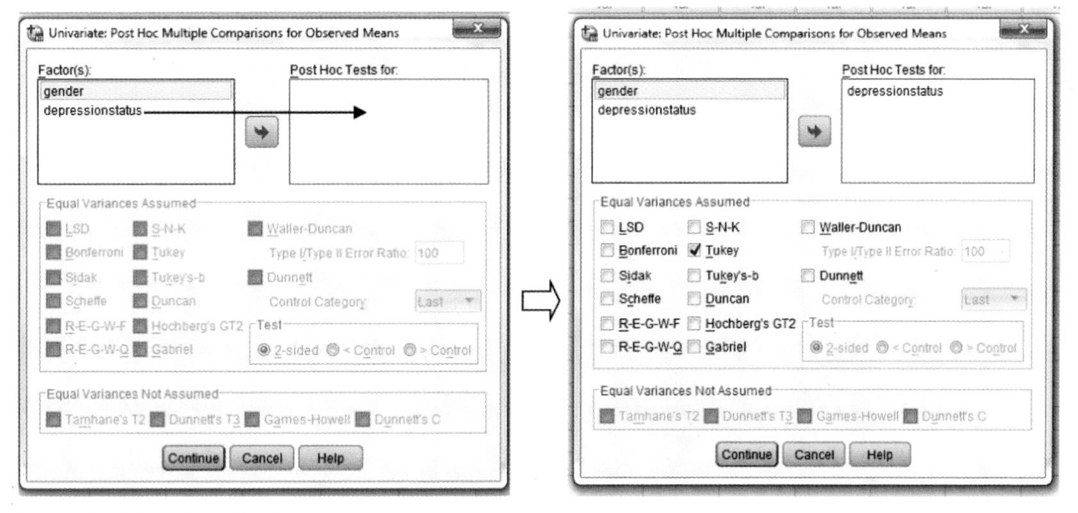

Figure 11.9 *Post hoc* selection

Interpretation of output

Figure 11.10 suggests that sleep satisfaction scores are higher (poorer) for women than for men, and that these scores worsen with increasing severity of depression. Women with severe depression appear to have particularly poorer sleep satisfaction scores than any other group, suggesting a possible interaction between gender and depression severity.

Gender	Depression status	Mean	Std. deviation	N
Male	None	47.80	9.689	15
	Mild	53.60	14.677	15
	Severe	59.53	11.363	15
	Total	53.64	12.766	45
Female	None	46.67	11.114	15
	Mild	58.40	14.913	15
	Severe	74.67	13.211	15
	Total	59.91	17.332	45
Total	None	47.23	10.261	30
	Mild	56.00	14.741	30
	Severe	67.10	14.346	30
	Total	56.78	15.460	90

Figure 11.10 Descriptive data

Figure 11.11 confirms that there was homogeneity of between-group variances. Earlier, we saw that this is an important consideration in analysing ANOVA data. As we had equal group sizes, we would probably have been OK here in any case.

F	df1	df2	Sig.
1.004	5	84	.421

Figure 11.11 Levene's test for equality of variances

Figure 11.12 shows that there was a significant main effect for gender (highlighted in red; $F(1, 84) = 5.529$, $p = .021$); we know from Figure 11.8 that sleep satisfaction scores were poorer for women. The 'Partial Eta Squared' data (highlighted in orange) will be used in effect analyses later. There was a significant main effect for 'depression status' (highlighted in blue; $F(2, 84) = 18.608$, $p < .001$). However, we cannot tell (yet) where the source of that difference lies; we need *post hoc* tests for that. There was also a significant interaction between gender and depression status, in respect of sleep satisfaction scores (highlighted in green; $F(2, 84) = 3.180$, $p = .047$). We will need to conduct further analyses to examine the exact nature of that interaction (see later).

Source	Type III sum of squares	df	Mean square	F	Sig.	Partial eta squared
Corrected model	7847.556[a]	5	1569.511	9.821	.000	.369
Intercept	290134.444	1	290134.444	1815.502	.000	.956
gender	883.600	1	883.600	**5.529**	**.021**	.062
depressionstatus	5947.489	2	2973.744	**18.608**	**.000**	.307
gender * depressionstatus	1016.467	2	508.233	**3.180**	**.047**	.070
Error	13424.000	84	159.810			
Total	311406.000	90				
Corrected total	21271.556	89				

Figure 11.12 ANOVA statistics

Figure 11.13 shows the Tukey *post hoc* outcome. Significant between-pair differences occur where significance ('Sig.') is less than .05 (also illustrated by an asterisk). Examples here are highlighted in red font. In this outcome, sleep satisfaction scores were significantly poorer for reporting 'severe depression' than for 'mild depression' ($p = .003$) and than 'no depression' ($p < .001$). Furthermore, sleep satisfaction was poorer for those reporting 'mild depression' than 'no depression' ($p = .023$).

(I) Depression status	(J) Depression status	Mean difference (I-J)	Std. error	Sig.	95% confidence interval Lower bound	Upper bound
None	Mild	−8.77*	3.264	.023	−16.55	−.98
	Severe	−19.87*	3.264	.000	−27.65	−12.08
Mild	None	**8.77***	3.264	**.023**	.98	16.55
	Severe	−11.10*	3.264	.003	−18.89	−3.31
Severe	None	**19.87***	3.264	**.000**	12.08	27.65
	Mild	**11.10***	3.264	**.003**	3.31	18.89

Figure 11.13 *Post hoc* statistics

Investigating source of significant interaction

Earlier, we learned that there are three ways in which we can examine the likely source of a significant interaction when we find one: we can produce a line graph; we can explore the data statistically, focusing on one group of an independent variable at a time (using the 'Split File' facility in SPSS); or we can perform something called simple effects (using SPSS syntax codes). I will demonstrate the first two methods here (as they are quite straightforward), but leave the simple effects method until later (despite the name, it is relatively complex).

Line graph

We can use the line graph that we created earlier (Figure 11.4) to perform a visual analysis of the potential source of interaction. This indicated that sleep satisfaction scores became poorer with increasing levels of depression severity, but there is evidence of gender differences here. Rates of increase in sleep satisfaction are quite steady for men; for women this increase is very much more dramatic. Gender differences in respect of sleep satisfaction scores may be apparent only where depression is most severe.

Split File method

While the line graph is useful, we need more objective statistical data. We can use SPSS to produce some additional tests, the type of which depends on the number of groups for each variable. In our case, we need to explore mean sleep satisfaction scores in respect of depression status, but for each gender group. Because there are three depression severity groups, we need to examine that using independent one-way ANOVAs. We perform this twice – once for males and once for females, – we have to split the data set to do this (we will see how to do that shortly). We also need to examine mean sleep satisfaction scores in respect of gender, at each level of depression severity (splitting the file once again). As gender has two groups, we need independent t-tests, which we do three times (once for each of the depression severity groups). To account for multiple comparisons, significance cut-off points should be adjusted according to the number of additional tests undertaken.

11.8 Take a closer look

Additional tests needed to locate source of interaction

Between-group interactions	Method for locating source
Sleep satisfaction scores across depression severity groups, but according to gender	2 × independent one-way ANOVAs 1. when gender = male 2. when gender = female
Sleep satisfaction scores across gender, but according to depression severity	3 × independent t-tests: 1. when depression = none 2. when depression = mild 3. when depression = severe

Sleep satisfaction vs. depression severity, according to gender

To investigate sleep satisfaction scores in respect of depression severity, according to gender groups, we need to split our data across the two gender groups. Once we have done that we can undertake two independent one-way ANOVAs. As it happens, because of the flexibility of the function we are about to execute, we only need to 'ask' for the test once.

Using the SPSS file Sleep satisfaction

Select **Data** → **Split File** → (in new window) select **Compare Groups** radio button → transfer **Gender** to **Groups Based on:** → click **OK**

Select **Analyze** → **Compare Means** → **One-Way ANOVA...** → (in new window) transfer **Sleep satisfaction** to **Dependent Variable** → transfer **Depression status** to **Factor** → click **Options...** → (in new window) select **Descriptive** → click **Continue** → click **OK**

Gender		N	Mean	Std. deviation	Std. error	95% confidence interval for mean		Minimum	Maximum
						Lower bound	Upper bound		
Male	None	15	47.80	9.689	2.502	42.43	53.17	33	67
	Mild	15	53.60	14.677	3.789	45.47	61.73	20	76
	Severe	15	59.53	11.363	2.934	53.24	65.83	42	76
	Total	45	53.64	12.766	1.903	49.81	57.48	20	76
Female	None	15	46.67	11.114	2.870	40.51	52.82	35	67
	Mild	15	58.40	14.913	3.851	50.14	66.66	37	86
	Severe	15	74.67	13.211	3.411	67.35	81.98	50	97
	Total	45	59.91	17.332	2.584	54.70	65.12	35	97

Figure 11.14 Descriptives (reported by gender)

We have undertaken two additional tests, so we divide the significance cut-off point by two. We will have a significant outcome only where $p < .025$. Figure 11.14 presents the mean scores and other descriptive data, by group. Figure 11.15 indicates that sleep satisfaction scores get significantly poorer across increasing levels of depression severity for women only: $F(2, 42) = 17.095$, $p < .001$; the effect for men is non-significant: $F(2, 42) = 3.533$, $p = .038$.

Gender		Sum of squares	df	Mean square	F	Sig.
Male	Between-groups	1032.578	2	516.289	3.533	.038
	Within-groups	6137.733	42	146.137		
	Total	7170.311	44			
Female	Between-groups	5931.378	2	2965.689	17.095	.000
	Within-groups	7286.267	42	173.483		
	Total	13217.644	44			

Figure 11.15 ANOVA outcome (reported by gender)

Sleep satisfaction vs. gender, according to depression severity

To investigate sleep satisfaction scores in respect of gender, according to depression severity, we need to split our data across the three depression groups. Once we have done that we can undertake 'three' independent t-tests.

Using the SPSS file Sleep satisfaction

Select **Data** → **Split File** → (in new window) select **Compare Groups** radio button → transfer **Depression status** to **Groups Based on:** → click **OK**

Select **Analyze** → **Compare Means** → **Independent Samples T Test...** → (in new window) transfer **Sleep satisfaction** to **Test Variable(s)** → transfer **Gender** to **Grouping Variable** → click **Define Groups** → (in new window) enter **1** in box for **Group 1** → enter **2** in box for **Group 2** → click **Continue** → click **OK**

We have three additional tests, so we divide the significance cut-off point by three; we will have a significant outcome only where $p < .016$. Figure 11.16 presents the mean scores and

Depression status		Gender	N	Mean	Std. deviation	Std. error mean
None	Sleep satisfaction scores	Male	15	47.80	9.689	2.502
		Female	15	46.67	11.114	2.870
Mild	Sleep satisfaction scores	Male	15	53.60	14.677	3.789
		Female	15	58.40	14.913	3.851
Severe	Sleep satisfaction scores	Male	15	59.53	11.363	2.934
		Female	15	74.67	13.211	3.411

Figure 11.16 Descriptive data (reported by depression severity)

Independent samples test

Depression status			Levene's test for equality of variances		t-test for equality of means						
										95% confidence interval of the difference	
			F	Sig.	t	df	Sig. (2-tailed)	Mean difference	Std. error difference	Lower	Upper
None	Sleep satisfaction scores	Equal variances assumed	.615	.440	.298	28	.768	1.133	3.807	−6.665	8.932
		Equal variances not assumed			.298	27.489	.768	1.133	3.807	−6.672	8.938
Mild	Sleep satisfaction scores	Equal variances assumed	1.226	.278	−.888	28	.382	−4.800	5.402	−15.866	6.266
		Equal variances not assumed			−.888	27.993	.382	−4.800	5.402	−15.867	6.267
Severe	Sleep satisfaction scores	Equal variances assumed	.040	.843	−3.364	28	.002	−15.133	4.499	−24.350	−5.917
		Equal variances not assumed			−3.364	27.388	.002	−15.133	4.499	−24.359	−5.908

Figure 11.17 Independent t-test (reported by depression severity) – truncated

other descriptive data, by group. Figure 11.17 shows that there is a significant difference in satisfaction scores only when the participants have severe depression: $t\ (28) = -3.364$, $p = .002$; gender differences are non-significant for mild, $t(28) = -0.888$, $p = .382$, and no depression, $t(28) = -0.298$, $p = .768$. Sleep satisfaction scores become significantly poorer with increasing depression severity only for women, and these scores are significantly poorer for women when they have severe depression.

You must remember to switch off the Split File facility, otherwise subsequent analyses will be incorrect.

Select **Data** → **Split File** → (in new window) select **Analyze all cases, do not create groups** radio button → click **OK**

Effect size and power

We can use G*Power to calculate effect size and power (see Chapter 4). This time, we need to do three analyses: one for each of the main effects and one for the interaction:

Open G*Power:

From **Test family** select **F tests**
From **Statistical test** select **ANOVA: Fixed effects, special, main effects and interaction**
From **Type of power analysis** select **Post hoc: Compute achieved – given α, sample size and effect size power**

Now we enter the **Input Parameters**:

Gender main effect

To calculate the **Effect size**, click on the **Determine** button (a new box appears).
In that new box, tick onradio button for Direct ➜ type **0.062** in the **Partial η^2** box (we get that from Figure 11.12, referring to the 'eta squared figure' for Gender, as highlighted in orange) ➜ click on **Calculate and transfer to main window**

Back in original display, for **α err prob** type **0.05** (the significance level) ➜ for **Total sample size** type **90** (the overall sample size) ➜ for **Numerator df** type **1** (we get the *df* from the Gender row in Figure 11.12) ➜ for **Number of groups** type **2** (Gender groups for male and female) ➜ click on **Calculate**

From this, we can observe two outcomes: Effect size (d) **0.26** (which is 'medium'); and Power (1-β err prob) **0.67** (which is a little short of our target of 0.80- see Chapter 4 to see why that is the optimal outcome).

Depression severity main effect

Follow procedure from above, in Determine box type **0.307** in the **Partial η^2** box ➜ click on **Calculate and transfer to main window** ➜ back in original display **α err prob** and **Total sample size** remain as above ➜ for **Numerator df** type **2** ➜ for **Number of groups** type **3** (Depression groups for None, Mild, and Severe) ➜ click on **Calculate**

From this, we can observe two outcomes: Effect size (d) **0.67** (which is large); and Power (1-β err prob) **1.00** (which is 'perfect').

Interaction

In Determine box type **0.070** in the **Partial η^2** box ➜ click on **Calculate and transfer to main window** ➜ back in original display **α err prob** and **Total sample size** remain as above for **Numerator df** type **2** ➜ for **Number of groups** type **6** (3 Depression groups for × 2 Gender groups) ➜ click on **Calculate**

From this, we can observe two outcomes: Effect size (d) **0.27** (which is medium); and Power (1-β err prob) **0.62** (which is also a little underpowered).

Writing up results

We should also present a narrative summary of those findings:

> An independent two-way ANOVA indicated that there was a significant main effect for gender in respect of sleep satisfaction scores ($F(1, 84) = 5.529$, $p = .021$; $d = 0.26$), and a significant main effect for depression severity ($F(2, 84) = 18.608$, $p < .001$; $d = 0.67$). Tukey *post hoc* analyses on the depression severity data showed that sleep satisfaction was poorer for those reporting severe depression compared with no depression ($p < .001$), and than for those reporting mild symptoms ($p = .003$). Furthermore, sleep satisfaction scores were poorer for those reporting mild depression than for those reporting none ($p = .023$). There was significant interaction between gender and depression severity ($F(2, 84) = 3.180$, $p = .047$; $d = .27$). Statistical and graphical analyses indicated that the interaction was illustrated by sleep satisfaction only becoming significantly poorer with increasing depression severity for women, and that sleep satisfaction scores were only poorer for women than men when they had severe depression.

Table 11.2 Sleep satisfaction scores according to HADS depression levels and gender

Sleep satisfaction scores									
				Depression status vs. gender					
	Main effects			Male			Female		
Depression status	Mean	SD	N	Mean	SD	N	Mean	SD	N
None	47.23	10.26	30	47.80	9.69	15	46.67	11.14	15
Mild	56.00	14.74	30	53.60	14.68	15	58.40	14.91	15
Severe	67.10	14.35	30	59.53	11.36	15	74.67	13.21	15
Gender									
Male	53.64	12.77	45						
Female	59.91	17.33	45						
Grand mean	56.78	15.46	90						

Chapter summary

In this chapter we have explored the independent multi-factorial ANOVA. At this point, it would be useful to revisit the learning objectives that we set at the beginning of the chapter.

You should now be able to:

- Recognise that we use an independent multi-factorial ANOVA to explore differences in mean scores from a single parametric dependent variable, across two or more between-group independent variables, each with two or more groups.
- Appreciate that, if a significant outcome is found across an independent variable that has three or more groups, we need additional tests to locate the source of the difference: planned contrasts can be used if we have made specific predictions about the outcomes between each group; *post hoc* tests must be applied in all other cases.
- Understand that main effects may be found for each of the independent variables, in respect of the dependent variable outcome, and that there may be interactions between those independent variables.
- Appreciate how to locate the source of interactions.
- Understand that the data should be interval or ratio, and be reasonably normally distributed. There should also be homogeneity of variances between the groups. Group membership must be exclusive; no person (or case) can appear in more than one group of a single independent variable.
- Calculate outcomes manually (using maths and equations).
- Perform analyses using SPSS, and know how to select the appropriate planned contrast or post hoc test.
- Examine significance of outcomes, and the source of significance by using additional independent t-tests and independent one-way ANOVAs where appropriate.
- Explore effect size and statistical power.
- Understand how to present the data and report the findings.

Research example

It might help you to see how independent multi-factorial ANOVA has been applied in published research. In this section you can read an overview of the following paper:

> Yuen, H.K. and Hanson, C. (2002). Body image and exercise in people with and without acquired mobility disability. *Disability and Rehabilitation*, 6: 289–296. DOI: http://dx.doi.org/10.1080/09638280110086477

If you would like to read the entire paper you can use the DOI reference provided to locate that (see Chapter 1 for instructions). In this research the authors investigated self-perceived body image in a group of 60 people according to whether they had a severe physical disability (or not) and whether they took active exercise. Physical disability was represented by participants who either had a spinal cord injury, or had undergone a lower-limb amputation (thus using a wheelchair). Collectively, they were referred to as the Acquired Mobility Disability (AMD) group. Thirty AMD adults were recruited along with 30 controls, matched for gender, age, ethnicity and exercise level. AMD and controls were further divided by the extent of their self-reported exercise: active vs. non-active. Due to prior matching, in both groups there were 17 active participants and 13 sedentary. Perceptions regarding body image were taken from the Multidimensional Body-Self Relations Questionnaire (MBSRQ; Cash 1990). The MBSRQ focuses on a range of body image and weight-related perceptions, including physical appearance, fitness, satisfaction with specific body parts (face, hair, torso, etc.) and pre-occupation with weight, diet, body fat and so on. All perceptions were measured on a Likert scale of 1–5, with higher scores representing greater satisfaction.

The results showed that there was a significant main effect for disability on MBSRQ scores for appearance orientation, $F(1, 56) = 10.44, p = .002$ (AMD group scored higher) and health evaluation, $F(1, 56) = 10.48, p = .002$ (able-bodied scored higher). There was also a significant main effect for exercise for appearance evaluation, $F(1, 56) = 10.75, p = .002$; fitness orientation, $F(1, 56) = 40.96, p < .001$; health orientation, $F(1, 56) = 9.31, p = .003$; body areas satisfaction, $F(1, 56) = 6.44, p = .014$; and fitness evaluation, $F(1, 56) = 4.41, p = .040$. Active participants scored higher on all of these sub-scales. There was a significant interaction between disability and exercise on MBSRQ scores, $F(1, 56) = 22.46, p, < .001$ (active able-bodied individuals scored higher on fitness perceptions than any other sub-group) and fitness orientation, $F(1, 56) = 7.84, p = .007$ (non-active able-bodied individuals felt less satisfied than anyone else). To clarify (as the authors do not), 'orientation' is the emphasis that someone puts on the importance of a factor (such as appearance, health or fitness); 'evaluation' is how people 'feel' about themselves on that factor. Despite some points of clarity, this paper demonstrates well the use of an independent multi-factorial ANOVA in research practice. However, the authors describe the statistical procedure as a '2 × 2 two-way ANOVA'. This is rather vague – as we will see in subsequent chapters, it is useful to indicate whether this is an independent, repeated-measures, or mixed multi-factorial ANOVA.

Extended learning task

You will find the data set associated with this task on the website that accompanies this book (available in SPSS and Excel format). You will also find the answers there.

Following what we have learned about independent multi-factorial ANOVA, answer these questions and conduct the analyses in SPSS and G*Power. If you do not have SPSS, do as much as you can with the Excel spreadsheet. For this exercise, we examine patient anxiety towards a forthcoming operation, which is measured on a scale of 0 (very anxious) to 100 (no anxiety). We will explore this in relation to whether the patient has a history of anxiety and in respect of the nature of information supplied to that patient.

Open the data set **Operation anxiety**

1. Describe the independent variables, including the groups that represent them.
2. What is the dependent variable?
3. Check for normal distribution across the groups.
4. Conduct an independent multi-factorial ANOVA.
5. Describe what the SPSS output shows.
6. Explain how you accounted for homogeneity of variance.
7. Describe the main effects and interactions.
8. Conduct appropriate additional analyses to indicate where the differences are (if there were any).
9. State the effect size and power, using G*Power.
10. Report the outcome as you would in the results section of a report.

Appendix to Chapter 11

Exploring simple effects

In the main text of this chapter we looked at how to explore the source of interaction between two independent variables, in respect of an outcome. We started by illustrating the interaction by way of a line graph (Figure 11.4). This is useful in that it provides some idea about where the potential source of interaction might lie. However, we need statistical measures to confirm that. We also learned how to use the 'Data Split File' facility in SPSS to explore differences in one independent variable at each level of the second independent variable. This method will probably suffice for most cases, if only to give us a fairly accurate guide. However, strictly speaking, that method does not fully account for the inter-relationship between the variables. To see how one independent variable operates at each level of another, taking into account these factors, we need something called 'simple effects'. In order to do these we need to use syntax, which is the coding language used by SPSS. This section is rather more advanced than what would normally be expected for an undergraduate student to understand – what we performed earlier should be more than enough for most needs. However, you may want to learn this, more advanced, method if you are really keen. Syntax cannot be obtained through the usual menus in SPSS – you have to write these out for yourself (sorry!).

11.9 Take a closer look

Using syntax

Syntax is the programming language that SPSS uses in performing the data analysis. Most of the time we do not use it directly, as SPSS converts the menu commands that we set into syntax. You can see what that syntax looks like, as it is usually shown in the output just prior to the main data analysis – it will look something like the example below. When we present the output tables in this book, the syntax lines are omitted because we do not usually need to see them. This syntax describes the procedure that we used to run the independent two-way ANOVA example we ran earlier:

```
UNIANOVA sleep BY gender depressionstatus
/METHOD=SSTYPE(3)
/INTERCEPT=INCLUDE
/POSTHOC=depressionstatus(TUKEY)
/PRINT=ETASQ HOMOGENEITY DESCRIPTIVE
/CRITERIA=ALPHA(.05)
/DESIGN=gender depressionstatus gender*depressionstatus.
```

You don't need to worry too much about what all of that means, but a brief overview might be useful. The first line describes the type of test used; 'UNIANOVA' suggests that it was a univariate ANOVA, which was explored using 'sleep' as the dependent variable, with 'gender' and 'depressionstatus' as the independent variables. The next two lines refer to some options that we took by default (sum of squares type and whether we included the intercept), so don't think about those too much as we did not cover it. The fourth line confirms that we selected Tukey as the *post hoc* test for 'depressionstatus'. The sixth line says that we selected estimates of effect size, homogeneity of variance and descriptive statistics when we set the Options. The next line is another default (we left the significance at p = .05). The final line confirms our main effects (Gender and Depression status) and the interaction (Gender vs. Depression status).

Finding the simple effects for our independent variables

We will now find the simple effects for HADS depression level on sleep satisfaction scores, across the levels of Gender, and those for Gender across the levels of HADS depression level. We will do this using syntax.

Open SPSS data set Sleep satisfaction

Select **File → New → Syntax** (you will see a blank syntax window, as shown in Figure 11.18).

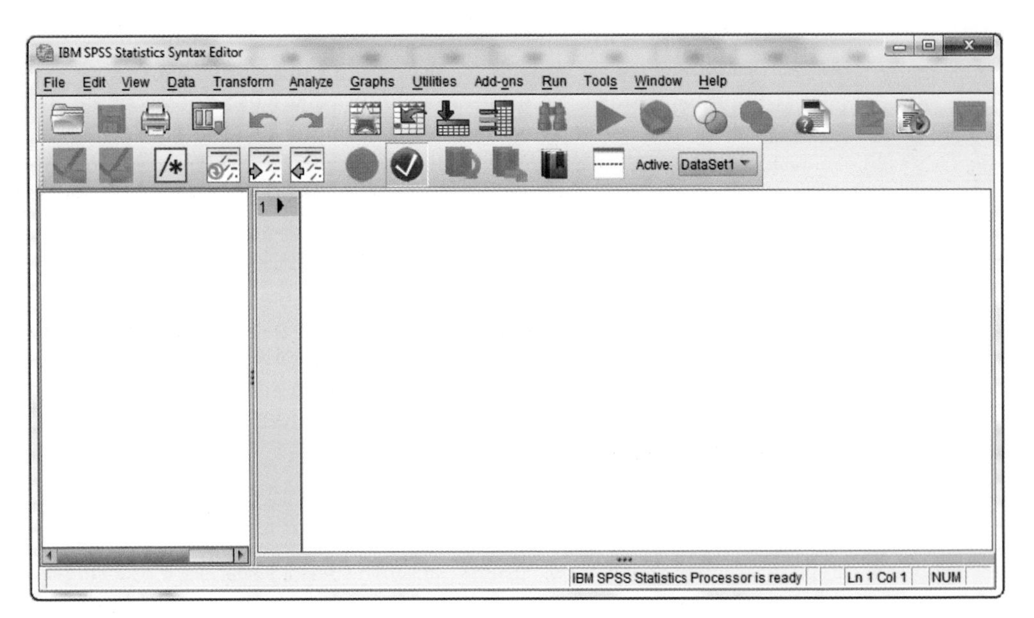

Figure 11.18 Blank syntax window

Now we insert the commands; the following must copied *exactly* into the command window:

```
MANOVA sleep BY gender (12) depressionstatus (1 3)
/DESIGN = gender WITHIN depressionstatus(1) gender WITHIN depressionstatus(2)
gender WITHIN depressionstatus(3)
/DESIGN = depressionstatus WITHIN gender(1) depressionstatus WITHIN gender(2)
/PRINT CELLINFO SIGNIF(UNIV MULT AVERF HF GG).
```

When that code is copied into the syntax window SPSS adds colour to help identify the elements. The revised syntax window is shown in Figure 11.19.

We should explain what those terms mean. MANOVA is the simple effect name (you don't need to know why). This is followed in the first line by the parameters: the dependent variable (sleep) and the two independent variables (gender) and (depressionstatus). The numbers in brackets indicate the lowest and highest nominal categories for each variable. The next two lines

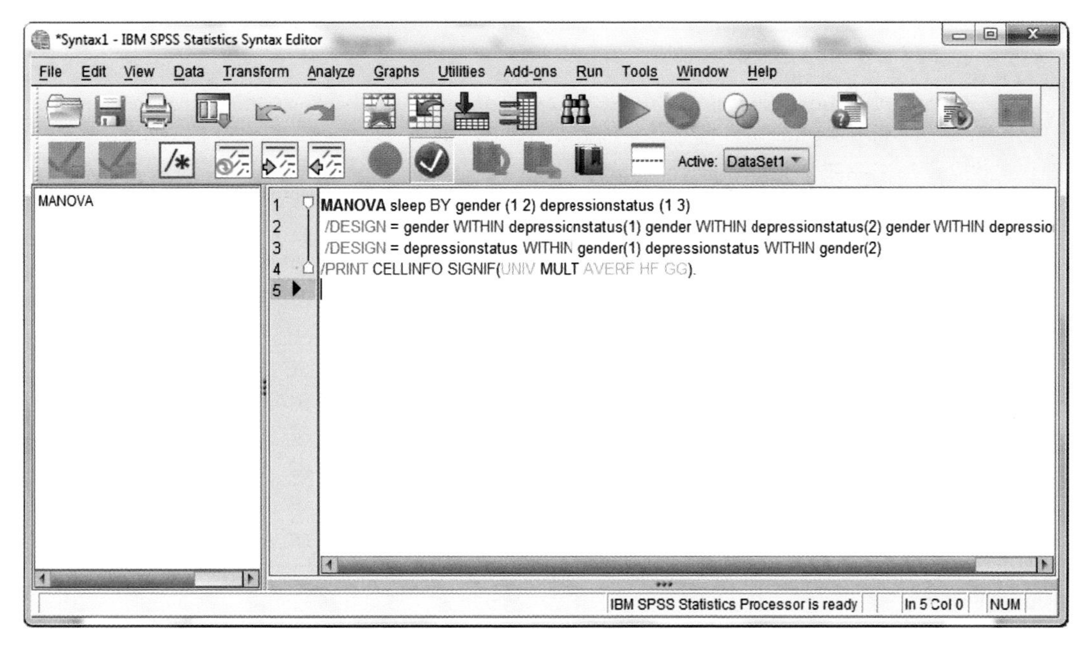

Figure 11.19 Completed syntax window

(**DESIGN**) present the command to find the simple effects: we look at 'gender' when 'depressionstatus' categories are 1, 2 and 3; then we examine 'depressionstatus' when 'gender' categories are 1 and 2. The PRINT line determines the output tables (don't worry about what that means). Once you have copied the text shown above into the command window, perform the following action:

You will get a series of output tables – those of most concern to us are shown in Figure 11.20.

These analyses show how to interpret the interaction that we found in the independent two-way ANOVA. The order in which the variables are presented in the 'Cell Means and Standard Deviations' will help us identify which effects the remaining output tables are referring to.

'Analysis of Variance – Design 1' explores the effect of 'gender' at each level of 'depressionstatus' in respect of 'sleep satisfaction scores'. It shows that we have a significant simple effect (model): $F(3, 89) = 2.81$, $p = .044$. It means that the interaction is partly found in the way that sleep satisfaction scores varied in respect of gender, but did so differently at each level of the depression severity independent variable. There was no difference in sleep satisfaction scores between men and women when 'depression status = 1' (no depression; $F(1, 86) = 0.04$, $p = .837$) or '2' (mild depression; $F(1, 86) = 0.77$, $p = .384$), but there was a significant difference across gender when 'depression status = 3' (severe depression; $F(1, 86) = 7.63$, $p = .007$). This supports what we saw earlier, but we can be confident that we have explored unique variation within the gender variable.

'Analysis of Variance – Design 2' explores the effect of 'depression severity' at each level of Gender (male vs. female), in respect of 'sleep satisfaction scores'. It shows that we have a significant simple effect (model): $F(4, 893) = 10.34$, $p < .001$. The order that gender was presented in the output is important here: males were shown first, so they are 'gender (1)' in this analysis. The output shows that the interaction is found in the way that sleep satisfaction scores varied in

```
Cell Means and Standard Deviations
Variable .. sleep Sleep satisfaction scores
     FACTOR           CODE              Mean    Std. Dev.   N          95%CI

    gender           Male
    depressi         None             47.800      9.689    15     42.434 - 53.166
    depressi         Mild             53.600     14.677    15     45.472 - 61.728
    depressi         Severe           59.533     11.363    15     53.241 - 65.826
    gender           Female
    depressi         None             46.667     11.114    15     40.512 - 52.821
    depressi         Mild             58.400     14.913    15     50.141 - 66.659
    depressi         Severe           74.667     13.211    15     67.351 - 81.983
    For entire sample                 56.778     15.460    90     53.540 - 60.016

* * * * * Analysis of Variance -- Design   1 * * * *

    Tests of Significance for sleep using UNIQUE sums of squares
    Source of Variation                   SS        DF      MS          F   Sig of F

    WITHIN+RESIDUAL                    19371.49     86    225.25

    GENDER WITHIN DEPRESSIONSTATUS(1)      9.63      1      9.63    .04      .837
    GENDER WITHIN DEPRESSIONSTATUS(2)    172.80      1    172.80    .77      .384
    GENDER WITHIN DEPRESSIONSTATUS(3)   1717.63      1   1717.63   7.63      .007

    (Model)                             1900.07      3    633.36   2.81      .044
    (Total)                            21271.56     89    239.01

    * * * * * Analysis of Variance -- Design   2 * * * *

    Tests of Significance for sleep using UNIQUE sums of squares
    Source of Variation                   SS        DF      MS      F   Sig of F

    WITHIN+RESIDUAL                    14307.60     85    168.32

    DEPRESSIONSTATUS WITHIN GENDER(1)   1032.58      2    516.29  3.07      .052
    DEPRESSIONSTATUS WITHIN GENDER(2)   5931.38      2   2965.69 17.62      .000

    (Model)                             6963.96      4   1740.99 10.34      .000
    (Total)                            21271.56     89    239.01

    * * * * * Analysis of Variance -- Design   3 * * * *

    Tests of Significance for sleep using UNIQUE sums of squares
    Source of Variation                   SS        DF      MS      F     Sig of F

    WITHIN CELLS                       13424.00     84    159.81
    gender                               883.60      1    883.60    5.53      .021
    depressionstatus                    5947.49      2   2973.74   18.61      .000
    gender BY depressionstatus          1016.47      2    508.23    3.18      .047

    (Model)                             7847.56      5   1569.51    9.82      .000
    (Total)                            21271.56     89    239.01
```

Figure 11.20 Simple effects output

respect of depression severity, but did so differently for males than females. There was no significant difference for sleep satisfaction scores across depression severity when 'gender = 1' (male; F (2, 85) = 3.07, p = .052), but there was a significant difference when 'gender = 2' (female; F(2, 85) = 17.62, p < .001). This supports what we saw earlier, but we can be confident that we have explored unique variation within the depression severity variable.

'Analysis of Variance – Design 3' confirms the overall ANOVA outcome that we saw in the main analyses.

12 REPEATED-MEASURES MULTI-FACTORIAL ANOVA

Learning objectives

By the end of this chapter you should be able to:

- Recognise when it is appropriate to use repeated-measures multi-factorial ANOVA
- Understand the theory, rationale, assumptions and restrictions associated with the test
- Calculate outcomes manually (using maths and equations)
- Perform analyses using SPSS, and explore outcomes, the source of significance and source of interaction
- Explore effect size and statistical power
- Understand how to present the data and report the findings

What is repeated-measures multi-factorial ANOVA?

A repeated-measures multi-factorial ANOVA examines outcomes from a single (parametric) dependent variable, across two or more categorical independent variables, each represented by two or more within-group conditions (all conducted across a single group). We saw what we meant by parametric data in Chapter 5. As we saw with independent multi-factorial ANOVA (Chapter 11), a feature of this test is that we can measure outcomes for each independent variable, and the extent that the independent variables 'interact' – an interaction occurs when the outcome across one independent variable differs across the groups or conditions of another independent variable. It is more than likely that you would choose to perform repeated-measures multi-factorial ANOVA *because* you had predicted that there would be an interaction. Locating the source of interactions is little more straightforward than it is for the between-group version.

Research questions for repeated-measures multi-factorial ANOVA

We can illustrate repeated-measures multi-factorial ANOVA by way of a series of research questions. We will need two different examples to reflect the two types of analysis we will demonstrate in this chapter. We need to perform slightly different methods according to the number of within-group conditions that make up the independent variables. In the first instance, we will explore outcomes where both independent variables have just two conditions. In the second scenario, we will see an example where one of those independent variables has three conditions, while the other has two.

In our first example, a group of cognitive psychology researchers, CALM, investigate factors that have an impact on memory. In this research, CALM explore how interference from competing modes of perception can compromise memory. If participants are presented stimuli visually, how well will they recall items if they are simultaneously presented with auditory information? To test this, the researchers examine participants' recall of words and numbers, in the presence of verbal or numeric interference. As a result, there are four scenarios, manipulating information that is presented on a computer screen:

1. Visually present words and audibly present the sound of someone chanting numbers.
2. Visually present words and audibly present the sound of someone saying contrasting words.
3. Visually present numbers and audibly present the sound of someone saying words.
4. Visually present numbers and audibly present someone chanting contrasting numbers.

All of these scenarios are presented to a single group of people. There are two within-group independent variables, each with two conditions – this will produce a 2 × 2 repeated-measures multi-factorial ANOVA. In essence, for each presentation (word or number) the interference is either different (number for word/word for number) or the same (word for word/number for number). Thirty minutes after each presentation, the participant has to write down as many of items that they could recall. CALM predict the similar-mode interferences to be associated with poorest recall.

12.1 Take a closer look
Variables for 2 × 2 repeated-measures multi-factorial ANOVA

Dependent variable:	Number of items recalled
Independent variable 1:	Presentation: word or number
Independent variable 2:	Interference: same or different

In the second research example, a group of higher education researchers, FUSS, seek to examine student preferences towards the type of lesson they receive and the expertise of the person who delivers that lesson. To measure student satisfaction with course content, the researchers manipulate those two variables. There are three types of lesson: interactive lecture, standard lecture or video. These lessons are presented by two types of lecturer: expert or novice. All students attend all of the scenarios, for which there are two within-group independent variables, one with the three conditions and the other with two conditions. In each scenario the students rate their satisfaction with the content. FUSS predict that students will report most satisfaction with the interactive lecture, compared with other lesson types. They also predict that students will prefer the expert lecturer. They further hypothesise that the expert lecturer will be preferred more strongly in the interactive lecture scenario than in any other condition. This research would be examined with a 3 × 2 repeated-measures multi-factorial ANOVA.

12.2 Take a closer look

Variables for 3 × 2 repeated-measures multi-factorial ANOVA

Dependent variable: Satisfaction with course content
Independent variable 1: Lecture type: interactive lecture, standard lecture or video
Independent variable 2: Lecturer expertise: expert or novice

Theory and rationale

Main effects and interactions

We first encountered main effects and interactions in Chapter 11. The outcome of a single independent variable is a 'main effect', while an 'interaction' occurs when there is a significant difference in the outcome across one independent variable at different conditions of another independent variable. The interaction effect should never be seen as a by-product of the analysis. Such an effect should have been predicted prior to selecting repeated-measures multi-factorial ANOVA In Chapter 11, we saw a series of graphical representations of main effects and interactions (you might like to look at those again). The nature of the variables in within-group analyses will differ for those graphs, but the principle is the same. We will see some graphical examples of within-group main effects and interactions later.

We could illustrate main effects and interactions in repeated-measures multi-factorial ANOVA, using the research example set by CALM. You will recall that they are examining how same vs. different perception modality interference might affect recall of items presented to participants. There are two within-group independent variables, each with two conditions. Here are some examples of what we might observe:

Possible main effect outcomes

Within-group independent variable 1: presentation

Participants might recall more words than numbers

Within-group independent variable 2: interference

Participants might recall more items when interference is in a different modality

Possible interaction

Within-group interaction: presentation vs. interference

Same-modality interference might be more pronounced for numbers than words

Identifying differences

The methods used to identify within-group differences for repeated-measures multi-factorial ANOVA are similar to the 'one-way' version (Chapter 10). The extent that response outcomes vary across conditions is illustrated by the variance. The overall variance is represented by the total sum of squares. In within-group studies, this is partitioned into within-participant and between-participant variance. The latter relates to how the scores vary across the sample, which we do not need in these analyses. We need to partition the within-participant variance into that which can be explained and that which cannot. Explained variance will be found in the model sum of squares. The unexplained (error) variance will be represented by the residual sum of squares. We need to partition the model sum of squares and residual sum of squares for each of the independent variables (to indicate main effects) and for the interaction(s). The model sum of squares is found from condition mean scores and sub-condition mean scores, in relation to the grand mean (the average number of items recalled regardless of condition). Residual sums of squares are found from case variances (we can see how this is all done in Box 12.3). The sums of squares are expressed in relation to degrees of freedom to find the mean squares; degrees of freedom (*df*) represent the number of values that are 'free to vary' in the calculation, while everything else is held constant (see Chapter 6). An F ratio is found by dividing the model mean square by the residual mean square; this is undertaken for each independent variable and for the interaction.

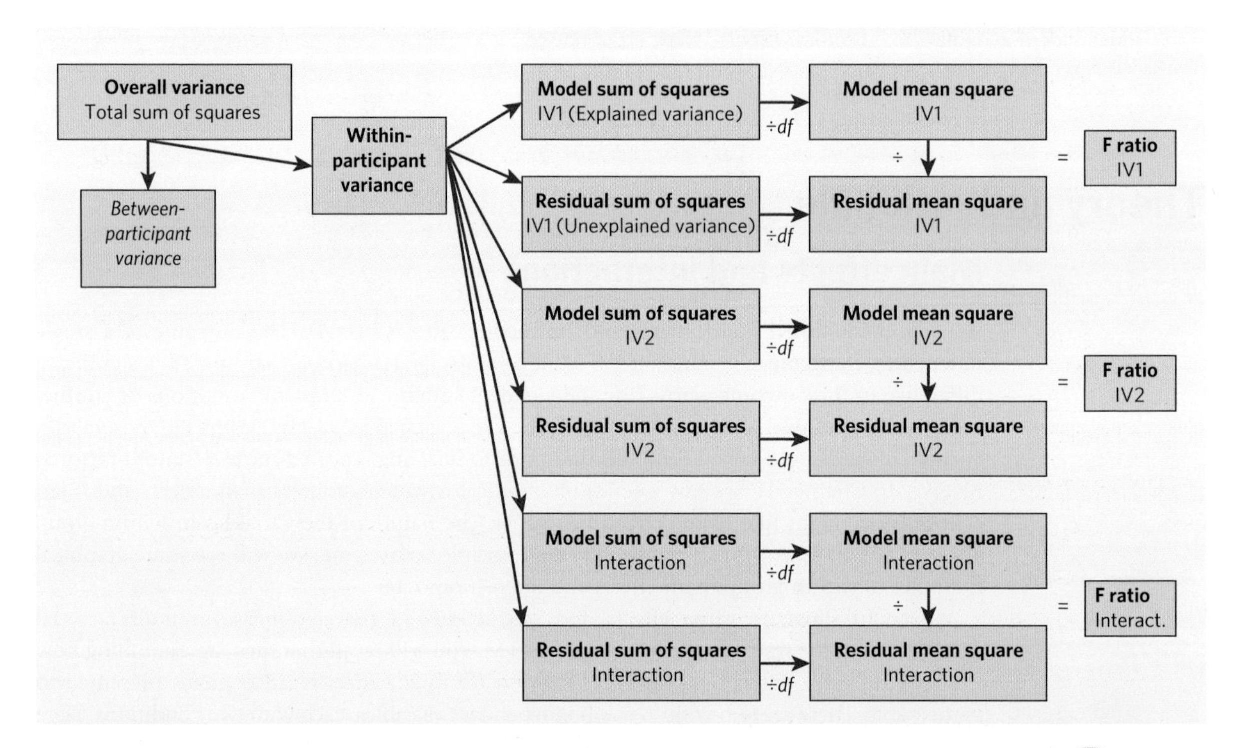

12.3 Calculating outcomes manually

Repeated-measures multi-factorial ANOVA calculation

To illustrate how we can calculate the outcome of a repeated-measures multi-factorial ANOVA, we will examine some data relevant to the CALM research question. The researchers are examining recall across two independent variables: item presentation (word or number) and interference (same or different). Participants are presented with items on a computer screen, whilst receiving simultaneous auditory information. The dependent variable is the number of items the participants can correctly recall 30 minutes after the presentation. The data are shown in Table 12.1. **You will find a Microsoft Excel spreadsheet associated with these calculations on the web page for this book.**

Table 12.1 No. of items recalled following numeric or verbal interference

	Word		**Number**		**Presentation mean**		**Interference mean**					
Partici-pant	**Diff**	**Same**	**Diff**	**Same**	**Word**	**Number**	**Diff**	**Same**	**Overall mean**	**Case variance All**	**Case variance P**	**Case variance I**
1	28	4	22	5	16.00	13.50	25.00	4.50	14.75	146.25	3.13	210.13
2	22	3	18	6	12.50	12.00	20.00	4.50	12.25	84.25	0.13	120.13
3	21	3	17	14	12.00	15.50	19.00	8.50	13.75	59.58	6.13	55.13
4	27	4	21	8	15.50	14.50	24.00	6.00	15.00	116.67	0.50	162.00
5	21	3	17	14	12.00	15.50	19.00	8.50	13.75	59.58	6.13	55.13
6	20	3	17	5	11.50	11.00	18.50	4.00	11.25	72.25	0.13	105.13
7	19	3	16	11	11.00	13.50	17.50	7.00	12.25	48.92	3.13	55.13
8	16	2	14	8	9.00	11.00	15.00	5.00	10.00	40.00	2.00	50.00
9	25	4	20	10	14.50	15.00	22.50	7.00	14.75	90.25	0.13	120.13
10	17	2	15	11	9.50	13.00	16.00	6.50	11.25	44.25	6.13	45.13
11	19	3	16	8	11.00	12.00	17.50	5.50	11.50	53.67	0.50	72.00
12	20	3	17	8	11.50	12.50	18.50	5.50	12.00	62.00	0.50	84.50
Mean	**21.25**	**3.08**	**17.50**	**9.00**	**12.17**	**13.25**	**19.38**	**6.04**	**Sum:**	**877.67**	**28.50**	**1134.50**

Grand mean: 12.71 **Grand variance:** 58.55 *(We saw how to calculate grand variance in Chapter 9)*

Total sum of squares (SS_T):

$SS_T = s^2_{grand}(N - 1)$ = grand variance × number of scores (48) *minus* 1 = 58.55 × 47 = **2751.92**

Within-participant sum of squares (SS_W):

SS_W = Case variance (total) × number of conditions (3) *minus* 1 = 877.67 × 3 = **2633.00**

We saw how to calculate case variance in Chapter 10; this is the case variance regardless of condition.

Between-participant sum of squares (SS_B): $SS_T - SS_W$ = 118.92

Model sum of squares (SS_M) $\Sigma n_k(\bar{x}_k - \bar{x}_{grand})^2$

SS_M includes the model (explained) variance and the residual (unexplained) variance. The model sum of squares needs to be partitioned into each IV and the interaction between the IVs.

SS_M Total

Deduct grand mean from sub-condition mean, square it, multiply by number of participants (12)

$= (12 \times (21.5 - 12.71)^2) + (12 \times (3.08 - 12.71)^2) + (12 \times (17.50 - 12.71)^2) + (12 \times (9.00 - 12.71)^2) = 2427.75$

Presentation condition SS_M P:

Deduct grand mean from presentation mean, square it, multiply by number of scores (24)

$= (12 \times (12.17 - 12.71)^2) + (12 \times (13.25 - 12.71)^2) =$ **14.08**

df for $SS_M P (df_M P)$ = (no. of conditions *minus* 1) = 2 − 1 = 1 (this is presentation numerator *df*)

Interference condition SS_M I:

Deduct grand mean from interference mean, square it, multiply by number of scores (24)

$= (24 \times (19.38 - 12.71)^2) + (24 \times (6.04 - 12.71)^2) =$ **2133.33**

df for $SS_M I$ ($df_M I$) = (no. of conditions *minus* 1) = 2 − 1 = 1 (this is interference numerator *df*)

Interaction:

SS_M PxI = whatever is left over from overall model sum of squares

SS_M Total − SS_M P − SS_M I = 2427.75 − 14.08 − 2133.33 = **280.34**
df for SS_M P × I ($df_M P_x I$) = df_M P × df_M I = 1

Residual sum of squares (SS_R)

SS_R represents the error, but we need to express this for each IV and the interaction between IVs. This is found from the case variance for each IV.

Presentation condition:

SS_R P = (Case variance (presentation) × no. of conditions) − SS_M P = (28.50 × 2) − 14.08 = **42.92**

df for SS_RP = no. of participants *minus* 1 = 12 − 1 = 11

Interference condition:

SS_R I = (Case variance (interference) × no. of conditions) − SS_MI = (1134.50 × 2) − 2133.33 = **135.67**

df for SS_RI = no. of participants *minus* 1 = 12 − 1 = 11

Interaction:

SS_R PxI = whatever is left over from overall variance

SS_W all − SS_M all − SS_R P − SS_R I = 2633.00 − 2427.75 − 42.92 − 135.67 = **26.67**
df for SS_R PxI = no. of participants *minus* 1 = 12 − 1 = 11

Mean squares

Mean squares = sum of squares ÷ *df*

Presentation mean square

Model mean square (MS_M P) = SS_M P ÷ df_M P = 14.08 ÷ 1 = 14.08

Residual mean square (MS_R P) = SS_R P ÷ df_R P = 42.92 ÷ 11 = 3.90

Interference mean square

Model mean square (MS_M I) = SS_M I ÷ df_M I = 2133.33 ÷ 1 = 2133.33

Residual mean square (MS_R I) = SS_R I ÷ df_R I = 135.67 ÷ 11 = 12.33

Interaction mean square

Model mean square (MS_M PxI) = SS_M PxI ÷ df_M PxI = 280.33 ÷ 1 = 280.33

Residual mean square (MS_R PxI) = SS_R PxI ÷ df_R PxI = 26.66 ÷ 11 = 2.42

F ratios

To calculate the F ratios, we divide the model mean squares by residual mean square.

Presentation = MS_M P ÷ MS_R P = 14.08 ÷ 3.90 = 3.61

Interference = MS_M I ÷ MS_R I = 2133.33 ÷ 12.33 = 172.97

Interaction = MS_M PxI ÷ MS_R PxI = 280.33 ÷ 2.42 = 115.64

Each F ratio can be compared with the relevant part of the F-distribution table (see Appendix 4) according to the *df* and significance level. In using that table, we need to refer to the cross-section of the numerator *df* (model sum degrees of freedom) and the denominator *df* (residual sum degrees of freedom). We express that as (numerator *df*, denominator *df*). In this example, because there are just two conditions across both of the independent variables, the numerator *df* is 1, while the denominator *df* is 11 (for each of the main effects and for the interaction).
The cut-off point for F at (1, 11) = 4.84; any F ratio that exceeds this is significant at $p < .05$.

Presentation: F = 3.61 → no significant difference in recall across presentation

Interference: F = 172.97 → significant difference in recall across interference

Interaction: F = 115.64 → significant interaction between presentation and interference

We can also use Excel to calculate the critical value of F and to provide the actual p value (see web resource for this book). In our example Presentation: $p = .084$ Interference: $p < .001$, Interaction: $p < .001$.

Locating the source of main effects

As we saw in Box 12.3, the F ratio determines whether the mean dependent variable scores differ significantly across the independent variable groups. If that does show a significant outcome, we need to report the source of that difference. If the independent variable is represented by two within-group conditions, we simply use the mean scores to illustrate this. If there are three or more conditions, additional tests are needed to locate the source of that difference. In the CALM research example there were two conditions for both independent variables. The ANOVA outcomes in Box 12.3 suggest that there was a significant effect across the interference conditions. If we refer to the mean scores in Table 12.1, we can conclude that significantly more words were recalled when the interference was in a different sensory modality (e.g. saw words and heard numbers) than when in the same modality (saw words and heard conflicting words). There was no main effect for the type of presentation, so we do not need to explore any further with that one.

In the second of our research examples, the FUSS group are exploring student satisfaction across two independent variables. The teacher expertise variable has two conditions; if a significant main effect is found, the mean score will inform us about the source of difference. The variable measuring the type of lesson has three conditions (interactive lecture, standard lecture and video); if that main effect is significant, we will need additional tests to locate the source of the difference. As we saw in Chapter 10, we can use planned contrasts *or post hoc* tests to explore differences between each pair of conditions. We can perform planned contrasts only if specific (one-tailed) predictions were made about the relationship between each pair of conditions. If a non-specific (two-tailed) prediction is made, *post hoc* tests should be undertaken. FUSS did make specific predictions, so we might be justified in employing planned contrasts in that instance. However, as we do not have a 'control group', we could only employ a non-orthogonal contrast (we explored the conventions of use in Chapter 10). This means that we would need to adjust the significance cut-off point by the number of additional tests run (to account for multiple comparisons). Since these planned contrasts involve quite a bit of work, it is probably more sensible to run *post hoc* tests in any case (where adjustments for multiple comparisons are included).

In repeated-measures multi-factorial ANOVA, the *post hoc* test options remain the same as they are for the 'one-way'version. There are fewer choices in SPSS for within-group ANOVAs than we find in between-group tests. Of the three choices that are available, the Bonferroni adjustment is generally the most preferred. It is suitably conservative (aiming to avoid too many Type I errors), while maintaining power (this reducing Type II errors). It also automatically accounts for multiple comparisons (we have discussed this at length in earlier chapters).

Locating the source of interaction

If there is a significant interaction between any pair of independent variables in respect of outcome scores, we need to locate its source. In the CALM research example, there appears to be an interaction between presentation and interference in respect of the number of words recalled. If we look at the mean scores in Table 12.1 we can see some evidence of 'crossover': more numbers were recalled than words in the same-modality condition, while more words were recalled than numbers in the different-modality condition. We could also plot a line graph – this would provide a visual analysis (we will see how to do that later). However, ultimately, we need formal statistical analyses. The methods for exploring interactions for repeated-measures multi-factorial ANOVA are rather less laborious than they are for independent multi-factorial ANOVA. To locate the source of within-group interactions, we simply run additional repeated-measures one-way ANOVAs for each variable (or we can use related t-tests to examine variables that have only two conditions). This simplicity is due to the fact that each 'condition' has its own column in SPSS, and we can analyse these directly. We will see how this is performed a little later.

Assumptions and restrictions

There are a number of assumptions that we need to satisfy when performing repeated-measures multi-factorial ANOVA. There must be at least two independent variables, each with at least two conditions, which must be measured across a single group. Furthermore, each person (or case) must be present in all conditions of all independent variables (see Chapter 10 to see why that is important). Outcomes are taken from one (parametric) dependent variable. The data should be interval or ratio, and be reasonably normally distributed. You can read more about parametric requirements in Chapter 5. However, remember that ANOVA is robust, so is able to withstand relatively minor violations. Serious problems with normal distribution may need more additional action, such as transformation (see Chapter 3). If all of that fails, we could run non-parametric tests, but there is no equivalent test for repeated-measures multi-factorial ANOVA. Furthermore, as we saw in Chapter 10, the usual arguments regarding the subjectivity of some ordinal data are reduced in within-group studies (because it is contained within the participants). If the data are clearly non-parametric, we could perform separate Wilcoxon signed-rank tests and/or Friedman's ANOVAs for each independent variable, but this would not allow us to examine the potential interaction between the variables.

Sphericity of within-group variances

If an independent variable has three or more conditions, we need to check that we have 'equality of within-group variances' across pairs of conditions. We call this measurement 'sphericity' (see Chapter 10). If there are two conditions, it means that we have only one pair of conditions, so there is nothing to compare (we assume sphericity). We can measure sphericity with Mauchly's test, which is produced automatically by SPSS in all repeated-measures ANOVAs. This test indicates whether those variances are significantly *different* between the pairs of conditions – to assume sphericity, we need the outcome to be non-significant. When we examine the CALM data, both independent variables have two conditions, so we can ignore Mauchly's outcome (but look at Figure 12.15 to see what that output looks like in this scenario).

When we explore the FUSS data, one of the independent variables has three conditions, so we will need to consult Mauchly's outcome. The result will determine which line of F ratio outcome we can report. If Mauchly's is non-significant, we can read from the 'sphericity assumed' line. If it is not, we need to refer to one of the other three lines of ANOVA output. Each of those adjusts the F ratio to account for the violation of sphericity. We looked at the arguments determining which line to choose in Chapter 10, so we will not repeat that here. Generally, most researchers select Greenhouse–Geisser or Huynh–Feldt outcomes (there is very little to choose between them). Field (2009) suggests that we take an average of those two outcomes if they differ markedly.

12.4 Take a closer look
Summary of assumptions and restrictions

- There must be at least two categorical independent variables
- Each IV must be categorical, with at least two conditions measured across one group
  - Each person (or case) must be present in all conditions
- There must be one numerical dependent variable
  - DV data should be interval or ratio, and reasonably normally distributed
- We need to account for sphericity of within-group variances

How SPSS performs repeated-measures multi-factorial ANOVA

We will perform two SPSS analyses, one where both independent variables have just two conditions. This will be a 2 × 2 repeated-measures multi-factorial ANOVA, where we will focus on the CALM data that we explored manually earlier. The second example will explore one independent variable that has three conditions, while the other variable has two. This will be an example of a 3 × 2 repeated-measures multi-factorial ANOVA – we will use the FUSS data.

2 × 2 repeated-measures two-way ANOVA

For this example, we will continue with the example from the CALM research data that we explored manually. You may recall that the researchers were investigating how recall can be compromised by presenting conflicting information across two different senses. Words or numbers are presented to a single group on a computer screen. This is the first independent variable (with two conditions: word or number). Simultaneously, sounds are presented through headphones. These are either different to the visual presentation (words spoken when numbers are presented, or numbers chanted when words are presented), or the sounds are the same as the presented stimulus and potentially interfere with recall (words are presented while different words are spoken or numbers are presented with conflicting spoken numbers). That 'interference' is the second independent variable (with two conditions: different or same). The dependent variable is the number of correctly recalled items (30 minutes after presentation). Those 'numbers' are clearly interval, but we will still need to check whether those data are normally distributed (see later).

12.5 Nuts and bolts
Setting up the data set in SPSS

When we create the SPSS data set for repeated-measures multi-factorial ANOVA, we need to set up columns for each of the 'scenarios' that the participants will encounter. The columns are the independent variables, while the scores are the dependent variable. They are all coded as 'continuous'.

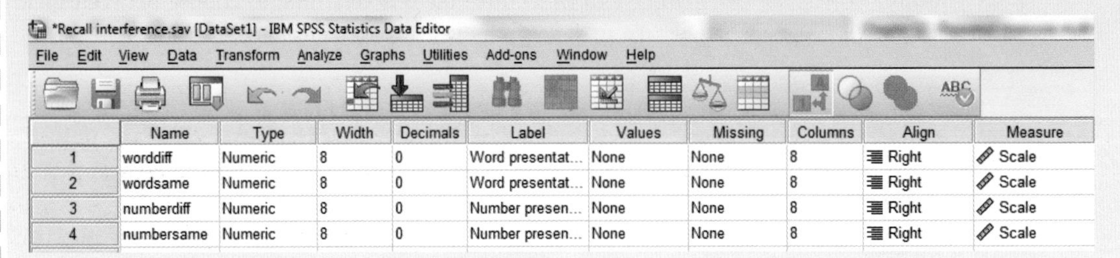

*Recall interference.sav [DataSet1] - IBM SPSS Statistics Data Editor

File Edit View Data Transform Analyze Graphs Utilities Add-ons Window Help

	Name	Type	Width	Decimals	Label	Values	Missing	Columns	Align	Measure
1	worddiff	Numeric	8	0	Word presentat...	None	None	8	Right	Scale
2	wordsame	Numeric	8	0	Word presentat...	None	None	8	Right	Scale
3	numberdiff	Numeric	8	0	Number presen...	None	None	8	Right	Scale
4	numbersame	Numeric	8	0	Number presen...	None	None	8	Right	Scale

Figure 12.1 Variable View for 'Recall interference' data

Figure 12.1 shows how the SPSS Variable View should be set up. Each of the four variable names represents the independent variables (word and number) and the conditions for each of them ('different' and 'same'). They are all 'continuous' variables, so we do not need to adjust anything in the Values column; the Measure column is set to Scale.

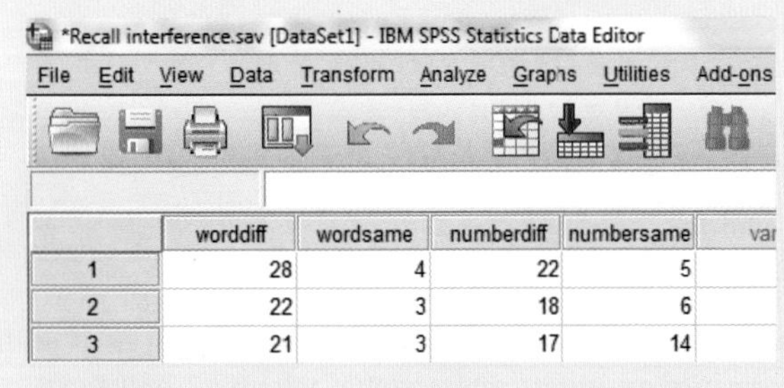

*Recall interference.sav [DataSet1] - IBM SPSS Statistics Data Editor

File Edit View Data Transform Analyze Graphs Utilities Add-ons

	worddiff	wordsame	numberdiff	numbersame	var
1	28	4	22	5	
2	22	3	18	6	
3	21	3	17	14	

Figure 12.2 Data View for 'Recall interference' data

Figure 12.2 illustrates how this will appear in the Data View. Each row represents a participant. When we enter the data we simply include the score for each participant in each condition.

Testing for normal distribution

As we have seen in previous chapters, we initially examine normal distribution with Kolmogorov-Smirnov and Shapiro-Wilk (KS/SW) tests (we saw how to do this for within-group studies in Chapter 8). If the outcome indicates that the data may not be normally distributed, we could additionally run z-score analyses of skew and kurtosis, or look to 'transform' the scores (see Chapter 3). We should look at the outcome from the KS/SW analyses (see Figure 12.3).

	Kolmogorov–Smirnov[a]			Shapiro–Wilk		
	Statistic	df	Sig.	Statistic	df	Sig.
Word presentation; different (number) interference	.193	12	.200*	.930	12	.384
Word presentation; same (word) interference	.300	12	.004	.809	12	.012
Number presentation; different (word) interference	.249	12	.038	.924	12	.323
Number presentation; same (number) interference	.211	12	.147	.915	12	.246

Figure 12.3 Kolmogorov-Smirnov/Shapiro-Wilk test for within-group conditions

Because we sample sizes of less than 50 (within all of the conditions) we should refer to the SW outcome. In Figure 12.3 three of the four levels show that the data appear to be normally distributed, because the significance ('Sig.) is greater than .05. There *may* be a problem with the 'word presentation/same interference' condition. We could perform additional z-scores analyses of skew and kurtosis (refer to Chapter 3 to see how that is done), but the overall outcome here is probably sufficient to proceed with the multi-factorial ANOVA. Remember, we are seeking *reasonable* normal distribution; ANOVA is pretty robust to minor violations.

Running repeated-measures multi-factorial ANOVA in SPSS

Using the SPSS file **Recall interference**
Select **Analyze → General Linear Model → Repeated Measures...** as shown in Figure 12.4

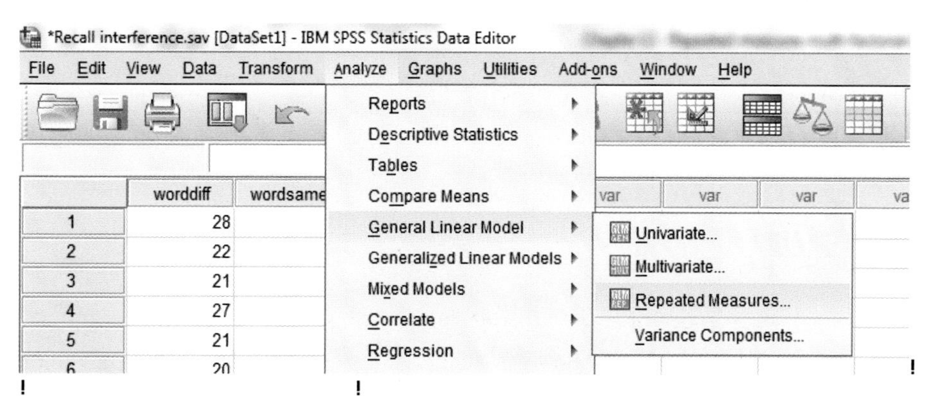

Figure 12.4 Repeated-measures multi-factorial ANOVA: procedure 1

In new window (see Figure 12.5), type **Presentation** in **Within-Subject Factor Name** box → type **2** for Number of Levels → click Add → type **Interference** in the **Within-Subject Factor Name** box → type **2** for **Number of Levels** → click **Add** → click **Define**

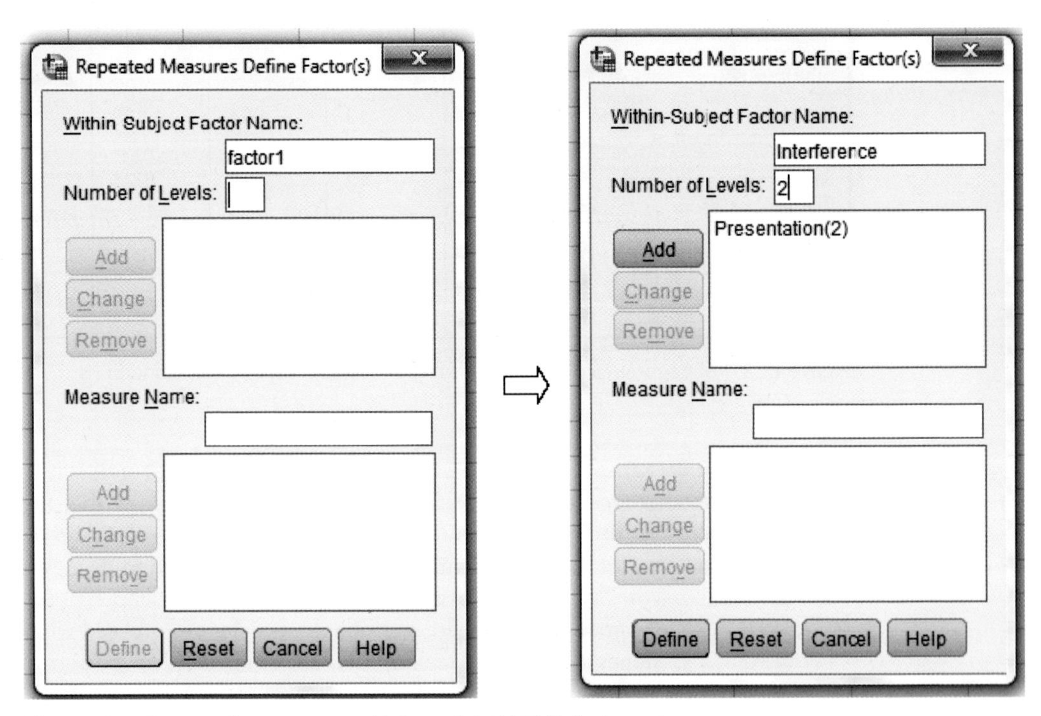

Figure 12.5 Repeated-measures multi-factorial ANOVA: factor entry

12.6 Nuts and bolts
Entering parameters in the correct order

It is very important that you enter the parameters in the correct order. We are given a clue to how we should do this in the window on the right-hand side of Figure 12.6 (_?_(1,1)... etc., see below).

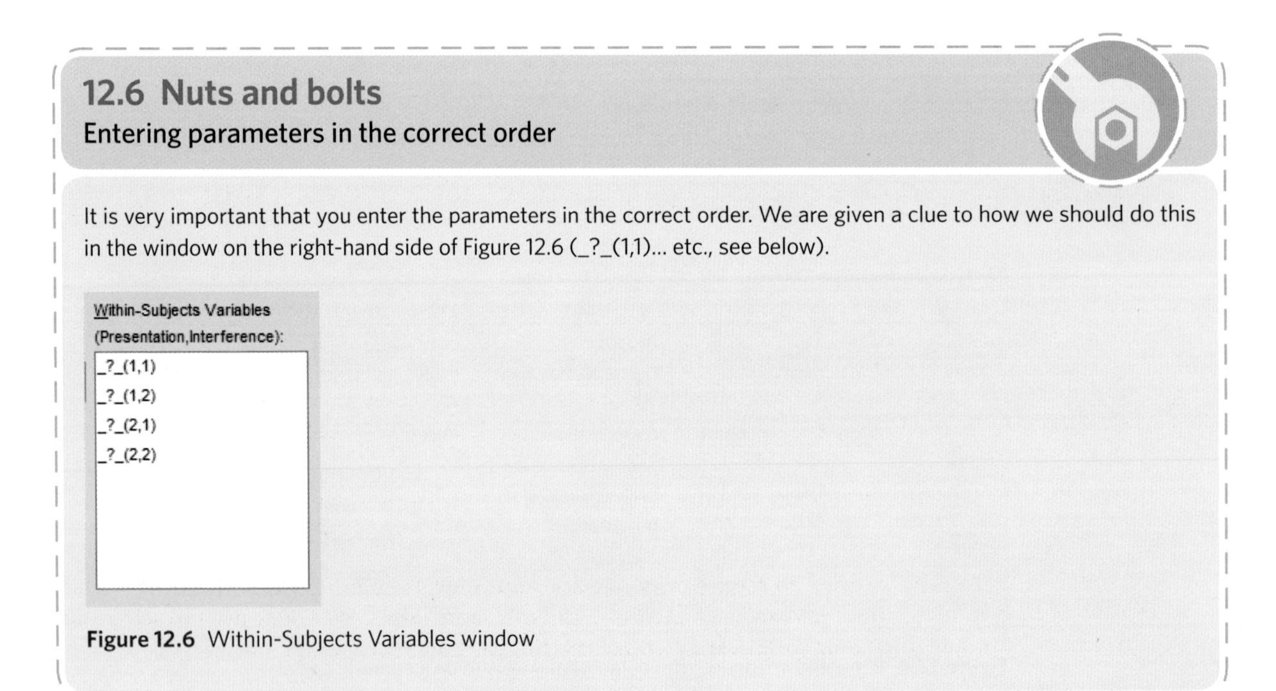

Figure 12.6 Within-Subjects Variables window

The numbers within the brackets indicate the independent variable conditions. So, (1,1) represents (condition 1 for independent variable 1, condition 1 for independent variable 2); (1,2) represents (condition 1 for independent variable 1, condition 2 for independent variable 2); (2,1) represents (condition 2 for independent variable 1, condition 1 for independent variable 2); and (2,2) represents (condition 2 for independent variable 1, condition 2 for independent variable 2).

In our example, the first independent variable is 'Presentation'; the conditions are (1) Word and (2) Number. The second independent variable is 'Interference'; the conditions are (1) Different and (2) Same. We can use that to guide us to enter the parameters in the correct order.

In new window (see Figure 12.7), transfer **worddiff** to **Within-Subjects Variables** to replace _?_(1,1) (by clicking on arrow, or by dragging the variable to that window) → transfer **wordsame** to_?_(1,2) → transfer **numberdiff** to _?_(2,1) → transfer **numbersame** to _?_(2,2) → click **Options...**

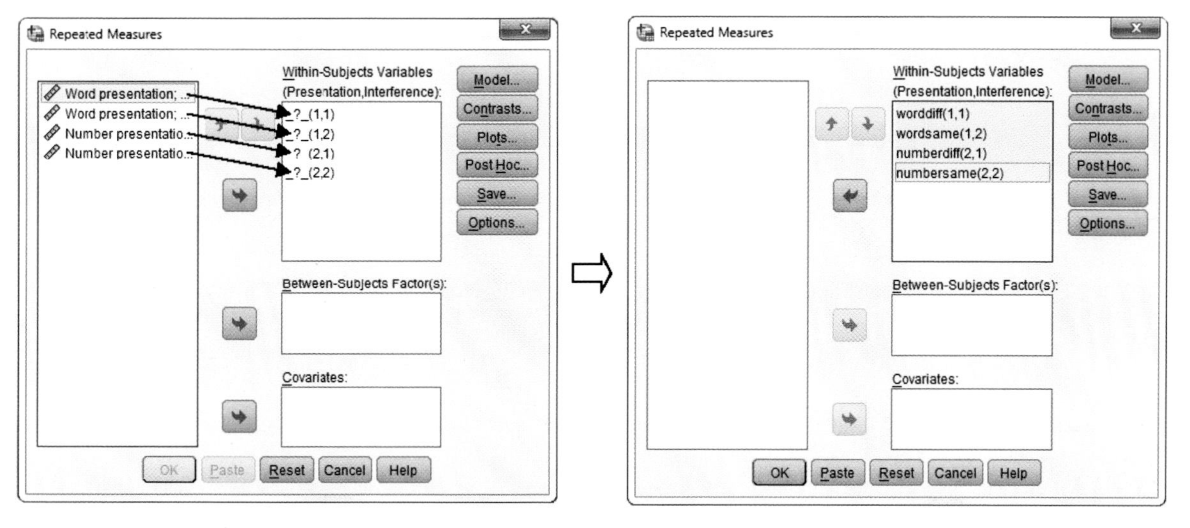

Figure 12.7 Repeated-measures multi-factorial ANOVA: parameter entry

In new window (see Figure 12.8), transfer **Presentation, Interference** and **Presentation*Interference** to **Display Means for:** → tick boxes under **Display** for **Descriptive statistics** and **Estimates of effect size** → click **Continue** → click **Plots...**

We do not want to select *post hoc* options this time as both our main effects have only two levels. If we had any variables with three or more levels, we would need to select this through the Compare Means function, selecting the Bonferroni option (see second example later).

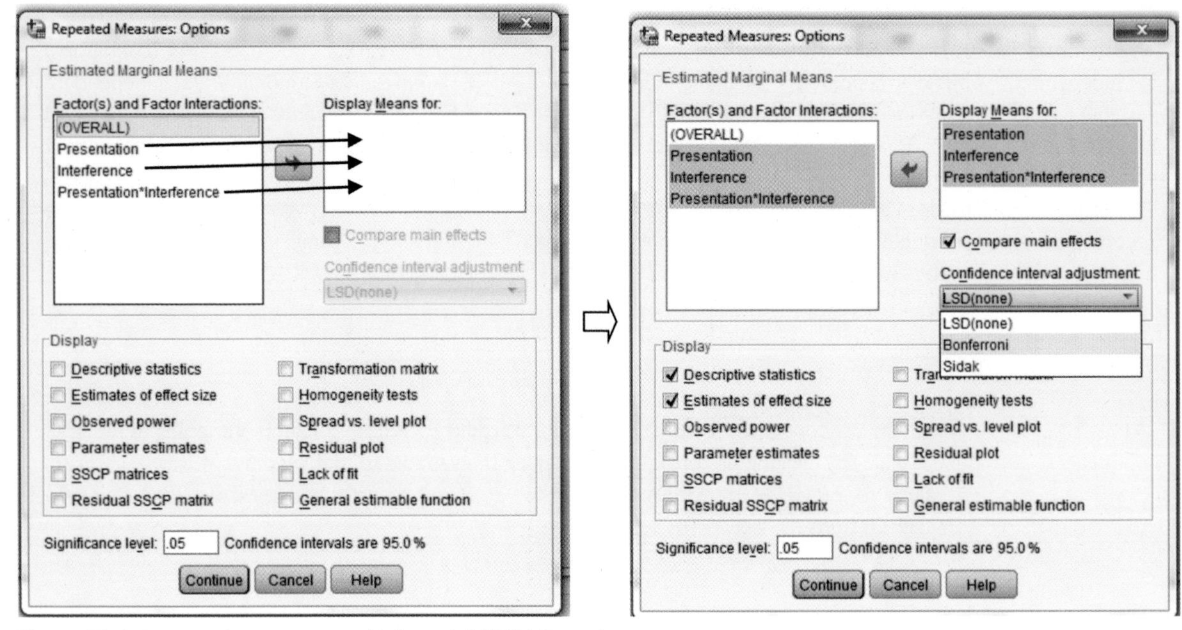

Figure 12.8 Repeated-measures multi-factorial ANOVA: options

In new window (see Figure 12.9), transfer **Presentation** to **Separate Lines:** ➜ transfer **Interference** to **Horizontal Axis:** ➜ click **Add** ➜ click **Continue** ➜ click **OK**

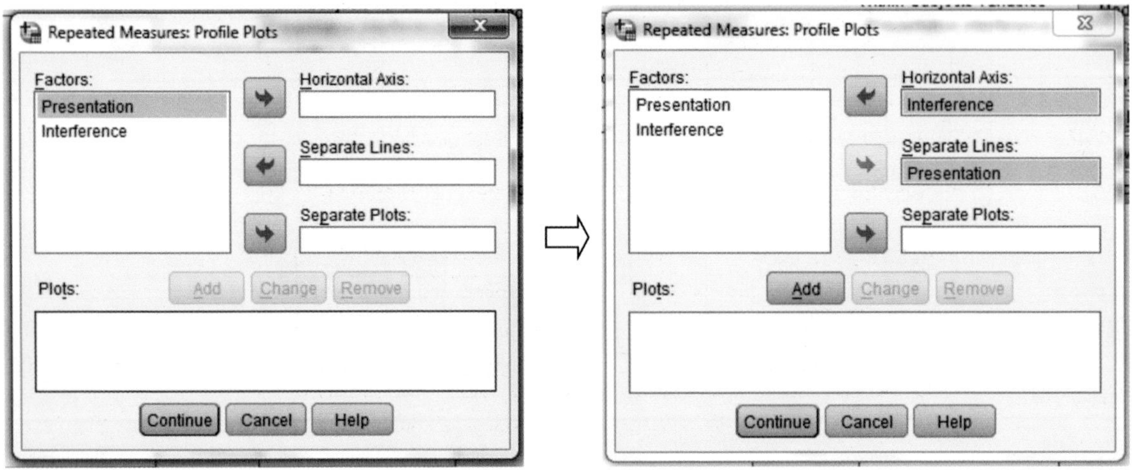

Figure 12.9 Repeated-measures multi-factorial ANOVA: plots

Interpretation of output

Figure 12.10 confirms the descriptive data It would appear that same-mode interference results in fewer words being recalled than with different-mode interference, but the way that these data are presented is not particularly helpful. We need to see the mean differences according to each main effect (independent variable) and for the interaction When we set up the SPSS parameters we asked for '**Estimated Marginal Means**' (see Figure 12.8). The data produced from that are more helpful, but we have an additional problem in that conditions are represented by numbers (see Figures 12.12–12.14). To help us here, we need an additional output (Figure 12.11) which can act as a 'key' for the remaining output.

	Mean	Std. deviation	N
Word presentation; different (number) interference	21.25	3.720	12
Word presentation; same (word) interference	3.08	.669	12
Number presentation; different (word) interference	17.50	2.393	12
Number presentation; same (number) interference	9.00	3.075	12

Figure 12.10 Descriptives for repeated-measures multi-factorial ANOVA

Presentation	Interference	Dependent variable
1	1	worddiff
	2	wordsame
2	1	numberdiff
	2	numbersame

Figure 12.11 Codebook for condition numbers

So, the 'Presentation' conditions are 1: word; and 2: number; while the 'Interference' conditions are 1: different; and 2: same.

Presentation			95% confidence interval	
	Mean	Std. error	Lower bound	Upper bound
1	12.167	.629	10.783	13.550
2	13.250	.467	12.223	14.277

Figure 12.12 Estimated marginal means: Presentation main effect

Recall appears to be a little better for numbers (2) than words (1).

Interference			95% confidence interval	
	Mean	Std. error	Lower bound	Upper bound
1	19.375	.881	17.435	21.315
2	6.042	.433	5.089	6.994

Figure 12.13 Estimated marginal means: Interference main effect

Recall appears to be considerably more depleted when the interference is in the same mode as the presentation (2) than it is for the different mode (1).

Presentation	Interference			95% confidence interval	
		Mean	Std. error	Lower bound	Upper bound
1	1	21.250	1.074	18.886	23.614
	2	3.083	.193	2.659	3.508
2	1	17.500	.691	15.979	19.021
	2	9.000	.888	7.046	10.954

Figure 12.14 Estimated marginal means: Interaction

Figure 12.14 is a little more difficult to follow. Differences in the Interference effect are shown according to each presentation. It would appear that the effect for interference is more dramatic for word presentation than it is for numbers.

Within–subjects effect	Mauchly' W	Approx. chi-square	df	Sig.	Epsilon[a] Greenhouse–Geisser	Huynh–Feldt	Lower bound
Presentation	1.000	.000	0	.	1.000	1.000	1.000
Interference	1.000	.000	0	.	1.000	1.000	1.000
Presentation * Interference	1.000	.000	0	.	1.000	1.000	1.000

Figure 12.15 Mauchly 's Test of Sphericity

At this point, if either of the independent variables had three or more conditions, we would need to consult the Mauchly's test outcome to determine which line of data we should read in the main ANOVA output. As both independent variables had two conditions, we can ignore that. However, we should refer briefly to that outcome on this occasion, so that we can see what that output looks like when there are only two conditions (see Figure 12.15). For each main effect and the interaction, we are told that Mauchly's W is the maximum possible (1.000), with a chi-square outcome of 0. This is highly non-significant (SPSS shows the significance here with a dot – but that is simply because it cannot be calculated). Mauchly's will always look like this when there are two conditions. But we already know that we can assume sphericity in any case.

Source		Type III sum of squares	df	Mean square	F	Sig.	Partial eta squared
Presentation	Sphericity assumed	14.083	1	14.083	**3.610**	**.084**	.247
	Greenhouse–Geisser	14.083	1.000	14.083	3.610	.084	.247
	Huynh–Feldt	14.083	1.000	14.083	3.610	.084	.247
	Lower bound	14.083	1.000	14.083	3.610	.084	.247
Error (presentation)	Sphericity assumed	42.917	11	3.902			
	Greenhouse–Geisser	42.917	11.000	3.902			
	Huynh–Feldt	42.917	11.000	3.902			
	Lower bound	42.917	11.000	3.902			
Interference	Sphericity assumed	2133.333	1	2133.333	**172.973**	**.000**	.940
	Greenhouse–Geisser	2133.333	1.000	2133.333	172.973	.000	.940
	Huynh–Feldt	2133.333	1.000	2133.333	172.973	.000	.940
	Lower bound	2133.333	1.000	2133.333	172.973	.000	.940
Error (interference)	Sphericity assumed	135.667	11	12.333			
	Greenhouse–Geisser	135.667	11.000	12.333			
	Huynh–Feldt	135.667	11.000	12.333			
	Lower bound	135.667	11.000	12.333			
Presentation * interference	Sphericity assumed	280.333	1	280.333	**115.638**	**.000**	.913
	Greenhouse–Geisser	280.333	1.000	280.333	115.638	.000	.913
	Huynh–Feldt	280.333	1.000	280.333	115.638	.000	.913
	Lower bound	280.333	1.000	280.333	115.638	.000	.913
Error (Presentation*Interference)	Sphericity assumed	26.667	11	2.424			
	Greenhouse–Geisser	26.667	11.000	2.424			
	Huynh–Feldt	26.667	11.000	2.424			
	Lower bound	26.667	11.000	2.424			

Figure 12.16 Tests of within-subjects effects

Figure 12.16 presents the significance outcome. There is a block of data for each independent variable: Presentation and Interference, and one for the interaction (Presentation*interference). Compare the sum of squares, mean square, and F ratio for each of these with what we found using manual calculations (Box 12.3). The 'Error' blocks present the residual sum of squares and residual mean square for each of the main effects and interaction We will read from the 'Sphericity Assumed' line for all of these analyses (for the reasons we gave earlier). The key information has been highlighted in coloured font to guide you a little more. Despite apparent differences shown in Figure 12.12, we do not have a significant main effect for Presentation (as highlighted in red): $F (1, 11) = 3.610$, $p = .084$. We will need the partial eta squared (η^2) data (highlighted in orange) when we look at effect size later. We do have a significant main effect for Interference (blue): $F (1, 11) = 172.973$, $p < .001$. Using the data from Figure 12.13, we know that recall is significantly poorer when the interference is in the same mode rather than in a different mode. We also have a significant interaction (green): $F (1, 11) = 115.638$, $p < .001$ – we will examine that shortly.

Source	Pr ...	Interference	Type III sum of squares	df	Mean square	F	Sig.	Partial eta squared
Presentation	Linear		14.083	1	14.083	3.610	.084	.247
Error(presentation)	Linear		42.917	11	3.902			
Interference		Linear	2133.333	1	2133.333	172.973	.000	.940
Error(interference)		Linear	135.667	11	12.333			
Presentation * Interference	Linear	Linear	280.333	1	280.333	115.638	.000	.913
Error (presentation*interference)	Linear	Linear	26.667	11	2.424			

Figure 12.17 Tests of within-subjects contrasts

Figure 12.17 is a little redundant on this occasion – it serves a purpose only when there are three or more conditions on an independent variable. This is because it describes how the data would be shown if drawn as a line graph. In most cases we want the data to be linear because that suggests that mean scores change in a consistent way across the conditions (we will see more about that when we examine 3 × 2 repeated-measures two-way ANOVA). In this case we have only two data points, so all of that is irrelevant and we can ignore it

Investigating source of significant interaction

As we said earlier, we need to locate the source of any significant interaction that we find. There are a few ways that this can be done. We could look at this graphically by way of a line graph and we could run additional repeated-measures one-way ANOVAs and/or related t-tests.

Graphical analysis

We requested a graph in SPSS earlier when we set up parameters for 'plots'. It is presented in Figure 12.18.

On the face of it, this graph is not easy to interpret (once again we have numbers where condition labels would be useful!). We could refer back to Figure 12.11 to help out but, better still, we can modify the graph to make it more meaningful. This is how we do that:

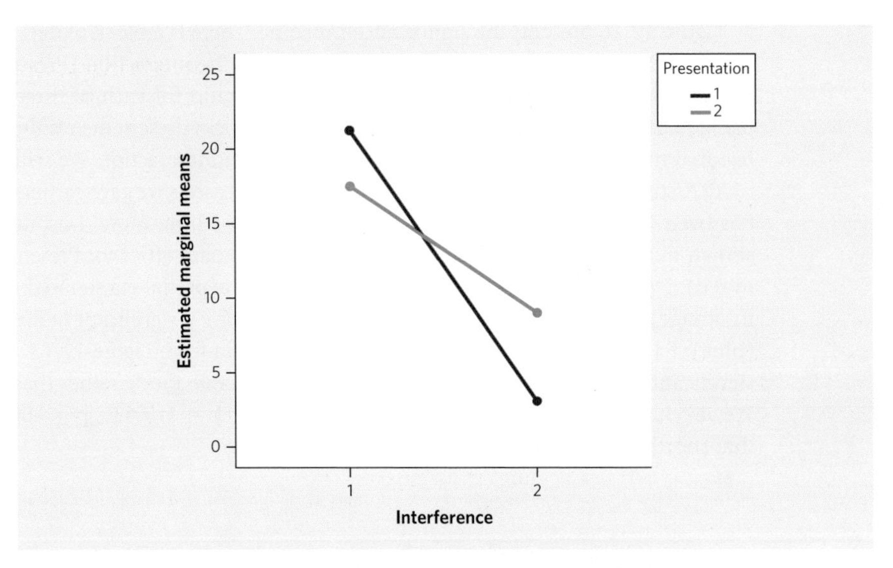

Figure 12.18 Line graph: Item recall by presentation and interference

Go to the graph in the SPSS output and double-click on it → a **Chart Editor** window will open (as shown in Figure 12.19)

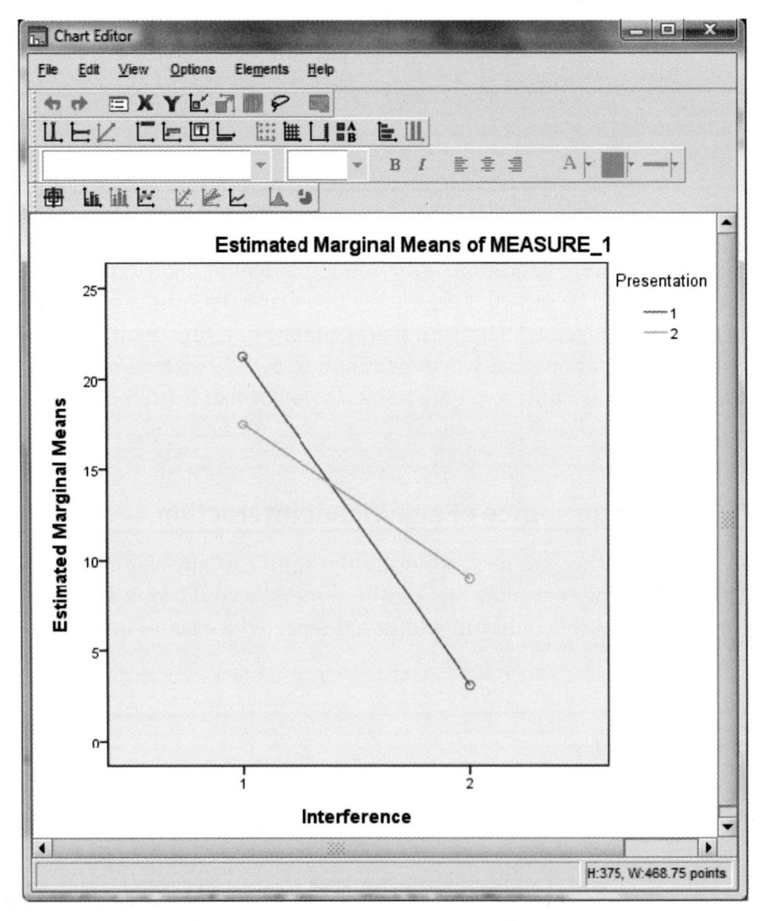

Figure 12.19 Chart editor window

Very carefully, click on the number **1** just below the word **Presentation** → if both numbers become highlighted (with an oval box) keep clicking until the number 1 only is surrounded by a square box (as shown in Figure 12.20)

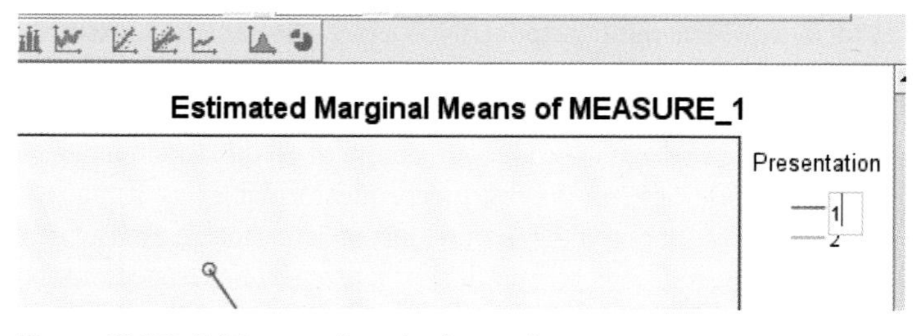

Figure 12.20 Editing numbers in chart editor

Overwrite number 1 with **Word** → then repeat what we did just now, this time for number 2 (still below Presentation) overwriting that with **Number** → go to the foot of the graph → click on the number **1** above **Interference** (initially both numbers 1 and 2 may become highlighted, but keep clicking until only the number 1 only is surrounded by a square box) → overwrite number 1 with **Different** → do the same for 2 , overwriting with **Same** → go to the left hand side of the graph and click twice on the **Estimated Marginal Means** label (with a short gap between clicks, rather than a fast double-click) → a new (horizontal) editing window will open → overwrite that label with **No of words recalled (mean)** → click anywhere away from that, and the new label will appear along the side of the graph

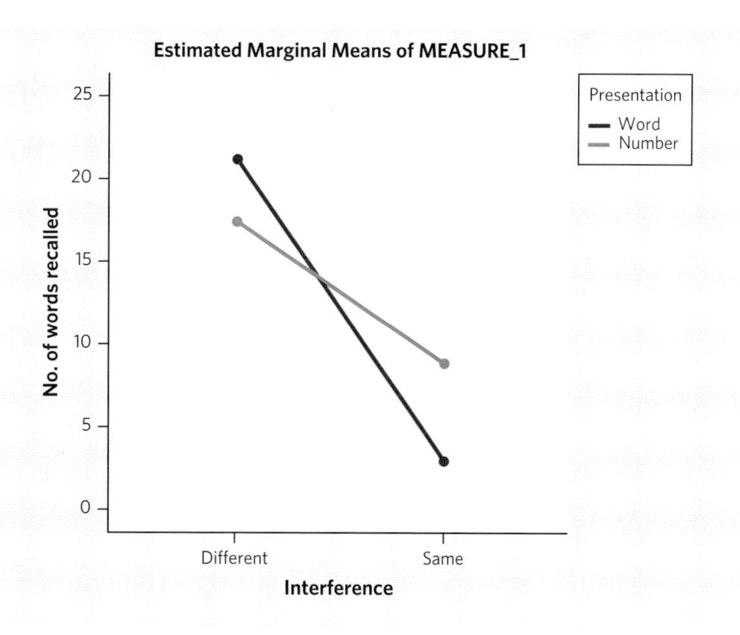

Figure 12.21 Line graph: Item recall by presentation and interference

Figure 12.21 is a much more sensible graph that we can actually interpret. The interaction appears to show that more words (blue line) are recalled than numbers (green line) when interference is in a different mode, while more numbers are recalled than words when the interference is in the same mode. Compare this graph to those that we saw in Chapter 11 (especially Figure 11.1g).

Statistical analysis

While the graph was fairly compelling in this instance, we really need statistical procedures to fully illustrate the nature of the interaction. The additional tests that we need to run are similar to those that we saw for independent multi-factorial ANOVA (see Chapter 11). However, we do not need to split the data set before performing them. The type of test we run depends on how many conditions we need to explore for each independent variable. Where there are two conditions (on the first independent variable) we undertake a related t-test for every condition on the second independent variable. If we have three or more conditions (on the first independent variable) we run a repeated-measures one-way ANOVA for every condition on the second independent variable. For our example, both independent variables have two conditions, so we will need two sets of two related t-tests (as shown in Box 12.7).

12.7 Take a closer look

Additional tests to locate source of interaction for 2 × 2 RM multi-factorial ANOVA

Here is a summary of the additional tests that we need to run to locate the source of interaction:

Within-group interactions	Method for locating source
Number of items recalled across 'Presentation' conditions, for each 'Interference' condition	2 × related t-tests: 1. word/same vs. number/same 2. word/different vs. number/different
Number of items recalled across 'Interference' conditions, for each 'Presentation' condition	2 × related t-tests: 1. same/word vs. different/word 2. same/number vs. different/number

Number of items recalled by Presentation conditions, at each level of Interference

Using the SPSS file **Recall interference**
Select **Analyze** → **Compare Means** → **Paired-Samples T Test...** → (in new window) transfer **wordsame** and **numbersame** to first line of **Paired Variables** → transfer **worddiff** and **numberdiff** to second line of **Paired Variables** → click **OK**

		Mean	N	Std. deviation	Std. error mean
Pair 1	Word presentation; same (word) interference	3.08	12	.669	.193
	Number presentation; same (number) interference	9.00	12	3.075	.888
Pair 2	Word presentation; different (number) interference	21.25	12	3.720	1.074
	Number presentation; different (word) interference	17.50	12	2.393	.691

Figure 12.22 Mean scores – presentation vs. word count, according to interference

		Paired differences					t	df	Sig. (2-tailed)
					95% confidence interval of the difference				
		Mean	Std. deviation	Std. error mean	Lower	Upper			
Pair 1	Word presentation; same (word) interference - number presentation; same (number) interference	−5.917	3.288	.949	−8.006	−3.828	−6.234	11	.000
Pair 2	Word presentation; different (number) interference - number presentation; different (word) interference	3.750	1.357	.392	2.888	4.612	9.574	11	.000

Figure 12.23 Related t-test outcome - presentation vs. word count, according to interference

Figures 12.22 and 12.23 show that recall was significantly more when numbers were recalled than words, when there was same-mode interference: $t(11) = -6.234$, $p < .001$, while more words were recalled than numbers when there was different-mode interference: $t(11) = 9.574$, $p < .001$. This shows very clear evidence of an interaction.

Number of items recalled by Interference conditions, at each level of Presentation

Using the SPSS file **Recall interference**

Select **Analyze** → **Compare Means** → **Paired-Samples T Test...** → (in new window) transfer **wordsame** and **worddiff** to first line of **Paired Variables** → transfer **numbersame** and **numberdiff** to second line of **Paired Variables** → click **OK**

		Mean	N	Std. deviation	Std. error mean
Pair 1	Word presentation; same (word) interference	3.08	12	.669	.193
	Word presentation; different (number) interference	21.25	12	3.720	1.074
Pair 2	Number presentation; same (number) interference	9.00	12	3.075	.888
	Number presentation; different (word) interference	17.50	12	2.393	.691

Figure 12.24 Mean scores - Interference vs. word count, according to presentation

		Paired differences					t	df	Sig. (2-tailed)
					95% confidence interval of the difference				
		Mean	Std. deviation	Std. error mean	Lower	Upper			
Pair 1	Word presentation; same (word) interference – word presentation; different (number) interference	−18.167	3.099	.895	−20.136	−16.197	−20.305	11	.000
Pair 2	Number presentation; same (number) interference – number presentation; different (word) interference	−8.500	4.462	1.288	−11.335	−5.665	−6.599	11	.000

Figure 12.25 Related t-tests outcome - interference vs. word count, according to presentation

Figures 12.24 and 12.25 tell us that recall was significantly poorer when a (conflicting) word was spoken with a visually presented word than when a number was spoken with the presented word: t (11) = −20.305, p < .001. Also, recall was significantly poorer when a (conflicting) number was spoken with a visually presented number than when a word was spoken with the presented number: t (11) = −6.599, p < .001. On the face of it, both outcomes are significant, so the difference between them is less clear. However, the effect in the number condition was much smaller than in the word condition (look at the t-scores).

Effect size and power

We can use G*Power to calculate effect size for us, and to show how much power our study had to detect the outcome. We need to do three analyses: one for each of the main effects and one for the interaction.

Open G*Power:

From **Test family** select **F tests**
From **Statistical test** select **ANOVA: Repeated measures, within-factors**
From **Type of power analysis** select **Post hoc: Compute achieved - given** α**, sample size and effect size power**

Now we enter the **Input Parameters**:

Presentation main effect

To calculate the **Effect size** , click on the **Determine** button (a new box appears).
In that new box, tick on radiobutton for Direct → type **0.247** in the **Partial** η^2 box (we get that from Figure 12.16 → click on **Calculate and transfer to main window**

Back in original display, for **α err prob** type **0.05** (the significance level) → for **Total sample size** type **12** (the overall sample size) → for **Number of groups** type **1** (we only had one sample group) → for **Number of repetitions** type **2** (word and number) → for **Corr among rep measures** type **0.5** (keep as default) → for **nonsphericity** type **1** (we get that from Figure 12.15 – Mauchly's W) → click on **Calculate**

<u>Effect size</u> (d) **0.57** (Good); and <u>Power</u> (1-β err prob) **0.95** (which is excellent, given that our target is 0.80).

Interference main effect

From Determine box type **0.940** in the **Partial** η^2 box → click on **Calculate and transfer to main window** → keep Number of repetitions as 2 → click on **Calculate**

Effect size (d) **3.96** (Very large); and Power (1-β err prob) **1.00** (Excellent).

Interaction

In Determine box type **0.913** in the **Partial** η^2 box → click on **Calculate and transfer to main window** → for **Number of repetitions** type **4** (2 presentation conditions × 2 interference conditions) → click on **Calculate**

<u>Effect size</u> (d) **3.24** (Very large); and <u>Power</u> (1-β err prob) **1.00** (Excellent).

Writing up results

Table 12.2 Mean number of items recalled

Condition	Mean	SD
Word presentation; number interference	21.25	3.72
Word presentation; word interference	3.08	0.69
Number presentation; word interference	17.50	2.39
Number presentation; number interference	9.00	3.08

We should also present a narrative summary of those findings:

Table 12.2 shows that item recall varied according to the nature of presentation (word or number) and the type of interference (different or same mode). A repeated-measures two-way ANOVA confirmed that there was no significant effect for presentation, $F(1, 11) = 3.610$, $p = .084$ ($d = .57$), but there was a significant effect for interference, $F(1, 11) = 172.973$, $p < .001$ ($d = 3.96$). Recall was significantly poorer when the interference was in the same mode rather than in a different mode. There was a significant interaction, $F(1, 11) = 115.638$, $p < .001$ ($d = 3.24$). Further investigation of the interaction showed that rates of recall were more dramatically reduced by same-mode interference (as opposed to different mode) for word presentation than for number.

3 × 2 repeated-measures two-way ANOVA

We will now explore a data set where we have one independent variable with three conditions and one with two conditions. We should explore what difference that makes to how we interpret the outcome and how we investigate the source of within-group differences and interactions. To illustrate this example, we will refer to the FUSS research question that we encountered earlier. FUSS are seeking to explore what type of lesson students prefer (interactive lecture, standard lecture or video) and who they would rather have deliver this (expert or novice lecturer). Those factors represent the two independent variables. The dependent variable is the satisfaction with course score – higher scores represent greater satisfaction.

Running tests in SPSS

Using the SPSS file **Lecture expert**

Select **Analyze** → **General Linear Model** → **Repeated Measures...** → (in new window) type **Lecture** in **Within-Subject Factor Name** → type **3** for **Number of Levels** → click **Add** → type **Expert** in **Within-Subject Factor Name** → type **2** for **Number of Levels** → click **Add** → click **Define** → (in next window) transfer **Interactive lecture - Expert** to **Within-Subjects Variables** to replace_?_(1,1) → transfer **Interactive lecture - Novice** to _?_(1,2) → transfer **Standard lecture - Expert** to _?_(2,1) → transfer **Standard lecture - Novice** to _?_(2,2) → transfer **Video - Expert** to _?_(3,1) → transfer **Video - Novice** to _?_(3,2) → click **Options...**

In addition to requesting descriptive data and estimated marginal means, we need to select a Bonferroni *post hoc* test to locate the source of difference across the 'Lecture' conditions (if there is a significant difference). We select both independent variables and the interaction for estimated marginal means because these give us better mean data than the descriptive statistics. The Bonferroni test is located within the 'Compare main effects' pull-down menu (see below) (we explained the rationale for our choice earlier).

Transfer **Lecture, Expert** and **Lecture*Expert** to **Display Means for:** ➜ tick **Compare main effects** box ➜ click pull-down arrow ➜ select **Bonferroni** (we need this for *post hoc* data this time, but only for the lecture type as that has three conditions) ➜ tick boxes under **Display** for **Descriptive statistics and Estimates of effect size** ➜ click **Continue** ➜ click **Plots...** ➜ (in next window) transfer **Lecture** to **Separate Lines:** ➜ transfer **Expert** to **Horizontal Axis:** ➜ click **Add** ➜ click **Continue** ➜ click **OK**

Interpretation of output

	Mean	Std. deviation	N
Interactive lecture – expert	56.00	11.137	72
Interactive lecture – novice	42.40	14.189	72
Standard lecture – expert	40.35	15.665	72
Standard lecture – novice	32.71	16.245	72
Video – expert	38.21	15.016	72
Video – novice	30.86	15.442	72

Figure 12.26 Descriptive statistics

Lecture	Expert	Dependent variable
1	1	ile
	2	iln
2	1	sle
	2	sln
3	1	ve
	2	vn

Figure 12.27 Codebook for condition numbers

Lecture	Mean	Std. error	95% confidence interval	
			Lower bound	Upper bound
1	49.201	1.197	46.815	51.588
2	36.528	1.742	33.054	40.002
3	34.535	1.664	31.218	37.852

Figure 12.28 Estimated marginal means: lecture main effect

Expert	Mean	Std. error	95% confidence interval	
			Lower bound	Upper bound
1	44.852	1.437	41.987	47.717
2	35.324	1.555	32.223	38.425

Figure 12.29 Estimated marginal means: expert main effect

Lecture	Expert	Mean	Std. error	95% confidence interval	
				Lower bound	Upper bound
1	1	56.000	1.312	53.383	58.617
	2	42.403	1.672	39.069	45.737
2	1	40.347	1.846	36.666	44.028
	2	32.708	1.914	28.891	36.526
3	1	38.208	1.770	34.680	41.737
	2	30.861	1.820	27.232	34.490

Figure 12.30 Estimated marginal means: Interaction

Figure 12.26 presents the descriptive data but, as we saw with the 2 × 2 analysis, the estimated marginal mean data shown in Figures 12.28–12.30 are probably more helpful. However, to interpret those, we need to understand what the 'numbers' mean for each condition. Figure 12.27 helps us with that. We can now see that the 'Lecture' conditions are 1: interactive, 2: standard and 3: video, while the 'Expert' conditions are 1: expert and 2: novice.

Figure 12.28 suggests that course satisfaction is much higher with the interactive lecture than with any other form of lesson. Figure 12.29 indicates that course satisfaction is higher when the lessons are presented by an expert rather than by a novice. Figure 12.30 suggests that there may be an interaction between lecture type and expertise: the difference in course satisfaction between expert and novice delivery appears to be greater within the interactive lecture, compared with other lesson types. However, we need to check all of this statistically.

Within-subjects effect	Mauchly's W	Approx. chi-square	df	Sig.	Epsilon[a]		
					Greenhouse–Geisser	Huynh–Feldt	Lower bound
Lecture	.004	386.645	2	.000	.501	.501	.500
Expert	1.000	.000	0	.	1.000	1.000	1.000
Lecture*Expert	.010	322.336	2	.000	.503	.503	.500

Figure 12.31 Mauchly's test of sphericity

We need to check sphericity on this occasion, as one of the independent variables (Lecture) has three conditions. This also affects the interaction, so we need to pay attention to that, too. The Expert variable has two conditions, so we do not need Mauchly's test, and can assume sphericity when we examine the output later (we presented the rationale for that argument earlier). For the Lecture variable, Mauchly's test is significant ('Sig.' < .05), so we should refer to one of the alternative significance outcomes in Figure 12.31. There are three options (see Chapter 10 to review some guidelines about which one to choose). We will select Huynh–Feldt for the Lecture main effect and the interaction (as Mauchly's test is significant for that, too).

Source		Type III sum of squares	df	Mean square	F	Sig.	Partial eta squared
Lecture	Sphericity assumed	18225.782	2	9112.891	66.767	.000	.485
	Greenhouse–Geisser	18225.782	1.002	18189.405	66.767	.000	.485
	Huynh–Feldt	18225.782	1.002	18187.848	**66.767**	**.000**	.485
	Lower bound	18225.782	1.000	18225.782	66.767	.000	.485
Error (lecture)	Sphericity assumed	19381.218	142	136.487			
	Greenhouse–Geisser	19381.218	71.142	272.430			
	Huynh–Feldt	19381.218	71.148	272.407			
	Lower bound	19381.218	71.000	272.975			
Expert	Sphericity assumed	9804.083	1	9804.083	**49.129**	**.000**	.409
	Greenhouse–Geisser	9804.083	1.000	9804.083	49.129	.000	.409
	Huynh–Feldt	9804.083	1.000	9804.083	49.129	.000	.409
	Lower bound	9804.083	1.000	9804.083	49.129	.000	.409
Error (expert)	Sphericity assumed	14168.583	71	199.558			
	Greenhouse–Geisser	14168.583	71.000	199.558			
	Huynh–Feldt	14168.583	71.000	199.558			
	Lower bound	14168.583	71.000	199.558			
Lecture * Expert	Sphericity assumed	895.792	2	447.896	15.671	.000	.181
	Greenhouse–Geisser	895.792	1.005	891.311	15.671	.000	.181
	Huynh–Feldt	895.792	1.005	891.120	15.671	.000	.181
	Lower bound	895.792	1.000	895.792	15.671	.000	.181
Error (lecture*expert)	Sphericity assumed	4058.542	142	28.581			
	Greenhouse–Geisser	4058.542	71.357	56.877			
	Huynh–Feldt	4058.542	71.372	56.864			
	Lower bound	4058.542	71.000	57.163			

Figure 12.32 Tests of within-subjects effects

Figure 12.32 shows that there are significant main effects for the Lecture variable (highlighted in red): $F(1.002, 71.148) = 66.767$, $p < .001$, and for the Expert variable (blue): $F(1, 71) = 49.129$, $p < .001$. There is also a significant interaction between Lecture and Expert in respect of 'course satisfaction' scores (green): $F(1.005, 71.372) = 15.671$, $p < .001$.

Exploring the source of main effects

Expert main effect

The significant effect for the 'expert' variable is straightforward as there are only two conditions. Using the mean data from Figure 12.29 and the statistics from Figure 12.32, we know that 'course satisfaction' is significantly higher when the lesson is presented by an expert than by a novice.

Lecture main effect

To explore the source of this difference we need to consult the Bonferroni *post hoc* test output.

(I) Lecture	(J) Lecture	Mean difference (I-J)	Std. error	Sig.[a]	95% confidence interval for difference[a]	
					Lower bound	Upper bound
1	2	12.674*	1.716	.000	8.467	16.880
	3	14.667*	1.654	.000	10.611	18.723
2	1	−12.674*	1.716	.000	−16.880	−8.467
	3	1.993*	.087	.000	1.780	2.207
3	1	−14.667*	1.654	.000	−18.723	−10.611
	2	−1.993*	.087	.000	−2.207	−1.780

Figure 12.33 Within-group *post hoc* test (lecture)

Figure 12.33 shows (with help from Figure 12.27) that 'course satisfaction' is significantly higher for the interactive lecture (1) than it is for the standard lecture (2) and video (3), and that satisfaction is significantly higher for the standard lecture (2) than it is for video (3).

Linear vs. quadratic outcome

Source	L ...	Expert	Type III sum of squares	df	Mean square	F	Sig.	Partial eta squared
Lecture	Linear		15488.000	1	15488.000	78.622	.000	.525
	Quadratic		2737.782	1	2737.782	36.032	.000	.337
Error(lecture)	Linear		13986.500	71	196.993			
	Quadratic		5394.718	71	75.982			
Expert		Linear	9804.083	1	9804.083	49.129	.000	.409
Error(expert)		Linear	14168.583	71	199.558			
Lecture*Expert	Linear	Linear	703.125	1	703.125	16.490	.000	.188
	Quadratic	Linear	192.667	1	192.667	13.266	.001	.157
Error(lecture*expert)	Linear	Linear	3027.375	71	42.639			
	Quadratic	Linear	1031.167	71	14.523			

Figure 12.34 Tests of within-subjects contrasts

Figure 12.34 shows the linearity of the outcome. This is relevant only for the Lecture condition as that has three conditions, because the Linear outcome is significant; if we presented the outcome in a line graph, there would be straight line between the conditions. This suggests that the scores increase incrementally across the conditions: 'course satisfaction' is getter better between 'video', 'standard lecture' and 'interactive lecture'. If the outcome was significantly 'quadratic', the line would be shown as a 'U' shape. This would suggest that 'course satisfaction scores fall between 'video' and 'standard lecture', but increase again for 'interactive lecture'. We would expect a linear outcome with our data.

Exploring the source of interaction

We need to use similar procedures to those we undertook with the CALM research data, focusing on line graphs and statistical analyses.

Graphical analysis

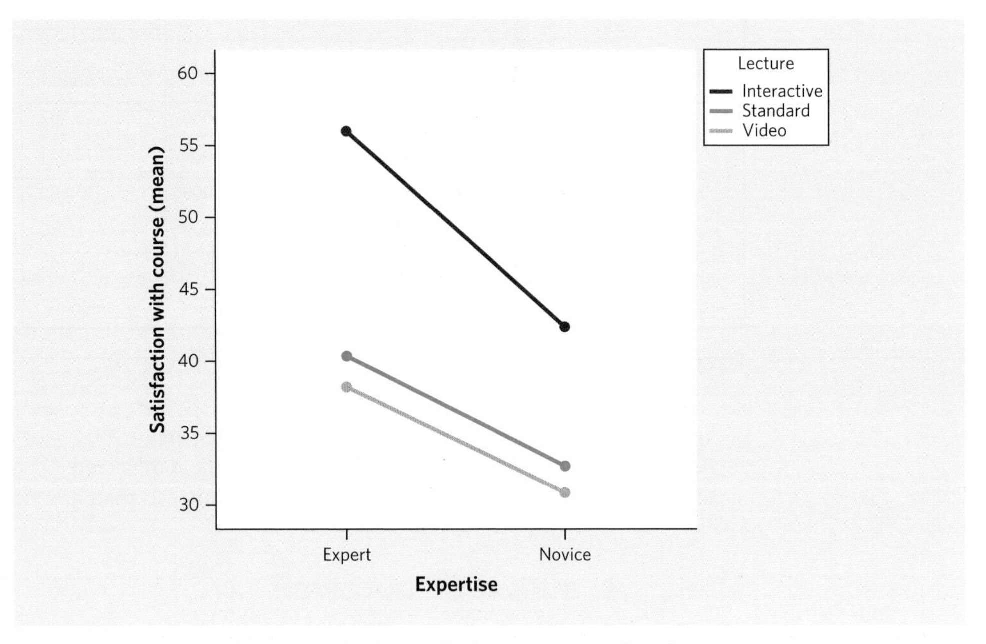

Figure 12.35 Line graph: course satisfaction by lecture type and expertise

Figure 12.35 has been amended using methods we saw earlier (see Figures 12.19–12.20). It suggests that 'course satisfaction' decreases between expert and novice presenters, but is more dramatic for 'interactive lecture' (blue line) than it is for 'standard lecture' (green) and 'video' (red).

Statistical analysis

For this example, one of the independent variables has three conditions, while the other has two conditions, so we will need some slightly different tests to the ones we performed earlier. To examine outcomes across the Lecture variable we need to undertake repeated-measures one-way ANOVAs. We need to do two of these: one for each level of the Expert variable. To examine the outcomes across the Expert variable we need three related t-tests, one for each level of the Lecture variable. Box 12.8 provides a summary.

12.8 Take a closer look

Additional tests to locate the source of interaction for 3 × 2 RM multi-factorial ANOVA

Within-group interactions	Method for locating source
Course satisfaction across 'Lecture type', for each 'expertise' condition	2 × repeated-measures one-way ANOVAs: 1. interactive/expert, standard/expert, video/expert 2. interactive/novice, standard/novice, video/novice
Course satisfaction across 'expertise', for each 'Lecture type' condition	3 × related t-tests: 1. expert/interactive vs. novice/interactive 2. expert/standard vs. novice/standard 3. expert/video vs. novice/video

Course satisfaction by Lecture type, at each level of expertise

Using the SPSS file **Lecture expert**

For Expert...

Select **Analyze** → **General Linear Model** → **Repeated Measures...** → (in new window) type **Lecture** in **Within-Subject Factor Name** box → type **3** for **Number of Levels** → click **Add** → click **Define** → (in next window) transfer**Interactive lecture - Expert** to **Within-Subjects Variables** to replace_?_(1) → transfer **Standard lecture - Expert** to _?_(2) → transfer **Video - Expert** to _?_(3) (we are only concerned with the ANOVA outcome for these analyses, so we do not need to do anything for **Options**) → click **OK**

For Novice...

As above, then click **Define** → (in next window, transfer all entries in **Within-Subjects Variables** back to variable list) → transfer **Interactive lecture - Novice** to _?_(1) → transfer **Standard lecture - Novice** _?_(2) → transfer **Video - Novice** _?_(3) → click **OK**

Source		Type III sum of squares	df	Mean square	F	Sig.	Partial eta squared
Expert	Sphericity assumed	13587.065	2	6793.532	91.264	.000	.562
	Greenhouse–Geisser	13587.065	1.002	13559.063	91.264	.000	.562
	Huynh–Feldt	13587.065	1.002	13557.865	91.264	.000	.562
	Lower bound	13587.065	1.000	13587.065	91.264	.000	.562
Error (expert)	Sphericity assumed	10570.269	142	74.439			
	Greenhouse–Geisser	10570.269	71.147	148.570			
	Huynh–Feldt	10570.269	71.153	148.557			
	Lower bound	10570.269	71.000	148.877			

Figure 12.36 Repeated-measures one-way ANOVA - satisfaction: lecture type vs. expert

Source		Type III sum of squares	df	Mean square	F	Sig.	Partial eta squared
Novice	Sphericity assumed	5534.509	2	2767.255	30.533	.000	.301
	Greenhouse–Geisser	5534.509	1.003	5516.355	30.533	.000	.301
	Huynh–Feldt	5534.509	1.003	5515.579	30.533	.000	.301
	Lower bound	5534.509	1.000	5534.509	30.533	.000	.301
Error (novice)	Sphericity assumed	12869.491	142	90.630			
	Greenhouse–Geisser	12869.491	71.234	180.666			
	Huynh–Feldt	12869.491	71.244	180.640			
	Lower bound	12869.491	71.000	181.260			

Figure 12.37 Repeated-measures one-way ANOVA - satisfaction: lecture type vs. novice

Figure 12.36 presents the descriptive data between the conditions. Both one-way ANOVA outcomes shown in Figure 12.37 indicate a significant difference across the lecture conditions so, on the face of it, show little difference. However, the 'effect' is greater for the Expert condition as the F ratio is higher ($F = 91.264$, $p < .001$) than it is for the Novice condition ($F = 30.533$, $p < .001$). Formal effect size calculations would also help clarify this.

Course satisfaction by Expert type, at each level of lecture type

Using the SPSS file **Recall interference**
Select **Analyze** → **Compare Means** → **Paired-Samples T Test...** → (in new window) transfer **Interactive lecture - Expert** and **Interactive lecture - Novice** to first line of **Paired Variables** → transfer **Standard lecture - Expert** and **Standard lecture - Novice** to second line of **Paired Variables** → transfer **Video - Expert** and **Video - Novice** to third line of **Paired Variables** → click **OK**

		Paired differences							
					95% confidence interval of the difference				
		Mean	Std. deviation	Std. error mean	Lower	Upper	t	df	Sig. (2-tailed)
Pair 1	Interactive lecture–Expert –Interactive lecture – Novice	13.297	15.327	1.782	9.746	16.848	7.463	73	.000
Pair 2	Standard lecture–Expert–Standard lecture–Novice	7.432	11.918	1.385	4.671	10.194	5.365	73	.000
Pair 3	Video–Expert–Video–Novice	7.347	11.439	1.348	4.659	10.035	5.450	71	.000

Figure 12.38 Related t-tests - course satisfaction: expert vs. novice (by lecture type)

Figure 12.38 shows that although all pairs show significant differences for course satisfaction between expert and novice conditions, t-scores suggest that difference appears to be higher for the interactive lecture (Pair 1) than it is for the standard lecture (Pair 2) and video (Pair 3). There appears to be very little difference in that outcome between pairs 2 and 3.

Chapter summary

In this chapter we have explored the repeated-measures multi-factorial ANOVA. At this point, it would be good to revisit the learning objectives that we set at the beginning of the chapter.

You should now be able to:

- Recognise that we use repeated-measures multi-factorial ANOVA to explore differences in mean scores from a single parametric dependent variable, across two or more within-group independent variables, each with two or more conditions.
- Appreciate that, if a significant outcome is found across an independent variable that has three or more conditions, we need additional tests to locate the source of the difference. Bonferroni *post hoc* tests are most commonly used to perform those analyses.
- Understand that main effects may be found for each of the independent variables, in respect of the dependent variable outcome, and that there may be interactions between those independent variables.
- Understand that the data should be interval or ratio, and be reasonably normally distributed. We need to account for sphericity of within-group variances when there are three or more conditions of an independent variable. These determine how we interpret the ANOVA outcome. Every person (or case) must appear in all conditions.

- Calculate outcomes manually (using maths and equations).
- Perform analyses using SPSS, including *post hoc* tests where appropriate.
- Examine significance of outcomes and the source of significance by using additional related t-tests and repeated-measures one-way ANOVAs where appropriate.
- Explore effect size and statistical power.
- Understand how to present the data and report the findings.

Research example

It might help you to see how repeated-measures multi-factorial ANOVA has been applied in published research. In this section you can read an overview of the following paper:

Petrilli, R.M., Roach, G.D., Dawson, D. and Lamond, N. (2006). The sleep, subjective fatigue, and sustained attention of commercial airline pilots during an international pattern. ***Chronobiology International, 23 (6):*** **1347 – 1362. DOI: http://dx.doi.org/10.1080/07420520601085925**

If you would like to read the entire paper you can use the DOI reference provided to locate that (see Chapter 1 for instructions).

In this research the authors investigated the serious problem of fatigue in long-haul airline pilots. Specifically, one group of 19 pilots were examined for mean sleep (in the previous 24 hours), self-rated fatigue and mean response speed (so, there were three separate dependent variables). In each repeated-measures two-way ANOVA, the mean dependent variable scores were examined in respect of two within-group independent variables: flight sector (where the aeroplane was going – Australia to Asia, Asia to Europe, Europe to Asia, and Asia to Australia) and state of flight (testing the dependent variables before and after the flight). Flight sectors varied in length of flight, number of time zones crossed, and whether a stop-over was included. Sleep was measured via 24-hour activity monitors that were attached to the pilots' wrists. Self-rated fatigue was measured from the Samn–Perelli Fatigue Checklist (Samn and Perelli, 1982). This is a Likert scale, which elicits perceptions of fatigue from 1 (fully alert, wide awake) to 7 (completely exhausted, unable to function effectively). Mean response speed was measured using a portable version of the psychomotor vigilance task (PVT). Response to stimuli is made by pressing a button using the dominant hand. Lower scores represent poorer response times. Self-rated fatigue scales and PVT tasks were completed five minutes before and after each flight.

Results – Sleep in previous 24 hours: There was no significant main effect for flight sector, $F(3,51) = 2.74, p = .06)$, but there was a significant main effect for stage of flight, $F(1,51) = 64.32$, $p < .001$. The amount of sleep obtained at the end of flights was lower than that received at the start (bear in mind that pilots were encouraged to sleep while off-duty during long-haul flights, and stop-over periods could be 1–2 days). There was also a significant interaction between flight sector and stage of flight, $F(3,51) = 4.79, p < .01$. Mean sleep was significantly lower for the end of the Europe – Asia and Asia – Australia flights than at the end of the Australia – Asia flights.

Self-rated fatigue: There was a significant main effect for flight sector, $F(3,24) = 4.95, p < .01$. Fatigue was higher at the end of flights than at the start. There was a significant main effect for stage of flight, $F(1,24) = 40.04, p < .001$. Fatigue ratings were lower (better) after the Europe – Asia flights than any other. (There was some confusion in the reporting of this outcome – if you read the full paper, did you spot it?) There was also a significant interaction between flight sector and stage of flight, $F(3,24) = 7.56, p < .01$; we were not told the source of that.

Mean response speed: There was no significant main effect for flight sector, $F(3,21) = 1.06$, $p = .39)$, but there was a significant main effect for stage of flight, $F(1,21) = 7.97, p < .05$. Response speed was lower at the end of flights than at the start. There was no significant interaction, $F(3,21) = 1.53, p = .24)$.

This paper provides a good example of how to report repeated-measures two-way ANOVA. However, you may have noticed some apparently odd differences in the presentation of the degrees of freedom (shown in brackets after 'F'). This was because the authors additionally employed fixed and random effects to their analyses. This is way beyond the scope of this book, so please do not concern yourself about that.

Extended learning task

You will find the data set associated with this task on the website that accompanies this book (available in SPSS and Excel format). You will also find the answers there.

Following what we have learned about repeated-measures multi-factorial ANOVA, answer these questions and conduct the analyses in SPSS and G*Power. If you do not have SPSS, do as much as you can with the Excel spreadsheet. In this example, we examine patient anxiety towards a forthcoming operation, which is measured on a scale of 0 (very anxious) to 100 (no anxiety). We suspect that patients will have their anxiety appeased to some extent if they are provided with information about the operation when they attend the preliminary outpatient appointment. Their anxiety may also depend on who is available to give them that information. The patients attend four appointments; at each appointment we manipulate the information given and who gives it. The information is either 'clear' or 'unclear' and it is provided to them by either a 'nurse' or a 'receptionist'.

Open the data set **Hospital anxiety**

1. Describe the independent variables and the conditions that define them.
2. What is the dependent variable?
3. Check for normal distribution across the conditions.
4. Conduct a repeated-measures multi-factorial ANOVA.
5. Describe what the SPSS output shows.
6. Explain how you accounted for sphericity (if it was necessary).
7. Describe main effects and interactions.
8. Conduct appropriate additional analyses to indicate where the differences are (if there were any).
9. State the effect size and power, using G*Power.
10. Report the outcome as you would in the results section of a report.

13 MIXED MULTI-FACTORIAL ANOVA

Learning objectives

By the end of this chapter you should be able to:

- Recognise when it is appropriate to use mixed multi-factorial ANOVA
- Understand the theory, rationale, assumptions and restrictions associated with the test
- Calculate main effects and interaction manually (using maths and equations)
- Perform analyses using SPSS, and explore outcomes, the source of significance and source of interaction
- Know how to measure effect size and power
- Understand how to present the data and report the findings

What is mixed multi-factorial ANOVA?

A mixed multi-factorial ANOVA explores mean dependent variable scores across one or more between-group independent variables (with at least two distinct groups) *and* one or more within-group independent variables (with at least two conditions). The dependent variable data should be parametric (we explored what that meant in Chapter 5, but we will extend that a little later). As with all multi-factorial ANOVAs, the central feature of the test is the way in which it examines interactions between independent variables.

Theory and rationale

Purpose of mixed multi-factorial ANOVA

The aim of a multi-factorial ANOVA is to explore an outcome across several factors at the same time. For the 'mixed' version, those factors are a combination of within-group conditions compared across independent groups. This can be very useful in research contexts. Quite often, we will need to examine outcomes over a series of stages (whether they be time points or several manipulations that the whole group experience), but we also want to see if those outcomes vary according to specific factors about the members of the group. For example, we might want to see how different types of teaching methods have an impact on a single group of students over time (that would be the within-group element). Such an examination might indicate which intervention *generally* works best. We might also want to investigate how learning performance varies by groups, such as gender, cultural background or social factors (that would be the between-group element). Such an investigation might tell us more about which intervention works best across different groups. In theory, there is no limit to how many within-group and between-group variables you can use, so long as you have at least one of each (although more complex examples are rather challenging!).

Research questions for mixed multi-factorial ANOVA

Throughout this chapter, we will use some research examples, from which we will develop a series of questions that we will explore using mixed multi-factorial ANOVA. We require several questions because we need to see different scenarios that reflect subtle changes in the analyses according to how many variables are used, and how many groups or conditions there are within each variable. The research examples focus on investigations carried out by the Greedy Pig Ice Cream Company to discover consumers' opinions on taste satisfaction towards their products. They have recently released two flavours of ice cream (chocolate and vanilla), which are available in various versions according to the fat content.

In the first trial, the company wants to explore the merit of releasing a 'half fat' version of its vanilla ice cream, alongside its 'full fat' and 'diet' versions. The team are sure that people generally prefer 'full fat' to 'diet', but they would like to know whether the 'half fat' version is sufficiently more popular than the diet version to warrant marketing it. They would still sell the diet version for all of their customers who are trying to lose weight! The company also wants additional information regarding how men and women differ in their satisfaction towards the ice cream. They know that women tend to have higher satisfaction than men, but does that vary according to fat content in the ice cream? In a second trial, the company wishes to discover how taste satisfaction differs between flavours of ice cream (chocolate and vanilla) and between the fat content (full fat and diet). The company expects ice creams with at least some fat to be more popular than the diet version, but they do not know which of the flavours are preferred, and how this might vary across fat content. Once again, they would like to find out how this varies by gender.

These scenarios are exactly the types of puzzle that mixed multi-factorial ANOVA seeks to resolve. In the first trial, we would need a relatively simple mixed (two-way) multi-factorial ANOVA (with a 3 × 2 design) to investigate differences in taste satisfaction (the dependent variable). It is a two-way ANOVA because there are two variables: fat content (within-group) and gender (between-group). It is a 3 × 2 design because there are three conditions on the within-group variable (full fat, half fat and diet) and two groups on the between-group variable (males and females). In the second trial, we would need a mixed (three-way) multi-factorial ANOVA (with a 2 × 2 × 2 design) to investigate differences in taste satisfaction. It is a three-way ANOVA because of the three variables: two within-groups (flavour and fat content) and one between-group (gender). It is a 2 × 2 × 2 design because there are two conditions on each of the within-group variables (chocolate vs. vanilla and full fat vs. diet) and two groups on the between-group variable (males vs. females). These scenarios are summarised in Box 13.1.

13.1 Take a closer look
Summary of mixed multi-factorial ANOVA research studies

Mixed multi-factorial (two-way) ANOVA (3 × 2)

- **Within-group IV:** fat content (three conditions: full fat, half fat and diet)
- **Between-group IV:** gender (two groups: men and women)
- **DV:** taste satisfaction scores

Mixed multi-factorial (three-way) ANOVA (2 × 2 × 2)

- **Within-group IV 1:** flavour (two conditions: chocolate and vanilla)
- **Within-group IV 2:** fat content (two conditions: full fat and diet)
- **Between-group IV:** gender (two groups: men and women)
- **DV:** taste satisfaction scores

13.2 Nuts and bolts
2 × 3 × 2 or 3 × 2 × 2? Which do I use?

Students often ask whether there are rules about what order to present the conditions and/or groups for mixed multi-factorial ANOVA. What goes first, within-group or between-group factors? The simple answer is that it does not matter a jot – it is a question of personal preference. However, it should be logical. Almost everyone has a different take on this. Some people describe the between-group variables first, followed by within-group (and present the numbers in that order to describe the levels); others do the reverse.

Main effects and interactions

As we saw in Chapters 11 and 12, an important factor of multi-factorial ANOVA is the description of main effects and interactions. A significant 'main effect' occurs if there are statistically significant differences in the dependent variable scores across the groups or conditions of an independent variable. There may be **between-group main effects** and/or **within-group main effects** (we will review some examples shortly). For each main effect, if there are three or more groups or conditions, we need *post hoc* tests to explore the source of the difference. An interaction occurs when there is a significant difference in the outcome across one independent variable at different groups or conditions of another independent variable. There is potentially at least one **within-between interaction** in every test, but there may be more depending on how many independent variables have been included. We will need to explore the source of any interaction that we find.

We can illustrate these points using examples from our research questions that we posed earlier. In the first trial, the Greedy Pig Ice Cream Company wanted to know whether taste satisfaction differs according to fat content ('full fat', 'half fat' and 'diet') and in respect of gender. Using a mixed multi-factorial ANOVA, we would explore two main effects one within-group (fat content) and one between-group (gender). We might find a significant difference in satisfaction scores across the fat content conditions (irrespective of gender). We would need a *post hoc* test to locate the source of that difference (because there are more than two conditions). That might show that 'full fat' ice cream is preferred over the 'half fat' and diet versions and that there is no difference between 'half fat' and 'diet'. We might also find that women report higher satisfaction for the ice cream than men, regardless of the fat content. The company also wants to know whether there are differences across the fat content conditions when accounting for gender – we can explore that with the interaction effects. We might find that it is only women who prefer 'full fat' ice cream over the 'half fat' and diet versions (men might show no difference in their preference). This would be an example of a within-between interaction. These scenarios are summarised in Box 13.3.

13.3 Take a closer look

Main effects and interactions in a mixed (two-way) multi-factorial ANOVA

The following scenarios present how 'taste satisfaction' scores might vary

Within-group main effect:	Differences in DV scores across conditions for one group Example: 'full fat' > 'half fat'; 'full fat' > 'diet'; 'half fat' = 'diet'
Between-group main effect:	Differences in dependent variable (DV) scores across distinct groups Example: Gender (women > men)
Within-between interaction:	Difference in the effect of any within-group IV at different levels of <u>any</u> between-group IV. Example: 'full fat' > 'half fat'; 'full fat' > 'diet' (but <u>only</u> for women)

In the second trial, the company wanted to investigate how taste satisfaction varied across flavours (chocolate and vanilla), fat content (full fat and diet), and by gender (males and females). We will explore three main effects, two within-groups ('flavour' and 'fat content') and one between-group (gender). None of the main effect analyses needs *post hoc* tests because there are only two conditions or groups. We might find that chocolate ice cream is preferred over vanilla (regardless of the fat content) and that full fat is preferred over the diet version (regardless of flavour) – both within-group main effect outcomes are expressed irrespective of the gender factor. We might additionally find that women report higher satisfaction for all ice cream than men, regardless of flavour or fat content. The interactions can be somewhat more complex in this example. Now we might have two within-between (two-way) interactions, in addition to one (two-way) within-group interaction, and we might have one three-way interaction. The two within-between interactions may occur between 'flavour' and 'gender' (chocolate flavour may be preferred over vanilla by men only) and 'fat content' (full fat might be preferred over the diet version by women only). The within-group interaction may operate between 'flavour' and 'fat content' (chocolate ice cream may be preferred over vanilla, but only for the full fat version). The three-way interaction might potentially occur between all three variables; diet/chocolate flavour may be preferred by men, while full fat/vanilla may be preferred by women. These scenarios are summarised in Box 13.4.

13.4 Take a closer look
Main effects and interactions in a mixed (three-way) multi-factorial ANOVA

The following scenarios present how 'taste satisfaction' scores might vary:

Within-group main effect 1:	Flavour (chocolate > vanilla)
Within-group main effect 2:	Fat content ('full fat' > diet)
Between-group main effect:	Gender (women > men)
Within-between interaction 1:	Chocolate > vanilla, but only for men
Within-between interaction 2:	'Full fat' > diet, but only for women
Within-group interaction:	Chocolate > vanilla, but only for 'full fat'
Three-way interaction:	Diet/chocolate flavour preferred by men; 'full fat'/vanilla by women

Establishing significant differences

To assess whether observed main effects and interactions are significant we use similar methods to those we saw with other multi-factorial ANOVAs (Chapters 11 and 12). However, this time we need to partition the overall variance (total sum of squares) into between-group, within-group and interaction sums of squares. Within each of those, we calculate model sums and residual sums of squares. We will see how that is calculated manually in Box 13.5. Model sums of squares are derived from group means or condition means in relation to the grand mean. Residual sums of squares are calculated from the variance across groups or conditions. Those outcomes are then expressed in relation to the relevant degrees of freedom (*df*); these represent the number of values that are 'free to vary' in the calculation, while everything else is held constant (see Chapter 6). From this we can calculate model and residual mean squares. For each main effect and interaction we calculate an F ratio, which is found by dividing the mean square by the residual square. Each F ratio is compared with cut-off points to determine which of the main effects and interactions are statistically significant.

13.5 Calculating outcomes manually
Mixed multi-factorial ANOVA calculation

We can illustrate how to calculate mixed multi-factorial ANOVA manually by using some data that reflect the first of our research questions, focusing on one within-group factor (fat content: 'full fat', 'half fat', and 'diet'), and one between-group element (gender). The outcome is 'taste satisfaction', with higher scores representing higher satisfaction. You will find a Microsoft Excel spreadsheet associated with these calculations on the web page for this book.

>

Table 13.1 Taste satisfaction scores towards ice cream, according to fat content and gender

		Fat content				
Subject	Gender	Full	Half	Diet	Case mean	Case variance
1	Male	66	64	55	61.67	34.33
2	Male	70	63	56	63.00	49.00
3	Male	68	67	63	66.00	7.00
4	Male	68	52	48	56.00	112.00
5	Male	61	62	57	60.00	7.00
6	Male	61	64	59	61.33	6.33
7	Male	68	65	55	62.67	46.33
8	Male	77	72	61	70.00	67.00
9	Male	69	56	53	59.33	72.33
10	Male	68	66	58	64.00	28.00
Condition means (male)		**67.60**	**63.10**	**56.50**	**62.40**	
		Condition variance (M)		**14.61**		
11	Female	68	70	71	69.67	2.33
12	Female	64	69	65	66.00	7.00
13	Female	76	77	68	73.67	24.33
14	Female	66	68	63	65.67	6.33
15	Female	68	67	61	65.33	14.33
16	Female	61	62	60	61.00	1.00
17	Female	63	71	65	66.33	17.33
18	Female	75	52	67	64.67	136.33
19	Female	73	71	69	71.00	4.00
20	Female	73	71	74	72.67	2.33
Condition means (female)		**68.70**	**67.80**	**66.30**	**67.60**	Total **644.67**
		Condition variance (F)		**15.95**		
Condition means (all)		**68.15**	**65.45**	**61.40**		

Grand mean: 65.00 **Grand variance:** 42.71

Grand mean = average of all scores; we saw how to calculate grand variance in Chapter 9 (but also see Excel spreadsheet).

Total sum of squares (SS_T):

$SS_T = s^2_{grand}(N - 1)$ = grand variance × number of scores (60) size *minus* 1 = 42.71 × 59 = 2520.00

Within-participant sum of squares (SS_W):

SS_W = Case variance (total) × number of conditions (3) *minus* 1 = 644.67 × 2 = 1289.33

We saw how to calculate case variance in Chapter 10.

Between-participant sum of squares (SS_B):

$SS_B = \sum n_k(\bar{x}_k - \bar{x}_{grand})^2$

Deduct grand mean from *each* case mean, square it, multiply by no. of conditions (3)

$SS_B = 3 \times [(34.33 - 65.00)^2 + (49.00 - 65.00)^2 + ...(2.33 - 65.00)^2] = 1230.67$

Within-between sum of squares (SS_{WB}):

$SS_{WB} = \sum n_k(\bar{x}_k - \bar{x}_{grand})^2$

Deduct grand mean from group condition mean, square it, multiply by no. of scores in conditions by group (10)

$SS_{WB} = 10 \times [(67.60 - 65.00)^2 + (63.10 - 65.00)^2 + (56.50 - 65.00)^2 + (68.70 - 65.00)^2 + (67.80 - 65.00)^2 + (66.30 - 65.00)^2] = 1058.40$

Between-group main effect

Model sum of squares $SS_M B \sum n_k(\bar{x}_k - \bar{x}_{grand})^2$

Deduct grand mean from group mean, square it, multiply by no. of people in each group (30)

$SS_M B = 30 \times [(62.40 - 65.00)^2 + (67.60 - 65.00)^2] = \mathbf{405.60}$

Degrees of freedom (*df*) for SS_M B (df_M B) = no. of groups *minus* 1 = 2 − 1 = 1

Residual sum of squares $SS_R B \sum s_k^2(n-1)$

Condition variance × (no. of people in each group by group [10] *minus* [10] − 1[= 9]) × no. of conditions (3) [= 27]

$SS_R B = (14.61 + 15.95) \times 27 = \mathbf{825.07}$

df for SS_R B(df_R B) = no. of groups × ([no. of people in each group by group *minus* 1] = 2 × 9 = 18

Within-group main effect

Model sum of squares ($SS_M W$) $\sum n_k(\bar{x}_k - \bar{x}_{grand})^2$

Deduct grand mean from condition mean, square it, multiply by no. of scores per condition (20)

$SS_M W = 20 \times [(68.15 - 65.00)^2 + (65.45 - 65.00)^2 + (61.40 - 65.00)^2] = \mathbf{461.70}$

df for SS_M W (df_MW) = no. of conditions *minus* 1 = 2

Residual sum of squares is calculated with interaction error.

Interaction

Model sum of squares ($SS_M BW$) calculated from whatever is left over from within-between sum of squares.

$SS_M BW = SS_{WB} - SS_M B - SS_M W = 1058.40 - 405.60 - 461.70 = \mathbf{191.10}$

df for SS_M BW (df_M BW) = df_MB × df_MW = 1 × 2 = 2

Within-Interaction error ($SS_R W$)

$SS_R W = SS_W - SS_M W - SS_M BW = 1289.33 - 461.70 - 191.10 = 636.53$

df for SS_R W(df_R W) = df_RB × df_MW = 18 × 2 = 36

Mean squares

Between: Model $\mathbf{MS_M B} = SS_M B \div df_M B = 405.60 \div 1 = \mathbf{405.60}$

Residual $MS_R B = SS_R B \div df_R B = 825.07 \div 18 = \mathbf{45.84}$

Within: Model $MS_MW = SS_MW \div df_MW = 461.70 \div 2 = 230.85$

Interaction: Model $(MS_MBW) = SS_MBW \div df_MW = 191.10 \div 2 = 95.55$

Within-Interaction Residual $MS_RW = SS_RW \div df_RW = 636.53 \div 36 = 17.68$

F ratios

Between $= MS_MB \div MS_RB = 405.60 \div 45.84 = \mathbf{8.85}$

Within $= MS_MW \div MS_RW = 230.85 \div 17.68 = \mathbf{13.06}$

Interaction $= MS_MBW \div MS_RW = 95.55 \div 17.68 = \mathbf{5.40}$

Each F ratio can be compared with the relevant part of the F-distribution table (see Appendix 4), according to the *df* and significance level. We saw how to use the table to find cut-off points in Box 12.3. The cut-off point for gender (1,18) = 3.006; F (gender) = 8.85 (significant, because greater than cut-off). Cut-off for fat content and interaction term (2,36) = 3.26; F (fat content) = 13.06 (significant); F (interaction) = 5.40 (significant).

We can also use Excel to calculate the critical value of F and to provide the actual p value. You can see how to do that on the web page for this book. In our example: Gender: $p = .008$ Fat content: $p < .001$ interaction: $p = .009$.

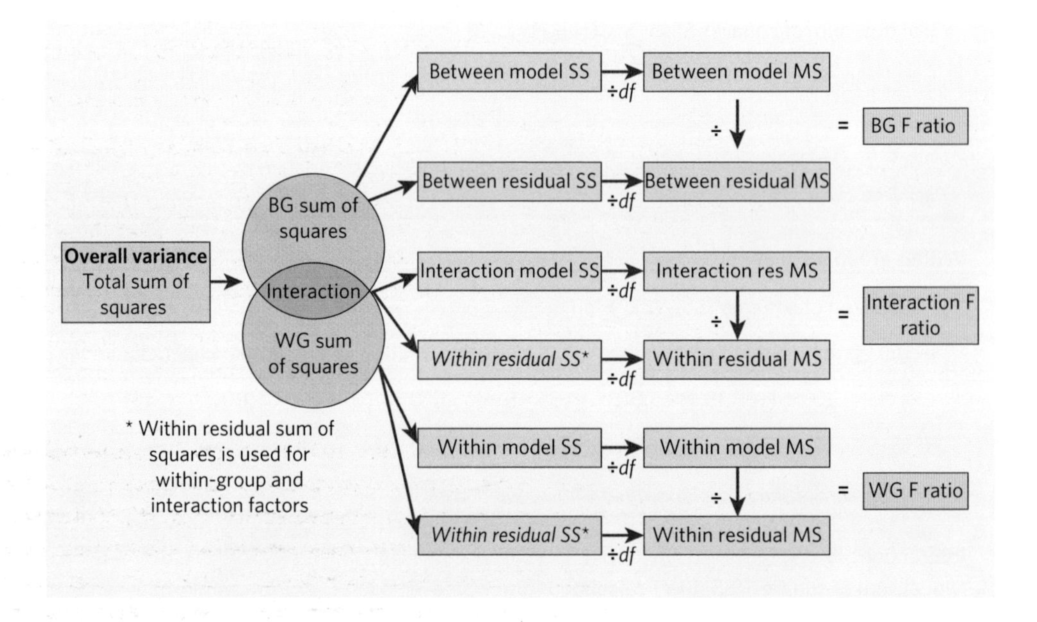

Locating the source of main effects

If we find a significant main effect, we need to report the source of that difference. As we have seen with previous ANOVA tests, we may need to do some more work to locate that. If the difference occurs where there are two groups or conditions, we can consult the mean scores to illustrate the source of that difference. In the example we explored manually just now, there was a significant effect for gender; the mean scores tell us that taste satisfaction was higher for women than it was for men. If a significant effect is found across three or more groups or conditions, we cannot use the mean scores, we need additional tests to locate the source of difference. In our example, there was a significant difference in satisfaction scores across fat content. So, we must perform additional analyses, in the form of either planned contrasts or *post hoc* tests. As we have seen in previous chapters, we can employ planned contrasts only if specific predictions have been made about the relationship between pairs of groups or conditions; otherwise we must use *post hoc* tests. A key difference between those analyses is the way in which they account for multiple comparisons. We explored the importance of considering multiple comparisons in Chapter 9. If we conduct several tests we increase the likelihood of finding a significant outcome by chance factors alone. In some circumstances we must make appropriate adjustments, or we run the risk of making Type I errors (see Chapter 4). The specific type of additional test that we perform depends on whether the independent variable is between- or within-groups.

Between-group main effects

If specific predictions were made about between-group outcomes and if one of the groups represents a control group, we can apply an orthogonal planned contrast. In that case, we do not need to adjust for multiple comparisons. If none of those groups is a control group, we must use a non-orthogonal planned contrast and we must adjust for multiple comparisons. We saw how to set contrast values and perform the tests in Chapter 9, so we will not repeat that here. If no specific predictions are made about between-group relationships, we must use a *post hoc* test. The conventions for choosing an appropriate test remain the same as we saw in Chapter 9 (in particular refer to Box 9.8). Those choices depend on whether there are equal group sizes and homogeneity of variance across the groups. The correct test is selected in SPSS via the Post Hoc menu of the repeated-measures procedures (as we will see later). Given the restrictions for orthogonal planned contrasts, it is likely that *post hoc* tests will be the best option in most cases.

Within-group main effects

As we saw in Chapter 10, there are no user-defined planned contrasts for repeated-measures analyses, only a series of standard contrasts. A 'Simple planned contrast' can be used, if specific predictions had been made about between-group and within-group relationships and if one of the between-group factors represented a control group. In that case, analyses can be undertaken without adjusting for multiple comparisons. 'Simple planned contrast' is the only orthogonal planned contrast available. Otherwise, a non-orthogonal contrast is needed. The most likely option we can use is the 'Repeated' contrast; adjustments for multiple comparisons must be made. We saw how to perform that in Chapter 10. If no specific predictions have been made about relationships between pairs of conditions, we should choose an appropriate *post hoc* test. These are not selected from the Post Hoc menu; that is for between-group analyses only. We obtain within-group *post hoc* tests via the Options menu, having indicated the independent variables that we would like to analyse. There are fewer choices of test – most researchers opt for the Bonferroni analysis (we will see how to do that later).

Locating the source of interaction

If we find a significant interaction, we need to illustrate that with graphs and/or statistics. Graphs can be requested within the Plots menu of SPSS and are very useful as an initial examination. However, this is viable only when we have two independent variables. We *could* run 3D graphs if we have three independent variables, but these are difficult to interpret. If we have more than three independent variables, graphs are simply not possible. Where available, it is often useful to look at graphs before exploring the statistics (as we will see later).

Statistical analyses can be pretty complex, but they are essential to confirm observations that we have undertaken visually (either with graphs or just looking at mean scores). Tests must be undertaken wherever there is an interaction between pairs of independent variables (there are procedures for more complicated interactions, but we will not deal with these here). The nature of the pair of variables being examined will determine which tests need to be employed. If at least one of those independent variables in the interaction pair represents between-group analyses, we will need to engage the Split File facility (we will see how later). In all cases, we must account for multiple comparisons. We can illustrate some typical scenarios using the research examples that we are focusing on in this chapter.

Research example 1: 3 × 2 mixed multi-factorial ANOVA

When we explored the Greedy Pig Ice Cream Company data earlier, we found a significant interaction between gender (between-group) and 'fat content' (within-group) in respect of taste satisfaction scores (dependent variable). We need to explore outcomes across gender at each of the fat content conditions, so we should perform three independent t-tests, with gender as the factor, and using the conditions 'full fat', 'half fat' and 'diet' as the Test Variable (we use those columns in SPSS as if they were dependent variables). We also need to examine outcomes across fat content but must do so for each gender group, so we should perform two repeated-measures one-way ANOVAs, with 'full fat', 'half fat' and 'diet' as the Within-Subjects Variables. We undertake two tests because we need to perform that test once for 'males' and then for 'females'. We will need to use the Split File facility to do this (because the groups are contained within a single column rather than in separate ones as we had for the first analysis). This is summarised in Box 13.6.

13.6 Take a closer look
Source of interaction: 3 × 2 mixed multi-factorial ANOVA

This how we set SPSS parameters for the example that we have just seen:

Table 13.2 Interaction scenarios

Analysis	Method
<u>Within-between interaction</u>	
Gender vs. fat content	3 × independent t-tests: taste satisfaction across gender, at each condition:[1] 1. for 'full fat' 2. for 'half fat' 3. for 'diet'
Fat content vs. gender	2 × repeated-measures one-way ANOVAs: satisfaction across fat content: 1. when gender = male 2. when gender = female[2].

[1] Using within-group columns for each fat content condition as the DVand gender as the IV
[2] Using 'fat content' conditions as 'within-subjects variables', but splitting file by gender

Research example 2: 2 × 2 × 2 mixed multi-factorial ANOVA

Our second research example extends the investigation in taste satisfaction towards the ice cream products by additionally focusing on flavour. Now there will be two within-group independent variables ('fat content' and 'flavour') and one between-group factor (gender). This time, all of the independent variables have just two levels: fat content ('full fat' vs. 'diet'), flavour ('chocolate' vs. 'vanilla') and gender (male vs. female). We will potentially have three two-way interactions: two within-between ('fat content vs. gender' and 'flavour vs. gender') and one within-group interaction ('fat content vs. flavour'). The series of additional tests that we will need to explore these potential interactions will be a little different to what we have just seen. It would make more sense to look at that in more depth when we analyse those data later.

However, we do need to clarify some conceptual points here, especially if you are about to explore some data that have two or more within-group factors in a mixed ANOVA. When we create SPSS data sets for such a scenario, we need to include a 'variable' for each permutation of the within-group conditions. In our example, there are two within-group variables, each with two conditions, so we create four 'variables' to account for the dependent variable score at each scenario: 'chocolate full-fat', 'chocolate diet', 'vanilla full fat' and 'vanilla diet'. When we compare the within-group variables with each other, we can use the procedures that we saw in Chapter 12. The problem comes when we need to compare each within-group variable with the between-group variable (gender). We need to compare satisfaction scores in respect of the flavour variable (chocolate vs. vanilla) separately for men and women (using the Split File facility). We need to do the same across the fat content variable ('full fat' vs. 'diet'). The variable columns do not directly reflect the within-group main effects (SPSS calculates estimated marginal means to report that outcome). So we need to create additional 'variables' to account for this, one for each main effect condition: chocolate, vanilla, full fat and diet. We will see how to do this later.

What extra tests do I need?

We have just seen two examples of how we might employ additional tests to investigate the source of interaction. Box 13.7 presents some scenarios that you might encounter for two-way interactions. The analysis of three-way interactions (or more) is probably a little complex for this humble book.

13.7 Take a closer look
Additional tests needed to locate the source of two-way interactions

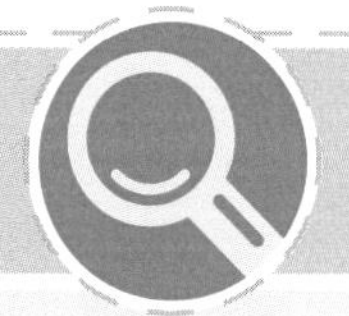

These examples should cover most of the scenarios that you are likely to encounter:

Table 13.3 Interaction scenarios

1st IV	2nd IV			
	Between-group		Within-group	
	2 groups	3+ groups	2 conditions	3+ conditions
Between-group with *n* groups	*n* × ITT[1]	*n* × IOWA[1]	*n* × RTT[1]	*n* × RMOWA[1]
Single within-group with *n* conditions	*n* × ITT[2]	*n* × IOWA[2]	*n* × RTT[3]	*n* × RMOWA[3]
One of two or more within-group IVs with *n* conditions	*n* × ITT[4]	*n* × IOWA[4]	*n* × RTT[3]	*n* × RMOWA[3]

ITT: independent t-test; IOWA: independent one-way ANOVA; RTT: related t-test; RMOWA: repeated-measures one-way ANOVA.

[1] SPSS data file will need to be split according to groups of first IV.
[2] Within-group conditions as DV, groups as IV.
[3] Using appropriate within-group conditions.
[4] Additional variables need to be created to reflect singular main effects.

Assumptions and restrictions

We need to satisfy several assumptions before performing mixed multi-factorial ANOVA. There must be at least two independent variables, at least one of which must be between-group and one within-group. All independent variables must have at least two groups or conditions. The dependent variable should be parametric, with interval data that are *reasonably* normally distributed (see Chapter 5). For this test, we need to investigate normal distribution for the dependent variable across the independent groups and over the within-group conditions (we shall see how later). However, bear in mind that we are looking for 'reasonable' normal distribution; ANOVA is quite robust to minor violations. More serious violations can be overcome with transformation (see Chapter 3). If that fails, we could consider bootstrapping (see Chapter 9). Questions concerning the use of ordinal data are less clear. It could be argued that our data (subjective 'taste satisfaction' scores) are ordinal. Strictly speaking, we cannot use such data, but many researchers do (partly because of the sheer wealth of information the tests can produce). We have considered arguments about this in previous chapters, so we will not repeat them here. There are no non-parametric equivalent tests for mixed multi-factorial ANOVA.

Homogeneity of between-group variance

We need to have equality of variance across groups and conditions. For the between-group portion of this mixed ANOVA, we must measure homogeneity via the Levene's test (we explored this in depth in Chapter 9). We need variance to be equal across the groups to avoid invalid interpretations of significance. This is a particular problem when there are unequal group sizes. Larger groups with variances that are larger than the smaller groups can reduce the likelihood of finding a significant outcome, while larger groups with smaller variances can inflate that chance. Group size differences and homogeneity of variance also have an impact on selecting an appropriate *post hoc* test. Violations of homogeneity of variance are not easy to address with mixed multi-factorial ANOVA. There are adjustments that we can make, such as those offered by Brown–Forsythe F or Welch's F. However, these are complex to perform manually and they are not available in SPSS for this test, unlike independent one-way ANOVA. If the group sizes are equal (or even quite similar) we probably do not need to worry too much. If there is greater inequality in the group sizes we could consider running the between-group variable alone in an independent one-way ANOVA, comparing with and without Brown–Forsythe F or Welch's F adjustments. If there is little difference then we are probably OK. If problems persist we may need to treat the observed outcome with extreme caution.

Sphericity of within-group variance

We also need to ensure that we have equal variance across pairs of within-group conditions – we call this sphericity (see Chapter 10). It is measured through Mauchly's test, which explores whether there is significant difference in variance between pairs of conditions, and assesses whether covariances are equal. Covariance measures the correlation between variables. For sphericity to be assumed we need Mauchly's test to be non-significant. However, this applies only when there are three or more conditions. If an independent variable has just two conditions, we can say that sphericity is assumed. When there are three or more conditions, if Mauchly's test is significant, sphericity cannot be assumed. When this happens it will affect the way in which we can interpret the main ANOVA outcome. We are presented with four lines of ANOVA outcome, each with a potentially different F ratio and significance outcome. If sphericity is assumed, we can select that line of output, otherwise, we must choose one of the other three lines. We discussed which line should be selected in Chapter 10, although most researchers opt for Greenhouse–Geisser or Huynh–Feldt. There is little to choose between them, but if there is a noticeable difference between those outcomes, it may be a good idea to take the average outcome (Field, 2009).

Homogeneity of variance-covariance matrices

We may also need to test for **homogeneity of variance-covariance matrices** (or inter-correlations), although this is more crucial when there are several dependent variables (we explore such examples in Chapter 14). However, homogeneity of variance-covariance matrices has some limited relevance in mixed multi-factorial ANOVA. We measure this outcome with **Box's M** test, which explores whether the correlation between the 'dependent variables' is significantly different between the groups. Although we have just the one dependent variable in mixed multi-factorial ANOVA, it could be argued that the range of within-group conditions performs like dependent variables when viewed against the between-group independent variable. In that sense, it is a measure of variances across the interaction. We should be concerned only if Box's M test is highly significant (at $p < .001$) and if there are unequal group sizes.

13.8 Take a closer look
Summary of assumptions and restrictions

- There must be at least two categorical independent variables
  - At least one between-group variable (with two or more distinct groups)
    - No person can appear in more than one group of each between-group variable
  - And at least one within-group variable (with two or more conditions across one sample)
    - Each person (or case) must be present in all conditions of all within-group variables
- DV data should be interval or ratio and reasonably normally distributed
- There should be between-group homogeneity of variances (particularly where there are unequal group sizes)
- We need to account for sphericity in the within-group variances
  - But only where the within-group independent variable has three or more conditions
- We may need to account for homogeneity of variance-covariance matrices

How SPSS performs mixed multi-factorial ANOVA

To perform mixed multi-factorial ANOVA in SPSS we use the repeated-measures methods that we have seen in Chapters 10 and 12 but add some between-group factors. We will run two tests, so that we can see some slightly different procedures and interpretation, according to the number of independent variables and the number of groups and/or conditions on those variables. Both tests focus on the research examples that we have been discussing throughout this chapter. In the first example, we will explore outcomes where we have one between-group independent variable (with two groups) and one within-group independent variable (with three conditions). In the second example we will examine one between-group independent variable (with two groups) and two within-group independent variables (each with two conditions).

Mixed ANOVA: one within-group IV vs. one between-group IV

In this first example, we will explore a simple two-way mixed multi-factorial ANOVA, using the first research question set by the Greedy Pig Ice Cream Company (see Box 13.9). We saw how to calculate this example manually in Box 13.5, now we will see how to perform the test in SPSS. In this analysis, 20 people (ten men and ten women) eat three versions of vanilla ice cream ('full fat', 'half fat' and 'diet') and report their satisfaction with the taste. The company wishes to know whether satisfaction differs across the fat content conditions, whether men and women vary in satisfaction, and whether the fat content preference differs across gender. Before we analyse this, we should remind ourselves about the dependent and independent variables with this quick summary:

Mixed multi-factorial (two-way) ANOVA (3×2)
Within-group independent variable: fat content (three conditions: 'full fat', 'half fat' and 'diet')
Between-group independent variable: gender (two groups: men and women)
Dependent variable: taste satisfaction scores

13.9 Nuts and bolts

Setting up the data set in SPSS

When we create the SPSS data set for this test, we need to account for between-group variables (where columns represent groups, defined by value labels) and within-group variables (where columns represent the dependent variable score according to each within-group condition).

Ice cream.sav [DataSet1] - IBM SPSS Statistics Data Editor

File Edit View Data Transform Analyze Graphs Utilities Add-ons Window Help

	Name	Type	Width	Decimals	Label	Values	Missing	Columns	Align	Measure
1	gender	Numeric	8	0	Gender	{1, Male}...	None	8	Right	Nominal
2	fullfat	Numeric	8	0	Full fat	None	None	8	Right	Scale
3	halffat	Numeric	8	0	Half fat	None	None	8	Right	Scale
4	diet	Numeric	8	0	Diet	None	None	8	Right	Scale

Figure 13.1 Variable View for 'Ice Cream' data

Figure 13.1 shows how the SPSS Variable View should be set up. The first variable is 'Gender'; this is the between-group independent variable. In the Values column, we include '1 = Male' and '2 = Female'; the Measure column is set to Nominal. The remaining variables ('Full fat', 'Half fat', and 'Diet') represent the within-group independent variables. We do not set up anything in the Values column; we set Measure to Scale.

*Ice cream.sav [DataSet1] - IBM SPSS Statistics Data Editor

File Edit View Data Transform Analyze Graphs Utilities Add-ons Win

5 :

	gender	fullfat	halffat	diet	var
1	Male	66	64	55	
2	Female	68	70	71	
3	Female	64	69	65	

Figure 13.2 Data View for 'Ice Cream' data

Figure 13.2 illustrates how this will appear in the Data View. It is the Data View that will be used to select the variables when performing this test. Each row represents a participant. When we enter the data for 'gender', we input 1 (to represent male) or 2 (to represent female); the 'gender' column will display the descriptive categories ('Male' or 'Female') or will show the value numbers, depending on how you choose to view the column. In the remaining columns ('fullfat'. 'halffat', and 'diet') we enter the actual satisfaction score (dependent variable) for that participant according to each within-group condition.

Testing for normal distribution

Although we have explored how to test for normal distribution on several occasions now, the method is slightly different, so we ought see how it is done. We need to perform a Kolmogorov–Smirnov/Shapiro – Wilk test to examine normal distribution for the taste satisfaction scores across the gender groups *and* over the within-group time points:

Open the SPSS file Ice cream

Select **Analyze** → **Descriptive Statistics** → **Explore** → (in new window) transfer **Full fat**, **Half fat** and **Diet** to **Dependent List** (by clicking on arrow, or by dragging variables to that window) → transfer **Gender** to **Factor List** → tick **Plots** radio button → click **Plots** box → (in new window) click **None** radio button (under **Boxplot**) → make sure that **Stem-and-leaf** and **Histogram** (under **Descriptive**) are unchecked → tick **Normality plots with tests** radio button → click **Continue** → click **OK**

	Gender	Kolmogorov-Smirnov[a]			Shapiro-Wilk		
		Statistic	df	Sig.	Statistic	df	Sig.
Full fat	Male	.235	10	.125	.877	10	.119
	Female	.192	10	.200*	.934	10	.485
Half fat	Male	.222	10	.176	.927	10	.421
	Female	.253	10	.070	.850	10	.058
Diet	Male	.161	10	.200*	.974	10	.926
	Female	.116	10	.200*	.979	10	.960

Figure 13.3 Kolmogorov-Smirnov test for taste satisfaction scores vs. gender

As group sizes are less than 50, we should refer to the Shapiro – Wilk outcome. We appear to have no problems with normal distribution here for all of the levels of the dependent variable (significance $> .05$) (see Chapter 3 for more information on normal distribution).

Running test in SPSS

Using the SPSS file Ice cream

Select **Analyze** → **General Linear Model** → **Repeated Measures...** (see Figure 12.4 for graphical example) → (in new window, as shown in Figure 13.4), type **Fat** in **Within-Subject Factor Name** box → type **3** for **Number of Levels** → click **Add** → click **Define**

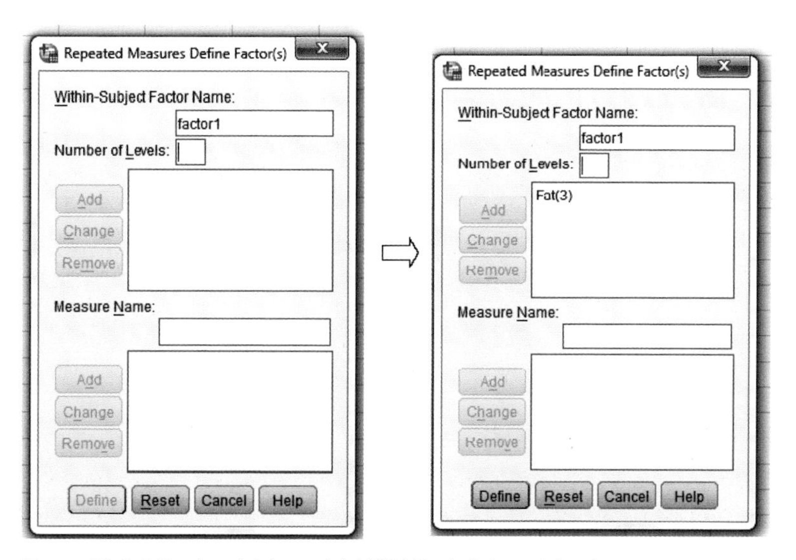

Figure 13.4 Mixed multi-factorial ANOVA: defining within factor

In new window (see Figure 13.5), transfer **Full fat** to **Within-Subjects Variables** to ?_(1,1) → transfer **Half fat** to ?_(1,2) → transfer **Diet** to ?_(1,3) → transfer **Gender** to **Between-Subjects factor** → click **Options...**

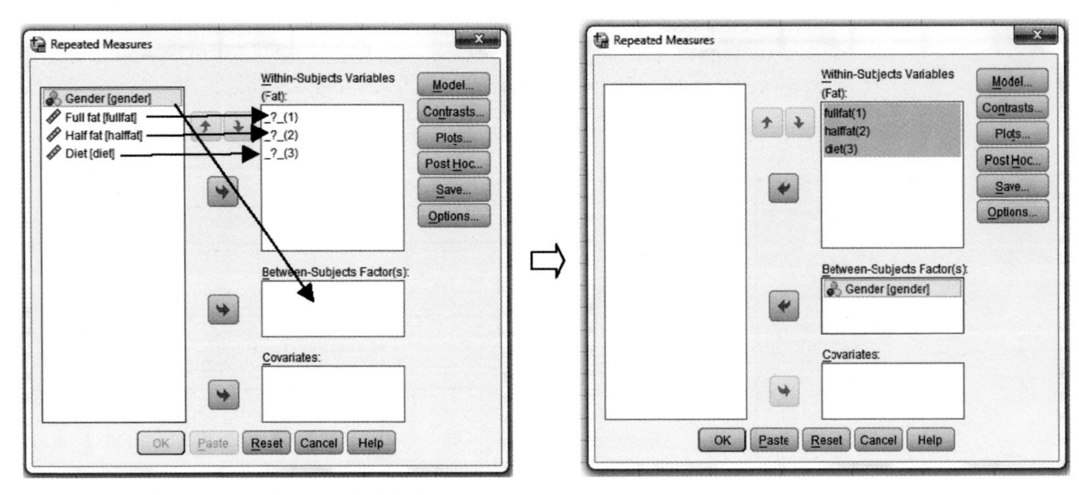

Figure 13.5 Mixed multi-factorial ANOVA: parameter entry

In new window (see Figure 13.6), transfer **Gender, Fat** and **Gender* Fat to Display Means for** → tick boxes under **Display** for **Descriptive statistics, Estimates of effect size and Homogeneity tests** → tick **Compare main effects** box → click pull-down arrow → select **Bonferroni** *(this will produce the within-group post hoc test)* → click **Continue** (it is at this point that we *could* set *post hoc* options for the between-group factor. In this case, we do not need to do that because Gender only has two groups. If the between-group factor had three or more groups we would now select the Post Hoc button, and follow the instructions shown in Chapter 11) → click **Plots** → transfer **Fat** to **Horizontal Axis:** → transfer **Gender** to **Separate Lines:** → click **Add** → click **Continue** → click **OK**

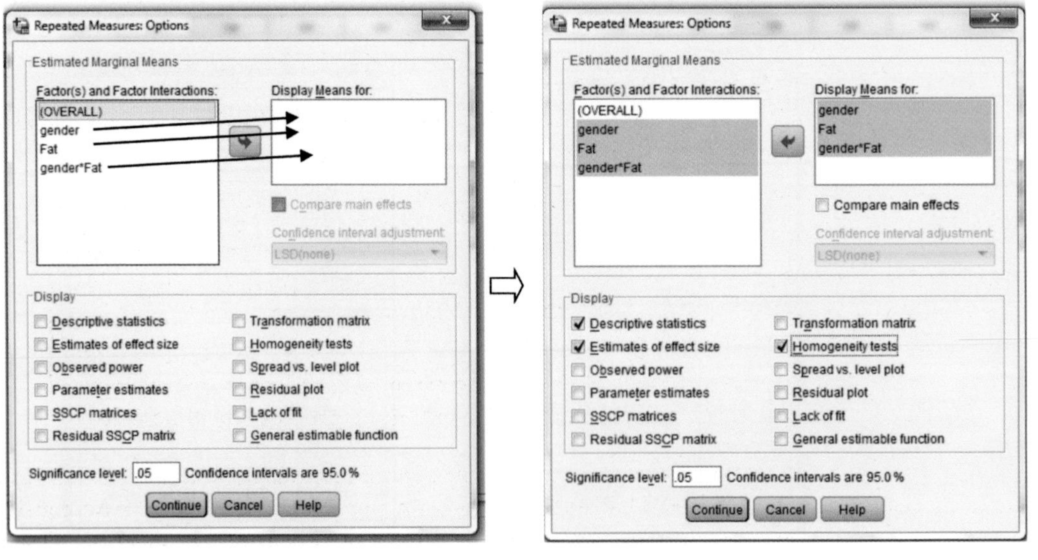

Figure 13.6 Mixed ANOVA: options

Interpretation of output

Before we look at the statistical output, it is often a good idea to look at a graphical representation of the outcome. This can give some idea of what we can expect; it gives us a 'feel' for the data. We can see potential main effects and interactions, which we can then seek to explore statistically afterwards. We asked for a graph when we set the Plots parameters just now. However, the initial graph produced by SPSS in repeated-measures analyses is not always too helpful, as it shows numbers where condition labels should be. We saw how to overcome that in Chapter 12 (in particular see Figures 12.18 and 12.19). The amended graph is shown in Figure 13.7.

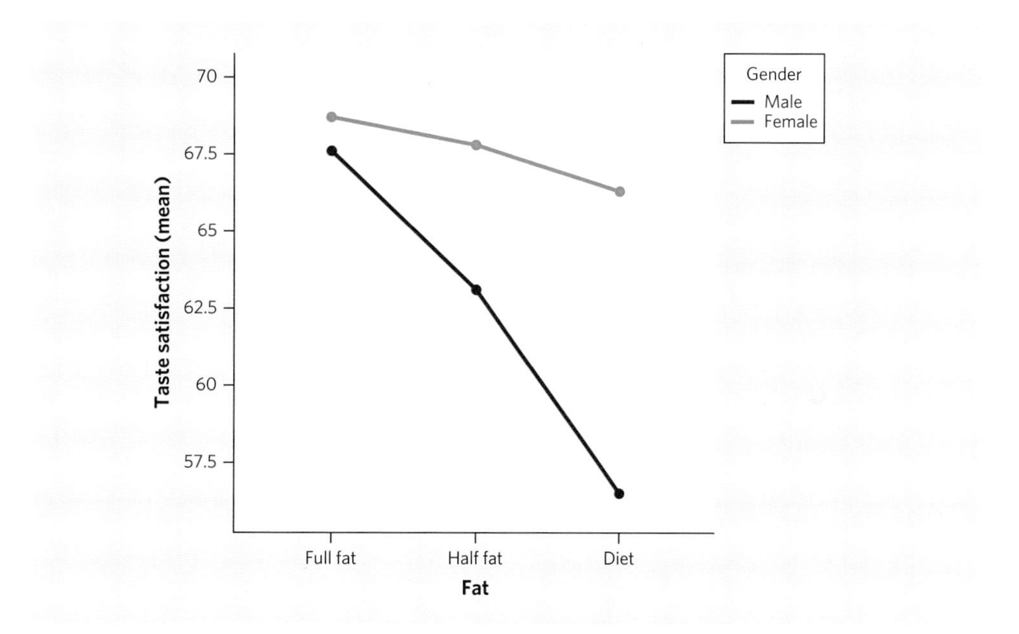

Figure 13.7 (Amended) line graph: Taste satisfaction by fat content and gender

Figure 13.7 appears to suggest that taste satisfaction is greater for the 'full fat' version of the ice cream than for 'half fat' and 'diet'. Overall, women (green line) have higher taste satisfaction than men (blue line), irrespective of the fat content of the ice cream. The decline in taste satisfaction across fat content appears to be much more dramatic for men than it is for women. But do the statistics support this observation?

Testing the assumptions

Before we examine statistical differences, we should check that we have met the conditions for homogeneity and sphericity.

	F	df1	df2	Sig.
Full fat	1.282	1	18	.272
Half fat	.096	1	18	.760
Diet	.124	1	18	.729

Figure 13.8 Homogeneity of (between-group) variances

Figure 13.8 shows the homogeneity tests for between-group variance at each fat content condition. Since none of these is significant, we can assume that we have equal between-group variances in taste satisfaction scores at each fat content condition (see Chapter 3 for more information on how to interpret homogeneity of variance tests).

Within-subjects effect					Epsilon[a]		
	Mauchly's W	Approx. chi-square	df	Sig.	Greenhouse–Geisser	Huynh–Feldt	Lower bound
Fat	.752	4.844	2	.089	.801	.916	.500

Figure 13.9 Sphericity of (within-group) variances

Figure 13.9 shows the outcome for sphericity of within-group variances. Since the significance ('Sig'.) is greater than .05, we can say that sphericity of within-group variance is assumed–we can read that line in the main ANOVA output. Had we violated that assumption, we would need to defer to one of the epsilon adjustments (such as Greenhouse–Geisser or Huynh–Feldt).

Box's M	11.876
F	1.617
df1	6
df2	2347.472
Sig.	.138

Figure 13.10 Homogeneity of variance-covariance matrices

Figure 13.10 shows Box's M test for homogeneity of variance-covariance matrices. This outcome measures whether the correlation in dependent variable scores, across the within-group conditions, remains constant for all groups. To meet this assumption, we need the significance to be greater than .001. It is, so we are OK.

Main effect for gender (between-group IV)

Gender			95% confidence interval	
	Mean	Std. error	Lower bound	Upper bound
Male	62.400	1.236	59.803	64.997
Female	67.600	1.236	65.003	70.197

Figure 13.11 Gender means

Figure 13.11 presents data on 'estimated marginal means' for gender. We are given those because we specifically requested them just now. Estimated marginal means are more useful than the usual 'Descriptive statistics' because they show mean scores for each main effect; traditional descriptive statistics tend to report outcomes only by group within each condition. Figure 13.11 *suggests* that women reported higher mean taste satisfaction scores than men.

Source	Type III sum of squares	df	Mean square	F	Sig.	Partial eta squared
Intercept	253500.000	1	253500.000	5530.462	.000	.997
Gender	405.600	1	405.600	**8.849**	**.008**	.330
Error	825.067	18	45.837			

Figure 13.12 ANOVA test for between-group differences

Figure 13.12 confirms that we have a significant between-group difference (highlighted in red), F (1, 18) = 8.849, p = .008. The outcome tells us that there is a significant main effect for gender (using mean scores in Figure 13.11, we know that 'taste satisfaction' scores are significantly higher for women). You might like to also check the 'mean square' and other data in Figure 13.12 with the calculations we did manually earlier. The partial eta squared figure (highlighted in orange) will be used for effect size calculations later.

Main effect for fat content (within-group IV)

Fat	Mean	Std. error	95% confidence interval	
			Lower bound	Upper bound
1	68.150	1.104	65.831	70.469
2	65.450	1.383	62.545	68.355
3	61.400	.964	59.375	63.425

Figure 13.13 Time-point condition means

Figure 13.13 presents the estimated marginal means for the fat content variable. SPSS is not very helpful when labelling these within-group differences, as it shows only the conditions as numbers. We need some additional output to help us here (shown in Figure 13.14). Using that, we can see that taste satisfaction scores *appear* to be highest for full fat (1) followed by half fat (2), with diet (3) apparently being the least preferred. Before we make any assumptions about that, we need to test whether the observed difference is significant.

Fat	Dependent variable
1	fullfat
2	halffat
3	diet

Figure 13.14 Within-subjects factors

Source		Type III sum of squares	df	Mean square	F	Sig.	Partial eta squared
Fat	Sphericity assumed	461.700	2	230.850	**13.056**	**.000**	.420
	Greenhouse–Geisser	461.700	1.603	288.091	13.056	.000	.420
	Huynh–Feldt	461.700	1.833	251.916	13.056	.000	.420
	Lower bound	461.700	1.000	461.700	13.056	.002	.420
Fat * gender	Sphericity assumed	191.100	2	95.550	5.404	.009	.231
	Greenhouse–Geisser	191.100	1.603	119.242	5.404	.015	.231
	Huynh–Feldt	191.100	1.833	104.269	5.404	.011	.231
	Lower bound	191.100	1.000	191.100	5.404	.032	.231
Error(fat)	Sphericity assumed	636.533	36	17.681			
	Greenhouse–Geisser	636.533	28.847	22.066			
	Huynh–Feldt	636.533	32.990	19.295			
	Lower bound	636.533	18.000	35.363			

Figure 13.15 ANOVA test for within-group differences (and interaction)

Figure 13.15 shows the statistical outcome for the within-group main effect and the interaction between that and the between-group main effect. We can see that there is a significant main effect for fat content (highlighted in blue), F (2, 36) = 13.056, p < .001. You might like to check that with the calculations we did manually earlier. We read from the 'Sphericity Assumed' line because we passed that test when we examined the assumptions earlier. Because we have three time points, we cannot tell from this where the difference lies within the main effect. To examine that we need the *post hoc* test (see Figure 13.16).

(I) Fat	(J) Fat	Mean difference (I-J)	Std. error	Sig.[a]	95% confidence interval for difference[a]	
					Lower bound	Upper bound
1	2	2.700	1.627	.343	−1.595	6.995
	3	6.750*	1.144	**.000**	3.730	9.770
2	1	−2.700	1.627	.343	−6.995	1.595
	3	4.050*	1.161	**.008**	.987	7.113
3	1	−6.750*	1.144	.000	−9.770	−3.730
	2	−4.050*	1.161	.008	−7.113	−.987

Figure 13.16 *Post hoc* test for within-group differences

The significant differences are highlighted in red. With help from Figure 13.14, we can see that taste satisfaction scores for 'full fat' are significantly higher than they are for 'diet' (p < .001) and that those for 'half fat' are significantly higher than they are for 'diet' (p = .008). There is no significant difference in taste satisfaction scores between 'full fat' and 'half fat' versions of the ice cream (see Chapter 10 for further guidance on interpretation of these tests).

Within-between interaction: gender vs. fat content

Gender	Fat	Mean	Std. error	95% confidence interval	
				Lower bound	Upper bound
Male	1	67.600	1.561	64.321	70.879
	2	63.100	1.956	58.991	67.209
	3	56.500	1.363	53.636	59.364
Female	1	68.700	1.561	65.421	71.979
	2	67.800	1.956	63.691	71.909
	3	66.300	1.363	63.436	69.164

Figure 13.17 Interaction means

Figure 13.17 presents some interesting patterns in taste satisfaction scores (we need Figure 13.14 once again to help decipher the conditions). If we look at gender for the fat point conditions, it would appear that differences in taste satisfaction scores, across the conditions, are much greater for men than for women. This suggests there probably is a within-between interaction, but we need to test that for significance. To do that, we refer back to Figure 13.15. This shows that we can see that we have a significant interaction (highlighted in green), F (2,36) = 5.404, p = .009. You might like to check that with those manual calculations from earlier, too.

Locating the source of interaction

Although the outcome from Figure 13.15 confirms that we have a significant interaction, we cannot be certain how that interaction is illustrated. Figure 13.7 gives us a pretty good idea, and this is supported by the estimated marginal means shown in Figure 13.17. Ultimately, we need some statistical analyses. There is no direct way to run *post hoc* tests on the interaction, so we need to run some additional tests (rather like we did in Chapters 11 and 12). In doing so, we must account for multiple comparisons to reduce the risk of making Type I errors (saying that we have an effect when there isn't one – see Chapter 4 for more information).

As we saw earlier, there are a number of ways to investigate the source of a significant interaction. In this case, we have a fairly straightforward two-way mixed ANOVA, with just one between-group variable (gender) and one within-group variable (fat content). If we refer to Box 13.6, we can see the type of additional tests that we should undertake. We need to explore the taste satisfaction scores in respect of gender, using three independent t-tests: one for each of the fat content conditions ('full fat', 'half fat' and 'diet'). We also need to examine taste satisfaction scores across the three 'fat content' conditions, using repeated-measures one-way ANOVA, which we do twice, once for men and then for women (using the Split File facility).

Gender vs. taste satisfaction scores, according to fat content:

Select **Analyze** → **Compare Means** → **Independent-Samples T Test...** → to **Grouping Variable** → click **Define Groups** → enter **1** in **Group 1** box → enter **2** in **Group 2** box → click **Continue** → click **OK**

	Gender	N	Mean	Std. deviation	Std. error mean
Full fat	Male	10	67.60	4.551	1.439
	Female	10	68.70	5.293	1.674
Half fat	Male	10	63.10	5.607	1.773
	Female	10	67.80	6.713	2.123
Diet	Male	10	56.50	4.223	1.335
	Female	10	66.30	4.398	1.391

Figure 13.18 Descriptive statistics

Independent samples test

		Levene's test for equality of variances		t-test for equality of means						
									95% confidence interval of the difference	
		F	Sig.	t	df	Sig. (2-tailed)	Mean difference	Std. error difference	Lower	Upper
Full fat	Equal variances assumed	1.282	.272	−.498	18	.624	−1.100	2.207	−5.737	3.537
	Equal variances not assumed			−.498	17.605	.624	−1.100	2.207	−5.745	3.545
Half fat	Equal variances assumed	.096	.760	−1.699	18	.106	−4.700	2.766	−10.511	1.111
	Equal variances not assumed			−1.699	17.446	.107	−4.700	2.766	−10.524	1.124
Diet	Equal variances assumed	.124	.729	−5.083	18	.000	−9.800	1.928	−13.851	−5.749
	Equal variances not assumed			−5.083	17.970	.000	−9.800	1.928	−13.851	−5.749

Figure 13.19 Independent t-test: gender vs. taste satisfaction scores (by fat content)

As we have three additional tests, we need to adjust for significance ($p < .05$) by dividing the usual cut-off point by 3 (so, $.05 \div 3 = .016$). We will have a significant difference only when $p < .016$. Figures 13.18 and 13.19 show that there is no significant difference in taste satisfaction scores in respect of gender, for 'Full fat' ($p = .624$) or 'Half fat' ($p = .106$), but there is a significant difference in taste satisfaction scores in respect of gender for 'diet', $t(18) = -5.083$, $p < .001$.

Taste satisfaction scores by fat content condition, according to gender:
To do this, we need to use the Split File facility that we encountered in Chapter 11. We will also need to adjust the significance cut-off point by the number of additional comparisons (two). We will have a significant difference only if $p < .025$.

Using the SPSS file Ice cream

Select **Data** → **Split File** → (in new window) select radio button for **Compare groups** → select Gender from list → click **Groups Based on:** arrow → click **OK**

Select **Analyze** → **General Linear Model** → **Repeated Measures...** → (in new window) type **Fat** in **Within-Subject Factor Name** box → type **3 in Number of Levels** → click **Add** → click **Define** → (in new window) transfer **Full fat, Half fat**, and **Diet** to **Within-Subjects Variables** → click **Options** (in next window) transfer **Fat** to **Display Means for:** → tick **Compare main effects** box → click pull-down arrow → click **Continue** → click **OK** (note that there should be NO entry in BG **Factor**)

Gender	Fat	Mean	Std. error	95% confidence interval Lower bound	Upper bound
Male	1	67.600	1.439	64.344	70.856
	2	63.100	1.773	59.089	67.111
	3	56.500	1.335	53.479	59.521
Female	1	68.700	1.674	64.914	72.486
	2	67.800	2.123	62.998	72.602
	3	66.300	1.391	63.154	69.446

Figure 13.20 Estimated marginal means

Gender	Within-subjects effect	Mauchly's W	Approx. chi-square	df	Sig.	Epsilon[a] Greenhouse–Geisser	Huynh–Feldt	Lower bound
Male	Fat	.508	5.421	2	.066	.670	.744	.500
Female	Fat	.545	4.863	2	.088	.687	.770	.500

Figure 13.21 Sphericity of (within-group) variances

We need to test sphericity again because the outcome may vary by gender, forcing us to interpret the statistics somewhat differently for men than we do for women. As it happens, Figure 13.21 tells us that Mauchly's test is not significant in either case, so we can claim that sphericity is assumed for both.

Gender	Source		Type III sum of squares	df	Mean square	F	Sig.	Partial eta squared
Male	Fat	Sphericity assumed	623.400	2	311.700	23.848	.000	.726
		Greenhouse–Geisser	623.400	1.340	465.114	23.848	.000	.726
		Huynh–Feldt	623.400	1.489	418.748	23.848	.000	.726
		Lower bound	623.400	1.000	623.400	23.848	.001	.726
	Error(fat)	Sphericity assumed	235.267	18	13.070			
		Greenhouse–Geisser	235.267	12.063	19.503			
		Huynh–Feldt	235.267	13.399	17.559			
		Lower bound	235.267	9.000	26.141			
Female	Fat	Sphericity assumed	29.400	2	14.700	.659	.529	.068
		Greenhouse–Geisser	29.400	1.374	21.396	.659	.479	.068
		Huynh–Feldt	29.400	1.540	19.095	.659	.494	.068
		Lower bound	29.400	1.000	29.400	.659	.438	.068
	Error(fat)	Sphericity assumed	401.267	18	22.293			
		Greenhouse–Geisser	401.267	12.367	32.446			
		Huynh–Feldt	401.267	13.857	28.958			
		Lower bound	401.267	9.000	44.585			

Figure 13.22 Repeated-measures one-way ANOVA for fat content (reported by gender)

Figures 13.20 and 13.22 tell us that we have a significant difference in taste satisfaction scores across the fat content for men, $F(2, 18) = 23.848$ $p < .001$, but not for women, $F(2, 18) = 0.659$, $p = .438$.

You must remember to switch off the Split File facility, otherwise subsequent analyses will be incorrect:

Using the SPSS file Ice Cream

Select **Data** → **Split File** → (in new window) select **Analyze all cases, do not create groups** radio button → click **OK**

Effect size and power

As we have seen in previous chapters, we can use G*Power to calculate effect size for us and to show how much power our study had to detect the outcome. We need to do three analyses: one each for the between-group main effect, within-group main effect and the interaction. Before we proceed with entering the data, we need to calculate the 'average' correlation across the variables (we will need this for the 'correlation between repeated measures' parameter shortly). We do not get that from any of the analyses that we have undertaken so far. We need to perform correlation analyses between the three conditions of the within-group variable and gender. We saw how to perform correlation in SPSS in Chapter 6. The output for this is shown in Figure 13.23.

		Gender	Full fat	Half fat	Diet
Gender	Pearson correlation	1	.117	.372	.768**
	Sig. (2-tailed)		.624	.106	.000
	N	20	20	20	20
Full fat	Pearson correlation	.117	1	.189	.340
	Sig. (2-tailed)	.624		.425	.142
	N	20	20	20	20
Half fat	Pearson correlation	.372	.189	1	.619**
	Sig. (2-tailed)	.106	.425		.004
	N	20	20	20	20
Diet	Pearson correlation	.768**	.340	.619**	1
	Sig. (2-tailed)	.000	.142	.004	
	N	20	20	20	20

Figure 13.23 Correlation between repeated-measures conditions

To find the 'correlation between repeated measures' in respect of these 'variables' we find the average correlation co-efficient from the repeated-measures factors shown in Figure 13.23 ([.189+ .340+ .619] ÷ 3 = .382).

Open G*Power:

From **Test family** select F tests

From **Type of power analysis** select Post hoc: Compute achieved-given α, sample size and effect size power

Gender (between-group) main effect

From **Statistical test** select ANOVA: Repeated measures, between factors
To calculate the **Effect size**, click on the **Determine** button (a new box appears) → under **Select procedure** choose Effect size from Variance
In box below, tick on radio button for Direct → type **0.330** in the **Partial η^2** box (we get that from Figure 13.12) → click on **Calculate and transfer to main window**
Back in original display, for α **err prob** type **0.05** (the significance level) → **Total sample size** type **20** (the overall sample size) → **Number of groups** type **2** (male and female) → **Number of measurements** type **3** (full fat, half fat, and diet) → **Corr among rep measures** type **0.382** (for the reasons we stated just now) → click on **Calculate**
Effect size (d) **0.70** (large); and Power (1-β err prob) **0.97** (strong-see Chapter 4).

Treatment (within-group) main effect

From **Statistical test** select ANOVA: Repeated measures, within factors
To calculate the **Effect size**, click on the **Determine** button
In that new box, tick on radio button for Direct → type **0.420** in the **Partial η^2** box (we get that from Figure 13.15) → click on **Calculate and transfer to main window**
Back in original display, for α **err prob** type **0.05** → **Total sample size** type **20** → **Number of groups** type **2** (as above) → **Number of repetitions** type **3** → **Corr among rep measures** type **0.382** → **nonsphericity** type **0.752** (Figure 13.9 – Mauchly's W) → click on **Calculate**
Effect size (d) **0.85** (large); Power (1-β err prob) **1.00** (perfect).

Interaction

From **Statistical test** select ANOVA: Repeated measures, within-between interaction
In Determine box type **0.231** in the **Partial η^2** box (Figure 13.15 again) → click on **Calculate**

and transfer to main window ➜ in the main window, all of the other data remain as above ➜ click on **Calculate**

Effect size (d) **0.55** (large); Power (1-β err prob) **1.00** (perfect).

Writing up results

Table 13.4 shows that taste satisfaction scores, towards vanilla ice cream, varied by gender and fat content. A mixed 3 × 2 multi-factorial ANOVA indicated a significant between-group difference for gender, $F(1, 18) = 8.849$, $p = .008$; $d = 0.70$. Women reported significantly greater satisfaction than men. There was a significant within-group difference for 'fat content', $F(2, 36) = 13.056$, $p < .001$; $d = 0.85$. Bonferroni *post hoc* tests indicated that significantly less satisfaction was reported for the 'diet' version than for 'full fat' ($p < .001$) and than for 'half fat' ($p = .008$). There was no difference between 'full fat' and 'half fat'. There was a significant interaction between gender and 'fat content', $F(2, 36) = 5.404$, $p = .009$; $d = 0.55$. Further examination (using independent t-tests and repeated-measures one-way ANOVAs) suggested that an interaction occurred because the main effect for 'fat content' was apparent for males only. Furthermore, although women reported significantly higher satisfaction than men, this was evident only for the diet version of the ice cream.

Table 13.4 Taste satisfaction scores by gender and fat content

	Satisfaction scores								
				Gender vs. fat content					
	Main effects			Male			Female		
Fat content	Mean	SE	N	Mean	SE	N	Mean	SE	N
Full fat	68.15	1.10	20	67.60	1.56	10	68.70	1.56	10
Half fat	65.45	1.38	20	63.10	1.96	10	67.80	1.96	10
Diet	61.40	0.96	20	56.50	1.36	10	66.30	1.36	10
Gender									
Male	62.40	1.24	10						
Female	67.60	1.24	10						

Mixed ANOVA: one between-group IV vs. two within-group IVs

To illustrate how we can explore a slightly more complex example of mixed multi-factorial ANOVA we will use the second of the research questions posed by the Greedy Pig Ice Cream Company. In this analysis, 20 people (ten men and ten women) eat four versions of ice cream (chocolate – full fat, chocolate – diet, vanilla – full fat, and vanilla – diet) and report their satisfaction regarding the taste. The company seeks to discover how taste satisfaction varies by flavour, 'fat content' and gender. Before we analyse this, we should remind ourselves about the dependent and independent variables with this quick summary:

Mixed multi-factorial (three-way) ANOVA (2 × 2 × 2)

- **Within-group independent variable 1:** flavour (two conditions: 'chocolate'and 'vanilla')
- **Within-group independent variable 2:** fat content (two conditions: 'full fat' and 'diet')
- **Between-group independent variable:** gender (two groups: men and women)
- **Dependent variable:** taste satisfaction scores

Setting up data set in SPSS

We saw how to create a data set, suitable for performing mixed multi-factorial ANOVA, in the two-way (3 × 2) example earlier (see Box 13.9). The procedure for setting up the next data set will be similar, so we need not repeat that. However, we may need to create some additional variables, should we find significant within-between interactions (for the reasons we stated earlier).

Running test in SPSS

Using the SPSS file Ice cream 2

Select **Analyze** → **General Linear Model** → **Repeated Measures...** → (in new window) type Flavour in **Within-Subject Factor Name** box → type **2** for **Number of Levels** → click **Add** → type Fat in **Within-Subject Factor Name** box → type **2** for **Number of Levels** → click **Add** → click **Define** → (in new window) transfer **Chocolate-full fat**, **Chocolate-diet**, **Vanilla-full fat** and **Vanilla-diet** to **Within-Subjects Variables** → transfer **Gender** to **Between-Subjects factor** → click **Options...**

In this next box, we set up the estimated marginal means (so that we can compare groups and conditions) and *would* set up within-group *post hoc* tests *if* we needed them (we do not, as there are only two conditions in each independent variable). We also do not need between-group *post hoc* tests (there are only two groups for gender). However, we can set up other options here. We will not attempt to ask for graphs, as we have three independent variables.

Transfer **gender**, **Flavour**, **Fat**, **gender*Flavour**, **gender*Fat**, **Flavour*Fat** and **gender*Flavour*Fat** to **Display Means for:** → tick boxes for **Descriptive statistics**, **Estimates of effect size** and **Homogeneity tests** → click **Continue** → click **OK**

Interpretation of output

Testing the assumptions

As usual, we should check that we have met the conditions for homogeneity and sphericity.

	F	df1	df2	Sig.
Chocolate – full fat	.777	1	18	.390
Chocolate – diet	.458	1	18	.507
Vanilla – full fat	.762	1	18	.394
Vanilla – diet	.097	1	18	.758

Figure 13.24 Homogeneity of (between-group) variances

Figure 13.24 shows the homogeneity tests for between-group variance for each within-group condition. Since none of these is significant, we can assume that we have equal between-group variances in satisfaction scores at each of the within-group conditions. We do not need to test for sphericity of within-group variances because there were only two conditions for each independent variable (see Chapter 10 for a further explanation of why this is not needed when there are only two conditions). We saw an example of what Mauchly's test looks like when we do have within-group variables with two conditions in Chapter 12 (see Figure 12.15).

Box's M	7.876
F	.594
df1	10
df2	1549.004
Sig.	.820

Figure 13.25 Homogeneity of variance-covariance matrices

Figure 13.25 confirms that we have homogeneity of variance-covariance matrices.

Between-group main effect: gender

			95% confidence interval	
Gender	Mean	Std. error	Lower bound	Upper bound
Male	75.350	.933	73.389	77.311
Female	74.275	.933	72.314	76.236

Figure 13.26 Gender means

Figure 13.26 shows that men *appear* to have reported (very slightly) higher satisfaction than women, but we need to test that for significance before we can make any assumptions.

Source	Type III sum of squares	df	Mean square	F	Sig.	Partial eta squared
Intercept	447752.813	1	447752.813	12847.488	.000	.999
Gender	23.113	1	23.113	.663	.426	.036
Error	627.325	18	34.851			

Figure 13.27 ANOVA test for between-group differences

Figure 13.27 confirms that we do not have a significant between-group difference, $F(1, 18) = .663$, $p = .426$; there is no significant main effect for gender. However, as we will see later, this is hiding a very interesting effect across the within-group conditions.

Within-group effects

We will need Figure 13.28 for several analyses of within-group main effects and interactions. The within-group main effects are highlighted in blue, within-group interactions in red, within-between interactions in green, and the three-way interaction in purple. We will refer to these

Source		Type III sum of squares	df	Mean square	F	Sig.	Partial eta squared
Flavour	Sphericity assumed	348.613	1	348.613	**17.800**	**.001**	.497
	Greenhouse–Geisser	348.613	1.000	348.613	17.800	.001	.497
	Huynh–Feldt	348.613	1.000	348.613	17.800	.001	.497
	Lower bound	348.613	1.000	348.613	17.800	.001	.497
Flavour * gender	Sphericity assumed	6003.113	1	6003.113	**306.520**	**.000**	.945
	Greenhouse–Geisser	6003.113	1.000	6003.113	306.520	.000	.945
	Huynh–Feldt	6003.113	1.000	6003.113	306.520	.000	.945
	Lower bound	6003.113	1.000	6003.113	306.520	.000	.945
Error(flavour)	Sphericity assumed	352.525	18	19.585			
	Greenhouse–Geisser	352.525	18.000	19.585			
	Huynh–Feldt	352.525	18.000	19.585			
	Lower bound	352.525	18.000	19.585			
Fat	Sphericity assumed	904.512	1	904.512	**28.703**	**.000**	.615
	Greenhouse–Geisser	904.512	1.000	904.512	28.703	.000	.615
	Huynh–Feldt	904.512	1.000	904.512	28.703	.000	.615
	Lower bound	904.512	1.000	904.512	28.703	.000	.615
Fat * gender	Sphericity assumed	6142.513	1	6142.513	**194.923**	**.000**	.915
	Greenhouse–Geisser	6142.513	1.000	6142.513	194.923	.000	.915
	Huynh–Feldt	6142.513	1.000	6142.513	194.923	.000	.915
	Lower bound	6142.513	1.000	6142.513	194.923	.000	.915
Error(fat)	Sphericity assumed	567.225	18	31.513			
	Greenhouse–Geisser	567.225	18.000	31.513			
	Huynh–Feldt	567.225	18.000	31.513			
	Lower bound	567.225	18.000	31.513			
Flavour * fat	Sphericity assumed	154.012	1	154.012	**5.933**	**.025**	.248
	Greenhouse–Geisser	154.012	1.000	154.012	5.933	.025	.248
	Huynh–Feldt	154.012	1.000	154.012	5.933	.025	.248
	Lower bound	154.012	1.000	154.012	5.933	.025	.248
Flavour * fat * gender	Sphericity assumed	78.013	1	78.013	**3.005**	**.100**	.143
	Greenhouse–Geisser	78.013	1.000	78.013	3.005	.100	.143
	Huynh–Feldt	78.013	1.000	78.013	3.005	.100	.143
	Lower bound	78.013	1.000	78.013	3.005	.100	.143
Error(flavour*fat)	Sphericity assumed	467.225	18	25.957			
	Greenhouse–Geisser	467.225	18.000	25.957			
	Huynh–Feldt	467.225	18.000	25.957			
	Lower bound	467.225	18.000	25.957			

Figure 13.28 Within-group effects and interactions

often throughout the following analyses. As we saw earlier, we need help identifying the within-group factors (see Figure 13.29).

Flavour	Fat	Dependent variable
1	1	chocfull
	2	chocdiet
2	1	vanfull
	2	vandiet

Figure 13.29 Within-group codes

Main effect for flavour (1st within-group IV)

			95% confidence interval	
flavour	Mean	Std. error	Lower bound	Upper bound
1	76.900	.735	75.356	78.444
2	72.725	.906	70.822	74.628

Figure 13.30 Flavour means

Figure 13.30 (with help from Figure 13.29) indicates that chocolate flavour appears to have received higher satisfaction scores than vanilla. To check whether that is significant, we need to refer to Figure 13.28. On this occasion we read from the block labelled 'flavour', which shows that we have a significant within-group effect, $F(1, 18) = 17.800$, $p = .001$. So we know that satisfaction for chocolate flavour is significantly higher than it is for vanilla.

Main effect for fat content (2nd within-group IV)

			95% confidence interval	
fat	Mean	Std. error	Lower bound	Upper bound
1	78.175	1.028	76.015	80.335
2	71.450	.776	69.820	73.080

Figure 13.31 Fat content condition mean scores

Figure 13.31 (with help from Figure 13.29) indicates that 'full fat' ice cream appears to have received higher satisfaction scores than 'diet'. The 'Fat block' section in Figure 13.28 shows that 'full fat'is significantly preferred over 'diet', $F(1, 18) = 28.703$, $p < .001$.

Interactions

Within-group interaction: Fat content × flavour

				95% confidence interval	
flavour	fat	Mean	Std. error	Lower bound	Upper bound
1	1	81.650	1.300	78.919	84.381
	2	72.150	.929	70.199	74.101
2	1	74.700	1.271	72.029	77.371
	2	70.750	1.194	68.241	73.259

Figure 13.32 Fat content vs. flavour interaction means

Figure 13.32 (with help from Figure 13.29) indicates that there appears to be a greater difference in satisfaction scores between 'full fat' and 'diet' versions for chocolate ice cream than for vanilla, with greater preference for 'full fat' in both cases. The 'Flavour*Fat' block in Figure 13.28 indicates that the interaction is significant, $F(1,18) = 5.933$, $p = .025$.

Within-between interaction 1: Flavour × gender

				95% confidence interval	
Gender	flavour	Mean	Std. error	Lower bound	Upper bound
Male	1	86.100	1.039	83.916	88.284
	2	64.600	1.281	61.908	67.292
Female	1	67.700	1.039	65.516	69.884
	2	80.850	1.281	78.158	83.542

Figure 13.33 Flavour vs. gender interaction means

You may recall that earlier we found no main effect for gender. However, when we look at Figure 13.33 we can see that this tells only half the story. From this output (with help from Figure 13.29), it would appear that men prefer chocolate flavour, while women prefer vanilla flavour. To confirm this, the 'Flavour*gender' block in Figure 13.28 shows that this interaction is significant, $F(1,18) = 306.520$, $p < .001$.

Within-between interaction 2: Fat content × gender

				95% confidence interval	
Gender	fat	Mean	Std. error	Lower bound	Upper bound
Male	1	69.950	1.454	66.895	73.005
	2	80.750	1.097	78.445	83.055
Female	1	86.400	1.454	83.345	89.455
	2	62.150	1.097	59.845	64.455

Figure 13.34 Fat content vs. gender interaction means

Figure 13.34 also suggests that there are gender differences in respect of preferences towards fat content, despite there being no main effect for gender – men appear to prefer the diet ice cream, while women prefer the full fat version. The 'Fat*gender' block in Figure 13.28 confirms that this interaction is significant, $F(1,18) = 194.923$, $p < .001$.

Three-way interaction: Fat content × flavour × gender

					95% confidence interval	
Gender	flavour	fat	Mean	Std. error	Lower bound	Upper bound
Male	1	1	81.100	1.839	77.237	84.963
		2	91.100	1.313	88.341	93.859
	2	1	58.800	1.798	55.022	62.578
		2	70.400	1.689	66.852	73.948
Female	1	1	82.200	1.839	78.337	86.063
		2	53.200	1.313	50.441	55.959
	2	1	90.600	1.798	86.822	94.378
		2	71.100	1.689	67.552	74.648

Figure 13.35 Three-way interaction means

When we examined the interaction between flavour and fat content, we found that the difference in satisfaction between 'full fat' and 'diet' ice cream was greater for chocolate flavour than it was for vanilla. Meanwhile, there was consistently greater preference for 'full fat'. Figure 13.35 suggests that the picture may be very different when we look at this by gender. For men, the difference in satisfaction scores between fat content versions was relatively similar between the flavours. What was (potentially) striking was that men preferred the diet version (as we saw in the interaction between fat and gender earlier). For women, there was an apparent larger difference in satisfaction scores between 'fat content' versions for vanilla flavour than there was for chocolate. This is in contrast to what we found for the overall sample when we examined the interaction between flavour and fat content. Interesting as all of that might seem, if we refer to the Flavour*Fat*gender block in Figure 13.28 we can see that there is not a significant interaction despite what we appear to see, $F(1,18) = 3.005$, $p = .100$. It is quite possible that we did not have enough participants to confirm this interaction.

Locating the source of interaction

Figure 13.28 shows that we have significant interactions for 'flavour vs. gender', 'fat content vs. gender' and 'flavour vs. fat content' (all in respect of taste satisfaction scores). We need to investigate the nature of these interactions. The estimated marginal means shown in Figures 13.32 to 13.34 provide some clues, but we need statistical confirmation. We saw a range of scenarios for the types of additional tests that we may need to perform in Box 13.7. In these analyses, we have two within-between interactions and one within-group interaction. Where the between-group variable is the first factor to be analysed, the remaining (within-group) variable must be investigated across the conditions, but using the Split File facility in SPSS to focus on one group at a time. In the second part of the within-between analyses, the examination is complicated by the fact that we have two within-group variables. This means that the range of within-group conditions has been set up in SPSS, but not the within-group main effects (we will address that shortly). The analysis of the within-group interaction is more straightforward because we can simply use the condition scenarios that are already in the SPSS data set (rather like we did in Chapter 12). All of the independent variables have two groups or conditions so, using the guidelines from Box 13.7, we can summarise the additional tests that we will need. In each analysis, we must adjust for multiple comparisons. As there are two tests in each case, we divide the usual significance cut-off point ($p < .05$) by 2. Significance will be met in any case only when $p < .025$. The additional tests are illustrated in Box 13.10.

13.10 Take a closer look
Source of interaction: 2 × 2 × 2 mixed multi-factorial ANOVA

We will need to conduct the following additional analyses to investigate the sources of the interaction that we observed in Figure 13.28:

Table 13.5 Additional tests needed

Analysis	Method
Within-between interactions	
Flavour vs. gender	2 × independent t-tests: taste satisfaction across gender for each flavour condition[1]: 1. for chocolate flavour 2. for vanilla flavour
	2 × related t-tests: taste satisfaction for 'chocolate vs. vanilla', by gender[2]: 1. when gender = male 2. when gender = female[3]
Fat content vs. gender	2 × independent t-tests: taste satisfaction across gender for each 'fat content' condition[1]: 1. for full fat 2. for diet
	2 × related t-tests: taste satisfaction for 'full fat vs. half fat', by gender[2]: 1. when gender = male 2. when gender = female[3]
Within-group interaction	
Flavour vs. fat content	2 × related t-tests: 1. chocolate/full fat vs. chocolate/diet 2. vanilla/full fat vs. vanilla/diet
Fat content vs. flavour	2 × related t-tests: 1. chocolate/full fat vs. vanilla/full fat 2. chocolate/diet vs. vanilla/diet

[1] Using within-group columns for each relevant condition as the DVand gender as the IV.
[2] New within-group variables need to be created (as shown in main text).
[3] Using the relevant conditions as 'within-subjects variables', but splitting file by gender.

Creating new (main effect) variables

So that we can analyse each of the within-group main effects across the gender groups, we need to 'create' the variables in SPSS. As it stands, the within-group columns currently present in our data set represent each of the condition scenarios that our group experienced (see Figure 13.36).

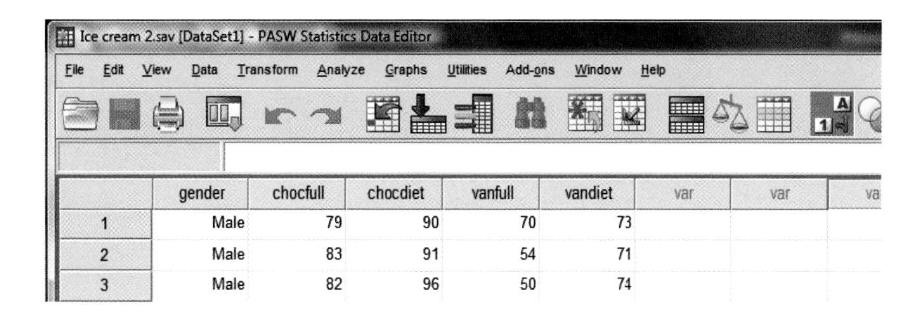

	gender	chocfull	chocdiet	vanfull	vandiet	var	var	va
1	Male	79	90	70	73			
2	Male	83	91	54	71			
3	Male	82	96	50	74			

Figure 13.36 Current variables in SPSS data set for Ice Cream 2

The current format of variables is good for analysing the main mixed ANOVA analyses and the within-group interaction ('chocfull' represents 'chocolate/full fat', 'chocdiet' for 'chocolate/diet', 'vanfull' for 'vanilla/full fat' and 'vandiet' for 'vanilla/diet'). However, to undertake the analyses of the within-between interaction, we need to see the outcomes for 'flavour' regardless of 'fat content' and the outcomes for 'fat content' regardless of flavour. We need to create variables for 'chocolate', 'vanilla', 'full fat' and 'diet' to do this (see later).

Using the SPSS file Ice cream 2

For 'Chocolate' condition

Select **Transform** → **Compute Variable...** → in **Target Variable** type **Chocolate** → click **()** (the brackets shown in the dashboard below **Numeric Expression** window) → transfer **Chocolate-full fat** to **Numeric Expression** → click + (*plus* sign in dashboard) → transfer **Chocolate-diet** to **Numeric Expression** (the string 'chocfull+chocdiet' will show within the brackets) → click to the right of the brackets → click / (forward slash in dashboard-to indicate 'divide by') → type **2** → click **OK**

For 'Vanilla' condition

Select **Transform** → **Compute Variable...** → delete all current entries in windows → in **Target Variable** type **Vanilla** → click **()** → transfer **Vanilla-full fat** to **Numeric Expression** → click + → transfer **Vanilla-diet** to **Numeric Expression** → click to the right of the brackets → click / → type **2** → click **OK**

For 'Full fat' condition

Select **Transform** → **Compute Variable...** → delete all current entries in windows → in **Target Variable** type **Fullfat** → click **()** → transfer **Chocolate-full fat** to **Numeric Expression** → click + → transfer **Vanilla-full fat** to **Numeric Expression** → click to the right of the brackets → click / → type **2** → click **OK**

For 'Diet' condition

Select **Transform** → **Compute Variable...** → delete all current entries in windows → in **Target Variable** type **Diet** → click **()** → transfer **Chocolate-diet** to **Numeric Expression** → click + → transfer **Vanilla-diet** to **Numeric Expression** → click to the right of the brackets → click / → type **2** → click **OK**

Go back to Data View of SPSS data set – the variables will now be presented as shown in Figure 13.37. We will use the original variable conditions for exploring the within-group interactions and the new variables for investigating the within-between interactions.

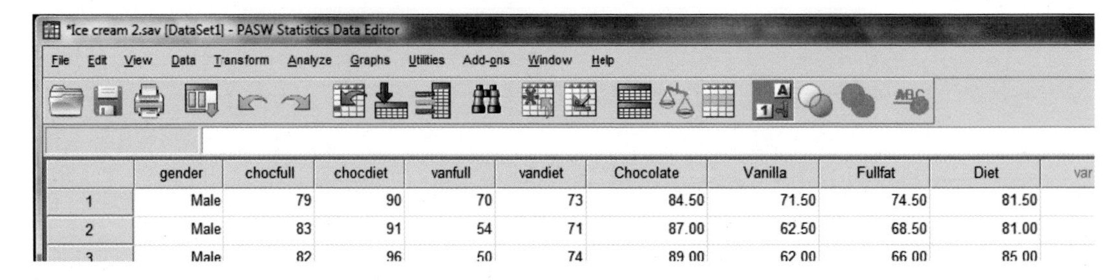

	gender	chocfull	chocdiet	vanfull	vandiet	Chocolate	Vanilla	Fullfat	Diet	var
1	Male	79	90	70	73	84.50	71.50	74.50	81.50	
2	Male	83	91	54	71	87.00	62.50	68.50	81.00	
3	Male	82	96	50	74	89.00	62.00	66.00	85.00	

Figure 13.37 Revised variables in SPSS data set for Ice Cream 2

Now we undertake the investigations of the interactions, using the analyses shown in Box 13.10.

Gender vs. taste satisfaction scores, according to flavour:

Using the SPSS file Ice cream 2

Select **Analyze** → **Compare Means** → **Independent-Samples T Test...** → (in new window) transfer **Chocolate** and **Vanilla** to **Test Variable List** → transfer **Gender** to **Grouping Variable** → click **Define Groups** → enter **0** in **Group 1** box → enter **1** in **Group 2** box → click **Continue** → click **OK**

	Gender	N	Mean	Std. deviation	Std. error mean
Chocolate	Male	10	86.1000	3.19548	1.01050
	Female	10	67.7000	3.37639	1.06771
Vanilla	Male	10	64.6000	3.95671	1.25122
	Female	10	80.8500	4.14360	1.31032

Figure 13.38 Descriptive statistics

Independent samples test

		Levene's test for equality of variances		t-test for equality of means						
									95% confidence interval of the difference	
		F	Sig.	t	df	Sig. (2-tailed)	Mean difference	Std. error difference	Lower	Upper
Chocolate	Equal variances assumed	.325	.576	12.516	18	.000	18.40000	1.47007	15.31149	21.48851
	Equal variances not assumed			12.516	17.946	.000	18.40000	1.47007	15.31082	21.48918
Vanilla	Equal variances assumed	.270	.610	−8.969	18	.000	−16.25000	1.81177	−20.05638	−12.44362
	Equal variances not assumed			−8.969	17.962	.000	−16.25000	1.81177	−20.05696	−12.44304

Figure 13.39 Independent t-test: gender vs. taste satisfaction scores (by flavour)–truncated

Figures 13.38 and 13.39 show that taste satisfaction was significantly higher for men than for women for chocolate flavour ice cream, $t\ (18) = 12.516$, $p < .001$ but taste satisfaction was significantly higher for women in respect of vanilla, t (18) = −8.969, p < .001. This is very clear evidence for the interaction.

Flavour vs. taste satisfaction scores, according to gender:

Using the SPSS file Ice cream 2

Select **Data** → **Split File** → (in new window) select **Compare groups** radio button → transfer **Gender** to **Groups Based on:** → click **OK**
Select **Analyze** → **Compare Means** → **Paired-Samples T Test...** → (in new window) transfer **Chocolate** and **Vanilla** to **Paired Variables** → click **OK**

Gender			Mean	N	Std. deviation	Std. error mean
Male	Pair 1	Chocolate	86.1000	10	3.19548	1.01050
		Vanilla	64.6000	10	3.95671	1.25122
Female	Pair 1	Chocolate	67.7000	10	3.37639	1.06771
		Vanilla	80.8500	10	4.14360	1.31032

Figure 13.40 Descriptive statistics

Gender			Paired differences					t	df	Sig. (2-tailed)
						95% confidence interval of the difference				
			Mean	Std. deviation	Std. error mean	Lower	Upper			
Male	Pair 1	Chocolate – Vanilla	21.50000	5.02770	1.58990	17.90340	25.09660	13.523	9	.000
Female	Pair 1	Chocolate – Vanilla	−13.15000	3.72715	1.17863	−15.81624	−10.48376	−11.157	9	.000

Figure 13.41 Related t-test: flavour vs. taste satisfaction scores (by gender)

Figures 13.40 and 13.41 show that chocolate flavour was significantly preferred over vanilla for men, t (9) = 13.523, p < .001, while vanilla was preferred by women, t (9) = −11.157, p < .001.

You must remember to switch off the Split File facility, otherwise subsequent analyses will be incorrect:

Using the SPSS file Ice cream 2

Select **Data** → **Split File** → (in new window) select **Analyze all cases, do not create groups** radio button → click **OK**

Get the idea now? Perhaps you could go ahead and complete the tasks for gender vs. fat content and flavour vs. fat content, as part of the exercises that you are set in these chapters (see later).

Chapter summary

In this chapter we examined mixed multi-factorial ANOVA. At this point, it would be good to revisit the learning objectives that we set at the beginning of the chapter.

You should now be able to:

- Recognise that we use mixed multi-factorial ANOVA to examine scores on a parametric dependent variable that are compared across two or more independent variables; at least one of these must be between-group and at least one within-group.
- Understand that the purpose of this test is to explore within-group and between-group main effects, and to examine interactions between them. As with all statistical tests, there are a number of assumptions and restrictions that need to be observed. All of the within-group conditions must be experienced by all of the members of a single group, and must contain at least two conditions. Between-group factors must be exclusive and independent, and include at least two groups. The dependent variable data should be at least interval, and be reasonably normally distributed. We saw how to examine normal distribution and interpret outcomes.
- Examine between-group homogeneity of variances using Levene's test, and homogeneity of variance-covariance matrices via Box's M test.
- Examine sphericity of within-group variances, using Mauchly's test. If sphericity is violated, we know that we should defer to alternative outcomes, as indicated by Huynh–Feldt or Greenhouse-Geisser adjustments.
- Calculate the outcome manually, using maths and equations.
- Perform different types of mixed multi-factorial ANOVA in SPSS, accounting for homogeneity of variance requirements, and setting the parameters to locate the source of main effects and interactions. Appropriate *post hoc* tests are needed when variables have three or more groups or conditions. If there are significant interactions, the source of those must also be explored, but significance cut-off points must be adjusted in proportion to the number of additional tests employed. In some cases, the data set will need to be subdivided into groups, using the 'Split File' facility in SPSS when exploring interactions involving between-group variables.
- Examine effect size and power, using G*Power software, across all main effects and interactions.
- Understand how to present the data, using appropriate tables, reporting the outcome in a series of sentences and correctly formatted statistical notation (such as $F(1, 18) = 28.703$, $p < .001$).

Research example

It might help you to see how mixed multi-factorial ANOVA has been applied in a research context. In this section you can read an overview of the following paper:

Mackinnon, A., Griffiths, K.M. and Christensen, H. (2008). Comparative randomised trial of online cognitive-behavioural therapy and an information website for depression: 12-month outcomes. *British Journal of Psychiatry, 192 (2)*: 130–134. DOI: http://dx.doi.org/10.1192/bjp.bp.106.032078

If you would like to read the entire paper you can use the DOI reference provided to locate that (see Chapter 1 for instructions).

In this research the authors investigated the effect of two forms of internet programs aimed at depressed patients, compared with a placebo control condition. One of the internet programs (MoodGYM) was based on cognitive-behavioural therapy (CBT). The second program (BluePages) was an educational site, guiding patients to treatments and therapies. The internet programs involved a number of weekly tasks and interactive exercises. The control condition involved a series of questions about patients' lifestyles. These three manipulations represented the between-group conditions; patients were randomly assigned to the groups. Outcome measures were determined by the Center for Epidemiologic Studies–Depression (CES-D) scale (Radoff, 1977). This measures psychological distress on a scale of 0–60, with higher scores representing poorer outcomes. The CES-D was used at baseline (pre-test), post-test, and at 6-month and 12-month follow-up points – those time points represented the within-group conditions. There were 525 patients recruited to the study.

The results indicated a significant between-group main effect for 'condition': F (2, 465.3) = 4.74, p = .009; *post hoc* tests were not reported between-groups, but were across some within-group conditions. There was a significant within-group effect for 'occasion': F (3, 379.3) = 48.45, p < .001; no *post hoc* tests were reported here either. There was a significant interaction between 'condition' and 'occasion': F (6, 379.2) = 2.90, p = .009. Planned contrasts were used to report the between-group effects at each time point (effectively reporting the source of interaction). At post-test, CES-D scores were significantly higher (poorer) for controls than for MoodGYM: F (1, 447.0) = 14.79, p < .001; *d* = .38 and BluePages: F (1, 439.7) = 8.13, p = .005; *d* = .29. There was no significant difference between MoodGYM and BluePages: F (1, 449.3) = 0.94, p = .332. At the 6-month follow-up, CES-D scores were significantly higher for controls than for MoodGYM: F (1, 405.5) = 4.49, p = .035; *d* = .27, but there was no significant difference between BluePages and controls: F (1, 396.6) = 2.80, p = .095; *d* = .21, or between MoodGYM and BluePages: F (1, 407.2) = 0.20, p = .652. At the 12-month follow-up, CES-D scores were significantly higher for controls than for MoodGYM: F (1, 388.7) = 4.09, p = .044; *d* = .27 and BluePages: F (1, 376.7) = 5.11, p = .024; there was no significant difference between MoodGYM and BluePages: F (1, 391.1) = 0.04, p = .849; *d* = .29.

This paper provides a good example of how to report mixed two-way ANOVA. Notice how the emphasis is on the interaction outcomes rather than on the main effect. Part of the point of conducting multi-factorial ANOVA is to explore interactions.

Extended learning task

You will find the data set associated with this task on the website that accompanies this book (available in SPSS and Excel format). You will also find the answers there.

1. Following what we have learned about mixed multi-factorial ANOVA, answer these questions and conduct the analyses in SPSS and G*Power. (If you do not have SPSS, do as much as you can with the Excel spreadsheet.) We will use a data set that examines a new form of treatment for depression in the community (patients attend GP practices rather than psychiatric outpatient clinics). The new treatment is 'care managed treatment', which is conducted via telephone by a trained nurse. This is compared with treatment as usual from the GP. To assess the impact of treatment we will examine quality of life perceptions, using the SF36 scale. This is scored from 0 to 100, with higher scores representing better perceptions. The perceptions are measured at baseline (before treatment), and again at treatment week 6 and treatment week 12.

Open the data set **Case managed depression**

1. Describe the independent variables.
   a. Specify the within-group independent variable/conditions.
   b. Specify the between-group independent variable/groups.
2. What is the dependent variable?
3. Check for normal distribution across the conditions and groups.
4. Conduct a mixed multi-factorial ANOVA.
5. Describe what the SPSS output shows.
6. Explain how you accounted for between-group homogeneity of variance.
7. Explain how you accounted for sphericity (if it was necessary).
8. Describe main effects and interactions.
9. Conduct appropriate additional analyses, to indicate the source of main effects and interactions.
10. Report the outcome as you would in the results section of a report.

14 MULTIVARIATE ANALYSES

Learning objectives

By the end of this chapter you should be able to:

- Recognise when it is appropriate to use multivariate analyses (MANOVA) and which test to use (traditional MANOVA or repeated-measures MANOVA)
- Understand the theory, rationale, assumptions and restrictions associated with the tests
- Calculate MANOVA outcomes manually (using maths and equations)
- Perform analyses using SPSS, and explore outcomes identifying the multivariate effects, univariate effects and interactions
- Know how to measure effect size and power
- Understand how to present the data and report the findings

What are multivariate analyses?

Multivariate analyses explore outcomes from *several* parametric dependent variables, across one or more independent variables (each with at least two distinct groups or conditions). This is quite different to anything we have seen so far. The statistical procedures examined in Chapters 7–13 differed in a number of respects, notably in the number and nature of the independent variables. However, these tests had one common aspect: they explored outcomes across a single dependent variable. With multivariate analyses there are at least two dependent variables.

Most commonly, we encounter these tests in the form of MANOVA (which is an acronym for **M**ultivariate **An**alysis **o**f **Va**riance) where the dependent variables outcomes relate to a single point in time. For example, we could investigate exam scores and coursework marks (two dependent variables) and explore how they vary according to three student groups (law, psychology and media – the between-group independent variable). The groups may differ significantly in respect of exam scores and with regard to coursework marks. Law students might do better in exams than coursework, while psychology students may perform better in their coursework than in exams; there may be no difference between exam and coursework results for media students. We will examine traditional MANOVA in the first part of this chapter.

We can also undertake multivariate analyses in within-group studies, using repeated-measures MANOVA. This is similar to what we have just seen, except that each of the dependent variables is measured over several time points. We can examine these outcomes with or without additional independent groups. For example, we could investigate the effect of a new antidepressant on a single group of depressed patients. We could measure mood ratings and time spent asleep on three occasions: at baseline (before treatment), and at weeks 4 and 8 after treatment. We could also explore these outcomes in respect of gender (as a between-group variable). We might find that men improve more rapidly than women on mood scores, but women experience greater improvements in sleep time. We will explore repeated-measures MANOVA later in the chapter.

What is MANOVA?

With MANOVA we examine two or more 'parametric' dependent variables across one or more between-group independent variable (we explored the criteria for parametric data in Chapter 5, although we will revisit this again shortly). Each dependent variable must represent a single set of scores from one time point (contrast that with the repeated-measures version). The scores across each dependent variable are explored across the groups of each of the independent variables. In theory, there is no upper limit to the number of dependent and independent variables that we can examine. However, it is not recommended that you use too many of either – multiple variables are very difficult to interpret (and may give your computer a hernia). We will focus on an example where there are two dependent variables and a single independent variable (sometimes called a one-way MANOVA).

14.1 Nuts and bolts

Multi-factorial ANOVA vs. multivariate ANOVA: what's the difference?

We have addressed this potential confusion in previous chapters, but it is worth reminding ourselves about the difference between these key terms.

Multi-factorial ANOVA: Where there are two or more independent variables (IVs)
Two-way ANOVA, three-way ANOVA: Describes the number of IVs in a multi-factorial ANOVA
Multivariate ANOVA: Where there are two or more dependent variables (as we have here)

Research question for MANOVA

Throughout this section, we will use a single research example to help us explore data with MANOVA. LAPS (Local Alliance of Pet Surgeries) are a group of vets. They would like to investigate whether the pets brought to their surgery suffer from the same kind of mental health problems as humans. They decide to measure evidence of anxiety and depression in three types of pets registered at the surgeries (dogs, cats and hamsters). The anxiety and depression scores are taken from a series of observations made by one of the vets, with regard to activity level, sociability, body posture, comfort in the presence of other animals and humans, etc. The vets expect dogs to be more depressed than other pets, and cats to be more anxious than other pets. Hamsters are expected to show no problems with anxiety (no fear of heights or enclosed spaces) or depression (quite at ease spending hours on utterly meaningless tasks, such as running around in balls and on wheels). However, hamsters may show a little more anxiety in the presence of cats.

14.2 Take a closer look
Summary of MANOVA research example

Between-group independent variable (IV): Type of pet (three groups: dogs, cats and hamsters)
Dependent variable (DV) 1: Anxiety scores
DV 2: Depression scores

Theory and rationale

Purpose of MANOVA

For each **MANOVA**, we explore the **multivariate effect** (how the independent variables have an impact upon the combination of dependent variables) and **univariate effects** (how the mean scores for each dependent variable differ across the independent variable groups). Within univariate effects, if we have several independent variables, we can explore interactions between them in respect of each dependent variable. MANOVA is a multivariate test – this means that we are exploring multiple dependent variables. It is quite easy to confuse the terms 'multivariate' and 'multi-factorial ANOVA', so we should resolve that here. If we consider the dependent variables represent scores that vary, we could call these 'variates'; meanwhile, we could think of independent variables in terms of 'factors' that may cause the dependent variable scores to vary. Therefore, 'multivariate' relates to many 'variates' and 'multi-factorial' to many 'factors'.

Why not run separate tests for each dependent variable?

There are a number of reasons why it is better to use MANOVA to explore outcomes for multiple dependent variables instead of running separate analyses. We *could* employ two independent one-way ANOVAs to explore how the pets differ on anxiety scores, and then in respect of depression scores. However, this would tell only half the story. The results might indicate that cats are more anxious than dogs and hamsters, and that dogs are more depressed than cats and hamsters. What that does not tell us is the strength of the relationship between the two dependent variables (we will see more about that later). MANOVA accounts for the correlation between the dependent variables. If it is too high we would reject the multivariate outcome. We cannot do that with separate ANOVA tests.

Multivariate and univariate outcomes

The multivariate outcome is also known as the MANOVA effect. This describes the effect of the independent variable(s) upon the combined dependent variables. In our example, we would measure how anxiety and depression scores (in combination) differ in respect of the observed pets: dogs, cats and hamsters. When performing MANOVA tests, we also need to explore the univariate outcome. This describes the effect of the independent variable(s) against each dependent variable separately. Using the LAPS research example, we would examine how anxiety scores vary between the pets, and then how depression scores vary between them. Once we find those univariate effects, we may also need to find the source of the differences. If the independent variable has two groups, that analysis is relatively straightforward: we just look at the mean scores. If there are three or more groups (as we do), we have a little more work to do. We will look at those scenarios later in the chapter.

Measuring variance

The procedure for partitioning univariate variance in MANOVA is similar to what we have seen in other ANOVA tests, but perhaps it is high time that we revisited that. In these tests we seek to see how numerical outcomes vary (as illustrated by a dependent variable). For example, in our research example, we have two dependent variables: depression scores and anxiety scores. We will focus on just one of those for now (anxiety scores). One of the vets uses an approved scale and observations to assess anxiety in each animal (based on a series of indicators). The assessment results in an anxiety score. Those scores will probably vary between 'patients' according to those observations. The extent to which those scores vary is measured by something called variance. The aim of our analyses is to examine how much of that variance we can explain – in this case how much variance is explained by differences between the animals (cat, dog or hamster). Of course, the scores might vary for other reasons that we have not accounted for, including random and chance factors. In any ANOVA the variance is assessed in terms of sums of squares (because variance is calculated from the squared differences between case scores, group means and the mean score for the entire sample). The overall variance is represented by the total sum of squares, explained variance by model squares, and the unexplained (error) variance by the residual squares.

In MANOVA there are several dependent variables, so it is a little more complex. The scores in *each* dependent variable will vary, so the variance for *each* total sum of squares will need to be partitioned into model and residual sums of squares. Within each dependent variable, if there is more than one between-group independent variable there will be a sum of squares for each of those (the main effects) and one for each of the interaction factors between them, plus the residual sum of squares. Those sums of squares need to be measured in terms of respective degrees of freedom – these represent the number of values that are free to vary in the calculation, while everything else is held constant (see Chapter 6). The sums of squares are divided by the relevant degrees of freedom (*df*) to find the respective model mean square and residual mean square. We then calculate the F ratio (for each main effect and interaction) by dividing the model mean square by the residual mean square. That F ratio is assessed against cut-off points (based on group and sample sizes) to determine statistical significance (see Chapter 9). Where significant main effects are represented by three or more groups, additional analyses are needed to locate the source of the difference (*post hoc* tests or planned contrasts, as we saw in Chapter 9). If there are significant interactions we will need to find the source of that too, just like we did with multi-factorial ANOVAs (see Chapter 11). This action is completed for *each* dependent variable – we call this process the univariate analysis.

However, because MANOVA explores several dependent variables (together), we also need to examine the multivariate outcome (the very essence of MANOVA). We need to partition the multivariate variance into explained (model) and unexplained (residual) portions. The methods needed to do this are complex (so see Box 14.3 for further guidance, in conjunction with the manual calculations shown at the end of this chapter). They have been shown separately (within Box 14.3) because some people might find what is written there very scary. You don't *need* to read that section, but it might help if you did. We will see how to perform these analyses in SPSS later.

14.3 Nuts and bolts
Partitioning multivariate variance

This section comes with a health warning: it is much more complex than most of the other information you will read in this book! To explore multivariate outcomes we need to refer to something called '**cross-product tests**'. These investigate the relationship between the dependent variables, partitioning that variance into model and residual cross-product. Total cross-products are calculated from differences between individual case scores and the grand mean for each dependent variable. **Model cross-products** are found from group means in relation to grand means. **Residual cross-products** are whatever is left.

Then the maths gets nasty! To proceed, we need to express sums of squares and cross-products in a series of matrices (where numbers are placed in rows and columns within brackets). This is undertaken for the model and error portions of the multivariate variance; they represent the equivalent of mean squares in univariate analyses. To get an initial F ratio, we divide the model cross **matrix** by the residual cross matrix. But we cannot do that directly, because you cannot divide matrices. Instead we multiply the model cross matrix by the inverse of the residual cross matrix (told you it was getting nasty!).

Now perhaps you can see why these manual calculations are safely tucked away at the end of this chapter. Even then, we are still not finished: we need to put all of that into some equations to find something called '**eigenvalues**'. Each type of eigenvalue is employed in a slight different way to find the final F ratio for the multivariate outcome (see end of this chapter).

Reporting multivariate outcome

When we have run the MANOVA analysis in SPSS, we are presented with several lines of multivariate outcome. Each line reports potentially different significance, so it is important that we select the correct one. There are four options: **Pillai's Trace**, **Wilks' Lambda**, **Hotelling's Trace** and **Roy's Largest Root**. Several factors determine which of these we can select. Hotelling's Trace should be used only when the independent variables are represented by two groups (we have three groups in our example). It is not as powerful as some of the alternative choices. Some sources suggest that Hotelling's T^2 is more powerful, but that option is not available in SPSS (although a conversion from Hotelling's Trace is quite straightforward but time consuming). Wilks' Lambda (λ) is used when the independent variable has more than two groups (so we could use that). It explores outcomes using a method similar to F ratio in univariate ANOVAs. Although popular, it is not considered to be as powerful as Pillai's Trace (which is often preferred for that reason). Pillai's Trace and Roy's Largest Root can be used with any number of independent variable groups (so these methods would be suitable for our research example). If the samples are of equal size, probably the most powerful option is Pillai's Trace (Bray and Maxwell, 1985), although this power is compromised when sample sizes are not equal *and* there are problems with equality of covariance (some books refer to this procedure as Pillai – Bartlett's test). We will know whether we have equal group sizes, but will not know about the covariance outcome until we have performed statistical analyses. Roy's Largest Root uses similar calculations as Pillai's Trace, but accounts for only the first factor in the analysis. It is not advised where there are platykurtic distributions, or where homogeneity of between-group variance is compromised. As we saw in Chapter 3, a platykurtic distribution is illustrated by a 'flattened' distribution of data, suggesting wide variation in scores.

14.4 Take a closer look
MANOVA: choosing the multivariate outcome - a summary

A number of factors determine which outcome we should use when measuring multivariate significance. This a brief summary of those points:

Pillai's Trace: Can be used for any number of groups, but is less viable when sample sizes in those groups are unequal and when there is unequal between-group variance
Hotelling's Trace: Can be used only when there are two groups (and may not be as powerful as Pillai's Trace). Hotelling's T^2 is more powerful, but it is not directly available from SPSS
Wilks' λ: More commonly used when the independent variable has more than two groups
Roy's Largest Root: Similar method to Pillai's Trace, but focuses on first factor. Not viable with platykurtic distributions

Reporting univariate outcome

There is much debate about how we should examine the univariate effect, subsequent to multivariate analyses. In most cases it is usually sufficient to simply treat each univariate portion as if it were an independent one-way ANOVA and run *post hoc* tests to explore the source of difference (where there are three or more groups). The rules for choosing the correct *post hoc* analysis remain the same as we saw with independent ANOVAs (see Chapter 9, Box 9.8). However, some statisticians argue that you cannot do this, particularly if the two dependent variables are highly correlated (see 'Assumptions and restrictions' below). They advocate **discriminant analysis** instead (we do not cover that in this book). In any case, univariate analyses should be undertaken only if the multivariate outcome is significant. Despite these warnings, we will explore subsequent univariate analysis so that you can see how to do them. Using our LAPS research data, we might find significant differences in depression and anxiety scores across the animal groups. Subsequent *post hoc* tests might indicate that cats are significantly more anxious than dogs and that dogs are significantly more depressed than cats. Hamsters might not differ from other pets on either outcome.

If we had more than one independent variable, we would also measure interactions between those independent variables (in respect of both outcome scores). For example, we could measure whether the type of food that pets eat (fresh food or dried packet food) had an impact on anxiety and depression scores. We would explore main effects for pet type and food type for each dependent variable (anxiety and depression scores in our example). In addition to main effects for anxiety and depression across pet category, we might find that depression scores are poorer when animals are given dried food, compared with fresh food, but that there are no differences in anxiety scores between fresh and dried food. Then we might explore interactions and find that depression scores are only poorer for dried food vs. fresh food for dogs but it makes no difference to cats and hamsters.

Assumptions and restrictions

There are a number of criteria that we should consider before performing MANOVA. There must be at least two parametric dependent variables, across which we measure differences in respect of one or more independent variable (each with at least two groups). To assume parametric requirements, the dependent variable data should be interval or ratio and should be reasonably normally distributed (we explored this in depth in Chapter 5). Platykurtic data can have a serious effect on multivariate outcomes (Brace *et al.*, 2006; Coakes and Steed, 2007), so we need to avoid that. As we saw earlier, if the distribution of data is platykurtic, it can also influence which measure we choose to report multivariate outcomes. We should check for outliers, as too many may reduce power, making it more difficult to find significant outcomes. In our research example, the dependent variables (depression and anxiety scores) are being undertaken by a single vet, using approved measurements so we can be confident that these data are interval.

There must be *some* correlation between the dependent variables (otherwise there will be no multivariate effect). However, that correlation should not be *too* strong. Ideally, the relationship between them should be no more than moderate where there is negative correlation (up to about $r = -.40$); positively correlated variables should range between .30 and .90 (Brace *et al.*, 2006). Tabachnick and Fidell (2007) argue that there is little sense in using MANOVA on dependent variables that effectively measure the same concept. Homogeneity of univariate between-group variance is important. We have seen how to measure this with Levene's test in previous chapters. Violations are particularly important where there are unequal group sizes (see Chapter 9).

We also need to account for homogeneity of multivariate variance-covariance matrices. We encountered this in Chapter 13, when we explored mixed multi-factorial ANOVA. However, it is of even greater importance in MANOVA. We can check this outcome with Box's M test. In addition to examining variance between the groups, this procedure investigates whether the correlation between the dependent variables differs significantly between the groups. We do not want that, as we need the correlation to be similar between those groups, although we have a problem only if Box's M test is very highly significant ($p < .001$). There are no real solutions to that, so violations can be bad news. Although in theory there is no limit to how many dependent variables we can examine in MANOVA, in reality we should keep this to a sensible minimum, otherwise analyses become too complex.

14.5 Take a closer look

Summary of assumptions and restrictions for MANOVA

- The independent variable(s) must be categorical, with at least two groups
- The dependent variable data must interval or ratio, and be reasonably normally distributed
- There should not be too many outliers
- There should be reasonable correlation between the dependent variables
  - Positive correlation should not exceed $r = .90$
  - Negative correlation should not exceed $r = -.40$
- There should be between-group homogeneity of variance (measured via Levene's test)
- Correlation between dependent variables should be equal between the groups
  - Box's M test of homogeneity of variance-covariance matrices examines this
- We should avoid having too many dependent variables

How SPSS performs MANOVA

To illustrate how we perform MANOVA in SPSS, we will refer to data that report outcomes based on the LAPS research question. You will recall that the vets are examining anxiety and depression ratings according to pet type: dogs, cats and hamsters. The ratings of anxiety and depression are undertaken by a single vet and range from 0 to 100 (with higher scores being poorer) – these data are clearly interval. That satisfies one part of parametric assumptions, but we should also check to see whether the data are normally distributed across the independent groups for each dependent variable (which we will do shortly). In the meantime, we should remind ourselves about the nature of the dependent and independent variables.

MANOVA variables

- **Between-group IV:** type of pet (three groups: dogs, cats and hamsters)
- **DV 1:** anxiety scores
- **DV 2:** depression scores

LAPS predicted that dogs would be more depressed than other animals, and cats more anxious than other pets.

14.6 Nuts and bolts

Setting up the data set in SPSS

When we create the SPSS data set for this test, we need to account for the between-group independent variable (where columns represent groups, defined by value labels - refer to Chapter 2 to see how to do that) and the two dependent variables (where columns represent the anxiety or depression score).

*Animals.sav [DataSet2] - IBM SPSS Statistics Data Editor

File Edit View Data Transform Analyze Graphs Utilities Add-ons Window Help

	Name	Type	Width	Decimals	Label	Values	Missing	Columns	Align	Measure
1	animal	Numeric	8	0	Animal	{1, Dog}...	None	8	Right	Nominal
2	anxiety	Numeric	8	0	Anxiety	None	None	8	Right	Scale
3	depression	Numeric	8	0	Depression	None	None	8	Right	Scale

Figure 14.1 Variable View for 'Animals' data

Figure 14.1 shows how the SPSS Variable View should be set up. The first variable is 'animal'; this is the between-group independent variable. In the Values column, we include '1 = Dog', '2 = Cat', and '3 = Hamster'; the Measure column is set to Nominal. Meanwhile, 'anxiety' and 'depression' represent the dependent variables. We do not set up anything in the Values column; we set Measure to Scale.

*Animals.sav [DataSet2] - IBM SPSS Statistics Data Editor

File Edit View Data Transform Analyze Graphs Utilities Add-ons Window Help

18 :

	animal	anxiety	depression	var	var	var	var	var	var
1	Dog	36	73						
2	Cat	80	48						
3	Hamster	50	67						
4	Hamster	28	50						

Figure 14.2 Data View for 'Animals' data

Figure 14.2 illustrates how this will appear in the Data View. Each row represents a pet. When we enter the data for 'animal', we input 1 (to represent dog), 2 (to represent cat) or 3 (to represent hamster); the 'animal' column will display the descriptive categories ('Dog', 'Cat' or 'Hamster') or will show the value numbers, depending on how you choose to view the column (using the Alpha Numeric button - see Chapter 2). In the remaining columns ('anxiety' and 'depression') we enter the actual score (dependent variable) for that pet according to 'anxiety' or 'depression'.

In previous chapters, by this stage we would have already explored outcomes manually. Since that is rather complex, that analysis is undertaken at the end of this chapter. However, it might be useful to see the data set before we perform the analyses (see Table 14.1).

Table 14.1 Measured levels of anxiety and depression in domestic animals

	Anxious			Depressed		
	Dogs	**Cats**	**Hamsters**	**Dogs**	**Cats**	**Hamsters**
	36	80	50	73	48	67
	48	93	28	87	48	50
	61	53	44	80	87	67
	42	53	44	62	42	50
	55	87	48	87	42	56
	42	60	67	67	42	56
	48	60	67	40	36	50
	48	98	50	90	61	49
	53	67	44	60	61	60
	48	93	80	93	42	48
Mean	**48.10**	**74.40**	**52.20**	**73.90**	**50.90**	**55.30**

Checking correlation

Before we run the main test, we need to check the magnitude of correlation between the dependent variables. This might be important if we do find a significant MANOVA effect. As we saw earlier, violating that assumption might cause us to question the validity of our findings. Furthermore, if there is reasonable correlation, we will be more confident that independent one-way ANOVAs are an appropriate way to measure subsequent univariate outcomes (we saw how to perform correlation in SPSS in Chapter 6).

Open the SPSS file Animals

Select **Analyze** → **Correlate** → select **Bivariate...** → transfer **Anxiety** and **Depression** to **Variables** (by clicking on arrow, or by dragging variables to that window) → tick boxes for **Pearson** and **Two-tailed** → click **OK**

		Anxiety	Depression
Anxiety	Pearson correlation	1	−.351
	Sig. (2-tailed)		.057
	N	30	30
Depression	Pearson correlation	−.351	1
	Sig. (2-tailed)	.057	
	N	30	30

Figure 14.3 Correlation between anxiety and depression

The correlation shown in Figure 14.3 is within acceptable limits for MANOVA outcomes. Although negative, it does not exceed r = −.400.

Testing for normal distribution

We should be pretty familiar with how we perform Kolmogorov-Smirnov/Shapiro-Wilk tests in SPSS by now, so we will not repeat those instructions (but do check previous chapters for guidance). On this occasion, we need to explore normal distribution for both dependent variables, across the independent variable groups. The outcome is shown in Figure 14.4.

	Animal	Kolmogorov–Smirnov[a]			Shapiro–Wilk		
		Statistic	df	Sig.	Statistic	df	Sig.
Anxiety	Dog	.206	10	.200*	.959	10	.773
	Cat	.192	10	.200*	.881	10	.134
	Hamster	.258	10	.057	.919	10	.349
Depression	Dog	.182	10	.200*	.922	10	.376
	Cat	.276	10	.030	.797	10	.014
	Hamster	.267	10	.041	.839	10	.043

Figure 14.4 Kolmogorov-Smirnov/Shapiro-Wilk test: anxiety and depression scores vs. animal type

As there are fewer than 50 animals in each group, we should refer to the Shapiro-Wilk outcome. Figure 14.4 shows somewhat inconsistent data, although we are probably OK to proceed. Always bear in mind that we are seeking *reasonable* normal distribution. We appear to have normal distribution in anxiety scores for all animal groups, but the position is less clear for the depression scores. The outcome for dogs is fine, and is too close to call for hamsters, but the cats data are potentially more of a problem. We could run additional z-score tests for skew and kurtosis, or we might consider transformation. Under the circumstances, that is probably a little extreme – most of the outcomes are acceptable.

Running MANOVA

Using the SPSS file Animals

Select **Analyze → General Linear Model → Multivariate...** as shown in Figure 14.5

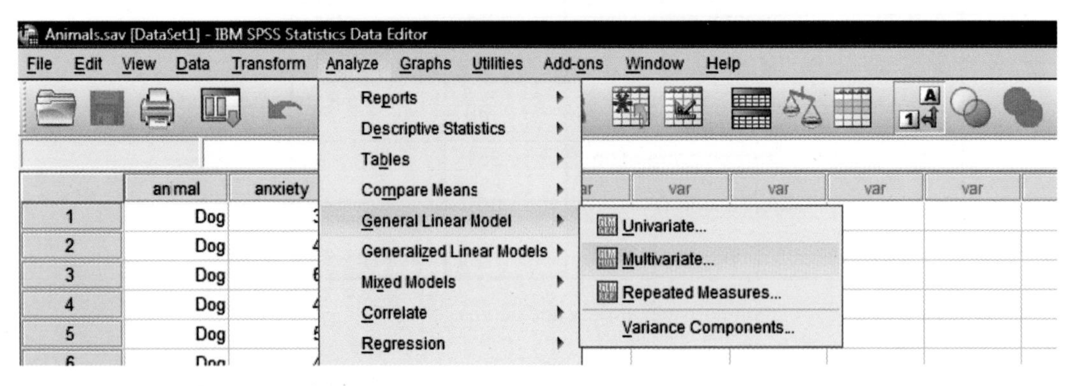

Figure 14.5 MANOVA: procedure 1

In new window (see Figure 14.6) transfer **Anxiety** and **Depression** to **Dependent Variables** → transfer **Animal** to **Fixed Factor(s)** → click on **Post Hoc...** (we need to set this because there are three groups for the independent variable)

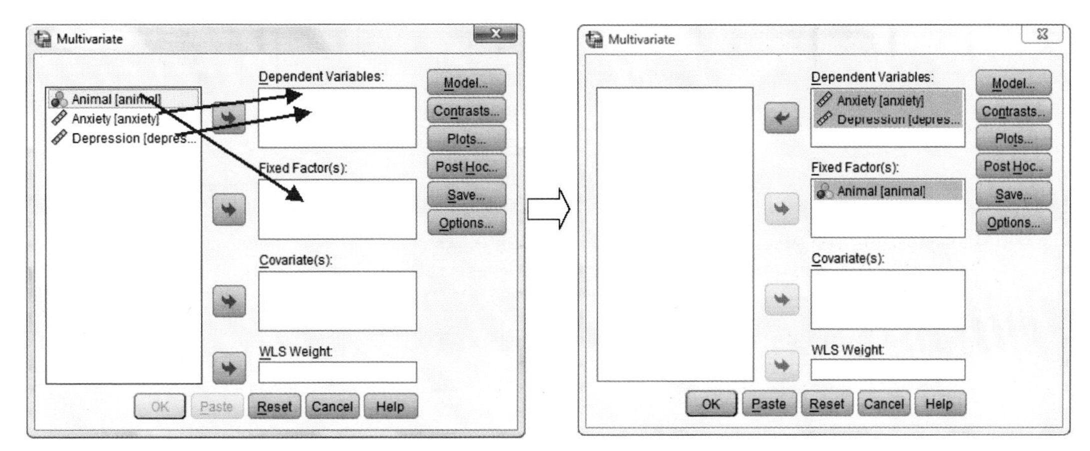

Figure 14.6 Variable selection

In new window (see Figure 14.7) transfer **Animal** to **Post Hoc Tests for** → tick boxes for **Tukey** (because we have equal group sizes) and **Games-Howell** (because we do not know whether we have between-group homogeneity of variance) → click **Continue** → click **Options...**

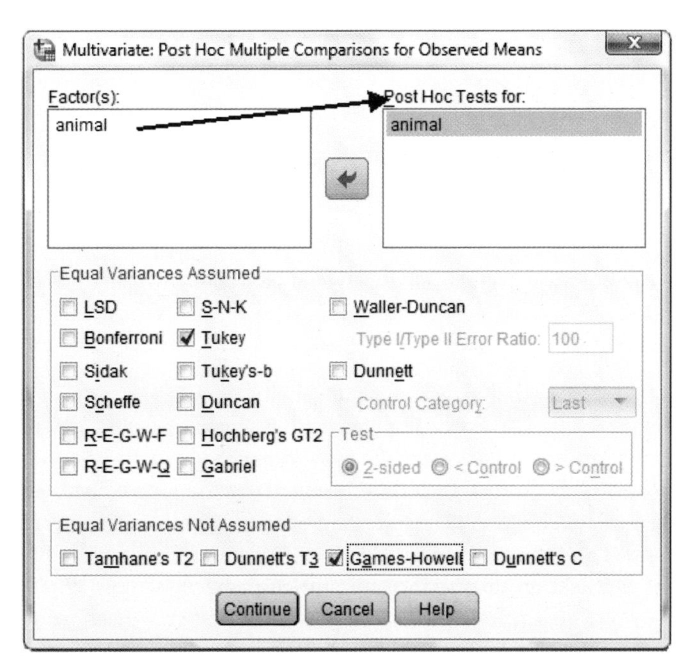

Figure 14.7 MANOVA: *post hoc* options

In new window (see Figure 14.8), tick boxes for **Descriptive statistics**, **Estimates of effect size** and **Homogeneity tests** → click **Continue** → click **OK**

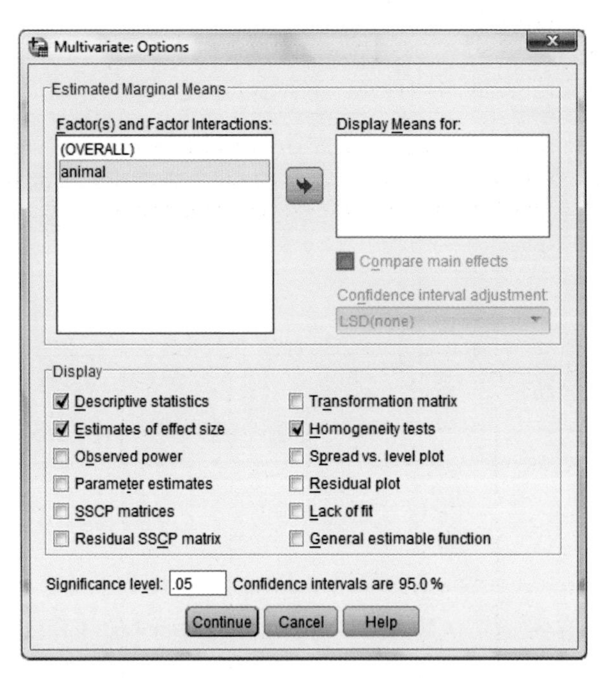

Figure 14.8 MANOVA: Statistics options

Interpretation of output

Checking assumptions

	F	df1	df2	Sig.
Anxiety	6.455	2	27	.005
Depression	2.518	2	27	.099

Figure 14.9 Levene's test for equality of variances

Figure 14.9 indicates that we have homogeneity of between-group variance for depression scores (significance > .05), but not for anxiety scores (significance < .05). There are some adjustments that we could undertake to address the violation of homogeneity in anxiety scores across the pet groups, including Brown–Forsythe F or Welch's F statistics. We encountered these tests in Chapter 9. However, these are too complex to perform manually for multivariate analyses and they are not available in SPSS when running MANOVA We could examine equality of between-group variance just for the anxiety scores. When we use independent one-way ANOVA to explore the univariate outcome, we can additionally employ Brown–Forsythe F and Welch's F tests (see later). This homogeneity of variance outcome will also affect how we interpret *post hoc* tests.

Box's M	13.813
F	2.055
df1	6
df2	18168.923
Sig.	.055

Figure 14.10 Box's M test for equality of variance-covariance matrices

Figure 14.10 shows that we can be satisfied that we have homogeneity of variance-variance-covariance matrices because the significance is greater than .001. It is important that the correlation between the dependent variables is equal across the groups – we can be satisfied that it is.

Multivariate outcome

	Animal	Mean	Std. deviation	N
Anxiety	Dog	48.10	7.141	10
	Cat	74.40	17.715	10
	Hamster	52.20	15.002	10
	Total	58.23	17.921	30
Depression	Dog	73.90	16.789	10
	Cat	50.90	15.140	10
	Hamster	55.30	7.258	10
	Total	60.03	16.666	30

Figure 14.11 Descriptive statistics

These initial statistics (presented in Figure 14.11) *suggest* that dogs are more anxious than cats and hamsters and that dogs are more depressed than cats and hamsters.

Effect		Value	F	Hypothesis df	Error df	Sig.	Partial eta squared
Intercept	Pillai's Trace	.978	575.667[a]	2.000	26.000	.000	.978
	Wilks' Lambda	.022	575.667[a]	2.000	26.000	.000	.978
	Hotelling's Trace	44.282	575.667[a]	2.000	26.000	.000	.978
	Roy's Largest Root	44.282	575.667[a]	2.000	26.000	.000	.978
animal	Pillai's Trace	.672	6.838	4.000	54.000	.000	.336
	Wilks' Lambda	**.407**	**7.387[a]**	4.000	52.000	**.000**	.362
	Hotelling's Trace	1.265	7.906	4.000	50.000	.000	.387
	Roy's Largest Root	1.086	14.661[b]	2.000	27.000	.000	.521

Figure 14.12 MANOVA statistics

Figure 14.12 presents four lines of data, each of which represents a calculation for multivariate significance (we are concerned only with the outcomes reported in the 'animal box'; we ignore 'Intercept'). We explored which of those options we should select earlier. On this occasion, we will choose Wilks' Lambda (λ) as we have three groups. That line of data is highlighted in red font here. We have a significant multivariate effect for the combined dependent variables of anxiety and depression in respect of the type of pet: $\lambda = 0.407$, $F(4, 52) = 7.387$, $p < .001$). We will use the Wilks 'λ' outcome (0.407) for effect size calculations later.

Univariate outcome

We can proceed with univariate and *post hoc* tests because the correlation was not too high between the dependent variables.

Source	Dependent variable	Type III sum of squares	df	Mean square	F	Sig.	Partial eta squared
Corrected model	Anxiety	4004.467[a]	2	2002.233	10.183	.001	.430
	Depression	2981.067[b]	2	1490.533	7.932	.002	.370
Intercept	Anxiety	101733.633	1	101733.633	517.397	.000	.950
	Depression	108120.033	1	108120.033	575.345	.000	.955
animal	Anxiety	4004.467	2	2002.233	**10.183**	**.001**	.430
	Depression	2981.067	2	1490.533	**7.932**	**.002**	.370
Error	Anxiety	5308.900	27	196.626			
	Depression	5073.900	27	187.922			
Total	Anxiety	111047.000	30				
	Depression	116175.000	30				
Corrected total	Anxiety	9313.367	29				
	Depression	8054.967	29				

Figure 14.13 Univariate statistics

Figure 14.13 suggests that both dependent variables differed significantly in respect of the independent variable (pet type): Anxiety (highlighted in blue font): $F(2, 27) = 10.183$, $p = .001$; Depression (green): $F(2, 27) = 7.932$, $p = .002$. We will use the partial eta squared data later when we explore effect size. As we know that we had a problem with the homogeneity of variance for anxiety scores across pet groups (Figure 14.9), we should examine those anxiety scores again, using an independent one-way ANOVA with Brown–Forsythe F and Welch's F adjustments.

Using the SPSS file Animals

Select **Analyze** → **Compare means** → **One-Way ANOVA** → (in new window) transfer **Anxiety** to **Dependent List** → transfer **Animal** to **Factor** → click **Options...** → tick boxes for **Brown-Forsythe** and **Welch** (we do not need any other options this time, because we are only checking the effect of Brown-Forsythe and Welch adjustments) → click **Continue** → click **OK**

Anxiety

	Sum of squares	df	Mean square	F	Sig.
Between-groups	4004.467	2	2002.233	10.183	.001
Within-groups	5308.900	27	196.626		
Total	9313.367	29			

Figure 14.14 Unadjusted ANOVA outcome

Anxiety

	Statistic[a]	df1	df2	Sig.
Welch	9.090	2	15.399	.002
Brown–Forsythe	10.183	2	20.638	.001

a. Asymptotically F distributed

Figure 14.15 Adjusted outcome for homogeneity of variance

Figure 14.14 confirms what we saw in Figure 14.13: unadjusted one-way ANOVA outcome, F (2, 27) = 10.183, p = .001. Figure 14.15 shows the revised outcome, adjusted by Brown–Forsythe F and Welch's F statistics. There is still a highly significant difference in anxiety scores across pet type, Welch: F (2, 15.399) = 9.090, p = .002. The violation of homogeneity of variance poses no threat to the validity of our results.

Post hoc analyses

Since we had three groups for our independent variable, we need *post hoc* tests to explore the source of the significant difference. Figure 14.16 presents the *post hoc* tests. As we saw in

Dependent variable		(I) Animal	(J) Animal	Mean difference (I-J)	Std. error	Sig.	95% confidence interval	
							Lower bound	Upper bound
Anxiety	Tukey HSD	Dog	Cat	–26.30*	6.271	.001	–41.85	–10.75
			Hamster	–4.10	6.271	.792	–19.65	11.45
		Cat	Dog	26.30*	6.271	.001	10.75	41.85
			Hamster	22.20*	6.271	.004	6.65	37.75
		Hamster	Dog	4.10	6.271	.792	–11.45	19.65
			Cat	–22.20*	6.271	.004	–37.75	–6.65
	Games–Howell	Dog	Cat	–26.30*	6.040	.003	–42.44	–10.16
			Hamster	–4.10	5.254	.721	–17.99	9.79
		Cat	Dog	**26.30***	6.040	**.003**	10.16	42.44
			Hamster	**22.20***	7.341	**.019**	3.42	40.98
		Hamster	Dog	4.10	5.254	.721	–9.79	17.99
			Cat	–22.20*	7.341	.019	–40.98	–3.42
Depression	Tukey HSD	Dog	Cat	**23.00***	6.131	**.002**	7.80	38.20
			Hamster	**18.60***	6.131	**.014**	3.40	33.80
		Cat	Dog	–23.00*	6.131	.002	–38.20	–7.80
			Hamster	–4.40	6.131	.755	–19.60	10.80
		Hamster	Dog	–18.60*	6.131	.014	–33.80	–3.40
			Cat	4.40	6.131	.755	–10.80	19.60
	Games–Howell	Dog	Cat	23.00*	7.149	.013	4.74	41.26
			Hamster	18.60*	5.784	.018	3.21	33.99
		Cat	Dog	–23.00*	7.149	.013	–41.26	–4.74
			Hamster	–4.40	5.309	.693	–18.43	9.63
		Hamster	Dog	–18.60*	5.784	.018	–33.99	–3.21
			Cat	4.40	5.309	.693	–9.63	18.43

Figure 14.16 *Post hoc* statistics

Chapter 9, there are a number of factors that determine which test we can use. One of those is homogeneity of variance. Earlier, we saw that anxiety scores did not have equal variances across pet type, which means that we should refer to the Games–Howell outcome for anxiety scores. This indicates that cats were significantly more anxious than dogs ($p = .003$) and hamsters ($p = .019$). There were equal variances for depression scores; since there were equal numbers of pets in each group, we can use the Tukey outcome. This shows that dogs are significantly more depressed than cats ($p = .002$) and hamsters ($p = .014$).

In summary, the multivariate analyses indicated that domestic pets differed significantly in respect of a combination of anxiety and depression scores; those dependent variables were not too highly correlated. Subsequent univariate analyses showed that there were significant effects for pet type in respect of the anxiety and (separately) in respect of depression scores. Tukey *post hoc* analyses suggested cats were significantly more anxious than dogs and hamsters, and that dogs were significantly more depressed than cats and hamsters.

Effect size and power

We can use G*Power to help us measure the effect size and statistical power outcomes from the results we found (see Chapter 4 for rationale and instructions). We have explored the rationale behind G*Power in several chapters now. On this occasion we can examine the outcome for each of the dependent variables and for the overall MANOVA effect.

Univariate effects:

From **Test family** select F tests

From **Statistical test** select ANOVA: Fixed effects, special, main effects and interaction

From **Type of power analysis** select Post hoc: Compute achieved – given α, sample size and effect size power

Now we enter the **Input Parameters**:

Anxiety DV

To calculate the **Effect size**, click on the **Determine** button (a new box appears).

In that new box, tick on radio button for Direct → type **0.430** in the **Partial η^2** box (we get that from Figure 14.13, referring to the 'eta squared figure' for animal, as highlighted in orange) → click on **Calculate and transfer to main window**

Back in original display, for **α err prob** type **0.05** (the significance level) → for **Total sample size** type **30** (overall sample size) → for **Numerator df** type **2** (we also get the *df* from Figure 14.13) → for **Number of groups** type **3** (Animal groups for dogs, cats, and hamsters) → for **Covariates** type **0** (we did not have any) → click on **Calculate**

From this, we can observe two outcomes: Effect size (d) **0.86** (very large); Power (1-β err prob) **0.99** (excellent- see Section 4.3).

Depression DV

Follow procedure from above: then, in **Determine** box type **0.370** in the **Partial η^2** box → click on **Calculate and transfer to main window** → back in original display **α err prob**, **Total sample size**, **Numerator df** and **Number of groups** remain as above → click on **Calculate**

Effect size (d) **0.77** (very large); Power (1-β err prob) **0.95** (excellent)

Multivariate effect:

From **Test family** select **F tests**

From **Statistical test** select **MANOVA: Global effects**

From **Type of power analysis** select **Post hoc: Compute achieved - given α, sample size and effect size power**

Now we enter the **Input Parameters**:

To calculate the **Effect size**, click on the **Determine** button (a new box appears).

In that new box, we are presented with a number of options for the multivariate statistic. The default is 'Pillai V', so we need to change that to reflect that we have used Wilks' Lambda:

Click on **Options** (in the main window) → click **Wilks U** radio button → click **OK** → we now have the Wilks U option in the right-hand window → type **0.407** in **Wilks U** (we get that from Figure 14.12) → click on **Calculate and transfer to main window**

Back in original display, for ***α* err prob** type **0.05** → for **Total sample size** type **30** → for **Number of groups** type **3** → for **Number of response variables** type **2** (the number of DVs we had) → click on **Calculate**

Effect size (d) **0.57** (large); Power (1-β err prob) **0.997** (excellent).

Writing up results

Table 14.2 Anxiety and depression scores by domestic pet type

Pet	N	Anxiety		Depression	
		Mean	SD	Mean	SD
Dog	10	48.10	7.14	73.90	16.79
Cat	10	74.40	17.71	50.90	15.14
Hamster	10	52.20	15.00	55.30	7.26

Perceptions of anxiety and depression were measured in three groups of domestic pet: dogs, cats and hamsters. MANOVA analyses confirmed that there was a significant multivariate effect: $\lambda = .407$, $F(4, 52) = 4.000$, $p < .001$, $d = 0.57$. Univariate independent one-way ANOVAs showed significant main effects for pet type in respect of anxiety: $F(2, 27) = 10.183$, $p = .001$, $d = 0.86$; and depression: $F(2, 27) = 7.932$, $p = .002$, $d = 0.77$. There was a minor violation in homogeneity of between-group variance for anxiety scores, but Brown–Forsythe F and Welch's F adjustments showed that this had no impact on the observed outcome. Games–Howell *post hoc* tests showed that cats were significantly more anxious than dogs ($p = .003$) and hamsters ($p = .019$), while Tukey analyses showed that dogs were more depressed than cats ($p = .002$) and hamsters ($p = .014$).

Presenting data graphically

We could also add a graph, as it often useful to see the relationship in a picture. However, we should not just replicate tabulated data in graphs for the hell of it – there should be a good rationale for doing so. We could use the drag and drop facility in SPSS to draw a bar chart (Figure 14.17):

Select **Graphs** → **Chart Builder** ... → (in new window) select **Bar** from list under **Choose from:** → drag **Simple Bar** graphic into empty chart preview area → select **Anxiety** and **Depression** together → drag both (at the same time) to **Y-Axis** box → transfer **Animal** to **X-Axis** box → to include error bars, tick box for **Display error bars** in **Element Properties** box (to right of main display box) → ensure that it states **95% confidence intervals** in the box below → click **Apply** (the error bars appear) → click **OK**

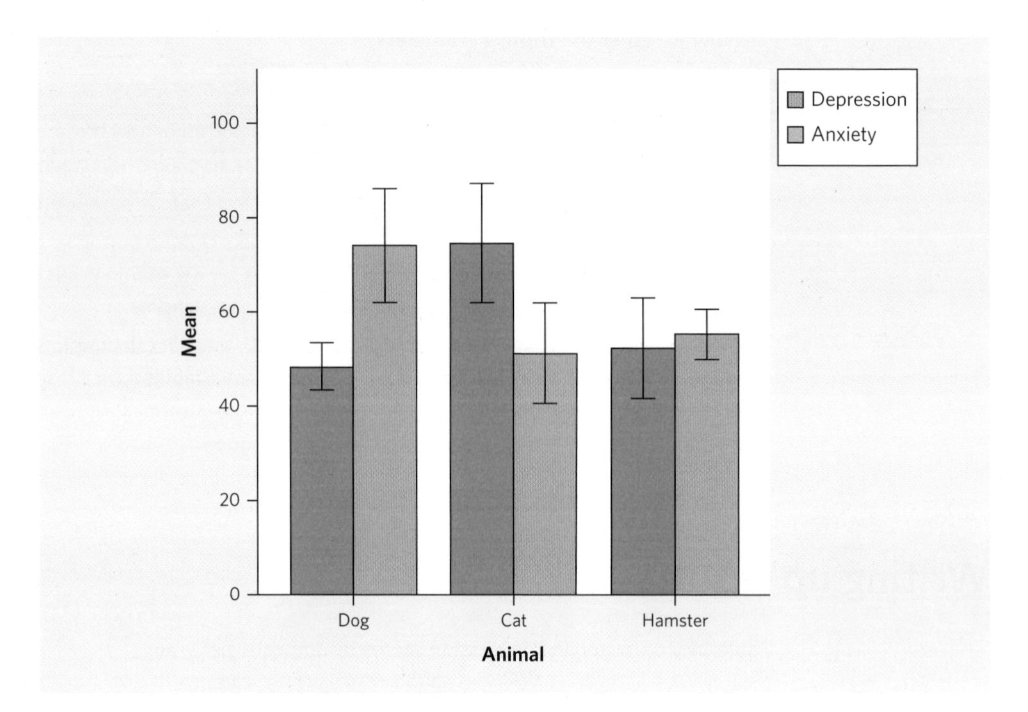

Figure 14.17 Completed bar chart

Repeated-measures MANOVA

Similar to traditional (between-group) MANOVA, the repeated-measures version simultaneously explores two or more dependent variables. However, this time those scores are measured over a series of within-group time points instead of the single-point measures we encountered earlier. For example, we could conduct a longitudinal study where we investigate body mass index *and* heart rate in a group of people at various times in their life, at ages 30, 40 and 50. Repeated-measures MANOVA is a multivariate test because we are measuring *two outcomes* at several time points (called 'trials'). Compare that to repeated-measures multi-factorial ANOVA, where we measure two or more *independent variables* at different time points, but for *one outcome*. In Chapter 12, we measured satisfaction with course content (the single dependent variable) according to two independent variables: the type of lesson received (interactive lecture, standard lecture or video) and expertise of the lecturer (expert or novice).

Repeated-measures MANOVA test is quite different. In the example we gave just now, body mass index and heart rate are the two dependent variables. The single within-group independent is the three time points (age). Despite those differences, we still perform these analyses in SPSS using the **general linear model** (GLM) repeated-measures function as we did in Chapters 10, 12 and 13. Only, the procedure is quite different (as we will see later). We can also add a between-group factor to repeated-measures MANOVA. We could extend that last example (measuring body mass index and heart rate at ages 30, 40 and 50) but additionally look at differences in those outcomes by the type of lifestyle reported by group members at the start of the study (sedentary, active or very active). Now we would have two dependent variables (body mass index and heart rate), one within-group independent variable (time point: ages 30, 40 and 50) and one between-group independent variable (lifestyle: sedentary, active or very active).

Research question for repeated-measures MANOVA

For these analyses we will extend the research question set by LAPS, the group of veterinary researchers that we encountered when we explored traditional MANOVA. They are still investigating anxiety and depression in different pets, only this time they have dropped the hamster analyses (as they showed no effects previously) and are focusing on cats and dogs. Also, they have decided to implement some therapies and diets for the animals with the aim of improving anxiety and depression To examine the success of these measures, anxiety and depression are measured twice for cats and dogs, at baseline (prior to treatment) and at four weeks after treatment. To explore this, we need to employ repeated-measures MANOVA with two dependent variables (anxiety and depression scores), based on ratings from 0–100, where higher scores are poorer (as before). There is one within-group measure, with two trials (the time points: baseline and week 4), and there is one between-group factor (pet type: cats and dogs). LAPS predict that outcomes will continue to show that cats are more anxious than dogs, while dogs are more depressed than cats. LAPS also predict that all animals will make an improvement, but do not offer an opinion on which group will improve more to each treatment.

14.7 Take a closer look
Summary of repeated-measures MANOVA research example

Between-group independent variable (IV): Type of pet (two groups: cats and dogs)
Within-group IV: Time point (two trials: baseline and four weeks post-treatment)
Dependent variable (DV) 1: Anxiety scores
DV 2: Depression scores

Theory and rationale

Multivariate outcome

Similar to traditional MANOVA, the multivariate outcome in repeated-measures MANOVA indicates whether there are significant differences in respect of the combined dependent variables across the independent variable (or variables). Although we will not even attempt to explore calculations manually, we still need to know about the partitioning of variance. Similar to traditional MANOVA, outcomes are calculated from cross-products and mean-square matrices, along with eigenvalue adjustments to find a series of F ratios. We are also presented with four (potentially different) F ratio outcomes: Pillai's Trace, Wilks' Lambda, Hotelling's Trace and Roy's Largest Root. The rationale for selection is the same as we summarised in Box 14.4.

Univariate outcome – main effects

Each dependent variable will have its own variance, which is partitioned into model sums of squares (explained variance) and residual sums of squares (unexplained 'error' variance) for each independent variable and (if appropriate) the interaction between the independent variables.

As we have seen before, all of this is analysed in relation to respective degrees of freedom. The resultant mean squares are used to produce an F ratio for each independent variable and interaction in respect of each dependent variable. This is much as we saw for traditional MANOVA, only we explore the outcome very differently. We need to use repeated-measures analyses to explore all univariate outcomes in this case (and mixed multi-factorial ANOVA is there between-group independent variables). If any of the independent variables have more than two groups or conditions, we will also need to explore the source of that main effect.

Locating the source of main effects

If significant main effects are represented by two groups or conditions, we can refer to the mean scores; if there are three or more groups or conditions, we need to do more work. The protocols for performing planned contrasts or *post hoc* tests are the same as we saw in univariate ANOVAs, so we will not repeat them here. Guidelines for between-group analyses are initially reviewed in Chapter 9, while within-group discussions begin in Chapter 10. For simplicity, we will focus on *post hoc* tests in these sections. In our example, the between-group independent variable (pet type) has two groups (dogs and cats). Should we find significant differences in anxiety or depression ratings, we can use the mean scores to describe those differences. The within-group independent variable has two trials (baseline and four weeks post-treatment), so we will not need to locate the source of any difference should we find one. Wherever there are significant differences across three or conditions, we would need Bonferroni *post hoc* analyses to indicate the source of the main effect.

Locating the source of interactions

Should we find an interaction between independent variables in respect of any dependent variable outcome, we need to look for the source of that. The methods needed to do this are similar to what we saw in Chapters 11–13 (when we explored multi-factorial ANOVAs). Interactions will be examined using a series of t-tests or one-way ANOVAs, depending on the nature of variables being measured. Where between-group independent variables are involved, the Split File facility in SPSS will need to be employed. A summary of these methods is shown in Chapter 13 (Box 13.7).

Assumptions and restrictions

The assumptions for repeated-measures MANOVA are pretty much as we saw earlier. We need to check normal distribution for each dependent variable, in respect of all independent variables, and account for outliers and kurtosis. We should also check that there is reasonable correlation between the dependent variables, and should avoid multicollinearity. If there are between-group independent variables, we need to check homogeneity of variances (via Levene's test). We also need to check that the correlation between the dependent variables is

14.8 Take a closer look
Summary of assumptions for repeated-measures MANOVA

As we saw in Box 14.5, plus...

- All within-group IVs (trials) must be measured across one group
    - Each person (or case) must be present in all conditions
- We need to account for sphericity of within-group variances

equal across independent variable groups (homogeneity of variance-covariance, via Box's M test). If that outcome is highly significant ($p < .001$), and there are unequal group sizes, we may not be able to trust the outcome. Since (by default) there will be at least one within-group independent variable, we will need to check sphericity of within-group variances (via Mauchly's test). The sphericity outcome will determine which line of univariate ANOVA outcome we read (much as we did with repeated-measures ANOVAs).

How SPSS performs repeated-measures MANOVA

To illustrate how we perform repeated-measures MANOVA in SPSS, we will refer to the second research question set by LAPS (the vet researchers). In these analyses, anxiety and depression ratings of 35 cats and 35 dogs are compared in respect of how they respond to treatment (therapy and food supplement). Baseline ratings of anxiety and depression are taken, which are repeated after four weeks of treatment. Ratings are made on a scale of 0–100, where higher scores are poorer. LAPS predict that outcomes will continue to show that cats are more anxious than dogs, while dogs are more depressed than cats. LAPS also predict that all animals will make an improvement, but cannot offer an opinion on which group will improve more to each treatment. The dependent and independent variables are summarised below:

MANOVA variables

- **Between-group IV:** Type of pet (two groups: cats and dogs)
- **Within-group IV:** Time point, with two trials (baseline and week 4)
- **DV 1:** Anxiety scores
- **DV 2:** Depression scores

14.9 Nuts and bolts
Setting up the data set in SPSS

Setting up the SPSS file for repeated-measures MANOVA is similar to earlier, except that we need to create a 'variable' column for each dependent variable condition and one for the independent variable.

Cats and dogs.sav [DataSet1] - IBM SPSS Statistics Data Editor

File Edit View Data Transform Analyze Graphs Utilities Add-ons Window Help

	Name	Type	Width	Decimals	Label	Values	Missing	Columns	Align	Measure
1	animal	Numeric	8	0	Animal	{1, Cat}...	None	12	Right	Nominal
2	anxbase	Numeric	8	0	Anxiety Baseline	None	-1	8	Right	Scale
3	anxwk4	Numeric	8	0	Anxiety week 4	None	-1	8	Right	Scale
4	depbase	Numeric	8	0	Depression Ba...	None	-1	8	Right	Scale
5	depwk4	Numeric	8	0	Depression we...	None	-1	8	Right	Scale

Figure 14.18 Variable View for 'Cats and dogs' data

As shown in Figure 14.18, we have a single column for the between-group independent variable (animal, with two groups set in the Values column: 1 = cat; 2 = dog), with Measure set to Nominal. The remaining variables represent the dependent variables: anxbase (anxiety at baseline), anxwk4 (anxiety at week 4), depbase (depression at baseline), depwk4 (depression at week 4). These are numerical outcomes, so we do not set up anything in the Values column and set Measure to Scale.

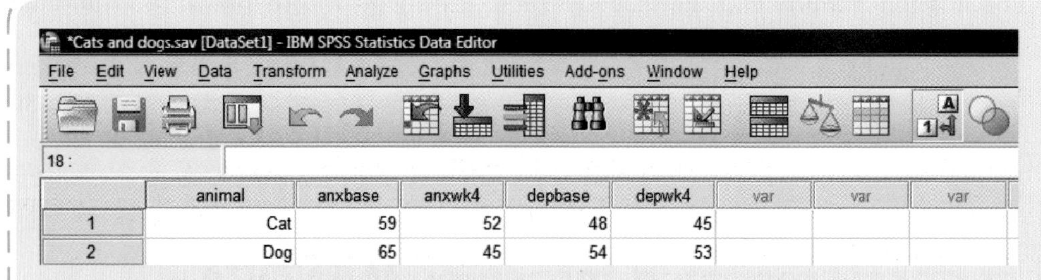

Figure 14.19 Data View for 'Cats and dogs' data

Figure 14.19 illustrates how this will appear in the Data View. As before, we will use this view to select the variables when performing this test. Each row represents a pet. When we enter the data for 'animal', we input 1 (to represent cat) or 2 (to represent dog); in the remaining columns we enter the actual score (dependent variable) for that pet at that condition.

The data set that we are using for these analyses is larger than the ones we are used to, so we cannot show the full list of data here. However, we will explore the mean scores and other descriptive data throughout our analyses.

Checking correlation

Reasonable correlation is one of the key assumptions of this test, so we ought to check that as we did earlier.

Open the SPSS file Cats and dogs

Select **Analyze** → **Correlate** → **Bivariate...** → transfer **Anxiety Baseline, Anxiety week 4, Depression Baseline** and **Depression week 4** to **Variables** → tick boxes for **Pearson** and **Two-tailed** → click **OK**

		Anxiety baseline	Anxiety week 4	Depression baseline	Depression week 4
Anxiety baseline	Pearson correlation	1	.488**	.491**	.401**
	Sig. (2-tailed)		.000	.000	.001
	N	70	70	70	70
Anxiety week 4	Pearson correlation	.488**	1	.465**	.582**
	Sig. (2-tailed)	.000		.000	.000
	N	70	70	70	70
Depression baseline	Pearson correlation	.491**	.465**	1	.893**
	Sig. (2-tailed)	.000	.000		.000
	N	70	70	70	70
Depression week 4	Pearson correlation	.401**	.582**	.893**	1
	Sig. (2-tailed)	.001	.000	.000	
	N	70	70	70	70

Figure 14.20 Correlation between dependent variables

Figure 14.20 shows that correlation across the dependent variables is acceptable (we need to focus only on the relationship between anxiety and depression measures).

Testing for normal distribution

As before, we need to check that the data are normally distributed. This time we need to explore outcomes for each of the dependent variables by within-group condition, across the animal groups:

		Kolmogorov–Smirnov[a]			Shapiro–Wilk		
	Animal	Statistic	df	Sig.	Statistic	df	Sig.
Anxiety baseline	Cat	.088	35	.200*	.968	35	.395
	Dog	.115	35	.200*	.946	35	.088
Anxiety week 4	Cat	.144	35	.062	.919	35	.014
	Dog	.092	35	.200*	.972	35	.506
Depression baseline	Cat	.164	35	.018	.943	35	.068
	Dog	.112	35	.200*	.950	35	.113
Depression week 4	Cat	.154	35	.035	.957	35	.190
	Dog	.101	35	.200*	.977	35	.666

Figure 14.21 Kolmogorov-Smirnov/Shapiro-Wilk test: anxiety and depression scores across time point, by animal type

As there are fewer than 50 animals in each group, we should refer to the Shapiro-Wilk outcome once more. Figure 14.21 shows that we can be satisfied that we have reasonable normal distribution in anxiety and depression scores, across the animal groups. The outcome for cats' anxiety at week 4 is potentially a problem, but given the overall picture we should be OK.

Running repeated-measures MANOVA

The method for performing repeated-measures MANOVA is different to what we did earlier. We do *not* use the GLM multivariate route, but build the analyses through repeated-measures methods:

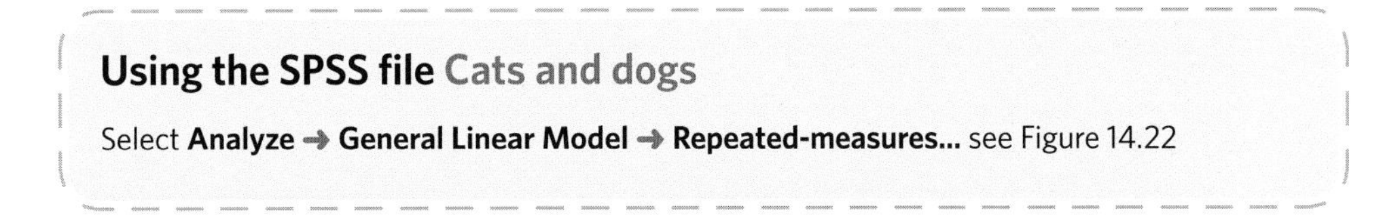

Using the SPSS file Cats and dogs

Select **Analyze → General Linear Model → Repeated-measures...** see Figure 14.22

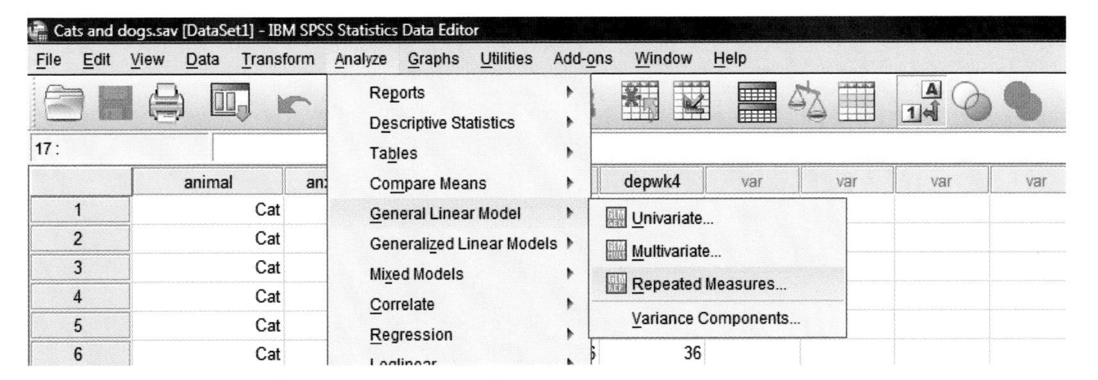

Figure 14.22 Repeated-measures MANOVA: procedure 1

In **Define Factors** window (see Figure 14.23), type **Weeks** into **Within-Subject Factor Name:** → type **2** into **Number of Levels:** → click **Add**

This sets up the within-group conditions for the analyses. You can call this what you like, but make it logical; 'weeks' makes sense because we are measuring across two time points. The number of levels is 2 because we have two time points; baseline and week 4.

Type **Anxiety** into **Measure Name:** → click **Add** → type **Depression** into **Measure Name:** → click **Add**

This defines the dependent variables. Again, call this what you want, but it makes sense to call our DVs 'anxiety' and 'depression'. The key thing is that we include a measure name for each DV that we have (in this case 2).

Click **Define**

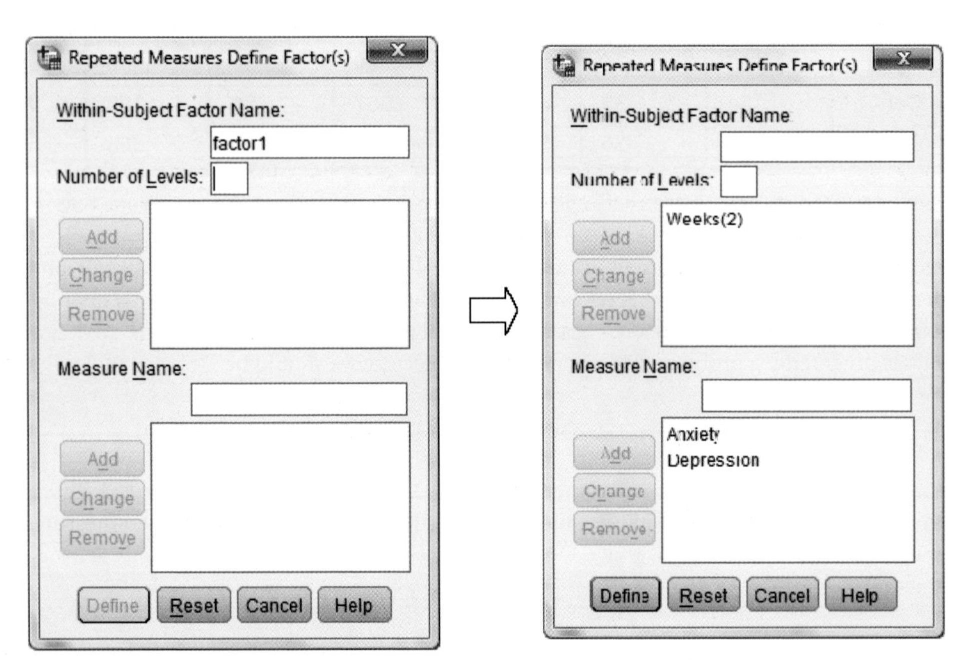

Figure 14.23 Define factors

In new window (see Figure 14.24), transfer **Anxiety Baseline** to **Within-Subjects Variables (Weeks)** to replace _?_ (1,Anxiety) → transfer **Anxiety Week 4** to **Within-Subjects Variables (Weeks)** to replace _?_ (2,Anxiety) → transfer **Depression Baseline** to **Within-Subjects Variables (Weeks)** to replace _?_ (1, Depression) → transfer **Depression week 4 to Within-Subjects Variables (Weeks)** to replace _?_ (2, Depression)

This sets up the within-group analyses. It is vital that this is undertaken in the correct order (which is why it helps to use logical names when defining the factors). In this case "1, Anxiety" is linked to "anxbase", "2, Anxiety" to "anxwk4", and so on.

Transfer **Animal** to **Between-Group Factor (s)** (to set up the between-group independent variable) → click **Options**

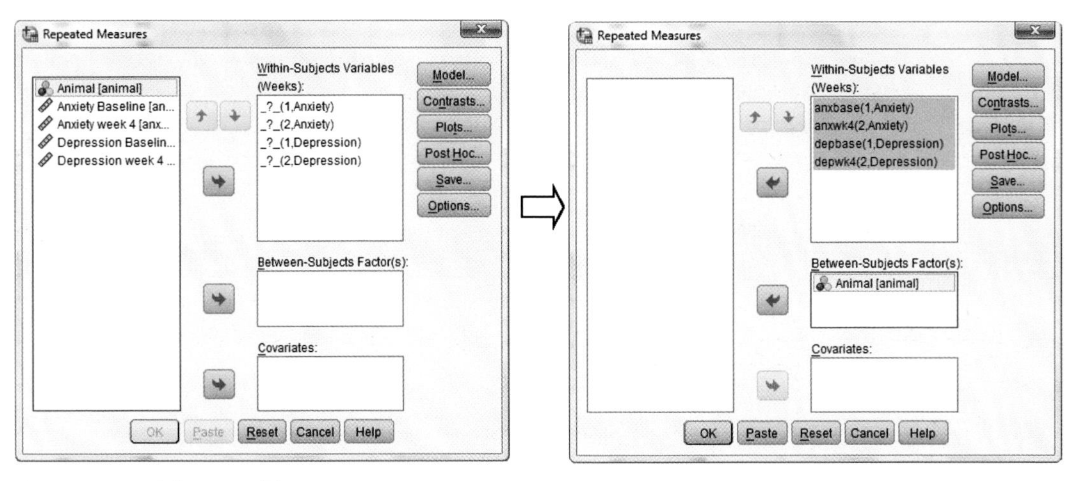

Figure 14.24 Select variables

In next window (See Figure 14.25), transfer **Animal, Weeks** and **Animal * Weeks** to **Display Means for:** ➜ tick boxes under **Display** for **Estimates of effect size** and **Homogeneity tests** ➜ click **Continue**

This sets up the univariate analyses of main effects and interactions. Both independent variables have two groups or conditions, so we do not need *post hoc* tests. If the within-group factor had three or more conditions, we would choose Bonferroni using the **Compare main effects** function (see Chapter 10). If the between-group factor had three or more groups, we would choose *post hoc* tests from the Post Hoc button back in the main menu (following instructions from Chapter 9) Click **Plots**

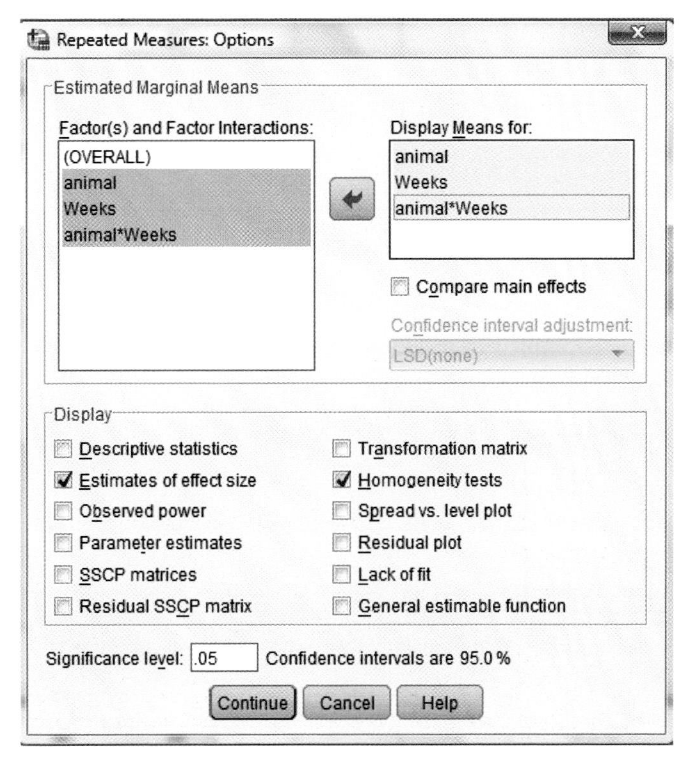

Figure 14.25 Options

In next window (see Figure 14.26), transfer **Weeks** to **Horizontal Axis:** → transfer **Animal** to **Separate Lines:** → click **Add** (this will give us some graphs that we can examine later) → click **Continue** → click **OK**

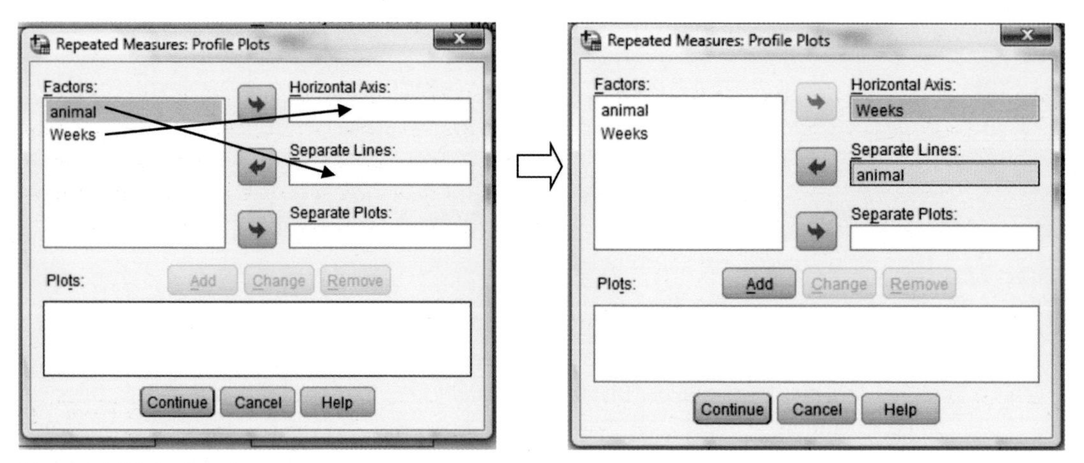

Figure 14.26 Profile plots

Interpretation of output

Checking assumptions

	F	df1	df2	Sig.
Anxiety baseline	.009	1	68	.923
Anxiety week 4	.099	1	68	.754
Depression baseline	.072	1	68	.789
Depression week 4	.264	1	68	.609

Figure 14.27 Levene's test for equality of variances

Figure 14.27 indicates that we have satisfied the assumption for between-group homogeneity of variance across animal groups for both dependent variables, at each condition (significance greater than .05).

Box's M	27.547
F	2.579
df1	10
df2	22106.773
Sig.	.004

Figure 14.28 Box's M test for equality of variance-covariance matrices

Figure 14.28 shows that we have also met the assumption of homogeneity of variance-covariance matrices (significance greater than .001).

Within-subjects effect	Measure	Mauchly's W	Approx. chi-square	df	Sig.	Epsilon[a]		
						Greenhouse–Geisser	Huynh–Feldt	Lower bound
Weeks	Anxiety	1.000	.000	0	.	1.000	1.000	1.000
	Depression	1.000	.000	0	.	1.000	1.000	1.000

Figure 14.29 Sphericity of within-group variance

Figure 14.29 presents the spherity outcome. We would need to check sphericity only if we had three or more within-group conditions. We had two, so we can ignore this output (but notice how outcomes are reported under these circumstances). We can state that sphericity is assumed, which will guide us to the correct line of univariate outcome later. When we have three or more conditions, we need to pay closer attention to the outcome (see Chapter 10).

Multivariate outcome

Effect			Value	F	Hypothesis df	Error df	Sig.	Partial eta squared
Between-subjects	Intercept	Pillai's Trace	.970	1082.734[a]	2.000	67.000	.000	.970
		Wilks' Lambda	.030	1082.734[a]	2.000	67.000	.000	.970
		Hotelling's Trace	32.320	1082.734[a]	2.000	67.000	.000	.970
		Roy's Largest Root	32.320	1082.734[a]	2.000	67.000	.000	.970
	animal	Pillai's Trace	.305	14.721[a]	2.000	67.000	.000	.305
		Wilks' Lambda	.695	14.721[a]	2.000	67.000	.000	.305
		Hotelling's Trace	.439	14.721[a]	2.000	67.000	.000	.305
		Roy's Largest Root	.439	14.721[a]	2.000	67.000	.000	.305
Within-subjects	Weeks	Pillai's Trace	.384	20.863[a]	2.000	67.000	.000	.384
		Wilks' Lambda	.616	20.863[a]	2.000	67.000	.000	.384
		Hotelling's Trace	.623	20.863[a]	2.000	67.000	.000	.384
		Roy's Largest Root	.623	20.863[a]	2.000	67.000	.000	.384
	Weeks * animal	Pillai's Trace	.104	3.894[a]	2.000	67.000	.025	.104
		Wilks' Lambda	.896	3.894[a]	2.000	67.000	.025	.104
		Hotelling's Trace	.116	3.894[a]	2.000	67.000	.025	.104
		Roy's Largest Root	.116	3.894[a]	2.000	67.000	.025	.104

Figure 14.30 Multivariate statistics

As we saw with traditional MANOVA, Figure 14.30 presents four lines of data for each outcome. The protocols for selecting the appropriate option remain as we saw earlier. There are two groups across our independent variable, so Pillai's Trace may be more suitable on this occasion. In these analyses we have a multivariate outcome across each independent variable and for the interaction between those independent variables. There is a significant multivariate effect for between-subjects (of the combined anxiety and depression scores) across animal group (regardless of time point): $V = .305$, $F(2, 67) = 14.721$, $p < .001$ (V is the sign we use to show the Pillai's Trace outcome; we will use the partial eta square outcome for effect size calculations later). There is also a significant multivariate effect across within-subjects time point (regardless of animal group): $V = .384$, $F(2, 67) = 20.863$, $p < .001$. We also have a significant multivariate effect across the interaction between animal group and time point: $V = .104$, $F(2, 67) = 3.894$, $p = .025$.

Univariate outcome

Between-group main effect

Measure	Animal	Mean	Std. error	95% confidence interval	
				Lower bound	Upper bound
Anxiety	Cat	43.786	1.336	41.120	46.451
	Dog	39.757	1.336	37.091	42.423
Depression	Cat	36.357	1.304	33.754	38.960
	Dog	40.429	1.304	37.826	43.031

Figure 14.31 Estimated marginal means

Source	Measure	Type III sum of squares	df	Mean square	F	Sig.	Partial eta squared
Intercept	Anxiety	244279.314	1	244279.314	1955.464	.000	.966
	Depression	206361.607	1	206361.607	1732.907	.000	.962
animal	Anxiety	568.029	1	568.029	4.547	.037	.063
	Depression	580.179	1	580.179	4.872	.031	.067
Error	Anxiety	8494.657	68	124.921			
	Depression	8097.714	68	119.084			

Figure 14.32 Between-group univariate ANOVA outcome

Figures 14.31 and 14.32 indicate that anxiety scores are significantly higher for cats than for dogs (regardless of time point), $F(1, 68), = 4.547, p = .037$; while depression scores are significantly higher for dogs than for cats, $F(1, 68), = 4.872, p = .031$.

Within-group main effect

Measure	Weeks	Mean	Std. error	95% confidence interval	
				Lower bound	Upper bound
Anxiety	1	44.843	1.078	42.692	46.994
	2	38.700	1.101	36.503	40.897
Depression	1	39.600	1.045	37.516	41.684
	2	37.186	.851	35.488	38.883

Figure 14.33 Estimated marginal means

Figures 14.33 and 14.34 indicate that anxiety scores are significantly higher at baseline than at week 4 (regardless of pet type), suggesting an improvement, $F(1, 68), = 32.026, p < .001$, and that depression scores also showed significant improvement, $F(1, 68), = 25.736, p < .001$.

Interaction

Figure 14.35 suggests a number of differences in improvement scores according to anxiety or depression when examined between cats and dogs. We already have seen that cats are more anxious than dogs, while dogs are more depressed than cats. It would also appear that there is a greater improvement in anxiety across time for cats than dogs, while improvements in depression scores (although *generally* higher for dogs) appear much the same for cats and dogs. The

Source	Measure		Type III sum of squares	df	Mean square	F	Sig.	Partial eta squared
Weeks	Anxiety	Sphericity assumed	1320.714	1	1320.714	32.026	.000	.320
		Greenhouse–Geisser	1320.714	1.000	1320.714	32.026	.000	.320
		Huynh–Feldt	1320.714	1.000	1320.714	32.026	.000	.320
		Lower bound	1320.714	1.000	1320.714	32.026	.000	.320
	Depression	Sphericity assumed	204.007	1	204.007	25.736	.000	.275
		Greenhouse–Geisser	204.007	1.000	204.007	25.736	.000	.275
		Huynh–Feldt	204.007	1.000	204.007	25.736	.000	.275
		Lower bound	204.007	1.000	204.007	25.736	.000	.275
Weeks * animal	Anxiety	Sphericity assumed	321.029	1	321.029	7.785	.007	.103
		Greenhouse–Geisser	321.029	1.000	321.029	7.785	.007	.103
		Huynh–Feldt	321.029	1.000	321.029	7.785	.007	.103
		Lower bound	321.029	1.000	321.029	7.785	.007	.103
	Depression	Sphericity assumed	14.464	1	14.464	1.825	.181	.026
		Greenhouse–Geisser	14.464	1.000	14.464	1.825	.181	.026
		Huynh–Feldt	14.464	1.000	14.464	1.825	.181	.026
		Lower bound	14.464	1.000	14.464	1.825	.181	.026
Error(weeks)	Anxiety	Sphericity assumed	2804.257	68	41.239			
		Greenhouse–Geisser	2804.257	68.000	41.239			
		Huynh–Feldt	2804.257	68.000	41.239			
		Lower bound	2804.257	68.000	41.239			
	Depression	Sphericity assumed	539.029	68	7.927			
		Greenhouse–Geisser	539.029	68.000	7.927			
		Huynh–Feldt	539.029	68.000	7.927			
		Lower bound	539.029	68.000	7.927			

Figure 14.34 Within-group univariate ANOVA outcome (and interaction)

					95% confidence interval	
Measure	Animal	Weeks	Mean	Std. error	Lower bound	Upper bound
Anxiety	Cat	1	48.371	1.524	45.330	51.413
		2	39.200	1.557	36.093	42.307
	Dog	1	41.314	1.524	38.273	44.356
		2	38.200	1.557	35.093	41.307
Depression	Cat	1	37.886	1.477	34.938	40.833
		2	34.829	1.203	32.428	37.229
	Dog	1	41.314	1.477	38.367	44.262
		2	39.543	1.203	37.143	41.943

Figure 14.35 Estimated marginal means

ANOVA outcome in Figure 14.34 shows that there was a significant interaction between weeks and pet type for anxiety scores, $F(1, 68) = 7.785$, $p = .007$, while there was no interaction between weeks and pet type for depression scores, $F(1, 68) = 1.825$, $p = .181$.

Graphical presentation of main effects and interaction

It would be useful to illustrate what we have just seen with some line graphs. We requested some graphs when we set the plot profiles in the SPSS commands. These are shown here, but have been adjusted to show more meaningful labels using the procedures we saw in Chapter 12 (in particular see Figures 12.18 and 12.20).

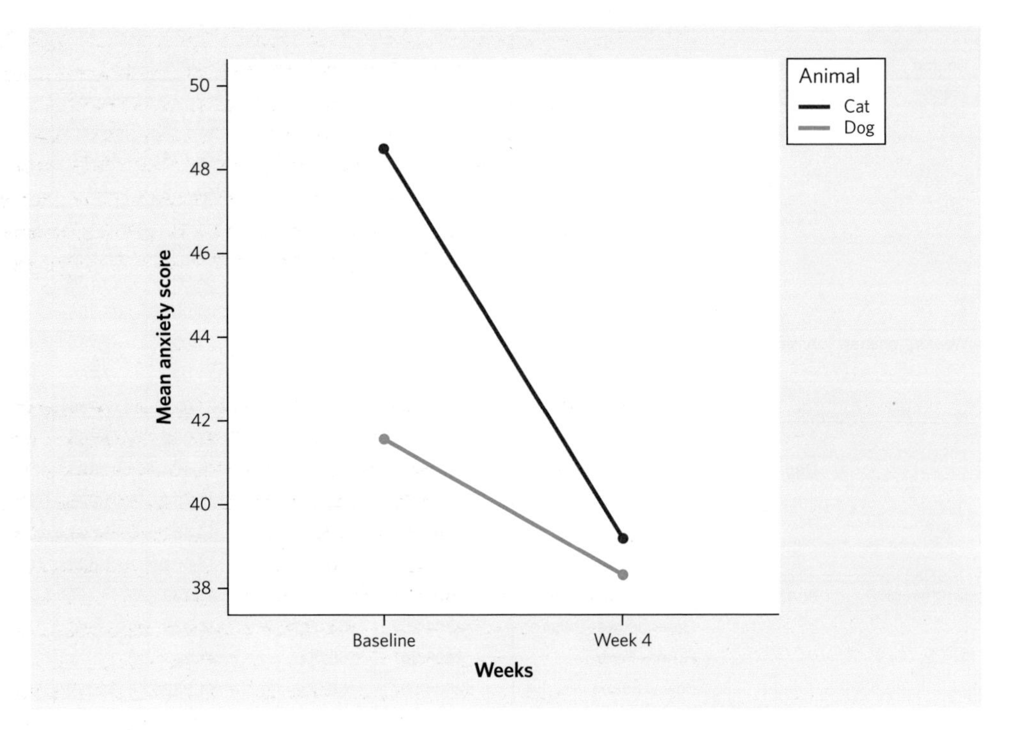

Figure 14.36 Line graph – anxiety scores across time points by animal type

Figure 14.36 shows that anxiety scores improved (reduced) more dramatically for cats than for dogs. The lines representing cats and dogs are not parallel, suggesting an interaction – this was supported by the statistical outcome in Figure 14.34.

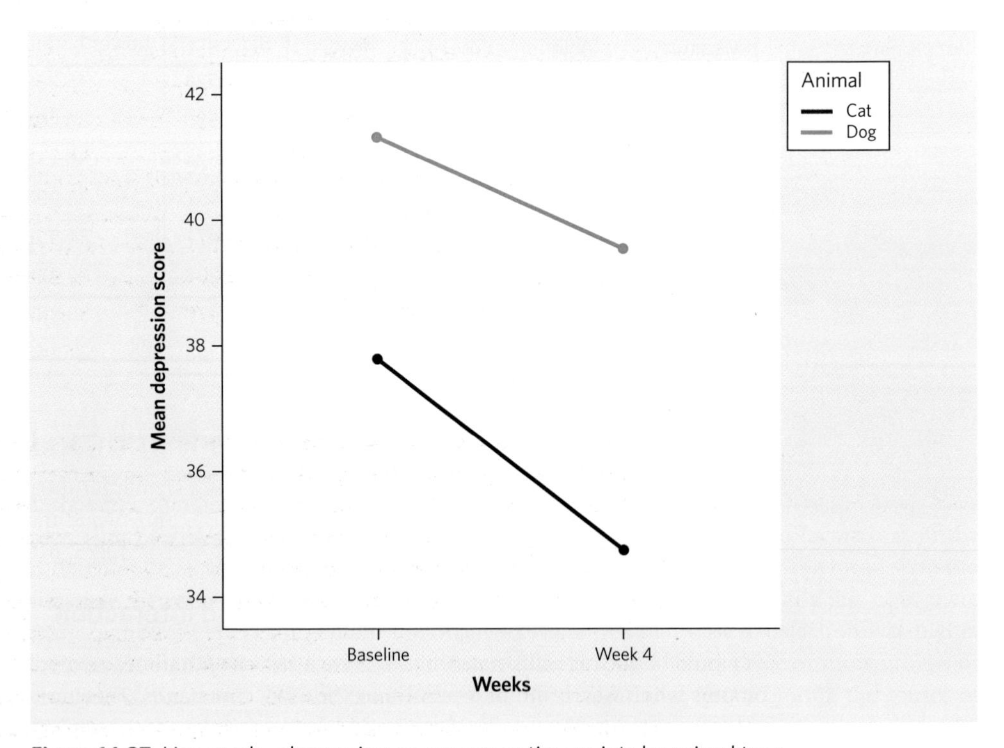

Figure 14.37 Line graph – depression scores across time points by animal type

Figure 14.37 shows that depression scores improved at much the same rate between cats and dogs. The lines representing cats and dogs are almost parallel, suggesting no interaction – this was also supported by the statistical outcome in Figure 14.34.

Finding the source of interaction

Whenever we find a significant interaction, we must explore the data further to illustrate the source of that interaction. This part of the analysis is the same as we saw for mixed multi-factorial ANOVA, so we will not repeat the finer detail regarding the sorts of tests that we need to perform. Box 13.7 in Chapter 13 presents an overview of those tests.

Anxiety scores - interaction between time points and animal type

Using guidelines from Box 13.7, we can apply this to what we need for the anxiety data. In each scenario we need two additional tests, so we should adjust significance cut-off points to account for multiple comparisons. Outcome will be significant only where $p < .025$ (usual cut off $.05 \div 2$).

14.10 Take a closer look

Source of interaction in anxiety scores between time points and animal type

Table 14.3 Tests needed to explore interaction

Analysis	Method
Animal type vs. time point	2 × independent t-tests: anxiety scores across animal type at each condition:[1] 1. for baseline 2. for week 4
Time point vs. animal type	2 × related t-tests:anxiety scores across time points, by animal type: 1. when animal = cat 2. when animal = dog[2]

[1] Using within-group columns for each time point condition as the DV and animal as the IV.
[2] Using 'time points' as 'within-subjects variables', but splitting file by animal type.

Animal type vs. anxiety ratings, according to time point:

Using the SPSS file Cats and dogs

Select **Analyze** → **Compare Means** → **Independent-Samples T Test...** → (in new window) transfer **Anxiety Baseline** and **Anxiety week 4** to **Test Variable List** → transfer **Animal** to **Grouping Variable** → click **Define Groups** → enter **1** in **Group 1** box → enter **2** in **Group 2** → click **Continue** → click **OK**

Figures 14.38 and 14.39 show that cats are significantly more anxious than dogs at baseline, $t(68) = 3.274$, $p = .002$ (well below the adjusted cut-off point). However, by week 4 (following treatment) there is no difference between cats and dogs in respect of anxiety scores, $t(68) = 0.454$, $p = 651$. That is certainly one explanation for the observed interaction.

	Animal	N	Mean	Std. deviation	Std. error mean
Anxiety baseline	Cat	35	48.37	8.981	1.518
	Dog	35	41.31	9.055	1.531
Anxiety week 4	Cat	35	39.20	9.634	1.628
	Dog	35	38.20	8.768	1.482

Figure 14.38 Descriptive statistics

Independent samples test

		Levene's test for equality of variances		t-test for equality of means						
									95% confidence interval of the difference	
		F	Sig.	t	df	Sig. (2-tailed)	Mean difference	Std. error difference	Lower	Upper
Anxiety baseline	Equal variances assumed	.009	.923	3.274	68	.002	7.057	2.156	2.756	11.359
	Equal variances not assumed			3.274	67.995	.002	7.057	2.156	2.756	11.359
Anxiety week 4	Equal variances assumed	.099	.754	.454	68	.651	1.000	2.202	−3.394	5.394
	Equal variances not assumed			.454	67.405	.651	1.000	2.202	−3.394	5.394

Figure 14.39 Independent t-test: animal type vs. anxiety scores (by time point)

Anxiety ratings across time point, according to animal type:

Using the SPSS file Cats and dogs

Select **Data** → **Split File** → (in new window) select **Compare groups** radio button → transfer Animal to **Groups Based on:** → click on **OK**

Select **Analyze** → **Compare means** → **Paired-Samples T Test...** → (in new window) transfer Anxiety Baseline and Anxiety week 4 to **Paired Variables** → click **OK**

Animal			Mean	N	Std. deviation	Std. error mean
Cat	Pair 1	Anxiety baseline	48.37	35	8.981	1.518
		Anxiety week 4	39.20	35	9.634	1.628
Dog	Pair 1	Anxiety baseline	41.31	35	9.055	1.531
		Anxiety week 4	38.20	35	8.768	1.482

Figure 14.40 Descriptive statistics

			Paired differences							
						95% confidence interval of the difference				
Animal			Mean	Std. deviation	Std. error mean	Lower	Upper	t	df	Sig. (2-tailed)
Cat	Pair 1	Anxiety Baseline–Anxiety week 4	9.171	11.092	1.875	5.361	12.982	4.892	34	.000
Dog	Pair 1	Anxiety Baseline–Anxiety week 4	3.114	6.475	1.095	.890	5.339	2.845	34	.007

Figure 14.41 Related t-test: anxiety scores across time point, by animal type

Figures 14.40 and 14.41 show that anxiety scores improved significantly between baseline and week 4 for cats, t (34) = 4.892, p < .001 (well below the adjusted cut-off point). Anxiety also improved significantly for dogs, t (34) = 2.845, p = .007 (but not nearly as much as for cats).

You must remember to switch off the Split File facility; otherwise subsequent analyses will be incorrect:

Select **Data** ➜ **Split File** ➜ (in new window) select **Analyze all cases, do not create groups** radio button ➜ click **OK**

Depression scores – interaction between time points and animal type

The interaction between time point and animal type in respect of depression scores was not significant, so we do not need to look any further (if we did it could be construed as 'fishing').

Effect size and power

Calculating effect size and achieved power for repeated-measures MANOVA is also a little different to what we undertook for traditional MANOVA. We start with univariate outcomes but, as we have one between-group factor and one within-group factor, those analyses are much as we did for mixed multi-factorial ANOVA. Before we proceed with entering the data, we need to find one further outcome: the 'average' correlation. We will need this for the 'correlation between repeated measures' parameter shortly. We have already examined the correlation between the conditions (see Figure 14.20), so we can use that to calculate 'average' correlation for the repeated measures (so, [.488 + .491 + .401 + .465 + .582 + .893] ÷ 6 = .557).

Univariate effects

From **Test family** select **F tests**

From **Type of power analysis** select **Post hoc: Compute achieved - given α, sample size and effect size power**

Between group:

From **Statistical test** select **ANOVA: Repeated measures, between factors**

For α **err prob** type **0.05** (significance level) ➜ **Total sample size** type **70** (overall sample size) ➜ **Number of groups** type **2** (cats and dogs) ➜ **Number of measurements** type **2** (baseline and week 4) ➜ for **Corr among rep measures** type **0.553** (as we saw just now)

To calculate the **Effect size**, click on the **Determine** button (a new box appears) ➜ under **Select procedure** choose **Effect size from Variance**

Anxiety

In box below, tick on radio button for Direct ➜ type **0.063** in the **Partial η^2** box (we get that from Figure 14.32) ➜ click on **Calculate and transfer to main window** ➜ back in original display ➜ click on **Calculate**

<u>Effect size</u> (d) **0.26** (medium); <u>Power</u> (1-β err prob) **0.68** (underpowered - see Chapter 4)

Depression

In box below, tick on radio button for Direct → type **0.067** in the **Partial η^2** box → click on **Calculate and transfer to main window** → back in original display → click on **Calculate**

Effect size (d) **0.27** (medium); Power (1-β err prob) **0.71** (underpowered)

Within-group:

From **Statistical test** select ANOVA: Repeated measures, within factors

For **α err prob** type **0.05** → **Total sample size** type **70** → **Number of groups** type **2** → **Number of repetitions** type **2** → **Corr among rep measures** type **0.553** → **nonsphericity** type **1** (see Figure 14.29 – Mauchly's W)

To calculate the **Effect size**, click on the **Determine** button

Anxiety

In new box, tick on radio button for Direct → type **0.320** in the **Partial η^2** box (we get that from Figure 14.34) → click on **Calculate and transfer to main window** → back in original display → click on **Calculate**

Effect size (d) **0.69** (large); Power (1-β err prob) **1.00** (perfect)

Depression

In new box, tick on radio button for Direct → type **0.275** in the **Partial η^2** box → click on **Calculate and transfer to main window** → back in original display → click on **Calculate**

Effect size (d) **0.62** (large); Power (1-β err prob) **1.00** (perfect)

Interaction:

From **Statistical test** select ANOVA: Repeated measures, within-between interaction

For **α err prob** type **0.05** → **Total sample size** type **70** → **Number of groups** type **2** → **Number of repetitions** type **2** → **Corr among rep measures** type **0.553** → **nonsphericity** type **1**

To calculate the **Effect size**, click on the **Determine** button

Anxiety

In new box, tick on radio button for Direct → type **0.103** in the **Partial η^2** box (Figure 14.34) → click on **Calculate and transfer to main window** → back in original display → click on **Calculate**

Effect size (d) **0.34** (medium); Power (1-β err prob) **1.00** (perfect)

Depression

In new box, tick on radio button for Direct → type **0.026** in the **Partial η^2** box → click on **Calculate and transfer to main window** → back in original display → click on **Calculate**

Effect size (d) **0.16** (small); Power (1-β err prob) **0.80** (strong)

Multivariate effects

From **Test family** select **F tests**

From **Type of power analysis** select Post hoc: Compute achieved – given α, sample size and effect size power

Between group:

From **Statistical test** select **MANOVA: Repeated measures, between factors**

For **α err prob** type **0.05** → **Total sample size** type **70** → **Number of groups** type **2** → **Number of measurements** type **2** → for **Corr among rep measures** type **0.553**

To calculate the **Effect size**, click on the **Determine** button → under **Select procedure** choose **Effect size from Variance**

In box below, tick on radio button for Direct → type **0.305** in the **Partial η^2** box (we get that from Figure 14.30) → click on **Calculate and transfer to main window** → back in original display → click on **Calculate**

Effect size (d) **0.66** (medium); Power (1-β err prob) **1.00** (perfect)

Within group:

From **Statistical test** select **MANOVA: Repeated measures, within factors**

For **α err prob** type **0.05** → **Total sample size** type **70** → **Number of groups** type **2** → **Number of repetitions** type **2** → **Corr among rep measures** type **0.553**

To calculate the **Effect size**, click on the **Determine** button

In new box, tick on radiobutton for Direct → type **0.384** in the **Partial η^2** box (Figure 14.30) → click on **Calculate and transfer to main window** → back in original display → click on **Calculate**

Effect size (d) **0.79** (large); Power (1-β err prob) **1.00** (perfect)

Interaction:

From **Statistical test** select **MANOVA: Repeated measures, within-between interaction**

For **α err prob** type **0.05** → **Total sample size** type **70** → **Number of groups** type **2** → **Number of repetitions** type **2**

To calculate the **Effect size**, click on the **Determine** button

In that new box, select **Effect size from criterion**

We are presented with a number of options for the multivariate statistic. The default is 'Pillai V', which is what we want → type **0.104** in **Pillai V** (Figure 14.30) → **Number of groups** type **2** → **Number of repetitions** type **2** → click on **Calculate and transfer to main window** → back in original display → click on **Calculate**

Effect size (d) **0.34** (medium); Power (1-β err prob) **0.79** (strong)

Writing up results

Perceptions of anxiety and depression were measured for cats and dogs at two time points: prior to treatment and four weeks after treatment (involving diet supplements and basic training). Repeated-measures MANOVA analyses confirmed that there were significant multivariate effects for animal group (V = .305, F (2, 67) = 14.721, p < .001, d = 0.69), treatment week (V =.384, F (2, 67) = 20.863, p < .001, d = 0.62) and the interaction between animal type and treatment week (V = .104, F (2, 67) = 3.894, p = .025, d = 0.34). Univariate between-group analyses showed that cats were significantly more anxious than dogs (F (1, 68), = 4.547, p = .037, d = 0.26), while dogs were more depressed than cats (F (1, 68), = 4.872, p = .031, d = 0.27). Within-group univariate analyses indicated that anxiety scores (F (1, 68), = 32.026, p < .001, d = 0.69) and depression scores (F (1, 68), = 25.736, p < .001, d = 0.62) were significantly

Table 14.4 Anxiety and depression scores by domestic pet type, across treatment time point

				Animal vs. week					
Anxiety		**Main effects**		**Cat**			**Dog**		
Week	Mean	SE	N	Mean	SE	N	Mean	SE	N
Baseline	44.84	1.08	70	48.37	1.52	35	41.31	1.52	35
Week 4	38.70	1.10	70	39.20	1.56	35	38.20	1.56	35
Animal									
Cat	43.79	1.34	35						
Dog	39.76	1.34	35						
Depression									
Week	Mean	SE	N	Mean	SE	N	Mean	SE	N
Baseline	39.60	1.05	70	37.89	1.48	35	41.31	1.47	35
Week 4	37.19	0.85	70	34.83	1.20	35	39.54	1.20	35
Animal									
Cat	36.36	1.30	35						
Dog	40.43	1.30	35						

improved between baseline and week 4 (irrespective of animal group). There was a significant interaction between animal type and treatment week for anxiety scores ($F(1, 68) = 7.785$, $p = .007$), but not for depression scores ($F(1, 68) = 1.825$, $p = .181$). Further analyses of the interaction for anxiety scores showed that while cats were significantly more anxious than dogs at baseline ($t(68) = 3.274$, $p = .002$), there was no difference between the groups by week 4 ($t(68) = 0.454$, $p = .651$). Improvements in anxiety scores were greater for cats than for dogs.

Chapter summary

In this chapter we have explored multivariate analyses, notably MANOVA and repeated-measures MANOVA. At this point, it would be good to revisit the learning objectives that we set at the beginning of the chapter.

You should now be able to:

- Recognise that we use (traditional) MANOVA to simultaneously examine several dependent variables (measured at a single time point) across one or more categorical independent variable. Meanwhile repeated-measures MANOVA explores several dependent variables at two or more time points (within-group); outcomes can be additionally measured across one or more between-group independent variable.
- Comprehend that multivariate analyses explore the overall effect on the combination of dependent variables, while univariate analyses examine main effects (and interactions) for each of the independent variables in relation to each of the dependent variables. Additional *post hoc* tests may be needed to explore the source of significant main effects. Further analyses may be needed to explore the source of interactions.

- Understand the assumptions and restrictions. For MANOVA we need parametric data (reasonable normal distribution and at least interval data), where there is reasonable correlation between the dependent variables. There should be homogeneity of between-group variance (where appropriate) and equality across variance-covariance matrices. There should also be sphericity of within-group variances.
- Perform analyses using SPSS, exploring multivariate effects, univariate effects, *post hoc* tests and analyses of interactions.
- Examine effect size and power, using G*Power software, across multivariate and univariate effects.
- Understand how to present the data, using appropriate tables, reporting the outcome in a series of sentences and correctly formatted statistical notation.

Research example (MANOVA)

It might help you to see how MANOVA has been applied in a research context. In this section you can read an overview of the following paper:

> Delisle, T.T., Werch, C.E., Wong, A.H., Bian, H. and Weiler, R. (2010). Relationship between frequency and intensity of physical activity and health behaviors of adolescents. *Journal of School Health, 80 (3)*: 134–140. DOI: http://dx.doi.org/10.1111/j.1746-1561.2009.00477.x

If you would like to read the entire paper you can use the DOI reference provided to locate that (see Chapter 1 for instructions).

In this research the authors examined the relationship between the frequency and intensity of physical activity in respect of several measures of health behaviour in US high school children. Some behaviours were likely to be detrimental to good health and others were likely to promote good health. Two separate analyses were undertaken: one that focused on vigorous physical activity (VPA) and one that explored moderate physical activity (MPA). We will report only the former here (you can read the paper to see more data).

Within VPA there was one independent variable, with three frequency groups: low (0–1 times per week), medium (2–4 times per week) and high (5 or more times per week). These groups were examined against health behaviours in four MANOVAs. Three reflected risky behaviours: alcohol consumption, cigarette smoking and taking marijuana. Each of those was reported across four dependent variables: length (for how long the behaviour had been performed), frequency (how often the behaviour was performed in the last month), quantity (the average monthly use) and heavy use (the number of days in past month where 'heavy use' was reported). One MANOVA analysis reported good health behaviours. This had three dependent variables: amounts consumed for fruit and vegetables, good carbohydrates and good fats. It would take too much time to explain here how each variable was measured, but you can read more about that in the paper. Data were collected from 822 11th- and 12th-grade high school students (in the USA, these youngsters are typically aged 16–17). The average age was 17; 56% of the sample was female.

The results were reported in a series of tables. There was no multivariate effect for alcohol: $F(8, 1614) = 0.95$, $p = .47$ (and no univariate effects). There was no multivariate effect for cigarettes: $F(8, 1598) = 1.35$, $p = .21$. However, there were significant univariate effects for frequency: $F(2, 812) = 3.59$, $p = .03$; and quantity of use: $F(2, 813) = 3.49$, $p = .03$. Tukey *post hoc* tests discovered lower frequency and quantity of cigarette use in young people partaking in high levels of VPA, compared with low levels. There was a significant multivariate effect for marijuana: $F(8, 1604) = 2.13$, $p = .03$. Subsequent univariate analyses indicated significant effects for frequency of use: $F(2, 810) = 2.99$, $p = .05$; and for heavy use: $F(2, 810) = 3.60$, $p = .03$. Once again, the *post hoc* tests suggested less detrimental use in high VPA vs. low VPA exercise behaviour. Finally, there was a highly significant multivariate effect for nutritional behaviour: $F(6, 1622) = 3.63$, $p = .001$. Univariate analyses showed significant effects for (good) carbohydrate intake: $F(2, 813) = 5.63$,

$p < .001$; and (good) fat consumption: $F(2, 814) = 10.68$, $p < .001$. The *post hoc* tests suggested better nutritional consumption among high VPA youngsters vs. low VPA.

This is a good example of a complex array of dependent and independent variables. Given the magnitude of the analyses, some effect size reporting would have been useful. Furthermore, some of the reporting of statistical notation did not comply with traditional standards. In particular, the authors reported high significance as '$p = .00$'; it is generally better to show this as '$p < .001$' (even in tables). The narrative reporting was more consistent.

Research example (repeated-measures MANOVA)

It might also help you to see how repeated-measures MANOVA has been applied in a research context:

> **Jerrott, S., Clark, S.E. and Fearon, I. (2010). Day treatment for disruptive behaviour disorders: can a short-term program be effective?** ***Journal of the Canadian Academy of Child & Adolescent Psychiatry, 19 (2)*****: 88–93. Web link (no DOI): http://www.cacap-acpea.org/en/cacap/Volume_19_Number_2_May_2010_s5.html?ID=581**

This research examined the effectiveness of a treatment programme for children with Disruptive Behaviour Disorder (DBD). This is a serious condition illustrated by aggression, hyperactivity, social problems and externalisation. Children with extreme behavioural problems are more likely to (later) engage in criminal behaviour, many need the services of educational specialists and they are often sent to residential care. Severe parental stress is common. In this study, 40 children with DBD (32 boys, 8 girls) were entered into a treatment programme. These were compared with 17 children who were on a waiting list for the programme. Children in treatment and waiting list groups did not differ on any behavioural measure prior to the study. Treatment involved several weeks of cognitive behavioural therapy (CBT) and parental training (see the paper for more detail). Measures for all groups were taken at baseline (or referral for waiting list) and four months after treatment (or post-referral for the waiting list group). Several measures were taken: The Child Behaviour Checklist (CBCL; Achenbach, 1991) was used to examine social problems, aggression and externalisation; the Conners' Parental Rating Scale Revised: Short Form (CPRS-R:S; Conners, 1997) was used to measure hyperactivity; the Eyberg Child Behaviour Index (ECBI; Eyberg and Pincus, 1999) was used to illustrate the intensity of behavioural problems; and the Parenting Stress Index (PSI; Abadin, 1995) was used to examine reported stress for the parents and for the child.

The results showed several differences between the groups, providing support for the treatment programme. There was a significant multivariate effect for combined outcomes across the groups: $F(5, 40) = 2.60$, $p = .04$. The authors actually reported this as follows: $F = 2.60$, $df = 5, 40$, $p = .04$. This is not incorrect per se, but it is not in accordance with standard conventions. The remainder of this summary will report outcomes as we have seen throughout this chapter, but do have a look at the paper to see how some reports differ in style. Also, the authors stated that Hotelling's T^2 was used for these multivariate analyses (presumably because the two groups had unequal sample sizes, making Pillai's Trace less viable). They did not report the T^2 value.

Univariate analyses showed that there were treatment effects across all of the outcomes: social problems, $F(1, 44) = 26.35$, $p < .001$; aggression, $F(1, 44) = 13.88$, $p = .001$; externalising, $F(1, 44) = 11.91$, $p = .001$; hyperactivity, $F(1, 44) = 21.90$, $p < .001$; and intensity, $F(1, 44) = 49.57$, $p < .001$. There was also a significant multivariate effect for the interaction between group and treatment: $F(5, 40) = 3.33$, $p = .013$. Only three univariate effects were significant for this effect: aggression, $F(1, 44) = 6.51$, $p = .014$; externalising, $F(1, 44) = 8.92$, $p = .005$; and intensity of behaviour, $F(1, 44) = 13.72$, $p = .001$. Tabulated data (not reported in the main text) suggested that treatment effects were significant in the treatment group only (presumably undertaken with related t-tests in respect of each group in turn). Independent t-tests showed that 'post-treatment' outcomes were significantly better for the treatment group than for the waiting list control on three measures: aggression, $t(53) = 2.61$, $p = .012$, $d = 0.79$; externalising, $t(53) = 3.41$, $p = .001$, *d* = 1.01; and intensity of behaviour, t (53) = 2.54, p = .014, *d* = 0.79. PSI measures were examined in a related t-test. Across the treatment group, reports of stress were significantly reduced from baseline to post-treatment for the child-related stress [t (33) = 5.76, p < .001] and parental stress [t (33) = 2.27, p = .03]. Neither effect was significant for the waiting list group.

Extended learning tasks

You will find the data set associated with these tasks on the website that accompanies this book (available in SPSS and Excel format). You will also find the answers there.

MANOVA

Following what we have learned about MANOVA, answer the following questions and conduct the analyses in SPSS and G*Power. (If you do not have SPSS, do as much as you can with the Excel spreadsheet.) In this example we examine how exercise levels may have an impact on subsequent depression and (independently) on quality of life perceptions. The depression and perceived quality of life scales are measured on a scale from 0-100; depression, 0 = severe, 100 = none; perceived quality of life, 0 =poor, 100 = good. There are nearly 350 participants in this study, so bear that in mind when making conclusions and drawing inferences from normal distribution measures.

Open the data set **Exercise, depression and QoL**

1. Which is the independent variable?
2. What are the independent variable groups?
3. Which are the dependent variables?
4. Conduct the MANOVA test.
   a. Describe how you have accounted for the assumptions of MANOVA.
   b. Describe what the SPSS output shows for the multivariate and univariate effects.
   c. Run *post hoc* analyses (if needed).
5. Describe the effect size and conduct power calculations, using G*Power.
6. Report the outcome as you would in the results section of a report.

Repeated-measures MANOVA

Following what we have learned about repeated-measures MANOVA, answer the following questions and conduct the analyses in SPSS and G*Power. (You will not be able to perform this test manually.) In this example we examine exam scores and coursework scores in a group of 60 students (30 male and 30 female) over three years of their degree course.

Open the SPSS data **Exams and coursework**

1. What is the between-group independent variable?
   a. State the groups.
2. What is the within-group independent variable?
   a. State the conditions.
3. Which are the dependent variables?
4. Conduct the MANOVA test.
   a. Describe how you have accounted for the assumptions of repeated-measures MANOVA.
   b. Describe what the SPSS output shows for the multivariate and univariate effects.
   c. Run *post hoc* analyses (if needed).
   d. Find the source of interaction (if there are any).
5. Describe the effect size and conduct power calculations, using G*Power.
6. Report the outcome as you would in the results section of a report.

Appendix to Chapter 14

Manual calculations for MANOVA

Table 14.5 presents some fictitious data that examine depression and anxiety among a group of 30 animals (10 dogs, 10 cats and 10 hamsters) which we will examine in respect of depression and anxiety. Which of our domestic friends are more likely to be depressed? Which are more likely to be anxious? If there is a pattern between the animals across anxiety, and depression, is that relationship independent of covariance between anxiety and depression? Please note that no animals were harmed during the making of this example. You will find an Excel spreadsheet associated with these calculations on the web page for this book. We saw these data tabulated in Table 14.1, but we should repeat this here (with added information on grand means and variance) so that we have data to refer to while undertaking our calculations.

Table 14.5 Measured levels of anxiety and depression in domestic animals

	Anxious			Depressed		
	Dogs	Cats	Hamsters	Dogs	Cats	Hamsters
	36	80	50	73	48	67
	48	93	28	87	48	50
	61	53	44	80	87	67
	42	53	44	62	42	50
	55	87	48	87	42	56
	42	60	67	67	42	56
	48	60	67	40	36	50
	48	98	50	90	61	49
	53	67	44	60	61	60
	48	93	80	93	42	48
Mean	**48.10**	**74.40**	**52.20**	**73.90**	**50.90**	**55.30**
Grand mean	*Anxious*	*58.23*		*Depressed*	*60.03*	
Variance	50.99	313.82	225.07	281.88	229.21	52.68
Grand variance	*Anxious*	*321.15*		*Depressed*	*277.76*	

Calculating the sum of squares and mean squares is the same as we have seen for other ANOVA models, which we will undertake for each of the dependent variables. The main difference this time is that we also need to perform analyses for variance between the dependent variables, which we explore with 'cross-products'. They are relatively simple to calculate, but the subsequent analysis of matrices is devilishly complex. We will take each dependent variable in turn, finding the sum of squares (total, model and residual), the mean squares of each, and the F ratio. This will be what we would have found had we undertaken two separate one-way ANOVA tests. We will then undertake the cross-products analysis to examine the multivariate (MANOVA) effect.

Anxious DV

Total sum of squares ($SS_{T\,ANX}$)

Formula for $SS_{T\,ANX} = \sum S^2_{grand}(n_k - 1)$ S^2_{grand} = grand variance (for anxious DV); $n = 10$

So, $SS_{T\,ANX} = 321.15 \times 9 = \mathbf{2890.36}$

Model sum of squares ($SS_{M\,ANX}$)

Formula for $SS_{MANX} = \sum n_k(\bar{x}_k - \bar{x}_{grand})^2$ $\bar{x}_{grand}$ = grand mean (for anxious); $n = 10$

(We take the grand mean from each group mean)

So, $SS_{MANX} = 10 \times (48.10 - 58.23)^2 + 10 \times (74.40 - 58.23)^2 + 10 \times$

$(52.20 - 58.23)^2 = \mathbf{4004.47}$

Degrees of freedom (*df*) = 3 IV groups *minus* 1so $df_{MANX} = 2$

Model mean square (MS_{MANX})

$MS_{MANX} = SS_{MANX} \div df_{MANX} = 4004.47 \div 2 = \mathbf{2002.23}$

Residual sum of squares (SS_{RANX})

Formula for $SS_{RANX} = \sum s^2_k(n_k - 1)$ S_k^2 = variance for each group (within anxious DV)

So, $SS_{RANX} = (50.99 \times 9) + (313.82 \times 9) + (225.07 \times 9) = \mathbf{5308.90}$

df = (30 animals *minus* 1) *minus* df_{MANX} so $df_{RANX} = 30 - 1 - 2 = 27$

Residual mean square (MS_{RANX})

$MS_{RANX} = SS_{RANX} \div df_{RANX} = 5308.90 \div 27 = \mathbf{196.63}$

F ratio $= MS_{MANX} \div MS_{RANX} = 2002.23 \div 196.63 = \mathbf{10.183}$

Depressed DV

Total sum of squares (SS_{TDEP})

Using the formula we saw earlier:

$SS_{T\,DEP} = 277.76 \times 9 = \mathbf{2499.82}$

Model sum of squares (SS_{MDEP})

Using the formula we saw earlier:

$SS_{MDEP} = 10 \times (73.90 - 60.03)^2 + 10 \times (50.90 - 60.03)^2 + 10 \times$

$(55.30 - 60.03)^2 = \mathbf{2981.07}$

$df_{MDEP} = 2$ (as it was for the Anxious DV)

Model mean square (MS_{MDEP})

$$MS_{MDEP} = SS_{MDEP} \div df_{MDEP} = 2981.07 \div 2 = \mathbf{1490.53}$$

Residual sum of squares (SS_{RDEP})

Using the formula we saw earlier:

$$SS_{RDEP} = (281.88 \times 9) + (229.21 \times 9) + (52.68 \times 9) = \mathbf{5073.90}$$

$$df_{RDEP} = 27 \qquad \text{(as it was for the Anxious DV)}$$

Residual mean square

$$MS_{RDEP} = SS_{RDEP} \div df_{RDEP} = 5073.90 \div 27 = \mathbf{187.92}$$

F ratio $= MS_{MDEP} \div MS_{RDEP} = 1490.53 \div 187.92 = \mathbf{7.932}$

<u>Cross-Products</u> (relationship <u>between</u> dependent variables)

Total cross-products (CP_T)

Formula for $CP_T = \sum((x_{ANX} - \bar{x}_{grand\ ANX}) \times (x_{DEP} - \bar{x}_{grand\ DEP}))$

(We take the grand mean from each case score, within each group, within each DV)

So, $CP_T = ((36 - 58.23) \times (73 - 60.03)) + ((80 - 58.23) \times (48 - 60.03)) +$

$((50 - 58.23) \times (67 - 60.03)) + \ldots((48 - 58.23) \times (93 - 60.03)) + ((93 - 58.23) \times$

$(42 - 60.03)) + ((80 - 58.23) \times (48 - 60.03)) = \mathbf{-3043.23}$

Model cross-products (CP_M)

Formula for $CP_M = \sum(n(x_{group\ ANX} - \bar{x}_{grand\ ANX}) \times (x_{group\ DEP} - \bar{x}_{grand\ DEP}))$; $n = \mathbf{10}$

(We take the grand mean from group mean, within each DV)

So, $CP_M = (10 \times ((48.10 - 58.23) \times (73.90 - 60.03))) + (10 \times ((74.40 - 58.23) \times$

$(50.90 - 60.03))) + (10 \times ((52.20 - 58.23) \times (55.30 - 60.03))) = \mathbf{-2596.13}$

Residual cross-products (CP_R)

$= CP_T - CP_M \qquad$ So, $CP_R = -3043.23 - (-2596.13) = \mathbf{-447.10}$

This is where it gets nasty. A matrix is a method of displaying the figures in a pattern of rows and columns. We need to produce two matrices: one for the model term and one for the error.

Model matrix (H)

$$H = \begin{pmatrix} SS_{MANX} & CP_M \\ CP_M & SS_{MDEP} \end{pmatrix}$$

So, substituting in what we calculated above:

$$H = \begin{pmatrix} 4004.47 & -2596.13 \\ -2596.13 & 2981.07 \end{pmatrix}$$

Error matrix (E)

$$E = \begin{pmatrix} SS_{RANX} & CP_R \\ CP_R & SS_{RDEP} \end{pmatrix}$$

So, substituting in what we calculated above:

$$E = \begin{pmatrix} 5308.90 & -447.10 \\ -447.10 & 5073.90 \end{pmatrix}$$

Effectively, what we have here with these two matrices is the model/residual mean squares. In normal circumstances, we would divide the model mean square by the residual mean square to get the F ratio, for the relationship between the dependent variables. Unfortunately you cannot divide one matrix by another – you have to multiply one by the inverse of the other.

To find the inverse of the error matrix (E), we first need to find two parameters: the 'minors' matrix of E and something called a determinant.

Minors matrix E (ME)

$$ME = \begin{pmatrix} SS_{RDEP} & -CP_R \\ -CP_R & SS_{RANX} \end{pmatrix}$$
$$= \begin{pmatrix} 5073.90 & 447.10 \\ 447.10 & 5308.90 \end{pmatrix}$$

Determinant E (DE)

$$DE = (SS_{RANX} \times SS_{RDEP}) - (CP_R \times CP_R)$$

$$= (5308.90 \times 5073.90) - (-447.10 \times -447.10) = 26736929.30$$

Inverse matrix E (E^{-1})

We divide the cells in ME by the determinant:

So: $5308.90 \div 26736929.30 = 0.000190$

$447.10 \div 26736929.30 = 0.000017$

$5073.90 \div 26736929.30 = 0.000199$

We put that into a matrix:

$$E^{-1} = \begin{pmatrix} 0.000190 & 0.000017 \\ 0.000017 & 0.000199 \end{pmatrix}$$

'Raw' F ratio

Now we can multiply H by E^{-1} which is the equivalent of dividing H by E. This is some way from our final answer, but is an integral part of it:

So $H \times E^{-1}$ (HE^{-1}) =

$$\begin{pmatrix} 4004.47 & -2596.13 \\ -2506.13 & 2981.07 \end{pmatrix} \times \begin{pmatrix} 0.000190 & 0.000017 \\ 0.000017 & 0.000199 \end{pmatrix} = \begin{pmatrix} A & B \\ C & D \end{pmatrix}$$

$$A = (4004.47 \times 0.000190) + (-2596.13 \times 0.000017) = 0.7165$$

$$B = (4004.47 \times 0.000017) + (-2596.13 \times 0.000199) = -0.4485$$

$$C = (-2596.13 \times 0.000190) + (2981.07 \times 0.000017) = -0.4428$$

$$D = (-2596.13 \times 0.000017) + (2981.07 \times 0.000199) = 0.5485$$

$$\text{So } HE^{-1} = \begin{pmatrix} A & B \\ C & D \end{pmatrix} = \begin{pmatrix} 0.7165 & -0.4485 \\ -0.4428 & 0.5485 \end{pmatrix}$$

Eigenvalues (λ)

Now we need to find something called 'eigenvalues', which we subsequently plot into a quadratic equation. This will give us a range of eigenvalues, which we examine according to various optional equations (but more of that later).

The first stage of this part is to multiply through HE^{-1} by λ and 0:

$$\text{So } \begin{pmatrix} 0.7165 & -0.4485 \\ -0.4428 & 0.5485 \end{pmatrix} - \begin{pmatrix} \lambda & 0 \\ 0 & \lambda \end{pmatrix} = \begin{pmatrix} F & G \\ H & I \end{pmatrix}$$

$$F = 0.7165 - \lambda$$

$$G = -0.4485 - 0 = -0.4485$$

$$H = -0.4428 - 0 = -0.4428$$

$$I = 0.5485 - \lambda$$

Put back in a matrix: $\begin{pmatrix} 0.7165 - \lambda & -0.4485 \\ -0.4428 & 0.5485 - \lambda \end{pmatrix}$

Now we need to describe that in the form of a quadratic equation, by multiplying that through:

$$= (0.7165 - \lambda) \times (0.5485 - \lambda) + (0.7165 \times 0.5485) - (-0.4485 \times -0.4428)$$

$$= \lambda^2 - .7165\lambda - .5485\lambda + .3930 - .1986$$

$$\text{or } \lambda^2 - 1.2650\lambda + .1944$$

Now we need to find the values of λ so that we can find our F ratios. To do that we need to change the order of the equation, so we make λ the subject. We need to use this equation to help us:

$$\lambda = \frac{-b \pm \sqrt{b^2 - 4ac}}{2a} \qquad \text{Where } a = 1;\ b = -1.2650;\ c = 0.1944$$

So, eigenvalues (λ) = 1.086027 or 0.179002

As we saw when we ran this test through SPSS, the MANOVA outcome produces four choices of test that determine the F ratio: Pillai-Bartlett Trace, Hotelling's Trace, Wilks' Lambda and Roy's Largest Root. We explored the relative benefits of each outcome earlier. This is how we calculate each of those outcomes:

Pillai–Bartlett Trace (V) (shown as Pillai's Trace in SPSS)

This test uses both eigenvalues in the following equation:

$$V = \sum_{i=1}^{s} \frac{\lambda_i}{1 + \lambda_i} \qquad \text{where } \lambda \text{ is each eigenvalue}$$

$$\text{So V} = \frac{1.086}{1 + 1.086} + \frac{.179}{1 + .179} = .672$$

Wilks' Lambda (Λ)

Multiplies total-to-error ratio across both eigenvalues in the following equation:

$$\Lambda = \prod_{i=1}^{s} \frac{1}{1 + \lambda_i}$$ where λ is each eigenvalue and Π is 'the product of' (the multiple)

$$\text{So } \Lambda = \frac{1}{1 + 1.086} \times \frac{1}{1 + .179} = 0.407$$

Hotelling's Trace (T^2)

Simply adds the two eigenvalues:

$$\text{So } T^2 = 1.086 + .179 = 1.265$$

Roy's Largest Root

Simply takes the first eigenvalue (1.086).

Each of those eigenvalues can be converted into an **F ratio**. The method is a little different for each one. For example, Wilks' Lamba is performed as follows:

$$\textbf{F (Wilks)} = \frac{1 - \Lambda^{1/s}}{df_n} \div \frac{\Lambda^{1/s}}{df_d}$$ From above, we know that $\Lambda = 0.407$

$$\text{Where s} = \sqrt{\frac{p^2q^2 - 4}{p^2 + q^2 - 5}}$$ p = no. of levels on IV (3); q = no. of DVs (2)

$$s = \sqrt{\frac{3^2 2^2 - 4}{3^2 + 2^2 - 5}} = 2$$ So, $\Lambda^{1/s} = \Lambda^{1/2} = \surd\Lambda$

df_M = no. of groups (3) − 1 = 2

df_n =numerator df = (p − 1) × q = 4

df_d = denominator df = ((group scores − 1) × (p × q)) − df_M = ((10 − 1) × 3 × 2) −2 = 52

$$\text{F (Wilks)} = \frac{1 - \sqrt{0.407}}{4} \div \frac{\sqrt{0.407}}{52} = 7.387$$

You could compare these outcomes to those we found using SPSS earlier.

We will not attempt to explore manual calculations for repeated-measures MANOVA.

15 ANALYSES OF COVARIANCE

Learning objectives

By the end of this chapter you should be able to:

- Recognise when it is appropriate to use analyses of covariance and which test to use: Analysis of Covariance (ANCOVA) or Multivariate Analysis of Covariance (MANCOVA)
- Understand the theory, rationale, assumptions and restrictions associated with the tests
- Calculate the outcomes manually for ANCOVA (using maths and equations)
- Perform analyses using SPSS and explore outcomes regarding covariate effect
- Know how to measure effect size and power
- Understand how to present the data and report the findings

What are analyses of covariance?

Analyses of covariance (ANCOVA) explore outcomes *after* accounting for other variables that may be related to that outcome. In any analysis, we are aiming to explain as much variance as possible, while controlling for as many additional factors as possible. When we have explored other ANOVA tests, we ultimately find an F ratio that describes how much variance we have explained in relation to how much remains unexplained. The 'unexplained' variance may be due to random or chance factors; we cannot measure these. Or it may be down to factors that we 'know about' but do not want to measure. Additional variables (not part of the main analysis) that may have an influence on the outcome are called '**covariates**'. Analyses of covariance can help us identify covariates and assess their impact on our outcome. If those covariates are not related to the explained variance, we can use ANCOVA to reduce the error variance (to provide a clearer picture of the original analysis). If the covariates are even partially related to the explained variance it means that some of that variance is shared. Depending on the extent of overlapping variance, the covariates may be 'confounding' the original outcome. We explore all of this in more depth later.

We will be exploring two types of analyses of covariance in this chapter: ANCOVA and Multivariate Analyses of Variance (MANCOVA). We will begin with ANCOVA, where a single dependent variable outcome is assessed across one or more independent variable, controlling for one or more covariate. After that, we will explore MANCOVA, which examines two or more dependent variable outcomes across one or more independent variables, controlling for one or more covariates. Both examples used here focus on between-group analyses. There are also analyses of covariance that can explore repeated-measures and mixed models. However, such endeavours are a little advanced for this book.

What is ANCOVA?

As we have just seen, ANCOVA is used to explore the impact of covariates on a single outcome. For example, when we explored independent one-way ANOVA (Chapter 9) we used an example that focused on how the number of lecture hours attended may differ between courses (law, psychology and media). Ultimately we found that there was a difference: $F(2, 27) = 12.687$, $p < .001$). The data set that we used showed only information on course type (the independent variable) and the number of hours spent in lectures (the dependent variable). In reality, our data sets will have many variables, some of which may covary with our measured outcome. In our example, we might also have measured age and the number of hours spent in fieldwork. It is possible that age is a factor in lecture attendance, regardless of the course attended – we might expect that older people are more likely to go to lectures. Also, some courses may be more field-related so hours spent doing that might increase at the expense of lecture attendance. ANCOVA seeks to examine how covariates influence the outcome. Ideally, ANCOVA will reduce the error variance and give us a clearer picture of the outcome, but it can also be used to explore the effect of a potential confounding variable. How we use ANCOVA depends on the extent to which the covariate is related to the **experimental effect** (the main effect that we are measuring between the independent variable and the dependent variable outcome). If it is related, the covariate will differ significantly across the independent variable.

For example, we might find that there is a relationship between age and the number of hours spent in lectures, but that the groups do not differ in respect of age. We can use that covariate in an effort to reduce the error variance (because the covariate is not related to the experimental effect). Meanwhile, we might observe that there is a correlation between the time spent in fieldwork and lectures, but also find that the groups differ significantly on the number of hours spent in fieldwork. We cannot use this covariate to reduce error variance, as it is not independent of the experimental effect. However, we could explore the extent that this covariate is 'interfering' with our original outcome. In this context, a number of outcomes may occur: a previously significant effect may no longer be significant (the covariate was entirely confounding the original outcome); a previously significant outcome is still significant, but the effect is reduced (the covariate was partially confounding the outcome); or a previously non-significant outcome is now significant (the original outcome was being masked by the covariate).

In all of these cases, ANCOVA provides a useful way to explore the problem of potentially confounding variables. It is often said that we should aim to control for all variables at the time of recruitment by matching participants for everything that we do not measure. In reality, that is very well, but recruitment is one of the hardest parts of research. Putting yet further restrictions on inclusion criteria only makes that harder still. It is also argued that we should randomise participants to experimental groups. This might be the best strategy, but there are cases where **randomisation** is neither practical nor ethical. Furthermore, randomising is no guarantee that we will have balanced groups on a range of variables. We can use ANCOVA to control for those variables.

15.1 Nuts and bolts
What is a covariate?

A covariate is any (additional) variable that is related to the outcome measure being examined (or has the potential to be related).

Research question for ANCOVA

Throughout this section, we will use a single research example to illustrate how we can use ANCOVA. The Healthy Lives Partnership (HeLP) is a group of psychologists focusing on sleep and fatigue and how these factors may affect mood. They decide to measure the extent that fatigue is a feature in depression. They collect data from 54 clients (27 who are currently depressed, 27 who are not depressed). They would have liked to match participants on other factors, such as current anxiety and the amount of sleep that they experience, but this was not feasible with such a small client list. Instead they explore these factors as covariates. Fatigue is measured from clinical observations leading to ratings for each client. Ratings are undertaken by one psychologist and range from 0–100, with higher scores representing greater fatigue. Anxiety is measured from an established rating scale – once again scores range from 0–100, with higher scores indicating greater anxiety. Total sleep time is measured in minutes and is based on sleep diaries. HeLP expect fatigue to be worse in depressed people. They also predict that total sleep time will have an impact on fatigue, but they are uncertain how much anxiety will affect outcomes.

15.2 Take a closer look
Summary of research example

ANCOVA variables
Dependent variable (DV): Fatigue
Independent variable (IV): Depression (two groups: yes or no)
Covariate (CV) 1: Anxiety scores
CV 2: Total sleep time

Theory and rationale

How ANCOVA can be used

ANCOVA can be employed in a wide range of contexts: we can 'filter out' error variance; we can explore pre-test vs. post-test effects; and we can control for variables statistically when we have not been able to do so physically.

Filtering out interference

As we saw earlier, when we find a significant difference, there is some error variance remaining (as shown in Figure 15.1). To reduce that variance still further, we can explore the effect of potential covariates using ANCOVA. If we have measured additional variables that were not part of the original experimental effect, we can investigate whether they contribute to the variance in our outcome measure. However, to reduce error variance, a covariate must be related to the outcome (the correlation between them must be *significant*) and must not be dependent on the independent variable (the covariate must not differ significantly in respect of the groups that represent the independent variable). If we satisfy those conditions, ANCOVA may provide us with a 'cleaner' picture of the experimental effect. Some sources refer to this action as removing 'noise' from the experiment (Tabachnick and Fidell, 2007); others call significant covariates 'nuisance variables' (Brace *et al.*, 2006).

In our research example, we are exploring whether fatigue is greater for depressed individuals. If that is indeed observed, we can explore additional variables to see whether the observed relationship can be strengthened through the application of covariates. In this case we are also examining anxiety and total sleep time. We might find that anxiety is related to fatigue, but does not differ according to the depression groups (at least in this sample). If that is the case, we could apply anxiety as a covariate and might expect to find that the relationship between fatigue and depression is stronger. We might also find that total sleep time is related to fatigue, but does not differ across the groups. If that is the case, we could apply total sleep as a covariate, too. However, if either of these variables is not related to fatigue they *cannot* be applied as a covariate (in any analysis). If we find that the covariates are related to fatigue, we must then check whether they differ in respect of the depression groups. If neither of them shows between group differences, we can apply *both* in an ANCOVA that seeks to reduce error variance. If either is related to depression groups, we can apply only the covariate that is not related. Any covariate that is related to the experimental effect must be examined separately to explore the extent of mediating or moderating effect (see below). This is because the variance found in the covariate is also shared with the explained variance in the experimental effect.

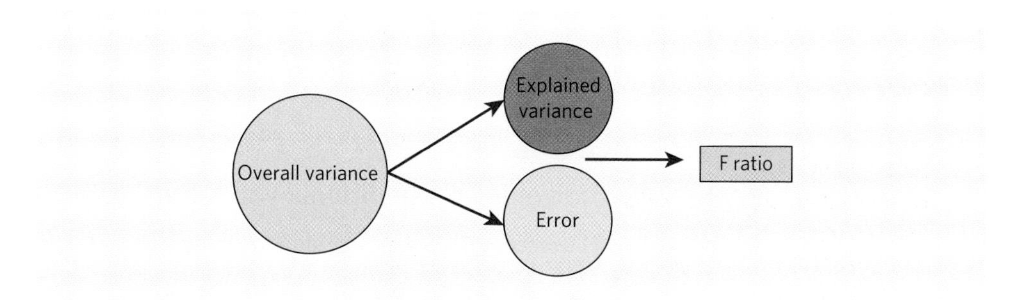

Figure 15.1 Main effect (prior to covariate)

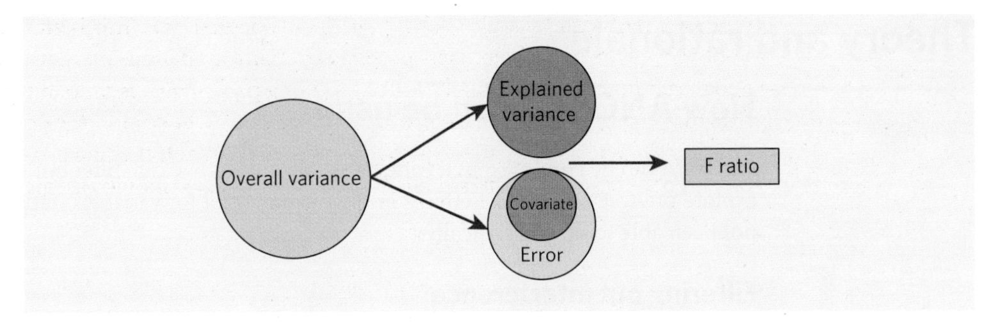

Figure 15.2 Independent covariate

Figure 15.2 shows that the covariate shares no variance with the explained variance. The action of the covariate will reduce the error variance and will increase the F ratio, producing a stronger effect.

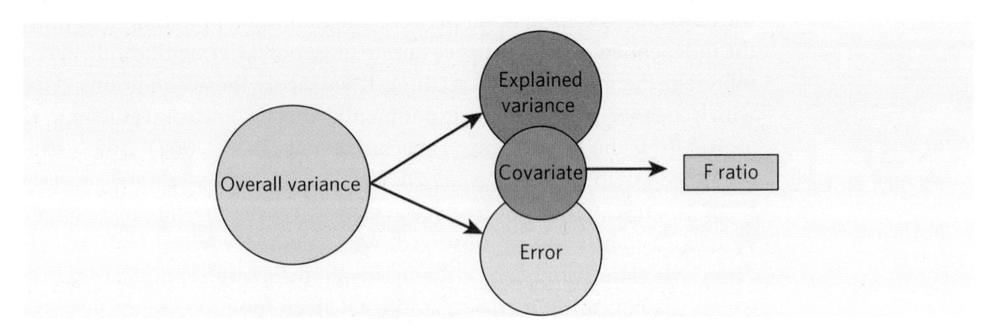

Figure 15.3 Dependent covariate

In Figure 15.3 the covariate shares some variance with the explained variance; there is a potential confounding effect on the outcome. The action of the covariate will reduce some error variance, but it will also reduce some of the explained variance. As a result, this will reduce the F ratio, producing a weaker effect.

Pre-test and post-test

Another common use of ANCOVA allows us to account for prior existing variables. The post-test data represent the main outcome measure, which is examined across independent variable groups. However, we may also need to account for the pre-test scores, in order to establish any pre-existing differences between groups. We can illustrate this with an extension to our research example. HeLP decide to investigate the effect of a sleep and exercise training programme to improve fatigue, and possibly reduce the risk for future depression. They randomly assign clients to two groups: one that receives the new training programme and one that receives basic literature on sleep and exercise. They predict that fatigue outcomes will show greater improvement in the training group, compared with the leaflet group. They may very well find that, but the result is a little meaningless if they do not account for fatigue prior to the study. For all they know, the people in the leaflet group always report poorer fatigue. To take care of that, HeLP should measure fatigue before (pre-test) and at the end of the study (post-test). Using ANCOVA, they could measure post-test fatigue scores across the groups, but control for pre-test fatigue scores.

Statistically controlling for additional variables

When we conduct between-group studies we should aim to control for as many additional variables as possible. If HeLP already knew that anxiety might interfere with their proposed investigation, they should probably try to ensure that people are matched for anxiety across the groups. For every depressed person they recruit from their client list, they should enlist a non-depressed person matched on anxiety measures. However, the researchers are also measuring total sleep time so, ideally, they should match participants on that, too. This is all very well, but HeLP do not have a very large client list – putting too many limitations on participant selection would make recruitment very difficult. Indeed, finding participants is probably the most challenging part of any research. In situations like this we can try to control for potentially confounding outcomes by introducing them as covariates in ANCOVA. If they are not related to the outcome measure, then we can exclude them. If they are related, we should explore what effect they have.

How ANCOVA is measured

The method for calculating ANCOVA uses similar methods to those we have seen in previous chapters. Outcomes are computed for the main effect, to assess how much variance in the dependent variable can be explained by differences across the independent variable groups. A similar calculation is made in respect of each covariate across the groups. In both cases, outcomes are examined using the methods that we saw for independent one-way ANOVA (see Chapter 9). In each analysis, the overall variance (total sum of squares) is partitioned into explained variance (model sum of squares) and error variance (residual sum of squares). Those sums of squares are then assessed with respect to the relative degrees of freedom. These represent the number of values that are 'free to vary' in the calculation, while everything else is held constant (see Chapter 6). From that outcome, mean squares are used to calculate the F ratio. We will have an F ratio for the main effect and each of the covariates. Statistical significance of main effect and covariates is assessed based on cut-off points in the F distribution (as we saw in Chapter 9).

To examine the effect of the covariate on the main outcome, we need to employ multivariate analyses similar to those that we saw with MANOVA (see Chapter 14). However, in addition to cross-products (variance across the combination of dependent variable and covariates), calculations are made in respect of cross-product partitions (equivalent to the mean squares of those cross-products). By assessing these across the model (explained) partitions and residual (error) partitions, we ultimately find another F ratio (which can be assessed for significance like any other F ratio). Exactly how that is calculated is rather complicated, so we will not discuss that here. If you would like to see how this is all done manually, there are some calculations at the end of this chapter. We will see how to perform the analyses in SPSS shortly.

Estimated marginal means

To illustrate the effect of a covariate we can refer to something called 'estimated marginal means' (or 'adjusted means'). We should use the research example that we referred to earlier in this chapter. Our researchers (HeLP) are examining how fatigue is reported across depressed and non-depressed groups. Using an independent one-way ANOVA, we might find that depressed people report significantly greater fatigue than those not depressed (we will use this in preference to an independent t-test so that we can compare outcomes before and after the application of a covariate). That outcome would have been calculated by comparing the mean scores (in relation to variance – as we saw in Chapter 9).

HeLP then decide to examine whether anxiety scores are interfering with that outcome. If we apply that as a covariate we may find that there is no longer a difference in fatigue scores across the depression groups; anxiety was confounding the outcome that we thought existed. The process of ANCOVA adjusts the mean dependent variable scores (for the outcome), across the independent variable, weighting those scores by the effect of the ANCOVA. Once that is done we get 'new' (adjusted) mean scores – shown by the estimated marginal means. When those adjusted mean scores are compared there is no longer a significant difference between them. This is illustrated still further in Box 15.3. We will see how to request estimated marginal means through SPSS a little later.

15.3 Take a closer look
Estimated marginal means

We can illustrate the effect of ANCOVA with estimated marginal means. We will examine mean fatigue scores across depression groups, and assess whether those scores are significantly different, before and after applying the covariate.

Table 15.1 Estimated marginal means/standard error (SE) before ANCOVA

	Fatigue scores		
Group	**N**	**Mean**	***SE***
Depressed	15	59.93	*3.89*
Not depressed	15	44.07	*3.89*

An independent one-way ANOVA shows that depressed people report significantly poorer fatigue than those not depressed, F (1, 28) = 8.305, p = .008.

Table 15.2 Estimated marginal means/standard error (SE) after ANCOVA

	Fatigue scores		
Group	**N**	**Mean**	***SE***
Depressed	15	56.21	*3.19*
Not depressed	15	47.79	*3.19*

After applying anxiety as a covariate, an independent one-way ANOVA shows that there was *no* significant difference in fatigue scores across the depression groups, F (1, 27) = 3.248, p = .083. Notice also how the gap between the groups in respect of estimated marginal mean scores in Table 15.2 is much closer than in Table 15.1. The effect of the ANCOVA has been to 'decrease' fatigue scores for the depressed group, and 'increase' them for the non-depressed group.

Assumptions and restrictions

There are several very important assumptions that we must address before employing ANCOVA (and before deciding how we can interpret the outcome). We have discussed two of these in some depth already. One of the key assumptions is that there must be reasonable correlation between the covariate and the dependent variable. Without that correlation ANCOVA cannot be conducted. However, we do not want the correlation to be too high either – something between 0.30 and 0.90 is ideal (correlation higher than that suggests that the covariate and dependent variable are measuring the same thing). We also need to check whether the covariate is dependent on the independent variable. In other words, do covariate outcomes vary across the groups? If this is the case, it does not violate ANCOVA, but it does mean that we will interpret outcomes differently. As we saw earlier, covariates that are not dependent on the independent variable can be used to reduce error variance in the main outcome. Any covariate that does vary across the groups must be examined separately to assess the extent that the variance in that covariate is shared with the explained variance in the experimental effect – there may be some confounding effect.

The covariate must be measured before any intervention. That makes sense in any case because the covariate is often an 'underlying' feature. For example, we could examine the effect of showing pictures of spiders to two groups: those who have a spider phobia and those who do not (the independent variable). We could measure stress in these participants immediately after viewing the pictures, perhaps with a heart rate monitor (the dependent variable). We might expect heart rate to be faster in the phobic group than on controls at that point. However, some people might have faster heart rates than others (irrespective of phobia)? To overcome that potential confounding variable, we could also measure heart rate as a baseline (pre-existing) condition; this could be the

15.4 Take a closer look
Homogeneity of regression slopes

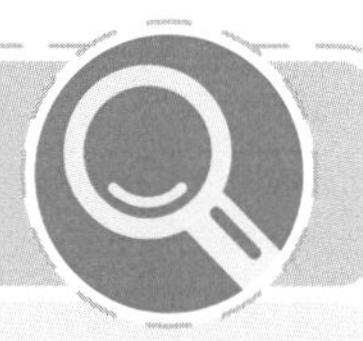

We can use some correlation examples to illustrate the importance of measuring homogeneity of regression slopes. In this scenario, we will explore the relationship between the dependent variable (fatigue) and the covariate (total sleep time) across depression groups (yes or no, the independent variable).

Table 15.3 Pearson correlation (*r*) between fatigue and total sleep time, across depression groups

	r	p
All	−.557	<.001
Depressed	−.490	.009
Non-depressed	−.555	.003

Table 15.3 shows that the correlation between fatigue and total sleep time *appears* to differ only slightly between the depressed and non-depressed groups. On the evidence of this, perhaps we might be confident that we have homogeneity of regression slopes. However, we can also explore that relationship using scatterplot graphs with 'regression' lines added.

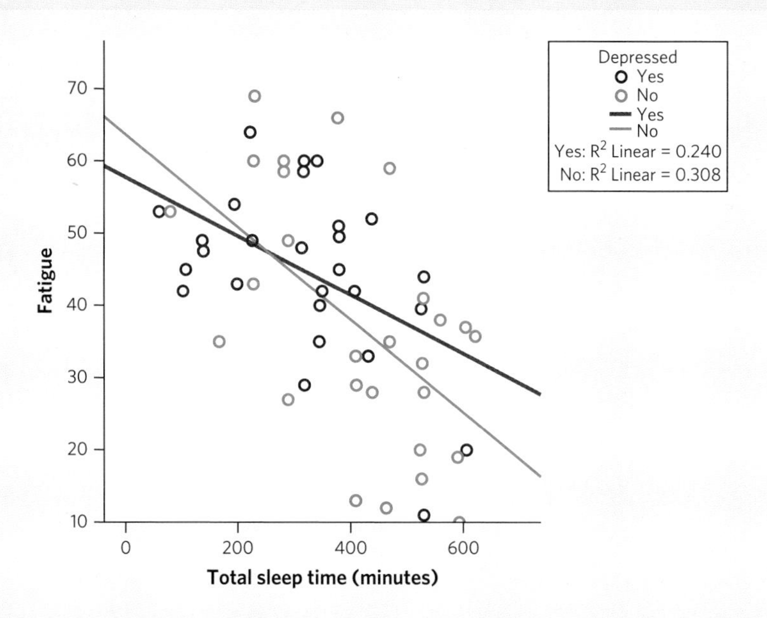

Figure 15.4 Scatterplot and regression lines

Figure 15.4 represents the regression slopes between fatigue and total sleep time – there is one each for the depressed group (blue line) and non-depressed group (green). To illustrate (total) homogeneity of regression slopes, those lines need to be parallel, but they clearly are not. As we saw in Chapter 11 (see Boxes 11.3 and 11.4), the 'crossover' of these lines *suggests* an interaction. We said that such an outcome may indicate that we have violated the requirement for homogeneity of regression slopes. The inconsistency in these two outcomes reinforces why we need something a little less subjective to formally measure that. We will see how to do that later.

covariate. But there would be little point measuring the covariate *after* the picture intervention. Where there is more than one covariate, they should not be highly correlated with each other. If the covariates are highly correlated we would be over-adjusting the ANCOVA effect.

Sample sizes are also important, not least because smaller samples will simply not generate enough power. Outcomes with a statistical power below 0.80 should be treated with caution (see Chapter 4 for more information on statistical power). There can also be problems if there are noticeably unequal samples, partly because the effect of controlling for the covariate is complicated by that inconsistency. There are ways to overcome this (see Tabachnick and Fidell, 2007, pp. 237–238), but it may not always be acceptable to do so. The covariates must be interval and be normally distributed across the independent variable groups. Normal distribution on the remaining dependent variables is less crucial, particularly in larger samples (Tabachnick and Fidell, 2007).

Homogeneity of regression slopes

One of the key assumptions of ANCOVA is the requirement for **homogeneity of regression slopes**. This assumes that the correlation between the covariate and dependent variable does not differ significantly across the independent variable groups. Earlier, we saw that the effect of running ANCOVA is illustrated by the change in the estimated marginal means. This outcome is determined by the correlation between the covariate and the dependent variable. If that correlation is skewed between the independent variable groups, the effect will be disproportionate. For that reason we must be able to demonstrate that we have homogeneity of regression slopes, otherwise we cannot trust the validity of the ANCOVA outcome. We will see how to measure that statistically later, but it would be useful to demonstrate that with an example, in Box 15.4.

15.5 Take a closer look

Summary of assumptions and restrictions

- There should be reasonable correlation between the covariate and dependent variable
- We must examine whether the covariate is dependent on the independent variable
- Covariates must be measured prior to interventions (independent variable)
- Covariates must be normally distributed
- The dependent variable *should* be normally distributed (across the groups)
- Sample sizes should be sufficient to ensure there is enough power to detect the hypothesis
  - Unequal sizes should be avoided
- There should be homogeneity of regression slopes between covariate and dependent variable
  - Across all groups of the independent variable

How SPSS performs ANCOVA

The manual calculations for ANCOVA are quite complex, so we will review them at the end of the chapter. In the meantime, we will explore how to perform the test in SPSS. For this example we should refer to the research question that HeLP set earlier. Data were collected from 54 people (27 depressed, 27 not depressed). Fatigue was measured from clinical observations from one psychologist, who rated people on a score of 0–100 (with higher scores representing greater fatigue). At the same time, the researchers collected data on current anxiety and total sleep time, which they believed might be covariates. Anxiety was measured on a rating scale of 0–100 (higher scores illustrating higher anxiety). Total sleep time was measured in minutes (average per night), as derived from a sleep diary completed over one week.

15.6 Nuts and bolts
Setting up the data set in SPSS

When we create the SPSS data set for ANCOVA, we need to set up one column for the dependent variable (which will have a continuous score), one for each covariate (also continuous) and one for each independent variable (which must be set up as categorical, with labels and values for groups).

Fatigue and depression.sav [DataSet1] - IBM SPSS Statistics Data Editor

File Edit View Data Transform Analyze Graphs Utilities Add-ons Window Help

	Name	Type	Width	Decimals	Label	Values	Missing	Columns	Align	Measure
1	Depressed	Numeric	8	0	Depressed	{1, Yes}...	-1	8	Right	Nominal
2	Fatigue	Numeric	8	0	Fatigue	None	-1	8	Right	Scale
3	Anxiety	Numeric	8	0	Anxiety score	None	-1	8	Right	Scale
4	Sleep	Numeric	8	0	Total sleep tim...	None	-1	8	Right	Scale

Figure 15.5 Variable View for 'Fatigue and depression' data

Figure 15.5 shows how the SPSS Variable View should be set up. The first variable is called 'depressed'; this is the independent variable. In the Values column, we include '1 = Depressed' and '2 = Not depressed'; the Measure column is set to Nominal. The second variable is 'Fatigue'; this is the dependent variable representing observed fatigue scores. The third variable is 'Anxiety'; this is the first covariate measuring ratings of anxiety. The fourth variable is 'Sleep'; this is the second covariate measuring average total sleep time across the week. For the dependent variable and covariates, we do not adjust anything in the Values column; the Measure column is set to Scale.

*Fatigue and depression.sav [DataSet1] - IBM SPSS Statistics Data Editor

File Edit View Data Transform Analyze Graphs Utilities Add-ons Window Help

	Depressed	Fatigue	Anxiety	Sleep	var	var	var
1	No	10	34	593			
2	Yes	11	15	530			
3	No	12	48	463			

Figure 15.6 Data View for 'Fatigue and depression' data

Figure 15.6 illustrates how this will appear in the Data View. Each row represents one of HeLP's clients. When we enter the data for 'Depression', we input 1 (to represent 'Depression') or 2 (to represent 'Not depressed'). Those columns will display the descriptive categories or the value numbers, depending on how you choose to view the column (you can change that using the Alpha Numeric button in the menu bar - see Chapter 2). Fatigue, anxiety and sleep outcomes are entered as whole numbers.

Prior tests for assumptions and restrictions

Before we run the ANCOVA analysis, there are several initial tests we need to conduct to ensure that we have not violated any assumptions.

Correlation between covariate and dependent variable

To begin with, we must test that there is reasonable correlation between the covariates and the dependent variable. If there is not, the covariate cannot be included in the analysis. We are looking for correlation between r = .30 and r = .90. We saw how to perform correlation in Chapter 6.

Open SPSS file Fatigue and depression

Select **Analyze** → **Correlate** → **Bivariate...** → transfer Fatigue, Anxiety and Sleep to **Variables** (by clicking on arrow, or by dragging variables to that window) → tick boxes for **Pearson** and **Two-tailed** → click **OK**

		Fatigue	Anxiety score	Total sleep time (minutes)
Fatigue	Pearson correlation	1	.344*	−.557**
	Sig. (2-tailed)		.011	.000
	N	54	54	54
Anxiety score	Pearson correlation	.344*	1	−.338*
	Sig. (2-tailed)	.011		.012
	N	54	54	54
Total sleep time (minutes)	Pearson correlation	−.557**	−.338*	1
	Sig. (2-tailed)	.000	.012	
	N	54	54	54

Figure 15.7 Correlation between covariates and dependent variable

Figure 15.7 shows that the correlation between the covariates ('Anxiety score' and 'Total sleep time') and the dependent variable (Fatigue) is reasonable – we have satisfied that requirement for both covariates.

Independence of covariate

Now we need to examine whether the covariates differ across the independent variable groups. As there are only two groups, we can use an independent t-test to examine this. We saw how to do that in Chapter 7.

Using the SPSS file Fatigue and depression

Select **Analyze** → **Compare Means** → **Independent-Samples T Test...** → transfer Anxiety and Total sleep time to **Test Variable(s)** → transfer Depressed to **Grouping Variable** → click **Define Groups** → enter 1 in **Group 1** box (for 'depressed') → enter 2 in **Group 2** (for 'not depressed') → click **Continue** → click **OK**

		Levene's test for equality of variances		t-test for equality of means						
									95% confidence interval of the difference	
		F	Sig.	t	df	Sig. (2-tailed)	Mean difference	Std. error difference	Lower	Upper
Anxiety score	Equal variances assumed	.215	.645	−.178	52	.860	−.889	4.999	−10.921	9.143
	Equal variances not assumed			−.178	51.635	.860	−.889	4.999	−10.922	9.145
Total sleep time (minutes)	Equal variances assumed	.293	.591	−2.305	52	.025	−91.667	39.767	−171.466	−11.868
	Equal variances not assumed			−2.305	51.924	.025	−91.667	39.767	−171.469	−11.865

Figure 15.8 Independent t-test outcome (covariate by independent variable group)

Figure 15.8 indicates that there is no significant difference in anxiety scores between depressed and non-depressed clients, t (52) = −0.178, p = .860. We can use that covariate to reduce error variance in the experimental outcome. However, there is a significant difference in total sleep time between the groups, t (52) = −2.305, p = .025, so we cannot use that covariate to reduce error variance. We should examine the extent to which total sleep time confounds the main outcome.

Testing for normal distribution

We need to examine normal distribution (particularly across the covariate) using the Kolmogorov-Smirnov/Shapiro-Wilk analyses. As there are fewer than 50 people in each group, we should refer to the Shapiro-Wilk outcome. Figure 15.9 shows that normal distribution appears to be fine for the dependent variable and covariates across the depression groups (significance is greater than .05).

		Kolmogorov–Smirnov[a]			Shapiro–Wilk		
	Depressed	Statistic	df	Sig.	Statistic	df	Sig.
Fatigue	Yes	.160	27	.075	.935	27	.089
	No	.119	27	.200*	.953	27	.254
Anxiety score	Yes	.100	27	.200*	.968	27	.542
	No	.106	27	.200*	.944	27	.156
Total sleep time (minutes)	Yes	.150	27	.124	.967	27	.513
	No	.143	27	.167	.938	27	.110

Figure 15.9 Kolmogorov-Smirnov/Shapiro-Wilk test for dependent variable and covariate (by group)

Testing for homogeneity of regression slopes

Earlier, we said that the correlation between the covariate and the dependent variable should not differ significantly across the independent variable groups. When we explored this, we did so with the help of some graphs, and we physically compared the correlation outcomes. However, we need to examine that objectively through statistical analyses. We need to assess whether the correlation (or more specifically the regression slopes) differs significantly between the groups (we will learn more about regression slopes in Chapter 16). To illustrate that, we build a 'custom model' in SPSS. Using that model, we do not want a significant interaction between the covariate and the dependent variable. In our example we are applying the covariates in separate analyses, so we will need to assess homogeneity of regression slopes for each of them.

Anxiety covariate

Using SPSS file Fatigue and depression

Select **Analyze → General Linear Model → Univariate. . .** as shown in Figure 15.10

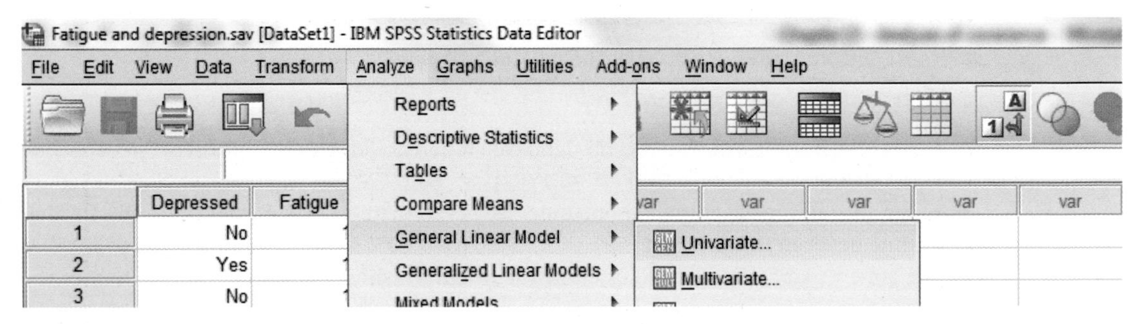

Figure 15.10 Homogeneity of regression slopes: procedure 1

In new window (see Figure 15.11), transfer **Fatigue** to **Dependent Variables** → transfer **Depressed** to **Fixed Factor(s)** → transfer **Anxiety** to **Covariate(s)** → click **Model**

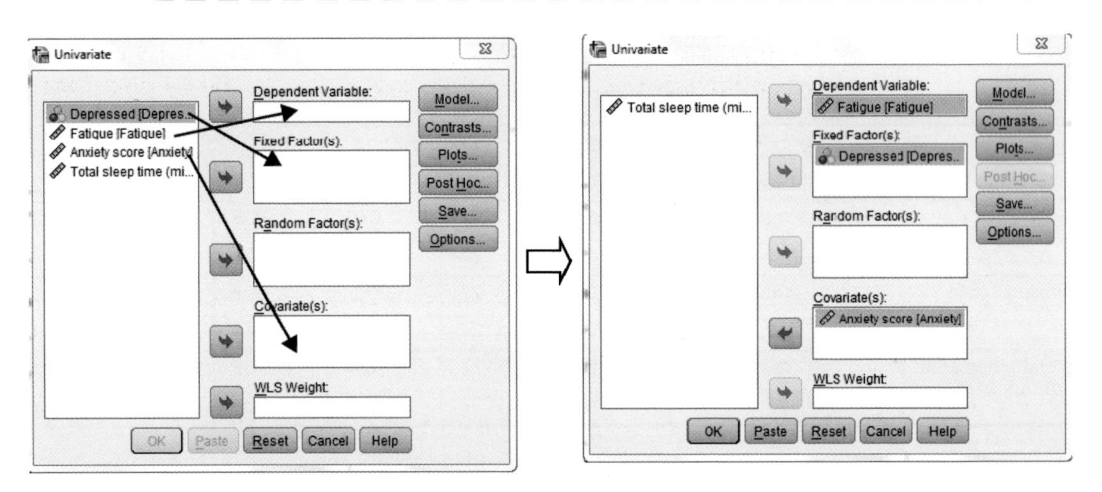

Figure 15.11 Setting parameters for homogeneity of regression slopes

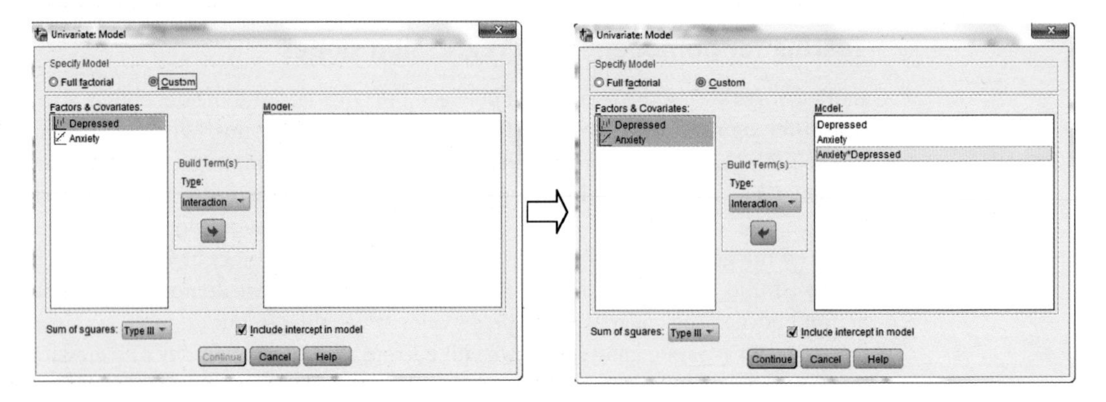

Figure 15.12 Testing homogeneity of regression: setting the model

In new window (see Figure 15.12) select **Custom** radio button → transfer **Depressed** and **Anxiety** (separately) to **Build Term(s)** → click on **Depressed** and **Anxiety** (so that both become highlighted) → transfer **Build Term(s)** → make sure that **Interaction** is selected in pull-down menu below **Build Terms** → click **Continue** → click **OK**

Source	Type III sum of squares	df	Mean square	F	Sig.
Corrected model	2890.817[a]	3	963.606	5.219	.003
Intercept	2262.838	1	2262.838	12.255	.001
Depressed	1045.035	1	1045.035	5.660	.021
Anxiety	1318.965	1	1318.965	7.143	.010
Depressed * Anxiety	652.443	1	652.443	**3.534**	**.066**
Error	9232.165	50	184.643		
Total	102979.000	54			
Corrected total	12122.981	53			

Figure 15.13 Homogeneity of regression slopes: Fatigue vs. Anxiety, by Depressed group

There is no significant interaction between 'Depressed' and 'Anxiety', $F(1, 50) = 3.534$, $p = .066$, which is precisely what we wanted (see Figure 15.13). If there had been a significant interaction it would mean that the requirement for homogeneity of regression slopes would have been violated. That would be a major problem for ANCOVA, as it would mean that the correlation between the covariate and dependent variable differs between the groups (see 'Assumptions and restrictions').

Total sleep time covariate

Using SPSS file Fatigue and depression

Select **Analyze** → **General Linear Model** → **Univariate. . .** → (in new window) transfer **Fatigue** to **Dependent Variables** → transfer **Depressed** to **Fixed Factor(s)** → transfer **Total sleep time** to **Covariate(s)** → choose *Custom Model* (as above) → transfer **Depressed** and **Sleep** (separately) to **Build Term(s)** → transfer combined **Depressed** and **Sleep** to **Build Term(s)** (you saw how just now) → click **Continue** → click **OK**

Source	Type III sum of squares	df	Mean square	F	Sig.
Corrected model	4000.503[a]	3	1333.501	8.209	.000
Intercept	26331.984	1	26331.984	162.093	.000
Depressed	63.012	1	63.012	.388	.536
Sleep	3042.881	1	3042.881	18.731	.000
Depressed * Sleep	152.466	1	152.466	**.939**	**.337**
Error	8122.478	50	162.450		
Total	102979.000	54			
Corrected total	12122.981	53			

Figure 15.14 Homogeneity of regression slopes: Fatigue vs. Sleep, by Depressed group

There is no significant interaction between 'Depressed' and 'Sleep', F (1, 50) = 0.939, p = .337 – we have satisfied the requirement for homogeneity of regression slopes here, too.

Before we proceed, we must remember to return the 'Model selection' back to its original state, otherwise future analyses will be reported incorrectly:

Select **Analyze** → **General Linear Model** → **Univariate...** → click **Model** → (in new window) select radio button for **Full factorial** → click **Continue** → (back in main window) click on **Total sleep time** in **Covariate(s)** → click on backwards arrow → click **OK** (delete any output produced by that action)

Running ANCOVA in SPSS

Main effect (before including covariate)

Before we undertake the ANCOVA analysis, we should explore the main effect for sleep quality scores, in respect of exercise group, *prior to* controlling for any covariate:

Using SPSS data set Fatigue and depression

Select **Analyze** → **General Linear Model** → **Univariate...** → (in new window) transfer Fatigue to **Dependent Variables** → transfer Depressed to **Fixed Factor(s)** → click **Continue** → click **Options** → (in new window) transfer Depressed to **Display Means for:** → tick boxes for **Descriptive statistics**, **Estimates of effect size** and **Homogeneity tests** → click **Continue** → click **OK**

Table 15.4 Estimated marginal means, prior to ANCOVA adjustment

	Fatigue scores		
Group	N	Mean	*SE*
Depressed	27	44.74	*2.85*
Not depressed	27	37.30	*2.85*

Table 15.4 suggests that it would *appear* that fatigue scores are poorer for the depressed group, but we need to check whether that difference is statistically significant.

Source	Type III sum of squares	df	Mean square	F	Sig.	Partial eta squared
Corrected model	748.167[a]	1	748.167	3.420	.070	.062
Intercept	90856.019	1	90856.019	415.349	.000	.889
Depressed	748.167	1	748.167	**3.420**	**.070**	.062
Error	11374.815	52	218.746			
Total	102979.000	54				
Corrected total	12122.981	53				

Figure 15.15 Univariate analysis, prior to ANCOVA adjustment

Figure 15.15 shows the main experimental effect for fatigue scores in respect of depression groups (before any covariate is applied). This output indicates that there was no significant difference in fatigue scores in respect of depression groups: F (1, 52) = 3.420, p = .070. However, this outcome certainly approached significance. If we were able to remove some error variance, perhaps the differences may be significant after all.

Applying a covariate to reduce error variance

When we examined the potential covariates earlier, we saw that total sleep time was dependent on depression status. Therefore, we cannot use that covariate to reduce the error variance. However, anxiety scores were independent of the depression groups, so that can be used in this context. We should explore what effect it had.

Using SPSS data set Fatigue and depression

Select **Analyze** → **General Linear Model** → **Univariate...** (the other settings will still be there from the last analysis; we just need to add the covariate) → (in new window) transfer Anxiety to **Covariate(s)** → click **OK**

Table 15.5 Estimated marginal means, after applying Anxiety scores as a covariate

	Fatigue scores		
Group	N	Mean	*SE*
Depressed	27	44.87	2.68
Not depressed	27	37.17	2.68

By applying anxiety as a covariate, Table 15.5 shows that the estimated marginal mean of fatigue scores for the depressed group has increased slightly, while this has decreased for the non-depressed group. Furthermore, the standard error has also decreased. All of this increases the likelihood that there is now a significant difference in fatigue scores between the groups.

Source	Type III sum of squares	df	Mean square	F	Sig.	Partial eta squared
Corrected model	2238.374[a]	2	1119.187	5.774	.005	.185
Intercept	2097.754	1	2097.754	10.823	.002	.175
Anxiety	1490.207	1	1490.207	7.689	.008	.131
Depressed	800.656	1	800.656	4.131	.047	.075
Error	9884.608	51	193.816			
Total	102979.000	54				
Corrected total	12122.981	53				

Figure 15.16 Effect of applying Anxiety scores as covariate

Figure 15.16 shows that the main effect for fatigue in respect of depression groups is significant when we apply anxiety scores as a covariate to reduce the error variance: F (1, 51) = 4.131, p = .047. The ANCOVA analysis has done exactly what we had hoped it would.

15.7 Nuts and bolts
Why has ANCOVA meant that the outcome is now significant?

To explain what has happened following the application of the covariate, we should look more closely at the error variance. For the unadjusted main effect (Figure 15.15), the residual mean square was 218.746. Once we applied the covariate (Figure 15.16) the residual mean square was 193.816. As we have seen throughout these chapters on ANOVA models, significance is determined by the F ratio, which is found by dividing the explained variance by the error variance. By reducing the error variance (residual mean square), we have increased the F ratio, increasing the likelihood of finding a significant outcome.

Effect size and power

We can use G*Power to show how effect size and statistical power are changed after we have controlled for a covariate in univariate analyses. We have used G*Power often in this book, but you can refer to Chapter 4 if you want a reminder about the rationale behind this program. We will start by looking at effect size and power of the unadjusted analysis, then repeat the process for adjusted effect.

From Test family select F tests

From **Statistical test** select ANCOVA: Fixed effects, special, main effects and interaction
From **Type of power analysis** select Post hoc: Compute achieved - given α, sample size and effect size power

Main (unadjusted) effect:

To calculate the **Effect size**, click on the **Determine** button (a new box appears).
In that new box, tick on **Direct** radio button → type 0.062 in **Partial η^2** box (we get that from 'Partial Eta Squared' in Figure 15.15) → click on **Calculate and transfer to main window** Back in original display, for α **err prob** type 0.05 (the significance level) → for **Total sample size** type **30** (the overall sample size) → for **Numerator df** type **1** (we get this from the highlighted row in Figure 15.15) → for **Number of groups** type **2** (Depression groups) → for **Covariates** type **0** → click on **Calculate**
From this, we can observe two outcomes: Effect size (d) **0.26** (which is medium); and Power (1-β err prob) **0.46** (which is considerably underpowered - see Chapter 4).

Adjusted effect (after controlling for anxiety scores):

To calculate the **Effect size**, click on the button (a new box appears).

In **Determine Effect Size** window, change Direct to **0.075** in the **Partial η^2** box (see Figure 15.16) → click on **Calculate and transfer to main window**

All of the other parameters (in the main box) remain the same, except **Covariates**, which should be changed to **1** → click on **Calculate**

From this, we can observe two outcomes: Effect size (d) **0.28** (which is still medium, but rather higher than before); and Power (1-β err prob) **0.53** (which is still underpowered, but a little better). We will explore this in more depth shortly.

Effect of (potentially) confounding covariate

Earlier, we saw that 'total sleep time' was dependent on depression groups, so could not be used as a covariate to reduce error variance. However, we can assess the potential confounding effect that total sleep time has on the main experimental outcome. You will recall that there was no significant difference in fatigue scores across the depression groups before we applied any covariate, $F(1, 52) = 3.420$, $p = .070$. We can examine the effect of controlling for total sleep time using ANCOVA in the same way as we did just now (but the effects are very different).

Using SPSS data set Fatigue and depression

Select **Analyze → General Linear Model → Univariate... →** (in new window) remove **Anxiety** from **Covariate(s)** box **→** replace with **Total sleep time → click OK**

Table 15.6 Estimated marginal means, after applying Total sleep time as a covariate

	Fatigue scores		
Group	N	Mean	*SE*
Depressed	27	42.32	2.52
Not depressed	27	39.72	2.52

By applying anxiety as a covariate, Table 15.6 shows that the estimated marginal mean of fatigue scores for the depressed group has decreased, while this has increased for the non-depressed group. This suggests that the difference between the groups is even more likely to be non-significant.

Source	Type III sum of squares	df	Mean square	F	Sig.	Partial eta squared
Corrected model	3848.037[a]	2	1924.019	11.858	.000	.317
Intercept	26180.086	1	26180.086	161.353	.000	.760
Sleep	3099.870	1	3099.870	19.105	.000	.273
Depressed	82.840	1	82.840	**.511**	**.478**	.010
Error	8274.944	51	162.254			
Total	102979.000	54				
Corrected total	12122.981	53				

Figure 15.17 Effect of applying Total sleep time as covariate

Figure 15.17 shows that the main effect for fatigue scores, in respect of depression groups, is even less significant, $F(1, 51) = 0.511$, $p = .478$. To explore what has happened we need to look at how the explained variance *and* error variance have been affected by the covariate. We can compare that to what we discussed in Box 15.7. If we compare Figure 15.15 with Figure 15.17, we can see a number of differences. Although the residual mean square has been reduced (218.746 to 162.254), and the error *df* has been reduced, we have an even smaller effect. To explain what has occurred, we need to look at the model mean square: this has been reduced from 748.167 to 82.840. The explained variance that we observed for fatigue across depression groups is *shared* with variance in total sleep time. It could be considered as a confounding effect, but it is more likely that the relationship between fatigue and total sleep time is so strong that they are difficult to prise apart. This is an important finding, as it tells us that HeLP might need to conduct another study, strictly matching depressed and non-depressed clients on total sleep time.

Writing up results

We should present the data in a table and discuss the statistical outcome (the descriptive data are taken from the SPSS output). We report the *actual* mean scores.

Table 15.7 Dependent variable (DV) and covariate (CV) scores, by depression group

	Depressed (N = 27)		Not depressed (N = 27)	
	Mean	SE	Mean	SE
Fatigue (DV)	44.74	2.85	37.30	2.85
Anxiety (CV)	62.56	3.38	63.44	3.68
Total sleep time, mins (CV)	320.11	27.58	411.78	28.65

Using independent one-way ANOVA, it was shown that there was no significant difference in fatigue scores between the depression groups, $F\ (1, 52) = 3.420$, $p = .070$. When anxiety scores were applied as a covariate in ANCOVA, it indicated that fatigue scores were significantly poorer for the depressed group, $F\ (1, 51) = 4.131$, $p = .047$. However, additional ANCOVA analyses suggested that variance in fatigue scores may be shared with variance in total sleep time, suggesting that they may be measuring the same construct.

MANCOVA: multivariate analysis of covariance

MANCOVA is used in similar circumstances to ANCOVA except that, now, we have several dependent variables (so it is also similar to MANOVA – see Chapter 14). We can use MANCOVA to examine two or more dependent variables simultaneously, in respect of one or more independent variables, but account for one or more covariates. This can be quite useful in studies where controlling for additional variables is a problem. Like ANCOVA, the effect of a covariate can serve to reduce error variance, but it can also be used to check that other variables are not confounding the observed outcome. As we will see shortly, there are many assumptions that we must attend to before undertaking a MANCOVA analysis.

Research question for MANCOVA

Throughout this section, we will refer to a research question that will illustrate how we can use MANCOVA to analyse data. PRAMS (Perinatal Research for Anxiety, Mood and Sleep) is a group of clinicians and academics who are exploring several projects focusing on maternal mental health. They are aware that maternal mood can be low during pregnancy and soon after giving birth. Sometimes this may lead to postnatal depression. From client observation, the researchers are aware that poor sleep can have an impact on mood. Based on this, PRAMS decide to examine sleep and mood across four study groups: pregnant women, mothers of infants aged less than three months, mothers with infants aged 3–6 months, and women without children. They recruit ten women to each group. Mood is measured from a series of questionnaires and observations, ultimately leading to a 'mood score' ranging from 0–250 (higher scores represent better mood). Total sleep time is taken from diaries and is measured in minutes. Because of the potential relationship between sleep and mood across these groups, the data could be analysed using MANOVA. However, PRAMS also know that mood and sleep may be affected by age and general health. Recruitment restrictions prevent the researchers from matching participants on these factors, so they decide to include them as covariates. Health scores are measured from clinician ratings, in a range of 0–150 (higher scores represent better general health).

15.8 Take a closer look
Summary of research example

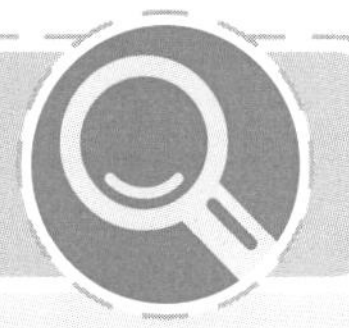

MANCOVA variables
DV 1: Mood
DV 2: Total sleep time
IV: Status (four groups: pregnant, with infant < 3 months, with infant 3–6 months, no children)
CV 1: Age
CV 2: General health

How MANCOVA is measured

The method for calculating outcomes is much the same as we saw for MANOVA and ANCOVA, so there is little point repeating all of that. Multivariate and covariate outcomes are measured from cross-product and cross-product partitions (see Chapter 14), while univariate outcomes are found from partitioning variance into the relevant sums of squares. Ultimately, each part of the analysis will be represented by F ratios, from which we can establish statistical significance.

Reporting multivariate outcome

As we saw when we explored MANOVA, we need to know which of the four multivariate outcomes to report from the SPSS output. For MANCOVA those rules still apply (see Box 14.4). The choice of significance test largely depends on the number of groups measured by the independent variables. When there are two groups, Pillai's Trace tends to be most favoured (even though it can be used with any number of groups). However, it is vulnerable to error if group sizes are unequal (and/or there is a lack of homogeneity of between-group variance). Hotelling's Trace can be used only when there are two groups, but it is considered less powerful than Pillai's Trace. Wilks' Lambda (λ) tends to be used where there are three or more groups. It is not considered to be as powerful as Pillai's Trace, but is nonetheless popular. As we have four groups represented by the independent variable we will choose Wilks' Lambda for our analyses.

Assumptions and restrictions

Before we can analyse our data we need to satisfy a number of assumptions. Effectively, we should observe the guidelines indicated for both MANOVA and ANCOVA tests. By definition, there must be at least two dependent variables. These should be parametric with data that are interval and reasonably normally distributed (although multivariate normality is quite robust to violations so long as the sample size exceeds 20, and ordinal data are frequently analysed through MANCOVA in published research). Covariates must be interval and normally distributed across the independent variable groups. If you need a reminder about data types and parametric requisites you could refer to Chapter 5. The dependent variables should not be dependent on each other; there are repeated-measures versions of MANCOVA, but we do not address these here. There must be at least *reasonable* correlation between the dependent variables (in Chapter 14, we said that the correlation co-efficient should be somewhere between $r = .30$ and .90). That correlation should not differ between the independent variable groups. We also encountered this in Chapter 14: it is measured through an examination of variance-covariance matrices via Box's M test (highly significant outcomes, where $p < .001$, should be avoided). Independent variables must be categorical.

There must be reasonable correlation between the dependent variables and the covariates, and there should not be between-group differences in respect of that (we referred to this as 'homogeneity of regression slopes' earlier in this chapter). There should also be homogeneity of between-group variance for each of the dependent variables. We assess this using Levene's test, as we have done elsewhere for (independent) ANOVA models. Independence of the covariates is less of a problem in MANCOVA in that they are frequently used to factor out shared variance in the outcome. Nevertheless, it is perhaps wise to check whether there are between-group differences in the covariates, and perform separate analyses for those that appear to confound the outcome from those that may help reduce error variance (we addressed this issue earlier). Outliers in the covariates should be avoided (although that will probably be taken care of when we measure normal distribution). Equal group sizes are preferable. Sample sizes should be large enough so that statistical power is not compromised.

How SPSS performs MANCOVA

We will now proceed to run MANCOVA through SPSS. To see how we can create a data set that is suitable for such analyses, refer to Box 15.6. The methods are the same, except that we need an additional dependent variable. You will recall that we are examining the PRAMS research, investigating sleep and mood factors in pregnancy, new mothers, and women without children. The dependent variables are 'total sleep time' (measured in minutes) and 'mood' (measured on a scale of 0–250, with higher scores representing better mood). The independent variable has four groups, representing pregnancy/mother status in the 40 women (pregnant, with child under three months, with child 3–6 months, and no children). The covariates are age and health status (measured on a scale of 0–150, with higher scores representing better health).

Prior tests for assumptions and restrictions

Before we can explore the main analyses, we need to check those assumptions and restrictions that we have just been exploring.

Correlation between covariates and dependent variables

We said that we need reasonable correlation between the dependent variables and between the covariates and the dependent variables. We also said that we should check that the correlation between the dependent variables does not differ significantly across the independent variable groups. We will assess that with the variance-covariance matrices outcome later.

		Health	Age	Total sleep time	Mood
Health	Pearson correlation	1	.192	.418**	–.008
	Sig. (2-tailed)		.235	.007	.961
	N	40	40	40	40
Age	Pearson correlation	.192	1	.185	.492**
	Sig. (2-tailed)	.235		.254	.001
	N	40	40	40	40
Total sleep time	Pearson correlation	.418**	.185	1	.548**
	Sig. (2-tailed)	.007	.254		.000
	N	40	40	40	40
Mood	Pearson correlation	–.008	.492**	.548**	1
	Sig. (2-tailed)	.961	.001	.000	
	N	40	40	40	40

Figure 15.18 Correlation between covariates and dependent variables

Open the SPSS file PND and sleep

Select **Analyze → Correlate → Bivariate... →** transfer Health, Age, Total sleep time and Mood to **Variables →** tick boxes for **Pearson** and **Two-tailed →** click **OK**

Figure 15.18 shows that the correlation between the covariates (health and age) and the dependent variables (mood and total sleep time) are mostly fine. Health is less strongly correlated with age and mood, but is significantly related to total sleep time, so should be included.

Independence of covariate

We need to examine whether the covariates differ across the independent variable groups. As there are four groups, we need to explore that with an independent one-way ANOVA. We saw how to do that in Chapter 9 (but we need only the main outcome, not any *post hoc* tests).

Using the SPSS file PND and sleep

Select **Analyze → Compare Means → One-Way ANOVA... →** transfer Health and Age to **Dependent List →** transfer Group to **Factor →** click **OK**

		Sum of squares	df	Mean square	F	Sig.
Health	Between-groups	202.400	3	67.467	.237	.870
	Within-groups	10256.000	36	284.889		
	Total	10458.400	39			
Age	Between-groups	20.275	3	6.758	.085	.968
	Within-groups	2875.700	36	79.881		
	Total	2895.975	39			

Figure 15.19 Independent one-way outcome (covariates by independent variable group)

Figure 15.19 indicates that there is no significant difference in health scores (F (3, 36) = 0.237, p = .870) or age (F (3, 36) = 0.085, p = .968) between the groups of women. These covariates may help reduce error variance.

Testing for normal distribution

We need to examine normal distribution (particularly across the covariate) using the KS/SW analyses (we will go straight to the outcome once again).

	group	Kolmogorov–Smirnov[a]			Shapiro–Wilk		
		Statistic	df	Sig.	Statistic	df	Sig.
Health	Pregnant	.160	10	.200*	.963	10	.820
	Infant < 3 months	.205	10	.200*	.898	10	.207
	Infant 3–6 months	.253	10	.068	.850	10	.058
	No children	.160	10	.200*	.939	10	.546
Age	Pregnant	.167	10	.200*	.915	10	.315
	Infant < 3 months	.165	10	.200*	.967	10	.859
	Infant 3–6 months	.213	10	.200*	.916	10	.321
	No children	.188	10	.200*	.875	10	.114
Total sleep time	Pregnant	.207	10	.200*	.919	10	.348
	Infant < 3 months	.151	10	.200*	.935	10	.494
	Infant 3–6 months	.199	10	.200*	.900	10	.219
	No children	.195	10	.200*	.906	10	.257
Mood	Pregnant	.141	10	.200*	.964	10	.833
	Infant < 3 months	.300	10	.011	.825	10	.029
	Infant 3–6 months	.198	10	.200*	.890	10	.169
	No children	.164	10	.200*	.939	10	.547

Figure 15.20 Normality tests for dependent variables and covariates (by group)

As there are fewer than 50 people in each group, we should refer to the Shapiro-Wilk outcome. Figure 15.20 shows that normal distribution appears to be fine for the dependent variables and covariates across the groups. Mood scores for women with an infant aged less than three months were potentially not normal, but given the other outcomes we can accept this as reasonable.

Testing for homogeneity of regression slopes

Now we check that the correlation between each covariate and the dependent variable does not differ across independent variable groups. We can examine this for both covariates together this time, as they were both shown to be independent of the group factor.

Using SPSS file PND and sleep

Select **Analyze** → **General Linear Model** → **Multivariate...** (as shown in Figure 14.5) → (in new window) transfer Total sleep time and Mood to **Dependent Variables** → transfer Group to **Fixed Factor(s)** → transfer Age and Health to **Covariate(s)** → click **Model** → (in new window) choose **Custom** → transfer group to **Build Term(s)** → transfer health to **Build Term(s)** → transfer age to **Build Term(s)** → transfer group and age (together) to **Build Term(s)** → transfer group and health (together) to **Build Term(s)** (make sure that **Interaction** is selected in pull-down menu below **Build Terms)** → click **Continue** → click **OK**

REMEMBER TO SWITCH THE MODEL BACK TO FULL FACTORIAL! (see page 376.)

Tests of between-subjects effects

Source	Dependent variable	Type III sum of squares	df	Mean square	F	Sig.
Corrected model	Total sleep time	89511.893[a]	11	8137.445	3.087	.008
	Mood	23977.669[b]	11	2179.788	4.073	.001
Intercept	Total sleep time	17107.539	1	17107.539	6.490	.017
	Mood	11276.690	1	11276.690	21.072	.000
group	Total sleep time	5094.026	3	1698.009	.644	.593
	Mood	2286.139	3	762.046	1.424	.257
health	Total sleep time	27461.515	1	27461.515	10.417	.003
	Mood	365.286	1	365.286	.683	.416
age	Total sleep time	1017.633	1	1017.633	.386	.539
	Mood	8387.166	1	8387.166	15.672	.000
group * age	Total sleep time	3073.630	3	1024.543	.389	**.762**
	Mood	443.120	3	147.707	.276	**.842**
group * health	Total sleep time	8590.598	3	2863.533	1.086	**.371**
	Mood	1036.597	3	345.532	.646	**.592**
Error	Total sleep time	73812.882	28	2636.174		
	Mood	14984.306	28	535.154		
Total	Total sleep time	5798079.000	40			
	Mood	1121701.000	40			
Corrected total	Total sleep time	163324.775	39			
	Mood	38961.975	39			

a. R Squared = .548 (Adjusted R Squared = .371)
b. R Squared = .615 (Adjusted R Squared = .464)

Figure 15.21 Homogeneity of regression slopes

To satisfy the assumption that correlation between the covariates and dependent variable do not differ across the groups of women, we need the interaction terms in this model to be non-significant. Figure 15.21 suggests that we have satisfied that assumption. There is no significant interaction between the covariates and independent variable for either dependent variable: 'age' covariate (shown by 'group*age') in respect of mood scores (p = .762) and total sleep time (p = .842); 'health' covariate (group*health) in respect of mood (p = .371) and total sleep time (p = .592).

Running MANCOVA in SPSS

Multivariate effect (before including covariates)

Before we see how to perform MANCOVA in SPSS, we should explore the multivariate and univariate outcomes without the covariates, along with estimated marginal means. This will give us something to compare with later. The initial data are derived from a simple MANOVA analysis (we saw how to perform that in SPSS in Chapter 14, so we will not repeat that here). For the moment, we are just exploring total sleep time and mood as dependent variables, with the women's status as the independent variable (Figure 15.22).

Dependent variable	group	Mean	Std. error	95% confidence interval Lower bound	Upper bound
Total sleep time	Pregnant	321.200	18.462	283.757	358.643
	Infant < 3 months	386.100	18.462	348.657	423.543
	Infant 3–6 months	403.200	18.462	365.757	440.643
	No children	390.800	18.462	353.357	428.243
Mood	Pregnant	137.400	8.862	119.428	155.372
	Infant < 3 months	166.400	8.862	148.428	184.372
	Infant 3–6 months	174.800	8.862	156.828	192.772
	No children	179.500	8.862	161.528	197.472

Figure 15.22 Estimated marginal means (prior to covariate inclusion)

Effect		Value	F	Hypothesis df	Error df	Sig.	Partial eta squared
Intercept	Pillai's Trace	.984	1053.999[a]	2.000	35.000	.000	.984
	Wilks' Lambda	.016	1053.999[a]	2.000	35.000	.000	.984
	Hotelling's Trace	60.228	1053.999[a]	2.000	35.000	.000	.984
	Roy's Largest Root	60.228	1053.999[a]	2.000	35.000	.000	.984
group	Pillai's Trace	.354	2.577	6.000	72.000	.026	.177
	Wilks' Lambda	.654	**2.764**[a]	**6.000**	**70.000**	**.018**	.192
	Hotelling's Trace	.519	2.942	6.000	68.000	.013	.206
	Roy's Largest Root	.497	5.967[b]	3.000	36.000	.002	.332

Figure 15.23 Multivariate outcome (prior to covariate inclusion)

Referring to the Wilks' Lambda outcome (under 'group') in Figure 15.23, we can report that we have a significant multivariate outcome (prior to covariate adjustment), in respect of total sleep time and mood across the women's status groups: $\lambda = .654$, $F(6, 70) = 2.764$, $p = .018$.

Source	Dependent variable	Type III sum of squares	df	Mean square	F	Sig.	Partial eta squared
Corrected model	Total sleep time	40621.075[a]	3	13540.358	3.973	.015	.249
	Mood	10691.075[b]	3	3563.692	4.538	.008	.274
Intercept	Total sleep time	5634754.225	1	5634754.225	1653.179	.000	.979
	Mood	1082739.025	1	1082739.025	1378.754	.000	.975
group	Total sleep time	40621.075	3	13540.358	**3.973**	**.015**	.249
	Mood	10691.075	3	3563.692	**4.538**	**.008**	.274
Error	Total sleep time	122703.700	36	3408.436			
	Mood	28270.900	36	785.303			
Total	Total sleep time	5798079.000	40				
	Mood	1121701.000	40				
Corrected total	Total sleep time	163324.775	39				
	Mood	38961.975	39				

Figure 15.24 Univariate outcome (prior to covariate inclusion)

Figure 15.24 indicates that there is a significant univariate outcome for total sleep time (F (3, 36) = 3.973, p = .015) and mood (F (3, 36) = 4.538, p = .008) across group status. If we were pursuing the MANOVA analysis, we would now need to explore the source of the main effects across the groups, using *post hoc* analyses. However, since we are focusing on MANCOVA in this section, we can leave that level of analysis until we apply the covariates.

Multivariate effect (after including covariates)

We will now see how to perform MANCOVA in SPSS, including the dependent variables (total sleep time and mood), the independent variable (group) and the covariates (age and health), using the PRAMS research data.

Using SPSS file PND and sleep

Select **Analyze** → **General Linear Model** → **Multivariate...** → (in new window) transfer Total sleep time and Mood to **Dependent Variables** → transfer Group to **Fixed Factor(s)** → transfer Age and Health to **Covariate(s)**

At this point we would normally select a relevant *post hoc* test to assess the source of any univariate effect. However, that option is not available through the usual **Post Hoc** route – we need to get this from the Options menu.

Click **Options** → (in new window) transfer group to **Display Means for:** → tick **Compare main effects** box → click on pull-down arrow under **Confidence interval adjustment** → select Bonferroni option (to produce *post hoc* outcomes) → tick boxes for **Descriptive statistics**, **Estimates of effect size**, and **Homogeneity tests** → click **Continue** → click **OK**

Interpreting output

Homogeneity checks

	F	df1	df2	Sig.
Total sleep time	1.738	3	36	.176
Mood	.582	3	36	.631

Figure 15.25 Homogeneity of between-group variances

Figure 15.25 indicates that we have homogeneity of between-group variances across the groups in respect of total sleep time and mood (significance is greater than .05 in both cases).

Box's M	7.375
F	.737
df1	9
df2	14851.910
Sig.	.676

Figure 15.26 Homogeneity of variance-covariance matrices

Figure 15.26 shows that the correlation between the dependent variables does not differ significantly between the women's groups.

Multivariate outcome (after applying covariates)

Dependent variable	group	Mean	Std. error	95% confidence interval	
				Lower bound	Upper bound
Total sleep time	Pregnant	317.397[a]	15.907	285.071	349.723
	Infant < 3 months	387.544[a]	15.886	355.259	419.829
	Infant 3–6 months	408.662[a]	15.958	376.231	441.094
	No children	387.697[a]	15.976	355.230	420.164
Mood	Pregnant	136.269[a]	7.109	121.822	150.716
	Infant < 3 months	165.566[a]	7.100	151.138	179.994
	Infant 3–6 months	173.970[a]	7.132	159.476	188.464
	No children	182.295[a]	7.140	167.785	196.805

a. Covariates appearing in the model are evaluated at the following values: Health = 110.80, Age = 28.28

Figure 15.27 Estimated marginal means (after applying covariates)

Effect		Value	F	Hypothesis df	Error df	Sig.	Partial eta squared
Intercept	Pillai's Trace	.421	11.996[a]	2.000	33.000	.000	.421
	Wilks' Lambda	.579	11.996[a]	2.000	33.000	.000	.421
	Hotelling's Trace	.727	11.996[a]	2.000	33.000	.000	.421
	Roy's Largest Root	.727	11.996[a]	2.000	33.000	.000	.421
health	Pillai's Trace	.363	9.419[a]	2.000	33.000	.001	.363
	Wilks' Lambda	.637	9.419[a]	2.000	33.000	.001	.363
	Hotelling's Trace	.571	9.419[a]	2.000	33.000	.001	.363
	Roy's Largest Root	.571	9.419[a]	2.000	33.000	.001	.363
age	Pillai's Trace	.416	11.732[a]	2.000	33.000	.000	.416
	Wilks' Lambda	.584	11.732[a]	2.000	33.000	.000	.416
	Hotelling's Trace	.711	11.732[a]	2.000	33.000	.000	.416
	Roy's Largest Root	.711	11.732[a]	2.000	33.000	.000	.416
group	Pillai's Trace	.543	4.221	6.000	68.000	.001	.271
	Wilks' Lambda	.499	**4.578[a]**	**6.000**	**66.000**	**.001**	.294
	Hotelling's Trace	.923	4.921	6.000	64.000	.000	.316
	Roy's Largest Root	.822	9.314[b]	3.000	34.000	.000	.451

Figure 15.28 Multivariate analyses (after applying covariates)

Figure 15.28 indicates that the multivariate outcome is much stronger subsequent to applying the covariates (compare this with the outcome shown in Figure 15.23); it would appear that the covariates have reduced some of the error variance. There is a highly significant multivariate effect across the groups for the combined dependent variables of total sleep time and mood: $\lambda = .499$, $F(6, 66) = 4.578$, $p = .001$. We refer to the 'group' line of data; the lines for 'age' and 'health' show the covariate effect. The effect on estimated marginal means can be assessed by comparing Figure 15.27 with Figure 15.23.

Univariate outcome (after applying covariates)

Figure 15.29 shows that univariate outcome is also much stronger after applying the covariates (compare outcome with Figure 15.24). Total sleep time ($F(3, 34) = 6.258$, $p = .002$) and mood scores ($F(3, 34) = 7.938$, $p < .001$) differ significantly across the groups of women.

Source	Dependent variable	Type III sum of squares	df	Mean square	F	Sig.	Partial eta squared
Corrected model	Total sleep time	77657.815[a]	5	15531.563	6.164	.000	.475
	Mood	21852.237[b]	5	4370.447	8.685	.000	.561
Intercept	Total sleep time	18466.690	1	18466.690	7.329	.011	.177
	Mood	12347.366	1	12347.366	24.536	.000	.419
health	Total sleep time	29920.293	1	29920.293	11.875	.002	.259
	Mood	322.489	1	322.489	.641	.429	.018
age	Total sleep time	2331.836	1	2331.836	.925	.343	.026
	Mood	11151.573	1	11151.573	22.160	.000	.395
group	Total sleep time	47306.388	3	15768.796	6.258	.002	.356
	Mood	11983.756	3	3994.585	7.938	.000	.412
Error	Total sleep time	85666.960	34	2519.616			
	Mood	17109.738	34	503.228			
Total	Total sleep time	5798079.000	40				
	Mood	1121701.000	40				
Corrected total	Total sleep time	163324.775	39				
	Mood	38961.975	39				

Figure 15.29 Univariate analyses

						95% confidence interval for difference[a]	
Dependent variable	(I) group	(J) group	Mean difference (I-J)	Std. error	Sig.[a]	Lower bound	Upper bound
Total sleep time	Pregnant	Infant < 3 months	−70.147*	22.491	.022	−133.157	−7.137
		Infant 3–6 months	−91.265*	22.585	.002	−154.537	−27.994
		No children	−70.300*	22.528	.022	−133.414	−7.186
	Infant < 3 months	Pregnant	70.147*	22.491	.022	7.137	133.157
		Infant 3–6 months	−21.118	22.475	1.000	−84.081	41.846
		No children	−.153	22.581	1.000	−63.415	63.109
	Infant 3–6 months	Pregnant	91.265*	22.585	.002	27.994	154.537
		Infant < 3 months	21.118	22.475	1.000	−41.846	84.081
		No children	20.965	22.691	1.000	−42.604	84.535
	No children	Pregnant	70.300*	22.528	.022	7.186	133.414
		Infant < 3 months	.153	22.581	1.000	−63.109	63.415
		Infant 3–6 months	−20.965	22.691	1.000	−84.535	42.604
Mood	Pregnant	Infant < 3 months	−29.297*	10.051	.038	−57.457	−1.138
		Infant 3–6 months	−37.701*	10.093	.004	−65.977	−9.425
		No children	−46.026*	10.068	.000	−74.232	−17.820
	Infant < 3 months	Pregnant	29.297*	10.051	.038	1.138	57.457
		Infant 3–6 months	−8.404	10.044	1.000	−36.543	19.735
		No children	−16.729	10.092	.639	−45.001	11.543
	Infant 3–6 months	Pregnant	37.701*	10.093	.004	9.425	65.977
		Infant < 3 months	8.404	10.044	1.000	−19.735	36.543
		No children	−8.325	10.141	1.000	−36.735	20.084
	No children	Pregnant	46.026*	10.068	.000	17.820	74.232
		Infant < 3 months	16.729	10.092	.639	−11.543	45.001
		Infant 3–6 months	8.325	10.141	1.000	−20.084	36.735

Figure 15.30 *Post hoc* (Bonferroni) outcomes

Locating source of main effect

Figure 15.30 indicates that pregnant women sleep for significantly shorter times and present poorer mood than all other women. There are no other between-group differences elsewhere for either dependent variable.

Effect size and power

Analyses of effect size and power for MANCOVA are currently not available in the G*Power software (and are too complex to undertake manually for this book).

Writing up results

Once again, we need to show some descriptive data in a table (Table 15.8) and discuss the statistical outcome.

Table 15.8 Total sleep time and mood scores by woman/child status

		Dependent variables			
		Total sleep time		Mood	
Status	N	Mean	SE	Mean	SE
Pregnant	10	321.20	18.46	137.40	8.86
Child < 3 months	10	386.10	18.46	166.40	8.86
Child 3–6 months	10	403.20	18.46	174.80	8.86
No children	10	390.80	18.46	179.50	8.86

Multivariate (MANOVA) analyses showed that pregnant women reported less sleep and poorer mood than other women who had young infants or were without children ($\lambda = .654$, $F(6, 70) = 2.764$, $p = .018$). These were confirmed across both dependent variables ($F(3, 36) = 3.973$, $p = .015$, and $F(3, 36) = 4.538$, $p = .008$ respectively). When age and general health were added as covariates in a MANCOVA these effects became even stronger ($\lambda = .499$, $F(6, 66) = 4.578$, $p = .001$). *Post hoc* (Bonferroni) analyses of the univariate outcomes (adjusted for age and health status) showed that pregnant women slept less than women with infants under three months ($p = .022$), women with infants aged 3–6 months ($p = .002$) and women without children ($p = .022$). Pregnant women also presented poorer mood than other women ($p = .038$, $p = .004$, and $p < .001$ respectively).

Chapter summary

In this chapter we have explored analyses of covariance, focusing on ANCOVA for univariate analyses and MANCOVA for multivariate outcomes. At this point, it would be good to revisit the learning objectives that we set at the beginning of the chapter.

You should now be able to:

- Recognise that we can use covariates to explore the effect that additional variables may have on an existing relationship between one or more dependent variable and one or more independent variable. The covariate may alter the strength of the original outcome, by reducing the error variance, or may illustrate the extent to which it may confound the original outcome.
- Appreciate that analyses of covariance can be employed in a wide range of contexts, including exploring moderator and mediator effects, measuring pre-test vs. post-test effects, and controlling for factors that we have not been able to match physically.
- Understand the assumptions and restrictions of ANCOVA: there must be one parametric dependent variable; there must be at least one independent variable; there must be at least one parametric covariate, which must be reasonably correlated with the dependent variable; correlation between the covariate(s) and the dependent variable should be equal across the independent variable groups; the covariate(s) should be reliable and consistent, and should be autonomous from the independent variable(s).
- Recognise the assumptions and restrictions of MANCOVA (in addition to those described for ANCOVA, the limitations for MANOVA also apply - see Chapter 14).
- Perform each test using SPSS.
- Examine effect size and power, using G*Power software, including showing the net effect of the covariate.
- Understand how to present the data, using appropriate tables, reporting the outcome in series of sentences and correctly formatted statistical notation.

Research examples

It might help you to see how analyses of variance have been applied in a research context. A separate published paper is reviewed for each of the ANCOVA and MANCOVA examples that we have just explored. If you would like to read the entire paper for either of these summaries you can use the DOI reference provided to locate that (see Chapter 1 for instructions).

ANCOVA

Ponizovsky, A.M., Grinshpoon, A., Levav, I. and Ritsner, M.S. (2003). Life satisfaction and suicidal attempts among persons with schizophrenia. ***Comprehensive Psychiatry, 44 (6):*** **442-447.DOI: http://dx.doi.org/10.1016/S0010-440X(03)00146-9**

In this research the authors examined the relationship between perceived quality of life (QoL) and suicide attempts in young people with schizophrenia. Outcome measures were examined across a range of QoL perceptions. These were related to the number of suicide attempts, and controlled for in respect of several current and historical measures of psychiatric illness. 227 Israeli patients with schizophrenia were recruited. The independent variable was represented by three groups, based on previous suicide attempts: non-attempters (124 patients), single attempters (75) and multiple attempters (28). The dependent variable was based on series of QoL perceptions drawn from the Quality of Life Enjoyment and Satisfaction Questionnaire (Q-LES-Q; Endicott *et al.*, 1993). The QoL domains included physical health, subjective feelings, leisure activities, social relationships, general activities, household duties, education/work, and satisfaction with medication. All items were scored on a scale of 1-5, with higher scores representing poorer perceptions. Additional variables were measured as potential covariates: current psychiatric symptoms were elicited from the Positive and Negative Syndrome Scale (PANSS; Kay *et al.*, 1987) and the Montgomery Asberg Depression

Rating Scale (MADRS; Montgomery and Asberg, 1979); historical evidence was based on previous psychiatric illness, suicidal behaviour, number and duration of psychiatric admissions, and demographic data (taken from patients' files).

Significant effects across the suicide groups were found for all QoL domains except physical health, $F = 1.3$, $p = .22$, and satisfaction with medication, $F = 1.1$, $p = .42$ (degrees of freedom for all outcomes = 2, 227). Significant outcomes included those for overall QoL, $F = 8.4$, $p = .002$; social relationships, $F = 8.9$, $p = .004$; and life satisfaction, $F = 4.0$, $p = .031$ (you can read the full outcomes in the paper). *Post hoc* analyses appeared to be undertaken with independent t-tests, although there was no indication about any adjustment for multiple comparisons. Where there were significant ANOVA effects, multiple attempters reported significantly poorer QoL than non-attempters in all cases; multiple and single attempters did not differ (except for social relationships). ANCOVAs were performed on all outcomes, controlling for current and historical psychiatric illness. Differences between multiple attempters and non-attempters remained for QoL scores, in respect of overall QoL, $F = 4.91$, $p = .03$; subjective feelings, $F = 6.80$, $p = .01$; and social relationships, $F = 3.73$, $p = .05$. Presumably, the other QoL domains were not significant, although this was not specifically mentioned.

MANCOVA

Alschuler, K.N. and Otis, J.D. (2012). Coping strategies and beliefs about pain in veterans with comorbid chronic pain and significant levels of posttraumatic stress disorder symptoms. *European Journal of Pain, 16* (2): 312–319.DOI: http://dx.doi.org/10.1016/j.ejpain.2011.06.010

In this research the authors investigated how former combat veterans dealt with chronic pain resulting from active service. Post-Traumatic Stress Disorder (PTSD) was confirmed via a formal diagnosis from the PTSD Checklist - Military Version (PCL-M; Weathers *et al.*, 1993). Rating was on the PCL-M range from 17–85 - a score above 50 indicates significant PTSD symptoms. A total of 194 war veterans were recruited to the study. Of those, 91 (47%) scored above the clinically relevant cut-off for PTSD. All veterans were assessed in respect of pain intensity via the McGill Pain Questionnaire (MPQ; Melzack, 1975). Outcomes (dependent variables) were examined in respect of attitude towards pain and coping styles. Attitudes towards pain were measured from the Survey of Pain Attitudes (SOPA; Jensen *et al.*, 1994). Questions focused on perceived control over the pain, disability as a result of the pain, avoidance of activity (to reduce further harm), emotion towards the pain, medications taken to deal with pain, the extent that the sufferer believed that others should respond with 'concern' towards their pain, and whether there was a 'cure' for the pain. Coping styles were examined using the Coping Strategies Questionnaire Revised (CSQ- R; Riley and Robinson, 1997). Questions focused on six domains. Two sub-scales explored coping strategies: cognitive style (such as positive self-talk) and distraction (such as thinking of something pleasant). Four domains explored maladaptive coping styles: catastrophising (thinking the worst), reinterpreting (such as pretending pain is not there), praying and hoping (the pain will just go away), and distancing (such as believing the pain is 'separate' to them). The researchers predicted that veterans (experiencing chronic pain) who showed greater evidence of PTSD would present more negative pain beliefs and would adopt more maladaptive coping styles (than those in the lower PCL-M group).

Although pain intensity (MPQ) scores did not differ between the PCL-M groups, MPQ scores were correlated with outcomes, so they were included as a covariate along with the age of the veterans. MANCOVA analyses (subsequent to controlling for pain intensity and age) showed a significant multivariate effect for pain beliefs and coping styles across the PTSD groups: $\lambda = .72$, $F = 4.90$, $p < .001$ (degrees of freedom were not reported). Univariate analyses with respect to attitudes towards pain indicated that PTSD veterans reported poorer pain control on the SOPA ($F = 13.41$, $p < .001$) and greater emotion ($F = 22.88$, $p < .001$) than those scoring less than 50 on PCL-M. For coping strategies, PTSD veterans reported significantly greater use of catastrophising on the CSQ-R ($F = 26.46$, $p < .001$).

Extended learning tasks

You will find the data sets associated with these tasks on the website that accompanies this book, (available in SPSS and Excel format). You will also find the answers there.

ANCOVA learning task

Following what we have learned about ANCOVA, answer the following questions and conduct the analyses in SPSS and G*Power. (If you do not have SPSS, do as much as you can with the Excel spreadsheet.) For this exercise, we will look at a fictitious example of treatment options for a group of patients. We will explore the effect of drug treatment, counselling or both on a measure of mood (which is measured on a scale from 0 (poor) to 100 (good), and is taken before and after the intervention). There are 72 participants in this study, with 24 randomly assigned to each of the treatment groups.

Open the data set **Mood and treatment**

1. Which is the independent variable (and describe the groups)?
2. What is the dependent variable?
3. What is the covariate?
4. What assumptions should we test for?
5. Conduct the ANCOVA test.
   a. Describe how you have accounted for the assumptions.
   b. Describe what the SPSS output shows, including pre- and post-treatment analyses.
   c. Describe the effect on estimated marginal means.
   d. Describe whether you needed to conduct *post hoc* analyses.
   i. Run them if they were needed.
6. Also show the effect size and conduct a power calculation, using G*Power.
7. Report the outcome as you would in the results section of a report.

MANCOVA learning task

Following what we have learned about MANCOVA, answer the following questions and conduct the analyses in SPSS and G*Power (you will not be able to perform this test manually). For this exercise, we will look at some data that explore the impact of two forms of treatment on anxiety and mood outcomes. The treatments are cognitive behavioural therapy (CBT) and medication. A group of 20 anxious patients are randomised into those treatment groups. Ratings of anxiety and depression are made by the clinician eight weeks after treatment. Both scales are scored in the range of 0–100, with higher scores representing poorer outcomes. To ensure that these outcomes are not related to prior anxiety, the anxiety ratings are also taken at baseline.

Open the SPSS data **CBT vs. drug**

1. Which is the independent variable (and describe the groups)?
2. What are the dependent variables?
3. What is the covariate?
4. What assumptions should we test for?
5. Conduct the MANCOVA test.
   a. Describe how you have accounted for the assumptions.
   b. Describe what the SPSS output shows, including pre- and post-treatment analyses.
   c. Describe the effect on estimated marginal means.
6. Report the outcome as you would in the results section of a report.

Appendix to Chapter 15
Mathematics behind (univariate) ANCOVA

Table 15.9 presents the raw data illustrating sleep quality perception scores (dependent variable), according to levels of exercise frequency (independent variable with three levels: frequent, infrequent and none). These were the data that we explored through SPSS earlier. The table also shows data for the age of each participant – this will be the covariate. You will find an Excel spreadsheet associated with these calculations on the web page for this book.

Table 15.9 Sleep quality scores (SQ), in respect of exercise frequency (adjusted for age)

Exercise frequency (IV)						
	Frequent		**Infrequent**		**None**	
	Age (CV)	**SQ (DV)**	**Age (CV)**	**SQ (DV)**	**Age (CV)**	**SQ (DV)**
	40	67	18	44	25	52
	47	80	24	44	14	46
	27	74	31	62	22	62
	27	37	21	39	22	46
	44	80	28	49	39	52
	30	62	21	39	34	52
	30	37	24	33	34	46
	49	83	24	56	25	36
	34	55	37	56	22	20
	37	86	24	39	25	26
Mean	**36.50**	**66.10**	**25.20**	**46.10**	**26.20**	**43.80**
Variance	**67.39**	**329.43**	**30.40**	**87.66**	**54.62**	**165.73**
Grand mean	*Age: 29.30*		*SQ: 52.00*			
Sum	*365*	*661*	*252*	*461*	*262*	*438*

CV = covariate (age); DV = dependent variable (sleep quality; SQ)

Response scores for the dependent variable and covariate are summed within independent variable group (we call these the sum of sums). The participant scores are multiplied across the dependent variable and covariate within each group (these are the sum of squared products). The average sums of squared products (total sums of squared products divided by the number of groups and the number of cases within the groups) are deducted from the average sum of sums (total sum of sums divided by the number of cases within the groups). This produces the model sum of cross-products (SP_M). The residual sum of cross-products is found by deducting the average sum of sums from the total sum of squared products.

Sum of sums SS

The 'Sum' columns for Age (x) and Sleep Quality (y) need to be calculated

$$\text{Age } SS_X: 365 + 252 + 262 = \mathbf{879}$$
$$\text{Sleep quality } SS_Y: 661 + 461 + 438 = \mathbf{1560}$$

Sum of squared products **SSP**

Calculated from the sum of each participant's DV score x CV score:

Exercise = Frequent SSP_F : (40 × 67) + (47 × 80)... + (37 × 86) = 25046 plus...
Exercise = Infrequent SSP_I : (18 × 44) + (24 × 44)... + (24 × 39) = 11924 plus...
Exercise = None SSP_N : (25 × 52) + (14 × 46)... + (25 × 26) = 11670 **SSP = 48640**

Main effect DV (Sleep quality; Y)

Model sum of squares ($SS_M Y$)

Formula for model sum of squares: $(SS_M Y) = \Sigma n_k(\bar{x}_k - \bar{x}_{grand})^2$

n_k = no. of items in group (10); $\bar{X}_k$ = group mean; $\bar{X}_{grand}$ = DV grand mean (52.00)

$$SS_M Y = (10 \times (66.10 - 52.00)^2) + (10 \times 46.10 - 52.00)^2) + (10 \times (43.80 - 52.00)^2) = \mathbf{3008.60}$$

Degrees of freedom (df): $df_M Y$ = no of groups (3) *minus* 1 = 2

Residual sum of squares ($SS_R Y$)

Formula for residual sum of squares: $(SS_{RY}) = \Sigma s_k^2(n_k - 1)$

s_k^2 = group variance; **n** = number of cases per group (10); **n** − 1 = 9

$$SS_R Y = (329.43 \times 9) + (87.66 \times 9) + (165.73 \times 9) = \mathbf{5245.40}$$

$df_R Y$ = (sample size *minus* 1) minus $df_M Y$ = 30 − 1 − 2 = 27

Mean squares

This is found by dividing sum of squares by the relevant degrees of freedom (df):

Model mean square **$MS_M Y$**: $SS_M Y \div df_M Y = 3008.60 \div 2 = \mathbf{1504.30}$

Residual mean square **$MS_R Y$**: $SS_R Y \div df_R Y = 5245.40 \div 27 = \mathbf{194.27}$

F ratio:

This is calculated from model mean square divided by residual mean square:

$$F_Y = \frac{MS_M Y}{MS_R Y} = 1504.30 \div 194.27 = \mathbf{7.74}$$

Main effect CV (Age; X)

Model sum of squares ($SS_M X$) Formulae as for DV

$$SS_M X = (10 \times (36.50 - 29.30)^2) + (10 \times (25.20 - 29.30)^2) + (10 \times (26.20 - 29.30)^2) = \mathbf{782.60}$$

$df_M X$ = no of groups (3) *minus* 1 = 2

Residual sum of squares ($SS_R Y$) Formulae as for DV

$$SS_R X = (67.39 \times 9) + (30.40 \times 9) + (54.62 \times 9) = \mathbf{1371.70}$$

$df_M X$(sample size *minus* 1) minus $df_M Y$ = 30 − 1 − 2 = 27

so df_{RX} = 29 − 2 = 27

Mean squares

Model mean square **$MS_M X$**: $SS_M X \div df_M X = 782.60 \div 2 = \mathbf{391.30}$

Residual mean square **$MS_R X$**: $SS_R X \div df_R X = 1371.70 \div 27 = \mathbf{50.80}$

F ratio:

$$F_X = \frac{MS_M X}{MS_R X} = 391.30 \div 50.80 = \mathbf{7.70}$$

Effect of CV on DV

Model sum of cross-products (SP_M)

Formula for model sum of squares: $$SP_M = \frac{\Sigma(Sum_k X \times Sum_k y)}{n} - \frac{SS_X \times SS_Y}{kn}$$

SS_X/SS_Y = Sum of sums (see earlier); k = no. of groups (3); n = no. of items per group (10)

$$SP_M = \frac{(365 \times 661) + (252 \times 461) + (262 \times 438)}{10} - \frac{879 \times 1560}{3 \times 10}$$

$$= 47219.30 - 45708.00 = \mathbf{1511.30}$$

Residual sum of cross products (SP_R)

Formula for residual sum of squares: SP_R = Sum of squared products $- \dfrac{\Sigma(\text{Sum}_k X \times \text{Sum } KY)}{n}$

$$SP_R = 48640 - 47219.30 \text{ (see } SP_M) = \mathbf{1420.70}$$

Model cross-product partitions (SS_M^P)

$$SS_M^P = SS_M Y - \left[\frac{(SP_M + SP_R)^2}{SS_M X + SS_R X}\right] - \frac{(SP_R)^2}{SS_R X} \text{ we know all of those products now:}$$

$$SS_M^P = 3008.60 - \left[\frac{(1511.30 + 1420.70)^2}{782.60 + 1371.70}\right] - \frac{1420.70^2}{1371.70} = \mathbf{489.60}$$

df_M^P = no of groups (3) *minus* 1 = 2

Residual cross-product partitions (SS_R^P)

$$SS_R^P = SS_R Y - \frac{SP_R^2}{SS_R X} = 5245.40 - \frac{1420.70^2}{1371.70} = \mathbf{3773.95}$$

df_M^P = Sample size *minus* no. of groups *minus* no. of covariates = 30 − 3 − 1 = 26

Mean cross-product partitions:

Mean cross-product partition (model): $MS_M^P = SS_M^P \div df_M^P = 489.60 \div 2 = \mathbf{244.80}$

Mean cross-product partition (residual): $MS_R^P = SS_R^P \div df_R^P = 3773.95 \div 26 = \mathbf{145.15}$

F ratio:

$$F^P = MS_M^P \div MS_R^P = 244.80 \div 145.15 = \mathbf{1.69}$$

We will not attempt to explore manual calculations for MANCOVA.

16 LINEAR AND MULTIPLE LINEAR REGRESSION

Learning objectives

By the end of this chapter you should be able to:

- Recognise when it is appropriate to use (simple) linear and multiple linear regression
- Understand the theory, rationale, assumptions and restrictions associated with each test
- Calculate the outcome manually (using maths and equations)
- Perform analyses using SPSS
- Know how to measure effect size and power
- Understand how to present the data and report the findings

What is linear regression?

Linear regression investigates relationships by examining the proportion of variance in a (numerical) outcome (dependent) variable that can be explained by one or more predictor (independent) variables. A numerical outcome is any observation that we can measure in terms of its magnitude. This might be income across a group of people, or it could be exam scores recorded for a group of students, or perhaps scores from a quality of life questionnaire (completed by several participants.) In all of those examples, it is quite likely that the outcome 'values' will fluctuate for one reason or another. Income may vary from person to person; student exam scores could differ across the class; quality of life scores might change according to participants. How much those scores vary across a single group will depend on a number of factors. Income may vary according to people's qualifications; exam scores might differ across the class according to the amount of revision a student did; quality of life scores might fluctuate between participants due to differences in their perceived physical health. Then again, scores in any of those examples might simply differ due to random factors. A linear regression model helps us determine what proportion of that variation is explained by factors that we have accounted for (such as qualifications, revision and perceived physical health) and the proportion that is unexplained (random factors or those which we have not accounted for).

What is simple linear regression?

Simple linear regression examines the proportion of variance in outcome scores that can be explained by a single predictor variable. For example, we could examine a group of participants in respect of quality of life scores as the **outcome variable** (from a questionnaire) and perceived physical health as the **predictor variable**. Note how we refer to 'outcome' and 'predictor' variables for 'dependent variable' and 'independent variable' respectively; these are more appropriate terms when referring to linear regression.

What is multiple linear regression?

Multiple linear regression examines the proportion of variance in outcome scores that can be explained by several predictor variables. This is perhaps a more realistic use of linear regression, since outcomes are usually the result of many causes. For example, quality of life scores may be explained by a whole series of factors in addition to perceived physical health, such as income, job satisfaction, relationship satisfaction and depression status. To examine this, we could use multiple linear regression to explore variance in quality of life scores, and investigate the proportion that is explained by those predicted factors, compared with that explained by random factors. In theory, there is no limit to how many predictors that we can include. However, as we will see later, there are restrictions on this according to sample sizes.

We will explore the various assumptions and restrictions regarding the type of variables that we can include later, but it is worth clarifying one important issue now. The outcome variable for linear regression must be represented by a numerical score. If we need to investigate a categorical outcome, such as the likelihood of passing an exam (yes or no), as predicted by a series of variables (such as the amount of revision undertaken, the number of lectures attended, etc.), we would undertake this with logistic regression (see Chapter 17).

Research questions for linear regression

To illustrate linear regression, we will use the example we referred to earlier regarding quality of life perceptions. In this scenario, the Centre for Healthy Independent Living and Learning (CHILL) would like to know what factors contribute to a perception of a good quality of life. They compose a questionnaire that attempts to capture those perceptions. Various questions are included, which are scored according to how positively the participant rates their quality of life. To investigate what factors might influence how people report quality of life, CHILL measure some additional factors. In an initial pilot study, in addition to the quality of scale, they ask the participants to rate how good they perceive their physical health to be. Later, in a larger study, CHILL recruit a new sample to complete quality of life questionnaires and ask them to rate their perceptions of physical health (as before), but also ask the participants to report their current income, state how satisfied they are with their job, indicate how satisfied they are in their relationship with their spouse or partner, and declare whether they are currently depressed (or not).

16.1 Take a closer look
Summary of linear regression examples

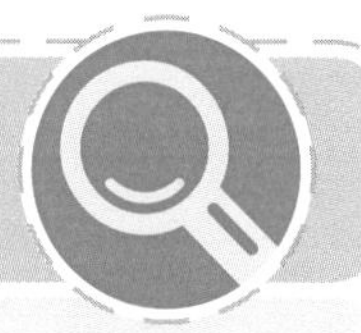

Simple linear regression

Outcome variable: Quality of life scores
Predictor: Perceived physical health

Multiple linear regression

Outcome variable: Quality of life scores
Predictors: Perceived physical health, income, job satisfaction, relationship satisfaction and depression status

Theory and rationale

Exploring linear regression models

Although the methods for calculating simple linear regression and multiple linear regression are similar, we will now divide this chapter to focus on each type. The potential use of these statistics is quite different and the rules and assumptions are very much stricter for multiple linear regression.

Simple linear regression

How simple linear regression works

The line of best fit

A key aim of linear regression is to find a ***model*** to illustrate how much of the outcome variable is explained by the predictor variable. Data points are drawn on a graph (called a scatterplot), through which we can draw a line that approximates the average of those points (we call this the line of best fit). That line is often referred to as the regression line – an example of just such a line is shown in Figure 16.1. It reflects the outcome from some data that seek to answer the first of the research questions set by the CHILL group. This line is described in terms of the gradient and where it crosses the Y axis (the vertical line to the left of the graph); some sources refer to this as the ***intercept***. The relationship between two variables is determined by the correlation between them. The gradient of the line illustrates how much the outcome variable changes for every unit change in the predictor variable. However, that predictor (significantly) contributes to the variance in the outcome only if that gradient is significantly greater than 0. However, that does *not* mean that the larger the gradient, the more likely it is to be significant. As we have seen throughout this book, significance is not determined by size alone.

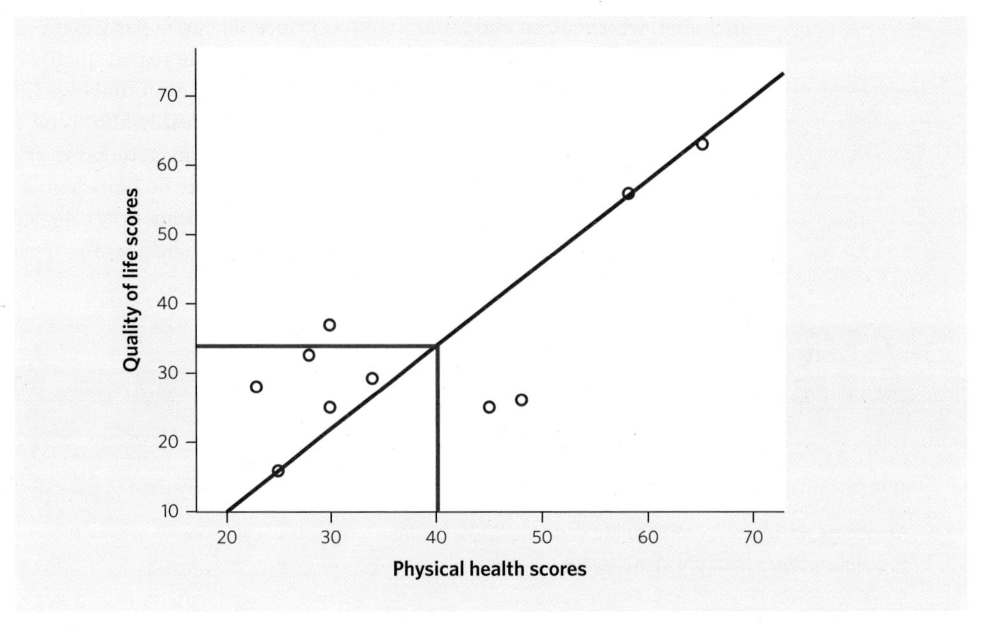

Figure 16.1 Scatterplot: quality of life perceptions vs. perceptions of physical health

It is important not to confuse the 'steepness' of the slope with the 'relationship' between the variables. It is quite possible for a steep line to represent a non-significant correlation, while a shallow line might relate to a highly significant correlation. This is because that gradient is dependent on the measurement scale used. As we saw just now, the significance of that gradient is determined by more than just the size.

We could use the line of best fit to estimate outcome scores from predictor values. For example, if we draw a vertical line from a 'Physical health' score of 40 to that line of best fit (as shown by the red line in Figure 16.1), and then a horizontal line from where it meets across to the vertical axis, we can estimate an outcome score of about 34. Of course, this is all rather subjective – we need an objective measure to formally predict outcome scores from predictor values. We do this with a simple linear regression equation.

16.2 Nuts and bolts

Simple linear regression equation

We can use an equation in simple linear regression to plot an outcome score from a predictor. You will not have to do this in the normal running of linear regression analyses through SPSS, but it will help you understand the process a little more if you see how it works.

Simple linear regression model equation: $Y_i = \beta_0 + \beta_1 X_i + \varepsilon_i$

Y = outcome variable score
i = the specific outcome score for participant (or case) 'i'
β_0 = constant (where the line crosses Y axis)
β_1 = gradient of line
X = predictor variable score
ε = error

In some sources you will see the gradient expressed as 'b_1' (in SPSS it is called B).

Using the equation shown in Box 16.2, once we know the parameters for 'β_0' and 'β_1' we can use values of X to find estimated values of Y. It is important to stress that these are *estimates* because we cannot account for the error until we have performed some more formal statistics. In Figure 16.1 the gradient ($\boldsymbol{\beta_1}$) is 0.776, while the constant ($\boldsymbol{\beta_0}$) is 3.863. We will see how to calculate the gradient and constant later but (for now) just take them as read for demonstration purposes. For example, if we took a physical health score of 40, we can use the linear regression equation to predict the outcome in quality of life scores:

$$Y = \beta_0 + \beta_1 X_1 = 3.863 + (0.776 \times 40) = 34.903$$

16.3 Nuts and bolts

Key terms in linear regression

Throughout this chapter we will encounter a number of terms that illustrate key aspects of a linear regression model.

Scatterplot: A series of data points that represents the predictor variable score and outcome value.
Best line of fit: The 'average' line drawn through the data points in a scatterplot
Gradient: The slope of the line. It illustrates how values in outcome scores change for every unit change in the predictor: a gradient significantly greater than 0 indicates that the predictor significantly contributes to the variance in outcome scores
Constant: Where the line crosses the Y axis (outcome variable)
Variance (R^2): The extent that the scores vary; the more that can be explained by the predictor the better.
'Success' of model: The extent that equation $Y_i = \beta_0 + \beta_1 X_i + \varepsilon_i$ can accurately predict an outcome value from any given predictor variable score.
ANOVA outcome (F): The success of the regression model; if F is significant, it suggests that the model is better at predicting outcome than some other (arbitrary) method.

Variance and correlation

Variance and correlation play a central role in linear regression. As we have seen in previous chapters, variance is the sum of the squared differences between each case score and the mean score in relation to the sample size. It describes the actual variation of the scores either side of the mean. Correlation is represented by *r*, multiple correlation by **R**, while variance is indicted by $\mathbf{R^2}$ – we will learn more about all of these later. We will see how to calculate R^2 in Box 16.5. In simple linear regression, we are concerned only with the correlation between the two variables in the model. In multiple linear regression, correlation becomes more complex. As we add variables to our model we need to employ semi-partial correlation instead. This is something that we first encountered in Chapter 6. However, it would make sense to deal with this when we explore multiple linear regression in more depth later in this chapter. For now, we will keep things simple.

The regression model

The 'success' of the linear regression *model* depends on how 'well' we can predict the outcome. Earlier, we used a regression model to predict an outcome score (quality of life perceptions) from a predictor value (physical health), the gradient of regression line and the constant. However, the full regression equation also includes an error value (which we do not know). Having predicted an outcome score of 34.903, we might subsequently find that the 'actual' outcome is 35. The difference (0.097) represents the error (or residual). The success of the regression model depends on how closely our predicted values match actual outcome. We will see how to quantify that success in Box 16.5, but it is ultimately measured by an F ratio outcome. If the F ratio is significant, it suggests that the model is better at predicting outcome than some random method (such as just using a mean score to predict the outcome).

Gradient of the slope

As we saw earlier, we can use the line of best fit to predict outcome. As we saw earlier, we could physically draw a line, as we did in Figure 16.1. Alternatively, we can use the (simple linear) equation that we saw in Box 16.2. In that equation, the gradient (β_1) tells us how outcome values change for every unit change in the predictor (this is shown as B in SPSS). However, we also need to know whether that predictor makes a meaningful contribution to the variance in the outcome scores. We do that by assessing whether the gradient is 'significantly greater than 0' (which SPSS calculates for us, using an independent t-test). If that indicates a significant outcome, it suggests that the gradient is significantly greater than 0 – it means that we can be confident that the predictor does indeed significantly contribute to the variance in outcome scores. In simple linear regression, this is somewhat academic as we have only one predictor variable. If we already know that we have a significant model, the predictor variable will significantly contribute to variance in the outcome. In multiple linear regression we have several lines; the gradients become more important (as we will see later). We will see how to calculate the significance of the gradient in Box 16.5.

Putting it all together

In simple linear regression we need to express three outcomes:

1. The amount of variance in the outcome variable explained by the model (R^2).
2. Whether that model is significantly better than using some other random method to predict outcome (via F ratio).
3. Whether the gradient is significantly greater than 0 (via an independent t-test).

Assumptions and restrictions

There are very few restrictions in the use of simple linear regression (unlike multiple linear regression, as we will see later). The outcome variable must be represented by continuous numerical scores. Strictly speaking, those values should be parametric (as we saw in Chapter 5, this means that they should be at interval and *reasonably* normally distributed). However, ordinal data are frequently used in linear regression models (the data used in our examples are probably ordinal). The predictor variable can be categorical or continuous. If the predictor is categorical it must be dichotomous (there must be only two groups, which must be given the values of 0 and 1, when setting up the value labels in SPSS). If a categorical variable has three or more groups, these should be recoded into several dichotomous dummy variables. For example, if the predictor variable represented ethnicity (British, Asian and African), you would need to set up three new variables: British (0 = yes, 1 = no), Asian (0 = yes, 1 = no) and African (0 = yes, 1 = no). However, the analysis would no longer be a simple linear regression as there would be several predictor variables; multiple linear analyses would be needed instead (see later).

16.4 Take a closer look
Summary of assumptions and restrictions

- Outcome variable must be represented by continuous numerical scores
  - Categorical outcomes can be measured using logistic regression (Chapter 17)
- Outcome variable scores should be *reasonably* normally distributed
- The predictor variable can be continuous or categorical
  - If categorical, this must be dichotomous (two groups)

Measuring outcomes in a simple linear regression model

We should now see how to examine those key criteria that we highlighted just now. We will see how to perform simple linear regression in SPSS shortly. In the meantime, we will explore how to undertake calculations manually.

16.5 Calculating outcomes manually
Simple linear regression calculation

To illustrate how to calculate simple linear regression manually, we will explore the first of the research questions that the CHILL group set earlier. The outcome variable (Y) is 'QoL scores', while the predictor variable (X) is 'physical health scores' (for both variables, higher scores indicate 'better' outcomes). The raw scores are presented in Table 16.1. **You will find a Microsoft Excel spreadsheet associated with these calculations on the web page for this book.**

Table 16.1 Quality of life (QoL) and physical health (PH) scores

	PH	QoL					
	X	Y	Y^1	$X - \bar{X}$	$(X - \bar{X})^2$	$Y^1 - Y$	$(Y^1 - Y)^2$
	58	56	48.85	19.4	376.36	7.15	51.12
	23	28	21.70	−15.6	243.36	6.30	39.69
	45	25	38.76	6.4	40.96	−13.76	189.34
	30	25	27.13	−8.6	73.96	−2.13	4.54
	25	16	23.25	−13.6	184.96	−7.25	52.56
	30	37	27.13	−8.6	73.96	9.87	97.42
	65	63	54.28	26.4	696.96	8.72	76.04
	28	33	25.58	−10.6	112.36	7.42	55.06
	48	26	41.09	9.4	88.36	−15.09	227.71
	34	29	30.23	−4.6	21.16	−1.23	1.51
Mean	**38.60**	**33.80**		$\Sigma(X - \bar{X})^2$	1912.40	$\Sigma(Y^1 - Y)^2$	794.98
St Dev	**14.58**	**14.70**					
St Error	**4.61**	**4.65**					

Correlation between X and Y

Formula for (Pearson's) correlation $r = \dfrac{\sum(x_i - \bar{x})(y_i - \bar{y})}{(N - 1)s_x s_y}$

x_i and y_i represent case scores for x and y; and are the respective mean scores
N = sample size (10); s = standard deviation
We use all case scores and mean scores from Table 16.1:

$$r = ((58 - 38.60) \times (56 - 33.80)) + ((23 - 38.60) \times (28 - 33.80)) + \ldots ((34 - 38.60) \times (29 - 33.80)) \div (9 \times 14.58 \times 14.70) = \mathbf{0.769}$$

Coordinates for line of best fit

We need to calculate the 'estimated' scores for Y^1 (we will need these for *model sum of squares* shortly).

Linear regression formula (without error term): $Y^1 = \beta_0 + \beta_1 X$

$$\beta_1 = r \times \left(\frac{\sigma Y}{\sigma X}\right) r = \text{ correlation; } \sigma = \text{ standard error } = 0.769 \times \left(\frac{4.65}{4.61}\right) = 0.776$$

So, now we know that $Y^1 = \beta_0 + (0.776)X$; if we use mean scores of X and Y:

$33.80 = \beta_0 + (0.776 \times 38.60) = \beta_0 + 29.95$: we rearrange that as: $\beta_0 = 33.80 - 29.95 = 3.85$

Now we can apply these to the equation to find values of Y^1 (as shown in Table 16.1):

$Y^1 = 3.85 + 0.776X$ (e.g. when $X = 58$: $Y^1 = 3.85 + (0.776 \times 58) = 48.85$)

Total sum of squares (SS_T) The formula for SS_T: $\sum(y_i - \bar{y})^2$

So, using the case scores for Y and the mean of Y (33.80):

$$SS_T = (56 - 33.80)^2 + (28 - 33.80)^2 + \ldots (29 - 33.80)^2 = \mathbf{1945.60}$$

Model sum of squares (SS_M) The formula for SS_M: $\sum(y_i^1 - \bar{y})^2$

So, using the case scores for Y^1 and the mean of Y:

$$SS_M = (48.85 - 33.80)^2 + (21.70 - 33.80)^2 + \ldots (30.23 - 33.80)^2 = \mathbf{1150.33}$$

SS_M degrees of freedom (df_M) = no. of predictors (in our case $df_M = 1$); this is the numerator *df*

Variance (R^2) As we saw earlier (R^2) $= \dfrac{SS_M}{SS_T} = 1150.33 \div 1945.60 = \mathbf{0.591}$

Residual sum of squares (SS_R): $SS_R = SS_T - SS_M = 1945.60 - 1150.33 = \mathbf{795.26}$

SS_R degrees of freedom (df_R) $= (N - 1) - df_M = (10 - 1) - 1 = 8$; this is the denominator *df*

F ratio This is found from: $\dfrac{MS_M}{MS_R}$ MS_M = mean model square; MS_R = residual mean square

$MS_M = SS_M \div df_M = 1150.33 \div 1 = \mathbf{1150.33}$
$MS_R = SS_R \div df_R = 795.26 \div 8 = \mathbf{99.41}$
F ratio $= 1150.33 \div 99.41 = \mathbf{11.57}$

We compare that F ratio to F-distribution tables (Appendix 4), according to numerator and denominator degrees of freedom (*df*), where $p = .05$. In this case, when numerator $df = 1$ and denominator $df = 8$, the cut-off point for F is 5.32. Our F ratio (11.57) is greater than that, so our model is significantly better at predicting outcome than using an arbitrary method such as the mean score.

We can also use Excel to calculate the critical value of F and to provide the actual p value. You can see how to do that on the web page for this book. In our example, $p = .009$. You can also perform the entire test in Excel.

Significance of the gradient

We need to find t, where $t = \beta_1 \div \sigma\beta_1$

We know that $\beta_1 = 776$; $\sigma\beta_1$ = standard error of gradient $\sigma\beta_1 = \dfrac{SEE}{\sqrt{\Sigma(x - \bar{x})^2}}$ where SEE = standard error of estimate: $\sqrt{\dfrac{\Sigma(y - y^1)^2}{n - k - 1}}$ where k = no. of predictors (1)

Using grey cells from Table 16.1, SEE $= \sqrt{\dfrac{794.98}{8}} = 9.97$ $\qquad \sigma\beta_1 = \dfrac{9.97}{\sqrt{1912.40}} = 0.228$

$$\text{So, } t = \beta_1 \div \sigma\beta_1 = 0.776 \div 0.228 = \mathbf{3.40}$$

We can also use Excel to calculate the critical value of t and to provide the actual p value. You can see how to do that on the web page for this book. In our case, $p = .009$ (two-tailed).

How SPSS performs simple linear regression

We can demonstrate how to perform simple linear regression in SPSS with the CHILL data that we used to calculate outcomes manually (see Box 16.5). You will recall that we are measuring quality of life scores as the outcome variable and perceived physical health scores as the predictor variable, in a group of ten participants (see Table 16.1). For both variables, higher scores indicate 'better' outcomes. We will assume that quality of life scores represent interval data, meeting parametric standards. However, we will need to check that the data are normally distributed.

16.6 Nuts and bolts

Setting up the data set in SPSS

When we create SPSS data sets for simple linear regression, we need to set up an outcome variable (where the column represents a continuous score) and a predictor variable (which can be continuous or categorical). In our example, the predictor variable is a numerical score. If the predictor had been represented by groups, we would use categories defined by value labels (e.g. 0 = male; 1 = female).

QoL and health.sav [DataSet1] - IBM SPSS Statistics Data Editor

File Edit View Data Transform Analyze Graphs Utilities Add-ons Window Help

	Name	Type	Width	Decimals	Label	Values	Missing	Columns	Align	Measure
1	QoL	Numeric	8	0	Quality of life s...	None	None	8	Right	Scale
2	physical	Numeric	8	0	Physical health...	None	None	8	Right	Scale
3	mood	Numeric	8	0	Mood	None	None	8	Right	Scale

Figure 16.2 Variable View for 'QoL and health' data

Figure 16.2 shows how the SPSS Variable View should be set up. The first variable is called 'physical'; this is the predictor variable, which represents 'Physical health scores'. The second variable is called 'QoL'; this is the outcome variable, which represents 'Quality of life scores'. Since both of these variables are represented by 'continuous' numerical scores, we do not need to adjust anything in the Values column and the Measure column is set to Scale.

QoL and health.sav [DataSet1] - IBM SPSS Statistics Data Editor

File Edit View Data Transform Analyze Graphs Utilities Add-ons Window Help

	QoL	physical	mood	var	var	var	var	var	var
1	56	58	48						
2	28	23	27						

Figure 16.3 Data View for 'QoL and health' data

Figure 16.3 illustrates how this will appear in Data View. This will be used to select the variables when performing this test. Each row represents a participant. When we enter the data, we simply input the relevant score for each participant in respect of the predictor variable and outcome variable.

Testing for normal distribution

To examine normal distribution, we investigate outcomes from Kolmogorov–Smirnov and Shapiro–Wilk tests (the sample size determines which outcome we should report). We only need to test the data from the outcome variable.

Open the SPSS file QoL and health

Select **Analyze** → **Descriptive Statistics** → **Explore** → (in new window) transfer Quality of life scores to **Dependent List** → select **Plots** radio button → click **Plots** box → (in new window) click **None** (under **Boxplot)** → make sure that **Stem-and-leaf** and **Histogram** (under **Descriptive**) are unchecked → tick **Normality plots with tests** radio button → click **Continue** → click **OK**

	Kolmogorov–Smirnov[a]			Shapiro–Wilk		
	Statistic	df	Sig.	Statistic	df	Sig.
Quality of life scores	.228	10	.150	.849	10	.056

Figure 16.4 Kolmogorov-Smirnov/Shapiro-Wilk test for quality of life scores

Because we have a sample size of ten (for each group) we should refer to the Shapiro–Wilk outcome (because the sample is smaller than 50). Figure 16.4 confirms that the outcome variable is probably normally distributed: $W(10) = .849$, $p = .056$. These tests investigate whether the data are significantly *different* from a normal distribution. Since 'Sig.' is greater than .05, we can be confident that is not the case.

Running test in SPSS

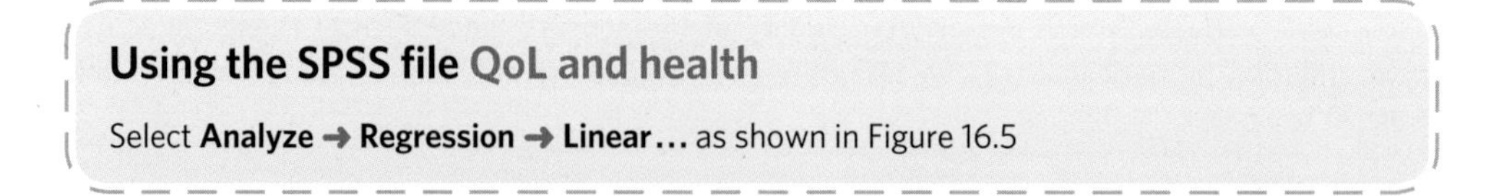

Using the SPSS file QoL and health
Select **Analyze** → **Regression** → **Linear...** as shown in Figure 16.5

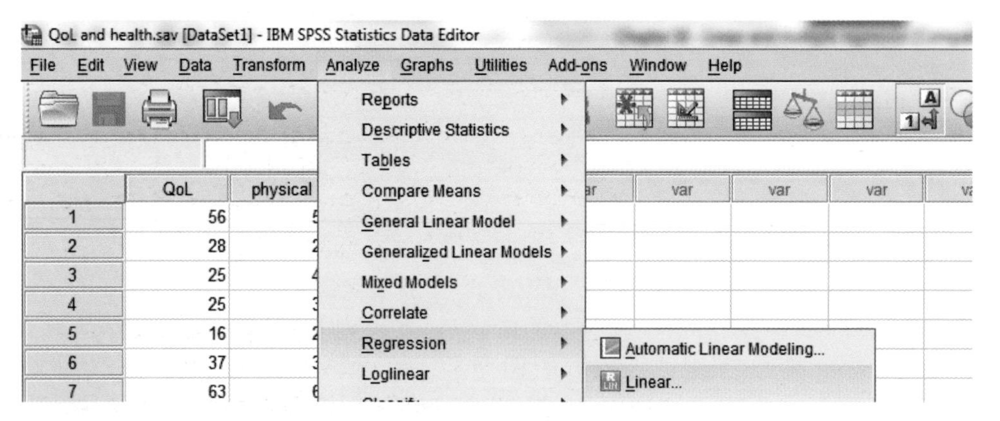

Figure 16.5 Linear regression: procedure 1

In new window (see Figure 16.6), transfer **Quality of life scores** to **Dependent:** ➜ transfer **Physical health scores** to **Independent(s)** ➜ click **OK**

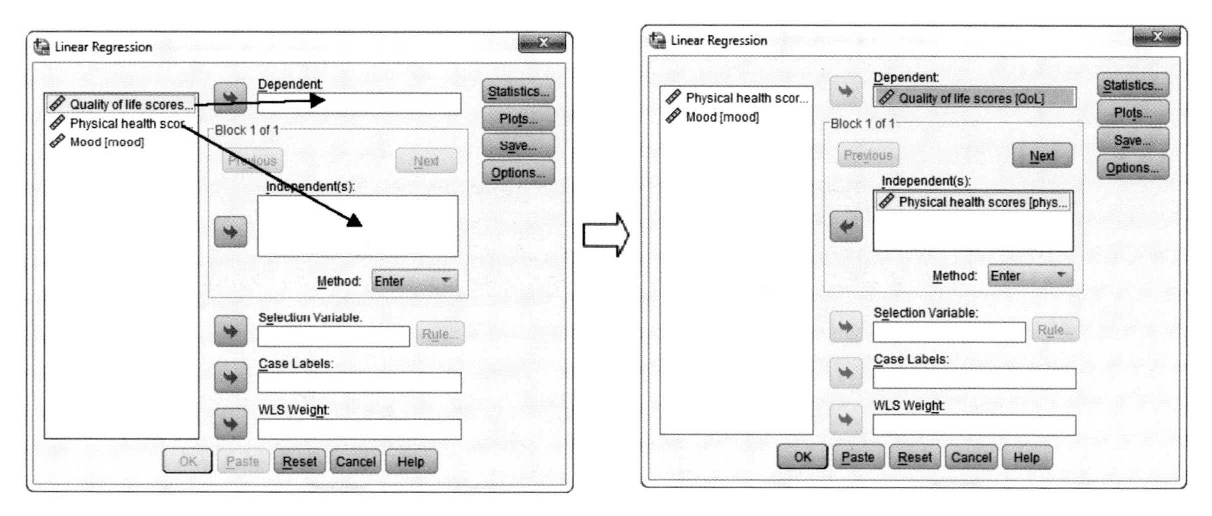

Figure 16.6 Linear regression: procedure 2

Interpretation of output

Figure 16.7 presents two key statistics of note. In simple linear regression, correlation between the variables is expressed by R (in our case, R = .769). Compare that to the manual calculations we performed earlier. The proportion of variance in the outcome variable (quality of life scores) that can be explained by the predictor variable is illustrated by R Square (R^2 = .591: compare that to the calculations in Box 16.5). To express the R^2 as a variance percentage, we multiply the outcome by 100 – 59.1% of all variance in quality of life scores can be explained by variations in reported physical health scores.

Figure 16.8 indicates that the regression model is significantly better at predicting outcome than some random method, F (1, 8) = 11.572, p = .009.

Model Summary

Model	R	R Square	Adjusted R square	Std. error of the estimate
1	.769[a]	.591	.540	9.970

a. Predictors: (constant), physical health scores

Figure 16.7 Linear regression: model summary

Model		Sum of squares	df	Mean square	F	Sig.
1	Regression	1150.325	1	1150.325	11.572	.009[a]
	Residual	795.275	8	99.409		
	Total	1945.600	9			

Figure 16.8 Significance of model

Coefficients[a]

Model		Unstandardised coefficients		Standardised coefficients	t	Sig.
		B	Std. error	Beta		
1	(Constant)	3.863	9.348		.413	.690
	Physical health scores	.776	.228	.769	3.402	.009

a. Dependent variable: quality of life scores

Figure 16.9 Model parameters

Figure 16.9 presents the data regarding the gradient of the linear regression line. The intercept reflects where the line crosses the Y axis, shown as the 'Constant' in SPSS (we know it as β_0 in the regression equation). In this case it is 3.863 (we calculated β_0 to be 3.85, which is close enough). The gradient of the line is shown in the 'Physical health scores' row, under 'B' in SPSS (this was β_1 in the regression equation). In this case, B = 0.776, with a standard error of 0.228 (as we calculated manually). It suggests that for every unit that physical health scores increase, quality of life scores increase by 0.776 of a point. The significance of that gradient is shown by the t score – in this case $t = 3.402$, $p = .009$. Therefore, physical health scores significantly contribute to the variance in quality of life scores.

Effect size and power

We can use G*Power to provide the effect size and statistical power for our linear regression outcome. (We have seen the importance of this in most of our statistical chapters so far, but especially see Chapter 4 for an explanation of the rationale.)

From **Test family** select **F tests**
From **Statistical test** select **Linear multiple regression: Fixed model, R^2 deviation from zero**
From **Type of power analysis** select **Post hoc: Compute achieved power - given α, sample size and effect size**

Now we enter the **Input Parameters**:

To calculate the **Effect size**, click on the **Determine** button (a new box appears).
In that new box, for **Squared multiple correlation p^2** type **0.591** (the R^2 figure) → click on **Calculate and transfer to main window**
Back in original display, for **α err prob** type **0.05** (the significance level) → for **Total sample size** type **10** → for **Number of predictors** type **1** → click on **Calculate**
Effect size (d) **1.44** (very strong); and Power (1-β err prob) **0.91** (very good - higher than the optimal target of 0.80 - see Chapter 4).

Writing up results

We should present the key data in a table and discuss the statistical outcome.

Table 16.2 Linear regression analysis of quality of life scores

Predictor variable	R^2	Adj. R^2	F	p	Constant	β_1	t	p (t)
Physical health scores	.591	.540	11.572	.009	3.863	.776	3.402	.009

Table 16.2 confirms that changes in reported physical health scores were significantly able to predict variance in quality of life scores. The linear regression model explained 59.1% of the overall variance in quality of life perceptions ($R^2 = .591$), which was found to significantly predict outcome, $F(1, 8) = 11.572$, $p = .009$, $d = 1.44$.

Multiple linear regression

Purpose of multiple linear regression: a reminder

Whereas simple linear regression explores the proportion of variance that can be explained in outcome scores by a single predictor, multiple linear regression investigates the proportion of variance that several predictors can explain. When we investigated simple linear regression, we referred to the first of the research questions set by the CHILL group at the beginning of this chapter, whereby we sought to establish how much of the variance in quality of life scores could be explained by reported physical health. In this section we will extend that by measuring how much of that variance (in quality of life scores) can be explained by participants' reports regarding their income, job satisfaction, relationship satisfaction and depression status, in addition to those perceptions about physical health.

How multiple linear regression works

Lines of best fit

Similar to simple linear regression, the line of best fit is very important. However, this time we have several lines (one for each predictor), each with its own gradient. Each line will illustrate the relationship between that predictor and the outcome variable. Because there are several lines, the regression equation becomes a little more complex (see Box 16.7). The constant will also vary each time an additional predictor is added. The gradient of each line will be used to determine how outcome scores change for a unit change in each predictor. Each gradient is also assessed to which predictors make a meaningful contribution to the outcome variance. A

16.7 Nuts and bolts
Multiple linear regression equation

To predict outcome scores in multiple linear regression, we need to account for several lines, each representing a predictor variable. As a result, the equation that we saw in Box 16.2 needs to be extended:

$$Y_i = \beta_0 + \beta_1 X_1 + \beta_2 X_2 + \ldots \beta_n X_n + \varepsilon_i$$

The first part of this equation ($\beta_0 + \beta_1 X_1$) and the error term (ε_i) are identical to simple linear regression - we still have the constant, or intercept, (β_0) and the first gradient ($\beta_1 X_1$). However, we also need a gradient for each of the remaining lines, hence ($\beta_n X_n$).

Once again, you will not have to calculate this (unless you want to) as SPSS does it for you, but it might aid your understanding of the processes needed to generate the multiple linear regression model.

predictor can be deemed to significantly contribute to the variance in outcome scores only if its gradient is significantly greater than zero (as we saw with simple linear regression).

Semi-partial correlation

Similar to simple linear regression, variance and correlation have a major role in determining outcome. However, we treat correlation very differently here. In Chapter 6, we briefly explored something called semi-partial correlation: this is central to multiple linear regression. It is also worth noting that the 'R' figure produced by SPSS for multiple linear regression actually relates to multiple correlation (unlike with simple linear regression where R is actually equivalent to the correlation between the two variables).

We should recap briefly here. In simple linear regression we explore the relationship between two variables. In that case, we can refer to the correlation between the predictor and outcome variable. However, in multiple linear regression, for every predictor that we add to the model, we explore the correlation between the predictor and outcome, while holding the remaining predictors constant (but only for the predictor variables, not the outcome). It might pay you to revisit the section in Chapter 6 referring to semi-partial correlation. A summary of the key points is shown in Box 16.8.

16.8 Nuts and bolts
Correlation, partial correlation and semi-partial correlation

We need to differentiate between different types of correlation when performing linear regression analyses. Here is a summary of those types:

Correlation: The standardised relationship between two variables.

Partial correlation: The relationship between two variables after controlling for a third variable that is held constant for both of the original variables

Semi-partial correlation: The relationship between two variables after controlling for a third variable that is held constant for only one of the original variables

We should put correlation and semi-partial correlation into context with linear regression by using the example that we explored in Chapter 6. There we explored the relationship between sleep quality perceptions and mood in a group of 98 participants, and found a moderately negative correlation: $r\,(96) = -.324$, $p = .001$. That seemed pretty straightforward until we wondered whether the observed relationship was affected by the participants' age. So, we employed partial correlation to examine the relationship between sleep quality perceptions and mood, while holding age constant for both variables. We found that correlation was now weaker and no longer significant: $r\,(95) = -.167$, $p = .051$. But then we wondered whether the relationship between sleep quality perceptions and mood was actually a relationship between sleep quality perceptions and age, and nothing to do with mood. To explore that, we undertook semi-partial correlation. We examined the relationship between sleep quality perceptions and mood, while holding age constant for just sleep quality perceptions. We found that the relationship was reduced further: $r\,(95) = -.128$, $p = .101$.

We can extend that point still further using the CHILL data that we explored for simple linear regression. This time we will take that a step further by exploring the implications of correlation and semi-partial correlation for linear regression. We were examining the relationship between perceptions of quality of life and physical health. Using those data, we can report the correlation between the variables:

Figure 16.10 indicates that we have a strong, positive (significant) correlation between quality of scores and physical health scores: $r\,(8) = .769$, $p = .005$. Earlier, we used a linear regression model to explore whether we could significantly explain the variance in quality of

Using the SPSS file QoL and health

Select **Analyze** ➔ **Correlate** ➔ **Bivariate...** ➔ transfer **Quality of life scores** and **Physical health scores** to **Variables** ➔ tick boxes for **Pearson** and **Two-tailed** ➔ click **OK**

		Quality of life scores	Physical health scores
Quality of life scores	Pearson correlation	1	.769**
	Sig. (1-tailed)		.005
	N	10	10
Physical health scores	Pearson correlation	.769**	1
	Sig. (1-tailed)	.005	
	N	10	10

Figure 16.10 Correlation between quality of life and physical health

scores from physical health scores. We saw that we explained 59.1% of the variance ($R^2 = .591$) in a significant regression model ($F(1, 8) = 11.572$, $p = .009$). Notice also how Figure 16.7 reported 'R' as .769. This is commensurate to the correlation coefficient (it always will be in simple linear regression). Furthermore, physical health scores significantly contributed to the variance in quality of life scores ($t = 3.402$, $p = .009$).

However, what if we suspected that quality of life scores had more to do with mood than physical health perceptions (despite the strength of the observed relationship)? To explore that suggestion, we can start by looking at the effect of employing a semi-partial correlation. We will examine the relationship between quality of life and physical health perceptions, while holding mood scores constant for physical health scores (only). Then we will look at the impact on the variance explained by the linear regression model:

Using the SPSS file QoL and health

Select **Analyze** ➔ **Regression** ➔ **Linear...** ➔ (in new window) transfer **Quality of life scores** to **Dependent** ➔ transfer **Physical health scores** and **Mood** to **Independent(s)** ➔ click **Statistics...** ➔ (in next window) tick boxes for **Estimates** , **Model fit** , and **Part and partial correlations** ➔ click **Continue** ➔ click **OK**

Actually, there are a lot more boxes that we need to tick for multiple linear regression, but we will explore that later.

		Unstandardised coefficients		Standardised coefficients			Correlations		
Model		B	Std. error	Beta	t	Sig.	Zero-order	Partial	Part
1	(Constant)	−7.845	8.668		−.905	.396			
	Physical health scores	.239	.279	.237	.858	.419	.769	.308	.151
	Mood	.967	.389	.689	2.488	.042	.872	.685	.438

a. Dependent variable: quality of life scores

Figure 16.11 Semi-partial correlation (and regression coefficients)

For the moment we will just focus on the final three columns of Figure 16.11 (the correlation data); we will come back to the regression coefficients later (but you may have already seen a pattern emerging). The semi-partial correlation between quality of life scores and physical health scores is considerably weaker than the initial correlation: $r = .151$ ($p = .419$). Now let's look at the regression outcome:

Model Summary

Model	R	R Square	Adjusted R square	Std. error of the estimate
1	.885[a]	.783	.721	7.765

a. Predictors: (constant), mood, physical health scores

Figure 16.12 Linear regression: model summary

Figure 16.12 indicates that 78.3% of variance in quality of life scores is explained in this model ($R^2 = .783$; actually we should refer to the adjusted R^2, but we will explain why later). We have explained more variance in the outcome than we did when we had only the one predictor. Notice also that 'R' does not relate to any observable correlation outcome from the reported data. This is because 'R' actually measures multiple correlation (whereas *r* measures standardised correlation); you don't need to know any more than that for now.

Model		Sum of squares	df	Mean square	F	Sig.
1	Regression	1523.579	2	761.790	12.636	.005[a]
	Residual	422.021	7	60.289		
	Total	1945.600	9			

Figure 16.13 Significance of model

Figure 16.13 indicates that the model significantly predicts the outcome variable, $F(2, 7) = 12.636$, $p = .005$. Now we can return to the regression co-efficients:

To explore the extent to which the predictor variables contribute to the overall variance in quality of life scores, we need to refer to the t-score data in Figure 16.14. In our initial analyses, physical health scores significantly contributed to quality of life scores (see Figure 16.9). Once we include mood into the model, physical health perceptions are seen to no longer contribute to variance in the outcome ($t = 0.858$, $p = .419$). This is despite the fact that there was a strong relationship between the predictor and the outcome. Meanwhile, mood scores do significantly predict variance in quality of life scores ($t = 2.488$, $p = .042$). This is precisely the kind of information that multiple linear regression seeks to uncover. There is a clear message here: just because there is a strong relationship between a predictor variable and outcome measure, it may not significantly contribute to the variance in that outcome once we explore other potential predictors.

		Unstandardised coefficients		Standardised coefficients			Correlations		
Model		B	Std. error	Beta	t	Sig.	Zero-order	Partial	Part
1	(Constant)	−7.845	8.668		−.905	.396			
	Physical health scores	.239	.279	.237	.858	.419	.769	.308	.151
	Mood	.967	.389	.689	2.488	.042	.872	.685	.438

a. Dependent variable: quality of life scores

Figure 16.14 Regression coefficients

However, we need to explore several more factors before we have the complete picture of what multiple linear regression can examine. To begin with, we need much larger samples than the example we have just explored – we used that to illustrate the point in a simple data set.

Variance and correlation

Another major difference with multiple linear regression relates to how we interpret the variance. Because we are using more variables, we need to report an *adjusted* variance, one that accounts for the number of variables used and the sample size. When we reported variance in simple linear regression, we used the term R^2. For multiple linear regression, we should report the adjusted R^2 (you can see how to calculate this at the end of this chapter.)

The impact of correlation needs to be more carefully considered with multiple linear regression, particularly the relationship between the predictor variables. While correlation between a single predictor and the outcome variable is important when evaluating how much variance has been explained, it is important that there is not too much correlation between predictors. We call this multi-collinearity. If we have too much, it can lead to inaccurate interferences about individual predictor variables. If two or more of these predictors are highly correlated to each other, their relationship may be so intertwined it may be impossible to distinguish between them.

The regression model

Once again, we need to examine whether the regression model is significantly better at predicting outcome than some other more random method, just as we did with simple linear regression. The success of the linear regression model depends on how closely our model predicts actual outcome. We can predict the outcome from the (extended) multiple regression model using the predictor values and gradients, and the constant. The difference between the predicted and actual outcome represents the error (or residuals).

We assess how closely we have modelled the actual outcome via the F ratio (just as we did with simple linear regression). We saw an example of how to calculate F ratio in Box 16.5. However, the route to get to this outcome in multiple linear regression is more complex (the manual calculations are shown at the end of this chapter). Although there are several predictor variables in these analyses, we still have only the one F ratio. If this is significant, it suggests that the model is better at predicting outcome than some random method.

Gradient of the slopes

Similar to simple linear regression, we need to assess the gradient of the slope between the predictor variable and the outcome variable. This is measured by the 'Beta' value (β_X in the multiple regression equation, but shown as 'B' in SPSS). However, when there are several predictors, we need to assess the gradient for each of the predictor variables included in the model. Each gradient is used to assess how outcome scores change for each unit change in that predictor. This is also examined to see whether it is significantly different from zero. As before, this is undertaken by means of an independent t-test. If that outcome is significant (for that predictor), it suggests that this predictor significantly contributes to the variance in the outcome scores. The calculations for that are more complex for multiple linear regression (you can see how this is done in the section at the end of the chapter).

Putting it all together

In multiple linear regression we need to express three outcomes – these are similar to simple linear regression, but with notable differences:

1. The amount of variance in the outcome variable explained by the model, indicated by the *adjusted* R^2 outcome.
2. Whether that model is significantly better at predicting outcome than some other random method (via F ratio).

3. Whether the gradient between *each* predictor variable and the outcome variable is significant, determined by the outcome from an independent t-test. A single predictor variable contributes to the variance in the outcome scores only if the gradient for that predictor is significantly greater than zero.

Assumptions and restrictions

16.9 Take a closer look
Summary of assumptions and restrictions

- There should be a normal distribution across the outcome variable
- Outliers need to be investigated as they may lead to Type II errors
- Avoid too many predictor variables in relation to the sample size
- Outcome variables must be continuous
- Predictor variables can be continuous or categorical
  - But categorical predictors must be dichotomous
    - And coded as 0 and 1 in SPSS values
- Outcome scores should be *reasonably* linear
  - Correlation should be at least $r = .30$,
  - But no higher than $r = .80$ (to avoid multi-collinearity)
- The residuals should not be correlated to each other (they should be independent)

There are many assumptions and restrictions that should be considered when conducting multiple linear regression, rather more than there are for simple linear regression. Many of these can be checked by selecting the appropriate parameters when running the statistic in SPSS. Some others will need to be run separately.

Parametric data

We should check that the outcome variable is normally distributed using the Kolmogorov–Smirnov test (it is unlikely that we would use an Shapiro–Wilk test as samples will generally need to be greater than 50, for reasons that will become clearer later). The outcome variable must be represented by continuous numerical scores. However, as we saw with simple linear regression, the predictor variable can be continuous (numerical) or categorical. If categorical, the predictor variable must be dichotomous (have two groups) and should be coded as 0 and 1 in the SPSS value labels facility (see later).

Outliers

Outliers are extreme scores in the data set. When we explored the regression line earlier (see Figure 16.1), we saw that some data points are very close to that line, while others are further away. If a data point is too far from the line of best fit, it could undermine the strength of the model. Outliers can overstate the error variance and can lead to Type II errors. As we saw in Chapter 4, a Type II error occurs when we falsely reject the experimental hypothesis. We can 'count' outliers in SPSS by referring to the '**standardised residuals**' output (we will see how to do that later). The output reports the values in a variable after they have been converted into z-scores (we saw what that meant in Chapter 3). Using what we know about normal distribution, we can determine how many outliers we would expect in a data set according to the sample size (see Box 16.10).

16.8 Nuts and bolts
Outlier cut-off points

In multiple linear regression, we set a limit for how many data points can exceed certain z-score values based on normal distribution. Too many outliers might reduce the power of the model.

Table 16.3 z-score limits

z-score cut-off point	Max limit	p value
1.96	5%	.05
2.58	1%	.01
3.29	0	.001

So, we would not expect more than 5% of our data to exceed a z-score of 1.96, no more than 1% to exceed a z-score of 2.58, and (ideally) none should exceed a z-score of 3.29.

Ratio of cases to predictors

The inclusion of 'too many' predictor variables should be avoided. As always, there is some debate about what is too many! Some books suggest that there should be at least ten times as many participants as predictors; others are more prescriptive. For example, Tabachnick and Fidell (2007) argue that there should be at least eight participants for every predictor, plus 50. We could write this simply as $N \geq 50 + 8m$ (where m is the number of predictors, and N is the sample size).

Correlation between outcome and predictor

Since linear regression depends on correlation, there must be at least moderate correlation between the outcome scores and predictor to justify running the analyses ($r = .30$ or higher).

Linearity

The relationship between outcome scores and the predictor must be **linear**. This is particularly important if the correlation is weak. We can examine that by way of a scatterplot (we will see how to do that later). For example, we might assume that there would be a relationship between income and quality of life perceptions, represented by a positive correlation. We would also expect that relationship to be linear. If we were to show this in a scatterplot the array of data would be in a straight line – as income increases quality of life perceptions improve. An example is shown in Figure 16.15. There is a clear linear trend here, shown by the cluster of data points from bottom left to top right, and further illustrated by the line of best fit.

However, they do say that money does not buy happiness! What if this were true? We may see that quality of life scores improve with increasing income up to a point. After that, quality of life perceptions may deteriorate with increasing income. If that were the case, the scatterplot displaying this relation may show what we call a 'quadratic' trend, where the cluster shows a distinct arched curve. An example is shown in Figure 16.16. This shows a very different trend to the one we saw before. There is a clear arch in the cluster of data, further illustrated by a quadratic line of best fit. We need to be especially careful to check linearity where correlation is weak. It is possible that it is masking a quadratic trend. If that were to be the case, it would impair the regression analysis.

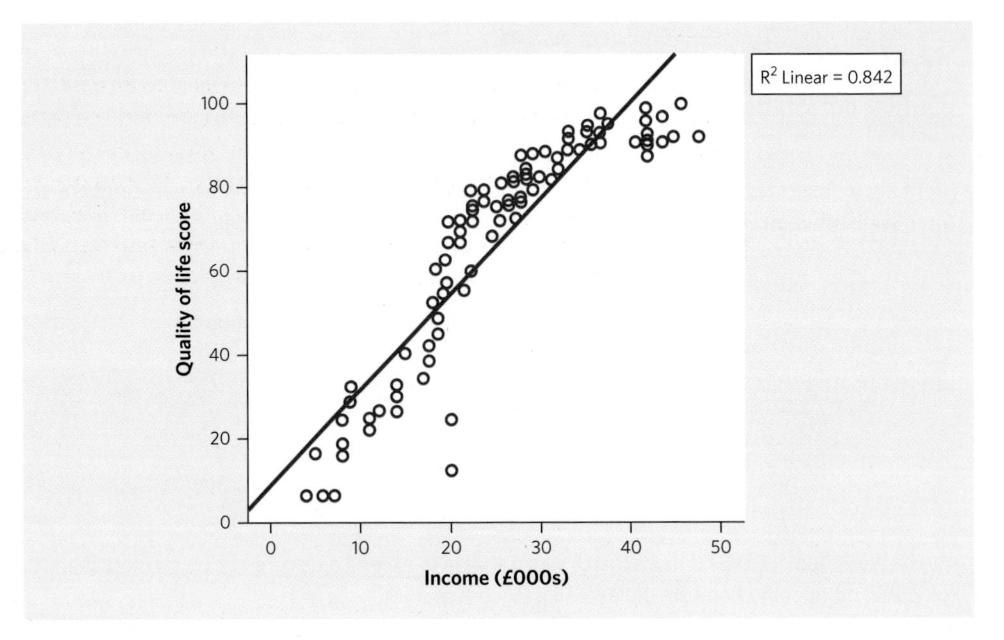

Figure 16.15 Scatterplot: quality of life perceptions vs. income (£000s)

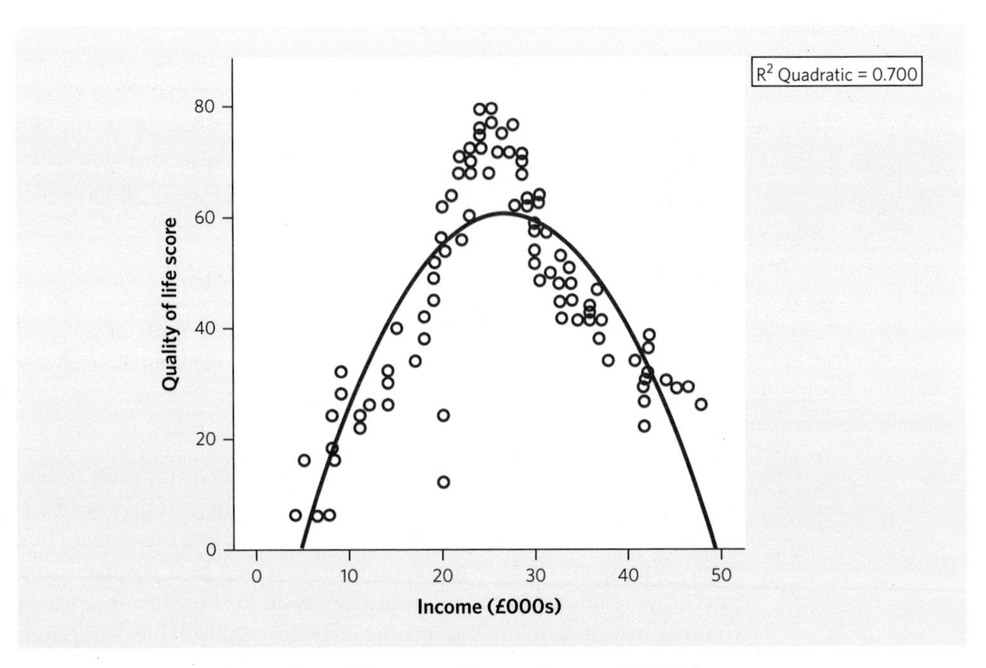

Figure 16.16 Scatterplot: quality of life perceptions vs. income (£000s)

Multi-collinearity

Continuing from the earlier point, the emphasis is on *reasonable* correlation – we do not need the correlation to be too high. Multi-collinearity occurs when two or more predictor variables are highly correlated. If two predictor variables were perfectly correlated with each other, it would make independent predictions impossible (because they are measuring the same thing). This can be measured in SPSS by selecting the collinearity option. We can also refer to the correlations tables that we have asked for. We need to be vigilant for correlations which exceed $r = .80$.

Independent errors

We need to make sure that the residuals are not correlated to each other (we saw what residuals are earlier). This can be examined by something called the **Durbin - Watson test**. We can ask for this test to be run when we set up the parameters for running the model in SPSS. As we will see later, this produces a statistic between 0 and 4: a score of 2 means that there is no correlation; a score of less than 2 indicates positive correlation; a score greater than 2 suggests negative correlation. Correct interpretation of that statistic can be complex, but you are fairly safe in rejecting an outcome that is less than 1 or greater than 3. Anything close to 2 is good.

16.11 Nuts and bolts
Methods of data entry in multiple linear regression

Enter: This option 'forces' all of the variables into the model simultaneously. Some researchers choose this option by default, unless they have good reason to believe that the variables should be entered in a particular order. Using the Enter method, we report the overall variance explained in the outcome variable by predictor variables (adjusted R^2), the significance of the model, and an assessment of the regression line gradients to illustrate which of them significantly contributes to the outcome variance. However, other researchers believe that the Enter method should be avoided - it is a 'saturated' model where all predictors are included, even when they have no independent influence on the outcome.

Hierarchical methods: These methods enter the predictor variables in pre-determined order. Select these with caution; you need to be able to justify clear evidence about why the variables should be entered in a specific order. Within hierarchical types there are three methods:

Forward: This method enters the variables starting with the strongest (the one with the highest correlation). If it significantly adds to the model it is retained; if it does not, it is excluded. This is repeated for each variable through to the weakest.

Backward: This method enters all of the predictors into the model, then proceeds to remove them, starting with the weakest. If the model is significantly improved by that removal, then the variable is left out; if it is not, the variable is re-included. This is repeated through to the strongest variable.

Stepwise: This method is similar to the forward method. This procedure adds each predictor variable to the model, assesses its relative contribution, retaining it if it significantly adds to the model. At that point, all of the other variables are assessed to see if they still significantly contribute, and are removed it they do not. This may seem like an attractive option, but you need to be able to demonstrate a clear rationale for choosing it. One such reason might be that you want to predict your outcome with the fewest possible predictors - we call this parsimony. Stepwise will do this because it will keep running the model until adding a predictor no longer significantly adds to the variance explained in the outcome. Once we have reported the initial model (adjusted R^2 variance, significance of model and strength of gradient), any additional predictor will be reported in terms of added variance (R^2 change), whether the model remains significant, and which of the included predictor variables significantly contribute to the outcome variance.

Measuring outcomes in a multiple linear regression model

When we examined simple linear regression, we explored the key outcome measures in the model by running a series of calculations manually. Those calculations are much more complex for multiple linear regression and would take up a lot of space here. For that reason, this is presented at the end of the chapter. In the meantime, we see how to perform this test in SPSS.

Methods of entering data in multiple linear regression

There are several ways in which we can enter, and examine, the predictors in multiple linear regression. The rationale behind which one we apply will depend on the nature of the investigation. We explore those options in Box 16.11. In short, although you will notice several options for entering data, there are technically only two types: forced (or simultaneous) methods and **hierarchical** methods. The most common forced entry type is '**Enter**', while '**Stepwise**' is frequently used as a hierarchical method. We will illustrate both of these data entry types in subsequent sections, using the same data set. Many texts prefer the Enter method as a first choice, (e.g. Brace *et al.* 2006, pp. 233–234), while some recommend that hierarchical (statistical) methods be reserved for model building, and the Enter should be retained for model testing (Tabachnick and Fidell, 2007, pp. 136–144). Whichever method you choose, it must be stressed that it would be wrong to simply think of every possible predictor and throw that into the model – there must be good prior evidence for including any variable. At the very least you need good theory to justify any choice.

How SPSS performs multiple linear regression

Multiple linear regression in SPSS: Enter method

For this analysis we will address the second of the research questions set by CHILL. You may recall that they wanted to explore how participants' quality of life perceptions might be predicted by reports regarding their physical health, income, job satisfaction, relationship satisfaction and depression status (we have five predictor variables). Quality of life perceptions are obtained from a questionnaire; higher scores represent 'better' perceptions. Perceptions of physical health and job/relationship satisfaction are taken from questionnaires (scored 0–100, with higher scores representing better perceptions). Income reflects annual salary (£000s per annum). The depression status predictor variable is categorical, indicating whether the participant is depressed (or not). The categories are indicated by value labels in SPSS, whereby 1 = 'yes' (participant is depressed) and 0 = 'no'. We said earlier that categorical variables in linear regression must be 'dichotomous' (they must have only two categories). However, so that SPSS knows that these are categorical, rather than numerical, we must use 0 and 1. They will then be recognised as binary terms and will be treated as categorical. We saw how to set up value labels in Chapter 2. There are 98 participants in the sample.

Is the sample large enough?

We should make sure that we have a sufficient sample for running this regression analysis in relation to the number of predictor variables. We are using five predictors so, using the guidance from Tabachnick and Fidell (2007) we saw earlier, we ensure that we have at least eight participants for every predictor, plus 50. We can express that as: $N \geq 50 + 8m$ (where m is the number of predictors and N is the sample size).

On this occasion: $50 + (8 \times 5) = 90$
Our sample is 98, so we can be confident that we have a large enough sample.

The remaining tests for checking the assumptions and restrictions of multiple linear regression need to be set up when we run the main analysis. We shall see how that is done now.

16.12 Nuts and bolts

Setting up the data set in SPSS

When we create the SPSS data set for multiple linear regression, we need to set up one column for the outcome variable (which will have a continuous score) and several columns for the predictor variables (which can be continuous or categorical). In this example, one of the predictors is categorical, so that column will need to represent two groups, defined by value labels.

Quality of life.sav [DataSet2] - IBM SPSS Statistics Data Editor

File Edit View Data Transform Analyze Graphs Utilities Add-ons Window Help

	Name	Type	Width	Decimals	Label	Values	Missing	Columns	Align	Measure
1	depressed	Numeric	8	0	Depressed	{0, No}...	9	8	Right	Nominal
2	relate	Numeric	8	0	Relationship sa...	None	-1	8	Right	Scale
3	job	Numeric	8	0	Job satisfaction	None	-1	8	Right	Scale
4	income	Numeric	8	0	Income	None	-1	8	Right	Scale
5	physical	Numeric	8	0	Physical health	None	-1	8	Right	Scale
6	QoL	Numeric	8	0	Quality of life s...	None	-1	8	Right	Scale

Figure 16.17 Variable View for 'Quality of life' data

Figure 16.17 shows how the SPSS Variable View should be set up. The first variable is called 'depressed'. This is the first of the predictors; this variable indicates whether the participant is depressed or not (it is a categorical variable). We can have categorical predictors, so long as they have only two groups (which must be coded with the values 0 and 1). In the Values column, we include '0 = No' and '1 = Yes'; the Measure column is set to Nominal. The next four variables ('relate', 'job', 'income' and 'physical') are also predictors, but they are represented by continuous scores. We do not need to adjust anything in the Values column; the Measure column is set to Scale. The final variable is 'QoL'; this is the outcome variable, which represents 'Quality of life scores'. Since this is also numerical, no adjustment is needed in the Values column; we also set Measure to Scale.

*Quality of life.sav [DataSet2] - IBM SPSS Statistics Data Editor

File Edit View Data Transform Analyze Graphs Utilities Add-ons Window Help

6 :

	depressed	relate	job	income	physical	QoL	var	var	var
1	Yes	39	19	20	32	34			
2	No	62	22	33	59	84			
3	Yes	44	21	33	36	50			
4	Yes	37	26	26	37	32			

Figure 16.18 Data View for 'Quality of life' data

Figure 16.18 illustrates how this will appear in the Data View. It is Data View that will be used to select the variables when performing this test. Each row represents a participant. When we enter the data for 'depressed', we input 0 (to represent 'no') or 1 (to represent 'yes'); the 'depression' column will display the descriptive categories ('No' or 'Yes') or will show the value numbers, depending on how you choose to view the column (you can change that using the Alpha Numeric button in the menu bar - see Chapter 2). For the remaining predictor variable columns ('relate', 'job', 'income' and 'physical'), and for the outcome variable column ('QoL'), we simply enter the relevant score for that participant.

Normal distribution

We should check that the outcome variable (quality of life score) is normally distributed. If you are particularly cautious, you should probably check that across the categorical variable groups for depressed vs. not depressed. We have seen how to run the Kolmogorov–Smirnov (KS) test on several occasions now (see earlier in this chapter for instance), so we will just report the outcome here.

	Kolmogorov–Smirnov[a]			Shapiro–Wilk		
	Statistic	df	Sig.	Statistic	df	Sig.
Quality of life score	.103	98	.013	.965	98	.010

Figure 16.19 Kolmogorov-Smirnov/Shapiro-Wilk test for quality of life scores

Figure 16.19 suggests that we *may* have a problem with normal distribution; the KS outcome is significant. We could check that further by calculating z-scores for skew and kurtosis. We have seen how to do that in previous chapters, so we shall not repeat that here (but do try it for yourself – you will see that we can be confident that the data are reasonably normally distributed).

Running Enter model in SPSS

Using the SPSS file Quality of life

Select **Analyze** → **Regression** → **Linear...** (as you were shown in Figure 16.5) → (in new window) transfer **Quality of life score** to **Dependent:** → transfer **Depressed**, **Relationship satisfaction**, **Job satisfaction**, **Income**, and **Physical health** to **Independent(s)** → select **Enter** in pull-down options for **Method** → click **Statistics** (as shown in Figure 16.20)

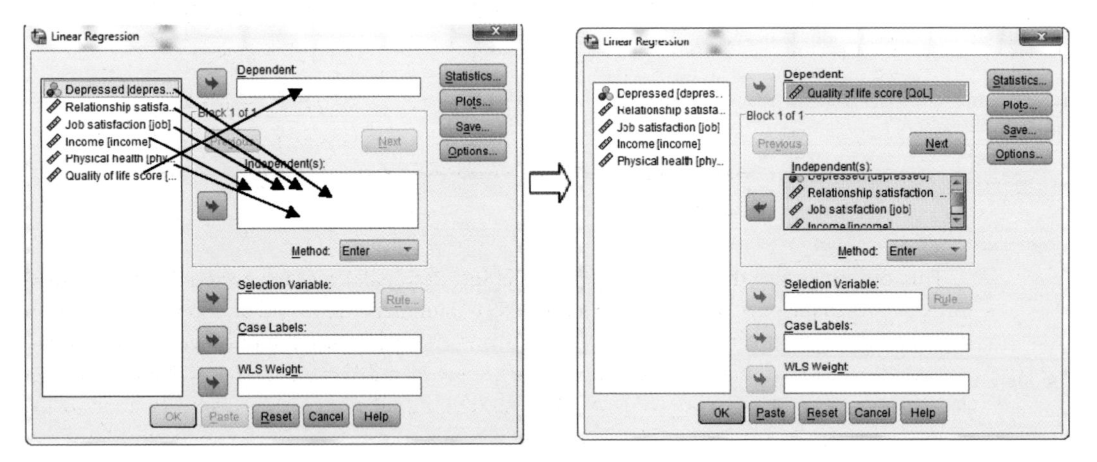

Figure 16.20 Multiple linear regression: selecting variables

In new window (see Figure 16.21), tick **Estimates** box for (under **Regression Coefficients)** → tick **Durbin-Watson** and **Casewise diagnostics** boxes (under **Residuals**) → tick **Outlier outside** radio button → set **standard deviations** to **2** (so that we can count how many z-scores are greater than that – we explained why that is important earlier) → tick **Model Fit**, **Part and partial correlations** and **Collinearity diagnostics** boxes → click **Continue** → (in original window) click **OK**

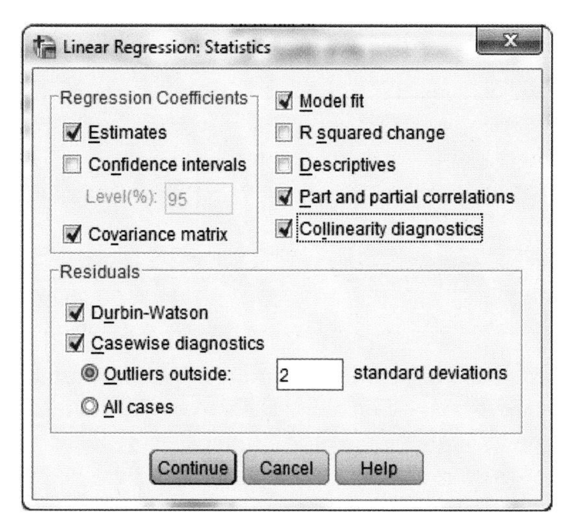

Figure 16.21 Multiple linear regression: statistical parameters

Interpretation of output

Checking assumptions and restrictions

Outliers

When we explored the assumptions and restrictions earlier, we saw a range of cut-off points determining the magnitude of z-scores that we could accept as a maximum. To assess this, we need to refer to the '**Casewise Diagnostics**' output (Figure 16.22).

Case number	Std. residual	Quality of life score	Predicted value	Residual
12	2.162	92	67.68	24.316
48	–2.300	18	43.88	–25.877
71	–2.061	8	31.18	–23.180
75	2.029	58	35.17	22.828

Figure 16.22 Casewise diagnostics

This output shows the number of z-scores (Std. Residuals) that have exceeded the defined limit (we set that to '2'). Had we chosen the SPSS default here, we would be shown only z-scores that exceed 3, which is not terribly helpful when we need to count the number of z-scores that exceed 1.96. We said that no more than 5% of z-scores should be higher than 1.96; in a data set of 98 that equates to five cases. We had four, so that's OK. We also said that no more than 1% should exceed 2.58 (that would be just one case in a sample of this size). We did not have any in any case, so that's good, too. If we did have 'too many' outliers, we could try to remove some (using techniques that we explored in Chapter 3).

Correlation

We said that we needed reasonable correlation between the variables (r = .30 to r = .80). It could be argued that we have a slight problem with the relationship between quality of life scores and physical health (r = .81), so we may need to be a little cautious about that one.

Linearity

Although we can be satisfied with the correlation outcome, we also said that we should check linearity, especially where there is evidence of a weak relationship. None of the outcomes in

Correlations

		Quality of life score	Depressed	Relationship satisfaction	Job satisfaction	Income (£000s)	Physical health
Pearson correlation	Quality of life score	1.000	−.702	.780	.473	.606	.814
	Depressed	−.702	1.000	−.627	−.411	−.472	−.643
	Relationship satisfaction	.780	−.627	1.000	.439	.425	.648
	Job satisfaction	.473	−.411	.439	1.000	.537	.643
	Income (£000s)	.606	−.472	.425	.537	1.000	.722
	Physical health	.814	−.643	.648	.643	.722	1.000

Figure 16.23 Correlation

Figure 16.23 poses a problem, but we will look further at the relationship between quality of life scores and job satisfaction, simply so we can see how we can examine linearity.

We saw how to request a scatterplot in SPSS, with a 'line of fit' in Chapter 6 (Correlation) – refer to Figures 6.9–6.11 for graphical representations.

Using the SPSS file Quality of life

Select **Graphs** ➜ **Chart Builder** ➜ (in new window) select **Scatter/dot** (from list under **Choose from:**) ➜ drag **Simple Scatter** graphic (top left corner) into **Chart Preview** window ➜ transfer **Quality of life score** to **Y-Axis** box (to left of new graph) ➜ transfer **Job satisfaction** to **X-Axis** box (under graph) ➜ click **OK**

There is evidence of a linear trend in Figure 16.24 (the cluster of data points is certainly not curved as you might find in a quadratic trend). However, to be certain, we could add the line of fit.

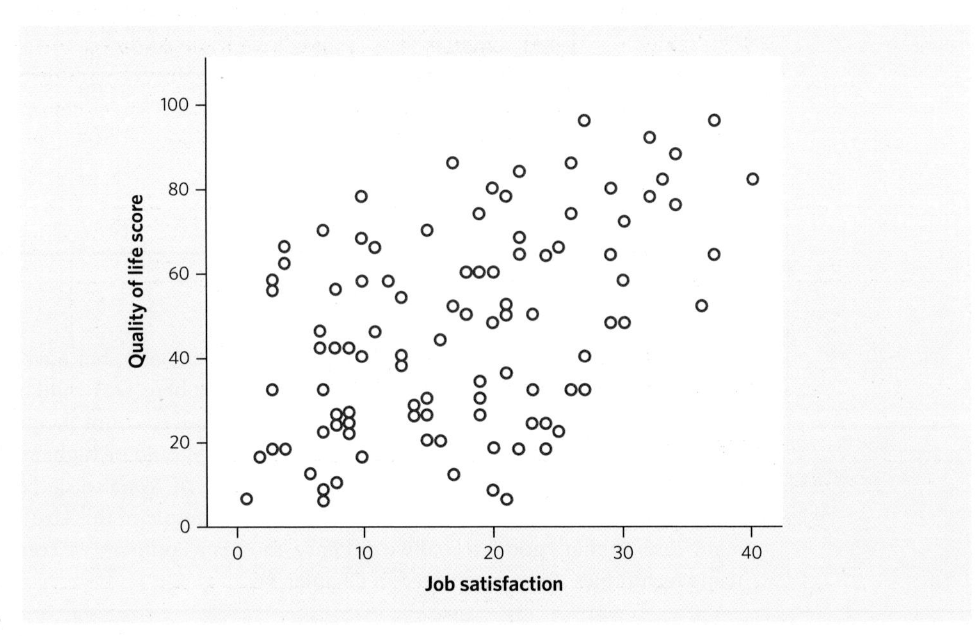

Figure 16.24 Scatterplot: quality of life scores vs. job satisfaction

In the SPSS output, double click on the graph (it will open in a new window, and will display some additional options) ➜ click on the icon '**Add Fit Line at Total**' (in the icons displayed above the graph) ➜ click on **Close** ➜ click on cross in top right hand corner of window showing adjusted graph

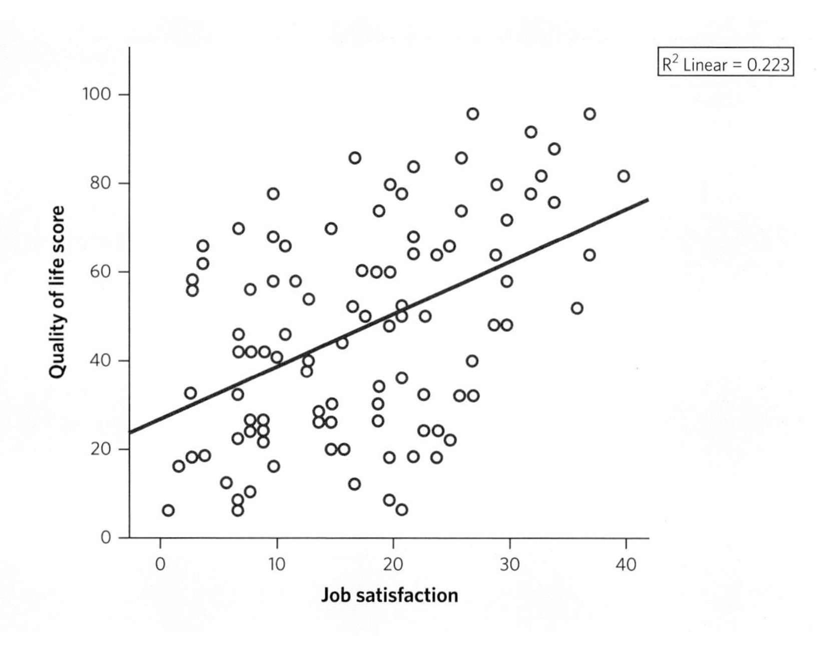

Figure 16.25 Scatterplot: quality of life scores vs. job satisfaction (with line of best fit)

Figure 16.25 confirms that we can be satisfied that the relationship is linear.

Multi-collinearity

Figure 16.26 presents the multi-collinearity data. What we are looking for here is that no single predictor is highly correlated with any of the computed dimensions; the variance proportions range from 0 to 1. Ideally, each predictor *should* be located against a different dimension. I have highlighted the highest variance for each predictor to help you see this. There may be a minor problem with 'Physical health' as the variance is .92 (slightly above the ideal maximum of .90). Also, 'Depressed' and 'Relationship satisfaction' are located on 'Dimension 6'; this is not ideal, but it is the only case. Overall, while not perfect, this is quite satisfactory.

Collinearity diagnostics[a]

Model	Dimension	Eigenvalue	Condition index	Variance proportions					
				(Constant)	Depressed	Relationship satisfaction	Job satisfaction	Income (£000s)	Physical health
1	1	5.210	1.000	.00	.00	.00	.00	.00	.00
	2	.541	3.104	.00	.21	.00	.01	.00	.01
	3	.113	6.797	.01	.02	.17	.74	.00	.00
	4	.077	8.241	.00	.00	.29	.19	.40	.05
	5	.037	11.805	.04	.19	.03	.05	.39	.92
	6	.023	15.175	.94	.57	.50	.00	.20	.01

a. Dependent variable: quality of life score

Figure 16.26 Collinearity diagnostics

Coefficients[a]

Model		Unstandardised coefficients		Standardised coefficients	t	Sig.	Correlations			Collinearity statistics	
		B	Std. error	Beta			Zero-order	Partial	Part	Tolerance	VIF
1	(Constant)	6.100	6.405		.952	.343					
	Depressed	−8.607	3.457	−.164	−2.490	.015	−.702	−.251	−.117	.509	1.966
	Relationship satisfaction	.520	.089	.387	5.823	.000	.780	.519	.273	.499	2.006
	Job satisfaction	−.286	.159	−.112	−1.798	.075	.473	−.184	−.084	.573	1.744
	Income (£000s)	.202	.158	.088	1.279	.204	.606	.132	.060	.465	2.149
	Physical health	.544	.105	.466	5.165	.000	.814	.474	.243	.271	3.690

a. Dependent variable: quality of life score

Figure 16.27 Collinearity statistics

Figure 16.27 shows another measure of multi-collinearity. We will need this output again later to analyse one of the key model outcomes, but, for the moment, we need to focus on the **Collinearity Statistics** (highlighted in orange font). To satisfy the criteria to avoid multi-collinearity, we need the '**Tolerance**' data not to be too close to zero. Scores below .1 are of serious concern; scores below .2 might cause some concern (Menard, 1995). We are comfortably above that, so we are good. The **VIF** figure performs a similar check on collinearity between the predictor variables and is the reciprocal of tolerance (VIF = 1 ÷ tolerance). VIF scores above 10 indicate a problem (Myers, 1990); we are well within those limits, too.

Independent errors

To satisfy this assumption, we need to show that there is no correlation between the residuals. To assess that, we need to refer to the Durbin–Watson outcome (Figure 16.28). Earlier, we said that the Durbin–Watson statistic is measured on a scale of 0 to 4, with 2 representing no correlation. We want to be as close to 2 as possible, and avoid figures below 1 and above 3. Our outcome is 1.906, which is quite close to 2, so that is encouraging. You should note that we will see this output again a little later, when we assess the main outcome.

Model summary[b]

Model	R	R square	Adjusted R square	Std. error of the estimate	Durbin-Watson
1	.893[a]	.797	.786	11.250	1.906

a. Predictors: (constant), physical health, depressed, job satisfaction, relationship satisfaction, income (£000s)
b. Dependent variable: quality of life score

Figure 16.28 Correlation between residuals

Checking model outcome

We saw this output table just now, but this time we are focusing on the main outcome. Figure 16.29 shows how much variance can be explained by the regression model. From this, we can see that 79.7% of variance in quality of life scores is explained by variations in the predictor variables included in the model (R^2 = .797). However, we should also report the adjusted R^2 (.786). The outcome (dependent) variable and predictors are confirmed in the footnotes.

Model summary[b]

Model	R	R Square	Adjusted R square	Std. error of the estimate	Durbin-Watson
1	.893[a]	.797	.786	11.250	1.906

a. Predictors: (constant), physical health, depressed, job satisfaction, relationship satisfaction, income (£000s)
b. Dependent variable: quality of life score

Figure 16.29 Explained variance

Figure 16.30 indicates that we have a highly significant model, $F(5, 92) = 72.310$, $p < .001$. The regression model is significantly better at predicting outcome than a random method. The outcome variable and predictors are confirmed once again in the footnotes.

ANOVA[b]

Model		Sum of squares	df	Mean square	F	Sig.
1	Regression	45755.400	5	9151.080	72.310	.000[a]
	Residual	11642.844	92	126.553		
	Total	57398.245	97			

a. Predictors: (constant), physical health, depressed, job satisfaction, relationship satisfaction, income (£000s)
b. Dependent variable: quality of life score

Figure 16.30 Significance of the model

Coefficients[a]

Model		Unstandardised coefficients		Standardised coefficients			Correlations			Collinearity statistics	
		B	Std. error	Beta	t	Sig.	Zero-order	Partial	Part	Tolerance	VIF
1	(Constant)	6.100	6.405		.952	.343					
	Depressed	−8.607	3.457	−.164	−2.490	.015	−.702	−.251	−.117	.509	1.966
	Relationship satisfaction	.520	.089	.387	5.823	.000	.780	.519	.273	.499	2.006
	Job satisfaction	−.286	.159	−.112	−1.798	.075	.473	−.184	−.084	.573	1.744
	Income (£000s)	.202	.158	.088	1.279	.204	.606	.132	.060	.465	2.149
	Physical health	.544	.105	.466	5.165	.000	.814	.474	.243	.271	3.690

a. Dependent variable: quality of life score

Figure 16.31 Regression parameters and predictor contribution

Figure 16.31 presents several pieces of key outcome. The 'Constant' figure (highlighted in green font) represents the first part of the regression equation, β_0 (6.100). The remaining data in that column report the regression gradients (β_x or 'B') for each predictor (highlighted in blue): Depressed (B = −8.607), Relationship satisfaction (B = 0.520), Job satisfaction (B = −0.286), Income (B = 0.202) and Physical health (B = 0.544). We explore what that actually means in Box 16.13. The t-test indicates which of those regression lines have a gradient that is significantly greater than 0 – these are the predictors that significantly contribute to the outcome variance. We have three: 'Depressed' ($t = -2.490$, $p = .015$), 'Relationship satisfaction' ($t = 5.823$, $p < .001$) and 'Physical health' ($t = 5.165$, $p < .001$). They all play an important part in determining quality of life perceptions. The gradients for job satisfaction and income are not significantly greater than zero. They do not contribute to the outcome variance (they are not important when we consider what factors affect quality of life perceptions).

Notice also the effect that the regression model has on the correlation between each predictor and the outcome. The 'zero order' correlations are moderate to high. By including all of the predictors into the model, we need to refer to the semi-partial correlation (under the 'Part' column in Figure 16.31). Initial correlations are compromised for all predictors, but none more so than those which ultimately do not significantly contribute to outcome variance.

Beta values and confidence intervals

You may have wondered why the gradients for 'job satisfaction' and 'income' were not significant. The gradient was expressed in the 'B' column in Figure 16.31; these represent 'Beta' values (β_n in the multiple linear regression equation). The Beta values were −0.286 and 0.202 respectively, which are by no means tiny. So why were they not significant (remember size does not always matter in linear regression)? The clue is in the 'standard error' outcome, also shown in Figure 16.31. We can use this to estimate confidence intervals. We encountered this measure in Chapter 4, when we explored how we determine whether an outcome is statistically significant. Confidence intervals represent the central spread of data (in this case Beta values). We usually report the central 95% of the data, as this equates to a significance cut-off of $p < .05$ (so we refer to them as 95% confidence intervals).

We can request confidence intervals when setting the parameters for running multiple linear regression in SPSS (the selection can be chosen in the 'Options' menu). By selecting that option, we will get a different 'coefficients' output to the one we saw for Figure 16.31 (see Figure 16.32).

As we saw just now, the Beta data (shown in red font in Figure 16.32) help us estimate how outcome scores change for each unit change in the predictor. The Beta value can be positive *or* negative depending on the direction of the relationship (it makes no difference to the significance). The 95% confidence interval data (blue font) indicate the range that the 'true' slope is likely to be found within. So, for relationship satisfaction, we have a Beta value of 0.520, and

16.13 Take a closer look
Interpreting gradients in multiple linear regression

In simple terms, the gradient of each predictor variable indicates the degree to which the unit value of the outcome variable increases according to every unit value increase in the predictor variable. In reality, it is a little more complex than that because that interpretation will vary depending on whether the predictor is categorical or continuous. To illustrate, we will examine the gradients for the predictor variables we have just encountered.

Table 16.4 Gradient interpretation

Predictor	Type	Gradient	Sig.	Interpretation
Depressed	Cat	−8.607	.015	QoL scores decrease (worsen) by 8.607 between categorisation of 'not depressed' (SPSS value code 0) and 'depressed' (1); this predictor significantly contributes to the variance in QoL scores.
Relationship satisfaction	Con	0.520	<.001	For every unit improvement in relationship satisfaction scores, QoL scores increase (improve) by .520; this predictor significantly contributes to the variance in QoL scores.
Job satisfaction	Con	−0.286	.075	It would *appear* that, for every unit improvement in job satisfaction scores, QoL scores decrease (worsen) by .286. However, this predictor does not significantly contribute to the variance in QoL scores, so has no predictive power.
Income	Con	0.202	.204	It would *appear* that, for every unit improvement in income, QoL scores increase (improve) by .202. However, this predictor does not significantly contribute to the variance in QoL scores; it has no predictive power.
Physical health	Con	0.544	<.001	For every unit improvement in physical health scores, QoL scores increase (improve) by .544; this predictor significantly contributes to the variance in QoL scores.

Key: Type: Cat = categorical predictor; Con = continuous predictor; QoL = quality of life

		Unstandardised coefficients		Standardised coefficients			95.0% confidence interval for B	
Model		B	Std. error	Beta	t	Sig.	Lower bound	Upper bound
1	(Constant)	6.100	6.405		.952	.343	−6.620	18.820
	Depressed	−8.607	3.457	−.164	−2.490	.015	−15.474	−1.741
	Relationship satisfaction	.520	.089	.387	5.823	.000	.343	.697
	Job satisfaction	−.286	.159	−.112	−1.798	.075	−.601	.030
	Income (£000s)	.202	.158	.088	1.279	.204	−.112	.516
	Physical health	.544	.105	.466	5.165	.000	.335	.754

a. Dependent variable: quality of life score

Figure 16.32 Beta values and confidence intervals

we can be 95% confident that the true slope of the regression line is in the range defined by 0.343 to 0.697. This range is determined by the standard error outcome. If that error is large, the 95% confidence interval range will be too wide. This will decrease the likelihood that the gradient is significant.

Now let's look at 'job satisfaction'. In this case, we can be 95% confident that the true slope of the regression line is in the range defined by −0.601 to +0.030. This means that at one end of the range the regression line describes a negative correlation, while at the other end it describes a positive one. This is not consistent: it must be one or the other. The gradient cannot be significant if the confidence intervals 'cross 0' (present a range that is negative to positive, or vice versa). You can probably see that the 95% confidence intervals for 'Income' illustrate a similar problem.

Effect size and power

The method for effect size and power is very similar to what we did for simple linear regression.

Open G*Power

From Test family select F tests

From **Statistical test** select **Linear multiple regression: Fixed model, R^2 deviation from zero)**
From **Type of power analysis** select **Post hoc: Compute achieved - given α, sample size and effect size power:**

Now we enter the **Input Parameters**

From the Tails box, select One

To calculate the **Effect size**, click on the **Determine** button (a new box appears).
In that new box, for **Squared multiple correlation ρ^2** type **0.786 → Calculate and transfer to main window**

Note that 'ρ' in G*Power (for squared multiple correlation) refers to the Greek letter 'rho'

Back in original display, for **α err prob** type **0.05** (the significance level) → for **Total sample size** type **98** → for **Number of predictors** type **5** → click on **Calculate**
Effect size(d) **3.67** (very strong); and Power(1-β err prob) **1.00** (excellent).

Writing up results

A multiple linear regression was undertaken to examine variance in quality of life scores. Five predictors were loaded into the model using the Enter method. Table 16.5 shows that the model was able to explain 79.7% of the sample outcome variance (Adj. $R^2 = .786$), which was found to significantly predict outcome, $F(5, 92) = 72.310$, $p < .001$. Three of the predictor variables

Table 16.5 Multiple linear regression analysis of quality of life scores

Predictor variable	R^2	Adj. R^2	F	p	Constant	Gradient	t	p
Model	.797	.786	72.310	<.001	6.100			
Depressed						−8.607	−2.490	.015
Relationship satisfaction						.520	5.823	<.001
Job satisfaction						−.286	−1.798	.075
Income (£000s)						.202	1.279	.204
Physical health						.544	5.165	<.001

significantly contributed to the model. Being depressed was related to poorer quality of life ($\beta = -8.607$, $t = -2.490$, $p = .015$), while increased relationship satisfaction ($\beta = 0.520$, $t = 5.823$, $p < .001$) and better physical health ($\beta = 0.544$, $t = 5.165$, $p < .001$) were significantly associated with improved quality of life scores. Two other predictor variables, job satisfaction and income, did not significantly contribute to variance. There was a very strong effect size ($d = 3.67$).

Multiple linear regression in SPSS: Stepwise method

We will now explore the same data set, but using the Stepwise method. As we saw in Box 16.11, this method can be a more economical way of predicting the outcome – we achieve our optimum outcome using the fewest possible predictor variables. SPSS will produce output reporting the key regression model factors, according to the addition of each predictor. The reported outcomes are the amount of explained variance (adjusted R^2), the significance of the model, and an analysis of the regression line gradients, in respect of which predictors significantly contribute to the outcome variance. SPSS will keep producing that outcome until the model can no longer significantly predict any more additional variance. The method for setting up the data set is the same as we saw earlier.

Examining assumptions and restrictions

We will briefly explore these as we progress through the main output, using the final model solution to determine how well they have been met. We need to ensure that we have a sufficient sample size, and that we have met linearity, multi-collinearity and independence of errors that we encountered earlier, but adjust that according to the number of predictor variables that are ultimately included. We will not repeat the rationale behind those assumptions, since we have explored that sufficiently in previous sections.

Running Stepwise model in SPSS

Using the SPSS file Quality of life

Select **Analyze** → **Regression** → **Linear...** (as you were shown in Figure 16.5) → (in new window) transfer **Quality of life score** to **Dependent:** → transfer **Depressed**, **Relationship satisfaction**, **Job satisfaction**, **Income**, and **Physical health** to **Independent(s)** → select **Stepwise** for **Method** → click **Statistics** → (in new window) tick **Estimates** box (under **Regression Coefficients**) → tick **Model Fit** , **R squared change** and **Part and partial correlations** boxes → click **Continue** → click **OK**

You also need to set parameters for testing assumptions and restrictions – as you were shown earlier

Interpretation of output

Figure 16.33 presents the variance that has been explained in the outcome at each stage of the model. The footnotes indicate which predictors were included at each point. Initially, 'Physical health' was included. At this stage, 66.2% of the (sample) variance was explained (Adj. $R^2 = .659$). In the second model, 'Relationship satisfaction' was added as a predictor. At this point, the model was able to predict 77.3% of the variance (Adj. $R^2 = .768$); an additional 11.0% of variance was explained (R^2 change $= .110$). In the third model, 'Depressed' was added as a predictor. At this stage, the model was able to predict 78.8% of the sample outcome variance (Adj. $R^2 = .781$); additional 1.5% of variance was explained (R^2 change $= .015$). This is the final model as no other predictors significantly contributed to the outcome.

Model summary

Model	R	R square	Adjusted R square	Std. error of the estimate	Change statistics				
					R square change	F change	df1	df2	Sig. F change
1	.814[a]	.662	.659	14.206	.662	188.426	1	96	.000
2	.879[b]	.773	.768	11.715	.110	46.163	1	95	.000
3	.888[c]	.788	.781	11.387	.015	6.557	1	94	.012

a. Predictors: (constant), physical health
b. Predictors: (constant), physical health, relationship satisfaction
c. Predictors: (constant), physical health, relationship satisfaction, depressed

Figure 16.33 Explained variance

Figure 16.34 shows the significance of the model at each stage of predictor inclusion. It makes sense that each of these is significant; once the inclusion of predictors no longer adds to significance, they are excluded. At the final stage, the model was still highly significant, $F(3, 94) = 116.233$, $p < .001$.

ANOVA[d]

Model		Sum of squares	df	Mean square	F	Sig.
1	Regression	38025.072	1	38025.072	188.426	.000[a]
	Residual	19373.173	96	201.804		
	Total	57398.245	97			
2	Regression	44360.450	2	22180.225	161.616	.000[b]
	Residual	13037.795	95	137.240		
	Total	57398.245	97			
3	Regression	45210.616	3	15070.205	116.233	.000[c]
	Residual	12187.629	94	129.656		
	Total	57398.245	97			

a. Predictors: (constant), physical health
b. Predictors: (constant), physical health, relationship satisfaction
c. Predictors: (constant), physical health, relationship satisfaction, depressed
d. Dependent variable: quality of life score

Figure 16.34 Significance of the models

Figure 16.35 presents key information regarding the regression gradients. By definition, each predictor will significantly contribute to the overall variance in the outcome, otherwise they would not have been included. In the final model the predictors 'Physical health'

Coefficients[a]

Model		Unstandardised coefficients		Standardized coefficients	t	Sig.	Correlations		
		B	Std. error	Beta			Zero-order	Partial	Part
1	(Constant)	7.090	3.213		2.206	.030			
	Physical health	.951	.069	.814	13.727	.000	.814	.814	.814
2	(Constant)	−5.634	3.245		−1.736	.086			
	Physical health	.621	.075	.531	8.279	.000	.814	.647	.405
	Relationship satisfaction	.586	.086	.436	6.794	.000	.780	.572	.332
3	(Constant)	7.692	6.085		1.264	.209			
	Physical health	.540	.079	.462	6.792	.000	.814	.574	.323
	Relationship satisfaction	.502	.090	.374	5.592	.000	.780	.500	.266
	Depressed	−8.948	3.494	−.170	−2.561	.012	−.702	−.255	−.122

a. Dependent variable: quality of life score

Figure 16.35 Regression parameters and predictor contribution

Excluded variables[d]

Model		Beta In	t	Sig.	Partial correlation	Collinearity statistics
						Tolerance
1	Depressed	−.305[a]	−4.279	.000	−.402	.587
	Relationship satisfaction	.436[a]	6.794	.000	.572	.580
	Job satisfaction	−.087[a]	−1.126	.263	−.115	.586
	Income (£000s)	.039[a]	.456	.649	.047	.479
2	Depressed	−.170[b]	−2.561	.012	−.255	.510
	Job satisfaction	−.103[b]	−1.633	.106	−.166	.585
	Income (£000s)	.079[b]	1.114	.268	.114	.476
3	Job satisfaction	−.100[c]	−1.628	.107	−.166	.585
	Income (£000s)	.071[c]	1.023	.309	.105	.475

a. Predictors in the model: (constant), physical health
b. Predictors in the model: (constant), physical health, relationship satisfaction
c. Predictors in the model: (constant), physical health, relationship satisfaction, depressed
d. Dependent variable: quality of life score

Figure 16.36 Excluded predictors

($\beta = 0.540, t = 6.792, p < .001$), 'Relationship satisfaction' ($\beta = 0.502, t = 5.592, p < .001$) and 'Depressed' ($\beta = -8.948$, $t = -2.561$, $p = .012$) all significantly contributed to the variance in quality of life scores.

Figure 16.36 confirms the variables excluded at each stage. In the initial models, some of the excluded predictors appear to contribute to the overall variance. They are subsequently included. In the final model, 'Job satisfaction' and 'Income' remain excluded, as they do not significantly predict the outcome.

Writing up results

We could report the outcome in much the same way as we did for the Enter version of multiple linear regression, but perhaps tweak the table a little to reflect the model stages.

Table 16.6 Multiple linear regression analysis of quality of life scores (n = 98)

Predictor variable	R^2	Adj. R^2	R^2/change	F	p	Gradient	t	p
Model	.788	.781		116.23	<.001			
Physical health			.662			.540	6.792	<.001
Relationship satisfaction			.110			.502	5.592	<.001
Depressed			.015			−8.948	−2.561	.012

A multiple linear regression was undertaken to examine variance in quality of life scores for 98 participants, using the Stepwise method. A significant model ($F(3, 94) = 116.23, p < .001$) predicted 78.8% of the sample outcome variance (Adj. R^2.781). Three predictors were entered into the model: better physical health ($\beta = 0.540$, $t = 6.792$, $p < .001$), increased relationship satisfaction ($\beta = 0.502$, $t = 5.592$, $p < .001$) and not being depressed ($\beta = -8.948$, $t = -2.561$, $p = .012$) were significantly associated with improved quality of life scores. Two other predictor variables (job satisfaction and income) were excluded from the model.

Chapter summary

In this chapter we have explored simple and multiple linear regression. At this point, it would be good to revisit the learning objectives that we set at the beginning of the chapter.

You should now be able to:

- Recognise that we use linear regression to explore the proportion of variance that we can 'explain' in outcome (dependent variable) scores, from a series of predictor (independent) variables. Simple linear regression employs one predictor variable; multiple linear regression uses several predictors.
- Understand that the purpose of linear regression is to build a 'model' that can be used to predict outcome more accurately than some arbitrary method (such as just using the mean score). Outcome variables must be numerical (and preferably normally distributed). Predictors can be numerical or categorical. If the predictor variable is categorical it must have only two groups and must be 'coded' as 0 and 1 when using the value labels in SPSS. These rules apply to simple and multiple linear regression. Beyond that multiple linear regression has a number of additional restrictions. There should be a limit on the number of outliers, otherwise the model might be compromised. While at least moderate correlation between predictor and outcome is needed, multi-collinearity between predictors should be avoided (it makes it harder to differentiate the effect of individual predictors). Correlation between residuals should also be avoided. There should be a sufficient sample size in relation to the number of predictors used.
- Calculate the outcome manually (using maths and equations). The methods for calculating outcome are very more complex for multiple linear regression than they are for simple linear regression.
- Perform analyses using SPSS, using the appropriate method. The procedures for simple linear regression are relatively straightforward. However, there is a range of methods that can be used for multiple linear regression. The main distinction lies in the way in which predictor variables are entered into the model. Unless evidence can be provided to justify the order in which the predictors should be entered, they should all be entered simultaneously using the 'Enter' method.
- Examine effect size and power using G*Power software. There are slight differences in the way that is performed between simple and multiple linear regression.
- Understand how to present the data, using appropriate tables, reporting the outcome in series of sentences and correctly formatted statistical notation. Three key statistics must be reported:

  1. The amount of variance explained (R^2 for simple linear regression; adjusted R^2 for multiple linear regression).
  2. The success (significance) of the model (e.g. $F(5, 92) = 72.310, p < .001$).
  3. Whether the gradient significantly contributes to variance in the outcome scores (for simple linear regression) or which predictor(s) significantly contribute to variance (for multiple linear regression). This is indicated by the independent t-test outcome.

Research example

It might help you to see how multiple linear regression has been applied in a research context (using Stepwise methods). In this section you can read an overview of the following paper:

> **Hopkins, R.O., Key, C.W., Suchyta, M.R., Weaver, L.K. and Orme, J.F. Jr. (2010). Risk factors for depression and anxiety in survivors of acute respiratory distress syndrome.** ***General Hospital Psychiatry, 32 (2)*** **: 147–155. DOI: http://dx.doi.org/10.1016/j.genhosppsych.2009.11.003**

If you would like to read the entire paper you can use the DOI reference provided to locate that (see Chapter 1 for instructions).

In this research the authors investigated the extent to which acute respiratory distress syndrome (a serious illness that restricts breathing) might pose a risk factor for developing depression and anxiety. Patients were assessed twice following hospital discharge, after one and two years. Although 120 patients were recruited, only 66 survived to take part at year 1 and 62 at year 2. To assess outcome, depression scores were measured using the Beck Depression Inventory (BDI: Beck, 1987), while anxiety scores were examined using the Beck Anxiety Inventory (BAI: Beck, 1993). Separate regression analyses were undertaken for six outcome variables: depression at year 1, at year 2 (excluding year 1 outcomes) and at year 2 (including year 1 outcomes). A series of demographic and illness-related variables was explored to investigate the effect on those outcomes (along with others that we do not explore here). Overall, eight predictor variables were included in each model (rather a lot given the sample size): age, gender, history of smoking, history of alcohol dependence, how long spent on breathing apparatus (measured by a variable called 'duration of mechanical ventilation'), a measure of breathing quality (measured by 'PaO_2/FiO_2'), a composite evaluation of cognitive performance (measured by 'presence of cognitive sequelae'), and an illness severity indicator (measured by 'APACHE II', which is short for 'Acute Physiological and Chronic Health Evaluation'). In the regression analyses, gradients were included when the significance was less than $p = .10$ (a somewhat optimistic interpretation).

A significant model was produced for all four outcomes: depression at year 1, $F(4, 59) = 4.73$, $p = .002$; depression at year 2 (excluding year 1 outcomes), $F(1, 57) = 5.85$, $p = .02$; depression at year 2 (including year 1 outcomes), $F(2, 56) = 46.20$, $p < .0001$; anxiety at year 1, $F(4, 59) = 5.087$, $p < .001$; anxiety at year 2 (excluding year 1 outcomes), $F(4, 54) = 5.18$, $p < .001$; anxiety at year 2 (including year 1 outcomes), $F(2, 556) = 19.75$, $p < .0001$ (it is not usual practice to report significance to this level – $p < .001$ is usually enough to show highly significant outcomes). I will report the remaining outcomes (variance and gradient evaluation) for only one of the models to save this summary from becoming too crowded. However, I am sure you get the picture (and you can read the paper yourself if you want to know more). For 'depression at year 1', 24.3% of the variance was explained (adjusted $R^2 = .191$). Four predictors significantly contributed to the outcome: history of alcohol dependence ($B = 7.74$, $t = 2.71$, $p = .009$), being female ($B = 4.44$, $t = 2.00$, $p = .05$), younger age ($B = -0.12$, $t = -1.77$, $p = .08$), and presence of cognitive sequelae ($B = 4.29$, $t = 1.96$, $p = .06$).

Extended learning task

You will find the data set associated with this task on the website that accompanies this book (available in SPSS and Excel format). You will also find the answers there.

Following what we have learned about multiple linear regression, answer these questions and conduct the analyses in SPSS and G*Power. (If you do not have SPSS, do as much as you can with the Excel spreadsheet.) For this exercise, we will look at a fictitious example of an exploration of factors that might contribute to overall satisfaction with health and fitness (the main outcome measure). This is scored on continuous scale from 0 (highly satisfied) to 100 (highly dissatisfied). Several other current measures may be associated with this main outcome. The 200 participants are assessed for current depression and psychosis, using clinical rating scales (providing a category of yes or no to both). The group were also asked to rate various aspects of their quality of life and health behaviours - each were rated on a score from 0 (poor) to 5 (good). These were degree of job stress, degree of stress, sleep quality, level of fatigue, frequency of exercise, quality of exercise, eating healthy food, and satisfaction with weight.

Open the data set healthy **Fitness**

1. Run this analysis, using 'Overall health and fitness' as the outcome variable, and the remaining variables as the predictors (using the 'Enter' method).
2. How much variance was explained by this model?
3. Was the model significant?
4. Which of the predictors were significant?
5. For the significant predictors, what does each value of 'B' tell you about the scores on the outcome variable?
6. Describe how each of the assumptions and restrictions were met.
7. Discuss the implication of any violations of assumptions.
8. Show the effect size and conduct a power calculation, using G*Power.
9. Write up the results with appropriate tables.

Appendix to Chapter 16

Calculating multiple linear regression manually

Earlier, we saw how to run multiple linear regression through SPSS. In this section we learn how we can calculate those outcomes manually. We will focus on a similar research design that we used to perform multiple regression analysis, but with a smaller sample and just three predictor variables (to aid demonstration). The outcome variable is still quality of life (QoL) scores, and the three predictor variables are physical health, job satisfaction and relationship satisfaction (shown as 'physical', 'job' and 'relate' in the following analyses). The raw scores are presented in Table 16.7, with additional calculations needed to examine the final outcome (as shown in Table 16.8). You will find an Excel spreadsheet associated with these calculations on the web page for this book.

Table 16.7 QoL scores, in respect of physical health, job satisfaction and relationship satisfaction

	Physical X_1	Job X_2	Relate X_3	QoL Y
	58	58	48	56
	23	40	44	28
	45	40	45	25
	30	24	35	25
	25	25	42	16
	30	37	36	37
	65	64	54	63
	28	22	22	33
	48	34	56	26
	34	26	44	29
Mean	38.60	37.00	42.60	33.80
St Dev	14.58	14.36	9.86	14.70
St Error	4.61	4.54	3.12	4.65

Table 16.8 Additional calculations

Y^1	$X_1 - \bar{X}_1$	$(X_1 - \bar{X}_1)^2$	$X_2 - \bar{X}_2$	$(X_2 - \bar{X}_2)^2$	$X_3 - \bar{X}_3$	$(X_3 - \bar{X}_3)^2$	$Y^1 - Y$	$(Y^1 - Y)^2$
55.73	19.4	376.36	21	441	5.4	29.16	−0.27	0.07
28.16	−15.6	243.36	3	9	1.4	1.96	0.16	0.03
37.35	6.4	40.96	3	9	2.4	5.76	12.35	152.52
24.96	−8.6	73.96	−13	169	−7.6	57.76	−0.04	0.00
18.27	−13.6	184.96	−12	144	−0.6	0.36	2.27	5.15
34.86	−8.6	73.96	0	0	−6.6	43.56	−2.14	4.58
59.31	26.4	696.96	27	729	11.4	129.96	−3.69	13.62
32.17	−10.6	112.36	−15	225	−20.6	424.36	−0.83	0.69
25.54	9.4	88.36	−3	9	13.4	179.56	−0.46	0.21
21.66	−4.6	21.16	−11	121	1.4	1.96	−7.34	53.88
		1912.40		1856.00		874.40		230.75

We will see how to calculate the values of Y^1 later.

Correlation between X_i and Y

Formula for (Pearson's) correlation earlier: $R = \frac{\sum(x_i - \bar{x})(y_i - \bar{y})}{(N-1)s_x s_y}$ S = standard deviation (Table 16.7)

Correlation between predictors and outcome (you will need to fill in the gaps . . .):

Physical vs. QoL: We already know that from simple linear regression $R = \mathbf{0.769}$

Job vs. QoL: $R = ((58 - 37.00) \times (56 - 33.80) + (40 - 37.00) \times (28 - 33.80) + \ldots (26 - 37.00) \times (29 - 33.80)) \div (9 \times 14.36 \times 14.70) = \mathbf{0.856}$

Relate vs. QoL: $R = ((48 - 42.60) \times (56 - 33.80) + (44 - 42.60) \times (28 - 33.80) + \ldots (44 - 42.60) \times (29 - 33.80)) \div (9 \times 9.86 \times 14.70) = \mathbf{0.295}$

Inter-correlation between predictors:

Physical vs. Job: $R = ((58 - 38.60) \times (58 - 37.00) + (23 - 38.60) \times (40 - 37.00) + \ldots (34 - 38.60) \times (26 - 37.00)) \div (9 \times 14.58 \times 14.36) = \mathbf{0.822}$

Physical vs. Relate: $R = ((58 - 38.60) \times (56 - 33.80) + (23 - 38.60) \times (28 - 33.80) + \ldots (34 - 38.60) \times (29 - 33.80)) \div (9 \times 14.58 \times 14.70) = \mathbf{0.671}$

Relate vs. Job: $R = ((56 - 33.80) \times (58 - 37.00) + (28 - 33.80) \times (40 - 37.00) + \ldots (29 - 33.80) \times (26 - 37.00)) \div (9 \times 14.70 \times 14.36) = \mathbf{0.622}$

We should put all of what we have just found in a table of correlations (see Table 16.9).

Table 16.9 Correlation and inter-correlation (Physical, Job and Relate vs. QoL)

	Physical X_1	Job X_2	Relate X_3	QoL Y
Physical	1	0.822	0.671	0.769
Job	0.822	1	0.622	0.856
Relate	0.671	0.622	1	0.295
QoL	0.769	0.856	0.295	1

At this point in simple linear regression, we used the correlation to calculate the gradient between the predictor and the outcome. It is not that simple in multiple linear regression. We need to conduct a separate calculation between each predictor and the outcome, but also need to account for the inter-correlation between the predictors. We do this by means of matrices of numbers. To account for the inter-correlation we take the red-font correlation co-efficients shown in Table 16.9 and express them as a matrix:

$$\textbf{Matrix of inter-correlation:} \begin{pmatrix} 1.000 & 0.822 & 0.671 \\ 0.822 & 1.000 & 0.622 \\ 0.671 & 0.622 & 1.000 \end{pmatrix}$$

We also need to express the correlation between the predictors and outcome as a matrix (using the blue-coloured correlation co-efficients).

$$\textbf{Matrix of correlation:} \begin{pmatrix} 0.769 \\ 0.856 \\ 0.295 \end{pmatrix}$$

To account for these inter-correlations, we need to divide the correlation between the predictors and outcome. However, we cannot divide one matrix by another: we must multiply one by the inverse of the other. We need to find the inverse of the inter-correlation matrix. The method to do that is not complex, but it is very tiresome and will take rather a lot of space. Instead, I will show you how you can find the inverse of a matrix using Microsoft Excel®.

Copy and paste inter-correlation matrix into cell B4 of a new spreadsheet (as shown in Figure 16.37).

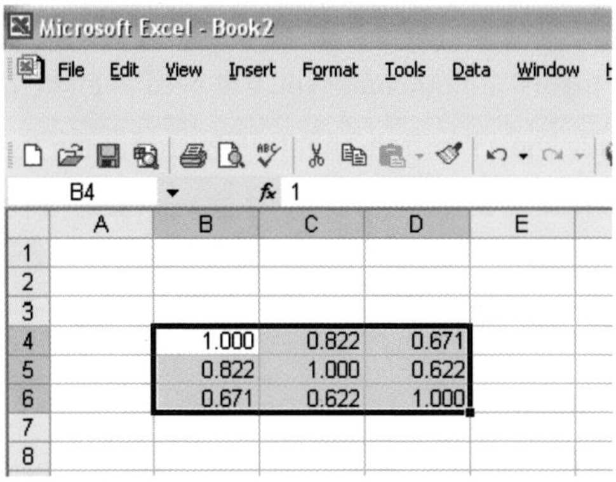

Figure 16.37 Finding inverse of matrix (step 1)

Now, in cell B8, type the following command: = minverse(B4:D6) → then tick the green arrow (as shown in Figure 16.38).

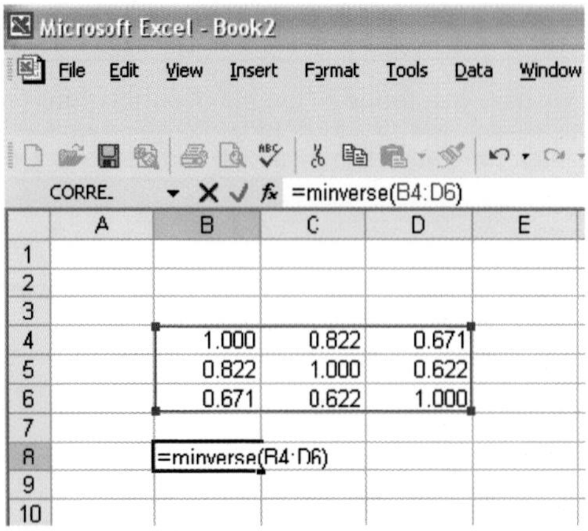

Figure 16.38 Finding inverse of matrix (step 2)

Now 'highlight' the cells B8 to D10 so that a 3 × 3 block is 'lit' with B8 in the top left-hand corner (see Figure 16.39). Now go to the Function window at the top of the spreadsheet and highlight over '=MINVERSE (B4:D6)'.

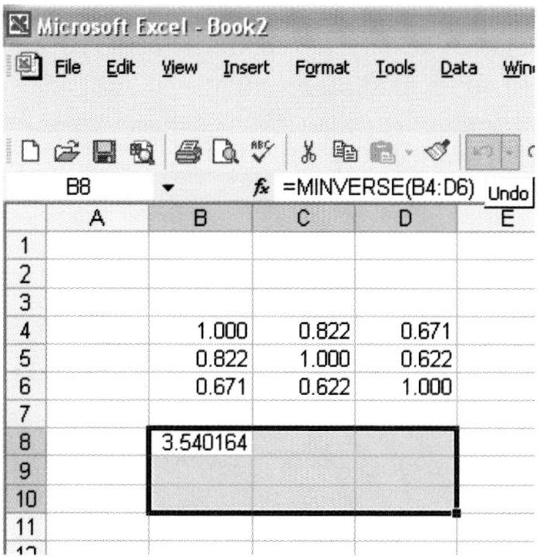

Figure 16.39 Finding inverse of matrix (step 3)

With that block still highlighted, press F2 and then press Crtl + Shift + Return together. You will see a new array, underneath the original one. This is your inverse matrix (see Figure 16.40).

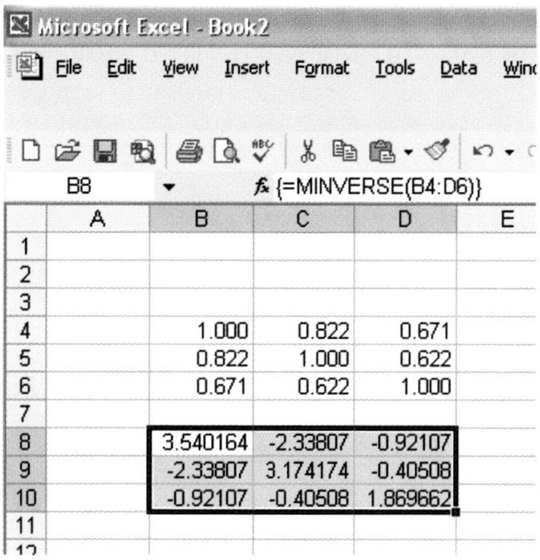

Figure 16.40 Finding inverse of matrix (step 4)

We can now use that inverse matrix to calculate the gradient between each predictor and the outcome variable:

$$\begin{pmatrix} 3.540 & -2.338 & -0.921 \\ -2.338 & 3.174 & -0.405 \\ -0.921 & -0.405 & 1.870 \end{pmatrix} \times \begin{pmatrix} 0.769 \\ 0.856 \\ 0.295 \end{pmatrix}$$

$$= \begin{pmatrix} (3.540 \times 0.769) + (-2.338 \times 0.856) + (-0.921 \times 0.295) \\ (-2.338 \times 0.769) + (3.174 \times 0.856) + (-0.405 \times 0.295) \\ (-0.921 \times 0.769) + (-0.405 \times 0.856) + (1.870 \times 0.295) \end{pmatrix}$$

$$= \begin{pmatrix} 0.448 \\ 0.800 \\ -0.503 \end{pmatrix} \text{ where the predictors are } \begin{pmatrix} X_1 \\ X_2 \\ X_3 \end{pmatrix}$$

To find the gradient, we multiply each of the figures in the final matrix by the respective ratio of standard error of Y by standard error of X_i:

Gradient Physical (X_1) vs. QoL $= 0.448 \times (4.65 \div 4.61) = \mathbf{0.452}$

Gradient Job (X_2) vs. QoL $= 0.800 \times (4.65 \div 4.54) = \mathbf{0.819}$

Gradient Relate (X_3) vs. QoL $= -0.503 \times (4.65 \div 3.12) = \mathbf{-0.750}$

Coordinates for best line(s) of fit

$$\text{Model regression line: } Y^1 = a + bX_1 + bX_2 + bX_3$$

We have the gradient for each predictor, so we can calculate the intercept (a):

$$Y_1 = a + (0.452)X_1 + (0.819)X_2 + (-0.750)X_3$$

So, if we use the mean of each predictor:

$$Y^1 = a + (0.452)(38.60) + (0.819)(37.00) + (-0.750)(42.60)$$
$$Y^1 = a + 15.80$$

If we use the average of Y:

$$33.80 = a + 15.80$$

So, $a = 33.80 - 15.80 = 18.00$ (actually, allowing for rounding, it is 17.987)

So now, Model regression line: $Y^1 = 17.987 + 0.452X\,1 + 0.819X_2 - 0.750X_3$

For example: $Y^1 = 17.987 + 0.452 \times (58) + 0.819 \times (58) - 0.750 \times (48) = 55.73$

We can put that value for Y^1 into Table 16.8.

Total sum of squares

This will be the same as our example for linear regression. It makes sense that the overall variation in the data will be the same (before we look at any potential model).

$$SS_T = 1945.60$$

Model sum of squares

Formula for total sum of squares (SS_M): $\sum (y_i^1 - \bar{y})^2$

So, $SS_M = (55.73 - 33.80)^2 + (28.16 - 33.80)^2 + \ldots (21.66 - 33.80)^2 = \mathbf{1714.77}$

Variance

$$R^2 = \frac{SS_M}{SS_T} = 1714.77 \div 1945.60 = 0.881$$

We also need adjusted R^2.

Adjusted $R^2 = 1 - (1 - R^2)\left(\frac{N - 1}{N - k - 1}\right)$ where N is sample size (10) and k is no. of predictors (3)

$$\text{Adjusted } R^2 = 1 - (1 - 0.881)\left(\frac{9}{6}\right) = \mathbf{0.822}$$

F ratio (success of model)

$$\text{F ratio} = \frac{MS_M}{MS_R}$$

We need to find model mean square and residual mean square.

First, we need to find the residual sum of squares (SS_R):

$$SS_R = SS_T - SS_M = 1945.60 - 714.77 = 230.83$$

The mean squares are found from the sum of squares divided by the degrees of freedom (*df*).

df for model (df_M) = number of predictors (3); *df* for error (df_R) = $(N - 1) - df_M = 6$

So $MS_M = SS_M \div df_M = 1714.77 \div 3 = 571.59$

$MS_R = SS_R \div df_R = 230.88 \div 6 = 38.48$

$\text{F ratio} = 571.59 \div 38.48 = 14.858$

In the F-distribution table, we would find the cut-off point for $df = 3, 6$ (when $p = .05$) is 4.76. Our F ratio (14.86) is greater than 4.76, so our model is significantly better than using the mean to predict outcome scores (QoL) from our predictor variables (physical health, job satisfaction and relationship satisfaction).

Significance of the gradients

A further check on the importance of the model is based on the t-test of the gradient; the calculations for this are much more complex, as it involves partial correlations. It is probably best that we leave it there!

You can see how to use Excel to run the entire multiple regression on the web page for this book.

17 LOGISTIC REGRESSION

Learning objectives

By the end of this chapter you should be able to:

- Recognise when it is appropriate to use logistic regression
- Understand the theory, rationale, assumptions and restrictions associated with the test
- Perform analyses using SPSS
- Understand how to present the data and report the findings

What is (binary) logistic regression?

Logistic regression predicts the likelihood of a categorical outcome (dependent variable) occurring, based on one or more predictors (independent variables). In **binary logistic regression** the outcome must be binomial – this means that there can be only two possible outcomes (such as 'yes or no', 'occurred or did not occur', etc.). For example, we could explore the likelihood of a diagnosis of major depressive disorder (yes or no). The likelihood for the outcome might be investigated in relation to a series of predictor variables – these can be numerical (continuous) or categorical. For example, the probability of a diagnosis of depression might be predicted by age, gender, sleep quality and self-esteem. The effect of the predictor upon outcome is described in terms of an '**odds ratio**'. This odds ratio represents a change in odds that results from a unit change in the predictor. So, in that sense, it is similar to the gradient in multiple linear regression. An odds ratio greater than 1 indicates a greater likelihood of the outcome occurring; if it is less than 1 that outcome is less likely to occur. An odds ratio of exactly 1 represents no change. For example, we might explore whether gender has an impact on the diagnosis of depression. If we found that female gender (as a predictor) had an odds ratio of 3.8, this would mean that being a woman increases the likelihood of having a diagnosis of depression by almost four times. Binary logistic regression explores a single outcome. Outcomes with more than two categories *can* be analysed, but multinomial logistic regression would be needed to do so (we do not explore that test in this chapter).

Research question for logistic regression

In this example, we explore some (fictitious) findings from a group of quality of life researchers known as the Centre for Healthy Independent Living and Learning (CHILL). They would like to know what factors contribute to a diagnosis of major depressive disorder (depression). They interviewed 200 participants and confirmed that 60 had a diagnosis of depression. The research group then asked the participants to complete some questionnaires that captured information on age, gender, sleep quality and self-esteem. CHILL sought to investigate the extent to which each of these variables poses a risk factor for depression.

17.1 Take a closer look
Summary of logistic regression example

Outcome variable: diagnosis of major depressive disorder (yes or no)
Predictors: age, gender, sleep quality and self-esteem

Theory and rationale

How logistic regression works

Similar to linear regression (Chapter 16), we express a logistic regression model in terms of an equation that reports how an outcome is explained by a constant, the gradient of each predictor (also known as the regression co-efficient) and the error term. The equation for any regression model must be linear (where $Y = \beta_0 + \beta_1 X_1$). However, logistic regression differs from linear regression in that the outcome variable is categorical rather than numerical. That outcome is expressed in binary terms (0 or 1), where 1 usually represents the 'positive' outcome. For example, if we are exploring a diagnosis of major depressive disorder as our outcome, 1 would indicate that depression was present, while 0 would signify no depression.

However, because the outcome is categorical, it cannot be linear, so we need to transform that outcome, using logarithmic transformation to make it linear (you can see more about logarithms in Box 17.3). This result of the transformation is illustrated in the logistic regression equation (see Box 17.2). Ultimately, the outcome is expressed in terms of a probability (or 'likelihood') of the outcome occurring. That probability is reflected in the equations shown in Box 17.2. For logistic regression there are two equations (as there are for linear regression): one where we have a single predictor variable and one for several predictors.

17.2 Nuts and bolts
Probability equations in logistic regression

There are two equations that express the probability of the outcome in logistic regression: one where there are single predictors and one when there are several:

$$\text{Single predictor: } P(Y) = \frac{1}{1 + e^{-(b_0 + b_1 x_1 + \varepsilon_i)}}$$

$$\text{Multiple predictors: } P(Y) = \frac{1}{1 + e^{-(b_0 + b_1 x_1 + b_2 x_2 + \ldots b_n x_n + \varepsilon_i)}}$$

where P = probability of Y (the outcome) equalling 1

On the face of it, these equations seem quite complex, but the expressions shown inside the brackets are actually the same as we saw for simple and multiple linear regression. As we saw in Chapter 16, b_0 is the constant (or intercept, where the regression line crosses the Y axis), b_n is the gradient (or *unstandardised* regression co-efficient) for the predictor variable x_n, while ε_i is the error term. Note that SPSS reports the gradient as 'B' in output tables. To convert the categorical outcome into a linear one, we need to employ logarithm transformation. We do this by using natural logarithms ($\log_e$). You can see more about logarithms in Box 17.3. The inverse function of $\log_e$ is e^x. In our case, 'x' is the linear regression equation.

Logistic regression outcomes

Log-likelihood

The outcome for logistic regression is determined by 'Y' (as shown in Box 17.2). That outcome will range between 0 and 1: values approaching 0 indicate that the outcome is unlikely, values approaching 1 suggest the outcome is probably likely. Once we know the constant (b_0) and the gradients for each predictor ($b_n x_n$), we can predict the 'expected' outcome. The only part we cannot estimate is the error value (ε_i). Hopefully, that will be small and the predicted model will closely match the actual outcome. The closer it is, the better the model fits the actual data

When we sought to estimate how well a linear regression model fitted the overall data, we used the F-ratio (see Chapter 16). In logistic regression we use likelihood ratios to estimate the goodness of fit. In simple terms, the initial likelihood could be found by expressing the probability of the outcome in relation to the number of people in the sample. If there are equal numbers of people in the two possible outcomes, the probability of any one outcome is 50:50, or 0.5. If there were two people in the sample, the likelihood ratio would be 0.5×0.5, which is 0.25. If we had a sample of 200 people (as we do in our example), the likelihood would be 0.5^{200}, which is a very small number: it represents $0.5 \times 0.5 \times 0.5\ldots$ 200 times! If you were to calculate this, the outcome would be 0.000…622, but with 60 zeros between the decimal point and the first digit! To make that easier to read, we write that as 6.22e–61. As it is so small, we find the natural logarithm ($\log_e$) for that number and multiply it by −2. This produces a figure that we can actually see (for example, $\log_e$ of 6.22e−61 is −138.629). We multiply that by −2 simply to get a positive number. The outcome is called **−2 log-likelihood** (or −2LL). As we will see later, our data do not have equal numbers in the outcome possibility, so the precise calculation is somewhat more complex than what we have just seen, but hopefully you get the general idea.

Furthermore, the logistic regression analysis calculates several log-likelihood outcomes. It does this through a process called iteration. Once the initial calculation has been performed, the iteration process refines the parameters, and calculates −2LL again. The aim is to reduce the outcome to as small a number as possible (because the smaller the number, the better the model fits the data). In theory, it could do this an infinite number of times, but we usually instruct the iteration process to stop when the 'improvement' in the −2LL outcome is less than .001. Ultimately, we are provided with two −2LL outcomes: one for the initial model (before we include the predictor variables) and one for the final model (after the inclusion of the predictors). We will address how to compare those outcomes in the next section.

17.3 Nuts and bolts

What are logarithms?

Logarithms were developed by John Napier in the 17th century. They are the inverse of a number expressed to a 'component' (such as 10^3). Where $Y = a^x$, 'a' represents the 'base' and 'x' signifies the 'component'. The inverse of $Y = a^x$ (the logarithm) is expressed as $x = \log(Y)_a$. Originally, (common) logarithms were expressed to base 10, so $10^3 = 1000$ (10 × 10 × 10). The common logarithm of 1000 to base 10 is therefore $\log(1000)_{10} = 3$. Logarithms were used to simplify calculations long before we had computers and hand-held calculators. Imagine we want to multiply 5.13 × 6.44. Easy to do these days, but not so easy without technology. The solution was found through logs. Every conceivable logarithm was published in log tables. We can use these tables to find $\log_{10}$ for 5.13 (0.710) and $\log_{10}$ for 6.44 (0.809). We add those outcomes (0.710 + 0.809 = 1.519) and we look up the inverse of the $\log_{10}$ for 1.519 in those same log tables. We would find that to be 33.037. You could confirm this with a calculator.

In logistic regression, we tend to use natural logarithms. These were developed by Leonhard Euler in the 18th century. In natural logs the base is called e (after Euler), where e = 2.71828182845904 (you don't need to know why). This logarithm is stated as $\log_e$; the inverse is e^x. The principles for analysis are exactly the same as we have just seen for common logs. We use natural logs in logistic regression for several reasons. First, as we saw earlier, we express the categorical outcome in a linear equation by using $\log_e$ and e^x. We saw the equation for logistic regression in Box 17.2. The 'regression' part of that equation was expressed as $e^{-b0+b1\times1+\varepsilon i}$. The 'minus' sign before the equation simply means '1 ÷ the equation'. For example, $4^{-1} = 1 \div 4 = 0.25$. We also use natural logs in logistic regression because the probability of outcome can be so small. For example, where ten people have an equal chance of an outcome, the likelihood is $0.5^{10} = 0.0000976$. This is not easy to work with, so we transform that to the natural $\log_e$ for 0.0000976, which is −6.931. In order that we can use a positive number, we multiply that outcome by −2 (13.862). We also explore logarithms as a method of transforming non-normally distributed data in Chapter 3.

Using log-likelihood ratios to assess the success of the regression model

To assess the 'success' of our model, we need to illustrate whether the predicted model is better at predicting outcome than some arbitrary (baseline) model or outcome. In linear regression, we use the mean outcome score as the baseline comparison – this is compared with predicted outcomes from the regression model (see Chapter 16). In logistic regression, we cannot use the mean score because our categorical outcome is a series of 1s and 0s. Instead, we use the 'most likely' outcome. If the measured event (such as the diagnosis of depression) occurred more often than it did not, we use a baseline indicator stating that the event occurred; if the frequency of 'no diagnosis of depression' is more prevalent, the baseline assumes that the event did not occur. The baseline indicator is represented by the logistic regression model with only the constant included. To assess the success of our model, we find the difference between '−2LL baseline' and '−2LL new', expressing this in terms of a chi-squared (X^2) outcome, as shown below (we will learn more about the properties of X^2 in Chapters 18 and 19).

$$\chi^2 = -2\text{LL (new)} \textit{ minus } - 2\text{LL (baseline)}$$

If the X^2 value is large and significant, we can say that our final model is better at predicting outcome than simply using the most common event (the initial model). We assess X^2 by comparing it with the cut-off point in chi-square distribution tables, according to the relevant degrees of freedom. In this instance, the degrees of freedom equal the number of predictor variables in the final model. If our X^2 value exceeds that cut-off point, we can say that the model is significantly better at predicting outcome than some arbitrary method (much like we can use the F ratio outcome in linear regression).

Gradient, correlation and variance (Wald statistic R and R^2)

Similar to linear regression, we can use correlation and variance in logistic regression to assess the success of the model, and refer to the gradient of the regression slope to indicate how changes in a predictor contribute to the outcome variable. However, measurement of correlation and variance are not quite so straightforward in logistic regression. Where there is one predictor, we measure correlation between that predictor and the outcome. Like any other correlation, it measures the strength of relationship between variables, and will range between −1 and 1 (see Chapter 6). A negative correlation with a predictor variable suggests that the outcome is less likely; a positive correlation suggests that it is more likely. When there are several predictors, rather like we found with multiple linear regression, partial correlation is used to explore the relationship between the predictor and the outcome. In both cases we use 'R' to signify that outcome (but must interpret them very differently). We can find R from something called a Wald statistic, which is measured from the unstandardised regression co-efficient (see Box 17.4).

Where R measures correlation (or partial correlation), R^2 estimates how much variance in the outcome has been explained by the predictor variable (just like linear regression). However, we cannot simply square R to calculate R^2. Instead, we can refer to two alternative statistics: Cox and Snell's R^2 and Nagelkerke's R^2. These outcomes are provided in the SPSS output (but you can see how they are calculated in Box 17.4). Cox and Snell's (1989) R^2 is based on the log-likelihood for the new and original model, along with the sample size. However, Nagelkerke (1991) argued that this calculation is limited, as it can never reach 1, and provided an alternative method. Both tend to be reported in statistical outcomes.

The Wald statistic examines the extent to which the gradient (unstandardised regression co-efficient) is greater than 0. It is equivalent to the t score in linear regression and assesses whether a predictor significantly contributes to predicting the outcome. It is calculated by dividing the gradient (b) by its standard error (this is done for each predictor):

$$\text{Wald} = \frac{b}{SE_b}$$

Effectively, the calculation for the Wald statistic provides a *z*-score, which we can use to determine significance (see Chapter 4). If the outcome is significant, it suggests that the predictor contributes to the outcome. However, some sources suggest that the Wald test should be used with caution where there are large regression co-efficients (Menard, 1995), and in smaller samples (Agresti, 1996). In such instances, the likelihood ratio test is thought to be more reliable (we saw how to calculate that earlier).

17.4 Nuts and bolts

Finding R and R^2 in logistic regression (using equations)

In logistic regression, R measures the correlation between the predictor and the outcome variable (where there is one predictor), or the partial correlation between each predictor and the outcome (where there are multiple predictors). We do not report R directly, but we can find it via the Wald statistic and log-likelihood:

$$R = \pm\sqrt{\frac{\text{Wald} - (2 \times \text{df})}{-2\text{LL (original)}}}$$

R^2 illustrates how variance in the outcome has been explained by the predictor. However, we cannot simply square R to find that, we must use the two following indicators:

$$\text{Cox and Snell's } R^2_{CS} = 1 - e^{\left[\frac{2}{n}(\text{LL(new)} - \text{LL(baseline)})\right]}$$

$$\text{Nagelkerke's } R^2_N = \frac{R^2_{CS}}{1 - e^{\left[\frac{2(\text{LL(baseline)})}{n}\right]}}$$

Both of these calculations provide an estimate of the variance explained by the predictor, often producing different outcomes (Nagelkerke's R^2 is less conservative).

Goodness of fit

We need to ensure that we have not lost too much data in our final model. We want the model to adequately 'fit' the data We can measure this by comparing the observed and expected frequencies. The observed outcomes are what actually occurred; the expected frequencies are what we predicted. We want these to be similar. In SPSS, we can request a procedure called the Hosmer and Lemeshow test. This test examines whether there is a significant difference between the observed and expected frequencies. We need a non-significant outcome to demonstrate **goodness of fit**. We will see more about that later when we analyse outcomes in SPSS.

Odds-ratios

We also need to find the logistic regression equivalent of gradient. Using SPSS, we can use the figure reported for 'Exp(B)'. As it happens, this outcome is even more useful in logistic regression as it can provide a direct assessment of odds ratios. It is a statement of likelihood that something will occur (we will learn more about odds ratios in Chapter 19). The odds ratio is found by dividing the odds of an event occurring by the odds that an event will not occur (you don't need to know exactly how, since SPSS does that for you). An odds ratio (OR) greater than 1 suggests that something is more likely to occur; an OR less than 1 indicates that it is less likely to happen Each predictor is assessed with regard to the odds ratio towards the outcome. The interpretation of the odds ratio is different depending on whether the predictor is categorical or continuous. Unlike linear regression, categorical predictors in logistic regression need not be dichotomous: they can have more than two categories. In our example, we have only one categorical predictor (gender). As this has two groups (male or female), it is probably better to use codes of 0 and 1. As we will see later, we tell SPSS which 'code' to treat as 'baseline'. In our example, we will direct SPSS to use 'females' (1) as baseline. If our analyses indicate that the Exp(B) outcome for gender (1) is 2.13, it suggests that women are more than twice as likely to be diagnosed with depression than men For continuous predictor variables, analysis of odds ratios is not so straightforward. In this case, the Exp(B) figure indicates the extent to which predicted odds ratios change in the outcome for every unit change in the predictor variable. For example, using our research example, we might find the Exp(B) outcome for age is 1.09. This means that, for every year that age increases, the likelihood for a diagnosis of depression increases by 1.09.

17.5 Take a closer look
Key elements in logistic regression

Log-likelihood:	Outcomes in logistic regression are expressed in terms of the likelihood that an event will occur. But these need to be converted via natural logarithms to provide a log-likelihood. We compare the log-likelihoods of the final and baseline models. The final model should be significantly smaller than baseline (examine via a chi-squared analysis): $\chi^2 = -2\text{LL}$ (new) *minus* -2LL (baseline).
Wald statistic:	Measures whether the regression co-efficient of a predictor significantly contributes to the outocme. It is calculated from the gradient (or regression co-efficient, b) and its standard error (SE): Wald $= b \div SE_b$.
Correlation (R):	Examines the strength of relationship between the predictor and outcome, as expressed by 'R'. When there are several predictors, partial correlation is used. It is calculated from the Wald statistic and the baseline log-likelihood (see Box 17.4).
Variance (R^2):	Indicates how much 'variation' in the outcome is explained by the model. Despite appearances, we cannot simply 'square' R. There are two alternative statistics that we can report: Cox and Snell's R^2 and Nagelkerke's R^2.
Goodness of fit:	Measures the extent that the final model still fits the data. This is assessed by comparing the observed and expected frequencies, via the Hosmer and Lemeshow test.
Odds ratio	Expresses the likelihood the outcome will occur according to changes in the predictor variable. An odds ratio (OR) greater than 1 indicates the event is more likely to occur; an OR less than 1 suggests that it is less likely.

Entering predictors into the model

Similar to multiple linear regression, there are choices about how to enter the predictor variables into the logistic regression model. The rationale behind those options is much the same as we saw in Box 16.11, so we will not extend that here. We will use the Enter method, where all the predictors are entered simultaneously. This will help illustrate how each predictor has contributed to the overall outcome. However, you may wish to enter the predictors in a hierarchy, so that you can specifically see the effect of each one. Within hierarchical methods, it is generally thought that '**backward stepwise**' is best. Forward stepwise method can falsely reject predictors if they have been 'suppressed' (significant only once another variable is held constant). It might increase the chance of Type II errors (where an experimental hypothesis is rejected when it should have been accepted – see Chapter 2). Among backward methods, the 'Backward: LR' tends to be used most because it uses likelihood ratio to determine significance. We will not deal with hierarchical methods in this book.

Categorical predictor variables

With logistic regression, if any predictor variable is categorical, we must tell SPSS that this is the case. There are several ways that we can do this. The most favoured (and default) option is the 'indicator' method, so we will focus on that. Also, it is possible for a categorical predictor to have more than two categories (unlike linear regression). For binary variables, we generally use a coding of 0 and 1 (we call this **dummy coding**). In such cases, a coding of 1 is taken as baseline. In our research example, gender is the only categorical predictor variable. When we explore our research example, we will set 'female' as the baseline because there is evidence that women are more likely to be depressed than men (Nolen-Hoeksema, 2001). In which case, we will code men as 0 and women as 1. When we perform logistic regression in SPSS, we use the 'Define Categorical Variables' box to determine which category should be treated as baseline. By default, SPSS takes the highest coded number as baseline (defined as 'last' in the SPSS parameters). If we had coded the groups as 1 and 2, SPSS would automatically use '2' as baseline. However, in dummy coding, '1' is effectively the 'low' number and '0' is 'high'. So, in our example, because we want SPSS to treat '1' (females) as baseline, we must ask for that to be looked at 'first'. We need to change the parameters to accommodate that (we will see how later).

In many cases, the coding for gender categories will not matter, but you will need to remember how you have coded the data (so that you interpret them correctly). On other occasions, logical coding may be more crucial. Let's say we have another categorical predictor variable in the form of current insomnia (yes or no). If we want to examine the extent to which a diagnosis of insomnia contributes to a diagnosis of depression, we would want to make 'insomnia = yes' the baseline and code that as 1 accordingly. Sometimes the categorical predictor might have more than two categories. For example, we could investigate how anxiety severity affects depression diagnosis. We might have three categories: none, mild and moderate (coded as 1, 2 and 3 respectively). When we set the categorical definition box in SPSS, we could choose to have 'no anxiety' as the reference (in which case we would select 'first' for the lowest coded number), or we could focus on severe anxiety (and select 'last' for the highest coded number).

Comparison of logistic regression to linear regression

When we looked at linear regression in Chapter 16, we explored how much variance in a numerical outcome variable could be explained by variations in one or more predictor variables. For example, we might investigate the extent that depression severity scores vary according to changes in age, gender, sleep quality and self-esteem. We might find that only variations in age and self-esteem significantly contribute to the variance in depression severity scores. In logistic regression we examine the extent that a categorical dichotomous outcome can be explained by individual predictor variables. For example, we could investigate the likelihood of receiving a diagnosis of depression as a result of those same predictors (age, gender, sleep quality and self-esteem). In logistic regression, instead of looking at how variations in predictors explain variance in the outcome score, we are investigating how those predictors change the likelihood of a categorical outcome. In our example, advancing age might increase the likelihood of depression (perhaps by a ratio of 1.5 for every five years' increase), while being female might increase the likelihood of depression fourfold. That likelihood can be expressed in terms of an odds ratio. We will explore odds ratios shortly, but will encounter them again in Chapter 19.

Assumptions and restrictions

There are fewer assumptions and restrictions for logistic regression compared with multiple linear regression. Normal distribution on the outcome variable does not apply because we dealing with categorical outcomes. By definition, the outcome variable must be categorical, with two possible outcomes, which must be coded as 0 and 1. The predictor variables can be continuous (numerical) or categorical. There are no restrictions on the number of categories for the predictor, or on the coding of those categories in SPSS (although they should be logical). Similar to linear regression, there must be reasonable linearity between predictors and the outcome. However, since the outcome is categorical, we need to transform it using **logarithms**. Therefore, it is probably more correct to say that there should be a linear relationship between each predictor and the log of the outcome variable. If there are several predictor variables, these should not be highly correlated with each other. As we saw in Chapter 16, if two predictor variables were perfectly correlated with each other, it would make independent predictions impossible (because they measure the same construct). We call this multi-collinearity.

17.6 Take a closer look
Summary of assumptions and restrictions

- The outcome variable must be categorical
  - For binary logistic regression, there must be only two outcome categories (coded as 0 and 1)
- Predictor variables can be continuous or categorical
  - They can have several categories
  - There are no restrictions on coding (but should be logical)
- The linear relationship between the predictor variable(s) and the 'outcome' should be reasonable

How SPSS performs logistic regression

For this analysis we will address the research question set by CHILL You may recall that we are examining data from 200 people, 60 of whom have a current diagnosis of major depressive disorder (depression). We are seeking to investigate the extent that four predictor variables (age, gender, sleep quality and self-esteem) predict a diagnosis of depression. Gender is the only categorical predictor; the remaining variables are continuous. Higher scores for sleep quality and self-esteem are 'better' scores.

Logistic regression model:
Outcome variable: diagnosis of major depressive disorder (yes or no)
Predictors: age, gender, sleep quality and self-esteem

17.7 Nuts and bolts
Setting up the data set in SPSS

When we create the SPSS data set for logistic regression, we need to set up one column for the outcome variable (which will need categorical coding), and several columns for the predictor variables (which can either be continuous or categorical). In the following section, we will be using the SPSS data set 'Depressed' to perform logistic regression.

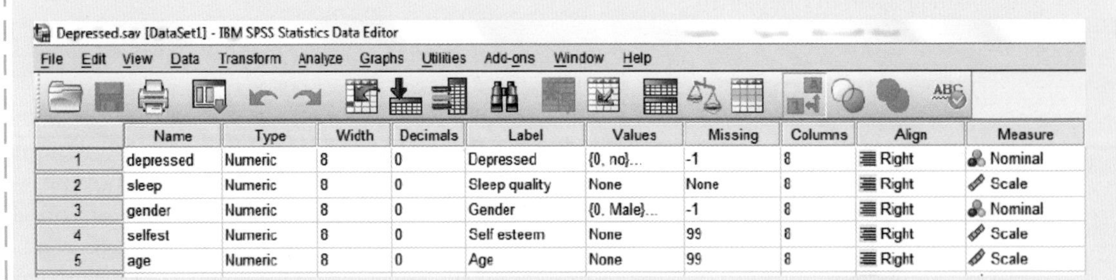

	Name	Type	Width	Decimals	Label	Values	Missing	Columns	Align	Measure
1	depressed	Numeric	8	0	Depressed	{0, no}...	-1	8	Right	Nominal
2	sleep	Numeric	8	0	Sleep quality	None	None	8	Right	Scale
3	gender	Numeric	8	0	Gender	{0, Male}...	-1	8	Right	Nominal
4	selfest	Numeric	8	0	Self esteem	None	99	8	Right	Scale
5	age	Numeric	8	0	Age	None	99	8	Right	Scale

Figure 17.1 Variable View for 'Depressed' data

Figure 17.1 shows how the SPSS Variable View should be set up (you should refer to Chapter 2 for more information on the procedures used to set the parameters). The first variable is called 'depressed'. This is the categorical outcome variable, which will be used to indicate whether the participant is depressed or not. In the Values column, we include '0 = No' and '1 = Yes'; the Measure column is set to Nominal. The next four variables ('Sleep quality', 'Gender', 'Self-esteem' and 'Age') are the predictors, all but 'Gender' are represented by continuous scores (so we do not need to adjust anything in the Values column; the Measure column is set to Scale).Gender is categorical; the Values column should be set as '0 = Male' and '1 = Female' (for the reasons we discussed earlier); the Measure column is set to Nominal.

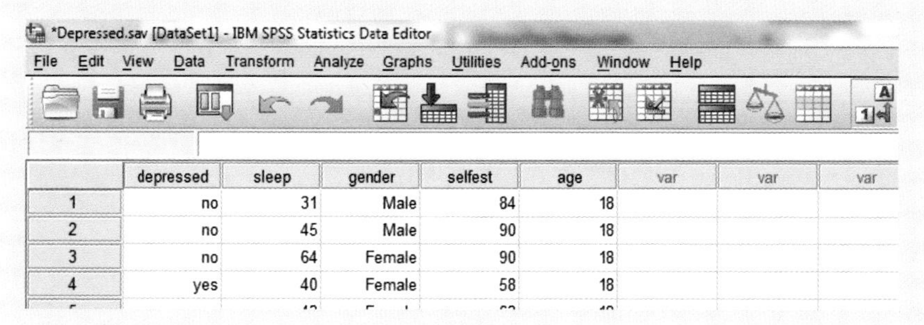

Figure 17.2 Data View for 'Depressed' data

Figure 17.2 illustrates how this will appear in Data View. Each row represents a participant. When we enter the data for 'depressed', we input 0 (to represent 'no') or 1 (to represent 'yes'); the 'depression' column will display the descriptive categories ('No' or 'Yes'). For the continuous predictor variables ('sleep', 'selfest', 'and 'age'), we simply enter the relevant score for that participant. For the gender variable, we need to enter 0 (to represent 'male') or 1 (to represent 'female').

Running the logistic regression model in SPSS

Using the SPSS file Depressed

Select **Analyze → Regression → Binary Logistic...** (as shown in Figure 17.3)

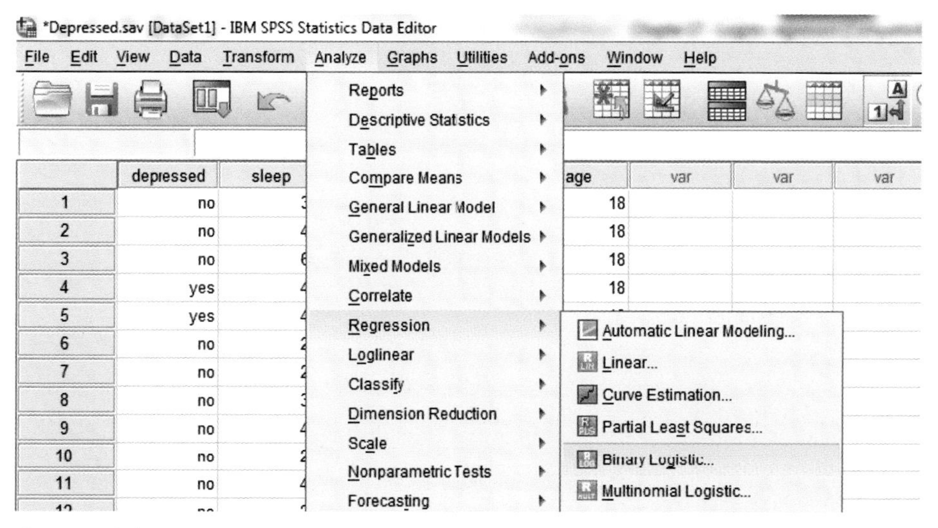

Figure 17.3 Logistic regression – step 1

In new window (see Figure 17.4), transfer **Depressed** to **Dependent:** → transfer **Sleep quality**, **Gender**, **Self esteem**, and **Age** to **Covariates:** → select **Enter** in pull-down options for **Method** → click **Categorical**

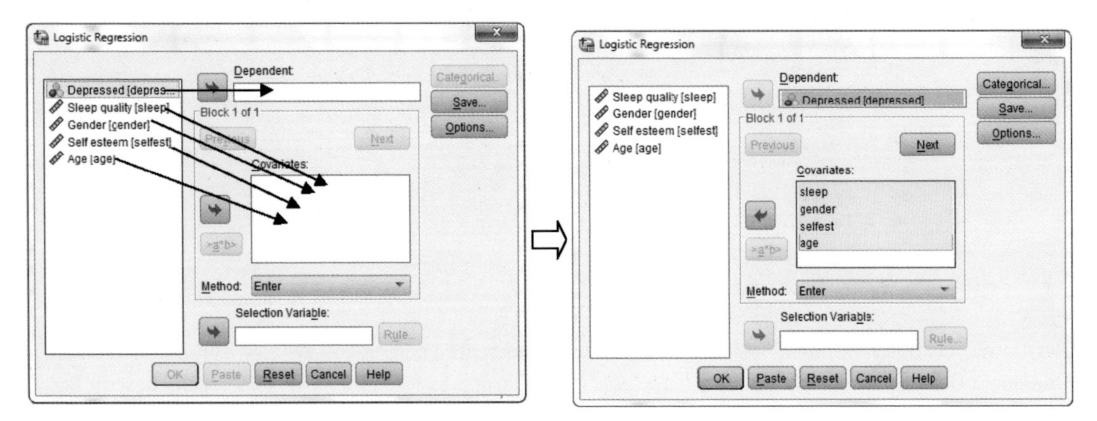

Figure 17.4 Logistic regression: choosing variables

In new window (see Figure 17.5), transfer **Gender** to **Categorical covariates** → select **Indicator** by **Contrast** (it will probably be set to that by default) → select **First** by **Reference Category** → click **Change** (we need to do this because we chose women to be the reference category, using dummy coding where 1 = female; therefore we want SPSS to look at '1' first) → click **Continue** → click **Options**

The choices made here for selecting the order of categorical predictor categories will change according to your circumstances (see earlier discussion)

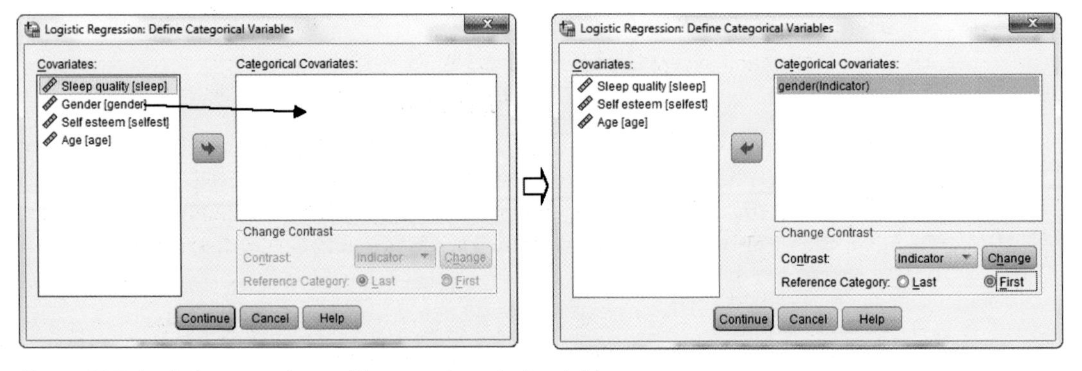

Figure 17.5 Logistic regression: setting up categorical variables

In new window (see Figure 17.6), tick boxes for **Classification plots, Hosmer-Lemeshow goodness-of-fit, Iteration history** and **CI for exp(B)** (set to 95%) under **Statistics and Plots** → click **Continue** → click **OK**

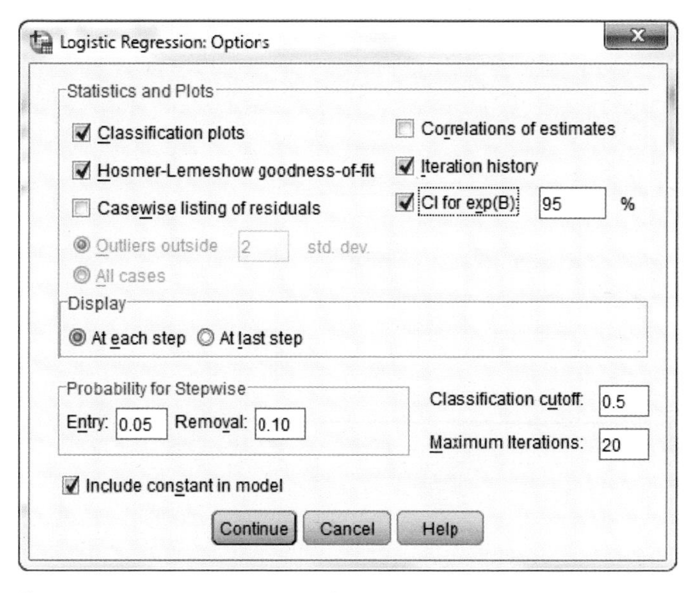

Figure 17.6 Logistic regression: Options

Interpretation of output

Coding confirmation

Figure 17.7 confirms that a positive diagnosis of depression is represented by a coding of 1.

Original value	Internal value
no	0
yes	1

Figure 17.7 Confirmation of dependent variable codes

Figure 17.8 confirms that women are the baseline category, since the parameter coding for Females is 1. This is also worth checking before you proceed with any further analyses.

			Parameter coding
		Frequency	(1)
Gender	Male	100	.000
	Female	100	1.000

Figure 17.8 Confirmation of categorical predictor variable codes

Baseline model (Block 0)

Figure 17.9 is of little use, other than indicating the log-likelihood (−2LL) of the baseline model, which we can compare to later.

Iteration		−2 Log likelihood	Coefficients
			Constant
Step 0	1	244.440	−.800
	2	244.346	−.847
	3	244.346	−.847

Figure 17.9 Initial iteration history

Figure 17.10 indicates how well the initial model predicts outcome. In this case (because there are more people without depression), the model predicts that no one is depressed. Since this occurs on 70% of occasions, this might seem quite good. However, the model misclassifies anyone who is actually depressed, thus correctly predicting 0% of positive diagnoses. Later, we will examine whether the final regression model is better at predicting outcome.

Observed			Predicted		
			Depressed		Percentage correct
			no	yes	
Step 0	Depressed	no	140	0	100.0
		yes	60	0	.0
	Overall percentage				70.0

Figure 17.10 Classification table for baseline model

Figure 17.11 shows that only the constant has been included in this initial model. It represents the outcome without any predictor variables.

		B	S.E.	Wald	df	Sig.	Exp(B)
Step 0	Constant	−.847	.154	30.152	1	.000	.429

Figure 17.11 Variables included in baseline model

Figure 17.12 shows the predictor variables, none of which has been included in the initial model. The key outcome here is represented by the 'Overall Statistics' data. It is highly significant, which suggests that the regression model would be significantly improved with the addition of at least one of the listed predictors. Had that outcome been non-significant, the regression model would not improve; it would be no better than simply using the baseline model to predict outcome. We are also given a clue to which predictors might significantly contribute to the final regression model. All but 'age' have significant outcomes, suggesting that they (and not age) have the potential to explain a diagnosis of depression. We will see whether that is the case shortly.

			Score	df	Sig.
Step 0	Variables	sleep	21.561	1	.000
		gender(1)	4.667	1	.031
		selfest	15.413	1	.000
		age	1.487	1	.223
	Overall statistics		37.367	4	.000

Figure 17.12 Variables excluded from baseline model

Final model (Block 1)

Figure 17.13 confirms that we used the Enter method. The final log-likelihood (−2LL) can be compared with the outcome in the baseline model. The term 'iteration' refers to the process that SPSS goes through, whereby the analysis is repeated until the best solution is found. The aim is to 'reduce' the log-likelihood to as small a figure as possible. In theory, this could go on infinitely, but the process stops once the change is too minimal to make further analyses worthwhile. In our case, the iterative process went through six steps before announcing the 'best' solution. The final figure for −2LL is shown as 194.783. This is somewhat reduced from 244.346 seen in the baseline model. This *suggests* that the final model is better at predicting whether someone has a diagnosis of depression than the initial model (which assumed that everyone was not depressed).

Iteration history[a,b,c,d]

Iteration		−2 Log likelihood	Coefficients				
			Constant	sleep	gender(1)	selfest	age
Step 1	1	205.502	2.430	−.036	.436	−.027	−.004
	2	195.783	4.263	−.065	.742	−.041	−.011
	3	194.797	4.986	−.078	.866	−.046	−.013
	4	194.783	5.077	−.079	.880	−.047	−.013
	5	194.783	5.079	−.079	.881	−.047	−.013
	6	194.783	5.079	−.079	.881	−.047	−.013

a. Method: Enter

b. Constant is included in the model

c. Initial−2 Log Likelihood: 244.346

d. Estimation terminated at iteration number 6 because parameter estimates changed by less than .001

Figure 17.13 Final iteration history

Having established that the final model *appears* to be better than the baseline model, because the log-likelihood has been reduced, we need to verify how much better it is. Figure 17.14 helps us with that analysis. The chi-squared figure represents the difference between the two log-likelihood figures, allowing for minor rounding differences (244.346 − 194.783 = 49.563). The outcome is compared with a chi-squared distribution for the relevant degrees of freedom (4). We can see that this is significant ($p < .001$), so we can be confident that the final model is significantly better at predicting outcome than the baseline model.

		Chi-square	df	Sig.
Step 1	Step	49.562	4	.000
	Block	49.562	4	.000
	Model	49.562	4	.000

Figure 17.14 Omnibus tests of model co-efficients

Figure 17.15 provides some information on how much variance has been explained by the final model. This outcome is similar to the R^2 figure we saw with linear regression in Chapter 16. Earlier, we said that we can use **Cox and Snell's R^2** and **Nagelkerke's R^2** to report variance. These two outcomes suggest that model explains between 21.9% and 31.1% of the variance.

Step	−2 Log likelihood	Cox & Snell R square	Nagelkerke R square
1	194.783[a]	.219	.311

Figure 17.15 Model summary

As we saw earlier, we need to ensure that our model adequately 'fits' the data. This is supported if the observed and expected frequencies are similar. SPSS produces a 'Contingency' table that illustrates this. We have not shown that here because all we need to know is provided by the **Hosmer** and **Lemeshow test**. Figure 17.16 reports this. This test examines whether there is a significant difference between the observed and expected frequencies. We don't want that to happen because we need them to be similar, not different. Because the outcome from the Hosmer and Lemeshow test is non-significant, we can be confident that we have adequate goodness of fit.

Step	Chi-square	df	Sig.
1	9.539	8	.299

Figure 17.16 Hosmer and Lemeshow goodness-of-fit

We can now revisit how well we are correctly predicting outcome in this final model. In Figure 17.10, we saw that the baseline model correctly predicted overall outcome on 70% of occasions. The final model shows some overall improvement, correctly predicting 75.5%. However, the sensitivity to successfully predicting positive diagnoses is considerably improved from 0% (in Figure 17.10) to 46.7% (in Figure 17.17).

Observed			Predicted		
			Depressed		Percentage correct
			no	yes	
Step 1	Depressed	no	123	17	87.9
		yes	32	28	46.7
	Overall percentage				75.5

Figure 17.17 Classification table for final model

		B	S.E.	Wald	df	Sig.	Exp(B)	95% C.I.for EXP(B)	
								Lower	Upper
Step 1[a]	sleep	−.079	.017	20.750	1	.000	.924	.893	.956
	gender(1)	.881	.365	5.813	1	.016	2.412	1.179	4.936
	selfest	−.047	.012	14.156	1	.000	.954	.932	.978
	age	−.013	.021	.380	1	.538	.987	.947	1.029
	Constant	5.079	1.444	12.367	1	.000	160.571		

Figure 17.18 Variables included in final model

Figure 17.18 probably shows the most important outcomes of all. The B column indicates the regression co-efficients for each predictor variable. Meanwhile, the **Wald statistic** and the respective significance indicate which of the predictor variables successfully predict a diagnosis of depression. However, it should be noted that SPSS reports the squared Wald statistic. Sleep quality, gender and self-esteem are significant ($p < .05$); age is not ($p = .538$). The **Exp(B)** column shows the odds ratios for each predictor. Values greater than 1 indicate a greater likelihood of depression diagnosis; values less than 1 signify reducing likelihood. We discussed the rationale for odds ratios earlier in the chapter. We analyse only those predictors that significantly predict the outcome. Higher scores for sleep quality and self-esteem indicate better scores. Since Figure 17.18 shows that the regression co-efficients are negative and the Exp(B) value is less than 1, increases in these scores represent decreasing likelihood for a diagnosis of depression: for every unit increase in sleep quality, the odds for depression decreases (OR .924; the 95% confidence intervals suggest that this is in the range of .893 to .956); and for every unit increase in self-esteem, the odds for depression decreases (OR .954; 95% CI: .932−.978). The final significant predictor is gender. Earlier, we selected 'females' to represent baseline (1). Therefore, Figure 17.18 indicates that women are more than twice as likely to be diagnosed with depression as men (OR 2.412; 95% CI: 1.179 −4.936).

Implication of results

It is worth stressing a point that we made about causality and correlation in Chapter 6. Even though our outcome has illustrated some important information about possible risk factors for depression, it does not mean that the predictors cause depression. Logistic regression employs correlation and partial correlation to examine variance in outcome according to a series of predictor variables. However, that correlation only measures relationships; it does not suggest cause and effect.

Checking assumptions

Before we can make any final statements about what we have just seen, we should check that we have not violated any of the assumptions and restrictions that we discussed earlier. We have left this until now because the procedures are slightly more complex and may have caused confusion had we tried to address this before we performed the main analyses.

Linearity

When we discussed assumptions and restrictions earlier, we said that we needed to demonstrate linearity between each predictor variable and the outcome. However, because that outcome is categorical, we need to express this in terms of the predictor and the log of the outcome. To assess linearity, we need to create a new variable that represents the interaction between the predictor variable and the natural log of that variable. This is how we do that:

Using the SPSS file Depressed

Select **Transform** → **Compute** (as you were shown in Figure 3.27, when we looked at transformation in Chapter 3) → In **Target Variable** type **Log Sleep** → click **Type & Label** button → (in new window) type **Log of sleep quality** → click **Continue** → select **Arithmetic** from **Function group** → scroll and select **Ln** from **Functions and Special Variables** list → click on 'up' arrow ('LN(?)' will appear in **Numeric Expression** window) → transfer **Sleep** to (**Numeric Expression** window should now read 'LN(sleep') → click **OK** (a new variable called LogSleep will appear in the data set)

Repeat the above for the other (continuous) predictor variables: self-esteem, and age, using similar variable names (LogSelfest, LogAge) and appropriate labels (we do not do this for the categorical predictor, gender)

Once this has been completed, we must re-run the logistic regression with these new interaction terms:

Select **Analyze** → **Regression** → **Binary Logistic...** (as shown in Figure 17.3) → transfer **Depressed** to **Dependent:** → transfer **Sleep quality**, **Gender**, **Self esteem**, and **Age** to **Covariates:** → then include the interaction terms: → select **Sleep quality** *and* **Log of Sleep quality** (at the same time), → click **>a*b>** by **Covariates:** → select **Self esteem** *and* **Log of self esteem** → click **>a*b>** by **Covariates:** → select **Age** *and* **Log of age** → click **>a*b>** by **Covariates:** → select **Enter** in pull-down options for **Method** → click **OK** (we do not need any of those options we used earlier)

On this occasion, we are concerned only with the logistic regression outcome that reports variables and interactions in the final model (we addressed the other outcomes in the earlier analysis) (Figure 17.19).

		B	S.E.	Wald	df	Sig.	Exp(B)
Step 1[a]	sleep	−.115	.418	.076	1	.783	.891
	selfest	−.406	.507	.641	1	.423	.666
	age	1.093	.549	3.964	1	.046	2.982
	LogSleep by sleep	.009	.089	.011	1	.915	1.010
	LogSelfest by selfest	.070	.098	.514	1	.474	1.072
	LogAge by age	−.237	.118	4.022	1	.045	.789
	Constant	.947	8.415	.013	1	.910	2.578

Figure 17.19 Variables and interactions included in final model

To satisfy the assumption of linearity, we do not want a significant outcome for the interaction terms. LogSleep by sleep ($p = .915$) and LogSelfest by selfest ($p = .474$) are fine; we can assume linearity between sleep quality and the outcome variable, and between self-esteem and the outcome. There is potentially more of a problem with LogAge by age, as this is significant ($p = .045$), suggesting that there may not be linearity between age and the outcome. As age was not found to be a predictor of depression diagnosis, we probably do not need to worry too much about that. In any case, it was pretty close to the significance cut-off point of p = .05, so we can make sensible allowances.

Multi-collinearity

Another assumption of logistic regression requires that we avoid multi-collinearity (when there are several predictors). This is similar to what we encountered in Chapter 16. We cannot use the functions of logistic regression in SPSS to test this, but we can use part of a multiple linear regression analysis instead. We only need to run the collinearity statistics for the main (untransformed) variables:

Using the SPSS file Depressed

Select **Analyze** → **Regression** → **Linear...** (as you were shown in Figure 16.5) → (in new window) transfer Depressed to **Dependent:** → transfer Sleep quality, Gender, Self esteem, and Age to **Independent(s)** → select **Enter** in pull-down options for **Method** → click **Statistics** → (in new window) tick box for **Collinearity diagnostics** (deselecting all other options) → click **Continue** → click **OK**

Model		Collinearity statistics	
		Tolerance	VIF
1	Sleep quality	.995	1.005
	Gender	.983	1.017
	Self-esteem	.860	1.163
	Age	.867	1.153

Figure 17.20 Collinearity statistics

As we saw in Chapter 16, to avoid multi-collinearity we need the 'Tolerance' data to be not too close to 0 (preferably not below .2) and the VIF figure not to exceed 10. Figure 17.10 suggests that we are fine on both accounts.

Model	Dimension	Eigenvalue	Condition index	Variance proportions				
				(Constant)	Sleep quality	Gender	Self-esteem	Age
1	1	4.351	1.000	.00	.01	.02	.00	.00
	2	.448	3.117	.00	.02	.93	.01	.00
	3	.105	6.434	.01	.82	.02	.01	.15
	4	.083	7.246	.00	.11	.01	.33	.27
	5	.013	18.259	.99	.05	.02	.65	.57

Figure 17.21 Collinearity diagnostics

We also explored this type of outcome in Chapter 16. The variance for each predictor is shared across the five dimensions. To avoid collinearity, we do not want the highest variance proportion for any variable to be located on the same dimension as another variable. Figure 17.21 shows that, for sleep quality and gender, this is fine: the highest variance proportion for those variables is located on dimensions 3 and 2 respectively. However, the highest variance proportions for self-esteem and age are both found on dimension 5. Once again, since age was not a significant predictor of outcome, we can probably ignore this. Indeed, the entire analysis could be run without the age variable.

Writing up results

Binary logistic regression was used to predict an outcome of major depressive order among 200 participants. The final model was able to explain between 21.9% and 31.1% of variance. The model was found to fit the data adequately (Hosmer and Lemeshow's $X^2 = 9.539$, $p = .299$), and was able to predict depression status (Omnibus X^2 (4) $= 49.562$, $p < .001$). Overall, the model was able to correctly predict 75.5% of all cases. Four predictors were included in the model, using the Enter method. Three of these successfully predicted depression status (squared Wald statistics are displayed in Table 17.1). Improvements in sleep quality and self-esteem were associated with decreased odds of depression (OR .924 and .954 respectively). Women were more than twice as likely as men to receive a diagnosis of depression (OR 2.412). Assumptions for linearity and multi-collinearity were satisfied.

Table 17.1 Logistic regression analysis of depression diagnosis (n = 200)

	Cox & Snell R^2	Nagelkerke R^2	HL X^2	sig	Wald²	*df*	p	Exp(B)
Model	.219	.311	9.539	.299				
Predictor variable:								
Sleep quality					20.750	1	< .001	.924
Gender (female = 1)					5.813	1	.016	2.412
Self esteem					14.156	1	< .001	.954
Age					.380	1	.538	.987
Constant					*12.367*	*1*	*< .001*	*160.571*

Key: HL - Hosmer and Lemeshow goodness of fit

Chapter summary

In this chapter we have explored (binary) logistic regression. At this point, it would be good to revisit the learning objectives that we set at the beginning of the chapter.

You should now be able to:

- Recognise that we use logistic regression to predict the outcome for a categorical variable, from one or more predictor variables.
- Understand that the purpose of binary logistic regression is to build a final model that is better at predicting outcome than simply using the most common occurrence. By definition, the dependent variable must be categorical, with only two categories of outcome; these must be coded as 0 and 1 in SPSS. Predictors can be numerical or categorical. Unlike linear regression, categorical predictors can have more than two categories, and there is no restriction on coding (although it should be logical). There should be linearity between each predictor and the outcome. Since the outcome is categorical, linearity has to be measured using log transformations. Multi-collinearity between multiple predictors should be avoided.

- Perform analyses using SPSS, using the appropriate method. Particular attention is needed regarding the method of 'entering' predictors into the model. Similar to multiple linear regression, unless evidence can be provided to justify the order in which the predictors should be entered, they should all be entered simultaneously using the Enter method.
- Understand how to present the data, using appropriate tables, reporting the outcome in series of sentences and correctly formatted statistical notation. Four key elements must be reported:
  - The amount of variance explained (using a combination of Cox and Snell's R^2 and Nagelkerke's R^2).
  - The success of the model. This is reported using several measures: Hosmer and Lemeshow's goodness of fit (which must produce a non-significant outcome), the ability to predict outcome (using the Omnibus χ^2, which should be high and significant) and the outcome classifications (stating the proportion of outcomes that were correctly predicted).
  - Which predictors were able to successfully predict outcome (indicated by the Wald statistics).
  - The odds ratios for those successful predictors in respect of the likelihood of outcome associated with that predictor.

Research example

It might help you to see how logistic regression has been applied in real research (using Stepwise methods). In this section you can read an overview of the following paper:

> **Tello, M.A., Jenckes, M., Gaver, J., Anderson, J.R., Moore, R.D. and Chander, G. (2010). Barriers to recommended gynecologic care in an urban United States HIV clinic. *Journal of Women's Health, 19 (8)*: 1511–1518. DOI: http://dx.doi.org/10.1089/jwh.2009.1670**

If you would like to read the entire paper you can use the DOI reference provided to locate that (see Chapter 1 for instructions).

In this research, the authors examined 200 women undergoing gynaecological treatment at an HIV clinic in Maryland, USA. The researchers sought to investigate two outcomes: whether the women missed gynaecology appointments and whether they were in receipt of a Papanicolaou (Pap) smear in the previous year. Previous evidence suggests that women with HIV are more likely to miss gynaecology appointments and fail to take a Pap smear, despite widespread availability. Both outcomes were explored in respect of the same predictor variables: age (< 40, 40–50, > 50), race (Caucasian, African American, other), education (not completed high school, at least completed high school), employment status (full-time or part-time, not working, disabled), dependent children in the household (none, at least one), CD4 count ($\leq$ 200 cells: CD4 is a protein found in human cells; a reduction in CD4 count is a marker for HIV-1 infection), HIV-1 (present or not), substance use in the past month (cocaine, heroin, amphetamine, marijuana, binged alcohol use), social support (very low/low, medium, high), depressive symptoms (none, moderate, severe), and intimate partner violence (IPV; yes or no). Social support was measured using the MOS Social Support Survey (Sherbourne and Stewart, 2002). Depressive symptoms were investigated using the Center for Epidemiologic Studies Short Depression Scale (CES-D 10; Andresen *et al*., 1994). IPV was examined using the Partner Violence Scale (MacMillan *et al*., 2006).

The results showed that 69% of the women missed at least one gynaecology appointment and 22% had no Pap smear in the past year. Using logistic regression, it was shown that missed appointments were predicted by moderate (odds ratio [OR] 3.1, 95% confidence interval [CI] 1.4–6.7) and severe depressive symptoms (OR 3.1, 95% CI 1.3–7.5) and past-month substance use (OR 2.3, 95% CI 1.0–5.3). Not having a Pap smear was associated with an education level of less than high school (OR 0.3, 95% CI 0.1–0.6). A combination of missed appointment and no Pap smear in the last year was predicted by moderate (OR 3.7, 95% CI 1.7–8.0) and severe (OR 4.4, 95% CI 1.7–11.1) depressive symptoms.

Although these odds ratios are useful in identifying risk factors for these women not attending gynaecological clinics, and/or not receiving a Pap smear, we are told nothing about the 'success' of the logistic regression model. We do not know how much variance was explained, nor do we know whether the model adequately fitted the data. We are also not informed about prediction rates for missing appointments and not receiving Pap smears. Without these data, we might question the validity of the results reported. It would have been useful had the authors included some additional data to address these points (as we saw in our procedures). Goodness of fit outcomes from the Hosmer and Lemeshow statistic would help assess the success of the model. Variance data might have been provided from Cox and Snell's R^2 and Nagelkerke's R^2. Classification data should have been included about observed and predicted rates of having a Pap smear.

Extended learning task

You will find the SPSS data associated with this task on the website that accompanies this book. You will also find the answers there.

Following what we have learned about logistic regression, answer these questions and conduct the analyses in SPSS and G*Power (there is no Microsoft 'Excel' alternative for this test; least not in this humble book). For this exercise, data are examined from 150 students regarding whether they passed an exam and what factors might predict success in the exam. The outcome variable is 'exam-pass' (yes or no); the predictors are the number of hours spent revising, current anxiety (based on a score of 0–100, with higher scores representing greater anxiety) and attendance at seminars (yes or no).

Open the SPSS data set **Exams**

1. Run this analysis, using the Enter method.
2. How much variance was explained by this model?
3. Does the model adequately fit the data?
4. How successfully is outcome predicted?
5. Which predictors successfully predicted outcome?
6. Describe the odds ratios for each of the significant predictors.
7. Describe how each of the assumptions and restrictions were met.
8. Write up the results with appropriate tables.

18 NON-PARAMETRIC TESTS

Learning objectives

By the end of this chapter you should be able to:

- Recognise a series of non-parametric tests and appreciate when it is appropriate to use each of them
- Understand the theory, rationale, assumptions and restrictions associated with each test
- Appreciate which parametric test each of those tests seeks to replicate
- Calculate the outcome manually (using maths and equations)
- Perform analyses using SPSS
- Know how to measure effect size and power
- Understand how to present the data and report the findings

Introduction

It may interest you to know that I spent a long time debating whether I should combine non-parametric tests into one chapter or present them as separate ones (or you may not care a jot). I was finally swayed towards the former because I realised I was repeating myself about the conventions regarding what constitutes parametric data and the effect that violations of those assumptions can have on the way we interpret outcome. A combined chapter means that I can get all of that theory and rationale out of the way before we proceed with the specific needs of each non-parametric test.

In the following sections we will take a fresh look at what happens when data fail to meet parametric criteria, extending what we explored in Chapter 5. We will compare the way in which outcomes are examined in parametric and non-parametric tests. We will also explore some common features of non-parametric tests, such as how to locate the source of difference when there are three or more groups or conditions (call it non-parametric *post hoc* tests, if you like). Once we have done that, we will get on with the specific procedures of each non-parametric test.

The focus here is those tests which specifically correspond to the parametric tests we explored in Chapters 7–10. We examine Mann–Whitney U (for independent t-test), Wilcoxon signed-rank (for related t-test), Kruskal–Wallis (for independent one-way ANOVA) and Friedman's ANOVA (for repeated-measures one-way ANOVA). Additional non-parametric tests, such as chi-squared tests and others for wholly categorical variables, will be explored in Chapter 19.

18.1 Take a closer look

A summary of statistical tests: study design and data type

Table 18.1 summarises some key univariate statistical tests and places them in the context of the number of independent variable groups/conditions they measure, whether they are used in between-group or within-group analyses, and in respect of parametric or non-parametric data.

Table 18.1 Statistical tests, according to design and data type

Groups or conditions	Between-groups		Within-groups	
	Parametric	Non-parametric	Parametric	Non-parametric
2	Independent t-test	Mann-Whitney U	Related t-test	Wilcoxon signed-rank
3+	Independent one-way ANOVA	Kruskal-Wallis	Repeated-measures one-way ANOVA	Friedman's ANOVA

Common issues in non-parametric tests

Parametric vs. non-parametric data: what's the big deal?

We explored some of the key features of parametric data in Chapter 5. These describe the extent to which we can trust the consistency of a range of numbers and whether those numbers can be compared with each other meaningfully. Parametric data should be *reasonably* normally distributed and must be represented by interval or ratio numbers. Although we have addressed these points before in this book, it will do no harm to go through it again – it may help to reinforce what we have learned. Normal distribution describes the extent to which the data are distributed either side of the mean (we saw how to measure that in Chapter 3). If the data are not normally distributed it may artificially inflate or deflate the mean score. Parametric tests rely on the mean score to determine outcome – if we cannot trust the mean score, we cannot trust the outcome from a test that uses that measure.

Interval data are objective ranges of numbers, where relative differences between numbers in the scale are meaningful. We can reliably say that 60° Fahrenheit is hotter than 30°. As a result, we can trust mean scores from interval data. Ratio data are an extension of interval data; numbers can be related to each other in terms of relative magnitude. Such data are equally suited for examining via the mean score. Ordinal data involve 'numbers', but it is probably meaningless to try to compare those numbers to each other. Attitude scales are a good example. These often ask participants to convey an opinion, perhaps on a scale of 1 to 10, where perhaps a score of 10 represents 'very strongly agree'. Since such ratings are prone to subjectivity, one person's score may be very different to someone else's. As a result, we have less faith in mean scores to measure ordinal data. We are more likely to rank the data and then make comparisons (as we will see later). However, it should also be noted that parametric tests are pretty robust and are able to withstand a degree of violation on these assumptions. The aim should be for *reasonable* normal distribution and a common-sense approach to determining what data are more likely to be ordinal.

18.2 Nuts and bolts
Examples of data types

If you are still a little confused about how to recognise different types of data, perhaps the examples in Table 18.2 might help.

Table 18.2 Definitions and examples of data

Type	Definition	Examples
Categorical	Distinct, non-numerical groups	Gender (male, female) Nationality (English, French, Welsh. . .)
Ordinal	'Numerical' data that can be ordered by rank, but differences between numbers in a scale may be meaningless	Rating scales (1 = strongly agree; 2 = agree; 3 = neither agree/disagree; 4 = disagree; 5 = strongly disagree) Race position (1st, 2nd, 3rd. . .)
Interval	Numerical data measurable in equal segments, but cannot be compared in relative magnitude (or ratio)	Time, age, income, temperature...
Ratio	Numerical (interval) data that can be compared in relative magnitude	Time, age, income. . .

How non-parametric tests assess outcome

Parametric tests use mean scores to determine whether there are differences in dependent variable outcomes across the independent variable. The mean score is the average of all of the scores. As we have just seen, data that are not normally distributed may 'bias' the mean score, while ordinal data may not have an 'objective' mean score. Either way, the mean score might be considered to be unreliable. Non-parametric tests do not rely on the mean score to determine outcome. Instead many of these tests rank the data and compare them between groups or conditions. Each test performs this ranking in a different way, so we will look at those methods when we explore the specific tests.

18.3 Nuts and bolts

How are non-parametric studies different?

Table 18.3 presents an overview of the fundamental differences between parametric and non-parametric tests. These focus on the features of the dependent variable (DV) data.

Table 18.3 Characteristics and features of parametric vs. non-parametric tests

	Parametric	Non-parametric
DV data distribution	Normal	Not normal*
DV data type	Interval or ratio	Ordinal*
Method of measurement	Assesses differences in mean DV scores between groups or conditions	Assesses differences in ranked DV scores between groups or conditions

*Non-parametric data may include normally distributed data if the scores are ordinal. Equally, the data may be interval or ratio, but may not be normally distributed.

Finding the source of differences

If we find that there is a significant difference between groups or conditions, we need to establish the source of that difference. If there are only two groups or conditions on the independent variable (as is the case with Mann–Whitney U and Wilcoxon signed-rank tests) we need do nothing further. We can simply refer to the descriptive data to tell us which group or condition is higher than the other (we will see how when we look at those tests specifically later on). However, if there are three or more groups or conditions (as we find with Kruskal–Wallis and Friedman's ANOVA) it is not so simple. We shall explore these methods in the respective sections for those tests.

Mann–Whitney U test

What does the Mann–Whitney U test do?

The Mann–Whitney U test explores differences in dependent variable scores between two distinct groups of a single (categorical) independent variable – it is the non-parametric equivalent of the independent t-test (see Chapter 7). We may have intended to explore an outcome using the t-test, but may have been prevented from doing so because the data are not parametric. It may be that the data are not normally distributed or because the scores used to measure outcomes are ordinal.

Research question for Mann–Whitney U

We met the research group MOANS in Chapter 6, when we learned about correlation. In this example, MOANS are seeking to examine perceived mood according to gender. They use a questionnaire that reports outcomes similar to a diagnosis of major depressive disorder. These assessments would normally be undertaken by a qualified clinician. However, MOANS are keen to investigate how people rate their own perception of depressive symptoms. Evidence suggests that depressive diagnoses are more predominant in women, so MOANS predict that women will report poorer 'depression scores' than men (so we have a one-tailed test). As the data focus is on self-report measures, it could be argued that the data are ordinal.

18.4 Take a closer look
Summary of Mann-Whitney U example

Dependent variable: self-rated depression scores
Independent variable: gender (male vs. female)

Theory and rationale

Using Mann-Whitney U instead of an independent t-test

In most cases, if we want to examine differences in outcome scores across two distinct groups, we would probably use an independent t-test. However, it is usually valid to do so only if the dependent variable data are parametric. Non-parametric tests, such as Mann–Whitney U, are

18.5 Nuts and bolts
Simple example of ranking for Mann-Whitney U test

The outcome in Mann-Whitney is derived from how dependent variable scores are ranked across the two groups. Table 18.4 provides a simple overview of how we calculate those ranks, using some pilot data from the MOANS research. The dependent variable is represented by self-rated 'depression' scores.

Table 18.4 Self-rated depression scores, by gender

Male scores	Rank	Female scores	Rank
65	10	25	3
56	8	20	2
52	7	48	6
34	5	62	9
27	4	18	1
Rank sum	**34**		**21**

We rank the scores from the smallest to the highest across the entire sample, regardless of group. A score of 18 is lowest, so that gets 'rank 1'. We carry on this process until we get to the highest score (65), which is allocated 'Rank 10' (we may also need to account for tied ranks, see Box 18.6). Then we assign the ranks to the respective (gender) groups. Those ranks are summed for each group. These rank totals form part of the analyses in a Mann-Whitney U calculation.

not confined by constraints. To establish the outcome with an independent t-test, we focus on the mean scores across the two groups. To explore differences with Mann–Whitney, the dependent variable scores are ranked in order of magnitude. Scores for the entire sample are ranked from lowest to highest. Those rankings are then apportioned to each group (see Box 18.5). We can then perform statistical analyses to determine whether the rankings are significantly different between the groups (as we will see later). We will use the MOANS research question to illustrate this.

Assumptions and restrictions

The assumptions of Mann–Whitney are very few. The Mann–Whitney U test must still examine data that are at least ordinal (they might be interval or ratio, but fail the parametric criteria because they are not normally distributed). This independent variable must be categorical and be represented by two distinct groups: no one can appear in more than one group at a time.

Establishing significant differences

We will use the MOANS research data to illustrate how we explore the magnitude of between-group differences in Mann–Whitney U. Self-rated depression scores are collected from 20 participants (ten men and ten women). The perceptions are taken from responses to a questionnaire, which are scored from 0–100 (higher scores represent poorer perceptions). Given previous evidence, MOANS predict that women will report higher depression scores than men.

18.6 Calculating outcomes manually
Mann-Whitney U calculation

Table 18.5 presents the MOANS data that we are examining. **You will find a Microsoft Excel spreadsheet associated with these calculations on the web page for this book.**

Table 18.5 Example data

Self-rated depression scores			
Male	**Rank**	**Female**	**Rank**
47	5.5	79	16.5
76	15	41	3
60	12	57	9.5
51	7	88	20
57	9.5	82	18.5
57	9.5	66	13
44	4	82	18.5
47	5.5	79	16.5
38	2	57	9.5
28	1	69	14
Rank sum	**71.0**		**139.0**

The first task is to rank the scores. We saw the basics of how we rank data in Box 18.5, but now we have some tied ranks to contend with. Once again, we start with the lowest score (in this case 28) and assign this rank number 1, the next highest score (38) is rank number 2, and so on. Tied numbers receive an average rank. For example, the score of 47 occurs twice, occupying the rank positions of 5 and 6. In this case we calculate the average of those ranks: $(5 + 6) = 11 \div 2 = 5.5$. This happens on several occasions in this data set (see Table 18.5). We then calculate the 'sum of ranks' for each group (shown in the respective columns in Table 18.5).

To assess whether the assigned ranks significantly differ between the groups, we need to find the U score. We calculate U from the highest sum of ranks and relate that to the sample size of the two groups, which is applied to the following equation:

$$U = (N_1 \times N_2) + \frac{N_1(N_1 + 1)}{2} - R_1$$

N_1 = sample size of group 1 (10); N_2 for group 2 (10); R_1 = largest sum of ranks (139.0)

$$\text{So, } U = (10 \times 10) + \left(\frac{10 \times (10 + 1)}{2}\right) - 139 = 16$$

To assess whether the U score is significant, we look this up in U tables (see Appendix 5). Where $N_1 = 10$, $N_2 = 10$, and $p = .05$, we find that the cut-off point for U is 23. Our U value is *less* than that, so it *is* significant. We can say that there is a significant difference between men and women in respect of depression scores (women showed the highest sum of ranks, so females score significantly higher on those depression scores).

How SPSS performs the Mann-Whitney U

We can perform Mann–Whitney U in SPSS. However, I do encourage you to try the manual calculations shown in Box 18.6 – you will learn so much more that way. To explore outcomes, we will examine the same data that we used in those manual calculations. The self-rated depression scores might be considered to be ordinal, which might explain why we would have chosen to employ Mann–Whitney U instead of an independent t-test. However, it also possible that the data were not normally distributed. We will not check that, as we have seen how to do that several times previously (however, you could refer to Chapter 7 to see how we would examine normal distribution in a scenario such as this one). Setting up the data set will be the same as it is for an independent t-test (see Box 7.8).

Running Mann-Whitney U in SPSS

Open the SPSS file Self-rated depression by gender

Select **Analyze → Nonparametric tests → Legacy dialogs → 2 Independent Samples . . .**(as shown in Figure 18.1)

Note that this procedure is based on SPSS version 19. If you have an older version than version 18, the procedure is slightly different (refer to Box 18.7 for guidance)

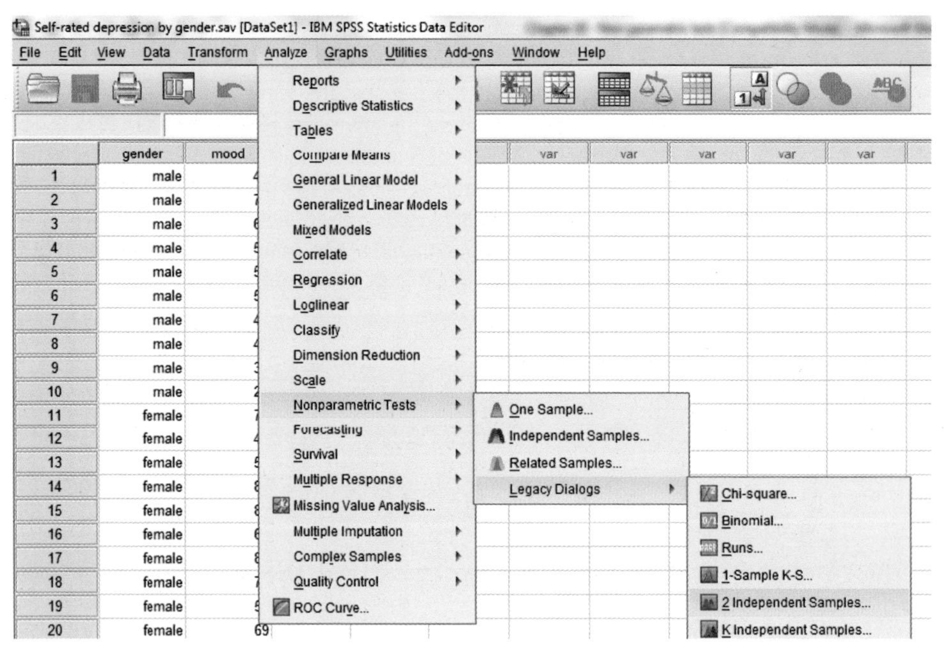

Figure 18.1 Mann-Whitney: procedure 1

In new window (see Figure 18.2), transfer **Self-rated depression** to **Test Variable List** → transfer **Gender** to **Grouping Variable** → click **Define Groups** → (in new window) enter **1** in **Group 1** → enter **2** in **Group 2** → click **Continue** → (back in original window) make sure that **Mann-Whitney U** test (only) is ticked (under **Test Type**) → click **OK**

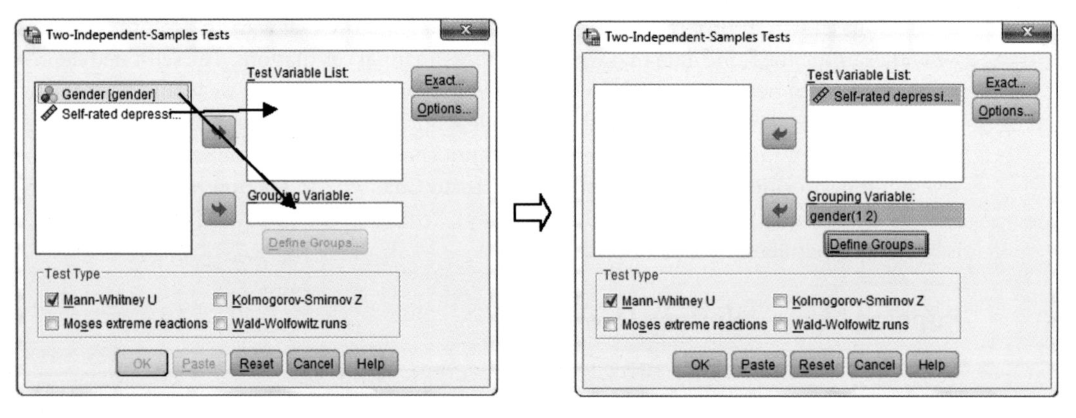

Figure 18.2 Mann-Whitney: procedure 2

You will also need some descriptive statistics to show the median scores and 95% confidence interval data between the groups for reporting purposes. The procedure has to be performed separately to the Mann–Whitney test because the 'Descriptive tests' facility (under Options) will only produce the median for the whole sample, which is not very useful.

Select **Analyze** → **Descriptive statistics** → **Explore . . .** → (in new window) transfer **Self-rated depression** to **Dependent List** → transfer **Gender** to **Factor List** → select **Statistics** radio button → click **OK**

18.7 Nuts and bolts

Running Mann-Whitney in SPSS prior to Version 18

The procedure for running the Mann-Whitney test that we saw just now is based on SPSS version 19. If your program is earlier than Version 18, the (initial) method is a little different:

Select **Analyze → Nonparametric Tests → 2 Independent Samples** as shown in Figure 18.3.

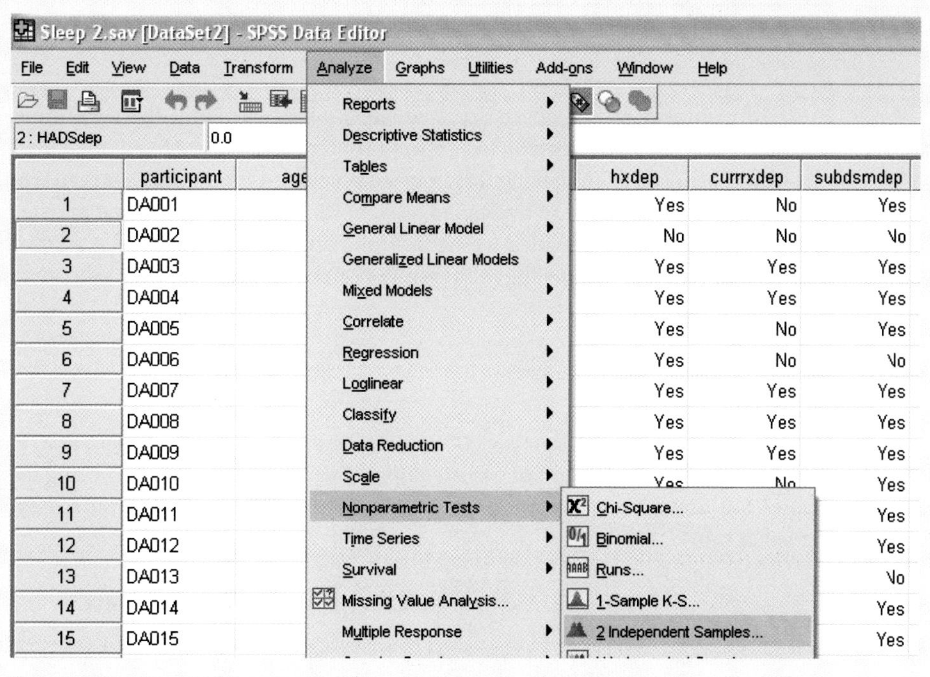

Figure 18.3 Mann-Whitney: procedure 1 (SPSS prior to version 18)

Interpretation of output

Figure 18.4 presents the differences in the ranks between depressed and not depressed.

	Gender	N	Mean rank	Sum of ranks
Depression scores	male	10	7.10	71.00
	female	10	13.90	139.00
	Total	20		

Figure 18.4 Confirmation of rank scores between the groups

Figure 18.5 confirms that we have a significant outcome, $U = 16.0$, $p = .009$. If we refer to the median scores (see Table 18.6), it is clear that females scored more poorly on the depression scores. The other statistics may also be useful, particularly the z-score (–2.583), which

	Depression scores
Mann–Whitney U	16.000
Wilcoxon W	71.000
Z	–2.583
Asymp. Sig. (2-tailed)	.010
Exact Sig. [2*(1-tailed Sig.)]	.009[a]

Figure 18.5 Mann-Whitney test statistic

is used for measuring the effect size (see later). SPSS calculates the significance based on a two-tailed test. If you have predicted a specific outcome (and so have a one-tailed hypothesis), you would be justified in halving the reported p-value to reflect that. You can read more about one-tailed tests, in relation to two-tailed tests, in Chapter 4. We also need the median and 95% confidence intervals data, which we get from the final output table that we got when we asked for descriptive statistics (not shown here). We can compare this to the outcome we calculated in Box 18.6.

Effect size

In previous chapters we have used G*Power to estimate Cohen's *d* effect size. We cannot use that in non-parametric tests, but we can employ Pearson's *r* effect size, by using the *z*-score from Figure 18.5. You simply divide the *z*-score by the square root of the sample size (you can ignore the *minus* sign):

$$r = \frac{Z}{\sqrt{n}} = \frac{2.583}{\sqrt{20}} = .578, \text{ which is a strong effect size.}$$

Writing up results

We can report the outcome, tabulating the data and describing the key points along with appropriate statistical notation.

Table 18.6 Depression rating scores by gender

	Median	95% CI
Male (n = 10)	49.50	41.10 - 59.90
Female (n = 10)	74.00	59.42 - 80.58

We would write this up in our results section as follows:

> Women reported poorer self-rated depression scores than men. A Mann-Whitney U test indicated that this difference was significant, with large effect size ($U = 16.0$; $N_1 = 10$; $N_2 = 10$; $p = .009$, $r = .58$).

Note how the results were presented. We need to report the 'U' outcome, but it is also useful to show the sample size (by group) indicated by 'N'. Note also how Table 18.6 presented information about the median and 95% confidence interval data; it did not show mean scores. We discovered earlier that the dependent variable data were not normally distributed. Therefore, the mean score might be skewed by outliers, so it would be misleading to include that in our results.

The median data are more useful in this instance. The 95% confidence intervals are even more useful because they show the range of scores for each group, but exclude the potential outliers (by definition they will be outside the range of 95% of the data). This will truly reflect the scope of scores shown by the two groups.

Presenting data graphically

Since this is a non-parametric test, we should display data according to median scores, not mean data However, we can still use bar charts to represent this, we simply need to make a few changes. The procedure is basically the same as for the independent t-test, so we can still use the drag and drop facility in SPSS. However, I will reiterate what I have said in previous chapters: never simply regurgitate data in graphs that have already been shown in tables, unless they show something novel that cannot be portrayed by numbers alone. I am showing you how to do graphs for demonstration purposes, in case you need them.

Full graphics for this procedure can be seen in Chapter 7 (see from Figure 7.9)

Select **Graphs** → **Chart Builder** . . . → (in new window) select **Bar** from list under **Choose from:** → drag **Simple Bar** graphic into Chart Preview area → transfer **Depression** to **Y-Axis** box → transfer **Gender** to **X-Axis** box

With the independent t-test we used the mean score to define our data; for Mann-Whitney we need the median score, so we need to change that parameter.

Go to **Element Properties** box (to right of main screen) → under **Statistic** (which will probably read **Mean**), click on pull-down arrow → select **Median** → tick box next to **Display error bars** (always good to have them) → click **Apply** → click **OK**

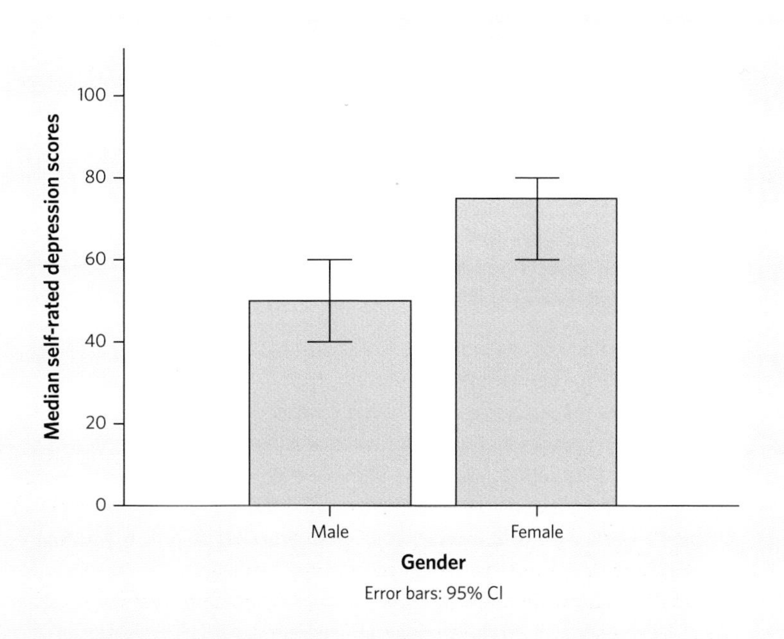

Figure 18.6 Completed bar chart: median self-rated depression scores by gender

18.8 Exercise
Mann-Whitney U mini-exercise

Try running an example yourself, using another data set. Once again, these data represent depression scores in respect of gender, but are from a large data set. Furthermore, the outcome data might be considered to be interval in this instance. However, tests suggest that the data are not normally distributed, hence the need to use Mann-Whitney U to explore between-group differences.

Open the dataset **QOL and Gender**

Run a Mann-Whitney test, with 'depression' as the dependent variable, and 'gender' as the independent variable. You will find the 'answers' for this task on the web page for this book.

Section summary

The Mann-Whitney U test is the non-parametric equivalent to the independent t-test. It explores the outcome of a single dependent variable, across two distinct groups of a categorical independent variable. We are more likely to use this test if the dependent variable data are not normally distributed, and/or those data are ordinal. Instead of comparing mean scores between two groups, the Mann-Whitney test ranks the scores in order of magnitude, then compares the ranks between the two groups.

18.9 Research example

Mann-Whitney U in practice

It might help you to see how Mann-Whitney U has been applied in published research. In this section you can read an overview of the following paper:

Maïano, C., Ninot, G., Stephan, Y., Morin, A.J.S., Florent, J.F. and Vallée, P. (2006). Geographic region effects on adolescent physical self: An exploratory study. *International Journal of Psychology, 41 (2)*: 73–84. DOI: http://dx.doi.org/10.1080/00207590544000004

If you would like to read the entire paper you can use the DOI reference provided to locate that (see Chapter 1 for instructions).

In this research the authors investigated adolescents' physical self-concept and self-esteem in respect of gender and geographical place of residence. The study examined 323 boys and 282 girls (aged 11–16) across two areas of France, comparing the warmer climates of Cagnes-Sur-Mer and Montpellier, to the colder Dunkerque and Nanterre. Physical self-perception and self-esteem were measured using the Physical Self-Perception Profile (PSPP; Fox and Corbin, 1989 – French translation). The authors also measured other self-concepts, but we will focus on these two for demonstration purposes. We are told that the dependent variable data were not normally distributed, which explains why Mann-Whitney U was used to examine the outcome.

The results indicated that boys had significantly higher (better) perceptions of physical self-concept ($U = 29883$, $N_1 = 323$; $N_2 = 282$; $p < .001$) and significantly higher self-esteem ($U = 40132$,

$N_1 = 323$; $N_2 = 282$; $p < .001$) than girls. Adolescents from the cooler northern areas had significantly better perceptions of physical self-concept ($U = 21962$, $N_1 = 323$; $N_2 = 282$; $p < .001$) and significantly higher self-esteem ($U = 17867$, $N_1 = 323$; $N_2 = 282$; $p < .001$) than those from the warmer southern climates. This study provides a useful example of how between-group study data can be explored through non-parametric methods. However, it is curious that the authors included mean scores in their tables when they had already told us that the data were not normally distributed - this is not good practice.

18.10 Exercise
Mann-Whitney U extended learning task

You will find the data set associated with this task on the companion website that accompanies this book (available in SPSS and Excel format). You will also find the answers there.

Learning task

Following what we have learned about Mann-Whitney U, answer the following questions and conduct the analyses in SPSS. (If you do not have SPSS, do as much as you can with the Excel spreadsheet). The fictitious data explored health satisfaction in 200 adults, which was compared in respect of the quality of their exercise. Health satisfaction is measured via ordinal scores. It might be expected that perceived health satisfaction would be higher (better) for those with a better quality of exercise.

Open the **Health satisfaction** data set

1. Why was a Mann-Whitney test needed to examine the data rather than an independent t-test?
2. Perform the Mann-Whitney U test and include all relevant descriptive statistics.
3. Describe what the SPSS output shows.
4. Calculate the effect size.
5. Report the outcome as you would in the results section of a report.

Wilcoxon signed-rank test

What does the Wilcoxon signed-rank test do?

The Wilcoxon signed-rank test explores differences in scores of a non-parametric dependent variable across two conditions of a single independent variable. It is the non-parametric equivalent of the related t-test (see Chapter 8). This is measured over one sample (all participants experience all conditions). As we saw with Mann–Whitney U, we may have intended to examine data with a related t-test but found the data to not comply with requirements for normal distribution. Alternatively, we may decide that the data are ordinal.

Research question for the Wilcoxon signed-rank test

A group of campaigners, WISE (Widening and Improving Student Experience), are seeking to investigate student anxiety during their first year at university. They give new students a questionnaire to complete at two time points during that first year: in freshers' week and at the end of the academic year. A series of eight questions explores perceptions of anxiety, focusing on issues such as self-esteem, financial worries, homesickness and other concerns. Each question is scored on a scale of 1 (very anxious) to 10 (very calm). These subjective responses might be considered to represent ordinal data, so might not be seen to be appropriate for parametric analyses. Therefore, WISE decide to examine outcomes using a Wilcoxon signed-rank test. The researchers expect there to be differences in reported anxiety at those time points, but do not specify when the students will be most anxious. Therefore, we have a two-tailed test.

18.11 Take a closer look

Summary of Wilcoxon signed-rank example

Dependent variable: self-rated anxiety scores

Independent variable: time point (freshers' week vs. last week of first academic year)

Theory and rationale

Using Wilcoxon signed-rank test instead of related t-test

When we explore an outcome across two within-group conditions (measured across a single group), we usually employ a related t-test. However, the legitimacy of doing so is compromised if the dependent variable data are not parametric. Non-parametric tests, such as Wilcoxon signed-rank, do not need to consider such restrictions. With a related t-test, outcomes are examined in relation to how the mean dependent variable scores differ over the two within-group conditions. Because we cannot rely on mean scores in non-parametric tests, Wilcoxon signed-rank test assesses differences according to how the scores are ranked. Dependent variable scores for each participant are examined across the two conditions. Those 'differences' are then ranked in order of magnitude (see Box 18.12). Then, each participant's ranked score is assigned a positive or a negative sign, according to which condition the highest score in. From this, all of the positive ranks are summed, followed by all of the negative ranks. Once we have done that, we can undertake statistical analyses to determine whether the rankings are significantly different between the conditions (see later). We will use the WISE research question to explore this.

18.12 Nuts and bolts

Simple example of ranking for Wilcoxon-signed rank test

The outcome in Wilcoxon signed-rank depends on how dependent variable scores differ across the conditions for each participant. Those differences are then ranked and those ranks apportioned to the 'direction of difference'. Table 18.7 provides a simple example of how we calculate those ranks, using some pilot data from the WISE research. The dependent variable is represented by self-rated 'anxiety' scores; the independent variable is the two time points.

Table 18.7 Student anxiety at freshers' week (FW) and end of year 1 (EY)

Participant	FW	EY	Diff	Rank	Sign	+	–
1	22	58	36	6	–		6
2	35	51	16	3	–		3
3	38	69	31	5	–		5
4	19	46	27	4	–		4
5	62	55	7	2	+	2	
6	30	31	1	1	–		1
					Rank sums	**2**	**19**

First, we calculate the difference in scores between conditions for each participant (e.g. for Participant 1, their self-reported anxiety at freshers' week (FW) was 22, while at the end of the academic year (EY) it was 58; a difference of 36). If there is no difference between the conditions, we give that a score of 0; that case is excluded from any further calculation. Where there is difference, the magnitude of difference is ranked (regardless of which score was higher for now), from the smallest (1) to the highest (36), across the entire sample regardless of condition. Then we assess the differences for direction. If the score 1 in Condition 1 (FW) is greater than Condition 2 (EY) we allocate a '+' sign, otherwise we put '–' (in this case it is the latter). The ranks are then apportioned to the relevant ± column; those columns are summed; the smallest sum (2 in this case) is applied to an equation that examines whether there is a significant difference between the conditions in respect of reaction times (see Box 18.13).

Assumptions and restrictions

The Wilcoxon signed-rank test must examine data that are at least ordinal (they might be interval or ratio, but if the data are not normally distributed, a non-parametric test might still be more appropriate). The independent variable must be categorical, with two within-group conditions (both measured across one single group). Every person (or case) must be present in both conditions.

Establishing significant differences

We will use the WISE research data to show how we calculate outcomes for Wilcoxon signed-rank. Self-rated anxiety scores are collected from 12 participants. As we saw earlier, responses are

18.13 Calculating outcomes manually
Wilcoxon signed-rank calculation

Table 18.8 presents the WISE data that we are examining. **You will find a Microsoft Excel spreadsheet associated with these calculations on the web page for this book.**

Table 18.8 Reported calmness at Freshers' week and after one full year of course

Participant	Freshers'	1 Year	Diff	Rank	Sign	+	–
1	16	37	21	4.5	–		4.5
2	18	18	0				
3	23	34	11	2	–		2
4	20	52	32	7.5	–		7.5
5	26	58	32	7.5	–		7.5
6	18	60	42	9.5	–		9.5
7	39	28	11	2	+	2	
8	13	34	21	4.5	–		4.5
9	16	16	0				
10	23	34	11	2	–		2
11	21	63	42	9.5	–		9.5
12	48	26	22	6	+	6	
					Sum of ranks	**8.0**	**47.0**

We saw how to rank data in Box 18.12, culminating in the sum of ranks shown in the final two columns of Table 18.8. In this example two cases should show no difference between conditions. As we indicated in Box 18.12, those cases are excluded from any further analysis. Once we have calculated the sum of ranks, we take the smallest sum (in this case 8.0), which we call '*T*' (some sources also refer to this as W, after Wilcoxon).

We also need to find the 'mean of T' ($\bar{T}$): $= \dfrac{n(n+1)}{4}$

n = the number of *ranked* participants (in our case 10 because we excluded 2)

And we need to find the standard error of T (SE_T): $= \sqrt{\dfrac{n(n+1)(2n+1)}{24}}$

$$\text{So, } \bar{T} = \frac{10 \times (10+1)}{4} = 27.5 \text{ and } SE_T = \sqrt{\frac{10 \times (10+1) \times (20+1)}{24}} = 9.811$$

To assess the outcome, we need to convert the difference in T to a z-score

$$z = \frac{T - \bar{T}}{SE_T} = \frac{8 - 27.5}{9.811} = -1.988$$

To assess whether the z-score is significant, we examine it in relation to scores from a normal distribution (see Chapter 3). From that, we know that any z-score that exceeds ±1.96 is significant at p = .05. Our z-score is −1.988. We can say that there is a significant difference in self-rated anxiety across the two conditions. Negative ranks were higher than positive ones, so 'anxiety' scores at end of Year 1 are higher (better).

in a Likert-scale format, providing a self-anxiety score of 1 (very anxious) to 10 (very calm) across eight questions; the overall scores range from 8 to 80. WISE state a non-directional hypothesis that there will be a difference in the anxiety ratings between the time points.

How SPSS performs the Wilcoxon signed-rank test

We can perform Wilcoxon signed-rank in SPSS (but do give the manual calculations a try). We will use the same WISE data that we explored manually just now. The anxiety ratings are obtained from a series of Likert scales. These are generally thought to produce ordinal data, so a non-parametric test is probably more appropriate. We may also be persuaded to use Wilcoxon signed-rank if the dependent variable data are not normally distributed. We will not run that here, but you can see how we tested for normal distribution in a similar context in Chapter 8, when we performed a related t-test. Setting up the data set in SPSS will be the same as we saw for the related t-test (see Box 8.7).

Running Wilcoxon signed-rank in SPSS

Open the SPSS file Student anxiety

Select **Analyze → Non-parametric tests → Legacy dialogs → 2 Related Samples . . .** (as shown in Figure 18.7)

This procedure is based on SPSS version 19. If you have an older version than version 18, the procedure is slightly different (see Box 18.7 for general guidance, but select '2 Related Samples' instead of '2 Independent Samples').

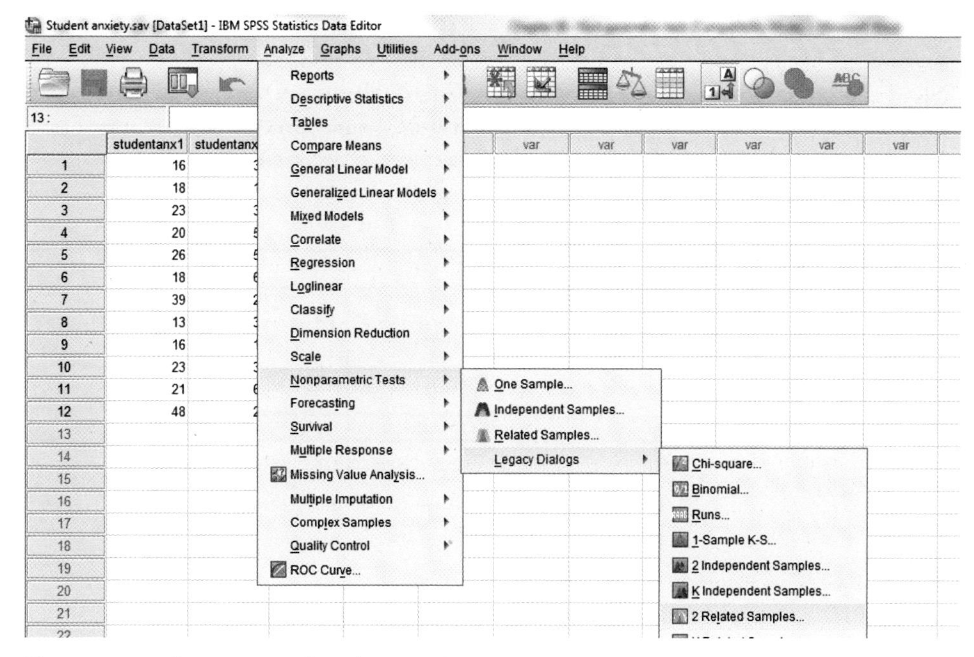

Figure 18.7 Wilcoxon signed-rank: procedure 1

In new window (see Figure 18.8), transfer **Student anxiety at end of Year 1** and **Student anxiety at freshers' week** to **Test Pairs** → make sure **Wilcoxon** (only) is ticked under **Test Type** → click **OK**

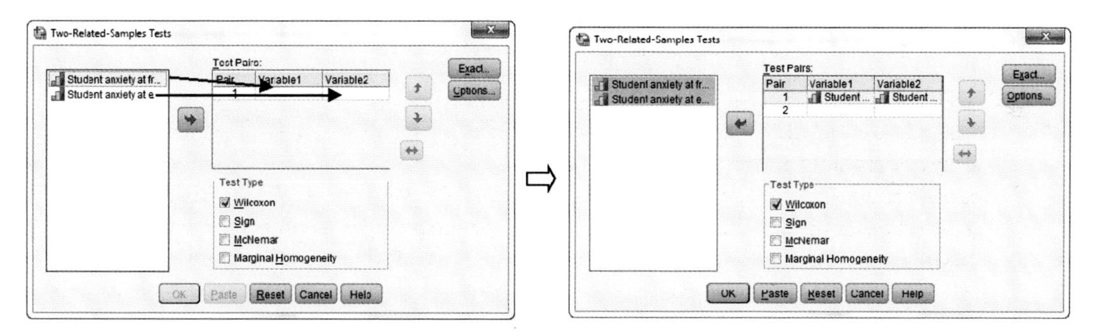

Figure 18.8 Wilcoxon signed-rank: procedure 2

You will also need some descriptive statistics to show the median scores and 95% confidence interval data across the conditions for reporting purposes. The procedure has to be done separately to the Wilcoxon test because the Descriptive tests option (under Options) will only produce the median for the whole sample, which is not very useful.

Select **Analyze** → **Descriptive statistics** → **Explore . . .** → (in new window) transfer **Student anxiety at end of Year 1** and **Student anxiety at freshers' week** to **Dependent List** → select **Statistics** radio button → click **OK**

Interpretation of output

Figure 18.9 presents the differences in the ranks between 'freshers' week' and 'end of year 1', in respect of anxiety. The smallest sum of ranks (8.00) is often reported as the Wilcoxon W statistic (and is the equivalent of *T* in our worked example in Box 18.13).

Ranks

		N	Mean rank	Sum of ranks
Student anxiety at freshers' week – Student anxiety at end of Year 1	Negative ranks	8[a]	5.88	47.00
	Positive ranks	2[b]	4.00	8.00
	Ties	2[c]		
	Total	12		

a. Student anxiety at freshers' week < Student anxiety at end of Year 1

b. Student anxiety at freshers' week > Student anxiety at end of Year 1

c. Student anxiety at freshers' week = Student anxiety at end of Year 1

Figure 18.9 Confirmation of rank scores across conditions

Figure 18.10 confirms that we have a significant outcome, $z = -1.997$, $p = .046$. We can use the median data (see Table 18.9) to tell us that anxiety scores are higher (more calm) at the end Year 1 (we asked for descriptive data when we set the test up earlier). SPSS calculates the significance based on a two-tailed test. If you have predicted a specific outcome (and so have a one-tailed hypothesis), you would be justified in halving the p-value to reflect that. We can compare the outcome to what we found in Box 18.13.

Test Statistics[b]

	Student anxiety at end of Year 1-student anxiety at freshers' week
Z	-1.997[a]
Asymp. Sig. (2-tailed)	.046

a. Based on negative ranks.

b. Wilcoxon Signed Ranks Test

Figure 18.10 Wilcoxon test statistic

Effect size

We cannot use G*Power for non-parametric tests, as we have done in previous chapters. However, we can use the *z*-score to approximate the effect size, using Pearson's *r* method. You simply divide the *z*-score by the square root of the sample size:

$$r = \frac{Z}{\sqrt{n}} = \frac{1.997}{\sqrt{12}} = .576, \text{ which is a strong effect size.}$$

Writing up results

We can report the outcome, tabulating the data and describing the key points along with appropriate statistical notation.

Table 18.9 Student anxiety: Freshers' week vs. end of Year 1 (n = 12)

	Median	95% CI
Freshers'	20.50	16.92 - 29.91
End year 1	34.00	28.05 - 48.62

We would write this up in our results section as follows:

> A Wilcoxon signed-rank test showed that students were significantly more relaxed (less anxious) by the end of their first year at university than they were in freshers' week: W = 8.00; $z = -1.997$, p = .046, with a strong effect size (r = .576).

Presenting data graphically

It would be useful to see how we might present some graphical data, but do bear in mind the protocols for including graphs and tables in your results section It is never good to include both, just for the sake of it, when they show the same thing. Graphs should be included only when they show something novel that the tables of data cannot. On this occasion, we look at how we can display some box plots, which are particularly useful for presenting median data.

Select **Graphs → Legacy Dialogs → Boxplot . . .** as shown in Figure 18.11

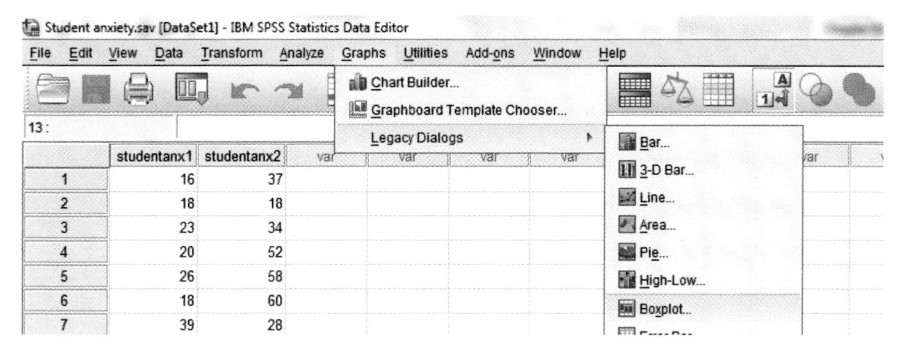

Figure 18.11 Creating a box plot- Step 1

In new window (see Figure 18.12), click **Simple → tick Summaries for separate variables** radio button **→** click **Define**

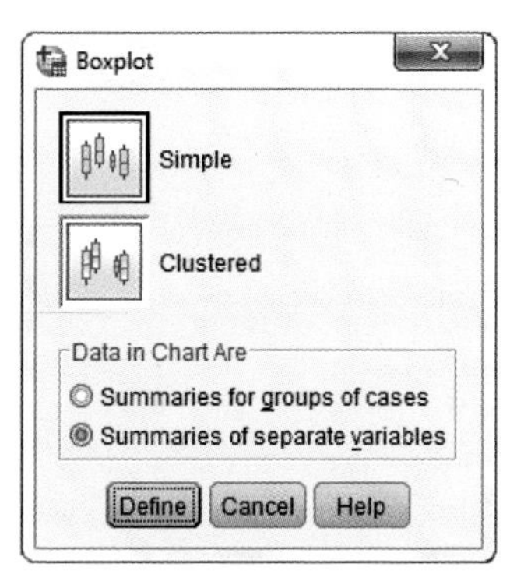

Figure 18.12 Creating a box plot- Step 2

In new window (see Figure 18.13) transfer **Student anxiety at end of Year 1** and **Student anxiety at freshers' week** to **Boxes Represent** → click **OK**

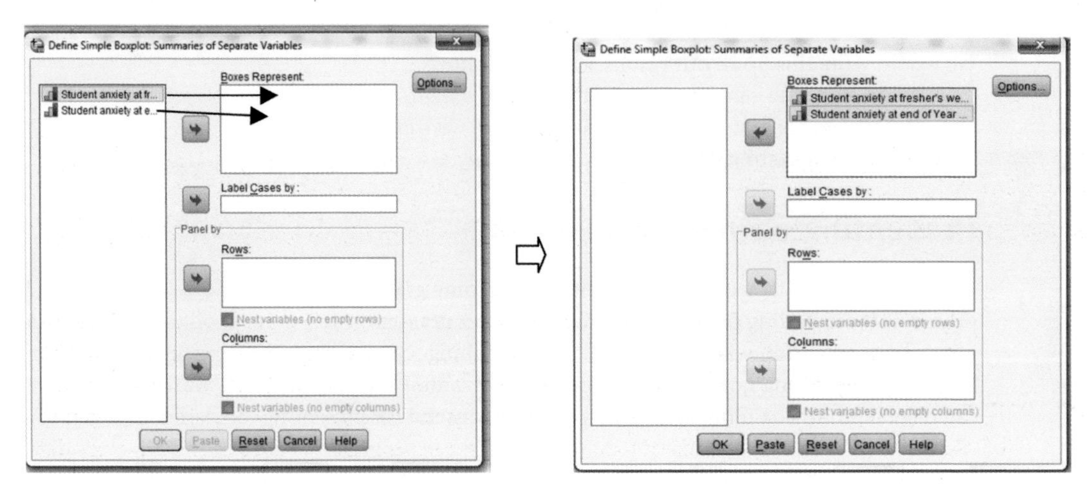

Figure 18.13 Creating a box plot- step 3

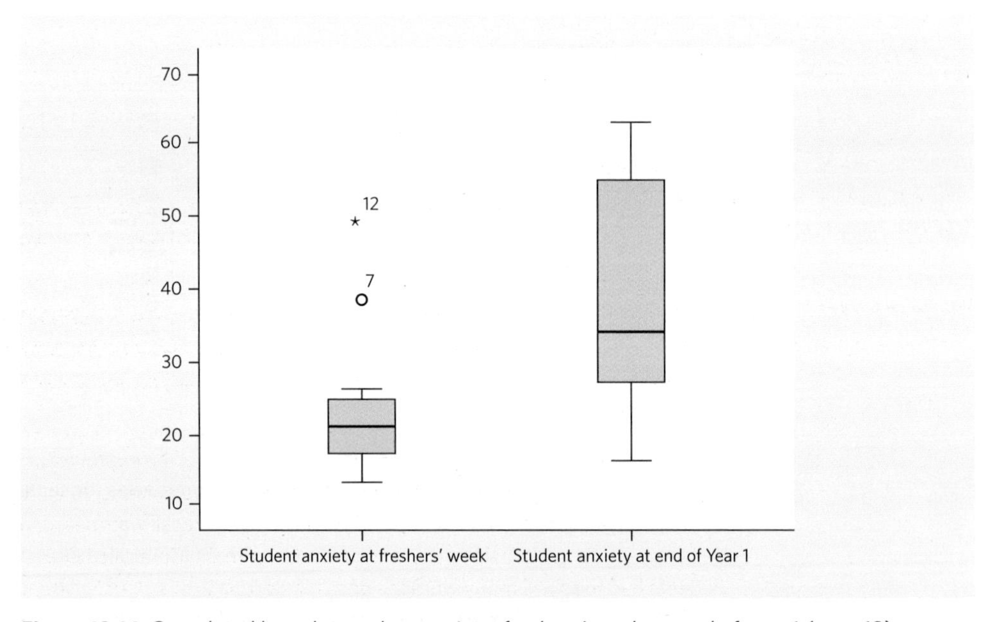

Figure 18.14 Completed box plot student anxiety: freshers' week vs. end of year 1 (n = 12)

18.14 Exercise
Wilcoxon signed-rank mini-exercise

In this example we explore the benefits of a new programme for treating depression via nurse-led therapy. Depression severity scores were measured using the Hospital Anxiety and Depression Scale (HADS) at baseline (prior to treatment) and at post-treatment week 12. Scores are self-rated from 1 to 4; higher HADS scores represent greater depression severity.

Open the data set **Case managed depression**

Run a Wilcoxon signed-rank test, with HADS baseline (depression) and HADS Week 12 (depression) as conditions. You will find the 'answers' for this task on the web page for this book.

Section summary

The Wilcoxon signed-rank test is the non-parametric equivalent to the related t-test. It explores the outcome of a single dependent variable, across two conditions of a categorical independent variable, measured over one group. We may need to use the Wilcoxon signed-rank test if the dependent variable data are not normally distributed and/or those data are ordinal. This test ranks the magnitude of differences across two conditions for each participant. Significant differences are established by examining the relative size of rank between the conditions.

18.15 Research example

Wilcoxon signed-rank in practice

It might help you to see how the Wilcoxon signed-rank test has been applied in published research. If you would like to read the entire paper you can use the DOI reference provided to locate that (see Chapter 1 for instructions).

Akdede, B.B.K., Alptekin, K., Kitiş, A., Arkar, H. and Akvardar, Y. (2005). Effects of quetiapine on cognitive functions in schizophrenia. ***Progress in Neuro-Psychopharmacology and Biological Psychiatry*****,** ***29*****: 233–238. DOI: http://dx.doi.org/10.1016/j.pnpbp.2004.11.005**

In this research the authors explored the extent that cognitive function in schizophrenia (verbal recall, word fluency, attention, motor skills and mental flexibility) can be improved with quetiapine treatment (an antipsychotic drug). This was given to 18 schizophrenic patients, in increasing doses. Cognitive function was measured across a number of domains. Word recall was measured with the Rey Auditory Verbal Learning Test. Verbal fluency was measured with the Controlled Oral Word Association Test.

Patients had to say as many words as possible starting with K, A or S. Attention was measured with the Digit Span test. Patients had to repeat numbers said to them, both forwards (as presented) and then in reverse order. Motor skills were measured with the Finger Tapping Test. Patients were measured on the speed that they were able to tap with the index finger of each hand. Scores represented the number of taps recorded for dominant and non-dominant hand. Mental flexibility was measured with the Trail Making Test. This draws on several cognitive skills, including sequencing and executive function. There were several other measures, but we will focus on these five for demonstration purposes.

The results measured baseline scores to those after eight weeks of treatment. There was no significant difference in verbal recall ($z = -0.34$, $p = .72$), or verbal fluency, ($z = -0.60$, $p = .55$). However, there were significant improvements in attention (Digit Span forwards; $z = -2.15$, $p = .03$), motor control (finger tapping, non-dominant hand; $z = -2.47$, $p = .01$) and sequencing tasks ($z = -2.40$, $p = .01$). This provides a useful example of how Wilcoxon signed-rank has been reported in a published study.

18.16 Exercise
Wilcoxon signed-rank extended learning task

You will find the data set associated with this task on the companion website that accompanies this book (available in SPSS and Excel format). You will also find the answers there.

Learning task

Following what we have learned about the Wilcoxon signed-rank test, answer the following questions and conduct the analyses in SPSS. (If you do not have SPSS, do as much as you can with the Excel spreadsheet.) This dataset explores participants' perceptions of body shape satisfaction before and after viewing images of slim models. To measure body shape satisfaction, a series of 15 questions was asked, scored in a Likert scale format, from 1 = very satisfied to 5 = very unsatisfied.

Open the **Body shape satisfaction** data set

1. Why was a Wilcoxon signed-rank test needed to examine the data rather than a related t-test?
2. Perform the Wilcoxon signed-rank test and include all relevant descriptive statistics.
3. Describe what the SPSS output shows.
4. Calculate the effect size.
5. Report the outcome as you would in the results section of a report.

Kruskal–Wallis test

What does the Kruskal–Wallis test do?

Kruskal–Wallis explores differences in scores of a non-parametric dependent variable between three or more groups of a single independent variable. It is the non-parametric equivalent of the independent one-way ANOVA (see Chapter 9).

Research question for Kruskal–Wallis

To help us explore outcomes in Kruskal–Wallis, we will look to some (fictitious) political research conducted by NAPS (National Alliance of Political Studies). They decide to investigate how attitudes towards the control of law and order vary between supporters of three political groups (socialist, liberal or conservative). The attitudes are examined based on responses to Likert-style questionnaires that elicit answers to a series of questions measured on a scale of 1 (in favour of greater control) to 7 (in favour of less control). As such, these scores would be considered to be ordinal and not suitable for parametric tests.

18.17 Take a closer look
Summary of Kruskal–Wallis example

Dependent variable: attitude scores towards control of law and order
Independent variable: political group (socialist, liberal or conservative)

Theory and rationale

Using Kruskal-Wallis instead of independent one-way ANOVA

In most cases, when we measure an outcome between three or more groups, we are likely to use an independent one-way ANOVA. However, you could be prevented from doing this if the dependent variable data are not parametric. Because we cannot rely on the validity of mean scores in non-parametric tests, Kruskal–Wallis ranks the dependent variable scores across the entire sample, then assigns those ranks to the relevant groups (see Box 18.18). We can undertake statistical analyses to determine whether the rankings are significantly different between the groups (as we will see later). We will use the NAPS data to illustrate this.

18.18 Nuts and bolts
Simple example of ranking for Kruskal-Wallis

The outcome in Kruskal-Wallis is calculated based on how dependent variable scores are ranked across three or more groups. Table 18.10 provides an overview of how we calculate those ranks, using pilot data from the NAPS research.

Table 18.10 Attitude scores towards control of law and order by political group

Socialist	Rank	Liberal	Rank	Conservative	Rank
48	4	73	12	50	5
61	7	42	3	70	9.5
35	1	38	2	75	13
57	6	70	9.5	86	15
64	8	71	11	79	14
Rank sum	**26**		**37.5**		**56.5**

We rank the scores from the smallest to highest across the entire sample, regardless of group. The score of 35 is lowest so that gets a rank of 1. A score of 86 is highest so gets the top rank of 15. Tied scores are averaged, just like they were for Mann-Whitney. For example, '70' is shared by two participants. Then we sum assigned ranks within each group (as shown in Table 18.10). Those rank sums are summed for each group. These rank totals form part of the analyses in a Kruskal-Wallis calculation.

Assumptions and restrictions

Compared with parametric tests there are very few assumptions and restrictions for Kruskal–Wallis. The only restriction for the dependent variable is that the data must be at least ordinal. It is possible that we might use interval or ratio data, but we may have decided to examine outcomes using a non-parametric test because the data are not normally distributed. The independent variable must be categorical and be represented by at least three distinct groups: no one can appear in more than one group at a time.

Establishing significant differences

To illustrate how we calculate between-group differences in Kruskal–Wallis, we will refer back to the NAPS research data. Attitudes towards the control of law and order are compared between supporters of three political groups (socialist, liberal or conservative). A Likert-style questionnaire is given to ten people from each group. The responses are measured on a scale of 1 (in favour of greater control) to 7 (in favour of less control). NAPS predict that there will be a difference in attitudes between the groups (but do not specify where those differences are likely to occur).

18.19 Calculating outcomes manually

Kruskal-Wallis calculation

Table 18.11 presents the NAPS data that we are examining. **You will find a Microsoft Excel spreadsheet associated with these calculations on the web page for this book.**

Table 18.11 Attitude scores towards law and order control, by political group

Socialist	Rank	Liberal	Rank	Conservative	Rank
45	21	42	16.5	39	12
30	4	39	12	36	9
42	16.5	45	21	33	7
45	21	42	16.5	33	7
51	27	48	25	42	16.5
39	12	45	21	33	7
39	12	45	21	30	4
57	29.5	54	28	27	2
48	25	57	29.5	24	1
48	25	39	12	30	4
Rank sum	**193.0**		**202.5**		**69.5**

We saw how to rank the scores in Box 18.18. Once we have completed the ranking, we calculate the rank sums for each group, as shown in Table 18.11. Then we apply those outcomes to the Kruskal-Wallis equation:

$$H = \frac{12}{N(N+1)} \sum_{i=1}^{k} \frac{R_i^2}{n_i} - 3(N+1)$$

N = total sample (30); $\sum$ = 'sum of'; R = group sum of ranks; i = group; n = group size (10 in each case)

$$\text{So, } H = \frac{12}{(30 \times 31)} \times \left(\frac{193.0^2}{10} + \frac{202.5^2}{10} + \frac{69.5^2}{10}\right) - (3 \times 31) = 14.21$$

Our outcome statistic (*H*) represents a χ^2 (chi-squared) score. We can assess the significance of this by examining χ^2 tables (see Appendix 6). The critical value for χ^2 at $df = 2$ (3 groups −1) at p = .05 is 5.99. Our χ^2 (14.21) is higher than that, so our (overall) difference is significant. This indicates that there is significant difference between the political groups in respect of attitudes towards law and order control. It does not tell us where the source of difference is (we need separate tests for that).

Finding the source of differences

As we have just seen, if we find that we have a significant between-group difference in Kruskal–Wallis, we still need to locate the source of the difference. Using the current example, we may know that attitude towards law and order appears to differ between political groups, but we do not know whether labour supporters have significantly less stringent opinions than conservative supporters (for example). This is a similar problem to what we encountered with independent one-way ANOVA (see Chapter 9). In that case, we could employ planned contrasts or *post hoc* tests (depending on whether specific hypotheses had been made about the outcomes between the groups). Following a significant Kruskal–Wallis outcome, there are a number of ways in which we can explore this. We *could* use a procedure called Jonckheere's Trend Test (we will not deal with that test here), or we can perform separate Mann–Whitney U tests for each pair of independent variable groups. However, we perform these additional tests only if there is a significant outcome from the Kruskal–Wallis analysis.

To run these additional Mann–Whitney U tests, we need to consider how we will deal with multiple comparisons before we interpret the outcome. As we saw in Chapter 9, the more tests we run, the more likely it is that we will find a significant difference (purely by chance factors alone). By running additional Mann–Whitney U tests after Kruskal–Wallis analysis, we may be increasing the likelihood of Type I errors. This occurs when we reject the null hypothesis when we should not have done so (see Chapter 4). If we make specific (one-tailed) predictions about between-group differences we are justified in using the standard cut-off point for significance ($p < .05$) for each pair of analyses. However, if we make only general (two-tailed) predictions about the overall outcome, we should adjust for multiple comparisons. We must divide the significance cut-off by the number additional analyses. In our example, NAPS did not make specific hypotheses. As there are three groups, we need to perform three Mann–Whitney U tests. We will need to divide the significance threshold by three. We will have a significant outcome only where $p < .016$ (see Box 18.20 for more details).

18.20 Nuts and bolts

Accounting for multiple comparisons in additional Mann-Whitney U tests

If we have established that there are significant between-group differences following Kruskal-Wallis analyses, we need to perform Mann-Whitney U tests for every pair of groups to locate the source of that difference. As we have just seen, if we have not made specific predictions about those outcomes, we must adjust the significance to account for multiple comparisons. We divide the threshold by the number of additional tests needed. Table 18.12 provides an overview of just some of the situations that we are likely to encounter.

Table 18.12 Number of Mann-Whitney U (MWU) tests required subsequent to Kruskal-Wallis tests

Groups	MWU tests	Sig. cut-off
3	3	.016
4	6	.008
5	10	.005
6	15	.003

For example, if Kruskal-Wallis analyses show a significant difference in outcomes between five groups, and we have made a non-specific (two-tailed) prediction, we will need to run ten Mann-Whitney U tests. For each test pair, the difference between the groups is significant only if $p < .005$.

How SPSS performs Kruskal-Wallis

We can perform Kruskal–Wallis in SPSS. To illustrate how we do that, we will examine the NAPS data that we saw in Box 18.19. Attitudes towards law and order (the dependent variable) are measured from a Likert scale. These data might be considered to be ordinal, hence the need for a non-parametric test to explore the outcome. We will not explore normal distribution, as we saw how to do that for this scenario in Chapter 9. Setting up the data set in SPSS will be similar to what we saw for independent one-way ANOVA (see Box 9.11).

Running Kruskal–Wallis in SPSS

Open the SPSS file Law and order

Select **Analyze → Non-parametric test → Legacy dialogs → K Independent Samples . . .** (as shown in Figure 18.15)

This procedure is based on SPSS version 19. If you have an older version than version 18, the procedure is slightly different (see Box 18.7 for general guidance, but select 'K Independent Samples' instead of '2 Independent Samples').

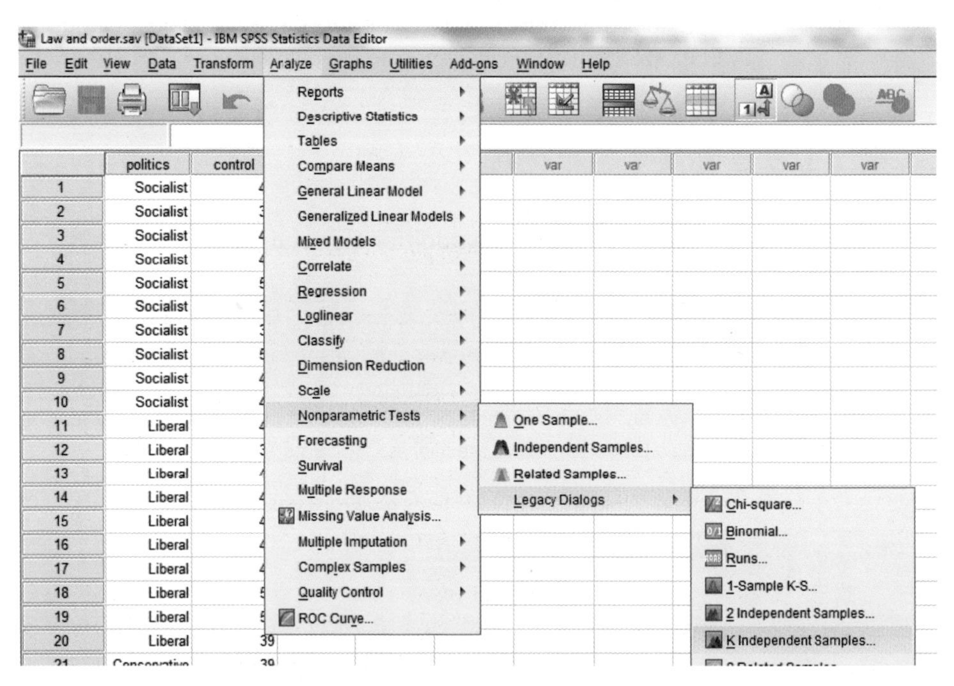

Figure 18.15 Kruskal–Wallis: procedure 1

In new window (see Figure 18.16), transfer **Strength of attitude towards control** to **Test Variable →** transfer **Political group** to **Grouping Variable →** click **Define Range →** (in new window) enter **1** for **Minimum** and **3** for **Maximum →** click **Continue →** (back in original window) make sure that **Kruskal–Wallis H** is (only) ticked (under **Test Type**) **→** click **OK**

You will also need some descriptive statistics to show the median scores and 95% confidence interval data between the groups for reporting purposes. The procedure has to be done separately to the Kruskal–Wallis test because the Descriptive tests option (under Options) will only produce the median for the whole sample, which is not very useful.

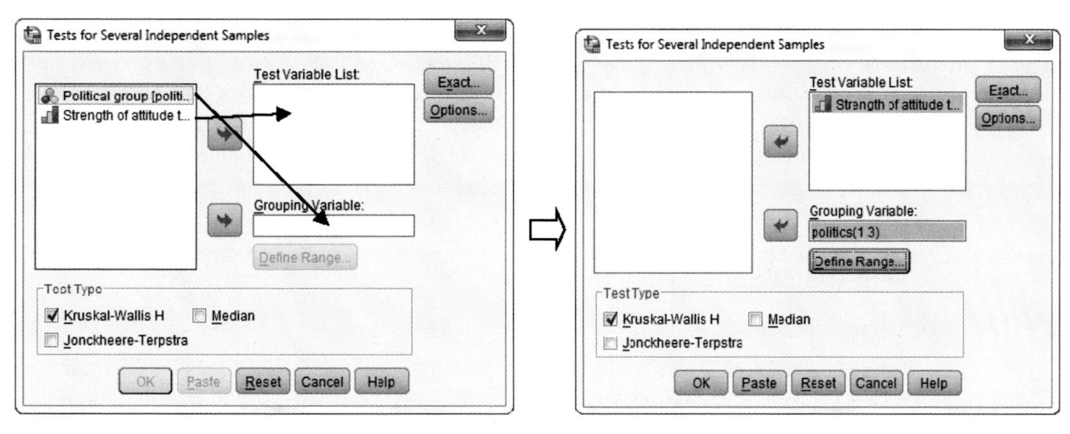

Figure 18.16 Kruskal-Wallis: procedure 2

Select **Analyze → Descriptive statistics → Explore... →** (in new window) transfer **Strength of attitude towards control** to **Dependent List →** transfer **Political group** to **Factor List → select Statistics** radio button **→** click **OK**

Interpretation of output

Figure 18.17 shows differences in ranked scores, concerning attitudes towards law and order, between political groups. Compare this with the outcome we calculated in Box 18.19.

	Political group	N	Mean rank
Strength of attitude towards control	Socialist	10	19.30
	Liberal	10	20.25
	Conservative	10	6.95
	Total	30	

Figure 18.17 Confirmation of rank scores between the groups

Figure 18.18 confirms that there is a significant difference in attitudes towards law and order control between the political groups, χ^2 (2) = 14.409, p = .001. You can compare this outcome to what we found when we calculated this by hand (see Box 18.19).

	Strength of attitude towards control
Chi-square	14.409
df	2
Asymp. Sig.	.001

Figure 18.18 Kruskal-Wallis test statistic

Locating the source of difference with Mann-Whitney U tests

The outcome reported in Figure 18.18 tells us only that there is a difference between the attitude scores overall. It does not tell us where the differences are according to each pair of analyses (socialist vs. liberal, socialist vs. conservative and liberal vs. conservative). We could refer to the median and 95% confidence interval of difference data, but this will be a numerical indication only – we need some statistical confirmation. We need to run three additional Mann–Whitney U tests to locate the difference, one for each pair of groups. Because we made no prediction about specific differences across the three groups, we need to adjust the significance cut-off point for the Mann–Whitney tests (see Box 18.20). Outcomes will be significant only if $p < .016$.

Using the SPSS file Law and order

Socialist vs. liberal:

Select **Analyze** → **Non-parametric test** → **Legacy dialogs** → **2 Independent Samples ...** → (in new window) transfer **Strength of attitude towards control** to **Test Variable** → transfer **Political group** to **Grouping Variable** → click **Define Groups** → (in new window) enter **1** for **Group 1** and **2** for **Group 2** → click **Continue** → click **OK**

Socialist vs. conservative:

As above, except . . . in **Define Groups** enter **1** for **Group 1** and **3** for **Group 2**

Liberal vs. conservative:

As above, except . . . in **Define Groups** enter **2** for **Group 1** and **3** for **Group 2**

Mann–Whitney U outcomes

Socialist vs. liberal

	Strength of attitude towards control
Mann–Whitney U	48.500
Wilcoxon W	103.500
Z	–.115
Asymp. Sig. (2-tailed)	.908
Exact Sig. [2*(1-tailed Sig.)]	.912[a]

Figure 18.19 Mann-Whitney test statistic

Figure 18.19 shows that there was no significant difference in attitudes between socialist and liberal groups, $z = -0.115$, $p = .908$ (we must assume a two-tailed test because this controls for tied scores).

Socialist vs. conservative

	Strength of attitude towards control
Mann–Whitney U	10.500
Wilcoxon W	65.500
Z	–3.003
Asymp. Sig. (2-tailed)	.003
Exact Sig. [2*(1-tailed Sig.)]	.002[a]

Figure 18.20 Mann-Whitney test statistic

Figure 18.20 shows that there was a significant difference in attitudes between socialist and conservative groups, $z = -3.003$, $p = .003$. We can use the median data from Table 18.13 to tell us that conservatives believe in stricter controls (in this cohort, using completely fictitious data). We obtained the median outcomes from the descriptive data that we asked for earlier.

	Strength of attitude towards control
Mann–Whitney U	4.000
Wilcoxon W	59.000
Z	−3.500
Asymp. Sig. (2-tailed)	.000
Exact Sig. [2*(1-tailed Sig.)]	.000[a]

Figure 18.21 Mann-Whitney test statistic

Liberal vs. conservative

Figure 18.21 shows that there was a significant difference in attitudes between liberal and conservative groups, $z = -3.500$, $p < .001$. The median data in Table 18.13 suggest that the conservative group hold more stringent views on law and order control than liberals.

Effect size

In non-parametric tests we need to use the z-score to estimate effect size (we cannot use G*Power, as we can for parametric tests). However, we are not given a z-score for Kruskal-Wallis, and we cannot *easily* convert the χ^2 statistic to a z-score. Instead, we can calculate the effect size for each of our Mann-Whitney tests. We divide the z-score (which we can get from the output tables) by the square root of the sample size:

$r = \frac{Z}{\sqrt{n}}$ (group sizes (n) relate to pairs of groups, so 20 for each pair of 10 students)

Socialist vs. liberal $= \frac{0.115}{\sqrt{20}}$ $r = .003$ (no effect)

Socialist vs. conservative $= \frac{3.003}{\sqrt{20}}$ $r = .671$ (strong effect)

Liberal vs. conservative $= \frac{3.500}{\sqrt{20}}$ $r = .783$ (strong effect)

Writing up results

Table 18.13 Attitude scores towards law and order control, by political group

Political group	Median	95% CI
Socialist (n = 10)	45.0	39.07 – 49.73
Liberal (n = 10)	45.0	41.33 – 49.87
Conservative (n = 10)	33.0	28.85 – 36.55

We would write this up in our results section as follows:

> A Kruskal-Wallis test indicated that there was a significant difference in attitudes towards law and order controls between political groups: $H(2) = 14.409$, $p = .001$. Subsequent Mann-Whitney U tests indicated that conservatives prefer stricter controls than socialists ($U = 18.5$ $N_1 = 10$; $N_2 = 10$; $p = .003$, $r = .67$) and liberals ($U = 4.0$; $N_1 = 10$; $N_2 = 10$; $p < .001$, $r = .78$).

Presenting data graphically

As we have seen in previous sections that explore non-parametric data, it would be good to look at some box plots, as these present median data quite nicely. With between-group data, SPSS has the useful facility whereby you can use a chart builder to design graphs more easily, without relying on having to understand the menus – you actually see what you are trying to create.

Select **Graphs** → **Chart Builder . . .** → (in new window) select **Boxplot** from list under **Choose from:** → drag **Simple Boxplot** graphic (in top left corner) to Chart Preview → trasnsfer- **Strength of attitude towards control** to **Y-Axis** box → transfer **Political group** to **X-Axis** box → click **OK**

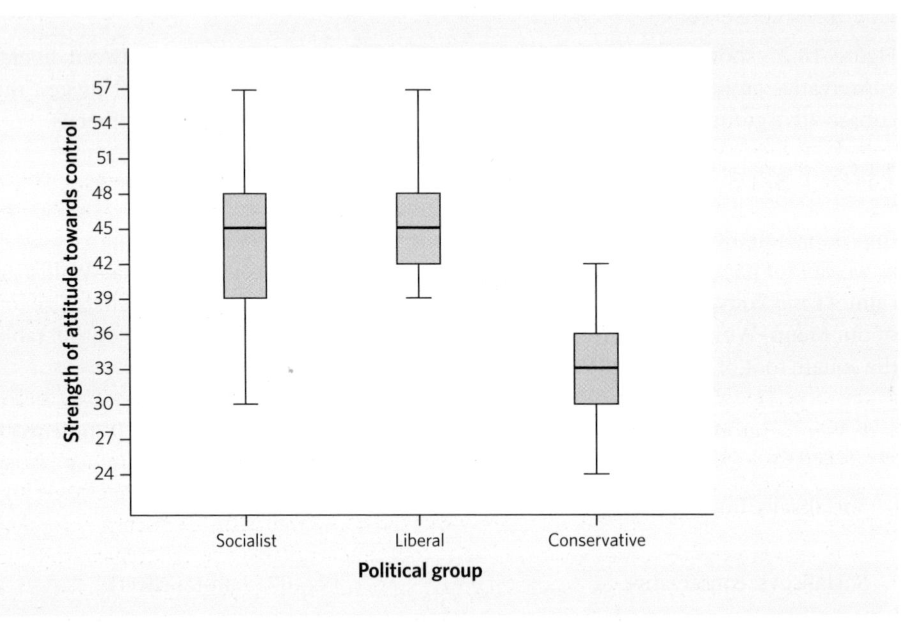

Figure 18.22 Completed box plot: attitude scores towards law and order control, by political group

18.21 Exercise
Kruskal-Wallis mini-exercise

Try running an example yourself using another data set (also located on the book's website). In this example we investigate whether attitudes towards health habits vary between three age groups: 18–25, 26–40, and 41 or over). Attitudes are measured on an ordinal Likert scale; higher scores represent more positive attitudes.

Open the data set **Health attitude**
Run a Kruskal-Wallis test, with Health attitude score as the dependent variable and Age as the independent variable. You will find the 'answers' for this task on the web page for this book.

Section summary

Kruskal-Wallis is the non-parametric equivalent to independent one-way ANOVA. It explores the outcome of a single dependent variable, across three or more distinct groups of a categorical independent variable. We are more likely to use this test if the dependent variable data are not normally distributed and/or those data are ordinal. Kruskal-Wallis examines outcomes by comparing how the scores are ranked across the groups (as opposed to comparing mean scores in an independent one-way ANOVA). Similar to ANOVA, Kruskal-Wallis indicates only whether there is a significant difference in dependent variable scores across groups. It does not tell us where those differences are located. Unlike ANOVA, there are no readily-available planned contrasts or *post hoc* tests that you can employ to investigate the source of difference. Instead, we have to run Mann-Whitney U tests for each pair of groups (adjusted for multiple comparisons if necessary).

18.22 Research example

Kruskal-Wallis in practice

It might help you to see how Kruskal-Wallis has been applied in published research. If you would like to read the entire paper you can use the DOI reference provided to locate that (see Chapter 1 for instructions).

> **Kieffer, J.M. and Hoogstraten, J. (2008). Linking oral health, general health, and quality of life. *European Journal of Oral Sciences, 116*: 445–450.DOI: http://dx.doi.org/10.1111/j.1600-0722.2008.00564.x**

In this research the authors examined oral health, general health and quality of life in a group of 118 psychology students in the Netherlands. Health-related quality of life (HRQoL) was measured with the RAND-36 (Hays and Morales, 2001). It examines a series of factors relating to physical and social functioning, physical and emotional roles, mental health, vitality, pain and general health. Final scores range from 0–100, with higher scores representing better HRQoL. Oral health-related quality of life (OHRQoL) was measured using the Oral Health Impact Profile (OHIP-49; Slade and Spencer, 1994). The scale focuses on functional limitations, physical pain, psychological discomfort, physical disability, psychological disability, social disability and handicap. Subjective reports from the RAND-36 and OHIP-49 can be divided into internal perceptions (limitations and discomfort) and external valuations (interpersonal and social experiences). A number of other factors were measured with additional scales, but we will focus on these two for demonstration purposes. Ultimately the combination of scales provided an overview of perceptions regarding self-rated general health (SRGH) and self-rated oral health (SROH). These were assessed based on the quality of perception, where 1 represented 'very good', 2 'good', 3 'fair' 4 'fairly poor' and 5 'indicated poor'. We are not specifically told why non-parametric tests were used. Normal distribution is not reported, so we can assume that the authors considered the data to be ordinal.

The results showed that there were significant differences in ratings across the SRGH components when assessing the number of symptoms: $H(2) = 12.1$, $p < .01$. Subsequent Mann-Whitney U tests indicated that the differences were between the categories 'very good' and 'fairly poor': $U = 122$, $p < .01$ and between the categories 'good' and 'fairly poor': $U = 604$, $p < .01$ (the higher the number of symptoms, the poorer the perceptions of general health). There were also significant differences in ratings across the SROH components, in relation to oral health symptoms: $H(2) = 14.8$, $p < .01$. Mann-Whitney tests found differences between 'very good' and 'fairly poor' categories: $U = 200$, $p < .01$ and those for 'good' and 'fairly poor': $U = 344$, $p < .01$ (the higher the number of oral symptoms, the poorer the perceptions of oral health). This study provides a well-reported example of how Kruskal-Wallis, and subsequently Mann-Whitney, are presented in a published study.

18.23 Exercise
Kruskal-Wallis extended learning task

You will find the data set associated with this task on the companion website that accompanies this book (available in SPSS and Excel format). You will also find the answers there.

Learning task

Following what we have learned about Kruskal-Wallis, answer the following questions and conduct the analyses in SPSS. (If you do not have SPSS, do as much as you can with the Excel spreadsheet.) In this data set we investigate students' satisfaction with their course. They are examined across three groups, according to how their performance is assessed: reports, essays or exams. Satisfaction is measured by a Likert questionnaire containing 16 questions, each of which has satisfaction rated from 1 (very poor) to 5 (very good).

Open the **Satisfaction with course** data set

1. Why was Kruskal-Wallis needed to examine the data rather than an independent one-way ANOVA?
2. Perform the Kruskal-Wallis test and include all relevant descriptive statistics.
3. Run an additional test to examine the source of difference, if needed.
4. Describe what the SPSS output shows.
5. Calculate the effect size.
6. Report the outcome as you would in the results section of a report.

Friedman's ANOVA

What does Friedman's ANOVA do?

Friedman's ANOVA explores differences in scores of a non-parametric dependent variable across three or more conditions of a single independent variable, measured over one sample (all participants experience all conditions). It is the non-parametric equivalent of the repeated-measures one-way ANOVA (see Chapter 10). On the face of it, the data used for examining outcomes in Friedman's ANOVA will look like those used for repeated-measures one-way ANOVA. However, we might find that the dependent variable data are not normally distributed, or decide that the data we are using represent ordinal numbers.

Research question for Friedman's ANOVA

A group of sleep researchers, SNORES (Sleep and Nocturnal Occurrences Research Group), decide to examine the benefit of a new drug to treat insomnia. To investigate this, the group run a trial with a single group of patients; they all receive the new drug 'Snooze'. To explore efficacy, SNORES examine self-rated sleep perceptions at three time points: at baseline (prior to treatment) and at treatment weeks 6 and 12. The participants complete a questionnaire about the quality of their sleep. The responses are scored on a scale of 0–100, with higher scores representing better perceptions. These subjective ratings of sleep *might* be considered to be ordinal,

18.24 Take a closer look
Summary of Friedman's ANOVA example

Dependent variable: self-rated sleep perception scores
Independent variable: time point (baseline, week 6 and week 12)

so might warrant analyses with non-parametric methods. SNORES expect differences, but make only non-specific (two-tailed) predictions about outcomes.

Theory and rationale

Using Friedman's instead of repeated-measures one-way ANOVA

When we examine outcomes across three or more within-group conditions (measured across a single group), we often do so with a repeated-measures one-way ANOVA. However, we should perform that test only if the data that we are examining are parametric. If the data are not parametric, we might not be able to trust the mean score (which is central to analyses with repeated-measures one-way ANOVA). Friedman's ANOVA calculates outcomes based on how scores are ranked across conditions rather than using the mean score (we will see how to undertake ranking in Box 18.25). The sum of those ranks, across the conditions, is applied to a calculation that determines whether there are significant differences between the conditions (as we will see later). We will use the SNORES research question to explore this.

18.25 Nuts and bolts
Simple example of ranking for Friedman's ANOVA

Ranking for Friedman's ANOVA is very similar to the methods used in the Wilcoxon signed-rank test, except that there are three or more conditions over which the ranking is performed. Once again, the sum of ranks (within each condition) plays a part in determining whether scores are significantly different between those conditions. Table 18.14 provides a simple example of how we calculate those ranks, using some pilot data from the SNORES research. The dependent variable is represented by self-rated 'sleep satisfaction' scores; the independent variable is the three time points (baseline, week 6 and week 12).

Table 18.14 Sleep satisfaction at baseline and treatment weeks

	Treatment week			**Rank**		
Participant	0	6	12	0	6	12
1	36	80	79	1	3	2
2	50	55	55	1	2.5	2.5
3	28	42	70	1	2	3
4	41	62	50	1	3	2
5	33	33	80	1.5	1.5	3
6	41	47	68	1	2	3
			Rank sums	**6.5**	**14**	**15.5**

For each participant, we rank their scores across the conditions, from the lowest to highest. For participant no. 1 the lowest scores was week 0, followed by weeks 12 and 6 (so the ranks scores are 1, 3 and 2). If scores are equal between conditions, the ranks are shared. This happens twice in our example. For participant no. 2 the two highest scores are equal. This would have represented ranks 2 and 3, so we share that: $(2 + 3) \div 2 = 2.5$. Once we have applied this to the entire sample, we sum the ranks for each condition (as shown in Table 18.14). These rank sums are applied to an equation, which assesses whether those ranks differ significantly between the conditions (see Box 18.26).

Assumptions and restrictions

Friedman's ANOVA must examine data that are at least ordinal (they might be interval or ratio but, if the data are not normally distributed, a non-parametric test might still be more appropriate). The independent variable must be categorical, with three or more within-group conditions (both measured across one single group). Every person (or case) must be present in all conditions.

Establishing significant differences

To explore outcomes for Friedman's ANOVA, we will use another example from the SNORES research question. Self-rated sleep satisfaction scores are collected from 11 patients. Each of these patients is given some new medication (Snooze) to help relieve their insomnia. Patients' ratings of sleep satisfaction are taken at baseline (week 0), and at post-treatment weeks 6 and 12. The degree of sleep satisfaction is measured on a scale of 0–100, with higher scores representing better perceptions. SNORES state a non-specific (two-tailed) hypothesis that there will be a difference in sleep satisfaction scores across the time points.

18.26 Calculating outcomes manually

Friedman's ANOVA calculation

Table 18.15 presents the SNORES data that we are examining. **You will find a Microsoft Excel spreadsheet associated with these calculations on the web page for this book.**

Table 18.15 Sleep perceptions, before and after treatment

Participant	Week 0	Week 6	Week 12	Rank 0	Rank 6	Rank 12
1	41	67	83	1	2	3
2	42	67	69	1	2	3
3	55	52	73	2	1	3
4	45	72	89	1	2	3
5	46	69	81	1	2	3
6	47	74	81	1	2	3
7	52	79	88	1	2	3
8	65	62	80	2	1	3
9	47	74	83	1	2	3
10	47	73	85	1	2	3
11	40	67	78	1	2	3
			Sum of ranks	**13**	**20**	**33**
			Mean rank	**1.18**	**1.82**	**3.00**

We saw how to rank scores across conditions in Box 18.25. We then calculate the rank sums for each condition, as shown in Table 18.15. Then we apply those outcomes to the equation for Friedman's ANOVA (F_r):

$$Fr = \left[\frac{12}{Nk(k+1)}\sum_{i=1}^{k} R_i^2\right] - 3N(k+1)$$

N = sample size (11); k = number of conditions (3); R = sum of ranks for each condition

$$So,\ F_r = \left[\frac{12}{11 \times 3 \times (4)} \times (13.0^2 + 20^2 + 33^2)\right] - (3 \times 11 \times (4)) = 18.73$$

F_r produces a value of χ^2 (chi square), which we compare within a χ^2 distribution table (see Appendix 6). The critical value for χ^2 at $df = 2$ (3 conditions *minus* 1) at p =.05 is 5.99. Our χ^2 is 18.73, which is greater, so our (overall) difference is significant. We need additional tests to examine the source of that difference.

Finding the source of difference

As we have just seen, if the outcome from Friedman's ANOVA is significant, it tells us only that there is a difference in scores across the conditions – it does not tell us where those differences are. Following the analysis of the SNORES data, we may find that sleep satisfaction ratings differ across the time points. However, at that point we do not know whether those ratings improve between baseline and treatment week 6 (for example). We saw a similar situation when we performed analyses on parametric data with repeated-measures one-way ANOVA (Chapter 10). We resolved that by performing Bonferroni *post hoc* tests. These are not suitable for Friedman's ANOVA Instead we *could* perform a procedure called Page's L Trend Test, but we will not explore that test in this book. More commonly, researchers employ separate Wilcoxon signed-rank tests for each pair of conditions (Baseline vs. Week 6, Baseline 0 vs. 12, and Week 6 vs. Week 12). Similar to Kruskal–Wallis, when we performed additional Mann–Whitney U tests, we need to consider how we should account for multiple comparisons. For the same reasons as we saw earlier, if we have made specific (one-tailed) predictions about the outcome between conditions, we do not need to make any adjustments. However, if we stated non-specific (two-tailed) hypotheses, we must divide the significance threshold by the number of additional tests needed. SNORES made only non-specific hypotheses in our example, so we must make adjustments. Because there are three conditions, if there is a significant Friedman's ANOVA outcome, we must perform three Wilcoxon signed-ranks tests. The significance threshold must be divided by three. We will have a significant outcome only where $p < .016$ (in a similar way to what we saw in Box 18.20).

How SPSS performs Friedman's ANOVA

We can perform Friedman's ANOVA in SPSS, examining the SNORES data that we used to calculate outcomes manually. Sleep satisfaction ratings (the dependent variable) are probably ordinal, hence the need for a non-parametric test to explore the outcome. We will not explore normal distribution as we saw how to do that for this scenario in Chapter 10. Setting up the data set in SPSS will be similar to what we saw for repeated-measures one-way ANOVA (see Box 10.8).

Running Friedman's ANOVA in SPSS

Open the SPSS file Sleep medication

Select **Analyze → Non-parametric test → Legacy dialogs → K related Samples . . .** (as shown in Figure 18.23)

This procedure is based on SPSS version 19. If you have an older version than version 18, the procedure is slightly different (see Box 18.7 for general guidance, but select 'K Related Samples' instead of '2 Independent Samples').

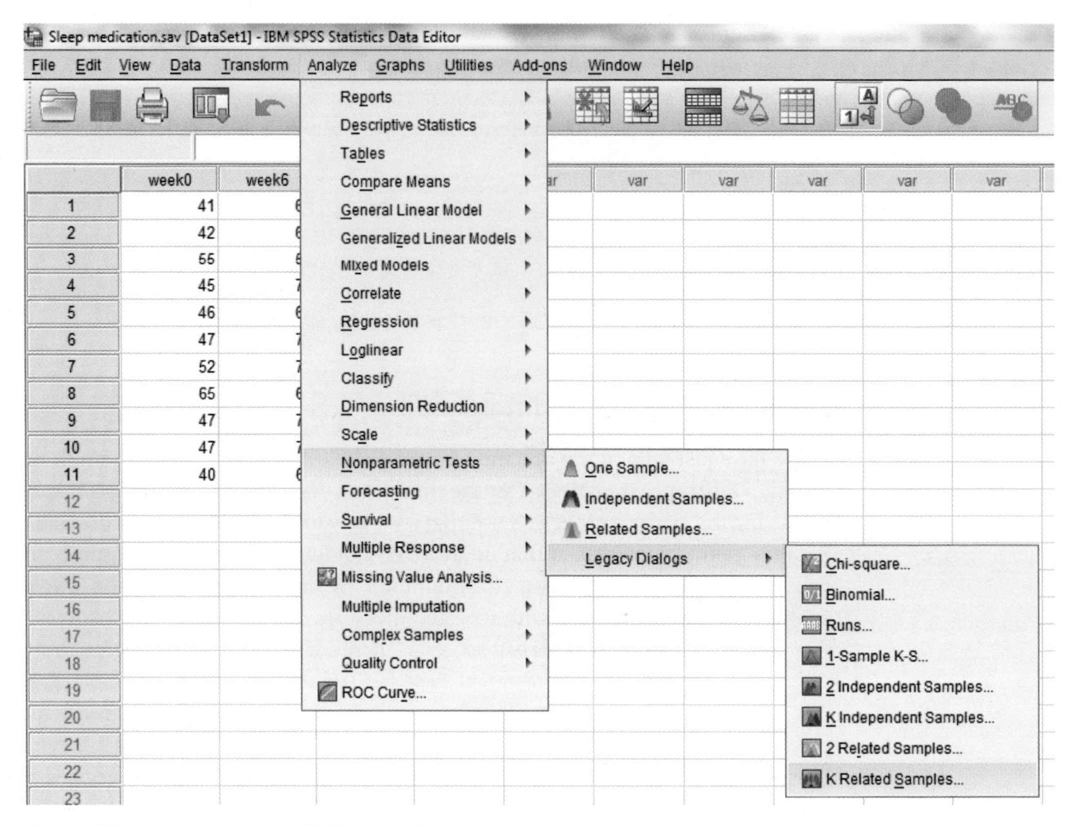

Figure 18.23 Friedman's ANOVA: procedure 1

In new window (see Figure 18.24), transfer **Sleep perceptions at baseline, Sleep perceptions at treatment week 6** and **Sleep perceptions at treatment week 12** to **Test Variables** ➜ make sure that **Friedman** is (only) ticked (under **Test Type**) ➜ click **OK**

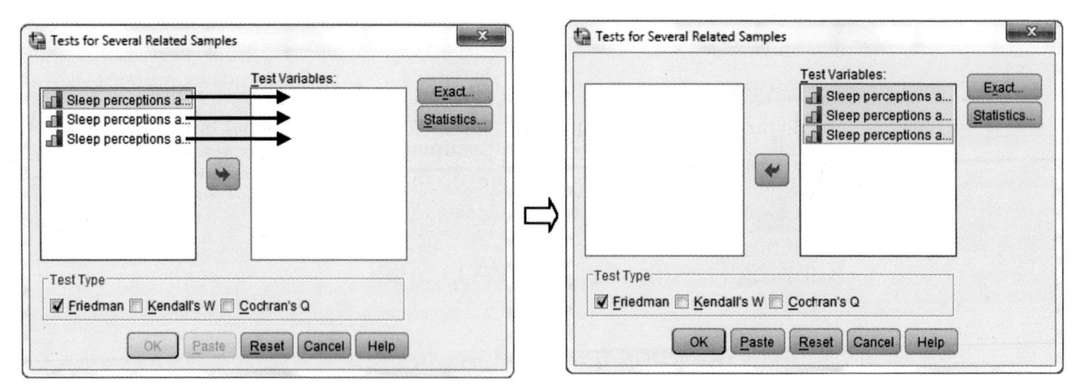

Figure 18.24 Friedman's ANOVA: procedure 2

You will also need some descriptive statistics to show the median scores and 95% confidence interval data between the groups for reporting purposes. The procedure has to be done separately to the Mann–Whitney test because the Descriptive tests option (under Options) will only produce the median for the whole sample, which is not very useful.

Select **Analyze → Descriptive statistics → Explore . . . →** (in new window) transfer **Sleep perceptions at baseline, Sleep perceptions at treatment week 6** and **Sleep perceptions at treatment week 12** to **Dependent List →** select **Statistics** radio button **→** click **OK**

Interpretation of output

Figure 18.25 shows differences in ranked scores concerning sleep satisfaction at each of the time points.

	Mean rank
Sleep perceptions at baseline	1.18
Sleep perceptions at treatment week 6	1.82
Sleep perceptions at treatment week 12	3.00

Figure 18.25 Confirmation of rank scores across conditions

Figure 18.26 confirms that we have a significant difference in sleep perceptions across the conditions, $\chi^2(2) = 18.727$, $p < .001$. We can compare this to what we calculated by hand in Box 18.26.

N	11
Chi-square	18.727
df	2
Asymp. Sig.	.000

Figure 18.26 Friedman's ANOVA test statistic

Finding the source of difference

As we learned earlier, the Friedman's ANOVA tells us only that there is a difference between the scores. It does not tell us where the differences are according to each pair of analyses. We need additional Wilcoxon signed-rank tests to locate the source of difference. In this case we have three conditions, so we need three extra tests. Also, as we saw earlier, we will need to make an adjustment to the significance cut-off, to account for multiple analyses (so the significance threshold is $p < .016$).

Using the SPSS file Sleep medication

Select **Analyze → Nonparametric Tests → 2 Related Samples . . . →** (in new window) transfer **Sleep perceptions at baseline** and **Sleep perceptions at week 6** to first line of **Test Pairs →** transfer **Sleep perceptions at baseline** and **Sleep perceptions at week 12** to second line of **Test Pairs →** transfer **Sleep perceptions at week 6** and **Sleep perceptions at week 12** to third line of **Test Pairs →** click **OK**

Wilcoxon signed-rank outcomes

Figure 18.27 shows the *z*-score and significance for each of the pairs of conditions. Remember, we have a significant difference only if $p < .016$ (because we had to adjust the cut-off point to account for multiple comparisons). There is a significant difference between each of the pairs. Using the median data we asked for (see Table 18.16), we can see that sleep perceptions were significantly improved between baseline and week 6 ($z = -2.697$, $p = .007$), between baseline and week 12 ($z = -2.937$, $p = .003$) and between week 6 and week 12 ($z = -2.937$, $p = .003$).

	Sleep perceptions at treatment week 6– sleep perceptions at baseline	Sleep perceptions at treatment week 12– sleep perceptions at baseline	Sleep perceptions at treatment week 12– sleep perceptions at treatment week 6
Z	–2.697[a]	–2.937[a]	–2.937[a]
Asymp. Sig. (2-tailed)	.007	.003	.003

Figure 18.27 Wilcoxon tests

Effect size

Similar to Kruskal–Wallis, we cannot use G*Power to calculate effect size for Friedman's ANOVA However, we can calculate Pearson's (*r*) effect size for each Wilcoxon signed-rank test that we ran to explore the source of difference. We divide the *z*-score by the square root of the sample size:

$$r = \frac{Z}{\sqrt{n}}$$ (the group size (n) is 22 on each occasion; we have two pairs of 11)

$$\text{Baseline vs. Week 6} = \frac{2.697}{\sqrt{22}} \quad r = .575 \text{ (strong effect)}$$

$$\text{Baseline vs. Week 12} = \frac{2.937}{\sqrt{22}} \quad r = .626 \text{ (strong effect)}$$

$$\text{Week 6 vs. Week 12} = \frac{2.937}{\sqrt{22}} \quad r = .626 \text{ (strong effect)}$$

Writing up results

We can report the outcome, tabulating the data and describing the key points along with appropriate statistical notation.

Table 18.16 Sleep perceptions, before and after treatment (n = 11)

Study week	Median	95% CI
Baseline	47.0	43.07 – 52.75
Treatment week 6	69.0	63.86 – 73.59
Treatment week 12	81.0	76.91 – 84.91

We would write this up in our results section as follows:

> Using Friedman's ANOVA, it was shown that there was a significant difference in sleep perception across treatment time points, χ^2 (2) = 18.727, P < .001. Subsequent Wilcoxon signed-rank

tests (with Bonferroni correction) indicated that sleep perceptions were significantly improved between baseline and week 6 (z = −2.697, p = .007), between baseline and week 12 (z = −2.937, p = .003) and between week 6 and week 12 (z = −2.937, p = .003).

Presenting data graphically

Box plots are a useful way of displaying non-parametric data. We can use SPSS to produce this for us.

Select **Graphs** → **Legacy Dialogs** → **Boxplot . . .** (as shown in Figure 18.11) → (in next window) click on **Simple** box → tick **Summaries for separate cases** radio button → click **Define** → (in next window) transfer **Sleep perceptions at baseline**, **Sleep perceptions at treatment week 6** and **Sleep perceptions at treatment week 12** to **Boxes Represent** → click **OK**

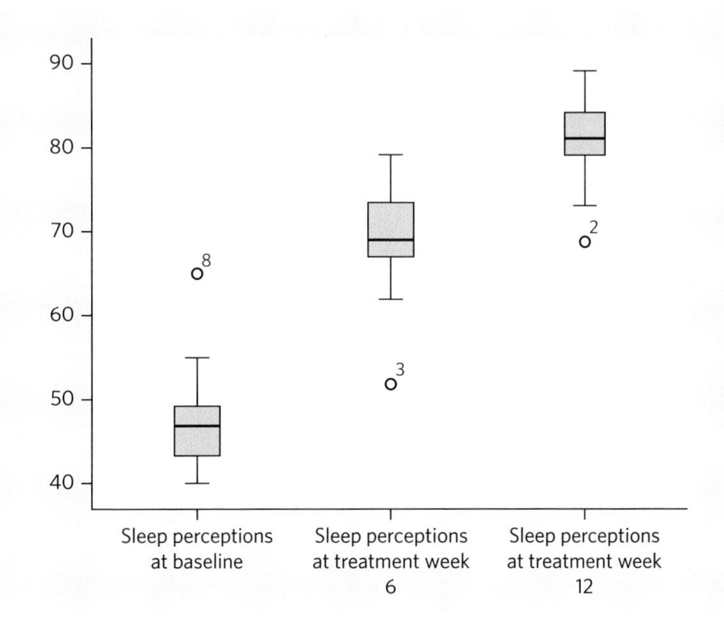

Figure 18.28 Box plot showing sleep perceptions across treatment time points

18.27 Exercise
Friedman's ANOVA mini-exercise

Try running an example yourself, using another data set (also located on the book's website). On this occasion we look at some data that examine a group of 50 people on their reported happiness levels, at different times of the week: Monday, Wednesday and Friday. Happiness was measured using a series of Likert-scale questions focusing on perceptions relating to work, relationships, general mood, tiredness, optimism, and so on. We will predict that there will be a difference in happiness across the days, but will not specify which will show the most joy (although I have a sneaking suspicion. . .).

Open the dataset **TGIF**

Run a Friedman's ANOVA test, with the days Monday, Wednesday and Friday as the conditions. You will find the 'answers' for this task on the web page for this book.

Section summary

Friedman's ANOVA is the non-parametric equivalent to repeated-measures one-way ANOVA. It is used where there is one within-group independent variable, with at least three conditions measured across a single group, measured in respect of one dependent variable. We are more likely to use this test if the dependent variable data are not normally distributed and/or those data are ordinal. Friedman's ANOVA examines outcomes by ranking dependent variable scores for each participant (or case) over the conditions, and explores whether those ranks differ significantly across the conditions for the entire sample. However, Friedman's ANOVA indicates only whether there are significant differences across the within-group conditions. To locate the source of difference, we need to run additional Wilcoxon signed-rank tests for each pair of conditions (adjusted for multiple comparisons where necessary).

18.28 Research example

Friedman's ANOVA in practice

It might help you to see how Friedman's ANOVA has been applied in published research. If you would like to read the entire paper you can use the DOI reference provided to locate that (see Chapter 1 for instructions).

Giaquinto, S. (2006). Death or improvement: the fate of highly disabled patients after stroke rehabilitation. *Clinical and Experimental Hypertension*, *28*: **357–364. DOI: http://dx.doi.org/10.1080/10641960600549629**

In this research the author explored functional improvements in stroke patients after discharge from hospital. Previous research had suggested that patients with the poorest post-stroke functioning showed little improvement over time. The author sought to demonstrate that this need not be the case, even for the most disabled patients who received rehabilitation prior to discharge. Functioning was measured via the Functional Independence Measure (FIM; Linacre *et al.*, 1994). FIM scores explore factors such as self-care, mobility and communication and are rated from 1 (need total assistance) through to 7 (patient has complete independence). FIM scores of less than 40 have previously predicted poor prognoses in stroke patients. FIM scores were measured at admission to hospital, upon discharge and one year after discharge. Initially, 176 stroke patients were recruited (with FIM scores of less than 40 at discharge). Only 89 of these were available at one-year follow-up (45 men, 44 women), so only they could be included in final analyses (supporting what we said about needing all participants to be present at all conditions). We were told that the dependent variable data were not normally distributed (hence the need for Friedman's ANOVA rather than a repeated-measures one-way ANOVA).

The results showed that median FIM scores for men were 24 at admission, 35 at discharge, and 66 at follow-up; for women median scores were 28, 37 and 63 respectively. Friedman's ANOVA showed a significant difference in FIM scores across the time point ($\chi^2 = 118.357$, $p < .001$). Degrees of freedom (*df*) were not presented; it is good practice to do so (*df* would be 2 in this case). Subsequent Wilcoxon signed-rank analyses indicated a significant difference between FIM scores at follow-up and admission ($z = -7.436$, $p < .001$), and between follow-up and discharge ($z = -7.357$, $p < .001$). We were not specifically told that these were indeed signed-rank analyses, but should assume that they were.

This paper provides a unique example of how Friedman's ANOVA, and subsequently Wilcoxon signed-rank tests, can be used in research.

18.29 Exercise

Friedman's extended learning task

You will find the data set associated with this task on the companion website that accompanies this book (available in SPSS and Excel format). You will also find the answers there.

Learning task

Following what we have learned about Friedman's ANOVA, answer the following questions and conduct the analyses in SPSS. (If you do not have SPSS, do as much as you can with the Excel spreadsheet.) In this data set we examine a group of 12 people whom we deprive of sleep in a laboratory and then test for perceptions of cognitive function. We ask our participants to rate how alert they feel on a Likert scale, rated from 1 (very alert) to 5 (feel like a zombie). We do this at three time points: after one hour, two hours, and three hours without sleep.

Open the **Sleep deprive** data set

1. Why was Friedman's ANOVA used rather than repeated-measures one-way ANOVA?
2. Perform Friedman's ANOVA and include all relevant descriptive statistics.
3. Run an additional test to examine the source of difference, if needed.
4. Describe what the SPSS output shows.
5. Calculate the effect size.
6. Report the outcome as you would in the results section of a report.

Chapter summary

In this chapter we have explored a range of non-parametric tests. At this point, it would be good to revisit the learning objectives that we set at the beginning of the chapter.

You should now be able to:

- Recognise when it is preferable to perform a non-parametric test instead of the parametric equivalent. Those parametric tests use the mean score to determine outcome. If the dependent variable data are not normally distributed, or if those data appear to be ordinal, we may not be able to trust that mean score. So, parametric tests may not be appropriate. Non-parametric tests examine outcomes based on how the scores are ranked, rather than depend on mean scores.
- Appreciate which non-parametric test is appropriate for each situation. All of the tests explored in this chapter examine data from a single (continuous) dependent variable. We use a Mann-Whitney U test (instead of an independent t-test) when the independent variable is represented by two distinct groups. We use a Wilcoxon signed-rank test (instead of the related t-test) when the independent variable is represented by two within-group categories. We use a Kruskal-Wallis test (instead of independent one-way ANOVA) when the independent variable is represented by three or more distinct groups. And we use Friedman's ANOVA (instead of repeated-measures one-way ANOVA) when the independent variable is represented by three or more within-group categories.

- Understand that the dependent variable must be represented by at least ordinal data. Interval or ratio data may be explored if they are not normally distributed. For between-group analyses, group membership must be exclusive (no person or case can appear in more than one group). For within-group analyses, every person or case must be present in every condition.
- Calculate the outcome manually (using maths and equations).
- Perform analyses using SPSS.
- Know how to measure effect size and power.
- Understand how to present the data and report the findings.

19 TESTS FOR CATEGORICAL VARIABLES

Learning objectives

By the end of this chapter you should be able to:

- Recognise when it is appropriate to use categorical tests
- Identify the appropriate test type in a range of contexts:
  - Chi-squared (χ^2) test, Yates' continuity correction, Fisher's exact test, layered χ^2 test, or loglinear analysis
- Understand the theory, rationale, assumptions and restrictions associated with each test
- Calculate the outcome by hand (using maths and equations)
- Perform analyses using SPSS
- Understand how to present the data and report the findings

What are tests for categorical variables?

Tests for categorical variables are used when all of the variables in the analysis are represented by categorical groups. Each variable is analysed according to 'frequencies' observed across the groups. The extent to which the distribution of frequencies is associated across the variables is then measured. For example, we might have two variables, mood status (depressed or not depressed) and gender (male and female). If we have 100 participants, we might find that 60 are depressed, while 40 are not depressed. All else being equal, we would expect the frequencies to be distributed in the same proportion across gender. Because of that, sometimes we illustrate those frequencies with 'percentages' to help us visualise what distribution we can expect. However, it must be stressed that the tests that we are about to explore base outcomes on the observed and expected frequencies, not the percentages (they are just added for illustration). Using the example we saw just now, we would find that 60% of the sample are depressed, while 40% are not. If we had 50 men and 50 women in our sample, we would expect that percentage distribution to remain the same. So, we would *expect* that 60% of men and 60% of women would be depressed (30 in each case) and 40% in both groups not to be depressed (20 in each case). In reality we might *observe* that 22 of men were depressed (44%) and 28 not depressed (56%), while 38 of women were depressed (76%) and 12 not depressed (24%). The 'expected' frequencies are compared with 'observed' frequencies. Statistical analyses examine whether that observed outcome is significantly different to the expected outcome.

Sometimes, we may have more than two variables. For example, we could examine the same 100 participants, but this time explore the same according to exercise frequency group (frequent, infrequent and none) and mood group (depressed vs. not), and then further still by gender (male vs. female). We might observe different mood outcomes across the exercise groups, which might differ still further across gender. The type of test we use will depend on three factors: how many variables we are exploring how many groups are represented by each variable and the number of cases being observed. We examine pairs of variables with tests such as Pearson's chi-squared, Yates' continuity correction and Fisher's exact test. If we have more than two variables, we might use a layered chi-squared or loglinear analysis. We will explain how to select the correct test a little later.

Research questions for tests for categorical variables

To illustrate tests for categorical variables, we should set some research questions. A group of sports psychologists, the Westchester Academy of Sports Psychologists (WASPS), decide to explore the relationship between exercise behaviour and mood, and to investigate whether these factors vary between men and women. WASPS recruit 100 participants, 45 men and 55 women. From a series of questionnaires and diagnostic interviews, they establish how many of the participants are currently depressed. Using records of attendance at the WASPS gym, they are able to establish how often the participants undertake exercise: frequently (two or more times a week), infrequently (once a week) and never. The researchers make three predictions:

1. Women are more likely to be depressed than men.
2. People who undertake regular exercise are less likely to be depressed.
3. Prediction 2 is likely to be more pronounced with women than men.

Strictly speaking, these tests of association permit only non-specific (two-tailed) hypothesis testing, although (as we will see later) there are ways in which we can observe how groups differ in the nature of that relationship.

19.1 Take a closer look
Summary of examples for tests of categorical variables

Two-variable example

Variable 1: Mood status - depressed vs. not depressed
Variable 2: Gender - male vs. female

Three-variable example

Variable 1: Mood status - depressed vs. not depressed
Variable 2: Exercise frequency group - frequent, infrequent, none
Variable 3: Gender - male vs. female

Theory and rationale

Measuring associations across two or more categorical variables

When we examine categorical data, we can display this in a table that shows these variables in a series of rows and columns – we call this a **cross-tabulation** (also known as a **contingency table**). The appearance of the cross-tabulation will depend on how many variables are being explored and how many groups are involved in those variables. The cross-tabulation is described in terms of the number of rows and columns, so a 2 × 2 cross-tabulation would have two rows and two columns, while a 2 × 3 table would have two rows and three columns (we always state the number of rows first). We should look at some examples before we proceed.

Table 19.1 2 × 2 cross-tabulation for depression vs. gender

	Gender	
Depressed	**Male**	**Female**
Yes	21	39
No	24	16

In the example shown in Table 19.1 we have two variables, gender and depression status: the gender variable has two groups (male and female); the depression status variable also has two groups (depressed and not depressed). It is an example of a 2 × 2 cross-tabulation. The numbers within the cells of this table represent the number of people who are depressed (or not) for men and then for women – we call this the 'observed frequencies'. In this example, the *proportion* of men who are depressed appear to be different to the proportion of depressed women; we shall see whether that is the case later.

Table 19.2 2 × 3 cross-tabulation for depression vs. exercise frequency

	Exercise frequency		
Depressed	**Frequent**	**Infrequent**	**None**
Yes	22	13	25
No	23	10	7

In the example shown in Table 19.2 we still have two variables, but this time one of the variables has three groups (exercise frequency: frequent, infrequent and none), while the depressed variable still has two groups. This is a 2 × 3 cross-tabulation because there are two rows and three columns. Looking at these observed frequencies, there seems little difference in proportions of frequent and infrequent exercise between those who are depressed or not. However, it would appear that people who do not exercise at all are more likely to be depressed than not.

Table 19.3 2 × 2 × 3 (layered) cross-tabulation for gender vs. depression vs. exercise frequency

Gender		Exercise frequency		
	Depression	Frequent	Infrequent	None
Male	Yes	6	5	10
	No	16	7	1
Female	Yes	16	8	15
	No	7	3	6

In the example shown in Table 19.3 we have three variables: gender, depression status and exercise frequency. It is a 2 × 2 × 3 cross-tabulation because there are two gender groups (male and female), two depression groups (depressed and not depressed) and three exercise frequency groups (frequent, infrequent and none). There are several ways to read this table: it seems that the proportion of depressed people is somewhat different between the exercise groups and the distribution appears to be somewhat differently expressed between men and women. Later in the chapter, we will explore tests that will help us determine whether the observed association is statistically significant.

How to determine significance in cross-tabulations

To examine whether the pattern of observed outcomes is significantly different between two variables, we need to explore observed and expected frequencies. This will illustrate whether the outcome differs sufficiently from what we would expect if there were no differences in outcomes between the groups.

Observed frequencies

In Tables 19.1 – 19.3 we saw what appeared to be differences between the observed frequencies of cases within groups for one variable across in relation to the groups of another variable. For example, in Table 19.1 it would seem that women in this sample are more likely to be depressed than men To help us establish whether associations are significant, we need to add row and column totals to the data we saw in Table 19.1. In Table 19.4, we can now see that 60 people are depressed, while 40 people are not depressed, and that 45 people in the group were male and 55 were female. We need to compare the information from the row and column totals to the more specific data

Table 19.4 Depression vs. gender (observed frequencies)

	Gender		
Depressed	Male	Female	Total
Yes	21	39	60
No	24	16	40
Total	45	55	100

shown in the centre of the cross-tabulation. We refer to these specific pieces of information as 'cells'. There are four cells in this example (depressed men, depressed women, non-depressed men and non-depressed women). It is these cells that we compare to the row and column totals.

Adding percentages

A useful option in the assessment of associations across categorical variables is the application of row or column percentages (it is better to choose one or other of these, rather than both, for reasons that will become clear later). This option does not affect the outcome; that can be calculated only from observed and expected frequencies, but it is a useful addition that very few statistics books give much credence to. In fact, if we only have percentage data, we have to know the sample and group sizes, so that we can convert this to frequency data. Without that the statistical analyses cannot be performed. The addition of percentages is particularly helpful when samples have unequal groups. We can illustrate this by extending the example that we have just seen. Let's say we had 28 men and 72 women in our sample. In that sample we observe that 17 of the 28 men, and 43 of the 72 women, are depressed, while 11 men and 29 women are not depressed. Now it is much more difficult to see whether there is an association between gender and depression. However, if we add percentages, that picture is much clearer. In this case, the 17 depressed men represent 60% of that group, while the 43 depressed women are also 60% of their group. Those rates reflect the overall percentage for depressed people, so there is no difference in the rates that depression is found among men and women.

SPSS will add those percentages for you, but it would be useful to see how that is calculated here. Those percentages can be presented across the rows and the columns, but it is probably wise to use only one of these (otherwise the tables can become crowded and confusing). In our example, we could calculate column percentages – these would illustrate the extent that depression status is distributed across the sample, and between men and women. However, we could equally calculate the percentage of depressed people who are male (vs. female) and compare that with the same outcome for non-depressed people. It does not matter which you do as it has no effect on the outcome (the statistical analyses are based on the frequency data, not the percentages). We will focus on column totals. To calculate the percentages for the *column* totals, we express each *row* total as a proportion of the overall sample. We will use the frequency data from Table 19.4:

Depressed (Yes): $(60 \div 100) \times 100 = 60.0\%$;

Not depressed (No): $(40 \div 100) \times 100 = 40.0\%$

Now we perform the calculations for the cells. To do this, we express the cell frequency by the total number of cases in that *row*:

Male: Depressed, $(21 \div 45) \times 100 = 46.7\%$; Not depressed, $(24 \div 45) \times 100 = 53.3\%$

Female: Depressed, $(39 \div 55) \times 100 = 70.9\%$; Not depressed, $(16 \div 55) \times 100 = 29.1\%$

We should now incorporate all of this into our display of observed frequencies (see Table 19.5). The data in black font are the observed frequencies; the data in red font are observed 'cell percentages', while the 'overall observed percentages' are shown in green font in the 'Total' column. Overall, we know that 60% of the participants are depressed, while 40% are not depressed. If men and women did not differ on depression status, it seems reasonable to expect that the total

Table 19.5 Depression vs. gender (observed frequencies/percentages)

	Gender		
Depressed	**Male**	**Female**	**Total**
Yes	21 46.7%	39 70.9%	**60** 60.0%
No	24 53.3%	16 29.1%	**40** 40.0%
Total	**45**	**55**	**100**

percentages for depression status would be the same according to gender: 60% of men and 60% of women would be depressed. Table 19.5 clearly shows that this is not the case.

Expected frequencies and percentages

Table 19.5 provides a much clearer picture of how depression status might vary by gender. However, as we have said, we cannot use the percentages to assess whether there is an association, we must compare the observed frequencies to expected frequencies. But how do we calculate expected frequencies? SPSS will provide those outcomes, but it would useful to see how it is done. To obtain these expected frequencies, for each of the cells we multiply the column total by the row total and divide by the overall sample size:

Depressed male: $(45 \times 60) \div 100 = 27$ Depressed female: $(55 \times 60) \div 100 = 33$

Non-depressed male: $(45 \times 40) \div 100 = 18$ Non-depressed female: $(55 \times 40) \div 100 = 22$

We can add these data to our table, which now shows observed and expected frequencies and percentages (see Table 19.6).

Table 19.6 Depression vs. gender (observed/expected frequencies/percentages)

Depressed		Gender		Total
		Male	Female	
Yes	Observed	**21** 46.7%	**39** 70.9%	**60** 60.0%
	Expected	*27* 60.0%	*33* 60.0%	
No	Observed	**24** 53.3%	**16** 29.1%	**40** 40.0%
	Expected	*18* 40.0%	*22* 40.0%	
	Total	**45**	**55**	**100**

The observed frequencies in Table 19.6 provide all the information we need now (presented in bold black font, with observed cell percentages in red). The expected frequencies are shown in black italics, while expected percentages are shown in green font. The expected percentages in the cells are the same as the overall percentages in the Total column. Now we can see a clearer picture of how observed and expected frequencies might be associated, with help from the observed and expected percentages. The statistical analyses will formally test that association, as we will see shortly.

Assumptions and restrictions

There are fewer restrictions on the use of tests for categorical variables than for most other statistical procedures. As we are examining purely categorical variables, these tests are non-parametric. We do not need to test the data for normal distribution. However, the categories that are used must be represented by distinct groups: no one can appear in more than one cell of cross-tabulation – this is perhaps the most crucial requirement of these tests. Furthermore, these tests can be used only for between-group studies; we cannot use within-group data. So, we could explore the frequency of people who are depressed according to gender; but we could not examine whether a single group of people were depressed, before and after undergoing cognitive therapy.

The remaining assumptions relate to the type of test that we can apply. Test selection will depend on the number of groups measured by the two variables. In most cases, the most common choice for examining categorical data is Pearson's chi-squared (χ^2) test. However, most statisticians argue that χ^2 can be used only in larger cross-tabulations, where at least one of the variables has three or more groups. Table 19.2 is a good example, as this is represented by a 2×3 cross-tabulation: there are two gender groups (male and female) and three exercise frequency groups (frequent, infrequent and none). If there are only two groups on both of the variables, we should use Yates' continuity correction.

With smaller samples, some statisticians argue that outcomes might be more vulnerable to error if we use these tests. For instance, Cochran (1954) argued that you cannot use Yates' continuity correction on contingency tables if more than 20% of cells have an expected frequency of less than 5, and none that is less than 1. It is argued in those cases that you should use Fisher's exact test instead. This is probably also relevant for χ^2 analyses, but many statistical packages (including SPSS) permit Fisher's test only for 2 × 2 tables. Others argue that the Cochran rule is too conservative and leads to too many Type II errors. Camilli and Hopkins (1978) state that it is OK to use χ^2 even with small expected frequency cell sizes if the overall sample size is greater than 20. In summary, it is probably not valid to use Pearson's χ^2 on 2 × 2 tables; otherwise we should employ Yates' continuity correction. We should consider using Fisher's test if more than 20% of the cells have an expected value of less than 5 (but only for 2 × 2 tables).

19.2 Take a closer look
Summary of assumptions and restrictions

- Variables must be categorical
- Data must be represented by frequencies
- Cases must be independent of each other
- Categorical tests can be applied to between-group studies only
- The type of test selected depends on number of groups of variable
    - And on number of cases in each cell

Which test to use

Each of the outcomes shown in Tables 19.1 – 19.3 would need to be explored with a different type of test, reflecting the number of variables and the number of groups represented by those variables. Pearson's chi-squared (χ^2) test is commonly used, but as we saw in the previous section, χ^2 should be used only when at least one of the variables has more than two groups. When there are only two groups on each variable we should use a test called Yates' continuity correction. Both tests are generally used with larger samples. In smaller samples (as defined in the last section) Fisher's exact test might be more appropriate. When there are more than two variables (as there were with the 2 × 2 × 3 cross-tabulation in Table 19.3), there are two tests that we could employ. A layered χ^2 will simply explore significant differences across one pair of variables, in respect of groups from a third variable. However, this will tell you nothing about the interaction between those variables. If interaction is important, you should use a procedure called loglinear analysis. We will explore all of these tests throughout the course of this chapter.

19.3 Take a closer look
Examining categorical variables

Table 19.7 Summary of tests used when exploring categorical variables

No. of variables	Cross-tabulation	Test
Two	2 × 2	Yates' continuity correction
	2 × 3, 3 × 2, 2 × 4...	Chi-squared (χ^2) test
	Small samples	Fisher's exact test
Three or more		Layered χ^2 test
	Interactions	Loglinear analysis

Measuring outcomes statistically

So far in this chapter we have been making somewhat superficial analyses of our data by looking at how observed frequencies and proportions *appear* to differ across rows and columns, in relation to expected frequencies and proportions. We need to test these outcomes statistically. Before we perform the tests in SPSS, we should have a quick look at how to calculate outcomes manually.

Categorical tests with two variables: Pearson's χ^2 test

Pearson's χ^2 is used to compare categorical outcomes from two variables, where at least one of those variables has three or more groups. It examines how closely an observed set of data matches expected frequencies. The χ^2 outcome is used to determine whether the observed outcome is significantly different to the expected frequencies (we see how to do that in Box 19.4). The null hypothesis states that there will be no difference.

19.4 Calculating outcomes manually
Pearson's χ^2 calculation

To calculate the outcome manually for χ^2 we can refer back to the research questions set by WASPS, the group of sports psychologists. We will compare how depression status (yes or no) varies according to exercise behaviour (frequent, infrequent or none). The data will be displayed in a 2 × 3 cross-tabulation (as shown in Table 19.8). Observed expected frequencies and percentages have been included.

Table 19.8 Depression status vs. exercise frequency (observed/expected frequencies/percentages)

		Exercise frequency			
Depression		**Frequent**	**Infrequent**	**None**	**Total**
Yes	Observed	**22** 48.9%	**13** 56.5%	**25** 78.1%	**60** 60.0%
	Expected	*27.0* 60.0%	*13.8* 60.0%	*19.2* 60.0%	
No	Observed	**23** 51.1%	**10** 43.5%	**7** 21.9%	**40** 40.0%
	Expected	*18.0* 40.0%	*9.2* 40.0%	*12.8* 40.0%	
	Total	**45**	**23**	**32**	**100**

The formula for χ^2 is:

$$\chi^2 = \sum \frac{(\text{Observed} - \text{Expected})^2}{\text{Expected}}$$

$$= \frac{(22-27.0)^2}{27.0} + \frac{(13-13.8)^2}{13.8} + \frac{(25-19.2)^2}{19.2} + \frac{(23-18.0)^2}{18.0} + \frac{(10-9.2)^2}{9.2} + \frac{(7-12.8)^2}{12.8} = 6.811$$

We compare our χ^2 to a cut-off point in chi-square tables, relative to the degrees of freedom (*df*)

df = (no. of rows *minus* 1) × (no. of columns *minus*1) = 2 × 1 = 2

Cut-off point in chi-square tables (see Appendix 6), where *df* = 2 (for p = .05) = 5.99

Our χ^2 = 6.811 is greater than 5.99, so we have a significant outcome.

There is a significant association between exercise frequency groups by depression status.

How SPSS performs Pearson's χ^2 (on a 2 × 3 cross-tabulation)

To demonstrate how to run Pearson's χ^2 in SPSS, we will refer to the WASPS research example again. We are exploring the association between exercise frequency and depression status, among 100 of WASPS' clients. They used attendance records at their gym to establish how often these clients undertake exercise, from which three groups are created, based on exercise frequency (frequent, infrequent and none). From a series of questionnaires and diagnostic interviews, WASPS establish how many of the participants are currently depressed. The researchers predict that people who take more exercise are less likely to be depressed. Before we analyse this, we should remind ourselves about the variables with this quick summary:

Pearson's χ^2 example

Variable 1: Exercise frequency group – frequent, infrequent, none

Variable 2: Mood status – depressed vs. not depressed

19.5 Nuts and bolts

Setting up the data set in SPSS

When we create the SPSS data set for testing categorical variables, we need to reflect that they are entirely categorical. This data set will measure all of the variables that we need to measure in this chapter, so we need something for gender, depression status and exercise frequency. In the following sections, we will be using the SPSS data set 'Exercise, mood and gender' to perform these tests.

Exercise mood and gender.sav [DataSet1] - IBM SPSS Statistics Data Editor

File Edit View Data Transform Analyze Graphs Utilities Add-ons Window Help

	Name	Type	Width	Decimals	Label	Values	Missing	Columns	Align	Measure
1	Gender	Numeric	8	0	Gender	{1, Male}...	None	8	Right	Nominal
2	Exercise	Numeric	8	0	Exercise	{1, Frequent...	None	8	Right	Nominal
3	Depressed	Numeric	8	0	Depressed	{1, Yes}...	None	8	Right	Nominal
4										

Figure 19.1 Variable View for 'Exercise, mood and gender' data

Figure 19.1 shows how the SPSS Variable View should be set up (you should refer to Chapter 2 for more information on these procedures). For all of the variables in this data set, the Measure column should be set to Nominal. The first variable is 'Gender'; we should set the Values as '1 = Male', and '2 = Female'. The second variable is 'Exercise; the values should be '1 = Frequent', '2 = Infrequent', and 3 = 'None'. The third variable is 'Depressed'; the values should be '1 = Yes' and '2 = No'.

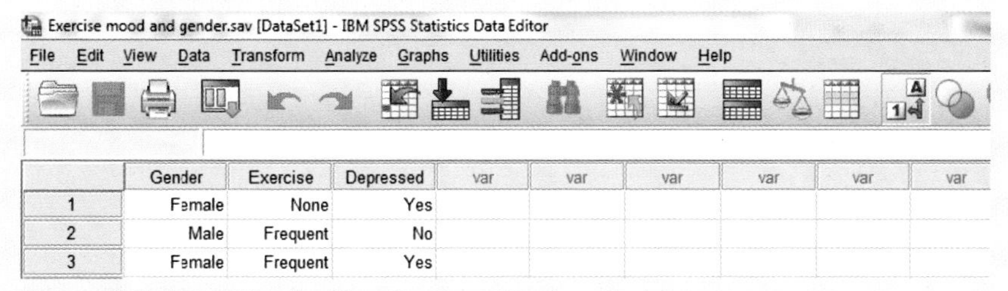

Exercise mood and gender.sav [DataSet1] - IBM SPSS Statistics Data Editor

File Edit View Data Transform Analyze Graphs Utilities Add-ons Window Help

	Gender	Exercise	Depressed	var	var	var	var	var	var
1	Female	None	Yes						
2	Male	Frequent	No						
3	Female	Frequent	Yes						

Figure 19.2 Data View for 'Exercise, mood and gender' data

Figure 19.2 illustrates how this will appear in Data View. It is Data View that will be used to select the variables when performing this test. Each row represents a participant. When we enter the data, we need to enter the value relevant to the group to which that participant belongs for each variable (as discussed just now). The columns will either display the descriptive categories, or will show the value numbers, depending on how you use the Alpha Numeric button in the menu bar (see Chapter 2).

Running Pearson's in SPSS

Open SPSS file **Exercise, mood and gender**

Select **Analyze** ➔ **Descriptive Statistics** ➔ **Crosstabs...** (as shown in Figure 19.3)

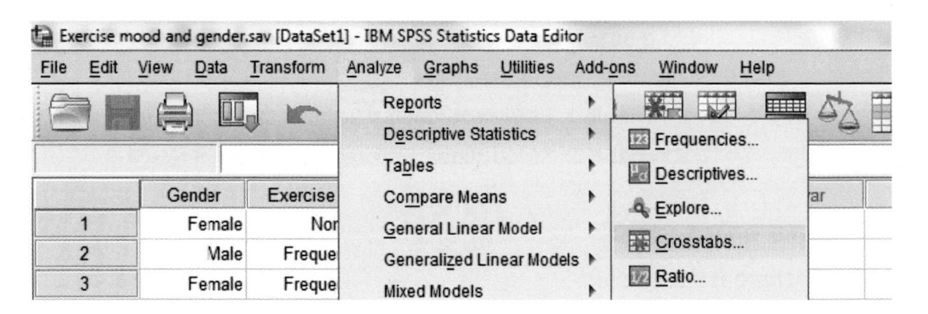

Figure 19.3 Pearson's χ^2 - step 1

In new window (see Figure 19.4), transfer **Depressed** to **Row(s)** ➔ transfer **Exercise** to **Column(s)** ➔ click **Statistics**

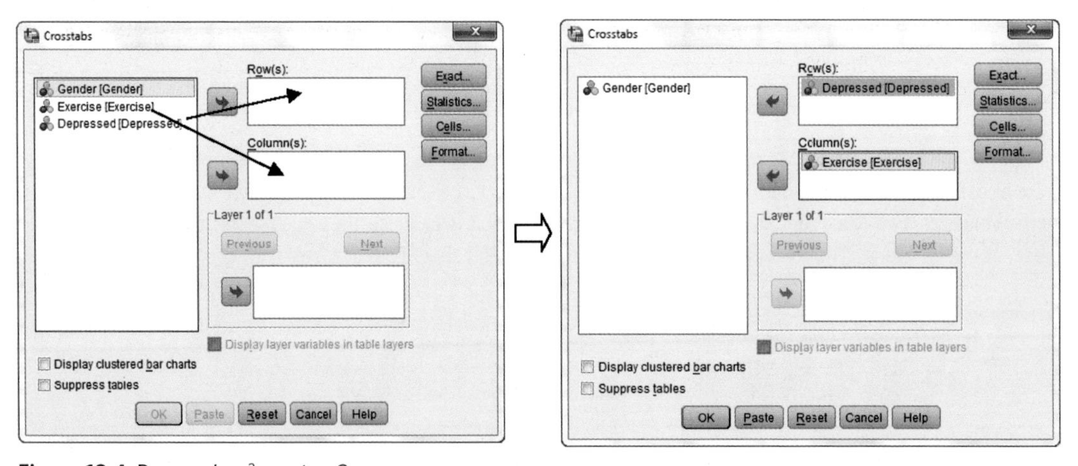

Figure 19.4 Pearson's χ^2 — step 2

In next window (see Figure 19.5), tick boxes for **Chi-square**, **Contingency coefficient** and **Phi and Cramer's V** ➔ click **Continue** ➔ (back in original window) click **Cells**

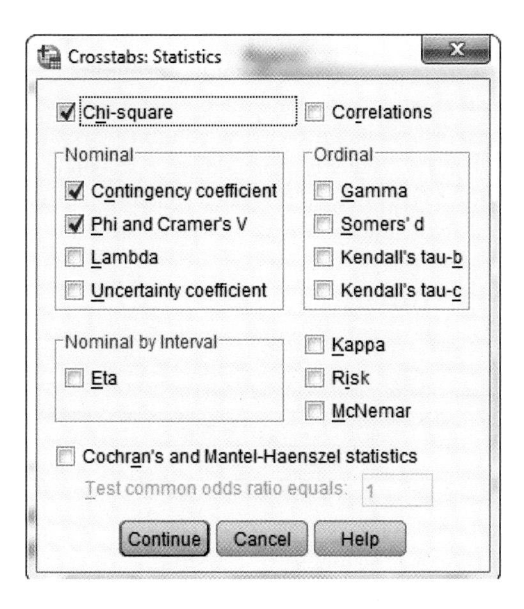

Figure 19.5 Pearson's χ^2 – Statistics options

These selections will provide the full range of statistical outcomes. We probably only need 'chi-squared' for a 2 × 3 cross-tabulation, but it is useful to ask for all of them – SPSS will guide you to the correct statistics. Phi and Cramer's are needed for information regarding variance (see later).

In next window (see Figure 19.6), tick boxes for **Observed** and **Expected** (under **Counts**) ➜ tick **Column** box (under **Percentages**) ➜ tick **Adjusted standardised** box (under **Residuals**) ➜ click **Continue** ➜ click **OK**

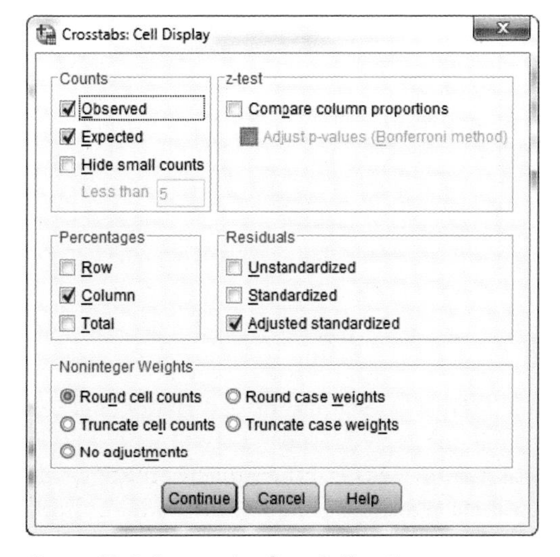

Figure 19.6 Pearson's χ^2 – Cell options

We have asked for a number of additional statistics in our cross-tabulation here. A basic table would just include observed frequencies (count), but that would not be very useful. We have included expected count so that we can compare the observed counts to expected frequencies (this is what the outcome statistic is based on). The column percentages are useful, too, for reasons that we explored earlier. The '**Adjusted standardised**' residuals are useful for highlighting where the biggest contribution to chi-squared is located (but also indicates potential sources of differences in the association across the variables).

Interpretation of χ^2 output

Depressed*Exercise cross-tabulation

			Exercise			Total
			Frequent	Infrequent	None	
Depressed	Yes	Count	22	13	25	60
		Expected count	27.0	13.8	19.2	60.0
		% within exercise	48.9%	56.5%	78.1%	60.0%
		Adjusted residual	−2.1	−.4	2.5	
	No	Count	23	10	7	40
		Expected count	18.0	9.2	12.8	40.0
		% within exercise	51.1%	43.5%	21.9%	40.0%
		Adjusted residual	2.1	.4	−2.5	
Total		Count	45	23	32	100
		Expected count	45.0	23.0	32.0	100.0
		% within exercise	100.0%	100.0%	100.0%	100.0%

Figure 19.7 Cross-tabulation

Figure 19.7 shows the cross-tabulation table for the variables 'Depressed' and 'Exercise'. You could compare this to the table we used in Box 19.4.

	Value	df	Asymp. Sig. (2-sided)
Pearson chi-square	6.811[a]	2	.033
Likelihood ratio	7.128	2	.028
Linear-by-linear association	6.389	1	.011
N of valid cases	100		

a. 0 cells (.0%) have expected count less than 5. The minimum expected count is 9.20

Figure 19.8 Outcome statistics (chi-squared tests)

Figure 19.8 confirms that there is a significant association between exercise behaviour according to depression status, χ^2 (2, N = 100) = 6.811, p = .033. However, this does not tell us the source of the association – we need the 'adjusted residual' data for that, looking for any values that are greater than 1.96. This is equivalent to the cut-off point for significance in a normal distribution (see Chapter 4). If we look at Figure 19.7, we can see that there are several cells where the adjusted residual is greater than 1.96. For those taking frequent exercise, the adjusted residual for 'not depressed' is 2.1, while the corresponding figure for 'depressed' is −2.1. This suggests that those undertaking frequent exercise are significantly less likely to be depressed. We can perform a similar analysis for those taking no exercise (where the adjusted residual is 2.5). This indicates that those taking no exercise are significantly more likely to be depressed.

SPSS provides a footnote here that we should pay attention to. It describes how many cells have an expected count of less than 5. If more than 20% of the cells show that, we need to use the Fisher's exact test rather than Pearson's χ^2. We have met that assumption. Indeed, had that outcome been violated, SPSS would automatically display the Fisher's test outcome (the absence of that report provides further evidence that we can use χ^2 to report our result).

Odds ratios

We can also express the observed association between exercise behaviour vs. depression status outcomes as an odds ratio. These express the relative likelihood of something occurring, where 'no difference' is represented by an odds ratio of 1. In a 2 × 3 cross-tabulation we need to look for odds ratios across pairs of variable cells. For example, we could compare 'depressed' vs. 'not depressed' in respect of 'frequent' and 'no' exercise. First we calculate the odds for being depressed when not taking exercise and then the odds for being depressed when undertaking frequent exercise. Finally, we divide the first calculation by the second, to obtain an 'odds ratio'.

Odds: depressed vs. not for no exercise (O1) = 25 ÷ (32-25) = 3.571

Odds: depressed vs. not for frequent exercise (O2) = 22 ÷ (45-22) = 0.956

Odds ratio = O1 ÷ O2 = 3.571 ÷ 0.956 = 3.734

This means that people who do not exercise are nearly four times more likely (OR 3.734) to be depressed than someone who undertakes regular exercise.

19.6 Take a closer look

More about odds ratios

The calculation we used to assess odds ratios just now might seem a little overly complex. For example, when we examined the odds of being depressed or not, when taking no exercise, we divided the cell count for 'depressed/no exercise' by the total number of people not taking exercise minus 'not depressed/no exercise'. Surely it would have been simpler just to divide the cell count for 'depressed/no exercise' by the cell count for 'not depressed/no exercise'? That would certainly give you the same answer, but that logic fails to recognise that there may be several rows and columns. For that reason we should always divide the measured cell count by the cell counts of all of the remaining cells in that row or column.

We could illustrate this by exploring the odds ratio for our last outcome in a slightly different way. Instead of focusing on the odds ratio for being depressed or not, in respect of not taking exercise, we could look at the odds ratio for frequent exercise vs. no exercise when someone is depressed, compared to someone who is not depressed. Using the data from Figure 19.7, we can see the following:

Odds: no exercise vs. frequent exercise for depressed (O1) = 25 ÷ (60 − 25) = 0.714
Odds: no exercise vs. frequent exercise for not depressed (O2) = 7 ÷ (40 − 7) = 0.212
Odds ratio = O1 ÷ O2 = 0.714 ÷ 0.212 = 3.367

This suggests that depressed people are nearly four times less likely (OR 3.367) to undertake any form of exercise than those not depressed.

		Value	Approx. Sig.
Nominal by nominal	Phi	.261	.033
	Cramer's V	.261	.033
	Contingency coefficient	.253	.033
N of valid cases		100	

Figure 19.9 Cramer's V

Explained variance

Figure 19.9 reports how much variance has been explained. In this case, **Cramer's V** (φ) = .261. If we square that and multiply that outcome by 100, we get 6.81. This means that 6.81% of the variation in depression status is accounted for by exercise behaviour.

Writing up results

We only need to show the basic cell details in our tabulated data, as shown in Table 19.9.

Table 19.9 Cross-tabulation of observed frequencies, exercise behaviour by depression status

	Exercise behaviour		
Depressed	Frequent	Infrequent	None
Yes	22	13	25
No	23	10	7

We would write this up in one of two ways:

> Pearson's χ^2 analyses indicated that people who do not exercise were significantly more likely to be depressed than someone who undertakes regular exercise, χ^2 (2) = 6.811, p = .033, φ = .261, odds ratio = 3.734.
>
> OR
>
> Pearson's χ^2 analyses indicated that depressed people are significantly less likely to undertake any form of exercise than those not depressed, χ^2 (2) = 6.811, p = .033, φ = .261, odds ratio = 3.367.

Categorical tests with two variables: Yates' continuity correction

Pearson's χ^2 is an appropriate test, so long as at least one of the variables has at least three groups or categories. If both variables are represented by two groups, Yates' continuity correction is probably the better option. The principle behind this test is the same as χ^2; the difference is that an adjustment is made to account for the number of groups. We still apply the χ^2 calculation, but deduct 0.5 when we examine the difference between observed and expected frequencies. The outcome is still compared with χ^2 distribution tables. If the outcome is significantly higher than the given cut-off point, we know that our observation is significantly different to the null hypothesis.

19.7 Calculating outcomes manually

Yates' continuity correction calculation

To calculate the outcome manually for Yates' continuity correction, we will use some more data from the WASPS research questions. We will compare how depression status (yes or no) varies according to gender (male or female); the data will be displayed in a 2 × 2 cross-tabulation (as shown in Table 19.10):

Table 19.10 Depression vs. gender (observed/expected frequencies/percentages)

Depressed		Gender		Total
		Male	Female	
Yes	Observed	**21** 46.7%	**39** 70.9%	60 60.0%
	Expected	*27* 60.0%	*33* 60.0%	
No	Observed	**24** 53.3%	**16** 29.1%	40 40.0%
	Expected	*18* 40.0%	*22* 40.0%	
	Total	**45**	**55**	**100**

Yates' continuity correction uses the Pearson's x^2 equation, but deducts 0.5 from the difference between observed and expected frequencies before squaring it:

$$\text{So, Yates}' = \sum \frac{((\text{Diff between Observed and Expected}) - 0.5)^2}{\text{Expected}}$$

We take the difference between observed and expected frequencies, ignoring the + or −
In this example, all of the differences between observed and expected frequencies come to 6
(21 − 27 = 6; 39 − 33 = 6; 24 − 18 = 6; 16 − 22 = 6)
We deduct 0.5 from this, and square it: $5.5^2 = 30.25$; we put that in top row of the equation:

$$\text{Yates}' = \frac{30.25}{27} + \frac{30.25}{33} + \frac{30.25}{18} + \frac{30.25}{22} = 5.093$$

df = (no. of rows *minus* 1) × (no. of columns *minus* 1) = 1 × 1 = 1
We look this up in chi-square tables: cut-off point for *df* = 1 is 3.84, where $p = .05$

Yates' = 5.093 > 3.84

There is a significant association between depression status by gender.

How SPSS performs Yates' continuity correction (on a 2 × 2 cross-tabulation)

To demonstrate how to run Yates' continuity correction, we will revisit the research question set by WASPS. You will recall that they were investigating exercise frequency among 100 of their clients (45 men and 55 women) and exploring the impact this may have on depression status. On this occasion, WASPS are focusing on gender and depression – they predict that women are more likely to be depressed than men. Before we analyse this, we should remind ourselves about the variables with this quick summary:

Yates' continuity correction example

Variable 1: Gender – male and female
Variable 2: Mood status – depressed vs. not depressed

Running Yates' continuity correction in SPSS

Using the SPSS file Exercise, mood and gender

Select **Analyze** → **Descriptive Statistics** → **Crosstabs...** (see Figure 19.3) → transfer **Depressed** to **Row(s)** → transfer **Gender** to **Column (s)** → click **Statistics** → (in new window) tick boxes for **Chi-square**, **Contingency coefficient** and **Phi and Cramer's V** → click **Continue** → click **Cells** → (in new window) tick boxes for **Observed** and **Expected** (under **Counts**) → tick **Column** box (under **Percentages**) → tick **Adjusted standardised** box (under **Residuals**) → click **Continue** → click **OK**

We saw the rationale for these selections earlier (for χ^2)

Interpretation of Yates' continuity correction output

Figure 19.10 shows the cross-tabulation table for the variables 'Depressed' and 'Gender'. Compare this to the table we used in Box 19.6.

Depressed*Gender cross-tabulation

			Gender		
			Male	Female	Total
Depressed	Yes	Count	21	39	60
		Expected count	27.0	33.0	60.0
		% within gender	46.7%	70.9%	60.0%
		Adjusted residual	–2.5	2.5	
	No	Count	24	16	40
		Expected count	18.0	22.0	40.0
		% within gender	53.3%	29.1%	40.0%
		Adjusted residual	2.5	–2.5	
Total		Count	45	55	100
		Expected count	45.0	55.0	100.0
		% within gender	100.0%	100.0%	100.0%

Figure 19.10 Cross-tabulation

Figure 19.11 confirms that there is a significant association between depression status and gender, Yates' $(1, N = 100) = 5.093$, $p = .024$. Note that, because we had a 2×2 cross-tabulation, we choose the line reading 'Continuity correction'. In fact, this statistic appears in the SPSS output only when we have just such a scenario. Unlike the outcome for Pearson's χ^2, we can immediately assess the source of association because we have only two pairs to compare (we do not need to refer to the adjusted residuals). The footnote confirms that we have satisfied the condition of not having too many cells with a count of less than 5.

Odds ratios

Similar to Pearson's χ^2, we can express the outcome as an odds ratio, although the calculations are simpler. On this occasion, we can divide one cell directly by another because there are only two rows and columns when using Yates' continuity correlation.

Odds: male vs. female for depressed (O1) $= 39 \div 21 = 1.857$
Odds: male vs. female for not depressed (O2) $= 16 \div 24 = 0.667$
Odds ratio $=$ O1 $\div$ O2 $= 1.857 \div 0.667 = 2.786$

Or we could calculate odds ratios for depression status within gender:

Odds: depressed vs. not for female (O1) = 39 ÷ 16 = 2.438
Odds: depressed vs. not for male (O2) = 21 ÷ 24 = 0.875
Odds ratio = O1 ÷ O2 = 2.438 ÷ 0.875 = 2.786

Either way, these odds ratios suggest that women are nearly three times more likely to be depressed than men.

	Value	df	Asymp. sig. (2-sided)	Exact sig. (2-sided)	Exact Sig. (1-sided)
Pearson chi-square	6.061[a]	1	.014		
Continuity correction[b]	5.093	1	.024		
Likelihood ratio	6.093	1	.014		
Fisher's exact test				.023	.012
Linear-by-linear association	6.000	1	.014		
N. of valid cases	100				

a. 0 cells (.0%) have expected count less than 5. The minimum expected count is 18.00

b. Computed only for a 2 x 2 table

Figure 19.11 Outcome statistics (chi-squared tests)

Explained variance

Figure 19.12 indicates that Cramer's V $\varphi = .246$. Using the calculations that we saw earlier, we can see that 6.05% of the variation in depression status is accounted for by gender ($.246^2 \times 100$).

		Value	Approx. Sig.
Nominal by nominal	Phi	–.246	.014
	Cramer's V	.246	.014
	Contingency coefficient	.239	.014
N. of valid cases		100	

Figure 19.12 Cramer's V

Writing up results

Table 19.11 Cross-tabulation of observed frequencies, depression status by gender

	Gender	
Depression	Male	Female
Yes	21	39
No	24	16

We would write this up as:

Yates' continuity correction analyses indicated that women are significantly more likely to be depressed than men, Yates' (1) = 5.093, p = .024, $\varphi = .246$, odds ratio = 2.786.

Fisher's exact test

In both of the last two examples, expected cell counts were within agreed parameters. As we saw earlier, if more than 20% of the cells in the cross-tabulation have an expected frequency of less than 5, or if any cell count is less than 1, we should use Fisher's exact test (but this is permitted only for 2 × 2 tables in SPSS). This test is not so much a method as an alternative choice of outcome. SPSS automatically provides the Fisher's exact test (in 2 × 2 tables) for us to refer to if the expected cell counts are too low (see Figure 19.11 for an example of how this is reported). If we find low expected cell counts in larger tables, we might need to combine groups.

Categorical tests with more than two variables

We have been exploring cases where we have two variables. Now we will look at what choices we have when there are more than two variables. We saw what a cross-tabulation might look like in this context in Table 19.3. For example, in addition to examining the association between exercise frequency and depression status, our research team (WASPS) might investigate the extent that might additionally vary according to gender. In these next sections we will examine two ways in which we can assess this. The first method that we will look at is a layered χ^2. This is the simpler of the two tests, but probably the least useful, because it will not allow analyses of interactions between variables. To explore interactions, we need to use a procedure called loglinear analysis. We will look at that test shortly.

Layered χ^2

By now, you hopefully know a little more about how we use observed and expected frequencies and proportions to assess associations between variables. In this section, we are adding a further variable, one that we 'layer' alongside the original two variables. We will use the data from Table 19.3, but extend that to include the observed and expected frequencies and percentages. You saw how to calculate these additional elements earlier in the chapter. In this example, we are exploring the association between depression status and exercise frequency, then examining that outcome further according to gender. However, as we will see, analyses from this are somewhat limited. In effect, all we are doing is exploring exercise frequency according to depression status for men and then repeating that for women.

Layered x^2 example

Variable 1: Mood status – depressed vs. not depressed
Variable 2: Exercise frequency group – frequent, infrequent, none
Variable 3: Gender – male vs. female

Table 19.12 Depression status vs. exercise frequency and gender (observed/expected frequencies/percentages)

Gender	Depression		Exercise frequency: Frequent	Infrequent	None	Total
Male	Yes	Observed	**6** 27.3%	**5** 41.7%	**10** 90.9%	**21** 46.7%
		Expected	*10.3* 46.7%	*5.6* 46.7%	*5.1* 46.7%	
	No	Observed	**16** 72.7%	**7** 58.3%	**1** 9.1%	**24** 53.3%
		Expected	*11.7* 53.3%	*6.4* 53.3%	*5.9* 53.3%	
		Total	**22**	**12**	**11**	**45**
Female	Yes	Observed	**16** 69.6%	**8** 72.7%	**15** 71.4%	**39** 70.9%
		Expected	*16.3* 70.9%	*7.8* 70.9%	*4.9* 70.9%	
	No	Observed	**7** 30.4%	**3** 27.3%	**6** 28.6%	**16** 29.1%
		Expected	*6.7* 29.1%	*3.2* 29.1%	*6.1* 29.1%	
		Total	**23**	**11**	**21**	**55**

There appear to be differences between men and women with regard to exercise frequency and depression status. Depressed men seem to be more likely to undertake no exercise at all, while non-depressed men appear to be more likely to undertake frequent exercise. The differences between observed and expected frequencies look quite large. Observed cell percentages (shown in red font) differ from column total percentages (green font). These may reflect differences between observed and expected frequencies. For women, differences between observed and expected frequencies appear smaller. However, we need to test all of this statistically.

Running layered χ^2 in SPSS

The procedure is the same as before, except that we add a layer to the variables that we include:

Using the SPSS file Exercise, mood and gender

Select **Analyze** → **Descriptive Statistics** → **Crosstabs...** (see Figure 19.3) → (in new window) transfer **Depressed** to **Row(s)** → transfer **Exercise** to **Column(s)** → transfer **Gender** to **Layer 1 of 1** → click **Statistics** → (in new window) tick boxes for **Chi-square**, **Contingency coefficient** and **Phi and Cramer's V** → click **Continue** → click **Cells** → (in new window) tick boxes for **Observed** and **Expected** (under **Counts**) → tick **Column** box (under **Percentages**) → tick **Adjusted standardised** box (under **Residuals**) → click **Continue** → click **OK**

Interpretation of output

Figure 19.13 presents the observed and expected frequencies, and relevant percentages. Figure 19.14 shows that there was a significant association between exercise frequency and depression status for men, χ^2 (2, N = 100) = 12.096, p = .002, but not for women, χ^2 (2, N = 100) = 0.041, p = .980. We can use the data in Figure 19.13 to illustrate the source of association for men, by referring to those cells where the adjusted residual is greater than 1.96. Using the methods we learned earlier, we can see that men who undertake frequent exercise are less likely to be depressed than not depressed, while men who take no exercise are *more* likely to be depressed.

Odds ratios

As before, we can express the outcome as an odds ratio, although it is necessary to do this only for men (the data for women showed no significant differences). We need to use the slightly more complex calculation because there are several rows and columns.

Odds: depressed vs. not for no exercise (O1) = 10 ÷ (11−10) = 10.000
Odds: depressed vs. not for frequent exercise (O2) = 6 ÷ (22−6) = 0.375
Odds ratio = O1 ÷ O2 = 10 ÷ 0.375 = 26.667

This means that men who do not exercise are nearly 27 times more likely (OR 26.667) to be depressed than someone who undertakes regular exercise, or...

Odds: no exercise vs. frequent exercise for depressed (O1) = 10 ÷ (21−10) = 0.909
Odds: no exercise vs. frequent exercise for not depressed (O2) = 1 ÷ (24−1) = 0.044
Odds ratio = O1 ÷ O2 = 0.909 ÷ 0.044 = 20.909

This suggests that depressed men are nearly 21 times less likely (OR 20.909) to undertake any form of exercise than those non-depressed men.

Depressed*Exercise*Gender cross-tabulation

Gender				Exercise Frequent	Infrequent	None	Total
Male	Depressed	Yes	Count	6	5	10	21
			Expected count	10.3	5.6	5.1	21.0
			% within exercise	27.3%	41.7%	90.9%	46.7%
			Adjusted residual	−2.6	−.4	3.4	
		No	Count	16	7	1	24
			Expected count	11.7	6.4	5.9	24.0
			% within exercise	72.7%	58.3%	9.1%	53.3%
			Adjusted residual	2.6	.4	−3.4	
	Total		Count	22	12	11	45
			Expected count	22.0	12.0	11.0	45.0
			% within exercise	100.0%	100.0%	100.0%	100.0%
Female	Depressed	Yes	Count	16	8	15	39
			Expected count	16.3	7.8	14.9	39.0
			% within exercise	69.6%	72.7%	71.4%	70.9%
			Adjusted residual	−.2	.1	.1	
		No	Count	7	3	6	16
			Expected count	6.7	3.2	6.1	16.0
			% within exercise	30.4%	27.3%	28.6%	29.1%
			Adjusted residual	.2	−.1	−.1	
	Total		Count	23	11	21	55
			Expected count	23.0	11.0	21.0	55.0
			% within exercise	100.0%	100.0%	100.0%	100.0%
Total	Depressed	Yes	Count	22	13	25	60
			Expected count	27.0	13.8	19.2	60.0
			% within exercise	48.9%	56.5%	78.1%	60.0%
			Adjusted residual	−2.1	−.4	2.5	
		No	Count	23	10	7	40
			Expected count	18.0	9.2	12.8	40.0
			% within exercise	51.1%	43.5%	21.9%	40.0%
			Adjusted residual	2.1	.4	−2.5	
	Total		Count	45	23	32	100
			Expected count	45.0	23.0	32.0	100.0
			% within exercise	100.0%	100.0%	100.0%	100.0%

Figure 19.13 Cross-tabulation

Explained variance

We can also see quite clear gender differences in respect of explained variance (see Figure 19.15). For men, Cramer's V (φ) = .518. This means that 26.8% of all variance in depression status in this sample is explained by exercise frequency. Meanwhile, for women, Cramer's V (φ) = .027 suggesting that only a very small amount of variance is explained (0.07%).

Loglinear analysis

Layered χ^2 has some benefits, but it does have its limitations. The main problem is that it only looks at superficial outcomes. If we want to extend what we have just seen, we need a statistical test that can explore more aspects of the relationship between the variables; loglinear analysis gives us

Gender		Value	df	Asymp. Sig. (2-sided)
Male	Pearson chi-square	12.096[a]	2	.002
	Likelihood ratio	13.399	2	.001
	Linear-by-linear association	10.812	1	.001
	N. of valid cases	45		
Female	Pearson chi-square	.041[b]	2	.980
	Likelihood ratio	.041	2	.980
	Linear-by-linear association	.019	1	.891
	N. of valid cases	55		

a. 0 cells (.0%) have expected count less than 5. The minimum expected count is 5.13.

b. 1 cells (16.7%) have expected count less than 5. The minimum expected count is 3.20.

Figure 19.14 Outcome statistics (chi-squared tests)

that flexibility. In a sense, loglinear analysis is the multivariate version of χ^2. This test will provide us with important information on associations and interactions. An association is represented by two-way relationships – it is a similar relationship to one that we see with correlation (see Chapter 6). An interaction illustrates three-way (or higher) relationships, comparable to what we saw in multi-factorial ANOVA (see Chapters 11 – 14), the difference being that we are dealing with categorical variables and frequency data rather than interval data Loglinear analysis provides an opportunity to examine cross-tabulations of categorical data that might otherwise be too complex to investigate in other ways. Perhaps more importantly you can use the analyses to explore the relative importance of the effect for each variable (and combination of variables) included.

Models in loglinear analysis

Loglinear analysis shares many features with linear regression (see Chapter 16). It provides a goodness of fit test that can be applied to the main effects, associations and interactions. In doing so, we are given a likelihood ratio statistic that can be used to assess how well the loglinear model represents the overall data (so shares similarities with logistic regression – see Chapter 17). The model produces a hierarchy that represents the main effects, associations and interactions. This is examined from the most complex (highest-order interactions) through to the lowest level main effects. This is quite similar to backward elimination in multiple linear regression.

Gender			Value	Approx. Sig.
Male	Nominal by nominal	Phi	.518	.002
		Cramer's V	.518	.002
		Contingency coefficient	.460	.002
	N. of valid cases		45	
Female	Nominal by nominal	Phi	.027	.980
		Cramer's V	.027	.980
		Contingency coefficient	.027	.980
	N. of valid cases		55	

Figure 19.15 Cramer's V

At this stage, it would probably help if we saw what these models and hierarchies mean. We can illustrate these by revisiting the case that we examined for layered χ^2. Our team of sports psychologists (WASPS) had set us the task of examining the effect that exercise might have on mood. We had three variables: exercise behaviour (frequent, infrequent and none) depression status (depressed vs. not depressed), and gender (male vs. female). Those are the three 'main effects' – we saw what we mean by main effects when we explored multi-factorial ANOVA (see Chapters 11–13). However, in loglinear analysis, we also have associations, which represent the relationship between pairs of variables. In this case we have three: 'exercise behaviour vs. depression status', 'exercise behaviour vs. gender' and 'depression status vs. gender'. We will also have a series of interactions – these explore the relationship between three or more variables. In this case we have just the one three-way interaction: 'exercise behaviour vs. depression status' vs. 'gender'. In other cases there may be yet further interactions. In our example, the three-way interaction is the highest and most complex part of the model. We call this the '**saturated model**'.

Removing parts of the model – finding the best fit

The role of loglinear analysis is to try to reduce that model into something simpler, without losing too much of the data (rather like regression). The analysis involves removing parts of the model and assessing how much difference it makes. We start with the highest level (in this case the three-way interaction). We compare that to the next (lower) level in the model, in this case the three two-way associations, by examining the difference between expected and observed frequencies. This is measured in terms of 'predictive power' or the 'goodness of fit'. The process starts with the most complex level, the saturated model in our example; this contains the three-way interaction. The goodness of fit for the saturated model is compared with data from the next level, discarding the saturated model. If the removal of the three-way interactions makes little difference, the more complex model is abandoned in favour of this lower level. We would then proceed to examine the associations. If removal makes a significant difference, the process stops (all we need to have explained is contained within the saturated model). In that case, there is no point analysing the lower orders as they are seen to be part of the higher order.

The effect of the difference is assessed in terms of a '**likelihood ratio**'. This is quite similar to the χ^2 statistic and Yates' continuity correction that we saw earlier. So the difference is examined by how much that likelihood ratio changes. If the change in likelihood ratio is significant it indicates that the removal of the term has made an important change to the fit of the data (so we stop). If it makes no difference, we move on to the two-way associations. In this case there are three associations (representing pairs of variables): 'exercise behaviour vs. depression status', 'exercise behaviour vs. gender' and 'depression status vs. gender'. If removal of any of those associations makes little difference to the likelihood ratio, they are abandoned in favour of the main effects. In our scenario, the 'exercise frequency vs. gender' association might be dropped to allow analysis of the main effects for 'exercise frequency' and 'gender'. If the removal of an association makes a difference to the likelihood ratio, we stop, as that explains what we need to know.

Assumptions and restrictions of loglinear analysis

For once there are very few assumptions and restrictions to concern us. Similar to χ^2, we need to account for low cell counts within the expected frequencies in the cross-tabulation. We should have none that is less than 1, and fewer than 20% of the cells should have expected frequencies of less than 5. Violations of these assumptions are more likely to decrease the chance of finding an effect (increasing Type II errors). There are ways that can adjust for such problems, such as collapsing (or combining) variables, but these methods are subject to several restrictions (you should consult advanced texts if you want to know more about that). However, there is a simpler way to overcome this problem – you can add an arbitrary figure to each cell (usually 0.5). In fact, SPSS does that by default when stating the saturated model in any case.

How SPSS performs loglinear analysis (on a 2 × 3 × 2 cross-tabulation)

To illustrate how we perform this test in SPSS, we return to our research question set by WASPS. The group is keen to explore the extent that exercise frequency has an impact on depression status, and whether there are any differences according to gender. They predict that those clients who undertake less exercise are more likely to be depressed than those who exercise frequently; women are more likely to be depressed than men; and there will be no difference in the frequency of exercise between men and women. Once again, we should remind ourselves about the variables with this quick summary:

Layered x^2 example

Variable 1: Mood status – depressed vs. not depressed

Variable 2: Exercise frequency group – frequent, infrequent, none

Variable 3: Gender – male vs. female

Before we proceed, we should check that we meet the restrictions on expected cell counts. We already have some information about this from our earlier analyses. Figure 19.13 shows that none of the cells in the cross-tabulation has an expected frequency of less than 1. Furthermore, only 1 cell (from a total of 12) contains a count of less than 5; this represents 8% of the cells. This means that we have not violated the restriction and will not need to adjust for small cell sizes.

Running loglinear analysis in SPSS

Using the SPSS file Exercise, mood and gender

Select **Analyze** → select **Loglinear** → select **Model Selection**... as shown in Figure 19.16

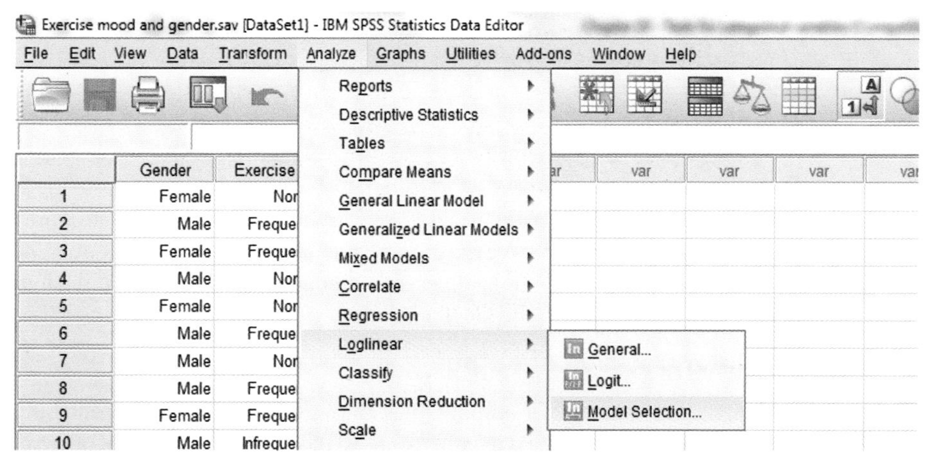

Figure 19.16 Loglinear analysis - step 1

In new window (see Figure 19.17), transfer **Gender**, **Exercise**, and **Depressed** to **Factor(s)** → in **Factor(s)** window, click **Gender(? ?)** → click **Define Range...** → (in new window) enter **1** in **Minimum** and **2** in **Maximum** → click **Continue** → click **Exercise(? ?)** → click **Define Range ...** → enter **1** in **Minimum** and **3** in **Maximum** → click **Continue** → click **Depressed(? ?)** → click **Define Range...** → enter **1** in **Minimum** and **2** in **Maximum** → click **Continue** → click **Options**

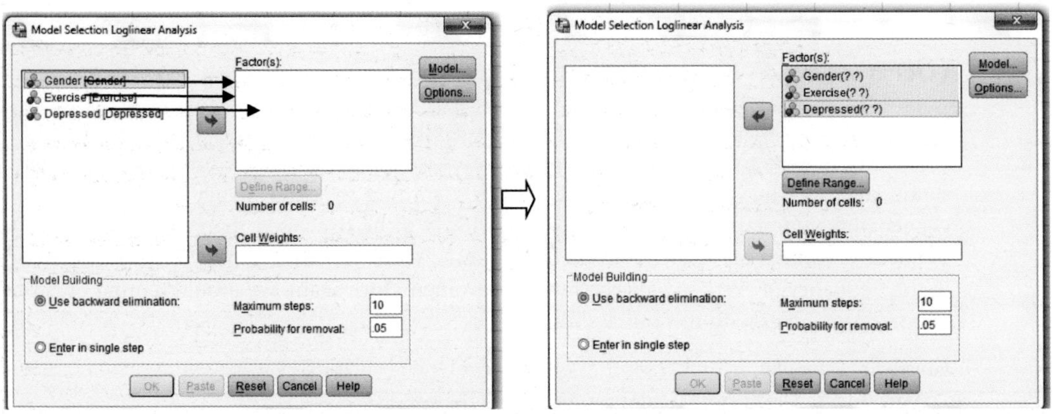

Figure 19.17 Loglinear analysis - step 2

In new window (see Figure 19.18), tick boxes for **Frequencies** and **Residuals** (under **Display**) → tick **Association table** box (under **Display for Saturated Model**) → change **Delta** (under **Model Criteria**) to **0** (see below) → click **Continue** → click **OK**

We have changed the 'Delta' figure to 0, because we did not need to adjust the observed frequencies in the saturated model on this occasion; we did not have too many cells with low expected frequency counts. On other occasions you might need to leave that as default, to allow for that adjustment.

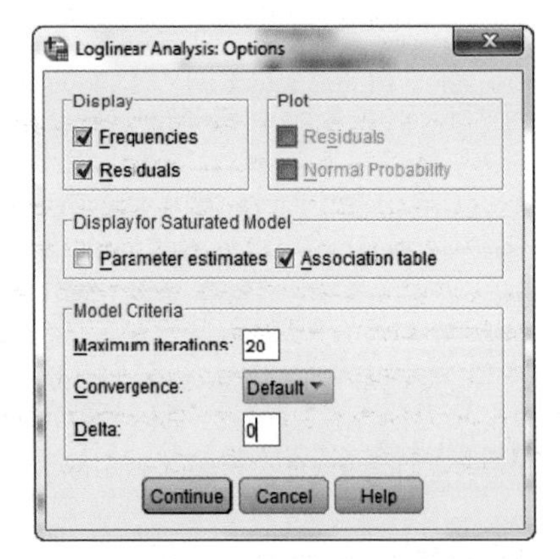

Figure 19.18 Loglinear analysis - step 3

Interpretation of output

SPSS produces a lot of output tables, but we are only really interested in some of them.

Figure 19.19 confirms that the saturated model presents identical observed and expected frequencies. If we reject the saturated model, for a simpler one, there will be differences between observed and expected frequencies. However, we will not want this difference to be too large, otherwise we would have lost too much data. We will look at this again with our final model shortly.

Figure 19.20 reports the Pearson statistic (equivalent to χ^2 that we saw earlier). To illustrate 'perfect fit' of the data, we need chi-square be very low, and non-significant. In our outcome, χ^2 is 0, with an infinite significance (note that when SPSS shows '.' for 'Sig', it means that the probability is '1', which is as high as it can get). We can be very confident that the saturated model represents perfect fit of the data. If χ^2 is large and significant (less than .05), it would mean that there is a poor goodness-of-fit. We will revisit this table again later, too.

Cell counts and residuals

Gender	Exercise	Depressed	Observed		Expected		Residuals	Std. Residuals
			Count[a]	%	Count	%		
Male	Frequent	Yes	6.000	6.0%	6.000	6.0%	.000	.000
		No	16.000	16.0%	16.000	16.0%	.000	.000
	Infrequent	Yes	5.000	5.0%	5.000	5.0%	.000	.000
		No	7.000	7.0%	7.000	7.0%	.000	.000
	None	Yes	10.000	10.0%	10.000	10.0%	.000	.000
		No	1.000	1.0%	1.000	1.0%	.000	.000
Female	Frequent	Yes	16.000	16.0%	16.000	16.0%	.000	.000
		No	7.000	7.0%	7.000	7.0%	.000	.000
	Infrequent	Yes	8.000	8.0%	8.000	8.0%	.000	.000
		No	3.000	3.0%	3.000	3.0%	.000	.000
	None	Yes	15.000	15.0%	15.000	15.0%	.000	.000
		No	6.000	6.0%	6.000	6.0%	.000	.000

a. For saturated models, .000 has been added to all observed cells

Figure 19.19 Saturated model observed and expected frequencies

Goodness-of-fit tests

	Chi-square	df	Sig.
Likelihood ratio	.000	0	.
Pearson	.000	0	.

Figure 19.20 Saturated model goodness-of-fit

Now we start examining the process of the backward elimination.

Figures 19.21 and 19.22 show pretty much the same thing, although the latter does it in slightly more detail (so we will focus on that). Starting with Step 0, the initial analysis shows the saturated model again. The second row in that step shows the effect of removing the highest-order term, the three-way interaction. In this case, there is a significant change in chi-square (7.579. $p = .023$).

K-Way and higher-order effects

	K	df	Likelihood ratio		Pearson		Number of iterations
			Chi-square	Sig.	Chi-square	Sig.	
K-way and higher-order effects[a]	1	11	34.131	.000	32.720	.001	0
	2	7	21.774	.003	20.997	.004	2
	3	2	7.579	.023	7.193	.027	3
K-way effects[b]	1	4	12.357	.015	11.723	.020	0
	2	5	14.195	.014	13.804	.017	0
	3	2	7.579	.023	7.193	.027	0

a. Tests that k-way and higher-order effects are zero.

b. Tests that k-way effects are zero.

Figure 19.21 Effect of removing terms in the model

Step summary

Step[a]		Effects	Chi-square[c]	df	Sig.	Number of iterations
0	Generating class[b]	Gender*Exercise*Depressed	.000	0	0	
	Deleted effect 1	Gender*Exercise*Depressed	7.579	2	.023	3
1	Generating class[b]	Gender*Exercise*Depressed	.000	0	0	

a. At each step, the effect with the largest significance level for the likelihood ratio change is deleted, provided the significance level is larger than .050.

b. Statistics are displayed for the best model at each step after step 0.

c. For 'deleted effect', this is the change in the chi-square after the effect is deleted from the model.

Figure 19.22 Backward elimination process (in more detail)

Therefore, the effect of removing the three-way effect would be significant, so we stop there and do not proceed to the lower-order effects. Step 1 confirms this by only reporting the outcome of the last step; it does not assess the outcome of removing the main effects. And there it ends. Had we rejected the saturated model, we would have needed to proceed to the next (lower) level. If you would like to see how that is done, you can refer to the second example for running loglinear analyses shortly.

Normally at this stage, having arrived at our 'final' model, we should check to see that this still represents a good fit of the data. There would be no point having a simpler model if we have lost too much of the data. As we accepted the first (saturated) model, we have lost nothing. This is confirmed in Figures 19.23 and 19.24 (they are identical to what we saw before we started – see Figures 19.19 and 19.20). This could be very different had we needed to reject the saturated model and defer to lower orders (see later section).

Although we have already found our optimal model, we can still trawl a little further to see where the effects are likely to be. When we set the SPSS parameters we asked for an association table, shown in Figure 19.25.

Cell counts and residuals

Gender	Exercise	Depressed	Observed		Expected		Residuals	Std. residuals
			Count	%	Count	%		
Male	Frequent	Yes	6.000	6.0%	6.000	6.0%	.000	.000
		No	16.000	16.0%	16.000	16.0%	.000	.000
	Infrequent	Yes	5.000	5.0%	5.000	5.0%	.000	.000
		No	7.000	7.0%	7.000	7.0%	.000	.000
	None	Yes	10.000	10.0%	10.000	10.0%	.000	.000
		No	1.000	1.0%	1.000	1.0%	.000	.000
Female	Frequent	Yes	16.000	16.0%	16.000	16.0%	.000	.000
		No	7.000	7.0%	7.000	7.0%	.000	.000
	Infrequent	Yes	8.000	8.0%	8.000	8.0%	.000	.000
		No	3.000	3.0%	3.000	3.0%	.000	.000
	None	Yes	15.000	15.0%	15.000	15.0%	.000	.000
		No	6.000	6.0%	6.000	6.0%	.000	.000

Figure 19.23 Final model observed and expected frequencies

Goodness-of-fit tests

	Chi-square	df	Sig.
Likelihood ratio	.000	0	.
Pearson	.000	0	.

Figure 19.24 Final model goodness-of-fit

Partial associations

Effect	df	Partial chi-square	Sig.	Number of iterations
Gender*Exercise	2	.974	.615	2
Gender*Depressed	1	4.825	.028	2
Exercise*Depressed	2	5.860	.053	2
Gender	1	1.002	.317	2
Exercise	2	7.328	.026	2
Depressed	1	4.027	.045	2

Figure 19.25 Overall effects

Figure 19.25 gives us an indication where the effects are. We already know that there was a three-way interaction. However, now we can look at the associations and main effects. There is no association between gender and exercise (χ^2 (2, N = 100) = .974, p = .615); exercise frequency does not appear to differ according to gender. There is a significant association between gender and depression status (χ^2 (2, N = 100) = 4.825, p = .028); from the earlier (traditional χ^2) analyses we can conclude that women are more likely to be depressed than men. There is a near-significant association between exercise and depression status (χ^2 (2, N = 100) = 5.860, p = .053). Our earlier investigations suggested that people who undertake no exercise are more likely to be depressed than those who take part in frequent exercise. This outcome is not so well supported using loglinear analysis, so we should probably treat that with caution.

The 'main effects' indicate which variables are the strongest predictors of outcome in the model. Using the chi-square data, it would appear that the strongest variable is 'exercise' (χ^2 = 7.328, p = .026), followed by 'depression status' (χ^2 = 4.027, p = .045). Gender does not appear to significantly contribute to the overall model (χ^2 (2, N = 100) = 1.002, p = .317).

Drawing it all together

All of this is very useful in telling us the relationship between the variables and the impact that they have on each other. SPSS gives us a great deal of information about interactions, associations and main effects. However, the output is less than forthcoming on the source of differences that we might see. Figure 19.23 helps a little with cell counts, but that is not as clear as it might be. To overcome that lack of detail, we could run a series of χ^2 tests for each pair of variables (rather like we did earlier in the chapter).

Writing up results

To account for what we have just found we should report this using a suitably adapted table of data (Table 19.13) (based on Figure 19.13) and some appropriate narrative.

We can use these data (along with the information we gleaned from simpler chi-squared analyses earlier, including odds ratio calculations) to write up the results as follows:

A three-way loglinear analysis was used to explore the relationship between exercise frequency (frequent, infrequent or none) in respect of depression status (depressed or not depressed) and

Table 19.13 Exercise frequency vs. depression status by gender

	Exercise behaviour						
	Frequent		Infrequent		None		All
	N	%	N	%	N	%	
Male							
Depressed	6	27.3	5	41.7	10	90.9	21
Not depressed	16	72.7	7	58.3	1	9.1	24
Female							
Depressed	16	69.6	8	72.7	15	71.4	39
Not depressed	7	30.4	3	27.3	6	28.6	16

according to gender. There was a significant three-way interaction, supporting the saturated model (χ^2 (2, N = 100) = 7.579, p = .023). Only one of the two-way associations was significant: gender vs. depression status (χ^2 (2, N = 100) = 4.825, p = .028); women were almost three times more likely to be depressed than men (odds ratio (OR) = 2.786. The association between exercise frequency and depression status neared significance (χ^2 (2, N = 100) = 5.860, p = .053); there was a tendency for depressed people to be more likely not to undertake any exercise (OR 3.367). There was no significant association between exercise frequency and gender (χ^2 (2, N = 100) = .974, p = .615). The main effect analyses indicated that exercise frequency was the strongest predictor of outcome (χ^2 (2, N = 100) = 7.328, p = .026), followed by depression status (χ^2 (2, N = 100) = 4.027, p = .045); gender did not appear to significantly contribute to the overall model (χ^2 (2, N = 100) = 1.002, p = .317). However, separate χ^2 analyses suggested that the effect of exercise frequency on depression status was stronger for men than for women; depressed men were almost 21 times less likely (OR 20.909) to undertake any form of exercise than those non-depressed men; no differences were found for women.

Loglinear analysis when saturated model is rejected

When we explored the three categorical variables from our research example with loglinear analysis earlier in the chapter, we were able to accept the 'saturated' model and did not need to proceed to the lower levels of the model. However, there will be instances where you need to do that, so it would be useful to see what to do when it does happen.

For this example we need a new set of data. In this scenario we are investigating a group of 180 patients in respect of three variables: self-reported sleep satisfaction (good or poor), whether the person is currently taking sleep medication (yes or no), and the current psychiatric diagnosis (major depressive disorder or generalised anxiety disorder – GAD). The procedure for running loglinear analysis in SPSS is the same as we saw earlier, so we will go straight to the output.

Figures 19.26 and 19.27 confirm what we have seen before about the saturated model – it is the 'perfect fit' between observed and expected frequencies, illustrated by a low, non-significant, Pearson outcome.

Starting at Step 0, the initial analysis shows the saturated model (see Figures 19.28 and 19.29). The second row in that step shows the effect of removing the highest-order term, the three-way interaction. In this case, there is very little change in chi-square (.001) and we can see that this is non-significant (p = .971). Therefore, the effect of change is minimal so, this time, we can proceed to the next lowest order, the three two-way associations. This is shown in Step 1. The first part of that step confirms the starting point (which is the same as the deleted effect in Step 0). The second part of Step 1 shows the effect of removing each of the associations. In all three cases chi-square is high and significant (diagnosis vs. sleep satisfaction: χ^2 = 11.817,

Cell counts and residuals

Diagnosis	Sleep satisfaction	Sleep medication	Observed		Expected		Residuals	Std. residuals
			Count[a]	%	Count	%		
Depressed	Good	Yes	18.000	10.0%	18.000	10.0%	.000	.000
		No	8.000	4.4%	8.000	4.4%	.000	.000
	Poor	Yes	32.000	17.8%	32.000	17.8%	.000	.000
		No	76.000	42.2%	76.000	42.2%	.000	.000
GAD	Good	Yes	22.000	12.2%	22.000	12.2%	.000	.000
		No	4.000	2.2%	4.000	2.2%	.000	.000
	Poor	Yes	10.000	5.6%	10.000	5.6%	.000	.000
		No	10.000	5.6%	10.000	5.6%	.000	.000

a. For saturated models, .000 has been added to all observed cells

Figure 19.26 Saturated model observed and expected frequencies

Goodness-of-fit tests

	Chi-square	df	Sig.
Likelihood ratio	.000	0	.
Pearson	.000	0	.

Figure 19.27 Saturated model goodness-of-fit

p = .001; diagnosis vs. sleep medication: $\chi^2 = 4.779$, p = .029; sleep medication vs. sleep satisfaction: $\chi^2 = 20.166$, p < .001). The effect of removing the associations from the model would be significant, so we stop there. Step 2 confirms this by only reporting the outcome of the last step; it does not assess the outcome of removing the main effects.

Figure 19.30 now shows that there is a difference between the observed and expected frequencies. To examine the extent of change we look at the residuals. In Figure 19.26 the residuals were 0; now they are a little more than that, but not much more. We do not want there to be too big a change here, otherwise our model has lost too much data in the process. Intuitively, we think that these changes are quite small, but we need Figure 19.31 to confirm that. In that output table we can see that chi-square is still small (.001) and that this is non-significant (p = .971), but then we saw the same data in Figure 19.29. Either way, it is clear that the final model is still a good fit to the original data.

K-Way and higher-order effects

	K	df	Likelihood ratio Chi-square	Likelihood ratio Sig.	Pearson Chi-square	Pearson Sig.	Number of iterations
K-way and higher-order effects[a]	1	7	135.737	.000	170.578	.000	0
	2	4	56.271	.000	70.820	.000	2
	3	1	.001	.971	.001	.971	4
K-way effects[b]	1	3	79.466	.000	99.758	.000	0
	2	3	56.270	.000	70.819	.000	0
	3	1	.001	.971	.001	.971	0

a. Tests that k-way and higher-order effects are zero

b. Tests that k-way effects are zero

Figure 19.28 Effect of removing terms in the model

Step summary

Step[a]			Effects	Chi-square[c]	df	Sig.	Number of iterations
0	Generating class[b]		DSMIV*sleeps at*sleepmeds	.000	0	.	
	Deleted effect	1	DSMIV*sleeps at*sleepmeds	.001	1	.971	4
1	Generating class[b]		DSMIV*sleeps at, DSMIV*sleep-meds, sleeps at*sleep-meds	.001	1	.971	
	Deleted effect	1	DSMIV*sleeps at	11.817	1	.001	2
		2	DSMIV*sleep-meds	4.779	1	.029	2
		3	sleeps at*sleep-meds	20.166	1	.000	2
2	Generating class[b]		DSMIV*sleeps at, DSMIV*sleep-meds, sleeps at*sleep-meds	.001	1	.971	

a. At each step, the effect with the largest significance level for the likelihood ratio change is deleted, provided the significance level is larger than .050

b. Statistics are displayed for the best model at each step after step 0

c. For 'deleted effect', this is the change in the chi-square after the effect is deleted from the model

Figure 19.29 Backward elimination process (in more detail)

Cell counts and residuals

Diagnosis	Sleep satisfaction	Sleep medication	Observed Count	Observed %	Expected Count	Expected %	Residuals	Std. residuals
Depressed	Good	Yes	18.000	10.0%	18.061	10.0%	−.061	−.014
		No	8.000	4.4%	7.968	4.4%	.032	.011
	Poor	Yes	32.000	17.8%	31.942	17.7%	.058	.010
		No	76.000	42.2%	76.028	42.2%	−.028	−.003
GAD	Good	Yes	22.000	12.2%	21.939	12.2%	.061	.013
		No	4.000	2.2%	4.032	2.2%	−.032	−.016
	Poor	Yes	10.000	5.6%	10.058	5.6%	−.058	−.018
		No	10.000	5.6%	9.972	5.5%	.028	.009

Figure 19.30 Final model observed and expected frequencies

Goodness-of-fit tests

	Chi-square	df	Sig.
Likelihood ratio	.001	1	.971
Pearson	.001	1	.971

Figure 19.31 Final model goodness-of-fit

Although we found our optimal model, we can still look a little further at the data to see where the effects are likely to be. Figure 19.32 suggests where the effects are. We already know that there was no three-way interaction, which is why the loglinear analysis proceeded to the two-associations. However, there are significant outcomes for all three associations: 'diagnosis vs. sleep satisfaction' ($\chi^2 = 11.817$, $p < .001$), 'diagnosis' vs. 'sleep medication' ($\chi^2 = 4.779$, $p = .029$), and 'sleep satisfaction' vs. 'sleep medication' ($\chi^2 = 20.166$,

Partial associations

Effect	df	Partial chi-square	Sig.	Number of iterations
DSMIV*sleepsat	1	11.817	.001	2
DSMIV*sleepmeds	1	4.779	.029	2
sleepsat*sleepmeds	1	20.166	.000	2
DSMIV	1	44.925	.000	2
sleepsat	1	33.118	.000	2
sleepmeds	1	1.424	.233	2

Figure 19.32 Overall effects

$p < .001$). We could run separate χ^2 tests to illustrate the exact nature of those associations. The final three rows explore the main effects. Once again these show us which variables are the strongest in the model. Diagnosis appears to be the strongest ($\chi^2 = 44.925$, $p < .001$), followed by sleep satisfaction ($\chi^2 = 33.118$, $p < .001$). Sleep medication does not appear to contribute at all ($\chi^2 = 1.424$, $p = .233$).

Chapter summary

In this chapter we have explored tests for categorical variables. At this point, it would be good to revisit the learning objectives that we set at the beginning of the chapter.
You should now be able to:

- Recognise that we use these tests when all of the variables are represented by categorical variables, each containing at least two distinct groups, the present frequency data.
- Understand that cross-tabulations are often used to display the data, incorporating observed and expected frequencies and percentages.
- Know how to calculate the outcomes for cross-tabulations.
- Appreciate when it is appropriate to apply each of the following tests:
  - Chi-squared (χ^2) test is used where there are two variables, where at least one of them has three or more groups (so long as assumptions for expected cell counts are met - see Fisher's exact test).
  - Yates' continuity correction is used where both variables have two groups (and subject to the assumption concerning expected cell counts).
  - Fisher's exact test is used instead of χ^2 or Yates' continuity correction, when more than 20% of the cells of an expected cell count of fewer than 5.
  - Layered χ^2 test is used where there are three or more variables (each with at least two distinct groups). In effect, this test simply presents traditional χ^2 tests across two of the variables, but does so for each group of a third variable. No interactions can be explored between those variables.

    Loglinear analysis extends layered χ^2 by exploring the data according to interactions between several (three or more) variables, associations between pairs of variables, and main effects across single variables.

- Understand that there are only a few assumptions and restrictions for these tests. These largely relate to expected cell counts.
- Manually calculate outcome for Pearson's χ^2 and Yates' continuity correction (using maths and equations).
- Perform analyses using SPSS.
- Understand how to present the data and report the findings.

Research example

It might help you to see how loglinear analysis has been reported in some published research. In this context you could read the following paper (an overview is provided below):

Bhattacherjee, A., Chau, N., Sierra, C.O., Legras, B., *et al.* (2003). Relationships of job and some individual characteristics to occupational injuries in employed people: a community-based study. *Journal of Occupational Health, 45 (6)*, 382–391.DOI: http://dx.doi.org/10.1539/joh.45.382

If you would like to read the entire paper you can use the DOI reference provided to locate that (see Chapter 1 for instructions).

In this research, the authors examined 2,562 French workers in respect of the frequency of industrial injuries that they experienced within a two-year period and the risk factors associated with those injuries. The annual incident rate of at least one industrial injury was 4.45% among the cohort. There were nine variables: gender (male or female); job type ('executives', 'intellectual professionals and teachers', 'manual labourers', 'office/administration employees', 'farmers, craftsmen and tradesmen', and' technicians or other'); age (29 or under, 30–39, 40–49, or 50 or over); body mass index (kg/m^2: 19 or under, 20–24, or 25 or over); smoking habit (current/ex-smokers or non-smokers); excess alcohol use (yes or no); regular psychotropic drug use (yes or no); presence of a disease (at least one disease or no disease); and occupational injury (presence or absence).

The loglinear analysis indicated that the saturated model was rejected. Ultimately, an unsaturated model including two-way associations (which the authors refer to as interactions) was accepted. This involved six of the factors: injury, age, gender, job type, disease and drugs. All associations between these factors were significant (in many cases, highly so), except gender vs. disease; all of the main effects were significant. In relation to occupational injury, the strongest risk factor was job type, followed by gender, psychotropic drug use, age and (finally) disease. In terms of odds ratios (OR) for experiencing at least one industrial injury, labourers (OR 6.40) and 'farmers, craftsmen, and tradesmen' (OR 6.40) were more than six times more likely than executives; 'employees' (OR 2.94) and 'technicians' (OR 3.14) were approximately three times more likely than 'executives, intellectual professionals and teachers'; men were twice as likely as women (OR 1.99); younger people (those aged 29 or younger) were nearly twice as likely as those aged 50 or older (OR 1.70); and those who had some form of disease were one-and-a-half times more likely than those who did not (OR 1.50).

This is quite a data-heavy study, but it illustrates the use of loglinear analysis quite well.

Extended learning task

You will find the SPSS data set associated with this task on the website that accompanies this book. (Excel spreadsheets are not very helpful in this instance.) You will also find the answers there.

Following what we have learned about Pearson's χ^2 and loglinear analyses, answer the following questions and conduct the analyses in SPSS. You will use one data set to explore both outcomes. The data reflect outcomes from a (fictitious) study of 129 women. The analyses focus on self-esteem (high or low), locus of control (internal or external) and hair colour (red, brown or blonde). It has been said that 'blondes have more fun'. Will this be reflected in higher self-esteem for this group compared with (say) redheads? Locus of control measures the extent that people believe that they have control over their own lives (internal) or they feel that life outcomes are forced upon them by others (external). We might expect people with low self-esteem to have an external locus of control, but is hair colour also involved in this relationship?

Open the SPSS data set **Hair colour**

1. Using an appropriate test, compare self-esteem and locus of control.
   a. Describe the pattern of data shown by the cross-tabulation.
   b. Present the statistical outcome and state whether this shows a significant difference between the variables.
   c. Calculate the odds ratio and variance for this outcome.
2. Using an appropriate test, compare hair colour and self-esteem.
   a. Describe the pattern of data shown by the cross-tabulation.
   b. Present the statistical outcome and state whether this shows a significant difference between the variables.
   c. Calculate the odds ratio and variance for this outcome.
3. Perform a layered chi-squared test using self-esteem and hair colour as the main analyses, and locus of control for the layer.
   a. Report how the relationship between self-esteem and hair colour differs according to locus of control.
4. Perform a loglinear analysis, using all of the variables.
   a. Describe the saturated model.
   b. Indicate whether the saturated model, or some other (lower, unsaturated) model, should be accepted.
   c. Describe all interactions, two-way associations and main effects.

20 FACTOR ANALYSIS

Learning objectives

By the end of this chapter you should be able to:

- Recognise when it is appropriate to use factor analysis
- Understand the different types of factor analysis (particularly principal components analysis)
- Be familiar with the stages of extracting factors in the process of principal components analysis
- Appreciate assumptions and restrictions associated with principal components analysis
- Perform analyses using SPSS
- Understand how to present the data and report the findings

What is factor analysis?

Factor analysis is a series of procedures that have two key purposes: data reduction and exploring theoretical structure. If we have a large questionnaire, with many questions, we may wish to 'reduce' that into sub-themes, to make analysis easier. On the other hand, we may choose to explore the structure of a questionnaire by examining its components. Typically, a questionnaire will explore one central theme (such as reported quality of life). However, within that theme, there may be several sub-themes (such as relationships, job satisfaction, physical health and mental well-being). Factor analysis seeks to explore the presence of those themes. There are several types of factor analysis but, most commonly, these are represented by two methods: **principal component analysis** and **principal axis factoring**. We explore the difference between these methods in more depth later. Whichever method is used, the outcome is explored by investigating correlation between responses to questions; those that are answered in a similar way to each other are grouped together into groups that we call 'factors '. It is the way that this is undertaken that differs between the methods. As we have seen in the regression chapters (16 and 17), when we assess correlation we also take account of variance. Principal component analysis analyses all the variance in the items, while principal axis factoring examines shared variance among the items (it estimates how much of the variability is due to common factors). We focus on principal components analysis in this chapter.

Research question for factor analysis

To illustrate factor analysis, we will use a research question that we will develop throughout the chapter. A group of researchers (the Mental Health Research Group; MHRG) are exploring what aspects might contribute to perceptions regarding quality of life and mood. MHRG devise a questionnaire containing 20 questions that they feel might define these perceptions. A full list of the questions can be seen in Table 20.2. The questions are written to explore four themes that could represent quality of life: mental well-being, relationships, job satisfaction and physical health. For example, 'I have frequent mood swings' (Question 1) might be measuring mental well-being; 'I am arguing with my partner a lot' (Question 6) could be tapping into perceptions concerning relationships; 'My mood is affecting my work' (Question 4) appears to reflect thoughts about work; and 'I feel dizzy and nauseous all the time' (Question 9) may be illustrating perceptions about physical health.

20.1 Take a closer look
Summary of factor analysis example

Primary theme of questionnaire: quality of life and mood perceptions
Possible sub-themes: relationships, job satisfaction, physical health and mental well-being

Theory and rationale

What is factor analysis used for?

Factor analysis is useful for a number of tasks. In a large questionnaire (perhaps about quality of life), several items may tap into separate sub-themes (such as relationships, job satisfaction, home life, social life, physical health, mental health, etc.). Factor analysis can help identify the presence of these sub-themes. In other cases, factor analysis can be used to identify psychological constructs from a scale, interview or questionnaire. Historically, factor analysis was originally used to identify 'traits' within personality scales. Most famously, this was undertaken in Eysenck's (1953) extraversion-introversion and neuroticism studies, and Cattell's (1966) 16-factor personality scale.

Factor analysis can be used to reduce the number of variables in a data set, to make it more manageable. Initially, every question is technically a variable. We may want to reduce that into something that represents a series of themes or constructs. Those constructs become the 'variables'; because we cannot actually 'see them', we call these '**latent variables**'. When we examined multiple linear regression (Chapter 16), we learned that we should not have too many predictor variables (as a ratio to the sample size). We also saw that we want to avoid predictors being too highly correlated with each other. Factor analysis can identify those variables that have high multi-collinearity and suggest those which can be combined. The reduced variable list can be re-analysed in regression. Whichever method is used, the end result is the same; a large set of variables is reduced to fewer (latent) variables.

Measuring validity

When we conduct research it is very important that we ask the right questions; otherwise we might get the wrong answer. Validity describes the extent that we are measuring what we claim to be. We explored the concept of validity in some depth in Chapter 5. Factor analysis examines something called **construct validity**, which is the degree to which a theory has been demonstrated in a test. For example, we could ask someone to report their IQ. If it is high we might claim that the person is intelligent. However, such an assumption might lack construct validity because we cannot be certain that IQ really does measure intelligence. Factor analysis measures construct validity by performing statistical analyses on the internal structure of a questionnaire. Part of that analysis involves assessing relationships between responses to different questions across the questionnaire; we will see more about that a little later. Earlier, we saw that our research group (MHRG) were looking to explore what constructs contribute to perceived quality of life. They have designed a questionnaire (with 20 questions) that they anticipate will uncover five themes: relationships, job satisfaction, home life, physical health and mental health. Through correlation, variance and mathematical 'rotation' principal components analysis examines the data and produces 'factors' that represent those latent variables. Hopefully, the located factors will be equivalent to the ones that MHRG predicted.

20.2 Take a closer look
What are latent variables?

Latent variables can be regarded as those that cannot be directly observed. They are often used in psychological research to represent constructs, such as perceptions. The latent variable may comprise several factors, often measured from responses to questionnaires or interviews. Mathematical models, such as factor analysis, explore the structure of questionnaires, and particularly the pattern of responses to those questions, to propose these latent variables. Each one should represent a facet of that questionnaire.

Methods for extracting factors

There are several 'versions' of factor analysis, each using a slightly different way to extract factors. Most of the arguments that differentiate these methods probably concern only the most dedicated statisticians. However, we should explore them briefly. All of the methods seek to explore the extent that responses vary. Each participant is likely to respond to questions differently to the next one. The amount that those responses differ is called 'variability'. The aim is to 'explain' as much of that variability as possible. Explained variability represents the response variation that we can confidently attribute to discernible patterns; anything left is 'error' variability.

Principal components analysis (PCA) explores the structure of questions within a questionnaire and seeks to locate any underlying latent variables. There are no specific hypotheses, but we would expect a large set of questions to be reliably reduced into a smaller set of factors. Principal axis factoring (PAF) is similar, in that it is also an exploratory method, but it uses a different method to find the latent variables. In doing so, PAF makes different assumptions about how the variables and extracted factors are related. PCA and PAF often analyse many items, but always within a single questionnaire. The analyses may produce several factors. Canonical factor analysis examines the relationship between two sets of variables. We might compare the factors produced from the assessed questionnaire against some other observable outcome. For example, the factors from the quality of life questionnaire might be compared to clinicians' ratings of the patient's mood. This method is also sometimes called **Rao's canonical factoring**. Other methods of factor extraction (such as image factoring, alpha factoring, least squares and maximum-likelihood) are also available in SPSS, but these are not used very often. If you are interested you should seek more information about them in more in-depth sources.

As we will see later, the variables in factor analysis should be 'numerical' (at least ordinal and preferably interval); we cannot use categorical data (you should refer to Chapter 5 if you need to remind yourself about what those terms mean). If we need to explore relationships between categorical (or nominal) variables, we can use **correspondence analysis**. For example, we could investigate two variables: UK location (England, Wales, Scotland and Northern Ireland) and haggis consumption (low, moderate and high). We might find that haggis consumption does not differ between people from England and Wales. Meanwhile, we may discover that high haggis consumption is associated with people from Scotland, but low consumption is associated with people from England. We saw similar methods of categorical data analysis in Chapter 19; correspondence analysis takes that a stage further. However, we do not explore that in this book.

20.3 Take a closer look

Summary of (most common) factoring methods

You may find the following table useful as a summary of methods we can use for factoring

Table 20.1 Factoring methods

Method	When used
Principal components analysis	Exploratory method, to locate themes (latent variables) from several (numerically rated) items in a single questionnaire. PCA analyses all the variance in the items.
Principal axis factoring	Also exploratory method, similar to PCA, except that PAF examines the common variance between items.
Canonical factor analysis	Examines relationship across two separate sources, such as factors from a questionnaire and another confirmatory source (such as clinical observation)
Correspondence analysis	Confirms relationship between factors of two categorical variables

Principal components analysis vs. principal axis factoring

The most popular methods of factor analysis are principal components analysis (PCA) and common factor analysis (principal axis factoring; PAF). The differences between PAF and PCA largely focus on the way the procedures deal with variability in the variables: PCA uses 'rotation' to maximise the amount of explained variance (by accounting for all of the variability in those variables); PAF assesses the extent to which the variability is due to common factors (otherwise known as communality). PAF uses mathematical models and is very strict in the way that unique and error variance is handled. PAF is often used to confirm hypotheses about the structure of a questionnaire. PCA is less cautious, where the aim is to reduce a large data set into more manageable factors, perhaps ahead of subsequent analyses (such as multiple linear regression). Often, this is conducted as part of the initial process of the research, perhaps to establish the validity of the questions that need to be asked to measure the constructs. Confirmatory analyses are likely much later in the process, and may be used to test theory generated by the factors or constructs. Hypothesis testing may be undertaken using advanced statistical tests, such as structural equation modelling. This chapter, like many introductory and moderately advanced sources, will focus on exploratory PCA.

Correlation in PCA

As we saw in Chapter 6, correlation explores the extent that two variables vary with each other. The strength of the relationship between those variables is measured on a scale of 0 (no correlation) to ± 1 (perfect correlation). Positive correlation suggests that, as the scores on one variable increase, the scores on the other variable also increase; we might observe this in the relationship between temperature and ice cream sales. Negative correlation indicates that, as the scores on one variable increase, the scores on the other variable decrease; we might see this in the relationship between temperature and the amount of clothes worn.

Correlation can also be used to examine the extent that people respond in a similar fashion between two questions. If they are similar, it could be argued that those questions tap into the same theme (or construct). For example, responses between the questions 'I have frequent mood swings' and 'I don't feel happy right now' will probably be quite similar; we might find a correlation of $r = .75$ between the responses to these questions. On the other hand, responses to the questions 'My mood is affecting my work' and 'I don't like Marmite' may be less similar; we might find a correlation of $r = .09$ in this instance.

In reality, patterns of responses may be a little less predictable than these examples. Throughout this chapter, we will illustrate principal components analysis by examining a questionnaire that uses 20 questions (set by our research group, MHRG) that measure several aspects of quality of life and mood. All of the questions in this questionnaire offer a response scale of 1 (definitely not true) to 5 (definitely true). The questionnaire has been completed by 586 participants.

Before we explore the sub-themes of this questionnaire, we should look at the correlation between the 20 questions, based on the responses from this sample (see Table 20.2). From that, we can see that answers to Question 1 (I have frequent mood swings) have a low negative correlation with Question 2 (I go home early because of headaches); $r = .141$. On the other hand the correlation between Question 1 and Question 4 (My mood is affecting my work) shows $r = .401$, indicating a moderate relationship. It would appear that the first two questions are not related, but the second two might be (both questions appear to be measuring mood perceptions).

Variance in factor analysis

In addition to correlation between two variables, it is also useful to measure how much variance is explained. That is done by simply squaring the correlation In our example, the correlation between answers to Questions 1 and 4 is .401; so the variance is $(0.401)^2 = 0.161$. This means that 16.1% of the variance in Question 4 is explained by variations in answers to Question 1; and vice versa.

How principal components analysis uses correlation

PCA loads variables onto factors according to patterns of correlation, grouping highly correlated items and ignoring the poorly correlated ones. When all of the variables are loaded onto a factor, a calculation is performed to illustrate the correlation between variables and this new factor; this is known as the **factor loading**. In reality, this is extremely complex, so you may be pleased

Table 20.2 Correlation matrix for all quality of life and mood questions

	1	2	3	4	5	6	7	8	9	10	11	12	13	14	15	16	17	18	19
1 I have frequent mood swings																			
2 I go home early because of headaches	.141																		
3 I have lost interest in my job	.330	.380																	
4 My mood is affecting my work	.401	.205	.375																
5 I feel like my thoughts are not my own	.382	.153	.295	.398															
6 I am arguing with my partner a lot	.176	.080	.212	.261	.266														
7 I am losing friends due to my mood	.279	.133	.347	.415	.322	.519													
8 I feel anxious	.384	.087	.277	.391	.298	.263	.358												
9 I feel dizzy and nauseous all the time	.131	.347	.352	.135	.143	.176	.121	.005											
10 I don't feel happy right now	.387	.187	.376	.391	.363	.344	.347	.609	.144										
11 I want to be alone all the time	.354	.142	.309	.302	.287	.449	.479	.342	.145	.434									
12 I am shutting people out	.336	.147	.366	.314	.265	.429	.473	.352	.124	.376	.451								
13 I am very introspective right now	.184	.173	.296	.338	.258	.319	.384	.275	.217	.296	.314	.361							
14 I feel unmotivated	.370	.126	.306	.415	.311	.288	.450	.612	.042	.570	.374	.406	.347						
15 I find it hard to talk to people	.317	.204	.286	.360	.329	.458	.485	.265	.171	.351	.506	.458	.235	.368					
16 I am taking more days off sick	.133	.240	.341	.150	.104	.132	.232	.137	.289	.160	.145	.220	.178	.098	.205				
17 I am not completing tasks at work	.161	.218	.299	.247	.146	.118	.172	.140	.190	.223	.141	.227	.158	.153	.202	.247			
18 I feel like my colleagues hate me	.313	.225	.413	.384	.298	.233	.476	.314	.132	.361	.375	.383	.189	.365	.431	.255	.412		
19 I feel that my health is suffering	.134	.258	.258	.115	.175	.215	.224	.099	.281	.176	.158	.210	.165	.114	.204	.251	.156	.147	
20 I always feel ill	−.037	.102	.188	.067	.045	.144	.090	.063	.247	.126	.092	.092	.071	.024	.103	.142	.031	.124	.261

to know that we will not explore how all of that works! However, if you would like to know more about that process you could refer to the very useful section provided by Dancey and Reidy (2008), pp. 419–423.

It is also worth noting that PCA bases the factor loading on the strength of the relationship; it does not matter whether that correlation is positive or negative. Quite often, reverse scoring is used in questionnaires (partly to dissuade participants going into 'response mode', simply ticking 'quite satisfied' for all questions, perhaps). This is a realistic reflection of how questionnaires are designed and PCA is able to deal with that (by ignoring the direction of relationship and focusing on the strength of that association). Other statistical tests (such as reliability analysis in Chapter 21) are not so helpful.

Finding factors (the hard way)

The mathematics involved in calculating factors, and the associated loading of correlation, is ludicrously complicated. I am not even going to attempt to explain it, mostly because it would mean that I would have to understand it too! In essence the calculations compare, multiply and otherwise manipulate matrices of data that represent the original correlations along with issues related to regression If you really want to know how to calculate the maths, you should refer to Tabachnick and Fidell (2007).

Factor analysis: some terminology

Although we might have dispensed with maths and formulae on this occasion, there is still a lot to learn about terminology before we interpret outcomes in SPSS.

Correlation matrix

The correlation matrix presents the magnitude of correlation between items. We saw an example in Table 20.2. We need at least reasonable correlation between the items (in excess of $r = .30$), otherwise there is no point in looking for relationships. However, we do not want too much either; we should avoid multi-collinearity (see assumptions and restrictions later).

Factor loading

Principal components analysis examines the correlation matrix to assess the relationship between groups of variables. Those items most strongly correlated are grouped into factors. The correlation between the item and its factor is called the factor loading. As we saw earlier, the squared factor loading equates to the explained variance between the item and its factor. The magnitude of the factor loading will guide us when we interpret the outcome. Once we have identified the factors, we explore explained variance through '**eigenvalues**' and 'communalities'.

Eigenvalues

Eigenvalues measure the amount of variance that has been explained by the factor. The higher that value the more important it is. A high eigenvalue explains much of the variance between the items and the factor; a low eigenvalue explains very little (so can be ignored). You do not need to know how the eigenvalue is calculated, but you do need to know the minimum value that we can usefully interpret. Although we should aim to explain as much as possible, we need to do so efficiently; we rarely explain all of the variance for all of the items. Ultimately, we must set a cut-off point that is the best compromise. As ever, there is much debate about where we should set the cut-off point. Kaiser (1960) recommends that we should only include eigenvalues in excess of 1 (which is now the commonly accepted cut-off point). Indeed, SPSS uses that criterion to assess the factors. However, Kaiser goes further to say that this cut-off point is valid only where there are less than 30 (initial) variables, and where the post-extraction communalities exceed

0.7 (see later). However, some statisticians find Kaiser's suggestion a little severe. For example, Jolliffe (2002) recommends that a cut-off point of 0.7–0.8 is appropriate.

Another useful guide to where we should set a cut-off point for eigenvalues is provided by something called a **scree plot**. Not only is this a good visual indicator, it is often used by default if we fail to meet Kaiser's criteria (mentioned just now). The scree plot is most useful in larger samples (exceeding 300 participants). It was devised by Cattell (1966); an example (using our data) is shown in Figure 20.1. The number of located factors is shown on the X axis (along the bottom), while the eigenvalues are presented on the Y axis (along the side). The scree plot provides a visual guide to a good cut-off point, which is judged to be where the line starts to 'level out'. More precisely, we call this the 'point of inflexion'; it indicates where the slope of the line changes dramatically. Initially, there is a very steep descent in this curve, followed by a shallower decline, before the line plateaus. In our example, this is probably around the fourth or fifth component here; that might suggest that variables could be reduced to four or five factors. The point at which the curve levels out could also be measured by its eigenvalue. This appears to be where we have an eigenvalue of 1, which reinforces the suggestion made by Kaiser.

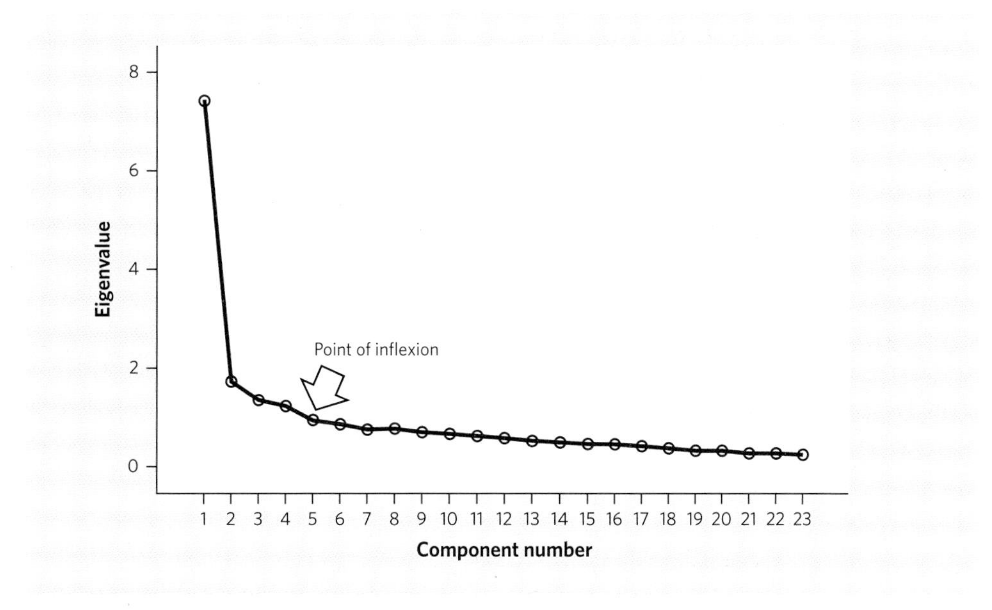

Figure 20.1 Scree plot of components in anxiety questions

Communality

Communality refers to the amount of variance that each variable shares with the other variables included in the analysis. Variance within an item will either be unique to that item (so it will share no variance with any other item across the questionnaire), or it will share some of that variance with others; we call this common variance. An item that has only unique variance will have a communality of 0; an item that shares all of its variance with others (it has none that is unique) has a communality of 1. Principal components analysis assumes that we start with a communality of 1. Once the factors have been extracted, communality can be calculated between the variable and the extracted factors; this is compared with the original communality. Before we start to reject factors (because they fall below the eigenvalue cut-off), we have a communality of 1 (all data are accounted for in the common variance); once we start excluding factors the communality is less than 1. While we want to maintain efficiency, we do not want to reject too many factors (so that we compromise communality). The closer this is to 1, the better. Kaiser (1960) recommends that (in samples of less than 300) extracted communalities should be at least 0.7.

20.4 Take a closer look
Summary of terms in principal components analysis

You may find the following summary useful for understanding the various terms that we use in principal components analysis

Table 20.3 Key terms

Term	Why it's important
Correlation matrix	Illustrates magnitude of correlation between items
Factor loading	Indicates magnitude of correlation between each item and its factor
Eigenvalues	Measures the amount of variance explained by the factor
Common variance	The variance within an item that is shared with other items in the questionnaire
Communality	Indicates the amount of common variance found within each item
Factor rotation	A series of statistical methods that help make more sense of extracted factors

Factor rotation

In principal components analysis the question items are examined to produce the first **factor extraction**; we call this the 'initial solution'. At this stage, it is more than likely that most of the items are loaded onto the first factor, and fewer onto the remaining factors. This does not tell us very much and makes interpretation difficult. We need to see if we can 'distribute' this a little more. All factor analysis methods do this through a statistical process called 'rotation'. If the data from the variables were plotted on a graph, **rotation** refers to turning the axes on a pivot point. The direction of 'turn' will depend on the type of rotation used. Initially, all of the factors are independent (they are not correlated to each other). In a two-dimensional example, the factors are at right angles to each other. In a more complex example, several factors will be at right angles in n-dimensional space (which is very difficult to conceptualise, let alone calculate). You do not need to know about the mathematics behind rotation, which is probably just as well. In essence rotation is achieved through matrix algebra and transformation; two factors need a 2×2 matrix, four factors require a 4×4 matrix, while ten factors need a 10×10 matrix (and so on). From that you can see how complicated it can get. The matrix is populated by the sines and cosines of the angle of rotation, which is then multiplied by the matrix of the original factors.

Rotation methods

There are several methods of rotation, although they are represented by just two types. **Orthogonal** rotation keeps the factors independent from each other by rotating them while maintaining the 90° angle between them. **Oblique** rotation draws the axes closer together (rotating them towards each other), increasing the likelihood that the factors will then be correlated to each other. If we have good theoretical reason to believe that the factors may be related to each other, then we can choose oblique rotation; we can assess the validity of that hypothesis in light of the outcome. However, if we have no reason to believe that the factors may be related, then we should choose orthogonal rotation. Most statistical sources recommend orthogonal methods by default.

20.5 Take a closer look
Rotation methods — a graphical view

It may help your understanding of the difference between orthogonal and oblique rotation methods by looking at these graphical representations.

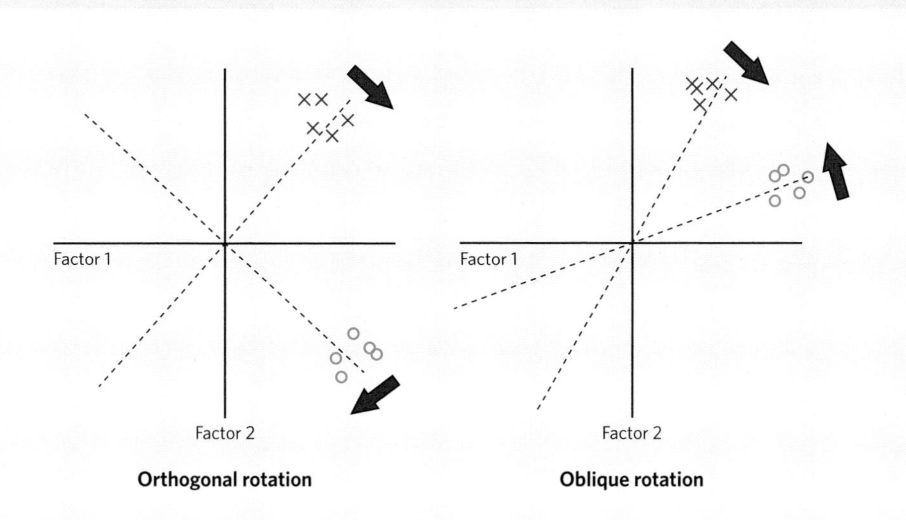

Figure 20.2 Rotation methods

Orthogonal rotation maintains the 90° angle between the factors (maintaining their independence); oblique rotation draws the factors towards each other, emphasising the correlation between them.

Once we have determined whether we can use oblique or orthogonal rotation methods, we still have a number of choices to make from within those categories. SPSS has five rotation options (or six if you include the option of 'no rotation'): three for orthogonal rotation (varimax, quartimax and equimax); and two for oblique rotation (direct oblimin and promax). For a review of rotation options, you should refer to Tabachnick and Fidell (2007, p. 639). With orthogonal rotation, varimax is often seen as the most sensible choice, as it tends to spread the loading of variables onto factors in a way that makes interpretation easier. For oblique rotation, direct oblimin is often regarded as the most sensible selection, while promax is normally reserved for very large samples.

Factor retention

Once we have employed the appropriate rotation, we are left with the 'final solution'. However, at that point there are yet more choices to make. Each factor will have a loading from every item to a greater or lesser extent. We cannot keep it like this, as it would make interpretation very tricky. We need to decide which factors to retain, and which we can discard. But what is the factor loading cut-off point? As ever, there are few rules, but many guidelines! Most of these suggestions depend on the significance of the loading, which will depend on the sample size. However, SPSS does not report the significance of this, so that's not very helpful. We can use guides from research that indicate the potential significance of loadings, according to sample size. Quite how that is calculated is unimportant, but a good guide is provided by Stevens (1992) – see Table 20.4.

Table 20.4 Guidelines for loading cut-off point, according to sample size

Sample size	Suggested loading cut-off point
50	.722
100	.512
200	.364
300	.298
600	.210
1000	.162

Some sources recommended a minimum of 75% variance should be accounted for (Dancey & Reidy, 2008) – this equates to a factor loading of .87 (variance is calculated by squaring the factor loading). However, it is probably a little harsh for the exploratory nature of principal components analysis. Many researchers choose .40 as the cut-off point in this context. By setting that threshold, it means that only factors with a loading in excess of .40 will be included. This has to be taken in context: a variable with a cut-off point of .40 explains only 16% of the variance in the factor; anything less than that may be meaningless. Remember, the ultimate aim is to explain as much variance, with as few factors, and as much communality as possible.

Assumptions and restrictions

There are a number of assumptions and restrictions to consider. The items contained in the questionnaire should be measured with at least ordinal data, although interval and ratio are preferred. Categorical data cannot be used with principal components analysis (correspondence analysis can be used in these cases). The data should be reasonably normally distributed. However, given that most samples are large by definition, this is less of a problem than it might be for other statistical procedures. Normal distribution is more of a problem if the final factors are used to conduct some other statistical analysis, such as an independent t-test (but then the rules for that test would apply anyway). We should avoid having too many outliers.

There should be at least reasonable linearity between variables; without correlation no factors will be found. At the same time, correlation should not be too high, otherwise we might have a problem with multi-collinearity. However, this is less of a problem for principal components analysis than it is for principal axis factoring. We can check correlation and multi-collinearity in several ways. We can request a correlation table when we set up the parameters for factor analysis in SPSS. From that, we can assess whether we have reasonable correlation; we only want a few co-efficients to be less than .30. To assess multi-collinearity we can check that there are only a very few correlations greater than .80. When we set up test parameters in SPSS, we can request something called a 'Determinant'; if that outcome is less than .00001, we may have a problem with multi-collinearity. Further checks can be undertaken by requesting outputs for the '**Kaiser-Meyer-Olkin** (KMO) Measure of Sampling Adequacy', '**Bartlett's test of sphericity**', 'anti-image' correlation, and 'reproduced' correlation. To ensure we have avoided multi-collinearity, the KMO test should be .500 at the very least (preferably at least .800). To ensure that we have good correlation, Bartlett's test should produce a highly significant chi-squared (χ^2) outcome. When we refer to the 'anti-image' correlation matrix, we need to check that we have exceeded a correlation of $r = .500$ for most of the items. The 'reproduced' correlation matrix can be used to check that we have not lost too much correlation following factor extraction; from that output we need that outcome to confirm that we have fewer than 50% 'non-redundant' residuals. We will explore these outcomes again later, when we analyse the output from an actual data set.

Sample sizes must be sufficient to cope with the rigours of factor analysis. However, there is much debate about what the minimum limit should be. Some sources say there should be at least 100 participants; others say it should be more like 200. Some argue that there should be more participants than items; Kline (1994) suggests at least a 2:1 ratio; others say that this should be more like 5:1. Comrey and Lee (1992) are more prescriptive: a sample of 50 is very poor; 100 is poor; 200 fair; 300 good; 500 good; and 1,000 is excellent (great if you can get the participants!). Comrey and Lee recommend having at least 300 cases, but they add that sample sizes can be lower if the factors are highly loaded (greater than .80; but that is rare). As usual, not everyone agrees: for instance, Guadagnoli and Velicer (1988) state that the sample size does not matter if there are four or more factors with a loading greater than .60. MacCallum, *et al.* (1999) refer to communalities as a determining factor. If the communality is greater than .6 the samples of around 100 may be adequate; if the communality is about .5, then there should be between 100 and 200 cases. In short, a sample size of 200 is a good target.

20.6 Take a closer look
Summary of assumptions and restrictions

- The measured items must be at least ordinal (preferably at least interval)
  - They should not be categorical
- Reasonable normal distribution is desirable (outliers should be avoided)
- There should be reasonable linearity between the items
  - But not so high that this might lead to multi-collinearity (particularly with principal axis factoring)
- Sample sizes should be sufficient (probably at least 200 cases or participants)

How SPSS performs principal components analysis

We will illustrate how to run principal components analysis through SPSS using the research example we explored earlier in this chapter. In that example, we explore the factor structure of a questionnaire containing 20 questions relating to quality of life and mood. A full list of the items used in this questionnaire can be seen in Table 20.2. All of those questions offer a response scale of 1 (definitely not true) to 5 (definitely true). The aim will be to investigate what sub-themes emerge from the questionnaire, which has been completed by 586 participants (a sample size that meets even the most stringent of requirements). We will not test for normal distribution, since that is less of a problem with principal components analysis. Before we analyse this, we should remind ourselves about the variables with this quick summary:

Principal components analysis
Primary theme of questionnaire: quality of life and mood perceptions
Possible sub-themes: relationships, job satisfaction, physical health and mental well-being

20.7 Nuts and bolts
Setting up the data set in SPSS

Creating the SPSS data set for principal components analysis is a lot more straightforward than it is for other tests; we simply need to set up a series of variables for continuous numerical scores.

Figure 20.3 shows how all of the variables are set as 'Ordinal' in the 'Measure' column. We do not need to enter any codes into the 'Value' column.

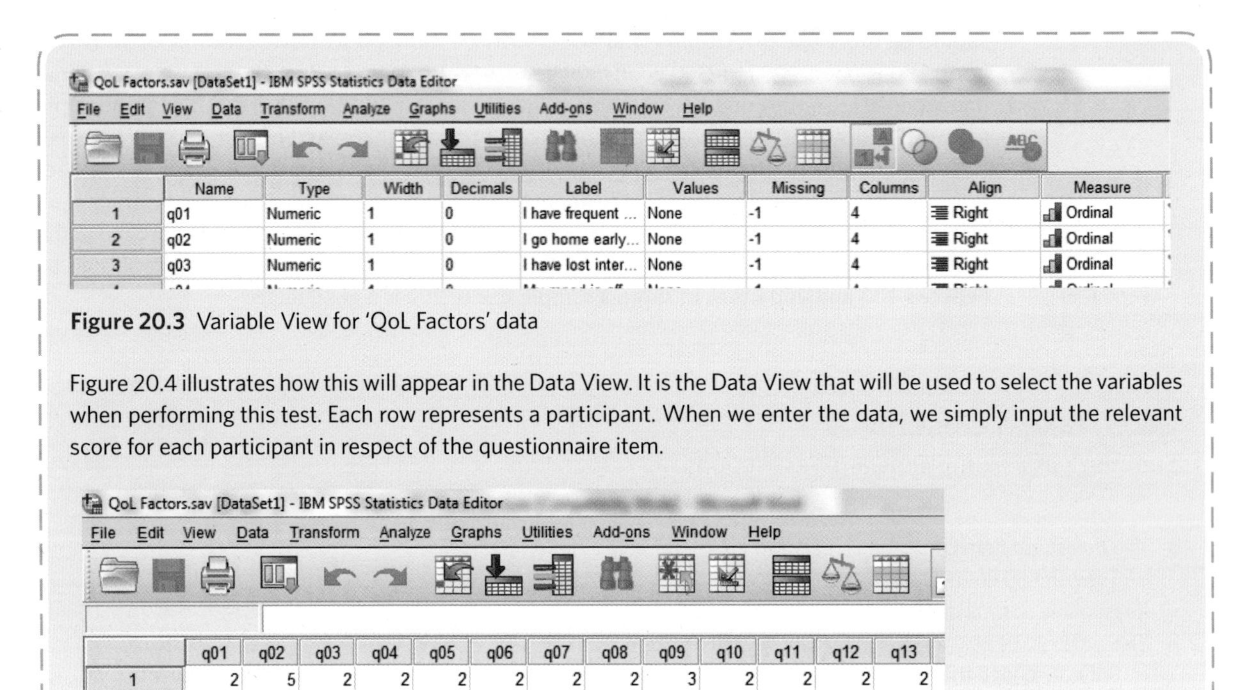

	Name	Type	Width	Decimals	Label	Values	Missing	Columns	Align	Measure
1	q01	Numeric	1	0	I have frequent ...	None	-1	4	Right	Ordinal
2	q02	Numeric	1	0	I go home early...	None	-1	4	Right	Ordinal
3	q03	Numeric	1	0	I have lost inter...	None	-1	4	Right	Ordinal

Figure 20.3 Variable View for 'QoL Factors' data

Figure 20.4 illustrates how this will appear in the Data View. It is the Data View that will be used to select the variables when performing this test. Each row represents a participant. When we enter the data, we simply input the relevant score for each participant in respect of the questionnaire item.

	q01	q02	q03	q04	q05	q06	q07	q08	q09	q10	q11	q12	q13
1	2	5	2	2	2	2	2	2	3	2	2	2	2
2	1	5	2	4	2	2	2	2	4	2	2	2	2
3	1	2	2	2	3	1	2	1	2	2	2	2	1
4	2	5	3	2	2	2	3	2	2	2	2	2	2

Figure 20.4 Data View for 'QoL Factors' data

Running tests in SPSS

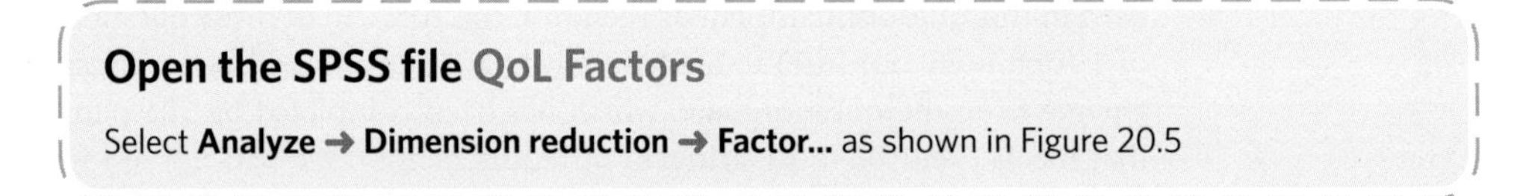

Open the SPSS file QoL Factors

Select **Analyze → Dimension reduction → Factor...** as shown in Figure 20.5

Figure 20.5 PCA: procedure 1

In new window (see Figure 20.6), transfer ALL of the variables to **Variables**

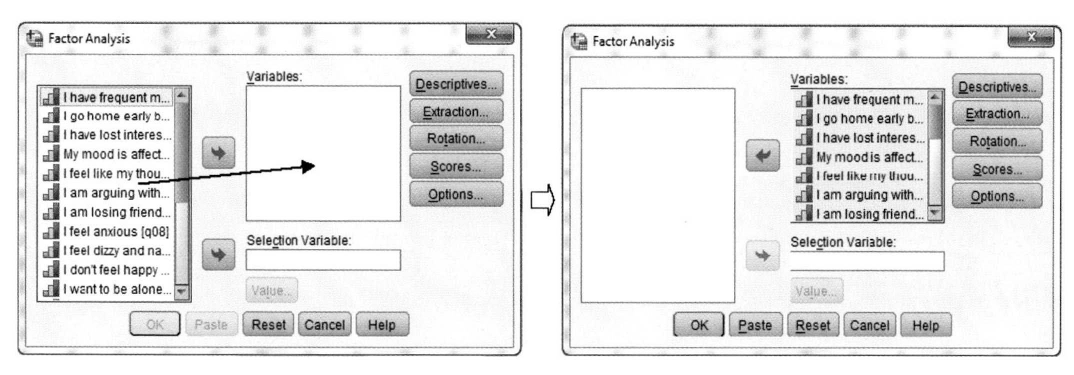

Figure 20.6 PCA: procedure 2

In new window (see Figure 20.7), tick boxes for **Initial solution** (useful for comparing to final solution), **Co-efficients** and **Significance levels**(to check initial correlation), **Determinant**, **KMO and Bartlett's test of sphericity** and **Anti-image** (to perform further checks on correlation and multi-collinearity) and tick **Reproduced** (to check that correlation has not been compromised after factor extraction) ➜ click **Continue** ➜ (back in main window) click **Extraction**

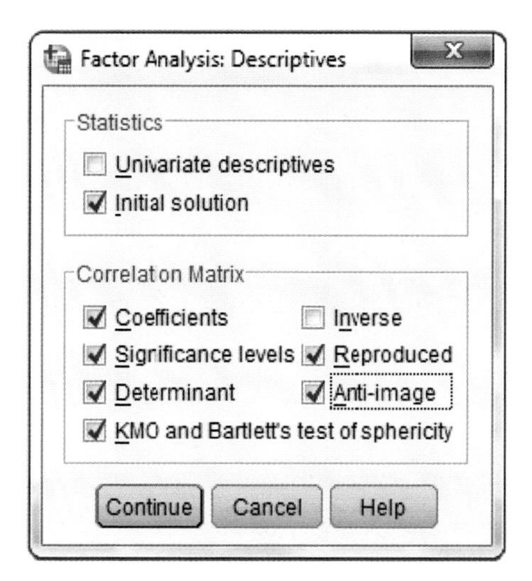

Figure 20.7 PCA: procedure 3

In new window (see Figure 20.8), select **Principal Components** for **Method** (use pull-down arrow to change if necessary) ➜ select radio button for **Correlation matrix** (you could choose Covariance matrix, it makes little difference on outcome) ➜ select radio button **Based on Eigenvalue** ➜ type **1** in box by **Eigenvalues greater than:** (to set up cut-off point, or set this to your required level — as discussed earlier) ➜ tick boxes for **Unrotated factor solution** (useful for comparing to final solution) and **Scree plot** (provides a graphical display of eigenvalues) ➜ click **Continue** ➜ (back in main window) click **Rotation**

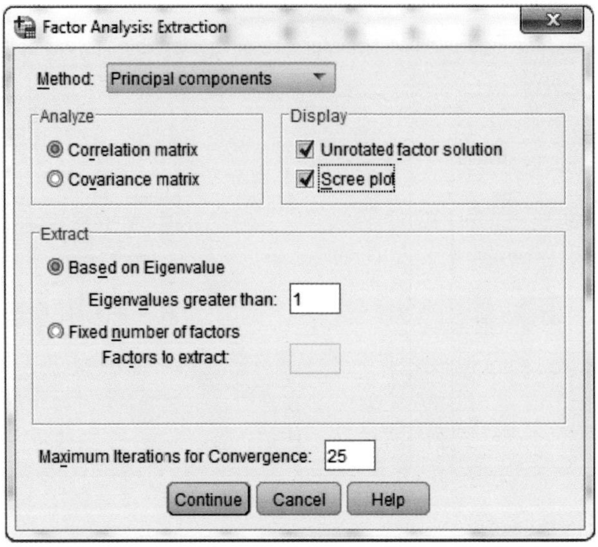

Figure 20.8 PCA: procedure 4

In new window (see Figure 20.9), tick radio button for **Varimax** (we discussed those options earlier) ➜ tick box for **Rotated solution** (we need this to identify the final factors) ➜ click **Continue** ➜ (back in main window) click **Options**

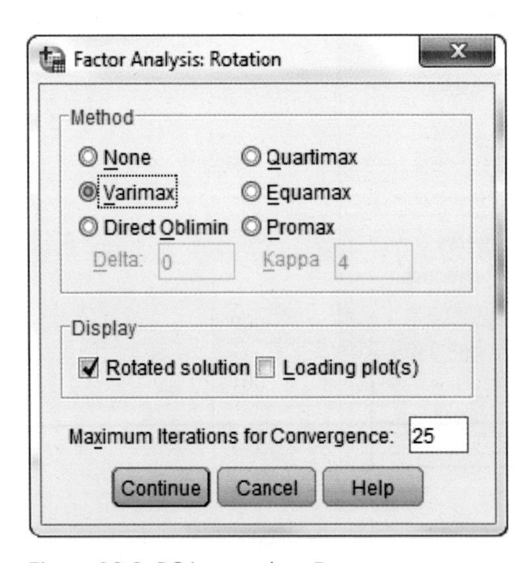

Figure 20.9 PCA procedure 5

In new window (see Figure 20.10), tick radio button for **Exclude cases listwise** (useful if there are any missing values) ➜ tick box for **Sorted by size** (makes interpretation of factors easier in final table) ➜ tick box for **Suppress small co-efficientsand** type **.40** in box next to **Absolute value below:** (this sets the cut-off for retaining factors, as we discussed earlier) click **Continue** ➜ click **OK**

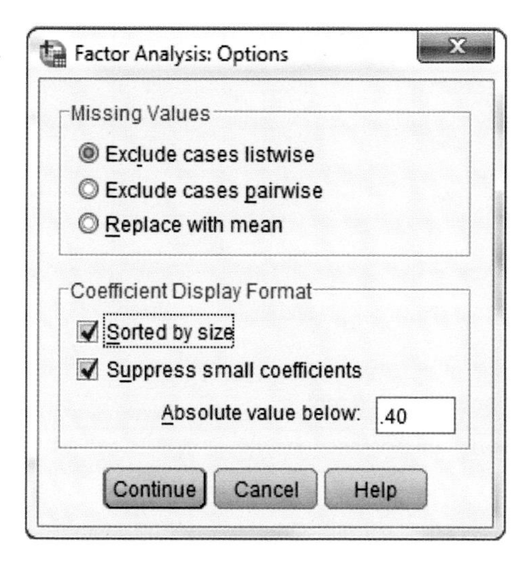

Figure 20.10 PCA: procedure 6

Interpretation of output

Checking assumptions: correlation and multi-collinearity

Before we examine the main outcome, we should check that we have satisfied the assumptions and restrictions. In particular, we should check that we have reasonable correlation and that we have avoided multi-collinearity. We presented the correlation for these data in Table 20.2. A scan through that tells us that we have pretty good correlation throughout; co-efficients are mostly in excess of .30 and none is above .80. A further check for multi-collinearity can be made by referring to the Determinant figure at the foot of the main correlation table that SPSS produces. This is too large to present here in full, but the relevant extract is shown in Figure 20.11.

	I find it hard to talk to people	.000
	I am taking more days off sick	.001
	I am not completing tasks at work	.000
	I feel like my colleagues hate me	.000
	I feel that my health is suffering	.001
	I always feel ill	.183

a. Determinant = .002

Figure 20.11 Determinant data for PCA (excerpt of correlation table)

We need not be too concerned about multi-collinearity with principal components analysis, but it is worth a look nonetheless. We said that we need the Determinant figure to exceed .00001. Figure 20.11 shows that we comfortably satisfy that criterion.

Figure 20.12 presents an extract of the relevant part of the anti-image correlation matrix (once again, it is too large to show all of it here). This is another check on multi-collinearity. The key focus here is the diagonal line where the numbers have been appended with a superscript 'a' (as shown in red font). These correlations reported across the 'diagonals' should be consistently above $r = .500$. Look at the full table in the SPSS output; you will see that they are.

	I always feel ill	.105	.034	–.058	–.020	.024
Anti-image correlation	I have frequent mood swings	.908[a]	.030	–.091	–.162	–.177
	I go home early because of headaches	.030	.862[a]	–.193	–.056	.000
	I have lost interest in my job	–.091	–.193	.921[a]	–.099	–.046
	My mood is affecting my work	–.162	–.056	–.099	.932[a]	–.156
	I feel like my thoughts are not my own	–.177	–.000	–.046	–.156	.936[a]

Figure 20.12 Anti-image correlation (extract)

KMO and Bartlett's test

Kaiser–Meyer–Olkin measure of sampling adequacy		.900
Bartlett's test of sphericity	Approx. chi-square	3657.185
	df	190
	Sig.	.000

Figure 20.13 Test for multi-collinearity

Figure 20.13 presents some further information on correlation and multi-collinearity. The Kaiser–Meyer–Olkin (KMO) Measure of Sampling Adequacy measures multi-collinearity. As we discussed earlier, we need this figure to be as high as possible in order to satisfy this assumption. Specifically, .500 is the minimum (but that would be very poor); anything above .800 is very good; and above .900 is excellent. Since we achieved the latter we can be confident that we can dismiss fears of multi-collinearity. The Bartlett's test confirms whether there is at least some good correlation between the variables. If there were no relationship between the variables, the X^2 score would be 0 and it would be non-significant. Our X^2 is very high and is highly significant ($p < .001$); there is considerable correlation between the variables, supporting the validity of factor analysis on this sample.

Checking assumptions: communalities

Figure 20.14 reports how factor extraction has impacted upon communality. Earlier, we said that principal components analysis starts with communality of 1. Once the factors have been extracted, communality may be compromised (but preferably not too much). The output shown here is only an extract again, but we can see that we have a column for the initial and extracted communalities. If we refer to the full version in the SPSS output, we can see that most of the extracted communalities are probably fine. However, none of them reaches Kaiser's (1960) target of 0.7; the average extracted communality is 0.52. This would be a problem had the sample size been fewer than 300; we had a sample of 586. But, we might have a problem with the question 'I am very introspective right now'; this has a very low extracted communality of 0.317.

	Initial	Extraction
I have frequent mood swings	1.000	.449
I go home early because of headaches	1.000	.453
I have lost interest in my job	1.000	.559
My mood is affecting my work	1.000	.473
I feel like my thoughts are not my own	1.000	.339

Figure 20.14 Communalities (extract)

Checking assumptions: correlation after factor extraction

Finally, we need to check that the process of extraction has not rendered our model meaningless. This is similar to the 'goodness of fit' measure in regression models – we want to explain as much as possible in the most efficient way. We focus on correlation, before and after factor extraction. We may have a problem if there is a large difference between those outcomes. We can check this by referring to the reproduced correlation matrix (we cannot show that here because it's too large). The top portion of that table shows the 'reproduced' correlations. This shows co-efficients after factor extraction. This outcome is deducted from the original correlation matrix (prior to extraction). The result is shown in 'residuals' in the lower portion of the reproduced matrix. We do not want these residuals to be too high. We could check that, but we can also refer to the 'non-redundant' residuals statistic beneath the table (see Figure 20.15). We do not want more than 50% of these residuals to be greater than .05. In our case this outcome is reported as being 41%, so we are fine.

	I am not completing tasks at work	−.095	−.137	−.108	−.051	−.039	.085
	I feel like my colleagues hate me	−.069	−.060	−.030	−.052	−.027	−.055
	I feel that my health is suffering	.065	−.065	−.084	−.014	.026	−.055
	I always feel ill	−.001	−.116	−.017	.059	−.027	−.083

Extraction method: principal components analysis

a. Reproduced communalities

b. Residuals are computed between observed and reproduced correlations. There are 79 (41.0%) nonredundant residuals with absolute values greater than 0.05

Figure 20.15 Reproduced correlation (and residual) matrix (truncated)

Main outcome

Figure 20.16 illustrates some key outcomes about variance, eigenvalues and factor loading. To begin with, all of the questions are allocated to a component (as shown in the first three columns). Then we examine the initial solution, where 'redundant' factors have been excluded. We now need to decide where to make the cut-off point. SPSS defaults to the Kaiser criterion (using an eigenvalue of 1). That might seem sensible, given the large sample size. In that case, we have four factors (as shown in Figure 20.16 – the factors that pass this test are shown in red font). However, we could equally have chosen to use Jolliffe's suggestion of using an eigenvalue of 0.7 − 0.8 as the cut-off. In that scenario we would have nine factors, although they would very difficult to interpret. We might also have chosen to base the factor extraction cut-off on the scree plot (see Figure 20.17). This is a very subjective measurement – the cut-off could be interpreted as being anything between two and five.

For consistency (and since we asked SPSS to use an eigenvalue of 1), we will go with the given output. To highlight the (initial) extracted factors more clearly, SPSS presents the loading of that initial factor plot in the next three columns (shown in blue in our Figure 20.16). At this stage, four factors have been extracted – overall these explain 52.353% of the variance. This is before the factors have been rotated, so most of the loading is on the first factor (31.087%), with the remaining (explained) variance spread over the other three factors. This is not the optimal loading, as too much loading is on the first factor (that is why we need rotation). The final three columns (highlighted in green) show the loading after (Varimax) rotation. The loading proportions have now changed – they are much more evenly spread: 17% is on the first factor, 15% on the second, 10% on the third and 10% on the fourth factor (we still have an overall variance of 52% explained).

Total variance explained

Component	Initial eigenvalues			Extraction sums of squared loadings			Rotation sums of squared loadings		
	Total	% of variance	Cumulative %	Total	% of variance	Cumulative %	Total	% of variance	Cumulative %
1	6.217	31.087	31.087	6.217	31.087	31.087	3.400	17.001	17.001
2	1.873	9.365	40.453	1.873	9.365	40.453	3.089	15.447	32.448
3	1.295	6.476	46.928	1.295	6.476	46.928	2.045	10.223	42.671
4	1.085	5.424	52.353	1.085	5.424	52.353	1.936	9.682	52.353
5	.949	4.745	57.097						
6	.892	4.459	61.556						
7	.797	3.984	65.540						
8	.764	3.821	69.361						
9	.734	3.670	73.031						
10	.692	3.458	76.490						
11	.641	3.205	79.694						
12	.572	2.858	82.553						
13	.543	2.715	85.267						
14	.534	2.669	87.937						
15	.510	2.550	90.486						
16	.457	2.285	92.771						
17	.408	2.039	94.810						
18	.376	1.881	96.691						
19	.343	1.713	98.405						
20	.319	1.595	100.000						

Extraction Method: principal components analysis

Figure 20.16 Variance and eigenvalues

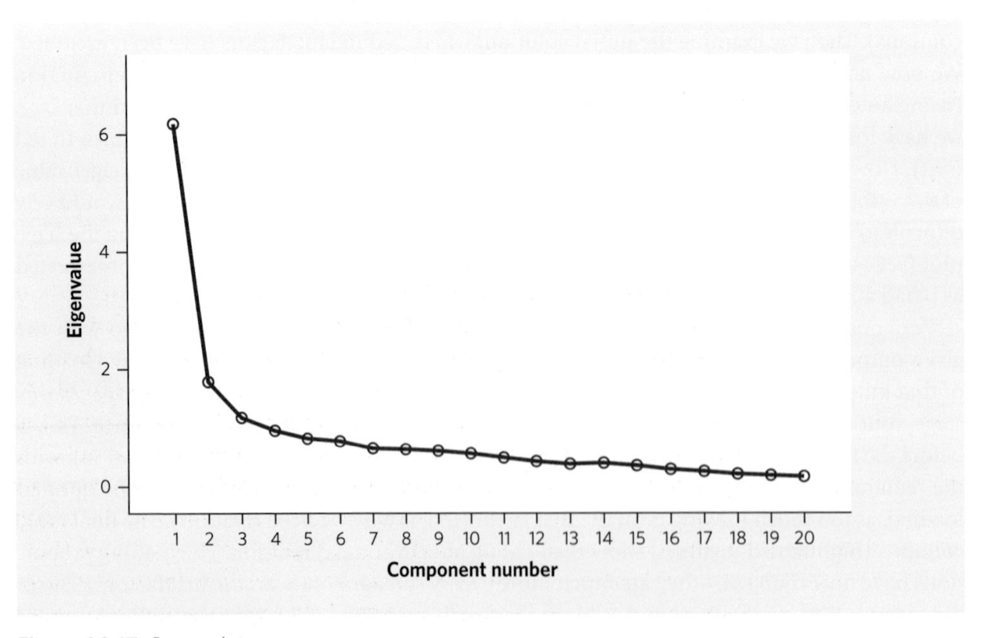

Figure 20.17 Scree plot

Component matrix[a]

	Component			
	1	2	3	4
I am losing friends due to my mood	.702			
I don't feel happy right now	.689			
I am shutting people out	.666			
I feel unmotivated	.663			
I find it hard to talk to people	.658			
I want to be alone all the time	.654			
I feel like my colleagues hate me	.640			
My mood is affecting my work	.632			
I have lost interest in my job	.626			
I feel anxious	.612			
I am arguing with my partner a lot	.572		.562	
I have frequent mood swings	.561			
I feel like my thoughts are not my own	.546			
I am very introspective right now	.528			
I feel dizzy and nauseous all the time		.628		
I go home early because of headaches		.504		
I am taking more days off sick		.469		
I feel that my health is suffering		.460		
I am not completing tasks at work				−.446
I always feel ill		.411		.427

Extraction method: principal components analysis

a. Four components extracted

Figure 20.18 Component matrix (prior to rotation)

Next we should look at the extracted component matrix as it appears on the first loading, prior to rotation (see Figure 20.18). This is useful to compare with the rotated matrix (Figure 20.19). We are shown only the factor loadings that are greater than .4 (because we set that limit earlier). Figure 20.18 confirms that most of the variables (14 out of 20) have been loaded onto the first factor, with very few across the other three factors (4, 0 and 2 respectively). This is not any better than before we did anything. So we should see how much better the solution is after rotation. Figure 20.19 confirms that we now have six variables on the first factor, five on the second, five on the third, and three on the fourth factor – that might be considered to be much better. Having said that, one variable (I am very introspective right now) has been omitted because it cannot be loaded onto any factor (as we saw earlier, it also has the lowest communality, so it clearly does not share much variance with other variables).

Rotated component matrix[a]

	Component			
	1	2	3	4
I feel anxious	.803			
I don't feel happy right now	.746			
I feel unmotivated	.744			
I have frequent mood swings	.603			
My mood is affecting my work	.570			
I feel like my thoughts are not my own	.505			
I am arguing with my partner a lot		.766		
I am losing friends due to my mood		.720		
I find it hard to talk to people		.689		
I want to be alone all the time		.658		
I am shutting people out		.624		
I am very introspective right now				
I am not completing tasks at work			.730	
I feel like my colleagues hate me			.584	
I have lost interest in my job			.495	
I go home early because of headaches			.483	.440
I am taking more days off sick			.482	
I always feel ill				.670
I feel dizzy and nauseous all the time				.669
I feel that my health is suffering				.627

Extraction method: principal components analysis.
Rotation method: varimax with Kaiser normalisation.

a. Rotation converged in eight iterations.

Figure 20.19 Component matrix (after rotation)

Making sense of the factors

Each variable can be attributed to one factor only. If a variable has been included on more than one, we select the highest loading to determine the factor it should be loaded on. If we look at Figure 20.19 we can see that this happens only once: we should declare that the question 'I go home early because of headaches' (initially loaded to factors 3 and 4) to be part of factor 3 (as that is where it has the highest loading).

However, although we achieved our aim of efficiently reducing our 20 questions into four factors, we still need to make sense about what those factors measure. We have already clarified the number of variables on each factor, but what does it mean? We cannot get SPSS to do this for us as we need to adopt some logical interpretation of the statistical outcome. We need to consider what theme might explain the combination of questions established by the factor loading. In Table 20.5 we can see the process of how we might do that. The process that we have used here may seem a little subjective; you may interpret the findings somewhat differently. But bear in mind that the statistical procedures 'created' those groupings – there is always some room for logical interpretation as well.

Table 20.5 Making sense of the factors

Factor	Question	Theme
1	I feel anxious I don't feel happy right now I feel unmotivated I have frequent mood swings My mood is affecting my work I feel like my thoughts are not my own	Could this factor be measuring mental well-being? Most of the questions appear to be exploring mood, anxiety, and psychotic symptoms. However, the question about 'mood affecting work' might be more suited to job-related issues (factor 3).
2	I am arguing with my partner a lot I am losing friends due to my mood I find it hard to talk to people I want to be alone all the time I am shutting people out	Could this factor be measuring relationship quality? The questions seem to tap into a range of perceptions about how well the respondent is getting on with other people, directly as a result of their mood
3	I am not completing tasks at work I feel like my colleagues hate me I have lost interest in my job I go home early because of headaches I am taking more days off sick	Could this factor be measuring job-related issues? Most of the questions appear to focus on perceptions about how the respondent is getting on at work. The final two questions are perhaps less clear; might they be better suited for physical health issues (see factor 4)?
4	I always feel ill I feel dizzy and nauseous all the time I feel that my health is suffering	These questions appear to be focusing on perceived physical health.

Writing up results

We should report our findings with a summary table and narrative. That table (Table 20.6) should probably focus on the final rotated solution

Table 20.6 Final rotated solution

	Factors			
	1	2	3	4
Variance explained (after rotation)	17.001	15.447	10.223	9.682
Initial eigenvalue	6.217	1.873	1.295	1.085
Variables (factor loadings)				
I am losing friends due to my mood	.702			
I don't feel happy right now	.689			
I am shutting people out	.666			
I feel unmotivated	.663			
I find it hard to talk to people	.658			
I want to be alone all the time	.654			
I feel like my colleagues hate me	.640			
My mood is affecting my work	.632			
I have lost interest in my job	.626			
I feel anxious	.612			
I am arguing with my partner a lot	.572		*.562*	
I have frequent mood swings	.561			

Table 20.6 Final rotated solution (*continued*)

	Factors			
	1	2	3	4
I feel like my thoughts are not my own	.546			
I am very introspective right now	.528			
I feel dizzy and nauseous all the time		.628		
I go home early because of headaches		.504		
I am taking more days off sick		.469		
I feel that my health is suffering		.460		
I am not completing tasks at work				−.446
I always feel ill		.411		.427

A questionnaire, using 20 questions to examine quality of life and mood, was answered by 586 participants. The aim was reduce those 20 questions into something more manageable. By using exploratory principal components analysis, with Varimax orthogonal rotation and an eigenvalue cut-off of 1.0, we were able to produce four factors that explained more than 52% of the data: mental well-being (six items, with 17% of explained variance), relationships (five items, 15%), work life (five items, 10%) and physical health (three items, 10%).

Chapter summary

In this chapter we have explored factor analysis; specifically focusing on the exploratory principal components analysis. At this point, it would be good to revisit the learning objectives that we set at the beginning of the chapter.

You should now be able to:

- Recognise that we use factor analysis to reduce large sets of items in a questionnaire into smaller, more meaningful, groups of similar themes or concepts.
- Understand that we can use different types of factor analysis according to the context. Principal axis factoring is used to confirm hypotheses about the predicted existence of factors, derived from a large set of questions. Principal components analysis (the main focus of this chapter) is used to explore potential themes that may exist within a larger group of items contained in a questionnaire.
- Comprehend that validity is a crucial element in research. This illustrates the extent to which we can demonstrate that we have actually measured what we are claiming to be. Factor analysis examines construct validity; the degree to which a theory has been supported by the items in test that is measuring it.
- Be familiar with the stages of extracting factors in the process of principal components analysis. Correlation is used to measure the strength of relationship between items. This information is used to identify potential groups of items that might become factors. Once found, the variance (correlation squared) is measured between each factor and its items, to produce the factor loading. The amount of variance explained there is measured by eigenvalues. Communality measures how much variance in one item is shared by other items in the questionnaire. We do not want to lose too much communality after factor extraction. Once we have an initial solution, the factors are statistically rotated to provide a more meaningful outcome. There are several methods of factor rotation; the correct one must be selected according to strict criteria.

- Be aware of the assumptions of principal components analysis, and the restrictions that must be met. The items being measured in the questionnaire must be at least ordinal. Reasonable normal distribution is preferred (but as larger data sets are generally required, this is not essential). There should be moderate correlation between most of the items (around r = .300 or higher); multicollinearity should be avoided (especially in principal axis factoring). Sample sizes need to be larger than for most other statistical tests (a sample of at least 200 is advised).
- Perform analyses using SPSS, using the appropriate method.
- Understand how to present the data and report the findings.

Research example

It might help you to see how principal components analysis has been applied in a research context. In this context you could read the following paper (an overview is provided below):

> **Sapin, C., Simeoni, M.C., El Khammar, M., Antoniotti, S and Auquier, P. (2005). Reliability and validity of the VSP-A, a health-related quality of life instrument for ill and healthy adolescents.** *Journal of Adolescent Health, 36 (4)*: 327–336. DOI: http://dx.doi.org/10.1016/j.jadohealth.2004.01.016

If you would like to read the entire paper you can use the DOI reference provided to locate that (see Chapter 1 for instructions).

In this paper the authors report the validation of a new generic self-report measure of adolescent health. Until that point there were no scales that captured the holistic nature of treating adolescent illness, nor one that reported the perceptions of those receiving care (at least not one that had been validated in France). To address that, the authors produced the VSP-A (Vécu et Santé Perçue de l'Adolescent, roughly translated to 'life and health perceptions of adolescents'). The questionnaire contained 37 questions that captured a range of health-related quality of life (HRQoL) perceptions, including mental health, body image, physical health, social relationships with peers, teachers, and family, and school performance. Each question was framed within a Likert scale, within a range of 1 (not at all/never) through to 5 (very much/always), focusing on the previous four weeks. Higher scores represented better HRQoL perceptions. Nearly 2,000 adolescents were given the VSP-A to complete: 1,758 were attending school, while 180 were hospitalised with either a medical, surgical or psychiatric condition. Previous examination of these perceptions was a lengthy process, involving several questionnaires. The aim was to establish a single questionnaire that could measure outcomes more quickly, but in a valid and reliable way.

The structure of the questionnaire was explored via principal components analysis, using Varimax rotation. The final outcome produced ten factors, which accounted for 74% of the overall variation (5% to 9% for each factor): Vitality (five items), psychological well-being (five items), relationships with friends (five items), leisure activities (four items), relationships with parents (four items), physical well-being (four items), relationships with teachers (three items), school performance (two items), body image (two items), and relationships with medical staff (three items). The authors then present data on the internal reliability of the questionnaire. This is a logical progression of what we have just learned. We need to be confident that the responses to the questions in each factor are consistent with each other. We will explore that in Chapter 21, so perhaps you could look at that chapter now to reinforce your learning.

The article that we have reviewed here is a good example of how principal components analysis has been applied in a research setting. However, it would have been useful to know a little information about overall linearity, communality and residual correlation.

Extended learning task

You will find the SPSS data associated with this task on the website that accompanies this book. You will also find the answers there.

Following what we have learned about principal components analysis, answer the following questions and conduct the analyses in SPSS. For this exercise, we will look at one of the initial reliability and validity tests on a quality of life and sleep questionnaire. It contained 38 questions that were intended to measure various aspects of perceived quality of life. The data were collected from 207 undergraduate students.

Open the SPSS data set **FA sleep**

1. Was the data set large enough?
2. Account for the other assumptions and restrictions of PCA.
3. Conduct the analysis in SPSS.
   a. Describe the extraction and rotation methods.
   b. How many factors were produced?
   c. How much variation did this explain?
   d. Describe the difference in loading distribution before and after rotation.
   e. Provide the factors with a name that reflects the questions loaded to it.

21 RELIABILITY ANALYSIS

Learning objectives

By the end of this chapter you should be able to:

- Recognise when it is appropriate to use reliability analysis
- Comprehend the importance of reliability in research
- Understand different types of reliability
- Appreciate assumptions and restrictions associated with principal components analysis
- Perform analyses using SPSS
- Understand how to present the data and report the findings

What is reliability analysis?

Reliability analysis examines consistency within responses across a group of items in a questionnaire. This might be in respect of all of the items in a questionnaire, but it is more likely to investigate sub-themes within that questionnaire. Reliability analysis is often seen as a logical follow-on from factor analysis (Chapter 20). In the previous chapter, we used principal components analysis (PCA) to locate four factors (or themes) from 20 questions contained within a single quality of life and mood questionnaire. Those factors were mental well-being, relationships, job and work issues, and physical health. We used PCA to explore the questionnaire to represent 'latent variables' to illustrate a series of themes. If you have not read that chapter, it is recommended that you do so now. Once we have established the presence of those themes, we would expect people to respond in a similar way on all items captured by each theme. Reliability analysis seeks to measure that consistency, on a scale of 0 to 1, where 1 is the most reliable outcome. We call this consistency '**internal reliability**'.

Research question for reliability analysis

In Chapter 20, we explored the research question set by the Mental Health Research Group (MHRG). They were seeking to investigate what aspects might contribute to perceptions regarding quality of life and mood. Having devised a questionnaire to measure that, we used PCA to detect four factors: mental well-being, relationships, job and work-related issues, and physical health. That analysis established the validity of the questionnaire, and the sub-themes within it. We are now set the task to confirm whether the questions included within those factors are answered consistently. This will confirm the reliability of those factors.

Theory and rationale

Why is reliability important?

Reliability is a crucial part of the research process. It examines how well we can trust the data, and is measured in respect of consistency and repeatability. We explored the concept of reliability in Chapter 5. Reliability is very important because it tells us how much we can depend on the outcome. Would we get the same outcome if we were to use these methods on a new data set? Would other researchers find what we do (using our methods)? Do participants respond to our questions consistently across time, and between contexts? Consistency can be examined over time (in respect of repeatability), between several researchers (in respect of reliable observations between them), for single researchers (in respect of their own consistency of ratings), and to measure the internal consistency of concepts within a questionnaire (to ensure that they appear to be measuring the same theme). We will look at all of these examples separately now.

Repeatability (test-retest reliability)

When we design data collection materials, such as questionnaires, we need to ensure that they are consistent over time. If they have been constructed correctly, the questions should be answered in the same way on repeated occasions. This may seem a little counterintuitive, in that some questionnaires are designed to measure concepts that change with circumstances – someone's mood will often vary day-to-day. However, all else being equal, if we measured the same person on separate occasions, where all circumstances remain constant, we should get an identical response each time. If that does not happen, it may indicate that there is a problem with the reliability of our question (perhaps it's too vague). We can test this with a procedure called **test-retest** correlation. For example, we could measure anxiety in a group of 50 people using a standard questionnaire. The questionnaire might have 20 items that ask the respondent how they react to stressful situations and how they feel about that. This would produce a score for each participant. Three weeks later we would give the same 50 people the same questionnaire to complete. We can compare the scores at the two time points using correlation (see Chapter 6). If the correlation co-efficient was around 0.70 or higher, we might feel that the questions were reliable over time (Shuttleworth, 2009); anything less than 0.50, we might have more reservations. We can base the assumptions on what we know about the strength of correlation. In Chapter 6, we stated that moderate correlation is represented by a co-efficient of 0.3–0.6, while anything above 0.7 is strong (Brace, *et al.* 2006). Reliability is an important factor in establishing validity of the constructs being measured, and provides greater strength to the results.

Inter-rater reliability

When studies involve several researchers rating a particular occurrence or behaviour, we need to make sure that those researchers are consistent with each other. Often, research involves observing people in natural settings, where the researchers need to record instances of certain behaviours. This could include the frequency and intensity of that behaviour. Since such ratings might be considered subjective, we need to ensure that all of the researchers are recording behaviour in the same way as each other. If they differ, we might feel less inclined to trust those observations. We could train the raters and then measure how consistent they are with each other. If they lack consistency, training might need to continue until that consistency is acceptable. Those researchers could also meet to compare ratings, and could agree on compromise where they differ greatly.

We can measure the consistency between observers with a process called **inter-rater reliability**. One of the most accepted forms of inter-rater reliability is **Cohen's kappa (κ)**. In this method, the number of agreed observations is calculated and compared with how many agreements would happen by chance in any case. The outcome can range from 0 to 1 (where 1 is perfect agreement); moderate concordance would be shown when κ is greater than 0.4, substantial agreement (probably the minimum target) is where κ is greater than 0.60 (Landis and Koch, 1977). Many sources recommend that it should be at least 0.70.

21.1 Calculating outcomes manually

The mathematics behind Cohen's kappa (κ)

You might find it useful to learn how to calculate Cohen's kappa; the formula for this is shown below:

$$\kappa = \frac{P_r - P_e}{1 - P_e}$$ where P_r is the proportion agreed; P_e is the proportion by chance factors

Now let's say that we have two coders rating the presence of a single behaviour, indicated by 'Yes' (it occurred) or 'No' (it did not). We find that our raters agree on 75% of occasions; we might feel that is quite good. However, we have not accounted for chance factors. There is a 50% probability that they agree by chance (because there are two options). We can apply that to our formula:

$$\kappa = \frac{0.75 - 0.50}{1 - 0.50} = 0.50$$

Such an outcome would be considered some way short of our target (assuming we want a minimum of 0.70). In fact, using this formula, we probably need more like 85% agreement to meet that target.

Intra-class correlation

Related to what we have just seen, some raters use scores (rather than categories) to assess the severity of an observation; calculating inter-rater reliability is more complex in this case. For instance, if the rating score ranges from 0–100, rater 'A' might on average score 86.2, while rater 'B' scores 78.9; we need to assess how consistent those ratings are with each other. Continuous scores like this can be examined using **intra-class correlation.** I do not propose to cover that here, but you might like to refer to McGraw and Wong (1996).

Intra-rater reliability

In the same way that we may need to measure that several raters are consistent with each other, we might also need to check that a single researcher makes consistent observations over several time points, or between observations. The methods used to measure **intra-rater reliability** are much the same as they are for inter-rater reliability, notably Cohen's kappa and intra-class correlation.

Internal consistency

We can also use reliability measures to examine whether our questionnaire possesses internal consistency. When we design a questionnaire we aim to ensure that the questions are answered consistently. In Chapter 20 we saw how we can use PCA to identify sub-themes within a questionnaire. We investigated a questionnaire that examined 20 questions relating to quality of life and mood. Following that process, we identified four sub-themes that appeared to be measuring mental well-being, relationships, job factors and physical health. To further examine that, we can assess the extent to which the questions within each sub-theme elicit consistent responses. Across the entire questionnaire, we might expect an individual to respond somewhat differently between certain questions, because they may be measuring various concepts. However, we would expect someone to answer questions within a theme in the same way. We can then compare that consistency across a group of people. Although this will vary to some extent, we would hope that the reliability is maintained acceptably throughout the group. There are a number of ways in which we can check internal consistency, but the most common are '**split half reliability**' and '**Cronbach's alpha**'(α).

In split half reliability, the variables are assessed by comparing one half of the group of questions to the other. Reliability is confirmed if the two halves of responses are highly correlated. This may sound attractive, but it is fraught with problems. How do we determine which questions to assign to each half? We could compare every possible combination of halves. This might be reasonable if there were just a few questions, but it would be wholly impractical if there were many questions (as there often are).

21.2 Take a closer look
Summary of reliability measurement methods

You may find the following summary useful for understanding a range of methods that can be used to measure reliability

Reliability type	What it does
Test-retest	Assesses consistency of responses to questionnaire items across time. Often referred to as repeatability
Inter-rater	Checks consistency of observational ratings between researchers. Often tested with Cohen's kappa
Intra-rater	Examines consistency of one researcher with regard to their ratings across time, or between observations
Internal consistency	Explores consistency of responses to questions within a specific theme, usually captured by several items on a questionnaire

A solution was provided by Cronbach (1951), who devised a calculation that performed that split half comparison on every possible permutation; this became known as Cronbach's α (you can see how this is calculated in Box 21.3). Cronbach's α is measured on a scale of 0 to 1, with 1 indicating perfect consistency. We are unlikely to find perfection, but we should aim for an acceptable level. Most sources consider values of 0.7 to 0.8 as being the minimum aim for acceptability. Kline (1999) was more specific; 0.8 is probably more suited for tests of IQ, while 0.7 is fine for other tests of ability, and the value can be as low as 0.6 for psychological constructs. Other statisticians urge caution with smaller sample sizes.

21.3 Calculating outcomes manually

The mathematics behind Cronbach's alpha (α)

The data in Table 21.1 represent a portion of the data set we will be examining later (those data include 586 participants, which would be rather difficult to summarise here). We will use this smaller portion to illustrate how the maths works (but you should note that we do not normally perform analyses on such small samples). The questions are taken from the quality of life and mood questionnaire that we have been using throughout these two chapters. From that, we will examine three questions that explore a sub-theme relating to mental well-being: "I feel anxious" (Question 8); "I don't feel happy right now" (Question 10); and "I feel unmotivated" (Question 14). We explore responses from 10 participants (cases). Each question elicits responses on a scale of 1 (definitely not true) to 5 (definitely true).

Table 21.1 Item analysis of mental well-being questions

	a	b	c								
Case	Q8	Q10	Q14		var a	var b	var c		cov ab/ba	cov ac/ca	cov bc/cb
1	2	3	2		0.09	0.16	0.49		−0.12	0.21	−0.28
2	2	2	3		0.09	0.36	0.09		0.18	−0.09	−0.18
3	1	2	1		1.69	0.36	2.89		0.78	2.21	1.02
4	2	1	2		0.09	2.56	0.49		0.48	0.21	1.12
5	3	1	3		0.49	2.56	0.09		−1.12	0.21	−0.48
6	2	3	3		0.09	0.16	0.09		−0.12	−0.09	0.12
7	4	3	3		2.89	0.16	0.09		0.68	0.51	0.12
8	1	2	2		1.69	0.36	0.49		0.78	0.91	0.42
9	2	5	4		0.09	5.76	1.69		−0.72	−0.39	3.12
10	4	4	4		2.89	1.96	1.69		2.38	2.21	1.82
Mean	**2.30**	**2.60**	**2.70**	*sum*	*10.10*	*14.40*	*8.10*	*sum*	*3.20*	*5.90*	*6.80*
				var	**1.122**	**1.600**	**0.900**	**cov**	**0.356**	**0.656**	**0.756**

We need to apply the data from Table 21.1 to the equation for Cronbach's α: $\frac{N^2\overline{Cov}}{\sum s^2_{item} + \sum Cov_{item}}$

$N = 3$ (no. of items); $\overline{Cov}$ = average covariance; s^2_{item} = item variance; Cov_{item} = item covariance

We need to calculate the variance for each question. We will demonstrate this for Q8:

We take the case score and deduct the mean Q8 score, and square it: $(2.00 - 2.30)^2 = 0.09$ (see 'var a')

We repeat that for each case, and sum all cases in 'var a'; we divide that sum by the no. of cases *minus* 1

$$(0.09 + 0.09 + \ldots 2.89) = 10.10 \div 9 = 1.122$$

Repeat that for all of the questions (see variance outcome under each column)

Next, we calculate the covariance between pairs of questions: ab, ba, ac, ca, ac, and cb; we will demonstrate with the covariance for Q8 vs. Q10 (ab):

We take the case score for Q8, and deduct Q8 mean; then do the same for Q10, and multiply them

$$(2.00 - 2.30) \times (3.00 - 2.60) = -0.30 \times .40 = -0.12$$

We repeat this for each case, and sum the cases; we divide that sum by the no. of cases *minus* 1

$$(-0.12 + 0.18 + \ldots 2.38) = 3.2 \div 9 = 0.356$$

Repeat that for all pairings. Remember to do this for 'ba' in addition to 'ab'; we need both (although only one calculation is shown in Table 21.1, to save space - see covariance outcome under each column)

Now, we apply those outcomes to the equation:

$$\text{Cronbach's } \alpha = \frac{3^2 \times ((1.122 + 1.600 + 0.900) \div 3)}{(1.122 + 1.600 + 0.900) + ((2 \times 0.356) + (2 \times 0.656) + (2 \times 0.756))} = 0.741$$

If this were replicated in the full data set, with all cases and items, we would be happy with that outcome.

Assumptions and restrictions

For once there are very few. The main issue relates to something called '**reverse scoring**' (see below), as it could seriously undermine the reliability of the outcome. We also need to be cautious of larger sample sizes, as these can over-inflate reliability.

Reverse scoring

When designing questionnaires, it is common practice to change the order of negative-to-positive response styles. This is particularly the case when asking the participant to rate an answer, such as 'definitely not true' through to 'definitely true', where scores of 1 to 5 are used to illustrate the range of answers. In our example, one of the questions in the mental well-being theme asks 'I don't feel happy right now'; a score of 1 would indicate that the respondent has said that they are feeling happy (it means that it is definitely not true that they . . . don't feel happy right now). However, we could have phrased the question 'I feel happy right now'; in that case, a score of 1 would suggest that the respondent is feeling unhappy. Reversing the polarity of questions is a good research method, as it helps prevent participants going into item-response mode, by ticking definitely true to everything; it makes them think more about their answer. The downside is that you need to watch out for that when conducting analyses.

In principal components analysis, reverse scoring is not a problem; we are only interested in the magnitude of correlation, not the direction. In reliability analysis it can be a problem; unadjusted it will skew the reliability outcomes. If your questionnaire employs reverse scoring, you will need to account for that prior to analysis. For that reason, it is probably wise not to score questions according to the wording of the response (1 = definitely not true, through to 5 = definitely true). Instead, it may be better to calculate a score according to positive or negative responses. So, irrespective of how the question is framed, higher scores could be assigned to more negative answers (for example). There is no reverse scoring in our example.

Key terms in reliability analysis

In addition to Cronbach's α, we can also report some other outcomes that will help assess the reliability of the questions being measured. While the alpha outcome will give us an overall indication of reliability for the entire set of items in a factor, we can also explore the relative contribution of each item.

Cronbach's alpha: the first, and probably most important, focus is the overall reliability of the factor; this is expressed in terms of the Cronbach's α score. Ranging from 0 to 1, we would want that to be as high as possible (usually at least 0.7, for reasons that we discussed earlier).

Item-total correlation: assesses the correlation between each item and the overall factor. The higher the correlation the better, but we would want this to be at least moderate ($\pm$ 0.3).

Squared multiple correlation: examines the multiple regression variance (R^2) score, treating the item as if it were an outcome variable in multiple linear regression, with the remaining items as predictor variables. We need this outcome to be reasonable too; perhaps at least 0.2 (see Chapter 16 for more information on linear regression).

Alpha if item removed: this is a recalculation of what Cronbach's α would be after a single item is removed from the group of questions being measured; if Cronbach's α increases dramatically, it indicates that the removed item was potentially compromising the reliability of the remaining items. If that were to happen, you might want to consider removing that item, or rewording it. Then you could collect more data and test the reliability again.

How SPSS performs reliability analysis

We will illustrate how to run reliability analysis through SPSS using the research example we explored earlier in this (and the previous) chapter. The aim is to explore the consistency of responses to groups of items within a quality of life and mood questionnaire. The groups of items represented the factors that we identified in Chapter 20: mental well-being, relationships, job and work-related issues, and physical health; the items can be seen in Table 21.2. The questions were answered by 586 participants, in a response mode of 1 (definitely not true) to 5 (definitely true). None of the questions was reverse scored. If you would like to see how to set up the data set for reliability analysis, refer to Box 20.5, as the same methods apply.

Table 21.2 Factor structure of quality of life and mood questionnaire

Factor	Description	Question	Q
1	Mental well-being	I feel anxious	8
		I don't feel happy right now	10
		I feel unmotivated	14
		I have frequent mood swings	1
		My mood is affecting my work	4
		I feel like my thoughts are not my own	5
2	Relationships	I am arguing with my partner a lot	6
		I am losing friends due to my mood	7
		I find it hard to talk to people	15
		I want to be alone all the time	11
		I am shutting people out	12
3	Job	I am not completing tasks at work	17
		I feel like my colleagues hate me	18
		I have lost interest in my job	3
		I go home early because of headaches	2
		I am taking more days off sick	16
4	Physical health	I feel that my health is suffering	19
		I always feel ill	20
		I feel dizzy and nauseous all the time	9

However, when we ran PCA in Chapter 20, we found that one question (Q13: I am very introspective right now) did not load onto any factor. It could be argued that this statement has more to do with 'relationships' than any of the other factors, so we will include it there for reliability analysis. It will be interesting to see how that question holds up.

Running reliability analysis in SPSS

To see how to create the SPSS data set for reliability analysis, refer to Box 20.7 in Chapter 20.

Open the SPSS file QoL factors

Select **Analyze → Scale → Reliability Analysis ...** as shown in Figure 21.1

Figure 21.1 Reliability analysis: – step 1

We need a separate analysis for each factor. For the first factor, we need those questions relating to 'mental well-being': see Table 21.2.

In new window (see Figure 21.2), transfer **I have frequent mood swings** (Q01), **My mood is affecting my work** (Q04), **I feel like my thoughts are not my own** (Q05), **I feel anxious** (Q08), **I don't feel happy right now** (Q10), and **I feel unmotivated** (Q14) to **Items: →** make sure that **Alpha** is selected by the **Model** box **→** type **Mental wellbeing** in the **Scale label** box **→** click **Statistics**

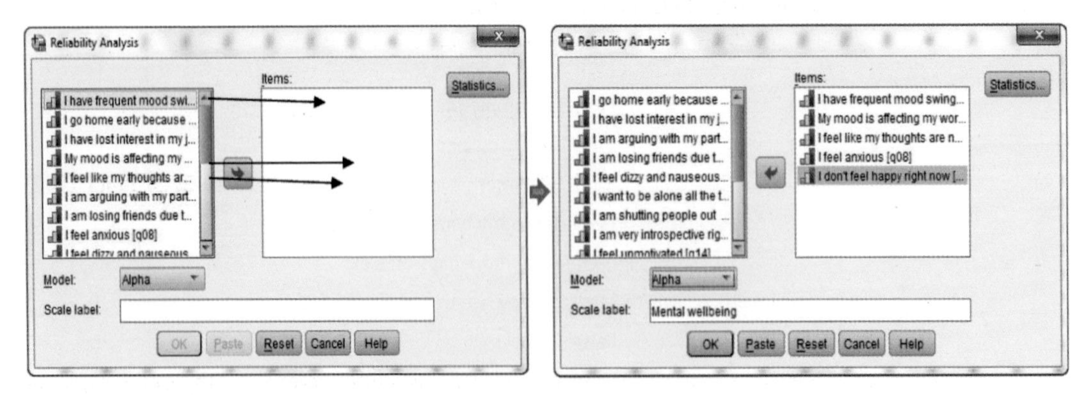

Figure 21.2 Reliability analysis: – step 2

In new window (see Figure 21.3), tick box for **Scale if item deleted** under **Descriptives for** → tick **Correlations** under **Inter-Item** → click **Continue** → (back in the previous window) click **OK**

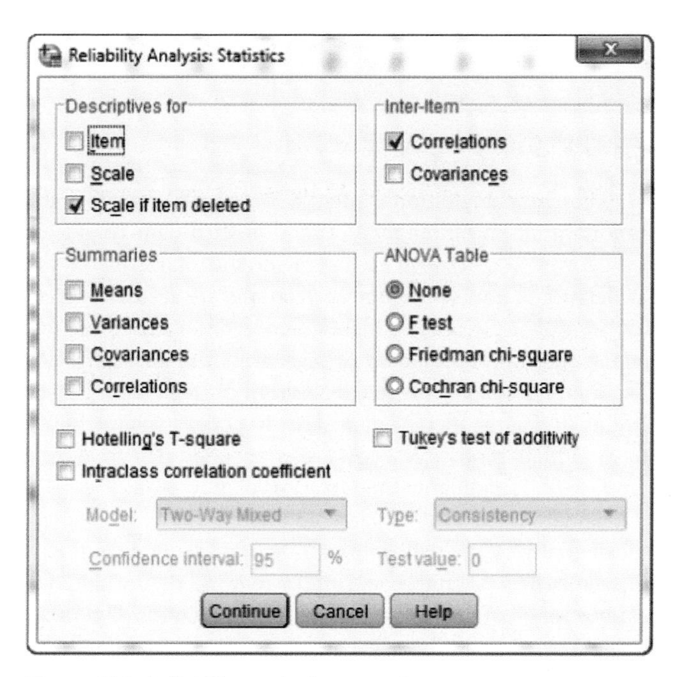

Figure 21.3 Reliability analysis: - step 3

Now we need to repeat that exercise for the remaining factors (you will need to remove the existing questions from **Items** prior to running each analysis). The settings in the **Statistics** menu stay the same:

On each occasion, start with:

Select **Analyze** → **Scale** → **Reliability Analysis** → then select the questions relevant to each factor

Transfer **I am arguing with my partner a lot** (Q06), **I am losing friends due to my mood** (Q07), **I want to be alone all the time** (Q11), **I am shutting people out** (Q12), **I am very introspective right now** (Q13), and **I find it hard to talk to people** (Q15) to **Items** → type **Relationships** in the **Scale label** box → click **OK**

Transfer **I go home early because of headaches** (Q02), **I have lost interest in my job** (Q03), **I am taking more days off sick** (Q16), **I am not completing tasks at work** (Q17), and **I feel like my colleagues hate me** (Q18) to **Items:** → type **Job** in the **Scale label** box → click **OK**

Transfer **I feel dizzy and nauseous all the time** (Q09), **I feel that my health is suffering** (Q19), and **I always feel ill** (Q20) → to **Items:** → type **Physical health** in the **Scale label** box → click **OK**

Interpretation of output

We will examine the output of each factor individually, although most of the initial explanation will be covered in the analysis of the first factor.

Mental well-being

Figure 21.4 confirms that there were six items in the analysis, and that Cronbach's alpha was .811. This is very good, given that we should aim for a minimum alpha of .70.

Cronbach's alpha	Cronbach's alpha based on standardised items	No.of items
.811	.812	6

Figure 21.4 Reliability statistics: mental well-being

Figure 21.5 presents the additional internal consistency outcomes; the final three columns are the most important in this output table. The 'Corrected Item-Total Correlation' (highlighted in red) shows the correlation between each item and the total score for that factor; we want at least moderate correlation here. Most of the correlations are at least moderate, and some are good. The 'Squared Multiple Correlation' (highlighted in blue) treats each item as an output variable, as if it were to be used in multiple linear regression; the remaining items are used as predictor variables (see Chapter 16 for more detail about linear regression). The output represents the R^2 value; the higher this is, the more variance that has been explained in that item. Once again, we want this to be as high as possible. Most are pretty good. The final column reports 'Cronbach's alpha if Item Deleted' (highlighted in green). We know from Figure 21.4 that Cronbach's α is .811. This final analysis examines what would happen to that if we removed the item. If alpha increases substantially by the removal of an item, you might want to consider doing just that. As it happens, alpha will not improve with the removal of any item here. Overall, we have very good reliability with all of the items contributing in an important manner.

	Scale mean if item deleted	Scale variance if item deleted	Corrected item – total correlation	Squared multiple correlation	Cronbach's alpha if item deleted
I have frequent mood swings	12.47	11.679	.521	.278	.792
My mood is affecting my work	12.06	11.010	.542	.302	.788
I feel like my thoughts are not my own	12.15	11.317	.467	.246	.806
I feel anxious	12.62	10.760	.636	.489	.767
I don't feel happy right now	12.61	10.621	.644	.465	.765
I feel unmotivated	12.35	10.702	.630	.461	.768

Figure 21.5 Item-total statistics: mental well-being

Relationships

The overall Cronbach's α is very good (Figure 21.6), showing that the six items possess high internal consistency (Figure 21.7).

Cronbach's alpha	Cronbach's alpha based on standardised items	No.of Items
.814	.814	6

Figure 21.6 Reliability statistics: relationships

	Scale mean if item deleted	Scale variance if item deleted	Corrected item – total correlation	Squared multiple correlation	Cronbach's alpha if item deleted
I am arguing with my partner a lot	13.53	13.716	.598	.368	.780
I am losing friends due to my mood	12.79	13.500	.651	.426	.767
I want to be alone all the time	13.31	14.664	.605	.378	.779
I am shutting people out	12.90	14.426	.595	.356	.781
I am very introspective right now	12.99	15.631	.425	.206	.815
I find it hard to talk to people	13.16	14.142	.587	.383	.782

Figure 21.7 Item-total statistics: relationships

The correlation between each item and the total factor score is moderate. The squared multiple regression is good for most items, although Question 13: 'I am very introspective right now', is on the low side (this was the item that had no factor loading, so it might explain why it performs less well here). However, when we look at how Cronbach's α might improve with the removal of an item it is clear that none of the items needs to be discarded. Once again, we have very good reliability with all of the items contributing in an important manner.

Job

The overall Cronbach's α is good (Figure 21.8), showing that the five items in that factor possess fairly high internal consistency (it is still pretty close to the conservative target of .70) (Figure 21.9).

Cronbach's alpha	Cronbach's alpha based on standardised items	No. of items
.683	.685	5

Figure 21.8 Reliability statistics: job

	Scale mean if item deleted	Scale variance if item deleted	Corrected item – total correlation	Squared multiple correlation	Cronbach's alpha if item deleted
I go home early because of headaches	13.74	9.105	.380	.167	.657
I have lost interest in my job	14.69	7.376	.529	.299	.589
I am taking more days off sick	14.48	7.919	.387	.156	.660
I am not completing tasks at work	14.55	8.204	.423	.208	.639
I feel like my colleagues hate me	14.96	8.059	.482	.269	.614

Figure 21.9 Item-total statistics: job

The correlation between each item and the total factor score is very good. The squared multiple regression is good for three of the five items; I go home early because of headaches, and I am taking more days off sick, are a little under 0.2 (the target we set earlier). Cronbach's α would not improve with the removal of any item. Once again, we have very good reliability with all of the items contributing in an important manner.

Physical health

This time, the overall Cronbach's α is poor (Figure 21.10) these three items do not possess high internal consistency. This often happens with a final factor (it accounted for the least variance in the factor analysis). Indeed, it may provide further evidence that the fourth factor should not have even been extracted in the first place.

Cronbach's alpha	Cronbach's alpha based on standardised items	No. of items
.514	.517	3

Figure 21.10 Reliability statistics: physical health

The data in this output (Figure 21.11) do not offer much help. The correlations are generally lower, but not completely lacking relationship. The squared multiple regression is generally poor throughout. Nevertheless, Cronbach's α would not benefit from the removal of any item. This factor simply possesses poor internal consistency.

	Scale mean if item deleted	Scale variance if item deleted	Corrected item – total correlation	Squared multiple correlation	Cronbach's alpha if item deleted
I feel dizzy and nauseous all the time	5.60	2.756	.333	.111	.414
I feel that my health is suffering	5.59	3.278	.344	.118	.390
I always feel ill	6.11	3.466	.316	.101	.434

Figure 21.11 Item-total statistics: physical health

Writing up results

We should report our findings as follows (there is no need for a table or graphs for this outcome):

> The four factors produced by the PCA were tested for reliability, using Cronbach's alpha (α). Factor 1 (mental well-being) showed very high internal consistency with an overall α of .811. Item-total correlations were generally at least moderate, the squared multiple regression generally confirmed that variance was moderately explained throughout. Cronbach's alpha would not benefit from the removal of any item. This level of internal consistency was also seen for Factor 2 (relationships; α = .814) and Factor 3 (job; α = .683). Internal consistency for the fourth factor (physical health) was not good (α = .514), showing low correlation and poor variation in the regression term. This was not improved by item removal.

Chapter summary

In this chapter we have explored reliability analysis. At this point, it would be good to revisit the learning objectives that we set at the beginning of the chapter.

You should now be able to:

- Recognise that we use reliability analysis to examine the consistency of responses to a group of items or questions. It is the next logical step from factor analysis, where the validity of themes and sub-themes has been established.
- Comprehend that reliability is an important factor in research. It confirms the consistency and repeatability of the methods used and the data gained from that research. In establishing reliability, we are adding to the validity of the constructs that we seek to measure.
- Understand different types of reliability. Repeatability of measures can be examined using test-retest reliability. Consistency of observational ratings between researchers can be explored using inter-rater reliability. Stability of observations from a single researcher can be investigated with intra-rater reliability. The internal consistency of responses to a group of items can be examined with split half reliability, but it is better analysed with Cronbach's alpha (and other measures associated with reliability analysis)
- Appreciate that there are very few assumptions and restrictions associated with reliability analysis. It is important that we account for reverse scoring and adjust if need be.
- Perform analyses using SPSS.
- Understand how to present the data and report the findings.

Research example

It might help you to see how principal components analysis has been applied in a research context. You could read the following paper (an overview is provided below):

Sapin, C., Simeoni, M.C., El Khammar, M., Antoniotti, S. and Auquier, P. (2005). Reliability and validity of the VSP-A, a health-related quality of life instrument for ill and healthy adolescents. *Journal of Adolescent Health, 36 (4):* 327–336. DOI: http://dx.doi.org/10.1016/j.jadohealth.2004.01.016

If you would like to read the entire paper you can use the DOI reference provided to locate that (see Chapter 1 for instructions).

We last saw this paper in Chapter 20, when we explored how the authors used principal components analysis to examine the factor structure of the VSP-A (Vécu et Santé Perçue de l'Adolescent – or, translated, the life and health perceptions of adolescents). From 37 questions, 10 factors were identified: Vitality (five items), psychological well-being (five items), relationships with friends (five items), leisure activities (four items), relationships with parents (four items), physical well-being (four items), relationships with teachers (three items), school performance (two items), body image (two items), and relationships with medical staff (three items). This paper also examines the internal consistency of those factors.

The results showed that all items possessed a minimum item-total correlation of 0.40 (so were at least moderate). The Cronbach's α for all factors exceeded 0.74, and no factor would benefit from the removal of any item. More specifically, Cronbach's α data were as follows: Vitality 0.84, psychological well-being 0.82, relationships with friends 0.81, leisure activities 0.81, relationships with parents 0.81, physical well-being 0.74, relationships with teachers 0.77, school performance 0.83, body image 0.85, and relationships with medical staff 0.86. This is a good example of how reliability analysis has been applied in a research setting.

Extended learning task

You will find the SPSS data associated with this task on the website that accompanies this book. You will also find the answers there.

Following what we have learned about reliability analysis, answer the following questions and conduct the analyses in SPSS. For this exercise, we extend what we explored in the exercise for PCA, where we investigated the factors present in a quality of life and sleep questionnaire. Now we should examine the internal consistency of those factors. To keep it simple, just focus on the first four factors produced by that outcome.

Open the SPSS data set **FA sleep**

1. For each factor report the outcome and implications:
   - **a.** Cronbach's alpha.
   - **b.** Item-total correlation.
   - **c.** Squared multiple regression.
   - **d.** Cronbach's alpha if item deleted.

Appendix 1
Normal distribution (z-score) table

Table A1.1 Probability of area under curve, to right of z-score

z	0	0.01	0.02	0.03	0.04	0.05	0.06	0.07	0.08	0.09
0	0.5000	0.5040	0.5080	0.5120	0.5160	0.5199	0.5239	0.5279	0.5319	0.5359
0.1	0.5398	0.5438	0.5478	0.5517	0.5557	0.5596	0.5636	0.5675	0.5714	0.5753
0.2	0.5793	0.5832	0.5871	0.5910	0.5948	0.5987	0.6026	0.6064	0.6103	0.6141
0.3	0.6179	0.6217	0.6255	0.6293	0.6331	0.6368	0.6406	0.6443	0.6480	0.6517
0.4	0.6554	0.6591	0.6628	0.6664	0.6700	0.6736	0.6772	0.6808	0.6844	0.6879
0.5	0.6915	0.6950	0.6985	0.7019	0.7054	0.7088	0.7123	0.7157	0.7190	0.7224
0.6	0.7257	0.7291	0.7324	0.7357	0.7389	0.7422	0.7454	0.7486	0.7517	0.7549
0.7	0.7580	0.7611	0.7642	0.7673	0.7704	0.7734	0.7764	0.7794	0.7823	0.7852
0.8	0.7881	0.7910	0.7939	0.7967	0.7995	0.8023	0.8051	0.8078	0.8106	0.8133
0.9	0.8159	0.8186	0.8212	0.8238	0.8264	0.8289	0.8315	0.8340	0.8365	0.8389
1.0	0.8413	0.8438	0.8461	0.8485	0.8508	0.8531	0.8554	0.8577	0.8599	0.8621
1.1	0.8643	0.8665	0.8686	0.8708	0.8729	0.8749	0.8770	0.8790	0.8810	0.8830
1.2	0.8849	0.8869	0.8888	0.8907	0.8925	0.8944	0.8962	0.8980	0.8997	0.9015
1.3	0.9032	0.9049	0.9066	0.9082	0.9099	0.9115	0.9131	0.9147	0.9162	0.9177
1.4	0.9192	0.9207	0.9222	0.9236	0.9251	0.9265	0.9279	0.9292	0.9306	0.9319
1.5	0.9332	0.9345	0.9357	0.9370	0.9382	0.9394	0.9406	0.9418	0.9429	0.9441
1.6	0.9452	0.9463	0.9474	0.9484	0.9495	0.9505	0.9515	0.9525	0.9535	0.9545
1.7	0.9554	0.9564	0.9573	0.9582	0.9591	0.9599	0.9608	0.9616	0.9625	0.9633
1.8	0.9641	0.9649	0.9656	0.9664	0.9671	0.9678	0.9686	0.9693	0.9699	0.9706
1.9	0.9713	0.9719	0.9726	0.9732	0.9738	0.9744	0.9750	0.9756	0.9761	0.9767
2.0	0.9772	0.9778	0.9783	0.9788	0.9793	0.9798	0.9803	0.9808	0.9812	0.9817
2.1	0.9821	0.9826	0.9830	0.9834	0.9838	0.9842	0.9846	0.9850	0.9854	0.9857
2.2	0.9861	0.9864	0.9868	0.9871	0.9875	0.9878	0.9881	0.9884	0.9887	0.9890
2.3	0.9893	0.9896	0.9898	0.9901	0.9904	0.9906	0.9909	0.9911	0.9913	0.9916
2.4	0.9918	0.9920	0.9922	0.9925	0.9927	0.9929	0.9931	0.9932	0.9934	0.9936
2.5	0.9938	0.9940	0.9941	0.9943	0.9945	0.9946	0.9948	0.9949	0.9951	0.9952
2.6	0.9953	0.9955	0.9956	0.9957	0.9959	0.9960	0.9961	0.9962	0.9963	0.9964
2.7	0.9965	0.9966	0.9967	0.9968	0.9969	0.9970	0.9971	0.9972	0.9973	0.9974
2.8	0.9974	0.9975	0.9976	0.9977	0.9977	0.9978	0.9979	0.9979	0.9980	0.9981
2.9	0.9981	0.9982	0.9982	0.9983	0.9984	0.9984	0.9985	0.9985	0.9986	0.9986
3.0	0.9987	0.9987	0.9987	0.9988	0.9988	0.9989	0.9989	0.9989	0.9990	0.9990
3.1	0.9990	0.9991	0.9991	0.9991	0.9992	0.9992	0.9992	0.9992	0.9993	0.9993

(Continued)

Table A1.1 (Continued)

z	0	0.01	0.02	0.03	0.04	0.05	0.06	0.07	0.08	0.09
3.2	0.9993	0.9993	0.9994	0.9994	0.9994	0.9994	0.9994	0.9995	0.9995	0.9995
3.3	0.9995	0.9995	0.9995	0.9996	0.9996	0.9996	0.9996	0.9996	0.9996	0.9997
3.4	0.9997	0.9997	0.9997	0.9997	0.9997	0.9997	0.9997	0.9997	0.9997	0.9998
3.5	0.9998	0.9998	0.9998	0.9998	0.9998	0.9998	0.9998	0.9998	0.9998	0.9998

Table A1.1 shows the probability of the area under the curve to the right of a given z-score. For example, we often use a z-score cut-off point of 1.96 to indicate the boundary of normal distribution We would expect data points beyond this to represent the upper 2.5% of a normally distributed data set, as shown by the right-hand segment in Figure A1.1 (assuming a two-tailed distribution). We can check this in the table by navigating down the first column to 1.9 and then across the row to 0.06. At a z-score of 1.96 the area under the curve to the right of this point is .9750 (or 97.5% when expressed as a percentage).

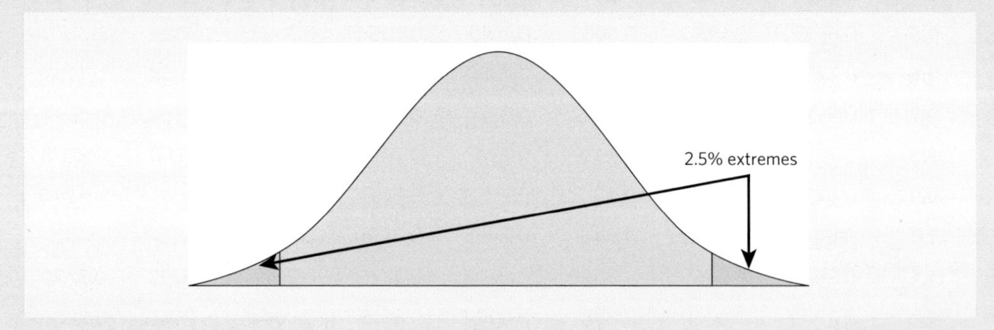

Figure A1.1 Two-tailed test

Table A1.2 Probability of area under curve, to left of z-score

z	0	0.01	0.02	0.03	0.04	0.05	0.06	0.07	0.08	0.09
−3.50	0.0002	0.0002	0.0002	0.0002	0.0002	0.0002	0.0002	0.0002	0.0002	0.0002
−3.40	0.0003	0.0003	0.0003	0.0003	0.0003	0.0003	0.0003	0.0003	0.0003	0.0002
−3.30	0.0005	0.0005	0.0005	0.0004	0.0004	0.0004	0.0004	0.0004	0.0004	0.0003
−3.20	0.0007	0.0007	0.0006	0.0006	0.0006	0.0006	0.0006	0.0005	0.0005	0.0005
−3.10	0.0010	0.0009	0.0009	0.0009	0.0008	0.0008	0.0008	0.0008	0.0007	0.0007
−3.00	0.0013	0.0013	0.0013	0.0012	0.0012	0.0011	0.0011	0.0011	0.0010	0.0010
−2.90	0.0019	0.0018	0.0018	0.0017	0.0016	0.0016	0.0015	0.0015	0.0014	0.0014
−2.80	0.0026	0.0025	0.0024	0.0023	0.0023	0.0022	0.0021	0.0021	0.0020	0.0019
−2.70	0.0035	0.0034	0.0033	0.0032	0.0031	0.0030	0.0029	0.0028	0.0027	0.0026
−2.60	0.0047	0.0045	0.0044	0.0043	0.0041	0.0040	0.0039	0.0038	0.0037	0.0036
−2.50	0.0062	0.0060	0.0059	0.0057	0.0055	0.0054	0.0052	0.0051	0.0049	0.0048
−2.40	0.0082	0.0080	0.0078	0.0075	0.0073	0.0071	0.0069	0.0068	0.0066	0.0064
−2.30	0.0107	0.0104	0.0102	0.0099	0.0096	0.0094	0.0091	0.0089	0.0087	0.0084
−2.20	0.0139	0.0136	0.0132	0.0129	0.0125	0.0122	0.0119	0.0116	0.0113	0.0110
−2.10	0.0179	0.0174	0.0170	0.0166	0.0162	0.0158	0.0154	0.0150	0.0146	0.0143

Table A1.2 (Continued)

z	0	0.01	0.02	0.03	0.04	0.05	0.06	0.07	0.08	0.09
−2.00	0.0228	0.0222	0.0217	0.0212	0.0207	0.0202	0.0197	0.0192	0.0188	0.0183
−1.90	0.0287	0.0281	0.0274	0.0268	0.0262	0.0256	0.0250	0.0244	0.0239	0.0233
−1.80	0.0359	0.0351	0.0344	0.0336	0.0329	0.0322	0.0314	0.0307	0.0301	0.0294
−1.70	0.0446	0.0436	0.0427	0.0418	0.0409	0.0401	0.0392	0.0384	0.0375	0.0367
−1.60	0.0548	0.0537	0.0526	0.0516	0.0505	0.0495	0.0485	0.0475	0.0465	0.0455
−1.50	0.0668	0.0655	0.0643	0.0630	0.0618	0.0606	0.0594	0.0582	0.0571	0.0559
−1.40	0.0808	0.0793	0.0778	0.0764	0.0749	0.0735	0.0721	0.0708	0.0694	0.0681
−1.30	0.0968	0.0951	0.0934	0.0918	0.0901	0.0885	0.0869	0.0853	0.0838	0.0823
−1.20	0.1151	0.1131	0.1112	0.1093	0.1075	0.1056	0.1038	0.1020	0.1003	0.0985
−1.10	0.1357	0.1335	0.1314	0.1292	0.1271	0.1251	0.1230	0.1210	0.1190	0.1170
−1.00	0.1587	0.1562	0.1539	0.1515	0.1492	0.1469	0.1446	0.1423	0.1401	0.1379
−0.90	0.1841	0.1814	0.1788	0.1762	0.1736	0.1711	0.1685	0.1660	0.1635	0.1611
−0.80	0.2119	0.2090	0.2061	0.2033	0.2005	0.1977	0.1949	0.1922	0.1894	0.1867
−0.70	0.2420	0.2389	0.2358	0.2327	0.2296	0.2266	0.2236	0.2206	0.2177	0.2148
−0.60	0.2743	0.2709	0.2676	0.2643	0.2611	0.2578	0.2546	0.2514	0.2483	0.2451
−0.50	0.3085	0.3050	0.3015	0.2981	0.2946	0.2912	0.2877	0.2843	0.2810	0.2776
−0.40	0.3446	0.3409	0.3372	0.3336	0.3300	0.3264	0.3228	0.3192	0.3156	0.3121
−0.30	0.3821	0.3783	0.3745	0.3707	0.3669	0.3632	0.3594	0.3557	0.3520	0.3483
−0.20	0.4207	0.4168	0.4129	0.4090	0.4052	0.4013	0.3974	0.3936	0.3897	0.3859
−0.10	0.4602	0.4562	0.4522	0.4483	0.4443	0.4404	0.4364	0.4325	0.4286	0.4247
0	0.5000	0.4960	0.4920	0.4880	0.4840	0.4801	0.4761	0.4721	0.4681	0.4641

Table A1.2 shows the probability of the area under the curve to the left of a given z-score. Using a similar example to earlier, we may wish to explore the cut-off point for scores that are below the lower end of what we would consider to be normally distributed. To do this, we would explore the area to the left of the curve for z-scores at the lower 2.5% of a normal distribution (as might be demonstrated by the left-hand segment in Figure A1.1). For example, if we use Table A1.2 to examine the z-score for −1.96 we see that probability under the curve is 0.0250 (which is 2.5% expressed as a percentage).

Appendix 2
t-distribution table

Table A2.1 t distribution for one- and two-tailed outcomes

probability (1-tail)	0.05	0.025	0.01	0.005	0.0025	0.001
probability (2-tail)	0.10	0.05	0.02	0.01	0.005	0.002
df						
1	6.314	12.706	31.821	63.657	127.321	318.309
2	2.920	4.303	6.965	9.925	14.089	22.327
3	2.353	3.182	4.541	5.841	7.453	10.215
4	2.132	2.776	3.747	4.604	5.598	7.173
5	2.015	2.571	3.365	4.032	4.773	5.893
6	1.943	2.447	3.143	3.707	4.317	5.208
7	1.895	2.365	2.998	3.499	4.029	4.785
8	1.860	2.306	2.896	3.355	3.833	4.501
9	1.833	2.262	2.821	3.250	3.690	4.297
10	1.812	2.228	2.764	3.169	3.581	4.144
11	1.796	2.201	2.718	3.106	3.497	4.025
12	1.782	2.179	2.681	3.055	3.428	3.930
13	1.771	2.160	2.650	3.012	3.372	3.852
14	1.761	2.145	2.624	2.977	3.326	3.787
15	1.753	2.131	2.602	2.947	3.286	3.733
16	1.746	2.120	2.583	2.921	3.252	3.686
17	1.740	2.110	2.567	2.898	3.222	3.646
18	1.734	2.101	2.552	2.878	3.197	3.610
19	1.729	2.093	2.539	2.861	3.174	3.579
20	1.725	2.086	2.528	2.845	3.153	3.552
21	1.721	2.080	2.518	2.831	3.135	3.527
22	1.717	2.074	2.508	2.819	3.119	3.505
23	1.714	2.069	2.500	2.807	3.104	3.485
24	1.711	2.064	2.492	2.797	3.091	3.467
25	1.708	2.060	2.485	2.787	3.078	3.450
26	1.706	2.056	2.479	2.779	3.067	3.435
27	1.703	2.052	2.473	2.771	3.057	3.421
28	1.701	2.048	2.467	2.763	3.047	3.408

Table A2.1 (Continued)

probability (1-tail)	0.05	0.025	0.01	0.005	0.0025	0.001
probability (2-tail)	0.10	0.05	0.02	0.01	0.005	0.002
df						
29	1.699	2.045	2.462	2.756	3.038	3.396
30	1.697	2.042	2.457	2.750	3.030	3.385
40	1.684	2.021	2.423	2.704	2.971	3.307
50	1.676	2.009	2.403	2.678	2.937	3.261
60	1.671	2.000	2.390	2.660	2.915	3.232
80	1.664	1.990	2.374	2.639	2.887	3.195
100	1.660	1.984	2.364	2.626	2.871	3.174
120	1.658	1.980	2.358	2.617	2.860	3.160
200	1.653	1.972	2.345	2.601	2.839	3.131

The values in Table A2.1 indicate critical values for t. These can be used to determine whether pairs of scores differ significantly. The reading of the table depends on whether a one-tailed or two-tailed prediction was employed, and upon the degrees of freedom (*df*) and the target probability level. For example, to find the critical value of t where $df = 18$, for a one-tailed outcome, and where the target probability is 0.05, we navigate along the columns until we reach 0.05 for a one-tailed test and down the rows until $df = 18$. We find that the 'cut-off' point is 1.734. A t score in excess of that will indicate that there are significant differences between the pairs of scores. Due to space restrictions, Table A2.1 displays only a restricted range of outcomes; a more extensive table of data is presented on the web page for this book. However, beyond 60 degrees of freedom, the cut-off points differ only slightly.

Appendix 3
r-distribution table

Table A3.1 r distribution for one-tailed and two-tailed outcomes

probability (1-tail)	0.05	0.025	0.01	0.005
probability (2-tail)	0.10	0.05	0.02	0.01
df				
1	0.988	0.997	1.000	1.000
2	0.900	0.950	0.980	0.990
3	0.805	0.878	0.934	0.959
4	0.729	0.811	0.882	0.917
5	0.669	0.754	0.833	0.875
6	0.621	0.707	0.789	0.834
7	0.582	0.666	0.750	0.798
8	0.549	0.632	0.715	0.765
9	0.521	0.602	0.685	0.735
10	0.497	0.576	0.658	0.708
11	0.476	0.553	0.634	0.684
12	0.458	0.532	0.612	0.661
13	0.441	0.514	0.592	0.641
14	0.426	0.497	0.574	0.623
15	0.412	0.482	0.558	0.606
16	0.400	0.468	0.543	0.590
17	0.389	0.456	0.529	0.575
18	0.378	0.444	0.516	0.561
19	0.369	0.433	0.503	0.549
20	0.360	0.423	0.492	0.537
21	0.352	0.413	0.482	0.526
22	0.344	0.404	0.472	0.515
23	0.337	0.396	0.462	0.505
24	0.330	0.388	0.453	0.496
25	0.323	0.381	0.445	0.487
26	0.317	0.374	0.437	0.479
27	0.311	0.367	0.430	0.471
28	0.306	0.361	0.423	0.463

Table A3.1 (Continued)

probability (1-tail)	0.05	0.025	0.01	0.005
probability (2-tail)	0.10	0.05	0.02	0.01
df				
29	0.301	0.355	0.416	0.456
30	0.296	0.349	0.409	0.449
35	0.275	0.325	0.381	0.418
40	0.257	0.304	0.358	0.393
45	0.243	0.288	0.338	0.372
50	0.231	0.273	0.322	0.354
60	0.211	0.250	0.295	0.325
70	0.195	0.232	0.274	0.302
80	0.183	0.217	0.257	0.283
90	0.173	0.205	0.242	0.267
100	0.164	0.195	0.230	0.254

The values in Table A3.1 indicate critical values for r. These can be used to determine whether a given correlation coefficient is statistically significant according to the relevant degrees of freedom (sample size minus 2). Cut-off points are provided in relation to the target probability for significance and are dependent on whether a one-tailed or two-tailed test was employed. For example, if we had a sample of 15 ($df = 13$), where we hypothesised that there would be a negative correlation between the two variables (one-tailed for a specific prediction), we might like to discover whether our observed outcome is significant, where $p < .05$. To find the critical value of r for this scenario, we navigate along the columns until we reach 0.05 for a one-tailed test and down the rows until $df = 13$. We find that the 'cut-off' point is 0.441. If the observed r value is greater than that cut-off point, the correlation co-efficient is significant.

Appendix 4
F-distribution table

Table A4.1 F distribution: cut-off values for F where probability = 0.05 (in right tail)

df1 / *df2*	1	2	3	4	5	6	7	8	9	10	20
1	161.448	199.500	215.707	224.583	230.162	233.986	236.768	238.883	240.543	241.882	248.013
2	18.513	19.000	19.164	19.247	19.296	19.330	19.353	19.371	19.385	19.396	19.446
3	10.128	9.552	9.277	9.117	9.013	8.941	8.887	8.845	8.812	8.786	8.660
4	7.709	6.944	6.591	6.388	6.256	6.163	6.094	6.041	5.999	5.964	5.803
5	6.608	5.786	5.409	5.192	5.050	4.950	4.876	4.818	4.772	4.735	4.558
6	5.987	5.143	4.757	4.534	4.387	4.284	4.207	4.147	4.099	4.060	3.874
7	5.591	4.737	4.347	4.120	3.972	3.866	3.787	3.726	3.677	3.637	3.445
8	5.318	4.459	4.066	3.838	3.687	3.581	3.500	3.438	3.388	3.347	3.150
9	5.117	4.256	3.863	3.633	3.482	3.374	3.293	3.230	3.179	3.137	2.936
10	4.965	4.103	3.708	3.478	3.326	3.217	3.135	3.072	3.020	2.978	2.774
11	4.844	3.982	3.587	3.357	3.204	3.095	3.012	2.948	2.896	2.854	2.646
12	4.747	3.885	3.490	3.259	3.106	2.996	2.913	2.849	2.796	2.753	2.544
13	4.667	3.806	3.411	3.179	3.025	2.915	2.832	2.767	2.714	2.671	2.459
14	4.600	3.739	3.344	3.112	2.958	2.848	2.764	2.699	2.646	2.602	2.388
15	4.543	3.682	3.287	3.056	2.901	2.790	2.707	2.641	2.588	2.544	2.328
16	4.494	3.634	3.239	3.007	2.852	2.741	2.657	2.591	2.538	2.494	2.276
17	4.451	3.592	3.197	2.965	2.810	2.699	2.614	2.548	2.494	2.450	2.230
18	4.414	3.555	3.160	2.928	2.773	2.661	2.577	2.510	2.456	2.412	2.191
19	4.381	3.522	3.127	2.895	2.740	2.628	2.544	2.477	2.423	2.378	2.155
20	4.351	3.493	3.098	2.866	2.711	2.599	2.514	2.447	2.393	2.348	2.124
21	4.325	3.467	3.072	2.840	2.685	2.573	2.488	2.420	2.366	2.321	2.096
22	4.301	3.443	3.049	2.817	2.661	2.549	2.464	2.397	2.342	2.297	2.071
23	4.279	3.422	3.028	2.796	2.640	2.528	2.442	2.375	2.320	2.275	2.048
24	4.260	3.403	3.009	2.776	2.621	2.508	2.423	2.355	2.300	2.255	2.027
25	4.242	3.385	2.991	2.759	2.603	2.490	2.405	2.337	2.282	2.236	2.007
26	4.225	3.369	2.975	2.743	2.587	2.474	2.388	2.321	2.265	2.220	1.990
27	4.210	3.354	2.960	2.728	2.572	2.459	2.373	2.305	2.250	2.204	1.974
28	4.196	3.340	2.947	2.714	2.558	2.445	2.359	2.291	2.236	2.190	1.959
29	4.183	3.328	2.934	2.701	2.545	2.432	2.346	2.278	2.223	2.177	1.945
30	4.171	3.316	2.922	2.690	2.534	2.421	2.334	2.266	2.211	2.165	1.932

Table A4.1 (Continued)

df1 / *df2*	1	2	3	4	5	6	7	8	9	10	20
40	4.085	3.232	2.839	2.606	2.449	2.336	2.249	2.180	2.124	2.077	1.839
60	4.001	3.150	2.758	2.525	2.368	2.254	2.167	2.097	2.040	1.993	1.748
120	3.920	3.072	2.680	2.447	2.290	2.175	2.087	2.016	1.959	1.910	1.659

The values in Table A4.1 indicate the critical values for F, where the probability for significance has been set at p = .05. To read this table, the columns (*df*1) represent the degrees of freedom in the numerator for the F ratio; the rows (*df*2) signify the denominator degrees of freedom. For example, say we have an ANOVA outcome of F (2, 27) = 12.68. We want to know if that F value is greater than the critical value. If it is, then we can say that the F ratio illustrates significant differences, where $p < .05$. To find the critical value, we look along the columns to *df* = 2 and down the rows to *df* = 27. We see that the critical value is 3.35. Our F value is greater than that, so it is significant. Space restrictions mean that only a limited range of values can be presented here, but you can see a fuller version on the web page for this book.

You may also need to check critical values of F for a more stringent significance cut-off, in which case you should refer to Table A4.2.

Table A4.2 F distribution: cut-off values for F where probability = 0.01 (in right tail)

df1 / *df2*	1	2	3	4	5	6	7	8	9	10	20
2	98.503	99.000	99.166	99.249	99.299	99.333	99.356	99.374	99.388	99.399	99.449
3	34.116	30.817	29.457	28.710	28.237	27.911	27.672	27.489	27.345	27.229	26.690
4	21.198	18.000	16.694	15.977	15.522	15.207	14.976	14.799	14.659	14.546	14.020
5	16.258	13.274	12.060	11.392	10.967	10.672	10.456	10.289	10.158	10.051	9.553
6	13.745	10.925	9.780	9.148	8.746	8.466	8.260	8.102	7.976	7.874	7.396
7	12.246	9.547	8.451	7.847	7.460	7.191	6.993	6.840	6.719	6.620	6.155
8	11.259	8.649	7.591	7.006	6.632	6.371	6.178	6.029	5.911	5.814	5.359
9	10.561	8.022	6.992	6.422	6.057	5.802	5.613	5.467	5.351	5.257	4.808
10	10.044	7.559	6.552	5.994	5.636	5.386	5.200	5.057	4.942	4.849	4.405
11	9.646	7.206	6.217	5.668	5.316	5.069	4.886	4.744	4.632	4.539	4.099
12	9.330	6.927	5.953	5.412	5.064	4.821	4.640	4.499	4.388	4.296	3.858
13	9.074	6.701	5.739	5.205	4.862	4.620	4.441	4.302	4.191	4.100	3.665
14	8.862	6.515	5.564	5.035	4.695	4.456	4.278	4.140	4.030	3.939	3.505
15	8.683	6.359	5.417	4.893	4.556	4.318	4.142	4.004	3.895	3.805	3.372
16	8.531	6.226	5.292	4.773	4.437	4.202	4.026	3.890	3.780	3.691	3.259
17	8.400	6.112	5.185	4.669	4.336	4.102	3.927	3.791	3.682	3.593	3.162
18	8.285	6.013	5.092	4.579	4.248	4.015	3.841	3.705	3.597	3.508	3.077
19	8.185	5.926	5.010	4.500	4.171	3.939	3.765	3.631	3.523	3.434	3.003
20	8.096	5.849	4.938	4.431	4.103	3.871	3.699	3.564	3.457	3.368	2.938
21	8.017	5.780	4.874	4.369	4.042	3.812	3.640	3.506	3.398	3.310	2.880
22	7.945	5.719	4.817	4.313	3.988	3.758	3.587	3.453	3.346	3.258	2.827

(Continued)

Table A4.2 F distribution: cut-off values for F where probability = 0.01 (in right tail) (Continued)

df1 / *df2*	1	2	3	4	5	6	7	8	9	10	20
23	7.881	5.664	4.765	4.264	3.939	3.710	3.539	3.406	3.299	3.211	2.781
24	7.823	5.614	4.718	4.218	3.895	3.667	3.496	3.363	3.256	3.168	2.738
25	7.770	5.568	4.675	4.177	3.855	3.627	3.457	3.324	3.217	3.129	2.699
26	7.721	5.526	4.637	4.140	3.818	3.591	3.421	3.288	3.182	3.094	2.664
27	7.677	5.488	4.601	4.106	3.785	3.558	3.388	3.256	3.149	3.062	2.632
28	7.636	5.453	4.568	4.074	3.754	3.528	3.358	3.226	3.120	3.032	2.602
29	7.598	5.420	4.538	4.045	3.725	3.499	3.330	3.198	3.092	3.005	2.574
30	7.562	5.390	4.510	4.018	3.699	3.473	3.304	3.173	3.067	2.979	2.549
40	7.314	5.179	4.313	3.828	3.514	3.291	3.124	2.993	2.888	2.801	2.369
60	7.077	4.977	4.126	3.649	3.339	3.119	2.953	2.823	2.718	2.632	2.198
120	6.851	4.787	3.949	3.480	3.174	2.956	2.792	2.663	2.559	2.472	2.035

Appendix 5
U-distribution table

Table A5.1 U distribution: cut-off values where one-tailed probability = 0.025; two-tailed probability = 0.05

N1 N2	4	5	6	7	8	9	10	11	12	13	14	15	16	17	18	19	20
4	0	1	2	3	4	4	5	6	7	8	9	10	11	11	12	13	13
5	1	2	3	5	6	7	8	9	11	12	13	14	15	17	18	19	20
6	2	3	5	6	8	10	11	13	14	16	17	19	21	22	24	25	27
7	3	5	6	8	10	12	14	16	18	20	22	24	26	28	30	32	34
8	4	6	8	10	13	15	17	19	22	24	26	29	31	34	36	38	41
9	4	7	10	12	15	17	20	23	26	28	31	34	37	39	42	45	48
10	5	8	11	14	17	20	23	26	29	33	36	39	42	45	48	52	55
11	6	9	13	16	19	23	26	30	33	37	40	44	47	51	55	58	62
12	7	11	14	18	22	26	29	33	37	41	45	49	53	57	61	65	69
13	8	12	16	20	24	28	33	37	41	45	50	54	59	63	67	72	76
14	9	13	17	22	26	31	36	40	45	50	55	59	64	67	74	78	83
15	10	14	19	24	29	34	39	44	49	54	59	64	70	75	80	85	90
16	11	15	21	26	31	37	42	47	53	59	64	70	75	81	86	92	98
17	11	17	22	28	34	39	45	51	57	63	67	75	81	87	93	99	105
18	12	18	24	30	36	42	48	55	61	67	74	80	86	93	99	106	112
19	13	19	25	32	38	45	52	58	65	72	78	85	92	99	106	113	119
20	14	20	27	34	41	48	55	62	69	76	83	90	98	105	112	119	127

The values in Table A5.1 indicate critical values for U, where the two-tailed probability for significance has been set at $p = .05$ (one-tailed $p = .025$). To read this table, the columns (N1) represent the smallest group size, with N2 the larger group size (they may often be the same). For example, to find the critical value of U when we have two groups of ten people we navigate along the columns to 10 and down the rows to 10. We find that the critical value is 23. If our (smallest) observed U value is *lower* than this cut-off point we can say that there is a significant difference in the outcomes between the groups.

Appendix 6
Chi-square (χ^2) distribution table

Table A6.1 χ^2 distribution: cut-off values according to probability and degrees of freedom (*df*)

p *df*	0.05	0.01	0.001
1	3.84	6.63	10.83
2	5.99	9.21	13.82
3	7.81	11.34	16.27
4	9.49	13.28	18.47
5	11.07	15.09	20.52
6	12.59	16.81	22.46
7	14.07	18.48	24.32
8	15.51	20.09	26.12
9	16.92	21.67	27.88
10	18.31	23.21	29.59
11	19.68	24.72	31.26
12	21.03	26.22	32.91
13	22.36	27.69	34.53
14	23.68	29.14	36.12
15	25.00	30.58	37.70
16	26.30	32.00	39.25
17	27.59	33.41	40.79
18	28.87	34.81	42.31
19	30.14	36.19	43.82
20	31.41	37.57	45.31
25	37.65	44.31	52.62
30	43.77	50.89	59.70
40	55.76	63.69	73.40
45	61.66	69.96	80.08
50	67.50	76.15	86.66
60	79.08	88.38	99.61
70	90.53	100.43	112.32

Table A6.1 (Continued)

p *df*	0.05	0.01	0.001
80	101.88	112.33	124.84
90	113.15	124.12	137.21
100	124.34	135.81	149.45

The values in Table A6.1 indicate critical values for χ^2, according to probability levels of 0.05, 0.01 and 0.001, in respect of the relevant degrees of freedom (this will vary according to the nature of the data being examined). For example, we may have an outcome that shows $\chi^2 = 6.811$, with 2 degrees of freedom. If we want to see if that outcome is statistically significant, where $p < .05$, we navigate along the columns to 0.05 and down the rows to $df = 2$. We find that the critical value for χ^2 is 5.99. Our outcome is greater than this, so it is significant. Space restrictions have limited the range of values that can be presented here, but you can see a fuller version on the web page for this book.

References

Abdin, R.R. (1995) *Parenting Stress Index, Third Edition: Professional Manual.* Florida: Psychological Assessment Resources.

Achenbach, T.M. (1991) *Manual for Child Behavior Checklist/4-18 and 1991 Profile.* Vermont: University of Vermont, Department of Psychiatry.

Agresti, A. (1996) *An Introduction to Categorical Data Analysis.* New York: John Wiley and Sons.

Andresen, E.M., Malmgren, J.A., Carter, W.B and Patrick, D.L. (1994) Screening for depression in well older adults: Evaluation of a short form of the CESD. *American Journal of Preventive Medicine.* 10: 77-84.

Beck, A.T. (1987) *Beck Depression Inventory: Manual.* San Antonio: The Psychological Corporation.

Beck, A. and Steer, R.A. (1993) *Beck Anxiety Inventory.* San Antonio: The Psychological Corporation.

Brace, N., Kemp, R. and Snelgar, R. (2006) *SPSS for Psychologists: A guide to data analysis using SPSS for Windows* (3rd edn). London: Routledge.

Bray, J.H. and Maxwell, S.E. (1985) *Multivariate Analysis of Variance.* Beverly Hills, CA: Sage Publications.

Bruera, E., Kuehn, N., Miller, M.J., Selmser, P. and Macmillan, K. (1991) The Edmonton Symptom Assesment System (ESAS): a simple method for the assessment of palliative care patients. *Journal of Palliative Care*, 7: 6-9.

Burns, W.C. (1996) *Spurious Correlations.* Retrieved March 2012 from **www.burns.com/wcbspurcorl.htm**

Camilli, G. and Hopkins, K.D. (1978) The applicability of chi-square to 2x2 contingency tables with small expected cell frequencies. *Psychological Bulletin,* 85 (1): 163-167.

Cash, T.F. (1990) Multidimensional body-self relations questionnaire, in Thompson, J.K. (ed.) *Body Image Disturbance: Assessment and Treatment.* New York: Pergamon Press, 125-129.

Cattell, R.B. (1966) The meaning and strategic use of factor analysis. In Cattell, R.B. (ed.) *Handbook of Multivariate Experimental Psychology*. Chicago: Rand McNally.

Cella, D.F., Tulsky, D.S., Gray G, *et al.* (1993) The Functional Assessment of Cancer Therapy scale: development and validation of the general measure. *Journal of Clinical Oncology* 11: 570-579.

Coakes, S.J. and Steed, L.G. (2007) *SPSS: Analysis without anguish: version 14.0 for Windows*. Brisbane: John Wiley & Sons Australia Ltd.

Cochran, W.G. (1954) Some methods for strengthening the common χ^2 tests. *Biometrics* 10 (4): 417-451.

Cohen, J. (1988) *Statistical Power Analysis for the Behavioral Sciences* (2nd edn) Hillsdale, NJ: Erlbaum.

Cohen, J. (1992). A power primer. *Psychological Bulletin*, 112 (1): 153-159.

Comrey, A.L. and Lee, H.B. (1992) *A First Course in Factor Analysis*. Hillsdale, NJ: Erlbaum.

Connors, C.K. (1997) *Connors' Parent Rating Scale - Revised: User's Manual.* North Tonwanda: Multi-Health Systems.

Coolican, H. (2009) *Research Methods and Statistics In Psychology* (5th edn). London: Hodder Education.

Cox, D.R. and Snell, E.J. (1989) *The Analysis of Binary Data* (2nd edn). London: Chapman and Hall.

Cronbach, L.J. (1951) Coefficient alpha and the internal structure of tests. *Psychometrika,* 16 (3): 297-334.

Dancey, C.P. and Reidy, J.G. (2008) *Statistics without Maths for Psychology* (4th edn). Harlow: Prentice Hall.

Eadie, W.T., Drijard, D., James, F.E., Roos, M. and Sadoulet, B. (1971) *Statistical Methods in Experimental Physics.* Amsterdam: North-Holland.

Endicott, J., Nee, J. and Harrison, W. (1993) Quality of life enjoyment and satisfaction questionnaire: A new measure. *Psychopharmacology Bulletin*, 29: 321-326.

Eyberg, S. and Pincus, D. (1999) *Eyberg Child Behavior Inventory and Sutter-Eyberg Student Behavior Inventory - Revised.* Florida: Psychological Assessment Resources.

Eysenck, H.J. (1953) *Uses and Abuses of Psychology.* Harmondsworth: Penguin.

Field, A. (2009) *Discovering Statistics using SPSS* (3rd edn). London: Sage.

Fox, K.R. and Corbin, C.B. (1989) The physical self-perception profile: Development and preliminary validation. *Journal of Sport and Exercise Psychology*, 11: 408-430.

Greenberg, L.M. and Walkman, I.D. (1993) Developmental normative data on the Test of Variables of Attention (TOVA). *Journal of Child Psychology and Psychiatry*, 34: 1019-1030.

Guadagnoli, E. and Velicer, W.F. (1988) Relation of sample size to the stability of component patterns. *Psychological Bulletin*, 103, 265-275.

Hays, R.D. and Morales, L.S. (2001) The RAND-36 measure of health-related quality of life. *Annals of Medicine,* 33: 350-357.

Howell, D.C. (2010) *Statistical Methods for Psychology* (8th edn). Belmont, CA: Cengage Wadsworth.

Jamieson, S. (2004) Likert scales: how to (ab)use them. *Medical Education,* 38 (12): 1212–1218.

Jensen, M.P., Turner, J.A., Romano, J.M. and Lawler, B.K. (1994) Relationship of pain-specific beliefs to chronic pain adjustment. *Pain,* 57: 301–309

Jolliffe, I.T. (2002) *Principal Component Analysis* (2nd edn). New York: Springer.

Kaiser, H.F. (1960) The application of electronic computers to factor analysis. *Educational and Psychological Measurement,* 20: 141–151.

Kay, S.R., Fiszbein, A. and Opler, L.A. (1987) The Positive and Negative Syndrome Scale (PANSS) for schizophrenia. *Schizophrenia Bulletin*, 13: 261–276.

Kline, P. (1994) *An Easy Guide to Factor Analysis.* New York: Routledge.

Kline, P. (1999) *Handbook of Psychological Testing.* London: Routledge.

Kovacs, M. (1981) Rating scales to assess depression in school-aged children. *Acta Paedopsychiatrica*, 46: 305–315.

Landis, J.R. and Koch, G.G. (1977) The measurement of observer agreement for categorical data. *Biometrics,* 33 (1): 159–174.

Linacre, J.M., Heinemann, A.W., Wright, B.D., Granger, C.V. and Hamilton, B.B. (1994) The structure and stability of the functional independence measure. *Archives of Physical Medicine and Rehabilitation.* 75: 127–132.

MacCallum, R.C., Widaman, K.F., Zhang, S.B. and Hong, S.H. (1999) Sample size in factor analysis. *Psychological Methods,* 4 (1): 84–99.

MacMillan, H.L., Wathen, C.N., Jamieson, E. et al. (2006) Approaches to screening for intimate partner violence in health care settings: A randomized trial. *Journal of the American Medical Assocation*, 296: 530-536.

McGraw, K.O. and Wong, S.P. (1996) Forming inferences about some intraclass correlation coefficients. *Psychological Methods,* 1 (1): 30–46.

Melzack, R. (1975) McGill pain questionnaire: Major properties and scoring methods. *Pain*, 1: 277–299.

Menard, S. (1995). *Applied Logistic Regression Analysis* (2nd edn). London: Sage.

Montgomery, S.A. and Asberg, M. (1979) A new depression scale designed to be sensitive to change. *British Journal of Psychiatry*, 134: 382–389.

Myers, R.H. (1990) *Classical and Modern Regression Application* (2nd edn). Boston: Duxbury Press.

Nagelkerke, N.J.D. (1991) A note on a general definition of the coefficient of determination. *Biometrika*, 78: 691–692.

Nolen-Hoeksema, S. (2001) Gender differences in depression. *Current Directions in Psychological Science*, 10 (5): 173–176.

Northwestern University (1997) *PROPHET StatGuide: Examining normality test results.* Retrieved March 2012 from **www.basic.northwestern.edu/statguidefiles/n-dist_exam_res.html**

Radloff, L.S. (1977) The CES-D Scale: A self-report depression scale for research in the general population. *Applied Psychological Measurement*, 1: 385–401.

Reips, U.-D. and Funke, F. (2008) Interval level measurement with visual analogue scales in Internet-based research: VAS Generator. *Behavior Research Methods*, 40 (3): 699–704.

Riley, J.L. and Robinson, M.E. (1997) The coping strategies questionnaire: Five factors or fiction? *Clinical Journal of Pain*, 13: 156–62.

Samn, S.W. and Perelli, L.P. (1982) Estimating aircraft fatigue: A technique with application airline operation. Brooks AFB, Tex: USAF School of Medicine. Technical report no. SAM_TR-82-21.

Sherbourne, C.D. and Stewart, A.L. (2002) The MOS support survey. *Social Science & Medicine*, 32: 705–714.

Shuttleworth, M. (2009) *Test—Retest Reliability*. Retrieved March 2012 from Experiment Resources: **www.experiment-resources.com/test-retest-reliability.html**

Slade, G.D. and Spencer, A.J. (1994) Development and evaluation of the Oral Health Impact Profile. *Community Dental Health*, 11: 3–11.

Spielberger, C. (1973) *Manual for the State-Trait Anxiety Inventory for Children.* Palo Alto, CA: Consulting Psychologists Press.

StatTrek.com (2012). *Normal Distribution Calculator: Online Statistical Table.* Retrieved March 2012 from **http://stattrek.com/online-calculator/normal.aspx**

Stevens, J.P. (1992) *Applied Multivariate Statistics for the Social Sciences* (2nd edn). Hillsdale, NJ: Erlbaum.

Tabachnick, B.G. and Fidell, L.S. (2007) *Using Multivariate Statistics* (5th edn). Boston, MA: Allyn and Bacon.

Vogt, P. (2005) *Dictionary of Statistics and Methodology: A Nontechnical Guide for the Social Sciences.* Thousand Oak, CA: Sage Publications.

Weathers, F.W., Litz, B.T., Herman, D.S., Huska, J.A. and Kean, T.M. (1993) The PTSD Checklist (PCL): Reliability, validity and diagnostic utility. In: Paper presented at the annual meeting of the international society for traumatic stress studies. San Antonio, TX.

Wilcox , R.R. (2005) *Introduction to Robust Estimation and Hypothesis Testing* (2nd edn). Burlington, MA: Elsevier.

Wolfram Alpha LLC (2011) *t-Score Calculator.* Retrieved March 2012 from **www.wolframalpha.com/entities/calculators/t-score_calculator/zn/5y/jy/**

Glossary

-2 log-likelihood (-2LL): A term used in logistic regression to determine the probability of a categorical outcome. Logistic regression outcomes are coded using the binary terms 0 and 1 (perhaps 1 = depressed; 0 = not depressed). The log-likelihood indicates how likely it is that a person or case will be coded as 1.

Adjusted standardised residuals: Indicates potential sources of differences in associations across categorical variables that have been examined by chi-squared (χ^2) analyses.

Alternative hypothesis: Often referred to as the experimental hypothesis. It is the prediction that suggests the experimental manipulation will have a significant effect on the outcome, or that an observed outcome is due to specific differences or associations.

Analysis of covariance (ANCOVA): A parametric statistical test that explores how much variance can be explained in a single (numerical) outcome that is explored across one or more (between-group) factor, while controlling for one or more additional variable.

Analysis of variance (ANOVA): A series of statistical tests that explore how much variance can be explained in a single (numerical) outcome, between one or more between-group factor, or across one or more within-group condition, or a mixture of between-group and within-group factors.

Association: See correlation

β_0 (intercept or constant): A measure in linear regression that indicates where the regression line crosses the y-axis.

B_x (gradient): A measure in linear regression that illustrates the slope of the regression line. The gradient indicates how outcome scores change for each unit change in a predictor. If the gradient is significantly different to 0, the predictor variable is seen to significantly contribute to variance in the outcome variable.

Backward stepwise: A hierarchical method of entering data into logistic regression.

Bar chart: A graphical method of presenting data, where bars represent a factor.

Bartlett's test of sphericity: A measure that examines correlation between variables, especially used in principal components analysis (a form of factor analysis).

Baseline: The pre-treatment, or pre-intervention, stage of a longitudinal study.

Between-group main effect: The effect that a single between-group independent variable has on the outcome (dependent variable). Usually applied in ANOVA tests, it is often measured from differences in mean outcome scores across the independent variable groups.

Between-group study: An examination of outcome (dependent variable) scores across independent groups or factors.

Between-groups t-test: Another name for the Independent t-test

Binary logistic regression: See Logistic regression

Biserial correlation: A test that examines association between one categorical variable and one continuous (numerical) variable. In contrast to Point-biserial correlation, the categorical variable *could* be viewed on a continuum (e.g. depression status; although this can be categorical (depressed or not depressed), there are also degrees of depression severity).

Bonferroni correction: A method to adjust the statistical significance of an outcome to account for multiple comparisons. This reduces the likelihood of making Type I errors. Also a method of *post hoc* analysis in ANOVA tests to examine the source of significance.

Box plot: A graphical method of presenting data, focusing on median and range data (also known as 'box and whisker plot').

Box's M test: A measure that examines homogeneity of variance-covariance matrices, most commonly used in MANOVA. In addition to checking that between-group variances are equal, this test also examines whether the correlation between the dependent variables is equal between the independent groups.

Brown-Forsythe F: A method of adjusting the F ratio in ANOVA when there is unequal variance between the groups (used alongside the very similar Welch's F).

Case mean: The average score across within-group conditions for a single case (or participant). Contrast with Condition mean.

Case variance: The extent to which scores vary, relative to the case mean.

Casewise diagnostics: A method of measuring outliers in linear regression.

Categorical data: Any data that are grouped by category, such as gender (the variable) which is (usually) categorised into two groups (men and women).

Cause-and-effect: The concept that an action or occurrence will lead to (or is directly responsible for) a specific outcome.

Central limit theorem: In probability theory, this states that samples larger than 30 will tend towards being normally distributed (subject to certain conditions - see Chapter 4). It also states that the mean of the sampling distribution will be equivalent to the mean of the population. The standard deviation of the population is the same as the average standard deviation of the sampling means; we find this by dividing the sample standard deviation by the square root of the sample size.

Chi-squared (χ^2) distribution: A range of numbers calculated from the 'sum of squares' of several independent normally distributed variables. It is used to predict outcomes when examining the association between categorical variables. It is also employed to test how well models of data fit observed outcomes. The distribution is used in a wide range of statistical tests to demonstrate the probability of statistical significance.

Chi-squared test: One of several applications of the chi-squared distribution. Commonly presented as χ^2, the outcome illustrates the association between two categorical variables. Generally used where at least one of the variables has three or more groups. If both variables have two groups Yate's continuity correction tends to be preferred.

Co-efficient: A measure of the strength and direction of the relationship (correlation) between two variables.

Cohen's kappa (κ): A measure to examine the extent of agreement between observations, notably used for assessing inter-rater reliability.

Collinearity statistics: Describes the extent that the relationship between variables can be illustrated graphically in a straight line (i.e. they are 'linear'). Two variables are perfectly collinear if there is an exact linear relationship between them.

Communality: A measure to show how much variance a variable shares with other variables. If it has unique variance, sharing none with other variables, it has communality of 0. If it shares all of its variance with other variables it has no unique variance and has a communality of 1. Communality is frequently used in factor analysis.

Condition mean: The average score across all cases in respect of a single within-group condition. Contrast with Case mean.

Conditions: Within-group factors in a single variable. Examples of a condition might be a time point in a longitudinal study, an intervention, or an experimental state. However, each condition must be performed by every person (or case) across a single group.

Confidence intervals: A range of values that is proposed to contain the true population value. The likelihood is measured in terms of probability (often 95%). The population value is usually based on the mean score of the sampling distribution.

Confidence intervals of difference: A measurement of probability where the range of values is represented by the difference in scores between two sets of values. Like traditional confidence intervals, the range is proposed to contain the actual mean difference between the groups. The likelihood that the range includes that value is also based on probability (usually 95%).

Confounding variable: Any variable that potentially has an effect on the outcome that was not expected and/or not directly measured (beyond the variable(s) that we predicted would have an effect).

Constant: See β_0

Construct validity: The extent to which a scale actually measures what we claim that it measures. If we design a questionnaire that is meant to measure fatigue, does it actually do that, or does it measure something quite different? One way to assess that is to examine the outcome alongside a similar scale that has already been shown to accurately measure the given construct.

Contingency table: A cross-tabulation of values from two or more categorical variables, presented in rows and columns. Each cell in the table represents the number of observations recorded for that outcome. These tables are often used to present data when assessing associations between the variables, which are examined statistically using chi-squared (or equivalent) tests.

Continuous data: Any range of data that have a numerical value. This excludes discrete data, such as categorical variables. Continuous data can be ordinal, interval or ratio.

Control group: The non-experimental group to which outcomes are compared.

Convenience sampling: A method of recruiting participants that uses whoever is available at a given time and place.

Correlation: A statistical test that measures the magnitude of the relationship between two variables. The precise method of calculation depends on the nature of the variables.

Correspondence analysis: A form of factor analysis, used to identify components within a scale. It is similar to principal components analysis, but is applied to categorical data rather than continuous data.

Counterbalancing: A method used in within-group studies to vary the order that the conditions are presented. This is performed to avoid practice and boredom effects.

Covariance: The average relationship between two variables.

Covariate: Any variable that is related to the (measured) outcome, or has the potential to be related to that outcome. Sometimes the covariate confounds the measured outcome. On other occasions, the covariate can help provide a clearer picture between the main predictor variables and the measured outcome.

Cox and Snell's R^2: One of two methods that are used to calculate how much variance that can be explained in a categorical outcome in logistic regression. Compare with Nagelkerke's R^2.

Cramer's V: A measure that calculates how much variance is explained between two categorical variables.

Critical value: The cut-off point in a distribution of scores, beyond which the probability of agreed statistical significance has been met, whereupon the null hypothesis can be rejected.

Cronbach's alpha (α): A measure that illustrates the degree of reliability of component items in a single construct, often used to measure consistency in a questionnaire.

Cross-products: A method of exploring the relationship between two dependent variables, often used in multivariate analyses.

Cross-sectional research: Studies that look at outcomes across large populations, often exploring differences between them. Usually, these focus on existing factors (as opposed to exploring the experimental effect of an intervention). Crucially, cross-sectional studies are performed at a single point in time (unlike longitudinal studies).

Cross-tabulation: Another name for a Contingency table.

Data Editor: The window in which all SPSS actions are performed, whether entering data or setting up variable parameters.

Data View: The SPSS window in which data are presented across columns that illustrate the variables and down rows to represent cases or participants. Contrast with Variable View.

Degrees of freedom: This describes the number of items that are free to vary, while one (or more) item is held constant, when examining differences or relationships. Those 'items' may be cases, variables, conditions, groups, or whatever - depending on what is being measured. This is used in many statistical tests to determine the cut-off point for significance. It might be the number of participants, less one (held constant) or the number of groups being examined (less one). Degrees of freedom are often shown as *df*.

Denominator: An extension of degrees of freedom (*df*) used in ANOVA statistical tests. Statistical significance is assessed from cut-off points in the F-distribution, relative to 'numerator' and 'denominator' degrees of freedom. In general, the numerator *df* relates to the number of groups or conditions (less one); the denominator *df* is usually found from the sample size (less one) minus the numerator *df*.

Dependent variable: The outcome, which is dependent on variations in the independent variable(s); for that reason, often referred to as the outcome variable.

Dichotomous (or binary) variable: A variable that only has two possible categories, such as depressed: yes or no.

Direct oblimin: An oblique (non-orthogonal) method of rotation in factor analysis (notably principal components analysis).

Discrete data: Specific values that are not part of a continuous range. Often used with categorical data to provide 'numbers' to represent categories, but can also represent a 'count' of numbers.

Discriminant analysis: A form of regression analysis, particularly useful with categorical dependent variables where there are more than two possible outcomes (thus excluding logistic regression). Discriminant analysis can also be used as a follow-up test in MANOVA to explore univariate outcomes.

DOI reference: A unique string of characters that can be entered into a web browser to instantly locate a published research article.

Dummy coding: A method of recoding variables with more than two categories into dichotomous (binary) ones. This may be necessary when applying multiple-category predictor variables to linear regression models. Dummy coding is also used in calculating 'planned contrast' outcomes to determine the source of significance in ANOVA where the variable has three or more groups or conditions.

Durbin-Watson test: An outcome in multiple linear regression that calculates the extent of correlation between the residuals. These 'residuals' represent the 'error' in the regression model; if the error values are highly correlated it will reduce the goodness of fit of the model.

Effect size: A measure of the actual difference or relationship (the 'effect'), usually quantified via Cohen's *d* or Pearson's *r* (correlation). The free software program G*Power is often useful here.

Eigenvalue: A measure of variance associated with factor analysis (most commonly) and in multivariate analyses. In factor analysis, the eigenvalue is used to determine how many factors to interpret after extraction.

Enter (method): A way of loading variables into regression models (by systematically including all of the variables at once).

Epsilon: The Greek letter (ε). It is used in repeated-measures ANOVAs to signify the outcome from adjusted tests of sphericity (to examine whether variances are equal across pairs of within-group conditions).

Equamax: An orthogonal method of rotation in factor analysis (notably principal components analysis).

Error: Anything that is not related to the experimental effect.

Error bars: A very useful addition to graphical representations of data (e.g. mean or median) that illustrates the spread either side of the central measures (often using 95% confidence intervals).

Estimated marginal means: The un-weighted mean scores, as presented (on request) in SPSS. These can be useful for reporting descriptive data when the mean scores across sub-groups or conditions are obscured by overall means. They are also useful for comparing outcomes before and after controlling for covariates.

Exp(B): A logistic regression outcome, reported by SPSS to indicate the odds (likelihood) of a categorical outcome, based on the contribution of a given predictor variable. If Exp(B) is less than 1, the predictor variable makes the outcome less likely; if it is greater than 1, then that outcome is more likely.

Experimental effect: See Effect size

Experimental group: The main group that is the focus of the experiment or study, often compared to a control group. For example, if a new drug is being tested on depressed patients, they are the experimental group. If we want to see how the outcomes compare in a non-depressed group we may recruit a control group.

Experimental hypothesis: See Alternative hypothesis

Explained variance: The extent that a statistical model can account for the variation in outcome scores by way of observable factors.This is most notably used in ANOVA and linear regression models. The more the variance can be explained, relative to random or error variance, the more likely it is that our model will be statistically significant.

Extraction: A process in factor analysis that determines the most important factors from a series of variables.

Extraneous variable: A variable, other than the experimental variables, that may have an influence on the outcome. We may need to control for these potential covariates, notably using ANCOVA.

F ratio: The proportion of the explained variance divided by the unexplained variance. Outcomes are compared across the F-distribution, relative to the numerator and denominator degrees of freedom, to determine how well the model fits the data. The greater the F ratio, the more likely it is that the data fits the linear regression model. Equally, the F ratio determines whether differences across groups or conditions are statistically significant.

Factor analysis: Strictly speaking, a *specific* method of exploring relationships between observable variables to model whether they form smaller groups of latent variables. The term is often used to describe similar techniques, such as principal components analysis.

Factor extraction: See Extraction

Factor loading: Describes the amount of variance explained between an observed variable and its latent variable in factor analysis.

Familywise error rate: The likelihood that a Type I error will be made when examining several outcomes across the same data set. When multiple comparisons are made, it is generally better to divide the cut-off point for statistical significance by the number of tests being conducted (that ask the same research question).

F-distribution table: The distribution of scores that describe values of F according to numerator and denominator degrees of freedom (see also F ratio).

First-order correlation: A simple correlation between two variables. Second-order correlations (and higher) describe the addition of additional variables in partial and semi-partial correlation.

Fisher's exact test: An alternative method (to the Chi-squared test and Yate's continuity correction) for calculating outcomes relating to associations between categorical variables. Tends to be used be used in smaller samples, where the chi-square distribution tends to be less reliable.

Friedman's ANOVA: A non-parametric test used in within-group analyses where there are three or more conditions. It is used as an alternative to Repeated-measures one-way ANOVA.

G*power: A free software program that runs power and effect size calculations.

Gabriel's: A *post hoc* analysis used to locate the source of significance in between-group ANOVA tests. It is particularly useful when there are unequal group sizes.

Games Howell: A *post hoc* analysis used to locate the source of significance in between-group ANOVA tests. It is particularly useful when homogeneity of variance has been violated.

General linear model (GLM): A method in SPSS for running ANOVA statistical analyses.

Goodness of fit: The extent to which the modelled data matches the overall data set it was taken from. Often that model is a prediction about outcomes; the goodness of fit describes how well that has been achieved.

Gradient: See B_x (gradient)

Grand mean: The average score of the entire sample regardless of groups or conditions.

Grand variance: The variance in outcome scores across the entire sample regardless of groups or conditions.

Greenhouse-Geisser: A method of calculating adjusted sphericity across pairs of conditions in repeated-measures ANOVA tests, used when Mauchly's test has been violated. A (plausible) alternative is the Huynh-Feldt test.

Group mean: The average outcome score across a single group.

Group variance: The variance in outcome scores across a single group.

Hierarchical regression: A method of entering variables into a regression model in a particular order.

Histogram: A way of presenting data graphically by way of bars to represent frequencies of scores across a series of points. Sometimes, a 'normal distribution' curve is added to the chart to demonstrate how 'symmetrically' the scores are distributed either side of the mean score.

Hochberg's GF2: A less common *post hoc* analysis used to locate the source of significance in between-group ANOVA tests. It is useful when there are unequal group sizes, similar to Gabriel's test, but tends only to be used where there are larger differences in group sizes.

Homogeneity of (between-group) variance: It is important that variances in outcome scores remain similar across the groups of the independent variable. This is usually examined with Levene's test. Relatively equal variances are particularly important where there are unequal group sizes.

Homogeneity of regression slopes: The assumption in ANCOVA that the correlation between the covariate and dependent variable should remain relatively constant across all levels (groups) of the independent variable.

Homogeneity of variance-covariance matrices: In multivariate analyses there is an additional assumption (beyond homogeneity of univariate between-group variances), the correlation between the dependent variables should remain relatively stable across the independent variable groups. It is often measured via Box's M test.

Hosmer and Lemeshow test: An outcome in logistic regression that indicates the proportion of variance than can still be explained (in the final model) relative to that found in the original model. It is a measure of success, in that we do not want to lose too much variance in establishing the final model. It is based on log-likelihood ratios.

Hotelling's Trace: One of several methods that can be used to report multivariate outcomes in MANOVA. The other options are Pillai's Trace, Wilks'Lambda, and Roy's Largest Root. Each method has its merits, according to the nature of the variables being explored. A fuller debate can be read in Chapter 14.

Huynh-Feldt: A method of calculating adjusted sphericity across pairs of conditions, used when Mauchly's test has been violated in repeated-measures ANOVA tests. A (plausible) alternative is the Greenhouse-Geisser test.

Hypothesis: A prediction about a research outcome, based on prior evidence and observation.

Independent errors: The assumption in linear regression that the residuals should not be highly correlated to each other.

Independent measures t-test: Another name for the Independent t-test

Independent multi-factorial ANOVA: A parametric statistical test that examines variance in a single dependent variable, in respect of two or more between-group independent variables.

Independent one-way ANOVA: A parametric statistical test that examines variance in a single dependent variable, in respect of one between-group independent variable.

Independent samples t-test: Another name for the Independent t-test

Independent t-test: A parametric statistical test that uses the t-distribution to explore differences in mean dependent variable scores across two distinct groups.

Independent two-way ANOVA: Another name for an Independent multi-factorial ANOVA (where there are two independent variables); this logic applies to an independent three-way ANOVA, and so on.

Independent variable: The variable that is 'manipulated' in experimental conditions to explore the effect on the dependent variable (outcome). It also applies to naturally occurring groups or conditions, across which differences in an outcome are examined.

Interaction: A term applied in multi-factorial ANOVA that illustrates how dependent variable scores over one independent variable differ across the groups or conditions of one or more additional independent variable; contrast this with the 'main effect'. For example, sleep satisfaction scores might be poorer for women than for men (main effect). Sleep satisfaction scores may also be poorer for people with depression than those without depression (another main effect). However, sleep satisfaction scores may be poorer for depressed women than non-depressed women, while the scores do not differ for men across the depression groups (a potential interaction). The observation needs to be confirmed statistically to fully demonstrate interaction.

Intercept: See β_0

Internal reliability: A measure of consistency within a scale. For example, if a series of questions in a quality of life questionnaire are supposed to measure health satisfaction, we would expect responses across those questions to be consistent. There are a number of ways to measure that, including split half reliability and formal reliability analyses (Cronbach's alpha).

Inter-rater reliability: A measure that examines the consistency of observations between different raters, using correlation analyses.

Interval data: Numerical scores that have an absolute value where the intervals between the numbers are consistent and objective. Age measured across a group of people is 'interval' because the gap between the ages of 10 and 15 is the same as it is between 20 and 25. Furthermore, the measure of age is consistent for whomever you measure it. Satisfaction ratings on a scale of 1 to 5 (where a score of 1 represents least satisfaction and 5 most satisfaction) are probably not interval. Someone may put greater emphasis on scores between 4 and 5 than they do between 1 and 2. Additionally, someone's rating of '4' may be very different to another person's rating (such subjective scores are more likely to be ordinal).

Intra-class correlation: A measure of consistency across similar constructs. One example might be to examine how two (full) siblings compare to each other across a series of traits. The outcome might be compared to observations across identical twins. Another example is where the numerically rated observations of two raters are compared (where the rating values have an identical range).

Intra-rater reliability: A measure that examines the consistency of observations for one rater across different observations, using correlation analyses.

Item-total correlation: A measure in reliability analyses that examines the correlation between one variable in a group of similar variables and the remaining variables in that group.

Kaiser-Meyer-Olkin (KMO): A test for multi-collinearity in factor analysis. Since the emergence or confirmation of factors depends on correlation, it is important the relationships between variables are at least reasonable (otherwise no factors will be found). However, it is also important that correlation is not too high, or differentiation becomes difficult. To ensure that multi-collinearity has been avoided, the KMO outcome should be greater than 0.5.

Kendall's Tau-b: A type of non-parametric correlation, where relationships are based on ranked scores. It is used in preference to Spearman's correlation where it is considered that there are too many tied scores.

Kolmogorov-Smirnov test: A method of examining normal distribution, often used in conjunction with the Shapiro-Wilk test. This test tends to be preferred with larger samples ($n > 50$). However, it should be used with caution in very large samples. Both tests report whether the outcome data are significantly different to a normal distribution. Since this outcome is not desirable, the outcome in these tests needs to be non-significant to show that the data are probably normally distributed.

Kruskal-Wallis: A non-parametric test used in between-group analyses where there are three or more groups. It is used as an alternative to Independent one-way ANOVA.

Kurtosis: An indicator of how data are spread, focusing on the 'peakedness' of the normal distribution curve. Ideally the kurtosis should be around '0' to reflect an optimum bell-shape curve (mesokurtic). A positive kurtosis (leptokurtic) suggests that the curve may be too peaked, where there may be too little variation in the data. A negative kurtosis (platykurtic) suggests that the curve may be too flat, where there may be too much variation in the data.

Latent variable: A variable that cannot be directly measured, but can be assumed to exist from its relationships with other (measurable) variables. Latent variables are often identified from factor analysis.

Layered chi-squared: A type of chi-squared test where associations between three or more categorical variables are measured.

Leptokurtic: See Kurtosis

Levene's test: A statistical method of examining homogeneity of variance between groups.

Likelihood: A measure of probability that assesses the chance of achieving a specific outcome.

Likelihood ratio: An indication of probability, expressed in terms of odds. For example, women may have a 2:1 likelihood ratio (to men) of being diagnosed with depression. This means that women are twice as likely to experience depression as men.

Likert scale: A questionnaire that requires respondents to report their answer on a numeric scale, or to choose from a series of potential answers that represent the strength of their response. These scales are often used to rate opinions, attitudes or satisfaction. An example might be asking students to rate their satisfaction of this book, on a scale of 1 to 10, where '10' represents 'most satisfied'.

Line graph: A graphical method of presenting data, where a line is drawn between data points to represent a trend. The data points might indicate groups or within-group time points or conditions.

Line of (best) fit: A type of line graph where the line is drawn between a cluster of data points, rather than specific ones, to represent an 'average trend' (see also Linear regression).

Linear regression: A statistical model that explores how much variance in a (continuous) outcome (dependent) variable can be explained by one or more predictor (independent) variable. Simple linear regression is used where there is one predictor; multiple linear regression is performed when there are several predictors. Relationships between the predictor and outcome variable are assessed in a series of lines of best fit. Correlation (particularly semi-partial correlation) and sums of squares (F ratio) are used to test the model.

Linear trend: See Linearity

Linearity: The extent the relationship between variables can be illustrated graphically in a straight line.

Logarithm: A method of presenting numbers in their 'base' form. The inverse of a logarithm is called the 'exponent'. If we know that $10^2 = 100$, the exponent in that function is '2'. Inversely, the logarithm of 100 to base 10 is 2. Most frequently, logarithms are expressed in relation to base 10 (the common logarithm), but any base can be used. The most common alternative is 'natural logarithm' (expressed to base 'e', for reasons that do not need to concern you for now). Logarithms were originally introduced to simplify calculations. In statistics, logarithms are used to transform a series of numbers into another form to allow alternative analyses.

Logistic regression: A statistical model that explores how much variance in a categorical outcome (dependent) variable can be explained by one or more predictor (independent) variable. The outcome is determined on likelihood ratios, where the chance of that outcome occurring is more likely, or less likely, when the predictor variable is added to the model. Most frequently, the dependent variable will have two possible outcomes (such as 'yes' or 'no'); this is called 'binary logistic regression'.

Log-likelihood: A statistic used in logistic regression to assess how well the model fits the data. The higher this outcome is, the poorer the fit.

Loglinear analysis: A statistical test that examines associations between three or more categorical variables. It is more sophisticated than layered chi-squared analyses, in that it can also explore interactions between those variables.

Longitudinal research: A study that examines outcomes across a single group over several time points. This is often used to measure progression from a baseline (pre-experimental) state through to post-intervention outcomes.

Lower-bound: A seldom used method of calculating adjusted sphericity across pairs of conditions, used when Mauchly's test has been violated in repeated-measures ANOVA tests. Greenhouse-Geisser and Huynh-Feldt tend to be preferred in these contexts.

Least squares difference (LSD): A seldom used *post hoc* test applied to find the source of difference in repeated-measures ANOVAs. Bonferroni adjustments tend to be used in most cases.

Main effect: A term applied in multi-factorial ANOVA that illustrates how dependent variable scores vary across a single independent variable (irrespective of the other independent variables). See 'Interaction' for further explanation.

MANCOVA: An acronym for Multivariate analysis of covariance.

Mann–Whitney U: A non-parametric test used in between-group analyses where there are two groups. It is used as an alternative to Independent t-test.

MANOVA: An acronym for Multivariate analysis.

MANOVA effect: The multivariate outcome in MANOVA (compare with univariate effect across single outcomes). See also Multivariate analysis.

Matrix: A way of presenting numbers in rows and columns, usually within brackets. For example: $\begin{pmatrix} 1 & 0 \\ 2 & 5 \end{pmatrix}$

Mauchly's test: A statistical measure that examines sphericity of within-group variance across pairs of conditions in repeated-measures ANOVA. The assumption of equal variances is violated if the outcome (Mauchly's W) is significant (because the test examines if the variances are different across the pairs of conditions). If Mauchly's test is non-significant, sphericity is said to be assumed. If it is violated, an adjustment is needed (usually employing Greenhouse-Geisser or Huynh-Feldt). The outcome determines which line to read in the SPSS output reporting F ratios.

Mean: The average score, which is found by dividing the sum of all the scores in a range by the number of scores in that range.

Mean difference: The average of the differences between two sets of scores.

Mean square: Part of the calculation of variance applied to many statistical tests, such as ANOVA and liner regression. It is the average sum of squares (overall variance) as apportioned across the experimental factors or conditions, or in respect of the error term.

Median: The 'middle' score in a range of numbers when they have been ordered from lowest to highest (or vice versa).

Mesokurtic: See Kurtosis

Mixed design: A study that includes both between-group and within-group variables in the same analyses (also known as 'mixed model').

Mixed multi-factorial ANOVA: A parametric statistical test that examines variance in a single dependent variable, in respect of two or more independent variables (in a mixed model design). At least one of the variables is explored between groups and at least one other is examined within-groups.

Mode: The most common number in a single data set.

Model cross-product: See Cross-products for full definition. Model cross-products refer to the proportion that can be attributed to the 'experimental' effect (as opposed to error or random effect).

Model mean square: See Mean square for full definition. Model mean square refers to the proportion that can be attributed to the 'experimental' effect (as opposed to error or random effect).

Model sum of squares: See Sum of squares for full definition. Model sum of squares refers to the proportion that can be attributed to the 'experimental' effect (as opposed to error or random effect).

Multi-collinearity: Where the linearity of two or more variables are very closely related. In many statistical tests, linearity and correlation play an important part in assessing the outcome. Where there are several variables being measured, if too many of them are strongly related to each other (they have multi-collinearity) they may effectively be measuring the same thing. This might reduce our efforts to find an actual effect.

Multi-factorial ANOVA: A series of parametric statistical tests that examine variance in a single outcome across two or more factors (independent variables). These tests can be applied in independent, repeated-measures, or mixed-model contexts.

Multiple comparisons: Where several analyses are conducted on a single data set, asking the same research question. This can pose a problem for statistical analyses, which depend on probability to determine outcome. The likelihood of finding a significant outcome that supports the hypothesis is increased simply by repeating the analyses. To overcome this, we can divide the agreed cut-off point for statistical significance by the number of tests that are being performed.

Multiple linear regression: See Linear regression

Multivariate analysis of covariance (MANCOVA): The multivariate version of Analysis of covariance. In this context it is a parametric statistical test that examines how much variance can be explained in two or more dependent (outcome) variables, explored across one or more (between-group) factor, while controlling for one or more additional variable.

Multivariate analysis (MANOVA): A parametric statistical test that examines how much variance can be explained in two or more dependent (outcome) variables, explored across one or more (between-group) factor. Within-group data in this context are explored with Repeated-measures MANOVA(even if the analyses also include between-group data).

Multivariate effect: The part of MANOVA that illustrates the effect of the independent variable(s) upon ALL of the outcome variables in combination.

Nagelkerke's R^2: One of two methods that are used to calculate how much variance that can be explained in a categorical outcome in logistic regression. Compare with Cox and Snell's R^2.

Negative correlation: A statistical test that measures the magnitude of the relationship between two variables whereby as the values in one variable increase, values in a second variable decrease (or vice versa).

Nominal variable: A variable that use numbers to represent names. Often used in SPSS to allocate numbers to categorical groups, such as gender where 1 = male and 2 = female.

Non-parametric: A term used to indicate that the data fail to meet one or more of the specific assumptions about the nature and dispersion of those data. It may be that the data are not normally distributed and/or that the 'numerical' values do not fit the requirements of interval data. In these cases statistical analyses (via non-parametric tests) tend to focus on ranked scores and/or median values rather than mean scores.

Normal distribution: A distribution of scores that assumes that data points are symmetrically dispersed around the mean. There are no outliers (skew = 0) and the peakedness of the distribution is mesokurtic (kurtosis = 0). One of the assumptions of parametric tests (such as t-tests, ANOVA, and linear regression) stipulates that the data should be reasonably normally distributed.

Null hypothesis: The default prediction that there will be 'no difference' or 'no relationship'. Contrast this with the Alternative (or experimental) hypothesis.

Numerator: See Denominator (degrees of freedom)

Oblique rotation: A method of rotation in factor analysis that assumes the factors are correlated to each other, and keeps them that way. The most common examples are direct oblimin and promax.

Observational research: A method of studying natural behaviour by observing it and recording instances, frequencies, and duration of behaviour in certain contexts. Naturalistic observation involves no intervention from the observers (such as watching children's behaviour in the playground). Participant observation occurs when the observer immerses into the group, to gain access to behaviours that might otherwise not be seen (such as observing secret societies).

Odds ratio: An expression of probability of outcomes between groups. Those probabilities can be presented in terms of 'odds'. For example, men may be three times more likely to watch football than not to do so (representing odds of '3' - or '3 divided by 1), while women may be twice as likely not to watch football than to do so (representing odds of '0.5' - or '1 divided by 2'). The 'odds ratio' is found by dividing one set of odds by the other: $3 \div 0.5 = 6$. This means that men are six times more likely to watch football than women are.

Omnibus outcome (ANOVA): The overall outcome, as determined by the F ratio, before exploring the source of difference (through *post hoc* tests or planned contrasts).

One-tailed hypothesis: A specific prediction about an outcome. Some examples: 1. Depression scores will be significantly poorer for women than men. 2. The sales of ice creams will increase as the temperature rises. Contrast this with a Two-tailed hypothesis.

Opportunity sample: Where a sample is drawn from people conveniently available from the immediate population. An example might be recruiting students from the coffee bar on campus.

Order effects: A series of possible outcomes that might occur in within-group studies as a result of repeating measures on the same group of people. An example might be learning or practice effects, where the participant becomes familiar with procedures and may anticipate outcomes.

Ordinal data: Where the 'values' of 'numerical' data have little meaning, other than perhaps that some numbers might be 'higher' than others, but where the magnitude of difference may be harder to quantify. Contrast this with interval data, where numerical differences are consistent and meaningful. For example, temperature and income are considered interval, because we can make meaningful inferences about differences in values in those ranges. Satisfaction scores are more likely to be ordinal, and not interval, as one person's rating of 1 (very satisfied) may be very different to someone else's rating. Ordinal data tend to be analysed in terms of how they are ranked, rather than the actual number. Indeed, another good example of ordinal data is race position - 1st, 2nd, 3rd, etc. Being first is not twice as good as being second.

Orthogonal rotation: A method of rotation in factor analysis that assumes the factors are not correlated to each other, and keeps them independent of each other. The most common examples are varimax, quartimax and equamax.

Outcome variable: Another name for the Dependent variable.

Outlier: An extreme score, likely to skew the distribution of the remaining scores, making inferences about the entire data set more difficult.

Output: The reports (often results) produced by SPSS when we run analyses or functions.

p value: A statement of the probability (p) that something happened by chance factors. It is the basis of most statistical analyses. A p value of '1' indicates that an outcome must have happened by chance; a p value of '0' means that there is no way it happened by chance. In reality, the p value will be somewhere in between and is expressed in decimal format. For example $p = .05$ (the common cut-off point for statistical significance) means that there is a 5% probability that the outcome occurred by chance.

Paired samples t-test: Another name for the Related t-test.

Parametric: A term used to indicate that the data meet both of the specific assumptions about the nature and dispersion of those data. Those data should be normally distributed and represented by interval numbers. In these cases statistical analyses (via parametric tests) tend to focus on mean scores and variance.

Partial correlation: An extension to traditional correlation whereby the relationship between two variables is measured, while controlling for additional variables, to see what effect that has on both of them.

Partial eta squared (η^2): A measure in ANOVA that examines the effect size of one specific factor in the analysis.

Path analysis: A complex form of statistical analyses that examines causality in tests similar to multiple linear regression and analyses of covariance. It is a specific form of Structural equation modelling, focusing on singular indicators.

Pearson's chi-squared: A more specific name for the Chi-squared test.

Pearson's correlation: A type of correlation used on parametric data, where outcomes are based on mean scores in each variable.

Pearson's r table: A distribution of values that determine probability outcomes for correlation.

Pillai's Trace: One of several methods that can be used to report multivariate outcomes in MANOVA. The other options are Hotelling's Trace, Wilks' Lambda, and Roy's Largest Root. Each method has its merits, according to the nature of the variables being explored. A fuller debate can be read in Chapter 14.

Planned contrasts: A group of statistical measures that can be used to examine the source of difference in ANOVA. Although more powerful than *Post hoc* tests, there are a number of restrictions on their use.

Platykurtic: See Kurtosis

Point-biserial correlation: A test that examines association between one categorical variable and one continuous (numerical) variable. In contrast to Biserial correlation, the categorical variable *is* strictly dichotomous (there can only be two possible outcomes, such as gender: male or female).

Population: A statistical term that refers to the specific group that is being measured. We can rarely measure every example of people or cases in that group, but we often use a sample (or several samples) that we feel best represents that population.

Positive correlation: A statistical test that measures the magnitude of the relationship between two variables whereby as the values in one variable increase, values in a second variable also increase (or vice versa).

***Post hoc* tests:** A group of statistical measures that can be used to examine the source of difference in ANOVA, most of which adjust for multiple comparisons.

Power: See Statistical power

Practice (boredom) effects: See Order effects

Predictor variable: Another name for the Independent variable, more traditionally used in regression analyses (so called because the variable predicts the outcome in the dependent variable).

Principal axis factoring: A form of factor analysis that is used to locate the presence of latent variables, focusing on the common variance between those variables.

Principal components analysis: A form of factor analysis that is used to locate the presence of latent variables, focusing on ALL of the variance between those variables.

Probability: A statistical term that seeks to explore the likelihood of things happening.

Promax: An oblique (non-orthogonal) method of rotation in factor analysis (notably principal components analysis).

Quadratic trend: In contrast to Linear trend, this represents a change in direction across data points. For example, patients may improve on depression scores between baseline and Week 4 of a drug trial, but may worsen thereafter. The quadratic trend line will have a 'U' shape.

Quartimax: An orthogonal method of rotation in factor analysis (notably principal components analysis).

Quasi-experiment: A study where the experimental effect cannot be directly manipulated for ethical or practical reasons. For example, true experiments involve randomly assigning participants to conditions (such as drug treatment group or placebo control group). On the other hand, it is not possible to randomly assign people to gender groups. Equally, if we wanted to explore the effect of child abuse on their development, we cannot ethically assign children to 'abuse' or 'non-abuse' groups. Instead, in quasi-experiments, we observe outcomes from pre-existing groups.

Quota sampling: A method of recruiting participants to a study into groups in the same proportions that they are believed to exist in the general population. For example, if we are investigating depression we might choose to recruit twice as many women as men, because (some) evidence suggests that women are two times more likely to be diagnosed with depression than men.

R: The symbol used to represent multiple correlation (between the predictor variables) in linear regression.

R^2: The symbol used to represent how much variance in the outcome scores is explained by the predictor variables in linear regression.

Random (sampling): A method used in research studies to recruit participants randomly. Contrary to popular misconception, it is NOT simply recruiting whoever is available at a given time and place (that would be opportunity or convenience sampling). Random number generators are used to dictate which potential participants are invited to take part (perhaps the 9th person who passes by).

Randomisation: A method used to (independently) allocate participants to groups in experimental studies without bias.

Ranked scores: A method used in non-parametric analyses to assign numbers to existing values that represent their relative order rather than their magnitude. The values are ordered from lowest to highest (or vice versa) and are assigned a number according to how that number is placed in that hierarchy.

Rao's canonical factoring: A method used in factor analysis whereby relationships are explored across two separate sources, such as latent factors from a questionnaire and another confirmatory source (such as clinical observation).

Ratio data: A specific form of interval data, where there is absolute '0' and where numbers can be compared to other numbers in that range in relative terms. For example, it could be said that someone who is aged 50 years is twice as old as someone aged 25. Meanwhile, temperature as measured by the Celsius scale (whilst interval) cannot be considered as ratio: a temperature of 30°C is not twice as hot as 15°C.

Regression line: See Line of (best) fit. In linear regression the gradient of the line is used to make inferences about how well the predictor variable contributes to the variance in the outcome.

Regression model: See Linear regression

REGWQ: A less common *post hoc* analysis used to locate the source of significance in between-group ANOVA tests. It should only be used where there are equal group sizes and homogeneity of variances across those groups. This method is more likely to be used when there are more than five groups.

Related samples t-test: Another name for the Related t-test.

Related t-test: A parametric statistical test that uses the t-distribution to explore differences in mean dependent variable scores across two within-group conditions across a single group.

Relationship: Another name for association or correlation.

Reliability: A measure of consistency in research methods and outcomes.

Reliability analysis: A statistical test that examines the internal consistency of themes within a questionnaire. The analyses include a series of measures (most notably Cronbach's alpha) that explore the relationship between items of each construct.

Repeated-measures: A method in within-group studies where all 'experimental' conditions are presented to a single group or sample.

Repeated-measures MANOVA: The within-group version of Multivariate analysis of variance (MANOVA), but which can also (additionally) contain between-group data.

Repeated-measures multi-factorial ANOVA: A parametric statistical test that examines variance in a single dependent variable, in respect of two or more within-group independent variables.

Repeated-measures one-way ANOVA: A parametric statistical test that examines variance in a single dependent variable, in respect of one within-group independent variable.

Repeated-measures t-test: Another name for the Related t-test.

Residual mean square: See Mean square for full definition. Residual mean square refers to the proportion that is attributed to the error or random effect (as opposed to 'experimental' effect).

Residual cross-product: See Cross-products for full definition. Residual cross-products refer to the proportion that can be attributed to the error or random effect (as opposed to 'experimental' effect).

Residual sum of squares: See Sum of squares for full definition. Residual sum of squares refers to the proportion that is attributed to the error or random effect (as opposed to 'experimental' effect).

Retrospective research: A form of quasi-experimental research that involves accessing previously collected data to make further analyses.

Reverse scoring: A method in questionnaire studies that shuffles the presentation of positively phrased and negatively phrased questions. The aim is to reduce the likelihood that respondents will answer questions without thinking carefully enough about their answers (often referred to as 'response mode').

Rotation: A method used in factor analysis to help the interpretation of factor loadings and ultimately the identification of latent variables.

Roy's Largest Root: One of several methods that can be used to report multivariate outcomes in MANOVA. The other options are Hotelling's Trace, Wilks' Lambda, and Pillai's Trace. Each method has its merits, according to the nature of the variables being explored. A fuller debate can be read in Chapter 14.

Saturated model: An initial model (often applied in regression and Loglinear analysis) that includes all of the variables (so it is a perfect fit). However, this may not tell us enough about specific relationships. To explore a simpler model, we need to remove parts of it until we find that optimum without losing too much information.

Scale data: A term (most commonly used in SPSS) to describe interval and ratio data.

Scatterplot: A graphical presentation of data where points indicate values across two variables. For example, one participant may score '25' on a sleep quality scale and '55' on a mood scale. The graph might plot sleep quality scores along the horizontal (x) axis and mood scores along the vertical (y) axis. In this case, the data point would be plotted 25 unit along the x axis and 55 up the y axis. This is repeated for all participants. The scatterplot can be used to visualise the correlation between the variables.

Scree plot: A graphical presentation that presents eigenvalues across each factor loading. It is used to help determine which factors should be retained to produce the most optimal model.

Second-order correlation: See First-order correlation

Semi-partial correlation: An extension to traditional correlation whereby the relationship between two variables is measured, while controlling for additional variables, to see what effect that has on just *one* of the original variables. Used in linear regression to explore the relationship between outcomes and predictor variables.

Shapiro-Wilk test: A method of examining normal distribution, often used in conjunction with the Kolmogorov-Smirnov test. This test tends to be preferred with smaller samples ($n < 50$). Both tests report whether the outcome data are significantly different to a normal distribution. Since, this outcome is not desirable, the outcome in these tests need to be non-significant to show that the data are probably normally distributed.

Sheffé: A common *post hoc* analysis used to locate the source of significance in between-group ANOVA tests. It should only be used where there are equal group sizes and homogeneity of variances across those groups.

Sidak: A seldom used *post hoc* test applied to find the source of difference in repeated-measures ANOVAs. Bonferroni adjustments tend to be used in most cases.

Simple contrast: One of several planned contrasts that can be used to locate the source of difference in ANOVA outcomes.

Simple effect: A term used in multi-factorial ANOVA that describes the effect of one independent variable on the outcome, at each level of additional independent variables.

Simple linear regression: See Linear regression

Skew: The extent to which a distribution of scores is biased by outliers. In a perfect situation, skew will equal 0. If there are extreme values at the lower end of the distribution there may be negative skew; if those outliers are at the higher end there may be positive skew. The value of the skew will determine whether the data are normally distributed, according to cut-off points relative to the sample size.

Spearman's correlation: A type of correlation used on non-parametric data, where outcomes are based on ranked scores in the variables.

Sphericity: A term usually applied to within-group studies to measure equality of variance across pairs of conditions. See also Mauchly's test.

Split half reliability: A method of assessing consistency of responses across several items in a single construct. Since this analysis is based on combining every possible pair of items, the test is constrained by the number of items it is comparing. Cronbach's alpha is a more robust way of doing this, and much easier to perform.

SPSS: A software program designed to perform statistical analyses. The acronym stands for 'Statistical Package for Social Sciences', although the program is applied to a much wider field these days (and the full name is rarely used anymore).

Spurious correlation: A relationship that only exists statistically, but is probably better explained by an association with other factors. There may be a very strong negative correlation between ice cream sales and the amount of clothes people wear. Despite the statistical relationship, it does not mean that eating ice cream forces people to remove their clothes. The correlation is spurious. The true relationship is between both variables and temperature.

Squared multiple correlation: One of the outcomes in reliability analysis that assess the internal consistency of items within a single construct found in a questionnaire.

Standard deviation: A statistical measure that examines the average variation of scores either side of the mean in a single sample.

Standard error: An estimate of the standard deviation in the population. Typically, it is found by dividing the standard deviation by the square root of the sample size.

Standard error of differences: This is similar to Standard error, except that it focuses on the distribution of differences between two populations. Typically, we will measure scores from two samples (or distinct groups within a sample) and examine the range of differences between them. From that, we will find a mean difference and a standard deviation of those differences.

Standardisation: A method of converting variable values into a standardised measure; this allows us to compare scores between several variables more directly. The most common form of conversion is known as a 'z' score. We can find this score in one of two ways: we can divide the score by the standard error of the range of scores from which it came; or we can deduct the score value from the mean of the scores in that distribution, and divide the outcome by the standard deviation of the scores in that distribution. Either way, it produces a new (z) score that resides within a distribution of scores, for which we know the mean is 0 and the standard deviation is 1.

Standardised residuals: A method of reporting the residual (error) values so that we can assess if they pose a threat to the significance of an outcome. If there is too much error variance our statistic may be compromised. Error values are standardised (see Standardisation), which we can assess with regard to how much they deviate from zero. Any standardised residuals that exceed ± 3.29 are likely to be true outliers that will skew our outcome. We may also be concerned about the distribution of our scores if more than 1% of the standardised residuals exceed ±2.58, and more than 5% exceed ±1.96.

Statistical notation: The format in which we report statistical outcome, using the relevant test statistic, degrees of freedom, and p (probability) value. Conventions vary for this and depend on the test being reported. For example, the APA-accredited method of reporting the outcome from an independent one-way ANOVA might be: $F(1, 84) = 5.529, p = .021$

Statistical power: An outcome that reports the ability to find the desired effect size. The optimal 'power' is usually 0.80; this reflects an 80% chance of not making a Type II error. We can estimate power uses program such as G*Power.

Statistical significance: The probability that the observed outcome occurred by chance. We usually set this at 5%: an outcome is statistically significant if there is a less than 5% probability that the outcome occurred by chance factors (often written as $p < .05$).

Stem-and-leaf plot: A graphical presentation of data that focuses on median and inter-quartile ranges, providing some illustration of the dispersion of data. These can be very useful as a visual guide to examining normal distribution.

Stepwise regression: A version of regression analysis where the variables are added to the model in hierarchies. Compare this to the Enter method, where all of the variables are added at the same time. Once a variable is included (based on semi-partial correlation) the remaining variables are assessed to see if they should be retained or removed.

Stratified sampling: Similar to Quota sampling, but the participants within the groups are randomly selected.

Structural equation modelling: A complex form of statistical analyses that examines causality in tests similar to multiple linear regression and factor analysis, focusing on multiple indicators. It is particularly useful for identifying latent variables.

Student's t-test: Another name for the Independent t-test.

Sum of squares: Part of the calculation of variance applied to many statistical tests, such as ANOVA and linear regression. It is the overall variance that can be apportioned across the experimental factors or conditions, or in respect of the error term.

Syntax: The programming language that SPSS uses.

Systematic variance: Another name for Explained variance.

Test-retest reliability: A method of examining the consistency of an outcome over time, usually applied to explore the properties of new questionnaires. The questionnaire is presented to the same group of people at two time points. Reliability is supported if there is strong correlation between the responses at the two time points.

Tolerance: A measure in multiple linear regression for examining multi-collinearity (which needs to be avoided). Outcomes below 0.1 present a serious concern, as it could increase the likelihood of making Type II errors.

Total sum of squares: Part of the calculation of variance applied to many statistical tests, such as ANOVA and linear regression. It is the overall variance in the outcome variable across the entire sample.

Transformation: A method of converting scores in one of several established processes to address a problem with a distribution of the data.

True experiment: A very specific type of research conducted under laboratory conditions, where all but the experimental conditions are controlled, and to which people or cases are randomly allocated to experimental groups.

t-score distribution: A distribution of scores that provide cut-off points for the independent t-test and related t-test to determine that differences between groups or conditions are significantly different to 0 (according to the relevant degrees of freedom). It is also applied to linear regression to examine whether the gradient between the predictor variable and outcome is significantly different to 0.

Tukey: A common *post hoc* analysis used to locate the source of significance in between-group ANOVA tests. It should only be used where there are equal group sizes and homogeneity of variances across those groups.

Two-tailed hypothesis: A non-specific prediction about an outcome. Some examples: 1. Depression scores will be significantly different between women and men. 2. There will be a correlation between the sales of ice creams and temperature. Contrast this with a One-tailed hypothesis.

Type I error: The experimental hypothesis has been accepted when it should not have been.

Type II error: The experimental hypothesis has been rejected when it should have been accepted.

Unexplained variance: The variance in the outcome values that cannot be explained by the experimental conditions or the elements that we are exploring (also known as error variance).

Univariate effect: Usually applied in the context of MANOVA. It is the part that illustrates the effect of the independent variable(s) upon EACH of the outcome variables (separately).

Unrelated t-test: Another name for the Independent t-test.

Unsystematic variance: Another name for Unexplained variance.

Validity: A term that covers a range of measures that examine whether something is measuring what we claim it to be. For example, we might believe that we are measuring mood in a group of participants, only to find that we were actually observing fatigue.

Variable: The entity that we are measuring, either as an outcome or as a factor that might influence that outcome.

Variable View: The SPSS window in which variable parameters are presented across columns that illustrate specific factors relating to variable (such as name, type, value labels, etc.). Each row represents a different variable. Contrast with Data View.

Variance: The extent to which scores vary. They may do so completely randomly (there is no explanation for why they vary) or they may vary due to specific factors. We aim to explain as much of that variance as possible from factors that we are observing.

Variance-covariance matrix: A matrix of numbers used primarily in multivariate analyses (see Homogeneity of variance-covariance matrices). It provides information about the variance within each variable and correlation between pairs of variables.

Varimax: The most common (orthogonal) method of rotation in factor analysis (notably principal components analysis).

VIF: A measure in multiple linear regression for examining multi-collinearity (which needs to be avoided); it stands for variance inflation factor. Outcomes above 10 present a serious concern, as it could increase the likelihood of making Type II errors.

Wald statistic: A measure used in logistic regression to assess whether a predictor variable significantly contributes to the 'variance' in the categorical outcome. It is based on values from a chi-squared distribution and is used to establish whether the gradient is significantly different to 0. It does this by dividing the gradient (the regression coefficient) by its standard error.

Welch's F: A method of adjusting the F ratio in ANOVA when there is unequal variance between the groups (used alongside the very similar Brown-Forsythe F).

Wilcoxon signed ranks: A non-parametric test used in within-group analyses where there are two conditions. It is used as an alternative to the Related t-test.

Wilks' Lambda: One of several methods that can be used to report multivariate outcomes in MANOVA. The other options are Hotelling's Trace, Roy's Largest Root and Pillai's Trace. Each method has its merits, according to the nature of the variables being explored. A fuller debate can be read in Chapter 14.

Within-between interaction: An interaction in a mixed multi-factorial ANOVA between a within-group independent variable and a between-group independent variable.

Within-group main effect: The effect that a single within-group independent variable has on the outcome (dependent variable). Usually applied in ANOVA tests, it is often measured from difference in mean outcome scores across the independent variable conditions (undertaken by the entire sample).

Within-group study: An examination of outcome (dependent variable) scores across conditions that have been experienced by the entire sample.

Within-group t-test: Another name for the Related t-test.

Yate's continuity correction: One of several applications of the chi-squared distribution. The outcome illustrates the association between two categorical variables. Generally used where both of the variables has two groups. If either variable has more than two groups (Pearson's) chi-squared test tends to be preferred.

Zero-order correlation: A simple correlation where there are two variables (and no covariates)

Z-score: See Standardisation

Index